Springer Proceedings in Mathematics & Statistics

Volume 229

Springer Proceedings in Mathematics & Statistics

This book series features volumes composed of selected contributions from workshops and conferences in all areas of current research in mathematics and statistics, including operation research and optimization. In addition to an overall evaluation of the interest, scientific quality, and timeliness of each proposal at the hands of the publisher, individual contributions are all refereed to the high quality standards of leading journals in the field. Thus, this series provides the research community with well-edited, authoritative reports on developments in the most exciting areas of mathematical and statistical research today.

More information about this series at http://www.springer.com/series/10533

Andreas Eberle · Martin Grothaus
Walter Hoh · Moritz Kassmann
Wilhelm Stannat · Gerald Trutnau
Editors

Stochastic Partial Differential Equations and Related Fields

In Honor of Michael Röckner
SPDERF, Bielefeld, Germany,
October 10–14, 2016

Springer

Editors
Andreas Eberle
Institute for Applied Mathematics
University of Bonn
Bonn
Germany

Moritz Kassmann
Faculty of Mathematics
Bielefeld University
Bielefeld
Germany

Martin Grothaus
Fachbereich Mathematik
Technische Universität Kaiserslautern
Kaiserslautern
Germany

Wilhelm Stannat
Institut für Mathematik
Technische Universität Berlin
Berlin
Germany

Walter Hoh
Faculty of Mathematics
Bielefeld University
Bielefeld
Germany

Gerald Trutnau
Department of Mathematical Sciences and Research Institute of Mathematics
Seoul National University
Seoul
South Korea

ISSN 2194-1009 ISSN 2194-1017 (electronic)
Springer Proceedings in Mathematics & Statistics
ISBN 978-3-030-09107-1 ISBN 978-3-319-74929-7 (eBook)
https://doi.org/10.1007/978-3-319-74929-7

Mathematics Subject Classification (2010): 60Hxx, 60Jxx, 31Cxx, 60Gxx, 35Kxx, 35Qxx, 47Hxx, 47Dxx, 35Jxx

Softcover re-print of the Hardcover 1st edition 2018

Printed on acid-free paper

This Springer imprint is published by the registered company Springer International Publishing AG part of Springer Nature
The registered company address is: Gewerbestrasse 11, 6330 Cham, Switzerland

Preface

On October 10–14, 2016, the Faculty of Mathematics at Bielefeld University hosted the international conference "Stochastic Partial Differential Equations and Related Fields." The goal of the conference was to bring together leading scientists and young researchers in order to present the current state of the art and directions for possible future developments in this exciting and rapidly developing field along with its many interactions.

The recent progress of the mathematical theory of stochastic partial differential equations is mainly driven by a growing interest in modeling and numerical approximation of stochastic processes on multiple space and time scales and by current groundbreaking progress in the mathematical rigorous analysis of stochastic dynamical systems at criticality provided by the theory of regularity structures and paracontrolled distributions. Their further investigation is very likely to influence the future development of stochastic analysis in the upcoming years, also with strong interactions with neighboring disciplines, in particular nonlinear partial differential equations.

More than 140 participants from all over the world, including many young researchers, postdocs, and Ph.D. students, participated in the conference. It consisted of 32 plenary talks by leading experts and 50 session talks given in three parallel sessions during the afternoons, covering major current developments in the field. During the days of the conference, the talks have focused in particular on the following areas:

- Monday, October 10: Theory of Dirichlet forms, potential theory, and geometry on metric spaces.
- Tuesday, October 11: Analysis of Kolmogorov operators, mild and variational solutions of stochastic partial differential equations.
- Wednesday, October 12: Numerical analysis of stochastic partial differential equations and analysis of stochastic partial differential equations with random coefficients, random transport equations.
- Thursday, October 13: Rough paths, singular stochastic partial differential equations, regularity structures, and the theory of paracontrolled distributions.

- Friday, October 14: Functional inequalities, Gibbs measures, and applications of stochastic partial differential equations, in particular to mathematical physics.

A special session of the conference was dedicated to celebrate the 60th birthday of Michael Röckner. His more than 30 years of research in the field of infinite dimensional stochastic analysis had a major impact on the development of the mathematical theory of stochastic partial differential equations up to its present state. It has inspired many young researchers to take up their research in this exciting field of mathematics.

Michael Röckner studied Mathematics and Physics at Bielefeld University (1976–1982). He received a Doktor degree in Mathematics from Bielefeld University in 1984 under the supervision of Sergio Albeverio. After spending the years 1984–1985 as a Visiting Fellow at Cornell University, he started a position as Lecturer at the University of Edinburgh in 1986, where he was promoted to Reader in 1989. In 1990, Michael Röckner returned to Germany and became a C3 Professor at the University of Bonn. Four years later, he accepted an offer from Bielefeld University for a Full Professorship (C4). In 2005, he became a Full Professor at Purdue University, W. Lafayette, Indiana, USA, but he returned to Bielefeld as Full Professor (W3) only one year later, where he is staying since then.

Michael Röckner has received several awards and recognitions, including the Heinz Maier-Leibnitz Prize (1989) and the Max Planck Research Award (1992), as well as offers for full professorships from the Universities of Bochum (1994), Leipzig (1998), Edinburgh (2005), and Bonn (2006). In 2014, he was awarded a "Specially-Invited-Professorship" by Jiangsu Normal University in Xuzhou. In 2017, he has received an honorary "Doctor of Science" degree from Swansea University and became a "Distinguished Visiting Professor" through the national "Innovative Talent Recruitment Program" at the Academy of Mathematics and System Science at the Chinese Academy of Sciences in Beijing. He received numerous grants from the German Research Foundation (DFG), the German Academic Exchange Service (DAAD), the Science and Engineering Research Council (SERC), the National Science Foundation (NSF), and the European Union. He served as Dean of the Department of Mathematics at Bielefeld University in the years 1997–1999, and from the year 2010 on until present.

During all of his academic career, Michael Röckner served the scientific community on several panels, including the DFG Commission for Collaborative Research Centers (2006–2011), the selection committee for Humboldt Research Awards (since 2012), and the panel "PE 1 Mathematics" of the European Research Council (2008–2013, Chairman 2015/2016, 2017/2018). He is currently also President of the German Mathematical Society (DMV).

Up to the present date, Michael Röckner supervised 21 Ph.D. students, 32 postdocs, and 7 Humboldt scholars. Many of them later became professors in Germany and abroad. Currently, he has published more than 260 research papers in scientific journals, more than 50 refereed contributions to proceedings, and he is the author of 8 monographs and the editor of 9 proceedings and special volumes.

The editors of this Festschrift asked the participants of the conference in Bielefeld to submit short research contributions related to the main topic of the conference. The aim was to present current research questions that can help especially young researchers but also mathematicians from different research areas to get an extensive overview on the current state of the art in the subject. The volume contains 34 short contributions to the following subject areas:

- Stochastic partial differential equations and regularity structures.
- Stochastic analysis including geometric aspects.
- Dirichlet forms, Markov processes, and potential theory.
- Applications including mathematical physics.

To honor Michael Röckner's contribution to the field, the editors also invited five colleagues to provide longer research surveys dedicated to one of Michael Röckner's research topics: Dirichlet forms and potential theory, analysis of Kolmogorov operators (in infinite dimensions), Fokker–Planck equations in Hilbert spaces, theory of variational solutions to stochastic partial differential equations and singular stochastic partial differential equations and their applications in mathematical physics.

With these 34 short and 5 longer contributions, we hope to provide a broad picture of the current state and open problems in this exciting research area.

Bonn, Germany — Andreas Eberle
Kaiserslautern, Germany — Martin Grothaus
Bielefeld, Germany — Walter Hoh
Bielefeld, Germany — Moritz Kassmann
Berlin, Germany — Wilhelm Stannat
Seoul, South Korea — Gerald Trutnau
October 2017

Acknowledgements

We would like to thank our colleagues that helped us in reviewing the contributions to this Festschrift.

We would also like to take this opportunity to thank all speakers and participants of the conference in Bielefeld for exciting talks and discussions, as well as the local organizers from Bielefeld for their excellent work. Financial support from the DFG and the CRC 701 for the organization of the conference is gratefully acknowledged.

Finally, we would like to thank also Marina Reizakis from Springer-Verlag for her support of this Festschrift.

Contents

Organization

Program Chair

Gerald Trutnau, Seoul National University, South Korea

Program Committee

Andreas Eberle, Universität Bonn, Germany
Martin Grothaus, TU Kaiserslautern, Germany
Walter Hoh, Universität Bielefeld, Germany
Moritz Kassmann, Universität Bielefeld, Germany
Wilhelm Stannat, TU Berlin, Germany
Gerald Trutnau, Seoul National University, South Korea

Stochastic Partial Differential Equations and Related Fields, Bielefeld, Germany, October 2016. Photographer: Christina Lux.

List of Participants

Sergio Albeverio, Universität Bonn
Markus Antoni, KIT, Karlsruhe
Sigurd Assing, the University of Warwick
Volker Bach, TU Braunschweig
Viorel Barbu, Romanian Academy, Iaşi
Wolf-Jürgen Beyn, Universität Bielefeld
Lucian Beznea, Institute of Mathematics of the Romanian Academy and University of Bucharest
Philippe Blanchard, Universität Bielefeld
Dirk Blömker, Universität Augsburg
Vladimir I. Bogachev, Moscow State University
Johannes Brasche, TU Clausthal
Zdzislaw Brzezniak, the University of York
Yana A. Butko, Universität des Saarlandes, Saarbrücken
Sandra Cerrai, University of Maryland, College Park
Jamil Chaker, Universität Bielefeld
Xiaopeng Chen, Shantou University
Zhen-Qing Chen, University of Washington, Seattle
Le Chen, the University of Kansas
Carsten Chong, TU München
Iulian Cimpean, Institute of Mathematics of the Romanian Academy, Bucharest
Petru A. Cioica-Licht, University of Otago, Dunedin
Sonja Cox, University of Amsterdam
Giuseppe Da Prato, Scuola Normale Superiore di Pisa
Robert Dalang, Ecole Polytechnique Fédérale de Lausanne
Lorenzo Dello Schiavo, Universität Bonn
Michael Demuth, TU Clausthal
Jean-Dominique Deuschel, TU Berlin
Martin Dieckmann, Universität Bielefeld
Joscha Diehl, MPI Leipzig
Andreas Dress, PICB Shanghai

Bruce K. Driver, University of California, San Diego
Jinqiao Duan, Illinois Institute of Technology, Chicago
Andreas Eberle, Universität Bonn
Shimaa Elesaely, TU Berlin
David Elworthy, the University of Warwick
Etienne Emmrich, TU Berlin
Nikita Evseev, Novosibirsk State University
Qianzhu Fan, the University of Manchester
Torben Fattler, TU Kaiserslautern
Hans Föllmer, HU Berlin
Martin Friesen, Universität Wuppertal
Peter Friz, TU and WIAS Berlin
Masatoshi Fukushima, Osaka University
Tadahisa Funaki, University of Tokyo
Barbara Gentz, Universität Bielefeld
Stefan Geiss, University of Jyväskylä
Máté Gerencsér, IST Austria, Klosterneuburg
Benjamin Gess, MPI Leipzig
Maria Gordina, University of Connecticut, Storrs
Friedrich Götze, Universität Bielefeld
Martin Grothaus, TU Kaiserslautern
Zhongkai Guo, South-Central University for Nationalities, Wuhan
Istvan Gyöngy, University of Edinburgh
Martin Hairer, the University of Warwick
Wolfhard Hansen, Universität Bielefeld
Erika Hausenblas, Montanuniversität Leoben
Michael Hinz, Universität Bielefeld
Antoine Hocquet, TU Berlin
Martina Hofmanova, TU Berlin
Walter Hoh, Universität Bielefeld
Fabian Hornung, KIT, Karlsruhe
Peter Imkeller, HU Berlin
Rahul Indoria, Universität Bielefeld
Arnulf Jentzen, ETH Zürich
Peng Jin, Bergische Universität Wuppertal
Kamran Kalbasi, the University of Warwick
Moritz Kassmann, Universität Bielefeld
Hiroshi Kawabi, Okayama University
Panki Kim, Seoul National University
Alexander Kolesnikov, Higher School of Economics, Moscow
Yuri Kondratiev, Universität Bielefeld
Yuri Kozitzky, Maria Curie-Skłodowska University, Lublin
Nicolai V. Krylov, University of Minnesota, Minneapolis
Peter Kuchling, Bielefeld University
Takashi Kumagai, Kyoto University

Eija Laukkarinen, University of Jyväskylä
Xue-Mei Li, the University of Warwick
Xiang-Dong Li, Academy of Mathematics and Systems Science, Beijing
Alessandra Lunardi, Università degli Studi di Parma
Terry Lyons, University of Oxford
Zhi-Ming Ma, Academy of Mathematics and Systems Science, Beijing
Ananta Kumar Majee, Universität Tübingen
Mateusz Majka, Universität Bonn
Carlo Marinelli, University College London
Bohdan Maslowski, Charles University, Prague
Prince Romeo Mensah, Heriot-Watt University, Edinburgh
Marvin Müller, ETH Zürich
Nora Müller, Universität Bielefeld
Alexandra Neamtu, FSU Jena
Alexander Nerlich, Universität Ulm
Julian Newman, Imperial College London
David Nualart, the University of Kansas, Lawrence
Hirofumi Osada, Kyushu University, Fukuoka
Youssef Ouknine, Université Cadi Ayyad, Marrakech
Ludger Overbeck, Universität Gießen
Etienne Pardoux, Aix-Marseille Université, Marseille
Tatyana Pasurek, Universität Bielefeld
Shige Peng, Shandong University
Nicolas Perkowski, HU Berlin
Marina Reizakis, Springer-Verlag, Heidelberg
Michael Röckner, Universität Bielefeld
Jasmin Röder, Universität Gießen
Barbara Rüdiger, Bergische Universität Wuppertal
Francesco Russo, ENSTA ParisTech, Palaiseau
Michael Scheutzow, TU Berlin
Torben Schröder, TU Berlin
Björn Schmalfuß, FSU Jena
Vitali Senin, TU Berlin
Stanislav Shaposhnikov, Moscow State University
Wilhelm Stannat, TU Berlin
Ludwig Streit, Universität Bielefeld
Karl-Theodor Sturm, Universität Bonn
Herry Suryawan, Universität Zürich
Masayoshi Takeda, Tohoku University, Sendai
Alexander Teplyaev, University of Connecticut, Storrs
Jonas Tölle, Aalto University, Helsinki
Gerald Trutnau, Seoul National University
Stefania Ugolini, Università degli Studi di Milano
Katharina von der Lühe, Universität Bielefeld
Robert Voßhall, TU Kaiserslautern

Feng-Yu Wang, Beijing Normal University and Swansea University
Hui Wang, Nanjing Normal University
Wei Wang, Nanjing University
Karolina Weber, TU Wien
Hendrik Weber, the University of Warwick
Lukas Wresch, Universität Bielefeld
Weina Wu, Universität Bielefeld
Xuetao Yang, Nanjing Normal University
Minoru Yoshida, Kanagawa University, Yokohama
Wen Yue, TU Wien
Lorenzo Zambotti, UPMC, Paris
Deng Zhang, Universität Bielefeld
Tusheng Zhang, the University of Manchester
Xicheng Zhang, Wuhan University
Anna Zhigun, TU Kaiserslautern
Quanxin Zhu, Universität Bielefeld
Rongchan Zhu, Beijing Institute of Technology and Universität Bielefeld
Xiangchan Zhu, Beijing Jiaotong University and Universität Bielefeld
Sebastian Zachrau, TU Berlin
Maria Ziemlanska, Leiden University

Part I
Longer Contributions

Stationary Fokker–Planck–Kolmogorov Equations

Vladimir I. Bogachev

Abstract We give a survey of results obtained over the last two decades on stationary Fokker–Planck–Kolmogorov equations with respect to measures, which are also called "double divergence form equations". The existence of densities and their integrability, continuity and weak differentiability are discussed. In case of equations on the whole space the existence and uniqueness of probability solutions are studied. A brief discussion of the infinite-dimensional case is included.

Keywords Stationary Fokker–Planck–Kolmogorov equation · Invariant measure Elliptic equation · Double divergence form equation

AMS Subject Classification Primary 35J15 · Secondary 35B65

1 Introduction

It is well known that any invariant (or stationary) measure μ of the diffusion process in a domain $\Omega \subset \mathbb{R}^d$ with generator of the form

$$Lu = L_{A,b}u = a^{ij}\partial_{x_i}\partial_{x_j}u + b^i\partial_{x_i}u = \text{trace}\,(AD^2u) + \langle b, \nabla u\rangle,$$

where $\langle\cdot,\cdot\rangle$ is the inner product in $\mathbb{R}^d$, a^{ij}, b^i are real Borel functions on Ω, $A(x) = (a^{ij}(x))_{i,j\le d}$ is a nonnegative definite matrix, $b = (b^1, \ldots, b^d)$, and throughout in such expressions we always mean the summation over repeated upper and

V. I. Bogachev (✉)
Lomonosov Moscow State University, Moscow, Russia
e-mail: vibogach@mail.ru

V. I. Bogachev
National Research University Higher School of Economics, Moscow, Russia

V. I. Bogachev
St.-Tikhon's Orthodox Humanitarian University, Moscow, Russia

A. Eberle et al. (eds.), *Stochastic Partial Differential Equations and Related Fields*, Springer Proceedings in Mathematics & Statistics 229,
https://doi.org/10.1007/978-3-319-74929-7_1

lower indices, satisfies the stationary Kolmogorov (or Fokker–Planck–Kolmogorov) equation

$$L^*\mu = 0, \tag{1}$$

which is understood in the sense of the integral identity

$$\int_\Omega Lu\,d\mu = 0 \quad \forall u \in C_0^\infty(\Omega),$$

where we assume that the coefficients a^{ij}, b^i are locally integrable with respect to the solution μ (which is automatically the case for locally bounded coefficients). This equation is meaningful and can be studied independently of diffusions (and under assumptions for which the existence of a diffusion connected with L might be unknown).

In terms of distributions equation (1) can be written in the form

$$\partial_{x_i}\partial_{x_j}(a^{ij}\cdot\mu) - \partial_{x_i}(b^i\cdot\mu) = 0. \tag{2}$$

Such equations are also called double divergence form equations.

A study of such equations goes back to the classical works of Fokker [35], Planck [49], and Kolmogorov [38–40]. Principal problems related to these equations concern the existence and uniqueness of solutions and the properties of solutions such as the existence of densities with respect to Lebesgue measure and continuity and differentiability of densities. These problems were posed already in [38] and in the process of later investigations they were precised and detailed in many directions. A new impetus to this line of research was given in the 1990s when intensive investigation of infinite-dimensional Fokker–Planck–Kolmogorov equations started (which was also influenced by some earlier work on non-symmetric Dirichlet forms, see [3]), which lead to new problems in the finite-dimensional case. One of such problems concerns equations with singular coefficients, i.e., coefficients that are locally integrable with respect to the considered solutions, but can fail to be locally integrable with respect to Lebesgue measure. It was shown in [15] (see also [14]) that even in this case under natural assumptions the solutions possess Sobolev class densities. An in-depth study of this phenomenon was undertaken in [10]. The present paper gives a concise account of the results obtained over the past twenty years in this direction with a particular emphasis on the work of Michael Röckner with his collaborators. Many of these results are connected (and were strongly influenced) by a parallel study of singular diffusions and diffusion operators, an important topic which is not touched here at all and which is also very popular in Röckner's group (see [31, 55] and the recent paper [50]). A thorough study of Fokker–Planck–Kolmogorov equations (elliptic and parabolic) has been recently presented in the monograph [13]. However, the present survey not only offers a simplified and more accessible exposition, but also includes some very recent achievements not covered by [13]. Proofs are given only for two new results and some illuminating examples.

If the matrix $A(x)$ is nondegenerate for all x, then every solution μ to Eq. (1) or (2) possesses a density ρ with respect to Lebesgue measure (see [10, 13] and Theorem 1 below), hence (1) can be equivalently written as the double divergence form equation

$$\partial_{x_i}\partial_{x_j}(a^{ij}\rho) - \partial_{x_i}(b^i\rho) = 0 \tag{3}$$

in the sense of distributions.

Since we have no assumptions about differentiability of A, Eq. (1) cannot be represented in the classic divergence form, however, (1) is exactly the form of the stationary Kolmogorov equation arising in many applications. If the matrix A is Sobolev differentiable, then this equation reduces to the well studied equation in the classic divergence form

$$\partial_{x_i}(a^{ij}\partial_{x_j}\rho + (\partial_{x_j}a^{ij} - b^i)\rho) = 0.$$

In case of less regular A, relatively little is known about properties of solutions, and also the number of papers on double divergence form equations is relatively small. Such equations are considered in papers [4, 5, 32–34, 41, 46–48, 53, 54], and also in the monograph [13], where some additional bibliography is given. In addition to connections with diffusion processes, the interest in equations of the form (3) is explained by the fact that such equations hold for the Green's function for the direct equation with respect to the second variable.

The principal distinction of double divergence form equations from equations of the usual divergence form and direct equations is that there is no increase of the regularity of solutions as compared to the regularity of A. For example, in the case of Hölder continuous A (and reasonable b), the solutions of the direct equation belong to the class C^2 and have Hölder continuous second derivatives, the solutions to the divergence form equation belong to the Sobolev class $W^{p,1}$, but solutions to some double divergence form equations even in the one-dimensional case will not be more regular than A itself. For example, if $b = 0$, then we obtain the equation $(A\rho)'' = 0$, whence $\rho = l/A$, where l is an affine function. Hence the regularity of ρ equals precisely the regularity of A. Say, if the matrix A is merely measurable, then so is the function ρ.

It is worth noting that even in case of smooth coefficients and smooth solutions, when our equation can be written as a usual second order equation

$$a^{ij}\partial_{x_i}\partial_{x_j}\rho + \text{first and zero order terms} = 0,$$

the structure of the lower order terms turns out to be rather special, which yields certain important properties of solutions to double divergence form equations not shared by general second order equations.

In this paper we consider locally finite Borel measures (possibly, signed) satisfying the Kolmogorov equation (1). It will be clear below that the case of signed measures does not reduce to that of nonnegative ones, and one has to deal with signed measures even in the problems concerned initially with probability solutions. We shall also consider more general equations

$$L^*_{A,b,c}\mu = 0, \tag{4}$$

similarly defined for the operator

$$L_{A,b,c}u = a^{ij}\partial_{x_i}\partial_{x_j}u + b^i\partial_{x_i}u + cu,$$

where c is a Borel function and in this case in the definition of Eq. (4) we require its local integrability with respect to the solution μ. As usual, the integrability with respect to a signed measure μ means the integrability with respect to its total variation $|\mu|$.

Throughout we always assume that $A(x) = (a^{ij}(x))_{i,j\le d}$ is a symmetric nonnegative definite matrix for every $x \in \Omega$.

Let $C_0^\infty(\Omega)$ denote the set of infinitely differentiable functions with compact support in an open set $\Omega \subset \mathbb{R}^d$. For a nonnegative Borel measure μ on Ω let $L^1_{loc}(\mu)$ denote the class of functions integrable with respect to μ on every compact set in Ω. Similarly we define $L^p_{loc}(\mu)$ for $p \in [1 < \infty)$. In case of the standard Lebesgue measure we write $L^p_{loc}(\Omega)$. Let $\|f\|_p$ denote the norm in L^p with respect to Lebesgue measure; in other cases we shall indicate the measure, for example, $\|f\|_{L^p(\mu)}$. The L^∞-norm will be denoted by $\|\cdot\|_\infty$.

The measure given by a density ξ with respect to the measure μ is denoted by $\xi \cdot \mu$.

Let $W^{p,r}(\Omega)$ denote the Sobolev class of functions on Ω belonging to $L^p(\Omega)$ along with their generalized partial derivatives up to order r. The Sobolev norm $\|f\|_{p,r}$ is given by the sum of the L^p-norms of the function and its generalized partial derivatives up to order r. Finally, $W^{p,1}_{loc}(\Omega)$ consists of all functions f such that $\zeta f \in W^{p,1}(\Omega)$ for all $\zeta \in C_0^\infty(\Omega)$.

2 The Case of a Non-differentiable Diffusion Matrix: Existence and Higher Integrability of Densities

The next result from [10] (see also [13, Theorem 1.5.2]) provides the existence of solution densities under minimal assumptions about the coefficients.

Theorem 1 *Suppose that a nonnegative locally finite Borel measure μ on a domain Ω satisfies the inequality*

$$\int_\Omega \mathrm{trace}(AD^2\varphi)\,d\mu \le C\Big(\sup_\Omega|\nabla\varphi| + \sup_\Omega|\varphi|\Big) \quad \forall\varphi \in C_0^\infty(\Omega)$$

with some number C. Then the measure $(\det A)^{1/d}\cdot\mu$ *has a density ρ with respect to Lebesgue measure and $\rho \in L^{d/(d-1)}_{loc}(\Omega)$. Therefore, if $1/\det A$ is locally bounded,* *then the measure μ itself has a density of this class. In particular, this is true for nonnegative solutions to the equation $L^*_{A,b}\mu = 0$.*

This result is sharp in the sense that one cannot omit the factor $(\det A)^{1/d}$ in front of μ and also it is impossible to increase the guaranteed order of integrability $d/(d-1)$ without additional conditions.

Example 1 Taking the density $\rho(x) = |x|^{1-d}|\ln|x||^{-\alpha}$, $\alpha > 1$, we obtain a solution to the equation $L^*_{A,b}\mu = 0$ with $A = \mathrm{I}$ and

$$b(x) = \frac{\nabla\rho(x)}{\rho(x)} = (1-d)|x|^{-2}x - \alpha\ln|x|\big|\ln|x|\big|^{-2}|x|^{-2}x,$$

and ρ belongs to $L^{d/(d-1)}(U)$ in the unit ball U, but does not belong to $L^p(U)$ with any $p > d/(d-1)$. Note that here we have $|b| \in L^r(\rho\,dx)$ only for $r = 1$.

It was shown in [34] (under the assumption of smoothness of the diffusion coefficients a^{ij}) that for the equation with zero drift the situation is somewhat different: the Green's function $G(x, y)$ for the direct equation satisfies Eq. (3) with respect to the variable y, i.e., the function $v(y) = G(x, y)$ for fixed x satisfies the equation $\partial_{y_i}\partial_{y_j}(a^{ij}v) = 0$, and, for any fixed $\lambda > 0$, the estimate $\lambda\cdot\mathrm{I} \le A \le \lambda^{-1}\cdot\mathrm{I}$ yields the existence of an exponent $q_\lambda > d/(d-1)$ for which the L^{q_λ}-norm of v is estimated with some constant depending on λ through the L^1-norm of v, but such a constant cannot be made independent of λ.

The existence of solution densities for signed solutions remains an open problem. We shall see below that densities exist in case of A of class *VMO*.

We recall that the class *VMO* consists of all locally integrable functions f on $\mathbb{R}^d$ such that there exists a continuous function ω (depending on f) on $[0, +\infty)$ with $\omega(0) = 0$, positive on $(0, +\infty)$, such that

$$\sup_{z\in\mathbb{R}^d, r<R}\int_{|x-z|<r}\int_{|y-z|<r}|f(x) - f(y)|\,dx\,dy \le \omega(R)$$

for all $R > 0$. This class contains all uniformly continuous functions, but also contains some locally unbounded functions. It is known that $W^{d,1}(\mathbb{R}^d) \subset VMO$. In case of functions on a domain, we define the class *VMO* just as the class of restrictions of functions from the global *VMO* to this domain.

In the next theorem we assume that the following condition is fulfilled:

(A1) $a^{ij} \in VMO$, $\quad\lambda_1\cdot\mathrm{I} \le A(x) \le \lambda_2\cdot\mathrm{I}$, where $\lambda_1, \lambda_2 > 0$ are constant.

Our assumptions about b and c are these:

(B1) $|b|, c \in L^q_{loc}(\Omega)$, where $q > d$.

We shall also consider an alternative condition

(B1μ) $|b|, c \in L^q_{loc}(|\mu|)$, where $q > d$.

A simple, but important example for applications where (B1μ) is fulfilled, but (B1) is not is given by a probability measure μ with density of class $\varrho \in C^\infty(\mathbb{R}^d)$ having zeros: every such measure satisfies the equation $\Delta\mu - \operatorname{div}(b\cdot\mu) = 0$ with the unit matrix A and drift

$$b(x) = \nabla\varrho(x)/\varrho(x),$$

defined by zero on the set of zeros of ϱ, because the integral of $\varrho\Delta\varphi$ equals the integral of $-\langle\nabla\varrho, \nabla\varphi\rangle = -\langle b, \nabla\varphi\rangle\varrho$. However, the drift b of such a form has points in a neighborhood of which it is not Lebesgue integrable. Say, if $\varrho(x) = cx^2e^{-x^2}$, then $b(x) = 2x^{-1} - 2x$. With respect to the measure μ itself this drift b can be integrable even to every power. The proof of the next theorem is given in [27].

Theorem 2 *Suppose that a signed measure μ satisfies Eq. (4), where the diffusion coefficient A satisfies conditions* (A1), *the coefficients b and c satisfy conditions* (B1) *or conditions* (B1μ). *Then it has a density $\varrho \in L^r_{loc}(\Omega)$ for all numbers $r \in [1, +\infty)$.*

In case where b and c are locally integrable to every power $p > 1$, we obtain the same integrability for ϱ. However, further improvement of integrability of b and c to local boundedness will not ensure the local boundedness of ϱ even for continuous A (which follows from the example in paper [5]). Nevertheless, one has the exponential integrability of the density.

Theorem 3 *If in the above theorem b and c are locally bounded, then for every ball U there exists a number $\delta > 0$, such that* $\exp(\delta|\varrho|) \in L^1(U)$.

Some additional conditions for raising the order of integrability of ϱ are obtained in [26, 27] in terms of eigenvalues of $A(x)$, but these conditions are more technical and do not give more than $d/(d-2)$ (see below).

In order to obtain the continuity of the solution density (or its local boundedness) the Dini condition on A is needed (the continuity of A is not enough). It would be interesting to find out whether the Dini condition can be further relaxed. In the one-dimensional case it is not needed.

We recall that the Dini condition for a modulus of continuity ω is expressed by

$$\int_0^1 \frac{\omega(t)}{t}\,dt < \infty. \tag{5}$$

This condition holds if $\omega(t) \le ct^\delta$ with $\delta > 0$ or even $\omega(t) \le c|\ln t|^{-1-\delta}$ for $t < 1$.

The role of the Dini condition in the following theorem and similar matters is explained by the integrability of the function $\omega(|x|)|x|^{-d}$ on the unit ball in $\mathbb{R}^d$, which reduces to (5) by passing to polar coordinates.

Theorem 4 *Suppose that on every ball A has a modulus of continuity satisfying the Dini condition. Let* $\det A > 0$ *everywhere and $|b|, c \in L^q_{loc}(\Omega)$, where $q > d$. Then the density ϱ of every solution (possibly, signed) to the equation $L^*_{A,b,c}\mu = 0$ has a continuous version.*

If, in addition, the matrix A is locally Hölder continuous of order $\delta \in (0, 1)$, then $\boldsymbol{\rho}$ has a density that is locally Hölder continuous of order δ.

It is shown in [27] that under the assumptions of Theorem 4, where now the modulus of continuity of A on the whole space satisfies Dini's condition, there holds the following integral representation of solutions obtained in [53, 54] for locally bounded b and c.

Let $\boldsymbol{\omega}_d$ be the area of the unit sphere in $\mathbb{R}^d$. Given $d > 2$, set

$$H(x, y) = \frac{1}{(d-2)\boldsymbol{\omega}_d}(\det A(y))^{-1/2}\langle A(y)^{-1}(x-y), x-y\rangle^{(2-d)/2}.$$

The function H, for each fixed y, serves as a fundamental solution for the equation with the constant diffusion matrix $A(y)$ and zero b and c, i.e., $L_{A(y),0,0}H(\cdot, y) = \delta_y$. Let us fix a function $\eta \in C_0^\infty(\mathbb{R}^d)$ with support in a ball U. The proof of the following given in [27] deals with the case $\eta(z) = 1$, but the general case is very similar.

Theorem 5 *Suppose that $L^*_{A,b,c}(\boldsymbol{\rho}\, dx) = 0$, where A satisfies the hypotheses of the previous theorem. Then the continuous version of $\boldsymbol{\rho}$ admits the following representation:*

$$\begin{aligned} -\eta(z)\boldsymbol{\rho}(z) = &\int_U \boldsymbol{\rho}(x)a^{ij}(z)H(x,z)\partial_{x_i}\partial_{x_j}\eta(x)\,dx \qquad (6)\\ &+2\int_U \boldsymbol{\rho}(x)a^{ij}(z)\partial_{x_i}H(x,z)\partial_{x_j}\eta(x)\,dx\\ &+\int_U \boldsymbol{\rho}(x)(a^{ij}(x)-a^{ij}(z))\partial_{x_i}\partial_{x_j}(\eta(x)H(x,z))\,dx\\ &+\int_U \boldsymbol{\rho}(x)b^i(x)\partial_{x_i}(\eta(x)H(x,z))\,dx + \int_U \boldsymbol{\rho}(x)c(x)\eta(x)H(x,z)\,dx. \end{aligned}$$

By using this representation one can prove the following sufficient condition for the global boundedness of $\boldsymbol{\rho}$ (previously only known in case of a Sobolev differentiable matrix A).

Theorem 6 *Suppose that a bounded measure μ satisfies the equation $L^*_{A,b,c}\mu = 0$, where A has a modulus of continuity on $\mathbb{R}^d$ satisfying Dini's condition, A and A^{-1} are uniformly bounded and $|b|, c \in L^q(|\mu|)$ with some $q > d$. Then μ possesses a uniformly bounded density.*

Proof We apply representation (6) and the following result from [54]: if nonnegative measurable functions u, K and f on $\mathbb{R}^d$ are such that

$$u, K \in L^1,\ f \in L^p + L^\infty,\ 1 < p \le +\infty,$$

i.e., $f = f_1 + f_2$, where $f_1 \in L^p$, $f_2 \in L^\infty$, and almost everywhere

$$u \le K * u + f,$$

then $u \in L^p + L^\infty$. We apply this result with $p = +\infty$ and $f = 2$. The representation of $\eta(z)\rho(z)$ will be used with a function η of the form $\eta(x) = \eta_0(z - x)$, where $\eta_0 \in C_0^\infty(\mathbb{R}^d)$ has support in the unit ball U and $\eta_0(0) = 1$. In that case in the left-hand side of the integral representation (6) we obtain $\rho(z)$. We observe that

$$|\partial_{x_i}\partial_{x_j} H(x,z)| \le C|x - z|^{-d}, \quad |\partial_{x_i} H(x,z)| \le C|x - z|^{1-d}.$$

Therefore,

$$|(a^{ij}(x) - a^{ij}(z))\partial_{x_i}\partial_{x_j}(\eta(x)H(x,z))| \le C\omega(|x - z|)|x - z|^{-d},$$

$$\int_{\mathbb{R}^d} |\rho(x)|\,|b^i(x)|\,|\partial_{x_i}(\eta(x)H(x,z))|\,dx \le \|b^i\|_{L^q(|\mu|)}\|\partial_{x_i}(\eta(x)H(x,z))\|_{L^{q/(q-1)}(|\mu|)}.$$

The second factor on the right is dominated by a constant multiplied by

$$\Bigl(\int_{\mathbb{R}^d} |x - z|^{(1-d)q/(q-1)}\,|\rho(x)|\,dx\Bigr)^{(q-1)/q} \le 1 + C\int_{\mathbb{R}^d} |x - z|^{(1-d)q/(q-1)}\,|\rho(x)|\,dx.$$

There is a similar estimate for the integral with the function c. Hence we obtain a bound on $|\rho(z)|$ of the form

$$|\rho(z)| \le C|\rho| * K(z) + 2,$$

where

$$K(x) = (|x|^{2-d} + |x|^{1-d} + |x|^{(1-d)q/(q-1)} + \omega(|x|)|x|^{-d})I_U(x).$$

Under our assumptions the function K is integrable over the whole space (equivalently, over U): the term with ω is taken care by Dini's condition, the term with $|x|$ to the power $(1 - d)q/(q - 1)$ is integrable since $q > d$, hence $q/(q - 1) > d/(d - 1)$, which yields that $(1 - d)q/(q - 1) > -d$, hence the desired integrability follows by passing to polar coordinates.

It is worth noting that the first results on the continuity of solutions to double divergence form equations were obtained by Sjögren [53, 54]: in the former paper the Hölder continuity of solutions was established in the case of the Hölder continuity of all coefficients and in the latter the continuity was proved in the case of locally bounded coefficients and the diffusion matrix satisfying the Dini condition. In [10] the Hölder continuity was established for drifts that are locally integrable to some power larger than d, but the matrix A was required to belong to the local Sobolev class $W^{p,1}_{loc}$ with $p > d$, which is stronger than the Hölder continuity. Even in the case of zero drift, merely the continuity (and nondegeneracy) of the diffusion matrix is not sufficient for the local boundedness of the density of a solution, as the known example from paper [5] shows.

Therefore, we have the following rather complete picture of integrability and continuity of densities of solutions to equation $L^*_{A,b}\mu = 0$ of the double divergence form in case of the matrix A of low regularity with locally bounded $1/\det A$.

- If $\mu \geq 0$, then a solution density exists and belongs to $L^{d/(d-1)}_{loc}$ without any other conditions (one only needs the inclusions $a^{ij}, b^i \in L^1_{loc}(\mu)$, no integrability of the coefficients with respect to Lebesgue measure is needed). However, the question about the existence of densities of signed solutions remains open.
- If we are given certain additional information about eigenvalues of $A(x)$ (see details in [27]), then we can raise the order up to $d/(d-2)$, but not more: even for $A = \mathrm{I}$ one cannot get $d/(d-2)$ if b is not integrable enough, as Example 2 below shows.
- If A is bounded and belongs to *VMO* on balls, say, is continuous, then also in the case of signed solutions there is a solution density $\boldsymbol{\rho}$ of class L^r_{loc} for every $r < d/(d-1)$, and if $|b| \in L^q_{loc}$ with some $q > d$, then this is true for all $r \in [1, +\infty)$. In case of locally bounded coefficients we have $\exp(\delta|\boldsymbol{\rho}|) \in L^1_{loc}$ with some $\delta > 0$. However, even for continuous A and $b = 0$ the density can be locally unbounded.
- If A has a modulus of continuity satisfying the Dini condition and $|b| \in L^q_{loc}$ with some $q > d$, then the density $\boldsymbol{\rho}$ has a continuous version.
- If A is locally Hölder continuous and $|b| \in L^q_{loc}$ with some $q > d$, then the density $\boldsymbol{\rho}$ is also locally Hölder continuous of the same order. If, in addition, the coefficients a^{ij} belong to the Sobolev class $W^{q,1}_{loc}$, then the density $\boldsymbol{\rho}$ also does.

In all these assertions the indicated bounds on orders of integrability of b and $\boldsymbol{\rho}$ are sharp.

Let us consider an example showing that even for the best possible identity diffusion matrix the order of integrability of the solution density can be low (strictly less than $d/(d-2)$) if the order of the integrability of the drift is low.

Example 2 The bounded measure μ with density $\boldsymbol{\rho}(x) = |x|^{2-d}$ in the unit ball $\boldsymbol{\Omega}$ in $\mathbb{R}^d$ with $d > 2$ satisfies the equation $L^*\mu = 0$ with the unit diffusion coefficient $A = \mathrm{I}$ and drift

$$b(x) = \frac{\nabla \boldsymbol{\rho}(x)}{\boldsymbol{\rho}(x)} = (2-d)|x|^{-2}x,$$

since $\nabla\boldsymbol{\rho}(x) = (2-d)|x|^{-d}x$, hence the integral of $\boldsymbol{\rho}\Delta\varphi$ equals the integral of $-\langle b, \nabla\varphi\rangle\boldsymbol{\rho}$ for all smooth compactly supported functions φ. Here the function $|b(x)| = (2-d)/|x|$ belongs to $L^r(\boldsymbol{\Omega})$ whenever $r < d$, but the function $\boldsymbol{\rho}^{d/(d-2)}$ is not integrable in a neighborhood of zero (we have $\boldsymbol{\rho} \in L^p(\boldsymbol{\Omega})$ for all $p < d/(d-2)$).

The case of a Sobolev differentiable matrix A is considered in the next section. We shall see there that the order d of local integrability of b is indeed critical for regular A.

3 The Case of a Sobolev Differentiable Diffusion Matrix

We now proceed to the case of Sobolev differentiable A.

Theorem 7 *Suppose that a signed measure μ satisfies Eq. (4), where the diffusion coefficient A satisfies the second condition in* (A1), *but the first condition is replaced with a stronger one: $a^{ij} \in W_{loc}^{p,1}(\boldsymbol{\Omega})$ with some $p > d$, and the coefficients b and c satisfy conditions* (B1) *or conditions* (B1μ). *Then μ has a density $\boldsymbol{\rho} \in W_{loc}^{p,1}(\boldsymbol{\Omega})$, hence this density admits a locally Hölder continuous version.*

For a proof, see [10] or [13, Chap. 1].

In case of a Sobolev differentiable density $\boldsymbol{\rho}$ our double divergence form equation can be rewritten as a classical divergence form equation

$$\partial_{x_i}(a^{ij}\partial_{x_j}\boldsymbol{\rho}) + \partial_{x_i}((\partial_{x_j}a^{ij} + b^i)\boldsymbol{\rho}) + c\boldsymbol{\rho} = 0.$$

For such equations there is a well-developed theory. In particular, if A and b are more regular, say, A is twice differentiable and b is once differentiable, then our equation can be rewritten in the classical direct form and studied in the framework of the classical theory (see [42, 43]). In particular, for infinitely differentiable A and b in case of nondegenerate A the solution density is also infinitely differentiable.

Let us also mention a result important for infinite-dimensional equations. Its main feature is that the hypotheses are dimension-free and the required integrability of b^i is much less than in the previous theorem. However, a natural local analog of this result is false.

Theorem 8 *Suppose that a probability measure μ on $\mathbb{R}^d$ satisfies the equation $L_{\mathrm{I},b}^*\mu = 0$ with $|b| \in L^2(\mu)$. Then μ has a density $\boldsymbol{\rho}$ such that $\sqrt{\boldsymbol{\rho}} \in W^{2,1}(\mathbb{R}^d)$ and*

$$\int_{\mathbb{R}^d} \frac{|\nabla\boldsymbol{\rho}|^2}{\boldsymbol{\rho}^2}\,\boldsymbol{\rho}\,dx \le \int_{\mathbb{R}^d} |b|^2\,d\mu.$$

In addition, the vector field $\nabla\boldsymbol{\rho}/\boldsymbol{\rho}$ is the orthogonal projection of b to the closure of gradients $\nabla\varphi$ of functions of class $C_0^\infty(\mathbb{R}^d)$ in the space $L^2(\mu, \mathbb{R}^d)$ of vector-valued mappings.

There is also an analogous result for more general equations $L_{A,b}^*\mu = 0$ with non-constant A. For proofs, see [9, 15] or [13, Chap. 3].

However, it is not known whether an analogous estimate holds for b in $L^p(\mu)$. Certain positive results in this direction employ restrictive additional hypotheses (see [11, 13, 48]). Let us mention a result that gives the global Sobolev regularity (see [13, Theorem 3.2.1]).

Theorem 9 *Suppose that a probability measure μ on $\mathbb{R}^d$ satisfies the equation $L_{A,b}^*\mu = 0$ with uniformly bounded A and A^{-1} such that A is Lipschizian on $\mathbb{R}^d$. Assume also that $|b| \in L^p(\mu)$ with some $p > d$. Then μ has a density $\boldsymbol{\rho} \in W^{p,1}(\mathbb{R}^d)$.*

This theorem is not trivial even for $A = \mathrm{I}$, but unfortunately the hypotheses that $p > d$ is not dimension-free.

Sufficient conditions for differentiability of solutions to stationary equations with respect to a parameter are obtained in [28].

4 Harnack's Inequality and Lower and Upper Bounds

We have seen in Theorem 6 that in case of a uniformly bounded A and A^{-1} with Dini's condition for A on the whole space and b in $L^p(|\mu|)$ with some $p > d$, the solution density is uniformly bounded. A closer look at the proofs enables one to get a more explicit uniform bound on $\|\rho\|_\infty$ in terms of the coefficients.

On the other hand, as concerned lower bounds, it is shown in [27] that if A satisfies Dini's condition and is nondegenerate and b is locally bounded, then the continuous version of a density of any nonnegative nonzero solution is positive and a variant of Harnack's inequality is established for it. In paper [45], the question was raised about the validity of Harnack's inequality for all nonzero nonnegative solutions and a positive answer was given in the case of identically zero low order terms. For more regular A, namely, of class $W^{p,1}_{loc}(\mathbb{R}^d)$ with $p > d$ (by the Sobolev embedding theorem, this condition is stronger than the Hölder continuity, hence is stronger than the Dini condition), this is true under a weaker assumption about b: it suffices that $b^i \in L^p_{loc}(\mathbb{R}^d)$.

It would be interesting to find minimal conditions on A under which the density ρ is locally separated from zero (i.e., $\inf_K \rho > 0$ on each compact set K). There is an example in [29] of a uniformly elliptic measurable matrix A and a nonzero nonnegative solution ρ to the equation $\partial_{x_i}\partial_{x_j}(a^{ij}\rho) = 0$ such that the function $1/\rho$ is not locally integrable on any ball (in this example ρ is the Green's function for the direct equation with a fixed value of the first variable).

For $A = \mathrm{I}$ the positivity of nonnegative solution densities (i.e., the local separation from zero in the sense explained above, which is equivalent to the absence of zeros for continuous nonnegative densities) is not ensured by conditions of the sort $|b| \in L^p_{loc}(\mu)$ (it is known that the local integrability of $|b|^{d+\varepsilon}$ with respect to Lebesgue measure implies positivity, see [10] or [13, Corollary 1.7.2]). However, it was shown in [1] that the stronger condition $\exp(\delta|b|) \in L^1_{loc}(\mu)$ with $\delta > 0$ yields that the continuous version of a nonzero nonnegative density is positive. Moreover, a lower bound on the density is obtained in [1] (see also [13]).

Under some additional assumptions, lower and upper bounds on densities of positive solutions are obtained in [11, 12, 18, 19, 48], and discussed in more detail in [13].

Let us mention a result that gives two-sided bounds on solution densities in terms of the equation coefficients.

Theorem 10 *Suppose that $A(x)$, $A(x)^{-1}$, $\nabla a^{ij}(x)$ are uniformly bounded and there are numbers $C, \beta > 0$ such that*

$$|b(x)| \le C|x|^\beta + C.$$

Then there is a number $K > 0$ such that

$$\varrho(x) \ge \varrho(0)\exp\{-K(1+|x|^{\beta+1})\}.$$

If, in addition,

$$\limsup_{|x|\to\infty} |x|^{-\beta-1}\langle b(x), x\rangle < 0,$$

then there are numbers $K_1, K_2 > 0$ such that

$$\exp\{-K_1(1+|x|^{\beta+1})\} \le \varrho(x) \le \exp\{-K_2(1+|x|^{\beta+1})\}.$$

Certainly, this two-sided estimate is not an asymptotic equivalence, but it is rather sharp. For example, the standard Gaussian density satisfies the equation with the drift $b(x) = -x$, which corresponds to $\beta = 1$.

5 Existence of Probability Solutions

Even for smooth coefficients equation (1) may fail to have nontrivial solutions on the whole space. For example, the simplest equation $\mu'' = 0$ on the real line has no non-zero solution of bounded variation: its solutions are given by densities that are affine functions. Locally bounded solutions exist under rather broad assumptions. For example, if $a^{ij} \in W^{p,1}_{loc}(\mathbb{R}^d)$ with some $p > d$, $1/\det A$ is locally bounded and $b^i \in L^p_{loc}(\mathbb{R}^d)$, then there is a positive continuous function ϱ of class $W^{p,1}_{loc}(\mathbb{R}^d)$ satisfying equation (3), see [13, 21].

The existence of solutions in the class of probability measures is ensured by the so-called Lyapunov functions. This condition goes back to Hasminskii's work [36] and was later extended to more general cases (see [13, 16]).

Theorem 11 *Suppose that $a^{ij} \in W^{p,1}_{loc}(\mathbb{R}^d)$ and $b^i \in L^p_{loc}(\mathbb{R}^d)$ with some $p > d$ and that $1/\det A$ is locally bounded. Assume, in addition, that there exist a continuous function $V \in W^{d,2}_{loc}(\mathbb{R}^d)$ and numbers $C, R > 0$ such that V is quasi-compact, i.e., the sets $\{V \le k\}$ are compact, and satisfies the estimate*

$$L_{A,b}V(x) \le -C \quad \textit{whenever} \quad |x| \ge R. \tag{7}$$

Then the equation $L^*_{A,b}\mu = 0$ has a solution in the class of probability measures.

We shall see in the next section that in this theorem there is only one solution in the class of probability measures.

A typical Lyapunov function which works in many examples is just the square of the norm. By using this function and its powers one can obtain results of the following type (see [16] and [13, Chap. 2]).

Corollary 1 *The assertion of the previous theorem is true if A is a locally Lipschitzian uniformly bounded mapping to the space of positive operators on $\mathbb{R}^d$ and $b = (b^i)$ is a mapping such that $b^i \in L^p_{loc}(\mathbb{R}^d)$, where $p > d$, and there exist positive numbers R and C such that*

$$\langle b(x), x\rangle \le -C - \sup_y \operatorname{trace} A(y) \quad \textit{whenever } |x| \ge R.$$

Corollary 2 *Let $A = (a^{ij})$ be a continuous mapping on $\mathbb{R}^d$ with values in the space of nonnegative symmetric linear operators on $\mathbb{R}^d$ and let b be a Borel vector field on $\mathbb{R}^d$. Suppose that there exists a quasi-compact function $V \in C^2(\mathbb{R}^d)$ with (7). Then the following assertions are true.*

(i) If b is continuous, then there exists a probability measure μ satisfying the equation $L^*_{A,b}\mu = 0$.
(ii) If $\det A > 0$ and b is locally bounded, then there exists a probability measure μ that has a density of class $L^{d/(d-1)}_{loc}(\mathbb{R}^d)$ and satisfies the equation $L^*_{A,b}\mu = 0$.

The following question remains open: can it happen that there are no solutions in the class of probability measures, but there is a non-zero signed solution of bounded variation on the whole space?

We now show that the existence of a quasi-compact Lyapunov function is not necessary for the existence of a unique probability solution. Let $d = 1$, $A = \mathrm{I}$ and

$$b(x) = -x + e^{x^2/2}.$$

Then $\rho(x) = e^{-x^2/2}$ is the unique probability solution of the equation $(\rho' - b\rho)' = 0$. However, there is no function $V \in W^{1,2}(\mathbb{R})$ such that $V \ge 0$, $\lim_{|x|\to\infty} V(x) = +\infty$ and for all sufficiently large x the inequality $L_{1,b}V(x) \le -\gamma$ holds with some number $\gamma > 0$. Indeed, suppose that for $|x| > R > 0$ we have $V''(x) + b(x)V'(x) \le -\gamma$. Let $x > R$ and

$$h(x) = \int_R^x \exp\biggl(\int_0^s b(t)\,dt\biggr)\,ds.$$

Then for $x > R$ we have

$$V'(x) \le \bigl(c - \gamma h(x)\bigr)/h'(x), \quad c = V'(R)h'(R),$$

which follows by the inequality $(V'h')' \le \gamma h'$, which holds by the equality $bh' = h''$ and the supposed inequality for V. We observe that $h(+\infty) = +\infty$, because we have

$b(x) \geq \delta e^{x^2/4}$ if $x > 0$. Hence there is a number $R_1 > R$ such that for all $x > R_1$ we obtain that $c - \gamma h(x) \leq 0$. Therefore, $V'(x) \leq 0$, which is impossible for a quasi-compact function.

A quasi-compact function V of class C^2 (or of class $W^{1,2}_{loc}(\mathbb{R}^d)$) for which

$$L_{A,b}V \leq -C$$

outside a ball is called a weak Lyapunov function. If we have

$$\lim_{|x|\to\infty} L_{A,b}V(x) = -\infty,$$

then V is called a strong Lyapunov function.

It is easy to give an example of a weak Lyapunov function that is not strong. For example, let $A = 1$, $b(x) = -x^3$, and $V(x) = |x|^{-1}$ if $|x| \geq 1$. We are unaware of examples of operators for which there is a weak Lyapunov function, but there is no strong one. If the diffusion coefficient A is bounded and an unbounded weak Lyapunov function V is Lipschitzian, then V^2 is a strong Lyapunov function, since

$$L_{A,b}(V^2) = 2VL_{A,b}V + 2\langle A\nabla V, \nabla V\rangle \leq -V|L_{A,b}V|$$

outside a ball.

Under rather broad assumptions the existence of a probability solution implies the existence of a Lyapunov function; see [22] and [13, Proposition 5.3.9] for the proof of the following result (the corresponding hypotheses are fulfilled, in particular, if $A = I$ and b has an at most linear growth).

Theorem 12 *Let μ be a probability solution of the equation $L^*_{A,b}\mu = 0$ on $\mathbb{R}^d$, where $a^{ij} \in C(\mathbb{R}^d) \cap W^{p,1}_{loc}(\mathbb{R}^d)$ with some $p > d$,* $\det A > 0$*, $b^i \in L^p_{loc}(\mathbb{R}^d)$, and*

$$\frac{\operatorname{trace} A(x)}{1+|x|^2}, \ \frac{|b(x)|}{1+|x|} \in L^1(\mu).$$

Then there is a Lyapunov function $V \in W^{p,2}_{loc}(\mathbb{R}^d)$ such that $V(x) \to +\infty$ and $L_{A,b}V(x) \to -\infty$ as $|x| \to \infty$ (hence V is a strong Lyapunov function).

We shall see in the next section that Lyapunov functions are connected also with uniqueness of probability solutions.

6 Uniqueness Problems

Since any multiple of a solution to a homogeneous equation is again a solution, the uniqueness problem must be posed in a reasonable class of solutions. Here we discuss uniqueness of probability solutions and uniqueness up to a multiple in the class of bounded solutions on the whole space.

In general, there is no uniqueness. For example, for $A = 1$ on the real line the probability densities $\rho_1(x) = x^2 e^{-x^2}$ and $\rho_2(x) = 2I_{[0,+\infty)}(x)\rho_1(x)$ satisfy the equation $\rho'' - (b\rho)' = 0$ with $b(x) = 2x^{-1} - 2x$; moreover, b is globally integrable with respect to both solutions.

One might hope that for smooth b the situation will be different. Indeed, on the real line this helps. However, for any $d > 1$ there is an infinitely differentiable mapping b such that the equation $L^*_{\mathrm{I},b}\mu = 0$ has several probability solutions (see [23, 24]), moreover, there are examples with infinitely many linearly independent probability solutions (see [51]). Let us mention one explicit example on the plane:

$$b_1(x, y) = -x - 2ye^{(x^2-y^2)/2}, \quad b_2(x, y) = -y - 2xe^{(y^2-x^2)/2}.$$

One can write explicitly two different probability solutions: one is the standard Gaussian density ρ and the other one is given by density

$$v(x, y) = [c\Phi(x) + c\Phi(y)]\rho(x, y),$$

where $\Phi'(s) = e^{-s^2/2}$. A detailed discussion of such examples is given also in [13].

Let us give sufficient conditions for uniqueness (see [13, Chap. 4]).

Theorem 13 *Suppose that* $a^{ij} \in W^{p,1}_{loc}(\mathbb{R}^d)$ *and* $b^i \in L^p_{loc}(\mathbb{R}^d)$ *with some* $p > d$ *and that* $1/\det A$ *is locally bounded. Assume, in addition, that there exist a number* C *and a positive function* $V \in C^2(\mathbb{R}^d)$ *such that* $\lim_{|x|\to\infty} V(x) = +\infty$ *and*

$$L_{A,b}V(x) \le CV(x) \quad \forall x \in \mathbb{R}^d.$$

Then the equation $L^*_{A,b}\mu = 0$ *can have at most one probability solution.*

Hence if $L_{A,b}V(x) \le -1$ outside a ball, there exists a unique probability solution.

In the next theorem a sufficient condition is expressed in terms of a given probability solution.

Theorem 14 *Let* μ *be a probability measure satisfying the equation* $L^*_{A,b}\mu = 0$, *where* A *and* b *satisfy the hypotheses of the previous theorem. Suppose that*

$$\frac{a^{ij}}{1 + |x|^2}, \ \frac{b^i}{1 + |x|} \in L^1(\mu).$$

Then μ *is the only probability solution to the above equation. In particular, this is the case if* $a^{ij}, b^i \in L^1(\mu)$.

Obviously, the additional condition in the last theorem is fulfilled if the coefficients are of linear growth (say, are uniformly continuous on the whole space).

It is not known whether in case $A = \mathrm{I}$ and smooth b the simplex of all probability solutions can be finite-dimensional not reducing to a singleton (in all known examples of non-uniqueness it is infinite-dimensional).

Uniqueness (up to a constant factor) in the class of solutions of bounded variation was studied in [7, 8, 20].

Some relation to uniqueness has the following problem: suppose that $A = \mathrm{I}$ and $b = \nabla\Phi$ with some smooth function Φ. If a probability solution exists, must it be of the form $ce^{-\Phi}$? Surprisingly enough, the answer is negative even in case $d = 1$ (when a probability solution is unique). Here is a simple counterexample:

$$\Phi(x) = -\frac{x^2}{2} + \int_0^x e^{s^2/2}\, ds.$$

The corresponding equation has a unique probability solution: the standard Gaussian density, but e^{Φ} is not integrable. If e^{Φ} is integrable, then the answer is positive. For other sufficient conditions, see [6, 13].

The uniqueness problem for solutions to stationary equations is closely connected with the uniqueness problem for invariant probability measures of semigroups. Let us recall that a bounded Borel measure μ on $\mathbb{R}^d$ is called invariant for a bounded linear operator T on the space $B_b(\mathbb{R}^d)$ of bounded Borel functions on $\mathbb{R}^d$ if

$$\int_{\mathbb{R}^d} Tf\, d\mu = \int_{\mathbb{R}^d} f\, d\mu \quad \forall\, f \in B_b(\mathbb{R}^d). \tag{8}$$

Similarly one defines invariance for a semigroup of operators T_t on $B_b(\mathbb{R}^d)$.

Under the assumptions of Theorem 11, according to which there exists a probability solution μ to the equation $L^*_{A,b}\mu = 0$ (this solution is unique by Theorem 13), there is a semigroup $\{T_t\}_{t\geq 0}$ of bounded operators on $B_b(\mathbb{R}^d)$ such that it extends to a strongly continuous semigroup on $L^1(\mu)$ whose generator extends $L_{A,b}$ on $C_0^\infty(\mathbb{R}^d)$. The latter means that the domain of the generator contains $C_0^\infty(\mathbb{R}^d)$ and the generator coincides with $L_{A,b}$ on this class. It is known that in case of a nondegenerate A of class $W^{p,1}_{loc}(\mathbb{R}^d)$ with $p > d$ and $b^i \in L^p_{loc}(\mathbb{R}^d)$, for every probability solution μ to the equation $L^*_{A,b}\mu = 0$ (even if it is not unique), there is always a strongly continuous sub-Markov semigroup $\{T_t\}^{\mu}_{t\geq 0}$ on $L^1(\mu)$ whose generator extends $L_{A,b}$ on $C_0^\infty(\mathbb{R}^d)$ (see [13, 23, 24]) and such that μ is sub-invariant with respect to the operators T_t^{μ}, i.e., in (8) there holds the inequality $\leq$ in place of the equality for $f \geq 0$. Moreover, μ is invariant with respect to this semigroup precisely when it is the only probability solution to the stationary equation (it is also the unique invariant probability measure for the semigroup). For $A = \mathrm{I}$ and smooth b the uniqueness of invariant probability measures for the semigroup was shown by Varadhan [56], who also posed the problem of generalizing his result to non-smooth drifts. A positive solution was obtained in [2], where the drift was assumed to be in $L^p_{loc}(\mathbb{R}^d)$ with $p > d$. The case of less regular coefficients A and b has been less studied so far. However, the aforementioned positivity and continuity of densities in case of A satisfying Dini's condition leads to the following result (note also that the case of Hölder continuous coefficients was considered in [44]).

Theorem 15 *Suppose that on every ball A has a modulus of continuity satisfying Dini's condition and is nondegenerate. Suppose also that b is locally bounded and there is a probability measure μ satisfying the equation $L^*_{A,b}\mu = 0$. Finally, assume that there is a Markov semigroup of operators T_t on $B_b(\mathbb{R}^d)$ that has μ as an invariant measure and extends to a strongly continuous semigroup on $L^1(\mu)$ whose generator extends $L_{A,b}$ on $C_0^\infty(\mathbb{R}^d)$. Then μ is the only absolutely continuous invariant probability measure for $\{T_t\}_{t\geq 0}$.*

Proof Let ν be another absolutely continuous probability measure invariant for $\{T_t\}_{t\geq 0}$. According to [13, Lemma 5.1.4 and Proposition 5.3.6], the measure $|\mu - \nu|$ is invariant for the semigroup and satisfies the equation $L^*_{A,b}|\mu - \nu| = 0$. In addition, ν also satisfies this equation. We know that under the stated assumptions the measures μ, ν and $|\mu - \nu|$ possess positive continuous densities with respect to Lebesgue measure. However, this is impossible for $|\mu - \nu|$. Indeed, if $\mu = \rho_\mu\, dx$, $\nu = \rho_\nu\, dx$, then $|\mu - \nu| = |\rho_\mu - \rho_\nu|$, but $|\rho_\mu - \rho_\nu|$ cannot be positive everywhere since $\rho_\mu - \rho_\nu$ is continuous and its integral over the space vanishes.

Under mild additional assumptions any invariant measure (possibly, signed) for $\{T_t\}_{t\geq 0}$ is absolutely continuous. However, some conditions are needed in order to connect this semigroup on $B_b(\mathbb{R}^d)$ and on $L^1(\mu)$: in the latter case different versions of $T_t f$ can be used. For example, the Ornstein–Uhlenbeck semigroup with the only invariant standard Gaussian measure γ can be redefined by letting $T_t f(0) = 0$ for all functions $f \in L^1(\gamma)$, which creates artificially "invariant" Dirac's measure at the origin (but this measure will not be invariant for the usual version of the semigroup).

7 The Infinite-Dimensional Case

The infinite-dimensional case, which in many respects stimulated the investigation of Fokker–Planck–Kolmogorov equations with growing coefficients, still remains much less studied. There are many interesting and challenging open problems in this area both in a rather abstract framework and for specific equations (for example, related to SPDEs). A more detailed discussion and further references can be found in [13, 17, 30]. Here we mention only one typical case: we take the so-called standard Gaussian measure γ on $\mathbb{R}^\infty$, which is the countable power of the standard Gaussian measure on the real line, and observe that it satisfies the equation $L_0^*\gamma = 0$ with the second order operator (the Ornstein–Uhlenbeck operator)

$$L_0 f = \sum [\partial^2_{x_i} f - x_i \partial_{x_i} f]$$

defined originally on the class $\mathcal{FC}$ of all cylindrical functions f of the form $f(x) = f_0(x_1, \dots, x_n)$, where $f_0 \in C_b^\infty(\mathbb{R}^n)$. The interpretation of the equation is similar to the finite-dimensional case:

$$\int L_0 f \, d\gamma = 0 \quad \forall f \in \mathscr{F}\mathscr{C}.$$

Now take $v = (v^i) \colon \mathbb{R}^\infty \to H = l^2$ and consider a first order perturbation

$$Lf = \sum [\partial_{x_i}^2 f - x_i \partial_{x_i} f + v^i \partial_{x_i} f], \quad f \in \mathscr{F}\mathscr{C},$$

i.e., $b(x) = -x + v(x)$, $b^i(x) = -x_i + v^i(x)$. As above, we can consider the equation

$$L^* \mu = 0$$

understood as the identity

$$\int Lf \, d\mu = 0 \quad \forall f \in \mathscr{F}\mathscr{C},$$

provided that $b^i \in L^1(\mu)$.

In the finite-dimensional case, if v is bounded, there is a unique probability solution to this equation. In addition, this solution has a Sobolev density.

The situation is partially similar in infinite-dimensions. It was proved by Shigekawa [52] that if $\|v\|_{l^2}$ is bounded or, more generally, belongs to $L^2(\gamma)$, then there is a probability solution μ to the equation $L^*\mu = 0$ and μ has a density with respect to γ belonging to some Sobolev class (defined similarly to the finite-dimensional case, but with γ in place of Lebesgue measure). Answering a question posed by Shigekawa, the following result was obtained in [15].

Theorem 16 *Let μ be a Borel probability measure such that x_i, $|v|_H \in L^2(\mu)$ and $L^*\mu = 0$. Then μ is absolutely continuous with respect to γ and $d\mu/d\gamma = \psi^2$, where ψ belongs to the Sobolev class $W^{2,1}(\gamma)$ with respect to γ.*

The analogous question in case $|v|_H, x_i \in L^1(\gamma)$ remains open.

Similarly one defines elliptic equations for measures associated with more general operators on $\mathscr{F}\mathscr{C}$ of the form

$$Lf = \sum_{i,j} a^{ij} \partial_{x_i} \partial_{x_j} f + \sum_i b^i \partial_{x_i} f$$

with Borel measurable functions a^{ij} and b^i. Now, in the definition of the equation $L^*\mu = 0$, one has to assume that $a^{ij}, b^i \in L^1(|\mu|)$.

Suppose that a probability measure μ satisfies $L^*\mu = 0$. Then its projection μ_n to $\mathbb{R}^n$ satisfies the equation $L^*_{A_n,b_n}\mu_n = 0$ whose coefficients a_n^{ij} and b_n^i are the conditional expectations of the original coefficients a^{ij} and b^i with respect to the projection π_n to $\mathbb{R}^n$, which follows immediately from the definition of the conditional expectation expressed equivalently by the identity

$$\int_{\mathbb{R}^n} a_n^{ij} f \, d\mu_n = \int_{\mathbb{R}^\infty} a^{ij} f \, d\mu, \quad f \in \mathcal{FC}.$$

By the property of conditioning, $a_n^{ij}, b_n^i \in L^1(\mu_n)$ once $a^{ij}, b^i \in L^1(\mu)$; more generally, $a_n^{ij}, b_n^i \in L^p(\mu_n)$ if $a^{ij}, b^i \in L^p(\mu)$. However, generally speaking, nothing else can be said about a_n^{ij}, b_n^i. For example, if $a^{ij} = \delta^{ij}$ and b^i are polynomials, then also $a_n^{ij} = \delta^{ij}$, but no smoothness is given for b_n^i. If b^i is bounded, then b_n^i has the same bound, but more general bounds are not inherited by conditional expectations. This is one of the reasons why we study finite-dimensional equations with drifts about which it is only known that they are integrable with respect to solutions, but not with respect to Lebesgue measure.

Example 3 Suppose that μ is a probability measure satisfying the equation $L^*\mu = 0$ with $a^{ij} = \delta^{ij}$ and $b^i \in L^p(\mu)$ for all $p < \infty$. Then the projections μ_n possess positive continuous densities belonging to all Sobolev classes $W^{p,1}(\mathbb{R}^n)$.

Stationary equations on finite- and infinite-dimensional manifolds are studied in [25].

I am most grateful to the anonymous referee for a thorough reading and very useful corrections and suggestions.

This work was supported by the RFBR Grant 17-01-00662 and the CRC 701 at Bielefeld University.

References

1. Agafontsev, B.V., Bogachev, V.I., Shaposhnikov, S.V.: A condition for the positivity of the density of an invariant measure. Dokl. Akad. Nauk **438**(3), 295–299 (2011) (in Russian); English transl. Dokl. Math. **83**(3), 332–336 (2011)
2. Albeverio, S., Bogachev, V., Röckner, M.: On uniqueness of invariant measures for finite- and infinite-dimensional diffusions. Commun. Pure Appl. Math. **52**, 325–362 (1999)
3. Albeverio, S., Röckner, M.: Classical Dirichlet forms on topological vector spaces - construction of an associated diffusion process. Probab. Theory Related Fields **83**, 405–434 (1989)
4. Bauman, P.: Positive solutions of elliptic equations in nondivergent form and their adjoints. Ark. Mat. **22**, 153–173 (1984)
5. Bauman, P.: Equivalence of the Green's functions for diffusion operators in R^n: a counterexample. Proc. Am. Math. Soc. **91**, 64–68 (1984)
6. Bogachev, V.I., Kirillov, A.I., Shaposhnikov, S.V.: Invariant measures of diffusions with gradient drift. Dokl. Ross. Akad. Nauk. **434**(6), 730–734 (2010) (in Russian); English transl. Dokl. Math. **82**(2), 790–793 (2010)
7. Bogachev, V.I., Kirillov, A.I., Shaposhnikov, S.V.: On probability and integrable solutions to the stationary Kolmogorov equation. Dokl. Ross. Akad. Nauk **438**(2), 154–159 (2011) (in Russian); English transl. Dokl. Math. **83**(3), 309–313 (2011)
8. Bogachev, V.I., Kirillov, A.I., Shaposhnikov, S.V.: Integrable solutions of the stationary Kolmogorov equation. Dokl. Ross. Akad. Nauk **444**(1), 11–16 (2012) (in Russian); English transl. Dokl. Math. **85**(3), 309–314 (2012)
9. Bogachev, V.I., Krylov, N.V., Röckner, M.: Regularity of invariant measures: the case of non-constant diffusion part. J. Funct. Anal. **138**, 223–242 (1996)

10. Bogachev, V.I., Krylov, N.V., Röckner, M.: On regularity of transition probabilities and invariant measures of singular diffusions under minimal conditions. Commun. Partial Differ. Equ. **26**(11–12), 2037–2080 (2001)
11. Bogachev, V.I., Krylov, N.V., Röckner, M.: Elliptic equations for measures: regularity and global bounds of densities. J. Math. Pures Appl. **85**(6), 743–757 (2006)
12. Bogachev, V.I., Krylov, N.V., Röckner, M.: Elliptic and parabolic equations for measures. Uspehi Matem. Nauk **64**(6), 5–116 (2009) (in Russian); English transl. Russian Math. Surv. **64**(6), 973–1078 (2009)
13. Bogachev, V.I., Krylov, N.V., Röckner, M., Shaposhnikov, S.V.: Fokker–Planck–Kolmogorov Equations. American Mathematical Society, Providence, Rhode Island (2015)
14. Bogachev, V.I., Röckner, M.: Hypoellipticity and invariant measures for infinite dimensional diffusions. C. R. Acad. Sci. Paris **318**, 553–558 (1994)
15. Bogachev, V.I., Röckner, M.: Regularity of invariant measures on finite and infinite dimensional spaces and applications. J. Funct. Anal. **133**, 168–223 (1995)
16. Bogachev, V.I., Röckner, M.: A generalization of Khasminskii's theorem on the existence of invariant measures for locally integrable drifts. Teor. Verojatn. i Primen. **45**(3), 417–436 (2000); correction: ibid. **46**(3), 600 (2001) (in Russian); English transl. Theory Probab. Appl. **45**(3), 363–378 (2000)
17. Bogachev, V.I., Röckner, M.: Elliptic equations for measures on infinite dimensional spaces and applications. Probab. Theory Related Fields **120**, 445–496 (2001)
18. Bogachev, V.I., Röckner, M., Shaposhnikov, S.V.: Positive densities of transition probabilities of diffusion processes. Teor. Verojatn. i Primen. **53**(2), 213–239 (2008) (in Russian); English transl. Theory Probab. Appl. **53**(2), 194–215 (2009)
19. Bogachev, V.I., Röckner, M., Shaposhnikov, S.V.: Lower estimates of densities of solutions of elliptic equations for measures. Dokl. Ross. Akad. Nauk **426**(2), 156–161 (2009) (in Russian); English transl. Dokl. Math. **79**(3), 329–334 (2009)
20. Bogachev, V.I., Röckner, M., Shaposhnikov, S.V.: On uniqueness problems related to elliptic equations for measures. J. Math. Sci. (New York) **176**(6), 759–773 (2011)
21. Bogachev, V.I., Röckner, M., Shaposhnikov, S.V.: On positive and probability solutions of the stationary Fokker–Planck–Kolmogorov equation. Dokl. Akad. Nauk **444**(3), 245–249 (2012) (in Russian); English transl. Dokl. Math. **85**(3), 350–354 (2012)
22. Bogachev, V.I., Röckner, M., Shaposhnikov, S.V.: On existence of Lyapunov functions for a stationary Kolmogorov equation with a probability solution. Dokl. Ross. Akad. Nauk **457**(2), 136–140 (2014) (in Russian); English transl. Dokl. Math. **90**(1), 424–428 (2014)
23. Bogachev, V.I., Röckner, M., Stannat, W.: Uniqueness of invariant measures and maximal dissipativity of diffusion operators on L^1. In: Infinite Dimensional Stochastic Analysis (Proceedings of the Colloquium, Amsterdam, 11–12 February, 1999); Clément, Ph., den Hollander, F., van Neerven, J., de Pagter, B. (eds.): Royal Netherlands Academy of Arts and Sciences, pp. 39–54. Amsterdam (2000)
24. Bogachev, V.I., Röckner, M., Stannat, W.: Uniqueness of solutions of elliptic equations and uniqueness of invariant measures of diffusions. Matem. Sbornik **193**(7), 3–36 (2002) (in Russian); English transl.: Sbornik Math. **193**(7), 945–976 (2002)
25. Bogachev, V.I., Röckner, M., Wang, F.-Y.: Elliptic equations for invariant measures on finite and infinite dimensional manifolds. J. Math. Pures Appl. **80**, 177–221 (2001)
26. Bogachev, V.I., Shaposhnikov, S.V.: Integrability and continuity of densities of stationary distributions of diffusions. (in Russian) Doklady Akademii Nauk **469**(1), 7–12 (2016); English transl. Doklady Math. **94**(1), 355–360 (2016)
27. Bogachev, V.I., Shaposhnikov, S.V.: Integrability and continuity of solutions to double divergence form equations. Annali di Matematica **196**(5), 1609–1635 (2017)
28. Bogachev, V.I., Shaposhnikov, S.V., Veretennikov, AYu.: Differentiability of solutions of stationary Fokker–Planck–Kolmogorov equations with respect to a parameter. Discret. Contin. Dyn. Syst. **36**(7), 3519–3543 (2016)
29. Cerutti, M.C.: Integrability of reciprocals of the Green's function for elliptic operators: counterexamples. Proc. Am. Math. Soc. **119**(1), 125–134 (1993)

30. Da Prato, G.: Kolmogorov Equations for Stochastic PDEs. Birkhäuser, Basel (2004)
31. Eberle, A.: Uniqueness and Non-uniqueness of Singular Diffusion Operators. Lecture notes in mathematics, vol. 1718. Springer, Berlin (1999)
32. Escauriaza, L.: Weak type-(1, 1) inequalities and regularity properties of adjoint and normalized adjoint solutions to linear nondivergence form operators with VMO coefficients. Duke Math. J. **74**, 177–201 (1994)
33. Escauriaza, L.: Bounds for the fundamental solution of elliptic and parabolic equations in nondivergence form. Commun. Partial Differ. Equ. **25**(5–6), 821–845 (2000)
34. Fabes, E.B., Stroock, D.W.: The L^p-integrability of Green's functions and fundamental solutions for elliptic and parabolic equations. Duke Math. J. **61**, 977–1016 (1984)
35. Fokker, A.D.: Die mittlere Energie rotierender elektrischer Dipole im Strahlungsfeld. Ann. Phys. **348**, 810–820 (1914)
36. Hasminskii, R.Z.: Ergodic properties of reccurent diffusion processes and stabilization of the solution of the Cauchy problem for parabolic equations. Teor. Verojatn. Primen. **5**(2), 196–214 (1960) (in Russian); English tranls. Theory Probab. Appl. **5**, 179–196 (1960)
37. Hino, M.: Existence of invariant measures for diffusion processes on a Wiener space. Osaka J. Math. **35**(3), 717–734 (1998)
38. Kolmogoroff, A.N.: Über die analytischen Methoden in der Wahrscheinlichkeitsrechnung. Math. Ann. **104**, 415–458 (1931)
39. Kolmogoroff, A.N.: Zur Theorie der stetigen zufälligen Prozesse. Math. Ann. **104**, 149–160 (1933)
40. Kolmogoroff, A.N.: Zur Umkehrbarkeit der statistischen Naturgesetze. Math. Ann. **113**, 766–772 (1937)
41. Krylov, N.V.: A certain estimate from the theory of stochastic integrals. Teor. Verojatnost. i Primenen. **16**, 446–457 (1971) (in Russian); English transl.: Theor. Probab. Appl. **16**, 438–448 (1971)
42. Krylov, N.V.: Lectures on Elliptic and Parabolic Equations in Hölder Spaces. American Mathematical Society, Rhode Island, Providence (1996)
43. Krylov, N.V.: Lectures on Elliptic and Parabolic Equations in Sobolev Spaces. American Mathematical Society, Rhode Island, Providence (2008)
44. Lorenzi, L., Bertoldi, M.: Analytical Methods for Markov Semigroups. Chapman & Hall, CRC, Boca Raton (2007)
45. Mamedov, F.I.: On the Harnack inequality for the equation formally adjoint to a linear elliptic differential equation. Sibir. Matem. Zh. **33**(5), 100–106 (1992) (in Russian); English tranls.: Siber. Math. J. **33**(5), 835–841 (1992)
46. Maz'ya, V., McOwen, R.: Asymptotics for solutions of elliptic equations in double divergence form. Commun. Partial Differ. Equ. **32**(1–3), 191–207 (2007)
47. McOwen, R.: On elliptic operators in nondivergence and in double divergence form. In: Analysis, Partial Differential Equations and Applications. Operator Theory: Advances and Applications, vol. 193, pp. 159–169. Birkhäuser Verlag, Basel (2009)
48. Metafune, G., Pallara, D., Rhandi, A.: Global properties of invariant measures. J. Funct. Anal. **223**, 396–424 (2005)
49. Planck, M.: Über einen Satz der statistischen Dynamik und seine Erweiterung in der Quantentheorie. Sitzungber. Preussischen Akad. Wissenschaften, 324–341 (1917)
50. Röckner, M., Shin, J., Trutnau, G.: Non-symmetric distorted Brownian motion: Strong solutions, strong Feller property and non-explosion results. Discret. Contin. Dyn. Syst. Ser. B **21**(9), 3219–3237 (2016)
51. Shaposhnikov, S.V.: On nonuniqueness of solutions to elliptic equations for probability measures. J. Funct. Anal. **254**(10), 2690–2705 (2008)
52. Shigekawa, I.: Existence of invariant measures of diffusions on an abstract Wiener space. Osaka J. Math. **24**(1), 37–59 (1987)
53. Sjögren, P.: On the adjoint of an elliptic linear differential operator and its potential theory. Ark. Mat. **11**, 153–165 (1973)

54. Sjögren, P.: Harmonic spaces associated with adjoints of linear elliptic operators. Annal. Inst. Fourier **25**(3–4), 509–518 (1975)
55. Stannat, W.: (Nonsymmetric) Dirichlet operators on L^1: existence, uniqueness and associated Markov processes. Ann. Sc. Norm. Sup. Pisa Cl. Sci. (4) **28**(1), 99–140 (1999)
56. Varadhan, S.R.S.: Lectures on Diffusion Problems and Partial Differential Equations. Tata Institute of Fundamental Research, Bombay (1980)

Liouville Property of Harmonic Functions of Finite Energy for Dirichlet Forms

Masatoshi Fukushima

Abstract A quasi-regular Dirichlet form is said to have a Liouville property if any associated harmonic function of finite energy is constant. We first examine this property for the energy form $\mathscr{E}^\rho$ on $\mathbb{R}^n$ generated by a positive function ρ. We next make a general consideration on a regular, strongly local and transient Dirichlet form $\mathscr{E}$ and an associated time changed symmetric diffusion process $\check{X}$ with finite lifetime. We show that $\check{X}$ always admits its one-point reflection $\check{X}^*$ at infinity by constructing the corresponding regular Dirichlet form. We then prove that, if $\mathscr{E}$ satisfies the Liouville property, a symmetric conservative diffusion extension Y of $\check{X}$ is unique up to a quasi-homeomorphism, and in fact, a quasi-homeomorphic image of Y equals the one-point reflection $\check{X}^*$ of $\check{X}$ at infinity.

Keywords Liouville property · Energy form · Strongly local transient Dirichlet form · One-point reflection at infinity · Symmetric extension

AMS 2010 Subject Classification Primary 60J50 · Secondary 31C25 · 60J60

1 Introduction

We consider a locally compact separable metric space E and a positive Radon measure m on E with full support. Given a regular Dirichlet form $(\mathscr{E}, \mathscr{F})$ on $L^2(E; m)$ with an associated Hunt process $X = (X_t, \zeta, \mathbf{P}_x)$ on E, let $\mathscr{F}_e$ and $\mathscr{F}^{\text{ref}}$ be its *extended Dirichlet space* and its *reflected Dirichlet space*, respectively. Then $\mathscr{F} \subset \mathscr{F}_e \subset \mathscr{F}^{\text{ref}}$ and the inner product $\mathscr{E}$ is extended from $\mathscr{F}$ to both spaces [6]. The notions of the extended and reflected Dirichlet spaces were introduced by Silverstein in [25, 26], respectively, in the same year 1974, but the latter notion was reformulated by Z.-Q. Chen [4] later in 1992 and further extended to a quasi-regular Dirichlet form in [6].

M. Fukushima (✉)
Branch of Mathematical Science, Osaka University, Toyonaka, Osaka 560-8531, Japan
e-mail: fuku2@mx5.canvas.ne.jp

A. Eberle et al. (eds.), *Stochastic Partial Differential Equations and Related Fields*, Springer Proceedings in Mathematics & Statistics 229,
https://doi.org/10.1007/978-3-319-74929-7_2

Define the linear subspace $\mathcal{H}^*$ of $\mathcal{F}^{\rm ref}$ by

$$\mathcal{H}^* = \{u \in \mathcal{F}^{\rm ref} : \mathcal{E}(u, v) = 0 \quad \text{for any} \quad v \in \mathcal{F}_e\}.$$

$\mathcal{H}^*$ is the collection of X-harmonic functions u on E of finite energy $\mathcal{E}(u, u)$. We will be concerned with a specific *Liouville property*

$$\dim(\mathcal{H}^*) = 1 \tag{1}$$

of the form $\mathcal{E}$ and its probabilistic significance.

We first give two general remarks on the Liouville property (1). A Borel function h on E is said to be *X-harmonic* if it is specified and finite up to quasi equivalence and if for every relatively compact open subset $G \subset E$, $\mathbf{E}_x[|h(X_{\tau_G})|] < \infty$ and $h(x) = \mathbf{E}_x[h(X_{\tau_G}]$ for q.e. $x \in E$, where τ_G denotes the first exit time from G. By the next proposition, we only need to consider the transient form $\mathcal{E}$ to study the Liouville property (1).

Proposition 1.1 (i) *If $\mathcal{E}$ is irreducible and recurrent, then $\mathcal{E}$ enjoys the property* (1). (ii) *If $\mathcal{E}$ is transient and if any bounded X-harmonic function on E is constant, then $\mathcal{E}$ enjoys the property* (1).

Proof (i) Suppose $(\mathcal{E}, \mathcal{F})$ is recurrent. Then $\mathcal{F}^{\rm ref} = \mathcal{F}_e$ by [6, Theorem 6.3.2]. Further $u \in \mathcal{F}_e$, $\mathcal{E}(u, u) = 0$ implies that the level set $\{x \in E : u(x) = c\}$ is invariant for each constant c by [6, Lemma 6.7.3]. Hence (1) follows from the irreducibility of $\mathcal{E}$.

The assertion (i) also follows from the identity $\mathcal{F}^{\rm ref} = \mathcal{F}_e$ and a Poincaré type inequality for $(\mathcal{F}_e, \mathcal{E})$ established in [17, Theorem 4.8.2] in the recurrent case, which requires an additional Sobolev type inequality holding for $(\mathcal{E}, \mathcal{F})$ however.
(ii) In view of [6, Remark 6.2.2], it holds under the transience of $\mathcal{E}$ that

$$\mathcal{H}^* = \{h = \mathbf{E}_\cdot[\varphi] : \varphi \in \mathbf{N}\},$$

for the space $\mathbf{N}$ of terminal random variables φ specified by [6, (6.2.1)]. For $\varphi \in \mathbf{N}$, let $\varphi_n = ((-n) \vee \varphi \wedge n$. Then $h_n(x) = \mathbf{E}_x[\varphi_n]$ is a bounded X-harmonic function and converges as $n \to \infty$ to h q.e. on E, yielding the assertion (ii). □

For an Euclidean domain $D \subset \mathbb{R}^n$, the *Beppo Levi space* and the *Sobolev space of order* $(1, 2)$ are defined, respectively, by

$$\mathrm{BL}(D) = \{u \in L^2_{\rm loc}(D) : |\nabla u| \in L^2(D)\}, \quad H^1(D) = \mathrm{BL}(D) \cap L^2(D). \tag{2}$$

$\mathbf{D}(u, v)$ will denote the Dirichlet integral $\int_D \nabla u(x) \cdot \nabla v(x) dx$ of $u, v \in \mathrm{BL}(D)$. The space $\mathrm{BL}(D)$ is just the space of Schwartz distributions whose first order derivatives are in $L^2(D)$. It was introduced and profoundly studied by Deny-Lions [12] following the preceding works by Beppo Levi [21], Nikodym [24] and Deny [10]. This space was one of the original sources of the notion of the *Dirichlet space* introduced by

Beurling-Deny [2] in 1959 which was basically free from the choice of the underlying symmetrizing measure. Later on, the space $\mathrm{BL}(D)$ was designated as $L_2^1(D)$ by Maz'ja [23] and studied in a more general context of the spaces $L_p^\ell(D)$ for $p > 0$ and integers ℓ. However the space $\mathrm{BL}(D)$ bears its own independent potential theoretic and probabilistic significances from the beginning. See [3, 10, 11, 13, 15] in this connection.

Now suppose a domain $D \subset \mathbb{R}^n$ is either of continuous boundary or an extendable domain relative to $H^1(D)$. The symmetric form $\mathcal{E}$ with $\mathcal{D}(\mathcal{E}) = \mathcal{F}$ defined by

$$\mathcal{E} = \frac{1}{2}\mathbf{D}, \qquad \mathcal{F} = H^1(D), \tag{3}$$

is then a regular strongly local irreducible Dirichlet form on $L^2(\overline{D})$ and the associated diffusion X on $\overline{D}$ is by definition the *reflecting Brownian motion* (*RBM* in abbreviation). The extended Dirichlet space of $\mathcal{E}$ is denoted by $H_e^1(D)$ and called the *extended Sobolev space of order* 1. $\mathrm{BL}(D)$ is nothing but the reflected Dirichlet space of this form $\mathcal{E}$ [6, p. 273]. The space $\mathcal{H}^* = \mathrm{BL}(D) \ominus H_e^1(D)$ consists of those functions on D with finite Dirichlet integral such that they are not only harmonic on D in the ordinary sense but also their quasi continuous versions are harmonic with respect to the RBM Z on $\overline{D}$.

It was shown in [5, Theorem 3.5] that $\mathcal{E}$ fulfills the Liouville property (1) when $D \subset \mathbb{R}^n$ is a uniform domain in the sense of [27]. On the other hand, it is demonstrated in [8] that $\dim(\mathcal{H}^*) = N$ when $n \geq 3$ and D is a Lipschitz domain with N number of *Liouville branches* in the specific sense formulated there.

In the simplest case that $D = \mathbb{R}^n$ the whole space, $\mathcal{H}^*$ is just the space of harmonic functions on $\mathbb{R}^n$ with finite Dirichlet integrals. Brelot [3] first observed that the property (1) is valid, namely, any harmonic function on $\mathbb{R}^n$ with finite Dirichlet integral is constant. See [17, Example 1.5.3] in this connection. A simple question arises:

(Q) Is the property (1) still valid for the whole space $\mathbb{R}^n$ and for more general Dirichlet forms than $\frac{1}{2}\mathbf{D}$?

In Sects. 2 and 3, we shall consider a measurable function $\rho(x)$ on $\mathbb{R}^n$ such that

$$0 < \lambda_\ell \leq \rho(x) \leq \Lambda_\ell < \infty, \quad \text{for every } x \in B_\ell := \{|x| < \ell\}, \quad \ell > 0. \tag{4}$$

for constants λ_ℓ, Λ_ℓ depending on $\ell > 0$, and the associated spaces $\mathcal{F}^\rho$, $\mathcal{G}^\rho$ and form $\mathbf{D}^\rho$ defined respectively by

$$\mathcal{F}^\rho = \{u \in L^2(\mathbb{R}^n; \rho dx) \ : \ |\nabla u| \in L^2(\mathbb{R}^n; \rho dx)\}, \tag{5}$$

$$\mathcal{G}^\rho = \{u \in L^2_{\mathrm{loc}}(\mathbb{R}^n) \ : \ |\nabla u| \in L^2(\mathbb{R}^n; \rho dx)\}, \tag{6}$$

$$\mathbf{D}^\rho(u, v) = \int_{\mathbb{R}^n} \nabla u(x) \cdot \nabla v(x)\, \rho(x)\, dx. \tag{7}$$

In the next section, we show that the *energy form* $\mathcal{E}^\rho = (\mathbf{D}^\rho, \mathcal{F}^\rho)$ is a regular strongly local irreducible Dirichlet form on $L^2(\mathbb{R}^n; \rho dx)$ and the *weighted Beppo Levi space* $(\mathcal{G}^\rho, \mathbf{D}^\rho)$ is the reflected Dirichlet space of the energy form $\mathcal{E}^\rho$. If the constants λ_ℓ, Λ_ℓ are independent of $\ell > 0$, then the energy form $\mathcal{E}^\rho$ admits $H_e^1(\mathbb{R}^n)$ and $\mathrm{BL}(\mathbb{R}^n)$ as its extended Dirichlet space and reflected Dirichlet space, respectively, so that the answer to the question **(Q)** is affirmative.

In Sect. 3, we give also an affirmative answer to **(Q)** for the energy form $\mathcal{E}^\rho$ when $n \geq 2$ and $\rho(x)$ is any positive C^∞-function depending only on the radial part r of the variable $x \in \mathbb{R}^n$. Presently we have no example of the energy form $\mathcal{E}^\rho$ on $\mathbb{R}^n$ for $n \geq 2$ violating the Liouville property (1).

In Sects. 4, 5 and 6, we shall make a general consideration on a regular, strongly local and transient Dirichlet form $(\mathcal{E}, \mathcal{F})$ on $L^2(E; m)$ and an associated diffusion process $X = (X_t, \zeta, \mathbf{P}_x)$ on E. X_t approaches to the point ∂ at infinity of E as $t \uparrow \zeta$ [6, Sect. 3.5]. But the lifetime ζ of X could be infinite and so, in place of X, we consider its time-changed process $\check{X} = (\check{X}_t, \check{\zeta}, \mathbf{P}_x)$ by means of its positive continuous additive functional whose Revuz measure ν is a finite measure on E charging no $\mathcal{E}$-polar set with full quasi-support. $\check{X}$ is ν-symmetric. As we see in Sect. 4, the lifetime $\check{\zeta}$ of $\check{X}$ is finite $\mathbf{P}_x$-a.s. for q.e. $x \in E$ and $\check{X}_t$ approaches to ∂ as $t \uparrow \check{\zeta}$.

Therefore the boundary problem of $\check{X}$ at ∂ looking for all possible Markovian extensions of $\check{X}$ beyond $\check{\zeta}$ makes perfect sense. A strong Markov process Y on a Lusin space $\widehat{E}$ is said to be an *extension* of $\check{X}$ if E is homeomorphically embedded into $\widehat{E}$ as an open subset, the part process of Y on E being killed upon leaving E is identical in law with $\check{X}$, and Y has no sojourn on $\widehat{E} \setminus E$, that is, Y spends zero Lebesgue amount of time on $\widehat{E} \setminus E$.

In Sect. 5, we show that the *time changed diffusion process* $\check{X}$ admits a ν-symmetric conservative diffusion extension $\check{X}^*$ from E to its one-point compactification $E^* = E \cup \{\partial\}$ by constructing a regular, strongly local, recurrent and irreducible Dirichlet form on $L^2(E^*; \nu)$, ν being extended to E^* by setting $\nu(\partial) = 0$. In accordance with [7], $\check{X}^*$ is called the *one point reflection* of $\check{X}$ at ∂.

Theorem 6.1 in Sect. 6 will state that, if $\mathcal{E}$ enjoys the Liouville property (1), then a ν-symmetric conservative diffusion extension Y of $\check{X}$ is unique and coincides with the one-point reflection $\check{X}^*$ of $\check{X}$ at ∂ up to a quasi-homeomorphism, namely, a quasi homeomorphic image of Y is identical with $\check{X}^*$, and furthermore the extended Dirichlet space of Y equals the reflected Dirichlet space $(\mathcal{F}^{\mathrm{ref}}, \mathcal{E}^{\mathrm{ref}})$ of $\mathcal{E}$ independently of the smooth measure ν employed in the time change.

The proof of Theorem 6.1 will make use of the following general observation. Owing to the works of S. Albeverio, Z.-M. Ma and M. Röckner [1, 22] and P.J. Fitzsimmons [14], the *quasi-regularity* of a Dirichlet form has been known to be not only a sufficient condition but also a necessary one for the existence of a properly associated right process. It is further shown by Z.-Q. Chen, Z.-M. Ma and M. Röckner [9] that a Dirichlet form is quasi-regular if and only if it is *quasi-homeomorphic* to a regular Dirichlet form on a locally compact separable metric space. On the other hand, it was known that, if two regular Dirichlet spaces are *equivalent* in the sense of [16] and [17, Appendix], then the equivalence can be induced by a certain quasi-homeomorphism of the underlying spaces. Hence the equivalence of two quasi-regular Dirichlet spaces can be induced by such a map as will be formulated in Theorem 6.2. In the specific setting of Theorem 6.1, the Dirichlet spaces of Y and $\check{X}^*$ are both quasi-regular and they can be shown to be equivalent if the Liouville property (1) is valid. Thus Theorem 6.1 follows from Theorem 6.2.

2 Energy Form $\mathscr{E}^\rho$ and Weighted Beppo Levi Space $\mathscr{G}^\rho$

For a fixed Borel function ρ on $\mathbb{R}^n$ satisfying (4), define $\mathscr{F}^\rho, \mathscr{G}^\rho, \mathbf{D}^\rho$ by (5), (6), (7), respectively. We put $\mathbf{D}_1^\rho(u,v) = \mathbf{D}^\rho(u,v) + \int_{\mathbb{R}^n} uv\rho dx, \ u, v \in \mathscr{F}^\rho$.

Proposition 2.1 *The energy form $\mathscr{E}^\rho = (\mathbf{D}^\rho, \mathscr{F}^\rho)$ is a regular strongly local and irreducible Dirichlet form on $L^2(\mathbb{R}^n; \rho dx)$.*

Proof Completeness: Suppose $\{u_k\} \subset \mathscr{F}^\rho$ is $\mathbf{D}_1^\rho$-Cauchy. There exists then $u \in L^2(\mathbb{R}^n; \rho dx)$ and u_k converges to u in $L^2(\mathbb{R}^n; \rho dx)$. For each $r > 0$, $\{u_k\big|_{B_r}\}$ is $\mathbf{D}$-Cauchy on B_r by (4) and so, $\partial_i u_k \ \to \ \partial_i u$ in $L^2(B_r), \ 1 \le i \le n$. One can find a subsequence $\{k_\ell\}$ such that

$$u_{k_\ell} \ \to \ u, \quad \partial_i u_{k_\ell} \ \to \ \partial_i u, \ 1 \le i \le n, \quad \text{a.e. on } \mathbb{R}^n, \quad \text{as } \ell \ \to \infty.$$

By Fatou's lemma

$$\mathbf{D}_1^\rho(u - u_m, u - u_m) \le \liminf_{\ell \to \infty} \mathbf{D}_1^\rho(u_{k_\ell} - u_m, u_{k_\ell} - u_m) \ \to \ 0, \ m \ \to \ \infty.$$

Regularity: Take any bounded $u \in \mathscr{F}^\rho$. For any $\varepsilon > 0$, we find $r > 0$ with

$$\int_{\mathbb{R}^n \setminus B_r} |\nabla u|^2 \rho dx < \varepsilon, \quad \int_{\mathbb{R}^n \setminus B_r} u^2 \rho dx < \varepsilon,$$

Choose $\varphi \in C_c^\infty(\mathbb{R}^n)$ with $\varphi = 1$ on B_r, $\varphi = 0$ on $\mathbb{R}^n \setminus B_{r+2}$, and $0 \le \varphi \le 1, \ 0 \le |\nabla \varphi| \le 1$ on $\mathbb{R}^n$. Then

$$\mathbf{D}_1^\rho(u - u\varphi, u - u\varphi) \le 2 \int_{\mathbb{R}^n \setminus B_r} (1 - \varphi(x))^2 |\nabla u(x)|^2 \rho(x) dx$$
$$+ 2 \int_{\mathbb{R}^n \setminus B_r} u(x)^2 |\nabla \varphi(x)|^2 \rho(x) dx + \int_{\mathbb{R}^n \setminus B_r} u^2 \rho dx \le 5\varepsilon.$$

Since $u\varphi \in H_0^1(B_{r+2})$, there exists $f \in C_c^\infty(B_{r+2})$ with

$$\int_{B_{r+2}} \left[|\nabla (u\varphi - f)|^2 + (u\varphi - f)^2 \right] dx < \varepsilon / \Lambda_{r+2}.$$

Hence, by taking (4) into account,

$$\mathbf{D}_1^\rho(u-f,u-f) \le 2\mathbf{D}_1^\rho(u-u\varphi,u-u\varphi)+2\mathbf{D}_1^\rho(u\varphi-f,u\varphi-f) < 12\varepsilon,$$

Markov property, strong locality and irreducibility follow from Theorem 3.1.1, Excercise 3.1.1 and Corollary 4.6.4 of [17], respectively. □

We consider the quotient space $\dot{\mathscr{G}}^\rho = \mathscr{G}^\rho/\mathscr{N}$ of $\mathscr{G}^\rho$ relative to the space $\mathscr{N}$ of constant functions on $\mathbb{R}^n$.

Lemma 2.2 *$\dot{\mathscr{G}}^\rho$ is a Hilbert space with inner product $\mathbf{D}^\rho$.*
If $u_k \in \dot{\mathscr{G}}^\rho$ is $\mathbf{D}^\rho$-convergent to $u \in \dot{\mathscr{G}}^\rho$ as $k \to \infty$, then there are constants c_k such that $u_k - c_k$ converges to u in $L^2_{\rm loc}(\mathbb{R}^n)$ as $k \to \infty$.

Proof We use the Poincaré inequality holding for each ball B_r, $r > 1$:

$$\int_{B_r} (u(x)-\langle u\rangle_1)^2 dx \le C_r \int_{B_r} |\nabla u(x)|^2 dx, \quad u \in H^1(B_r), \tag{8}$$

where $\langle u\rangle_1 = |B_1|^{-1}\int_{B_1} u(x)dx$ and C_r is some positive constant (cf. [19, (7.45)]).

Let $\{u_k\} \subset \dot{\mathscr{G}}^\rho$ be $\mathbf{D}^\rho$-Cauchy. There exist then $f_i \in L^2(\mathbb{R}^n;\rho dx)$ such that $\partial_i u_k \to f_i$ in $L^2(\mathbb{R}^n;\rho dx)$ and hence in $L^2(B_r)$ as $k \to \infty$, $1 \le i \le n$, $r > 1$. Set $c_k = \langle u_k\rangle_1$. By (8), $\{u_k - c_k\}$ is $L^2(B_r)$-Cauchy for each $r > 1$. Let $u \in L^2_{\rm loc}(\mathbb{R}^n)$ be the limit function. Then, for any $\varphi \in C_c^\infty(\mathbb{R}^n)$,

$$\int_{\mathbb{R}^n} f_i\varphi dx = \lim_{k\to\infty}\int_{\mathbb{R}^n}\partial_i(u_k-c_k)\cdot\varphi dx = -\int_{\mathbb{R}^n}(u_k-c_k)\partial_i\varphi dx = -\int_{\mathbb{R}^n} u\partial_i\varphi dx,$$

so that $f_i = \partial_i u$, $1 \le i \le n$. □

Let $(\mathscr{F}_e^\rho, \mathscr{E}^\rho)$ be the extended Dirichlet space of the energy form $\mathscr{E}^\rho$.

Corollary 2.3 *$\mathscr{F}_e^\rho \subset \mathscr{G}^\rho$ and $\mathscr{E}^\rho(u,u) = \mathbf{D}^\rho(u,u)$, $u \in \mathscr{F}_e^\rho$.*

Proof For $u \in \mathscr{F}_e^\rho$, there exist $u_k \in \mathscr{F}^\rho$, $k \ge 1$, which is $\mathbf{D}^\rho$-Cauchy and converge to u a.e. as $k \to \infty$ according to the definition [17, Sect 1.5]. By Lemma 2.2, $\{u_k\}$ is $\mathbf{D}^\rho$-convergent to some $v \in \mathscr{G}^\rho$ and, for some constants c_k and a subsequence $\{k_\ell\}$, $u_{k_\ell} + c_{k_\ell} \to v$ a.e. as $\ell \to \infty$. Hence $c = \lim_{\ell\to\infty} c_{k_\ell}$ exists and $u = v - c$ so that $u \in \mathscr{G}^\rho$ and $\{u_k\}$ is $\mathbf{D}^\rho$-convergent to u. □

For a general regular strongly local Dirichlet form $\mathscr{E} = (\mathscr{E}, \mathscr{F})$ on $L^2(E;m)$, the energy measure $\mu_{\langle u\rangle}$ is well defined for $u \in \mathscr{F}_{\rm loc}$ and, according to [6, Theorem 6.2.13], the reflected Dirichlet space of $\mathscr{E}$ can be defined by

$$\begin{cases} \mathscr{F}^{\rm ref} = \{u : \text{ finite } m\text{–a.e.on } E,\ \tau_k u \in \mathscr{F}_{\rm loc},\ \sup_k \mu_{\langle \tau_k u\rangle}(E) < \infty\} \\ \mathscr{E}^{\rm ref}(u,u) = \lim_{k\to\infty}\mu_{\langle \tau_k u\rangle}(E), \quad u \in \mathscr{F}^{\rm ref}, \end{cases} \tag{9}$$

where $\tau_k u(x) = (-k) \vee u(x) \wedge k$.

Let $(\mathscr{F}^{\rho,\mathrm{ref}}, \mathscr{E}^{\rho,\mathrm{ref}})$ be the reflected Dirichlet space of the energy form $\mathscr{E}^\rho$. By virtue of [6, Theorem 4.3.11], the energy measure of a bounded $u \in \mathscr{F}^\rho$ is given by $\mu_{\langle u\rangle}(dx) = |\nabla u|^2 \rho dx$. Thus we have from (9)

$$\begin{cases} \mathscr{F}^{\rho,\mathrm{ref}} = \{u : \text{ finite a.e.on } \mathbb{R}^n, \ \tau_k u \in H^1_{\mathrm{loc}}(\mathbb{R}^n), \ \sup_k \int_{\mathbb{R}^n} |\nabla \tau_k u|^2 \rho dx < \infty\} \\ \mathscr{E}^{\rho,\mathrm{ref}}(u,u) = \lim_{k\to\infty} \int_{\mathbb{R}^n} |\nabla \tau_k u|^2 \rho dx, \quad u \in \mathscr{F}^{\rho,\mathrm{ref}}. \end{cases} \tag{10}$$

Proposition 2.4 *It holds that*

$$\mathscr{F}^{\rho,\mathrm{ref}} = \mathscr{G}^\rho, \qquad \mathscr{E}^{\rho,\mathrm{ref}} = \mathbf{D}^\rho.$$

Proof From (10), we obviously have $\mathscr{G}^\rho \subset \mathscr{F}^{\rho,\mathrm{ref}}$ and $\mathscr{E}^{\rho,\mathrm{ref}}(u,u) = \mathbf{D}^\rho(u,u)$ for any $u \in \mathscr{G}^\rho$.

It remains to show that $\mathscr{F}^{\rho,\mathrm{ref}} \subset \mathscr{G}^\rho$, For any $u \in \mathscr{F}^{\rho,\mathrm{ref}}$, we see by Banach-Saks theorem (cf. [6, Theorem A.4.1]) that the Césaro mean sequence $\{f_\ell, \ell \geq 1\}$ of a suitable subsequence of $\{\tau_k u, k \geq 1\} \subset \mathscr{G}^\rho$ is $\mathbf{D}^\rho$-Cauchy and converges pointwise to u. By Lemma 2.2, there exist constants c_ℓ and $w \in \mathscr{G}^\rho$ such that $f_\ell - c_\ell \to w$ in $L^2(B_r)$ as $\ell \to \infty$ for each $r > 1$. Choose a subsequence $\{\ell_p\}$ such that $f_{\ell_p} - c_{\ell_p} \to w$ a.e. on $\mathbb{R}^n$ as $p \to \infty$. Then $c = \lim_{p\to\infty} c_{\ell_p}$ exists and $u = w + c \in \mathscr{G}^\rho$. □

3 Liouville Property of Rotation Invariant Energy Forms on $\mathbb{R}^n$

Theorem 3.1 *Let $\rho(x) = \eta(|x|)$, $x \in \mathbb{R}^n$, for a positive C^∞-function η on $[0,\infty)$ such that η is constant on $[0,\varepsilon)$ for some $\varepsilon > 0$.*
Then the energy form $\mathscr{E}^\rho$ satisfies the Liouville property (1) *when $n \geq 2$.*
When $n = 1$, dim$(\mathscr{H}^*) = 2$ in transient case.

Proof In view of Propositions 1.1 and 2.1, it suffices to consider only the transient case in order to verify the Liouville property (1). According to Theorem 1.6.7 in the first edition of [17], $\mathscr{E}^\rho$ is transient if and only if

$$\int_1^\infty \frac{1}{\eta(r)r^{n-1}} dr < \infty. \tag{11}$$

In what follows, we assume that η satisfies condition (11).

It then follows from $1/r = (r^{n-3}\eta(r))^{1/2}(\eta(r)r^{n-1})^{-1/2}$ and the Schwarz inequality that

$$\int_1^\infty r^{n-3}\eta(r)dr = \infty. \tag{12}$$

We use the polar coordinate

$$x_1 = r\cos\theta_1,\ x_2 = r\sin\theta_1\cos\theta_2,\ x_3 = r\sin\theta_1\sin\theta_2\cos\theta_3, \dots,$$

$$x_{n-1} = r\sin\theta_1\sin\theta_2\dots\sin\theta_{n-2}\cos\theta_{n-1},\ x_n = r\sin\theta_1\sin\theta_2\dots\sin\theta_{n-2}\sin\theta_{n-1}.$$

Then, for $u, v \in C_c^1(\mathbb{R}^n)$,

$$\begin{aligned} &\mathbf{D}^\rho(u,v) \\ &= \int_{(0,\infty)\times(0,\pi]^{n-2}\times[0,2\pi]} \left[u_r v_r + \frac{u_{\theta_1}v_{\theta_1}}{r^2} + \frac{u_{\theta_2}v_{\theta_2}}{r^2\sin^2\theta_1} + \dots + \frac{u_{\theta_{n-1}}v_{\theta_{n-1}}}{r^2\sin^2\theta_1\dots\sin^2\theta_{n-2}}\right] \\ &\qquad \times \eta(r)r^{n-1}\sin^{n-2}\theta_1\dots\sin\theta_{n-2}drd\theta_1\dots d\theta_{n-1}. \end{aligned} \tag{13}$$

For a C^∞-function u on $\mathbb{R}^n$, we denote by $I_\eta(u,u)$ the value of the integral of the right hand side of (13) for $v = u$.

By Proposition 2.4, the reflected Dirichlet space of $\mathscr{E}^\rho$ is given by $(\mathscr{G}^\rho, \mathbf{D}^\rho)$ where

$$\mathscr{G}^\rho = \{u \in L^2_{\rm loc}(\mathbb{R}^n) : \int_{\mathbb{R}^n} |\nabla u(x)|^2\eta(|x|)dx < \infty\}.$$

Since the extended Dirichlet space $(\mathscr{F}_e^\rho, \mathbf{D}^\rho)$ of the transient energy form $\mathscr{E}^\rho$ is a real Hilbert space possessing $C_c^\infty(\mathbb{R}^n)$ as its core, we have

$$\mathscr{H}^* = \{u \in \mathscr{G}^\rho : \mathbf{D}^\rho(u,v) = 0 \quad \text{for every} \quad v \in C_c^\infty(\mathbb{R}^n)\}. \tag{14}$$

By noting that $\rho(x) = \eta(|x|)$ is a C^∞-function on $\mathbb{R}^n$, we let

$$Lu(x) = \Delta u(x) + \nabla\log\rho(x)\cdot\nabla u(x), \quad x \in \mathbb{R}^n. \tag{15}$$

We say that u is $\mathscr{E}^\rho$*-harmonic* if

$$u \in C^\infty(\mathbb{R}^n), \quad Lu(x) = 0,\ x \in \mathbb{R}^n.$$

Equation (14) then implies that $u \in \mathscr{H}^*$ if and only if

$$u \text{ is } \mathscr{E}^\rho\text{-harmonic and} \quad I_\eta(u,u) < \infty. \tag{16}$$

It also follows from (13) that u is $\mathscr{E}^\rho$-harmonic if and only if

$$u \in C^\infty(\mathbb{R}^n),\quad \mathscr{L}u(r,\theta_1,\dots,\theta_{n-1}) = 0,\ (r,\theta_1,\dots,\theta_{n-1}) \in (0,\infty)\times(0,\pi)^{n-2}\times[0,2\pi], \tag{17}$$

where

$$
\begin{aligned}
&\mathscr{L}u(r,\theta_1,\ldots,\theta_{n-1}) \qquad (18)\\
&= \frac{1}{r^{n-1}}(u_r\cdot\eta(r)r^{n-1})_r + \frac{\eta(r)}{r^2\sin^{n-2}\theta_1}(u_{\theta_1}\sin^{n-2}\theta_1)_{\theta_1} + \frac{\eta(r)}{r^2\sin^2\theta_1\sin^{n-3}\theta_2}(u_{\theta_2}\sin^{n-3}\theta_2)_{\theta_2}\\
&+\ldots+\frac{\eta(r)}{r^2\sin^2\theta_1\ldots\sin^2\theta_{n-3}\sin\theta_{n-2}}(u_{\theta_{n-2}}\sin\theta_{n-2})_{\theta_{n-2}} + \frac{\eta(r)}{r^2\sin^2\theta_1\ldots\sin^2\theta_{n-2}}(u_{\theta_{n-1}})_{\theta_{n-1}}
\end{aligned}
$$

Now take any function $u\in\mathscr{H}^*$. We first claim that

$$u_{\theta_{n-1}} = 0. \tag{19}$$

Put $w = u_{\theta_{n-1}}$. Then $w\in C^\infty(\mathbb{R}^n)$ because $w(x) = -u_{x_{n-1}}(x)x_n + u_{x_n}(x)x_{n-1},\ x\in\mathbb{R}^n$. Due to the expression (18) of $\mathscr{L}$, we also have $\mathscr{L}w = (\mathscr{L}u)_{\theta_{n-1}} = 0$ on $(0,\infty)\times(0,\pi)^{n-2}\times[0,2\pi]$. Thus w satisfies (17) so that w is $\mathscr{E}^\rho$-harmonic.

For $B_r = \{x\in\mathbb{R}^n;\ |x|<r\}$ and the uniform probability measure $\Pi(d\xi)$ on ∂B_1, w therefore admits the Poisson integral formula

$$w(x) = \int_{\partial B_1} K_r(x, r\xi)w(r\xi)\Pi(d\xi).\quad x\in B_r, \tag{20}$$

where $K_r(x, r\xi)$ is the Poisson kernel for B_r with respect to L, which is known to be continuous in $(x,\xi)\in B_r\times\partial B_1$ (cf. [20]). We also note that $K_r(0, r\xi) = 1$ for any $\xi\in\partial B_1$ by the rotation invariance of L around the origin 0.

Fix $a>0$. It then holds For any $r>a$ that

$$K_r(x, r\xi_2) = \int_{\partial B_a} K_a(x, a\xi_1)K_r(a\xi_1, r\xi_2)\Pi(d\xi_1),\quad x\in B_a,\ \xi_2\in\partial B_1.$$

Hence, if we let $\sup_{x\in B_{a/2},\xi_1\in\partial B_1} K_a(x, a\xi_1) = C_a < \infty$, then, for $x\in B_{a/2},\ \xi_2\in\partial B_1$,

$$K_r(x, r\xi_2) \le C_a\int_{\partial B_1} K_r(a\xi_1, r\xi_2)\Pi(d\xi_1) = C_aK_r(0, r\xi_2) = C_a,$$

and it follows from (20) that

$$|w(x)| \le C_a\int_{\partial B_1} |w(r\xi)|\Pi(d\xi),\quad x\in B_{a/2},\quad r>a.$$

Recall that $w = u_{\theta_{n-1}}$. We denote by σ_n the area of ∂B_1. We multiply the both hand side of the above inequality by $r^{n-3}\eta(r)$, integrate in r from a to R, apply the Schwarz inequality and finally use the expression (13) to get

$$|u_{\theta_{n-1}}(x)| \le \frac{C_a}{\sqrt{\sigma_n}}\left[\int_a^R r^{n-3}\eta(r)dr\right]^{-1/2} \cdot \sqrt{I_\eta(u,u)}, \quad x \in B_{a/2},$$

which tends to 0 as $R \to \infty$ by (12) and (16). Since $a > 0$ is arbitrary, we arrive at (19). Here the finite energy property for $u \in \mathcal{G}^\rho$ is crucially utilized.

It also holds that

$$u_{\theta_k} = 0 \quad \text{for any} \quad 1 \le k \le n-2. \tag{21}$$

In fact, if we let $\xi_i = \frac{x_i}{r}$, $1 \le i \le n$, $\xi = (\xi_1, \dots, \xi_n) \in \partial B_1$, then θ_k, $1 \le k \le n-2$, is an angle of two n-vectors $\xi^{(k)} = (\underbrace{0, \dots 0}_{k-1}, \xi_k, \dots, \xi_n)$, $\quad \mathbf{e}_k = (\underbrace{0, \dots 0}_{k-1}, 1, 0, \dots, 0)$. Let O be an orthogonal matrix whose $(n-1)$-th and n-th row vectors are $\mathbf{e}_k$ and $\widehat{\mathbf{e}} = |\xi^{(k+1)}|^{-1} \cdot \xi^{(k+1)}$, respectively. We make the orthogonal transformation $\mathbf{y} = O\mathbf{x}$. Then θ_k equals an angle of two vectors on the (y_{n-1}, y_n)-plane in the new coordinate system $\mathbf{y}$ and (21) follows from (19) as the expression (7) of $\mathbf{D}^\rho(u, v)$ with $\rho(x) = \eta(|x|)$ remains valid for $\mathbf{y}$ in place of $\mathbf{x}$.

Thus u depends only on r and, in terms of a scale function $ds(r) = \dfrac{dr}{\eta(r)r^{n-1}}$ on $(0, \infty)$, (13) and (18) are reduced, respectively, to

$$I_\eta(u,u) = \sigma_n \int_0^\infty \left(\frac{du(r)}{ds(r)}\right)^2 ds(r), \qquad \mathcal{L}u(r) = \frac{1}{r^{n-1}}\frac{d}{dr} \cdot \frac{du(r)}{ds(r)}.$$

By (17), $\mathcal{L}u = 0$ so that $\frac{du(r)}{ds(r)}$ equals a constant C and $I_\eta(u,u) = \sigma_n C^2 s(0,\infty)$.

When $n \ge 2$, $s(0,\infty) = \infty$ and we get $C = 0$ from (16), yielding that u is constant. When $n = 1$ and $\rho(x) = \eta(|x|)$, $x \in \mathbb{R}$, for a positive continuous function η on $[0, \infty)$ with $\int_1^\infty \eta(x)^{-1}dx < \infty$, $u = c_1 + c_2 s$, $c_1, c_2 \in \mathbb{R}$, for $s(x) = \int_0^x \eta(|\xi|)^{-1}d\xi$, $x \in \mathbb{R}$. □

The condition for η to be a positive constant near 0 is just for simplicity and it can be weakened appropriately.

4 Strongly Local Transient Dirichlet Form $\mathcal{E}$ and a Time Change $\check{X}$ of the Associated Diffusion

Let $(E, m, \mathcal{E}, \mathcal{F})$ and $(\mathcal{F}_e, \mathcal{F}^{\text{ref}})$ be as is stated in the beginning of Introduction. Once for all, we assume that the Dirichlet form $(\mathcal{E}, \mathcal{F})$ on $L^2(E; m)$ is transient and strongly local. Let $X = (X_t, \zeta, \mathbf{P}_x)$ be the associated diffusion process on E.

The lifetime ζ of X can be finite or infinite. Since X admits no killing inside E, we get from [6, Theorem 3.5.2 and Corollary 3.5.3]

$$\mathbf{P}_x(\lim_{t\to\zeta} X_t = \partial) = 1 \quad \text{q.e.} \quad x \in E, \tag{22}$$

$$\mathbf{P}_x(\lim_{t\to\zeta} u(X_t) = 0) = 1 \quad \text{q.e.} \quad x \in E, \tag{23}$$

where ∂ is the point at infinity of E and u is any quasi continuous function belonging to the extended Dirichlet space $\mathscr{F}_e$.

In the remainder of this paper, we fix an arbitrary positive finite measure ν on E charging no $\mathscr{E}$-polar set such the the quasi-support of ν equals E. Let A be the positive continuous additive functional of X with Revuz measure ν. A typical example of such a measure ν is $\nu(dx) = f(x)m(dx)$ for a strictly positive Borel function f on E with $\int_E f dm < \infty$ and, in this case, $A_t = \int_0^{t\wedge\zeta} f(X_s)ds,\ t \geq 0$.

Let $\check{X} = (\check{X}_t, \check{\zeta}, \mathbf{P}_x)$ be the time changed process of X by means of A:

$$\check{X}_t = X_{\tau_t}, \quad \tau_t = \inf\{s : A_s > t\}, \qquad \check{\zeta} = A_\zeta.$$

$\check{X}$ is a diffusion process on E symmetric with respect to the measure ν and the Dirichlet form $\check{\mathscr{E}} = (\check{\mathscr{E}}, \check{\mathscr{F}})$ of $\check{X}$ on $L^2(E;\nu)$ is given by

$$\check{\mathscr{E}} = \mathscr{E}, \qquad \check{\mathscr{F}} = \mathscr{F}_e \cap L^2(E;\nu), \tag{24}$$

which is strongly local and regular [6, p. 183].

Proposition 4.1 (i) *It holds that*

$$\mathbf{P}_x(\check{\zeta} < \infty,\ \lim_{t\uparrow\check{\zeta}} \check{X}_t = \partial) = \mathbf{P}_x(\check{\zeta} < \infty) = 1 \ \textit{for q.e. } x \in E. \tag{25}$$

(ii) *The extended and reflected Dirichlet spaces of* $(\check{\mathscr{E}}, \check{\mathscr{F}})$ *equal* $(\mathscr{F}_e, \mathscr{E})$ *and* $(\mathscr{F}^{\text{ref}}, \mathscr{E}^{\text{ref}})$*, respectively.*

Proof (i) It suffices to show that

$$\mathbf{P}_x(A_\infty < \infty) = 1 \qquad \text{for q.e. } x \in E. \tag{26}$$

Take a strictly positive Borel function h on E with $\int_E h(x)m(dx)dx < \infty$.. By the transience of X and [6, Proposition 2.1.3], $Rh(x) < \infty$ for m-a.e. $x \in E$ where R is the 0-order resolvent of X. For integer $\ell \geq 1$, let $\Lambda_\ell = \{x \in E : Rh(x) \leq \ell\}$. Then $R((I_{\Lambda_\ell} h)(x) \leq \ell$ for any $x \in E$.

From [6, (4.1.3)], we have for each $\ell \geq 1$

$$\int_{\Lambda_\ell} \mathbf{E}_x[A_\infty] h(x)m(dx) = \langle R(I_{\Lambda_\ell} h), \nu\rangle \leq \ell\nu(E) < \infty.$$

As $m(E \setminus (\cup_{\ell=1}^\infty \Lambda_\ell)) = 0$, it follows that $\mathbf{E}_x[A_\infty] < \infty$ m-a.e. $x \in E$ and hence q.e. $x \in E$ by [6, Theorem A.2.13 (v)], yielding (26).

(ii) is a consequence of the invariance of the extended and reflected Dirichlet spaces under a time change by means of a fully supported positive Radon measure charging no $\mathscr{E}$-polar set [6, Corollary 5.2.12 and Proposition 6.4.6]. □

Since the lifetime $\check{\zeta}$ of the time changed diffusion $\check{X}$ is finite $\mathbf{P}_x$-a.s. for q.e. $x \in E$ by the above proposition, the boundary problem concerning possible Markovian extensions of $\check{X}$ beyond its lifetime $\check{\zeta}$ makes a perfect sense. For different choices of ν, the diffusions $\check{X}$ share a common geometric structure related each other only by time changes. So the study of the boundary problem for $\check{X}$ as we shall engage in the next two sections is a good way to make a closer look at the behavior of the diffusion process X around ∂.

5 One-Point Reflection of $\check{X}$ at ∂

Denote by E^* the one-point compactification $E \cup \{\partial\}$ of E. The measure ν is extended from E to E^* by setting $\nu(\{\partial\}) = 0$. In this section, we construct a ν-symmetric conservative diffusion extension of $\check{X}$ from E to E^* by constructing a regular strongly local Dirichlet form on $L^2(E^*; \nu)$. Note that $L^2(E^*; \nu)$ can be identified with $L^2(E; \nu)$.

Recall that the reflected Dirichlet space $(\mathcal{F}^{\rm ref}, \mathcal{E}^{\rm ref})$ of the regular strongly local Dirichlet form $\mathcal{E} = (\mathcal{E}, \mathcal{F})$ on $L^2(E; m)$ is given by (9). On account of the property [6, Theorem 4.3.10] of the energy measure $\mu_{\langle u \rangle}$ for $u \in \mathcal{F}_{\rm loc}$, this means that

$$1 \in \mathcal{F}^{\rm ref}, \qquad \mathcal{E}^{\rm ref}(1, 1) = 0. \tag{27}$$

Furthermore $\mathcal{F}_e$ does not contain a non-zero constant function because of the transience of $\mathcal{E}$. In what follows, every function in the space $\mathcal{F}_e$ is taken to be $\mathcal{E}$-quasi-continuous.

Let us define

$$\begin{cases} \mathcal{F}_e^* = \{u + c : u \in \mathcal{F}_e,\ c \in \mathbb{R}\}, \\ \mathcal{E}^*(u_1 + c_1, u_2 + c_2) = \mathcal{E}(u_1, u_2),\ u_i \in \mathcal{F}_e,\ c_i \in \mathbb{R},\ i = 1, 2. \end{cases} \tag{28}$$

$(\mathcal{F}_e^*, \mathcal{E}^*)$ is a subspace of $(\mathcal{F}^{\rm ref}, \mathcal{E}^{\rm ref})$.

Theorem 5.1 (i) *Define* $\check{\mathcal{F}}^* = \mathcal{F}_e^* \cap L^2(E; \nu)$. *The form* $\check{\mathcal{E}}^* = (\mathcal{E}^*, \check{\mathcal{F}}^*)$ *is then a regular strongly local Dirichlet form on* $L^2(E^*; \nu)$.
(ii) *The extended Dirichlet space of* $\check{\mathcal{E}}^*$ *equals* $(\mathcal{F}_e^*, \mathcal{E}^*)$. $\check{\mathcal{E}}^*$ *is recurrent.*
(iii) *Let* $\check{X}^* = (\check{X}_t^*, \mathbf{P}_x^*)$ *be the diffusion process on* E^* *associated with* $\check{\mathcal{E}}^*$. *The part of* $\check{X}^*$ *on* E *being killed upon hitting* ∂ *is then identical in law with the time changed diffusion* $\check{X}$. $\check{X}^*$ *is conservative and irreducible.*

Proof (i) *Completeness*. Suppose $w_n = u_n + c_n \in \check{\mathcal{F}}^*$, $n \geq 1$, are $\mathcal{E}_1^*$-Cauchy. Then $\{u_n\} \subset \mathcal{F}_e$ is an $\mathcal{E}$-Cauchy sequence. Due to the transience, it is $\mathcal{E}$-convergent to some $u \in \mathcal{F}_e$ and some subsequence $\{u_{n_k}\}$ of it converges to u q.e. on E in view of [6, Theorem 2.3.2]. $\{w_n\}$ is $L^2(E; \nu)$-convergent to some $w \in L^2(E : \nu)$ and a subsequence $\{w_{n'_k}\}$ of $\{w_{n_k}\}$ converges to w ν-a.e. on E. Hence $\lim_{k\to\infty} c_{n'_k} = c$ exists and $w = u + c$. Hence $\{w_n\}$ is $\mathcal{E}_1^*$-convergent to $w \in \check{\mathcal{F}}^*$.

Markov property. This follows from $(0 \vee w) \wedge 1 = [(-c) \vee u] \wedge (1-c) + c$ for $w = u + c \in \mathcal{F}_e^*$.
Regularity. For any bounded $w = u + c \in \check{\mathcal{F}}_e^*$, choose $u_k \in \mathcal{F}_e \cap C_c(E)$ $\mathcal{E}$-converging to u. We may assume that $\{u_k\}$ are uniformly bounded by [17, Theorem 1.4.2]. By [6, Theorem 2.3.2], a subsequence $\{\widetilde{u}_k\}$ of $\{u_k\}$ converges q.e. to u. Since $\int_E u_k^2 d\nu$ is uniformly bounded, the Césaro mean $\{f_k\}$ of a suitable subsequence of $\{\widetilde{u}_k\}$ converges to $f \in L^2(E; \nu)$. Since f_k converges to u ν-a.e., $f = u$. Then $f_k + c \in \check{\mathcal{F}}_e^* \cap C(E^*)$ is $\check{\mathcal{E}}_1^*$-convergent to $u + c = w$.
Strong locality. Suppose, for $w_i = u_i + c_i,\ u_i \in \mathcal{F}_e \cap C_c(E),\ i = 1, 2$, that w_1 is constant in a neighborhood of Supp(w_2). When $c_2 = 0$, $\mathcal{E}^*(w_1, w_2) = 0$ by the strong locality of $\mathcal{E}$. When $c_2 \neq 0$, the set $U = E^* \setminus \text{Supp}(w_2)$ is either empty or non-empty relatively compact open subset of E. In the former case, $\mathcal{E}^*(w_1, w_2) = 0$. In the latter case, $u_2 = -c_2$ on U, while Supp$(u_1) \subset U$ and $\mathcal{E}^*(w_1, w_2) = \mathcal{E}(u_1, u_2) = 0$ again. Therefore $\mathcal{E}^*$ is strongly local on account of [17, Theorem 3.1.2].
(ii) The inclusion $\subset$ can be shown by using [6, Theorem 3.2.3]. Conversely, for any $u \in \mathcal{F}_e^*$, its truncations are in $\check{\mathcal{F}}^*$ and convergent to u pointwise and in $\mathcal{E}^*$. Hence u is in the extended Dirichlet space of $\check{\mathcal{E}}^*$. Since $1 \in \mathcal{F}_e^*$ and $\mathcal{E}^*(1, 1) = 0$, $\check{\mathcal{E}}^*$ is recurrent.
(iii) By virtue of [17, Theorem 4.4.3], the part of $\check{\mathcal{E}}^*$ on E admits $\mathcal{F}_e \cap C_c(E)$ as its core. So it coincides with the Dirichlet form (24) on $L^2(E; \nu)$ associated with $\check{X}$. Hence $\check{X}^*$ is an extension of $\check{X}$.

Since $\check{\mathcal{E}}^*$ is recurrent, $\check{X}^*$ is recurrent and in particular conservative.

In order to verify the irreducibility of $\check{X}^*$, the resolvents of $\check{X}^*$, $\check{X}$ are denoted by R_α^*, R_α, respectively, and we let $u_\alpha(x) = \mathbf{E}_x[e^{-\zeta}],\ x \in E$. $(f.g)$ will stand for the integral $\int_E fgd\nu$. By the strong Markov property of $\check{X}^*$ at the hitting time of ∂, we have for any bounded Borel function f on E,

$$R_\alpha^* f(x) = R_\alpha f(x) + u_\alpha(x)\ R_\alpha^* f(\partial), \quad x \in E. \tag{29}$$

By (25), $u_\alpha(x) > 0$ for q.e. $x \in E$ and $1 - u_\alpha(x) = \alpha R_\alpha 1(x),\ x \in E$. By integrating the both hand sides of (29) by ν, we thus get $R_\alpha^* f(\partial) = \dfrac{(u_\alpha, f)}{\alpha(u_\alpha, 1)}$. Hence it follows from (29) that

$$(I_A, R_\alpha^* I_B) \geq \frac{(u_\alpha, I_A)(u_\alpha, I_B)}{\alpha(u_\alpha, 1)} > 0,$$

for any Borel sets $A, B \subset E$ with positive ν-measure, yielding the irreducibility of $\check{X}^*$. □

In accordance with [7], we call $\check{X}^*$ the *one-point reflection of* $\check{X}$ *at* ∂.

This theorem is a generalization of Theorem 3.2 in [18] where a stronger assumption of a Poincaré inequality for $\mathcal{E}$ was made. The first construction of such a one-point reflection at ∂ goes back to [15, Sect. 8] where $\check{X} = X$ and X was the absorbing Brownian motion on an arbitrary bounded domain of $\mathbb{R}^n$.

As has been presented in [18] and [6, Sects. 7.5 and 7.6], there is an alternative quite different way to construct a one-point reflection $\check{X}^*$ of $\check{X}$ at ∂ by using a Poisson point process of excursions of $\check{X}$ around ∂, which makes the structure of the constructed process $\check{X}^*$ more transparent but requires a certain regularity condition on the resolvent of $\check{X}$ in the construction. Notice that $\check{X}^*$ becomes irreducible, while $\check{X}$ may not be. See Example 6.2 in [18] in this connection.

In the next section, we shall show that, if $\mathscr{E}$ satisfies the Liouville property (1), then any symmetric conservative diffusion extension of $\check{X}$ must equal $\check{X}^*$ up to a quasi-homeomorphism.

6 Liouville Property of $\mathscr{E}$ and Uniqueness of a Symmetric Conservative Diffusion Extension of $\check{X}$

Let $\widehat{E}$ be a Lusin space into which E is homeomorpically embedded as an open subset. The measure ν on E is extended to $\widehat{E}$ by setting $\nu(\widehat{E} \setminus E) = 0$. Let $Y = (Y_t, \zeta^Y, \mathbf{P}_x^Y)$ be any ν-symmetric conservative diffusion process on $\widehat{E}$ whose part process on E being killed upon leaving E is identical in law with $\check{X}$. We denote by $(\mathscr{E}^Y, \mathscr{F}^Y)$ the Dirichlet form of Y on $L^2(\widehat{E}; \nu)$. We call Y a *ν-symmetric conservative diffusion extension* of $\check{X}$.

Theorem 6.1 *Suppose $\mathscr{E}$ satisfies the Liouville property* (1). *Then we have the following:*
(i) *As Dirichlet forms on $L^2(E, \nu)$,*

$$(\mathscr{E}^Y, \mathscr{F}^Y) = (\mathscr{E}^*, \check{\mathscr{F}}^*). \tag{30}$$

(ii) *The extended Dirichlet space of $(\mathscr{E}^Y, \mathscr{F}^Y)$ equals $(\mathscr{F}_e^*, \mathscr{E}^*) = (\mathscr{F}^{\text{ref}}, \mathscr{E}^{\text{ref}})$.*
(iii) *Y under $\mathbf{P}_{g\cdot\nu}^Y$ and $\check{X}^*$ under $\mathbf{P}_{g\cdot\nu}^*$ share the same finite dimensional distributions for any non-negative $g \in L^2(E; \nu)$.*
(iv) *A quasi-homeomorphic image of Y in the specific sense described in Theorem 6.2 below is identical with $\check{X}^*$.*

Proof of Theorem 6.1 **(i), (ii), (iii)** We use basic results due to Albeverio-Ma-Röckner [1, 22], Fitzsimmons [14] and Chen-Ma-Röckner [9] being formulated in [6]: $\mathscr{E}^Y$ is a quasi-regular Dirichlet form on $L^2(\widehat{E}; \nu)$ and Y is properly associated with it [6, Theorem 1.5.3]. By considering the image by the quasi-homeomorphsim j in [6, Theorem 3.1.13], we can therefore assume that $\widehat{E}$ is a locally compact separable metric space, ν is a fully supported positive Radon measure on $\widehat{E}$, $(\mathscr{E}^Y, \mathscr{F}^Y)$ is a regular Dirichlet form on $L^2(\widehat{E}; \nu)$ and Y is an associated diffusion Hunt process on $\widehat{E}$. E is now quasi-open and hence q.e. finely open in $\widehat{E}$.

Since $\check{X}$ is the part on E of Y, we can use [6, Theorem 3.3.8] to characterize its Dirichlet form (24) as

$$\check{\mathscr{F}} = \{u \in \mathscr{F}^Y : u = 0 \quad \text{q.e. on} \quad \widehat{E} \setminus E\}, \quad \check{\mathscr{E}} = \mathscr{E}^Y \text{ on } \check{\mathscr{F}} \times \check{\mathscr{F}},$$

where every function in $\mathscr{F}^Y$ is taken to be quasi continuous. This means that $\check{\mathscr{F}}$ is an ideal of $\mathscr{F}^Y$; if $u \in \check{\mathscr{F}}_b,\ v \in \mathscr{F}_b^Y$, then $uv \in \check{\mathscr{F}}_b$, in other words, $\mathscr{F}^Y$ is a Silverstein extension of $\check{\mathscr{F}}$.

We can then invoke [6, Theorem 6.6.9] about the maximality of the reflected Dirichlet space of $\check{\mathscr{E}}$ which equals $(\mathscr{F}^{\text{ref}}, \mathscr{E}^{\text{ref}})$ by virtue of Proposition 4.1 (ii) Thus we have

$$\mathscr{F}^Y \subset \mathscr{F}_a^{\text{ref}} \ (= \mathscr{F}^{\text{ref}} \cap L^2(E; \nu)),$$

But under the present assumption (1).

$$\mathscr{F}^{\text{ref}} = \mathscr{F}_e^*, \quad \mathscr{E}^{\text{ref}} = \mathscr{E}^*, \quad \mathscr{F}_a^{\text{ref}} = \check{\mathscr{F}}^*, \tag{31}$$

so that $\mathscr{F}^Y \subset \check{\mathscr{F}}^*$. As Y is assumed to be conservative while $\check{X}$ has a finite lifetime in view of Proposition 4.1 (i), $\check{\mathscr{F}}$ is a proper subspace of $\mathscr{F}^Y$. Hence we must have the identity $\mathscr{F}^Y = \check{\mathscr{F}}^*$. Since Y is a diffusion with no killing inside $\widehat{E}$, the regular Dirichlet form $(\mathscr{E}^Y, \mathscr{F}^Y)$ is strongly local so that $\mathscr{E}^Y(1, 1) = 0$. Consequently, for $w = u + c,\ u \in \check{\mathscr{F}},\ \mathscr{E}^Y(w, w) = \mathscr{E}^Y(u, u) = \mathscr{E}(u, u) = \mathscr{E}^*(w, w)$, yielding (30).

The assertion (ii) follows from (30), (31) and Theorem 5.1 (ii).

By (30), Y and $\check{X}$ generate the same strongly continuous Markovian semigroup on $L^2(E; \nu)$ yielding the assertion (iii). □

Here we give one remark on the above proof. Let $(\mathscr{E}, \mathscr{F})$ be a quasi-regular Dirichlet form and κ be the killing measure in the Beurling-Deny representation of $\mathscr{E}$. Theorem 6.6.9 in the book [6] by Z.-Q.Chen and the present author states that, among all of Silverstein extensions of $(\mathscr{E}, \mathscr{F})$, its reflected Dirichlet space is the maximal one. Actually this statement holds true under the condition that

$$\kappa = 0. \tag{32}$$

But the condition (32) is missing in that statement of [6]. We would like to take this opportunity to correct it by requiring the condition (32). Of course, (32) is fulfilled by the present strongly local Dirichlet form $(\check{\mathscr{E}}, \check{\mathscr{F}})$. In this connection, see also the proof of Theorem 7.1.6 in [6].

In accordance with [17, A.4], we say that a quadruplet $(E, m, \mathscr{E}, \mathscr{F})$ is a *Dirichlet space* if E is a Hausdorff topological space with a countable base, m is a σ-finite positive Borel measure on E and $\mathscr{E}$ with domain $\mathscr{F}$ is a Dirichlet form on $L^2(E; m)$. The inner product in $L^2(E; m)$ is denoted by $(\cdot, \cdot)_E$. For a given Dirichlet space $(E, m, \mathscr{E}, \mathscr{F})$, the notions of an *$\mathscr{E}$-nest*, an *$\mathscr{E}$-polar set*, an *$\mathscr{E}$-quasi-continuous numercal function* and '*$\mathscr{E}$-quasi-everywhere*' ('$\mathscr{E}$-q.e.' in abbreviation) are defined as in [6, Definition 1.2.12]. The *quasi-regularity* of the Dirichlet space is defined just as in [6, Definition 1.3.8]. We note that the space $\mathscr{F}_b = \mathscr{F} \cap L^\infty(E; m)$ is an algebra.

Given two Dirichlet spaces

$$(E, m, \mathscr{E}, \mathscr{F}), \qquad (\tilde{E}, \tilde{m}, \tilde{\mathscr{E}}, \tilde{\mathscr{F}}), \tag{33}$$

we call them *equivalent* if there is an algebraic isomorphism Φ from $\mathscr{F}_b$ onto $\tilde{\mathscr{F}}_b$ preserving three kinds of metrics: for $u \in \mathscr{F}_b$

$$\|u\|_\infty = \|\Phi u\|_\infty, \ (u, u)_E = (\Phi u, \Phi u)_{\tilde{E}}, \ \mathscr{E}(u, u) = \tilde{\mathscr{E}}(\Phi u, \Phi u).$$

One of the two equivalent Dirichlet spaces is called a *representation* of the other.

The underlying spaces E, $\tilde{E}$ of two Dirichlet spaces (33) are said to be *quasi-homeomorphic* if there exist $\mathscr{E}$-nest $\{F_n\}$, $\tilde{\mathscr{E}}$-nest $\{\tilde{F}_n\}$ and a one to one mapping q from $E_0 = \cup_{n=1}^\infty F_n$ onto $\tilde{E}_0 = \cup_{n=1}^\infty \tilde{F}_n$ such that the restriction of q to each F_n is a homeomorphism onto $\tilde{F}_n$. $\{F_n\}$, $\{\tilde{F}_n\}$ are called the *nests attached to the quasi-homemorphism* q. Any quasi-homeomorphism is quasi-notion-preserving.

We say that the equivalence Φ of two Dirichlet spaces (33) *is induced by a quasi-homeomorphism* q of the underlying spaces if

$$\Phi u(\tilde{x}) = u(q^{-1}(\tilde{x})), \quad u \in \mathscr{F}_b, \quad \tilde{m}\text{—a.e. } \tilde{x}.$$

Then $\tilde{m}$ is the image measure of m by q and $(\tilde{\mathscr{E}}, \tilde{\mathscr{F}})$ is the image Dirichlet form of $(\mathscr{E}, \mathscr{F})$ by q.

Theorem 6.2 *Assume that two Dirichlet spaces* (33) *are quasi-regular and that they are equivalent. Let* $X = (X_t, \mathbb{P}_x)$ *(resp.* $\tilde{X} = (\tilde{X}_t, \tilde{\mathbb{P}}_{\tilde{x}})$*) be an* m*-symmetric right process on* E *(resp. an* $\tilde{m}$*-symmetric right process on* $\tilde{E}$*) properly associated with* $(\mathscr{E}, \mathscr{F})$ *on* $L^2(E; m)$ *(resp.* $(\tilde{\mathscr{E}}, \tilde{\mathscr{F}})$ *on* $L^2(\tilde{E}; \tilde{m})$*). Then the equivalence is induced by a quasi-homeomorphism* q *with attached nests* $\{F_n\}$, $\{\tilde{F}_n\}$ *such that* $\tilde{X}$ *is the image of* X *by* q *in the following sense: there exist an* m*-inessential Borel subset* N *of* E *containing* $\cap_{n=1}^\infty F_n^c$ *and an* $\tilde{m}$*-inessential Borel subset* $\tilde{N}$ *of* $\tilde{E}$ *containing* $\cap_{n=1}^\infty \tilde{F}_n^c$ *so that* q *is one to one from* $E \backslash N$ *onto* $\tilde{E} \backslash \tilde{N}$ *and*

$$\tilde{X}_t = q(X_t), \qquad \tilde{\mathbb{P}}_{\tilde{x}} = \mathbb{P}_{q^{-1}\tilde{x}}, \qquad \tilde{x} \in \tilde{E} \backslash \tilde{N}. \tag{34}$$

A proof of this theorem can be carried out as is explained at the end of Introduction. See Appendix of [8] for the details.

Proof of Theorem 6.1 **(iv)** By Theorem 6.1 (i), the two Dirichlet spaces

$$(\widehat{E}, \nu, \mathscr{E}^Y, \mathscr{F}^Y), \qquad (E^*, \nu, \mathscr{E}^*, \check{\mathscr{F}}^*)$$

are equivalent in the above sense by the identity map Φ from $\mathscr{F}_b^Y$ onto $\check{\mathscr{F}}_b^*$. Hence Theorem 6.1 (iv) follows from Theorem 6.2.

To be more precise, there exist a quasi-homeomorphism q with attached $\mathscr{E}^Y$-nest $\{F_n\}$ on $\widehat{E}$ and $\mathscr{E}^*$-nest $\{\widetilde{F}_n\}$ on E^*, a ν-inessential Borel set N with $\cap_{n=1}^\infty F_n^c \subset N \subset \widehat{E}$

for Y and a ν-inessential Borel set $\widetilde{N}$ with $\cap_{n=1}^{\infty} \widetilde{F}_n^c \subset \widetilde{N} \subset E^*$ for $\check{X}^*$ such that q is one to one from $\widehat{E} \setminus N$ onto $E^* \setminus \tilde{N}$ and

$$\check{X}_t^* = q(Y_t), \qquad \tilde{\mathbf{P}}_{\tilde{x}}^* = \mathbf{P}_{q^{-1}\tilde{x}}^Y, \qquad \tilde{x} \in E^* \setminus \tilde{N}. \qquad \square$$

We note that the third assertion (iii) of Theorem 6.1 follows from the fourth one (iv) because the above map q preserves the ν-measure.

An analogous theorem to Theorem 6.1 has appeared in [5, Theorem 3.4] for the reflecting X on the closure of an Euclidean domain.

Acknowledgements I am grateful to Zhen-Qing Chen and Kazuhiro Kuwae for valuable discussions on the present subject. Indeed I owe the present proof of Proposition prop1.1 to their private communications.

References

1. Albeverio, S., Ma, Z.-M., Röckner, M.: Quasi-regular Dirichlet forms and Markov processes. J. Func, Anal. **111**, 118–154 (1993)
2. Beurling, A., Deny, J.: Dirichlet spaces. Proc. Nat. Acad. Sci. U.S.A. **45**, 208–215 (1959)
3. Brelot, M.: Etude et extensions du principe de Dirichlet. Ann. Inst. Fourier **5**, 371–419 (1953/54)
4. Chen, Z.-Q.: On reflected Dirichlet spaces. Probab. Theory Relat. Fields **94**, 135–162 (1992)
5. Chen, Z.-Q., Fukushima, M.: On unique extension of time changed reflecting Brownian motions. Ann. Inst. Henri Poincaré Probab. Statist. **45**, 864–875 (2009)
6. Chen, Z.-Q., Fukushima, M.: Symmetric Markov Processes, Time Change and Boundary Theory. Princeton University Press, Princeton (2011)
7. Chen, Z.-Q., Fukushima, M.: One-point reflections. Stoch. Process Appl. **125**, 1368–1393 (2015)
8. Chen, Z.-Q., Fukushima, M.: Reflections at infinity of time changed RBMs on a domain with Liouville branches. J. Math. Soc. Jpn. (To appear)
9. Chen, Z.-Q., Ma, Z.-M., Röckner, M.: Quasi-homeomorphisms of Dirichlet forms. Nagoya Math. J. **136**, 1–15 (1994)
10. Deny, J.: Les potentiels d'énergie finie. Acta Math. **82**, 107–183 (1950)
11. Deny, J.: Methods Hilbertiennes en théorie du potentiel. In: Potential Theory, pp. 121–120. Centro Internazional Mathematico Estivo, Edizioni Cremonese, Roma (1970)
12. Deny, J., Lions, J.L.: Les espaces du type de Beppo Levi. Ann. Inst. Fourier **5**, 305–370 (1953/54)
13. Doob, J.L.: Boundary properties of functions with finite Dirichlet integrals. Ann. Inst. Fourier **12**, 574–621 (1962)
14. Fitzsimmons, P.J.: On the quasi-regularity of semi-Dirichlet forms. Potential Analysis **15**, 151–185 (2001)
15. Fukushima, M.: On boundary conditions for multi-dimensional Brownian motions with symmetric resolvent densities. J. Math. Soc. Jpn. **21**, 58–93 (1969)
16. Fukushima, M.: Dirichlet spaces and strong Markov processes. Trans. Am. Math. Soc. **162**, 185–224 (1971)
17. Fukushima, M., Oshima, Y., Takeda, M.: Dirichlet Forms and Symmetric Markov Processes. de Gruyter, Berlin (1994). 2nd revised Edition (2010)
18. Fukushima, M., Tanaka, H.: Poisson point processes attached to symmetric diffusions. Ann. Inst. Henri Poincaré Probab. Statist. **41**, 419–459 (2005)
19. Gilbarg, D., Trudinger, N.S.: Elliptic Partial Differential Equations of Second Order. Springer, Berlin (1998)

20. Itô, S.: Fundamental solutions of parabolic differential equations and boundary value problems. Jpn. J. Math. **27**, 65–102 (1957)
21. Levi, B.: Sul principio di Dirichlet. Rend. Palermo **22**, 293–359 (1906)
22. Ma, Z.-M., Röckner, M.: Introduction to the Theory of (Non-symmetric) Dirichlet Forms. Springer, Berlin (1992)
23. Maz'ja, V.G.: Sobolev Spaces. Springer, Berlin (1985)
24. Nikodym, O.: Sur une classe de fonctions considérées dans le problème de Dirichlet. Fund. Math. **21**, 129–150 (1933)
25. Silverstein, M.L.: Symmetric Markov Processes. Lecture Notes in Mathematics, vol. 426. Springer, Berlin (1974)
26. Silverstein, M.L.: The reflected Dirichlet space. Ill. J. Math. **18**, 310–355 (1974)
27. Väisälä, J.: Uniform domains. Tohoku Math. J **40**, 101–118 (1988)

Regularization and Well-Posedness by Noise for Ordinary and Partial Differential Equations

Benjamin Gess

Dedicated to Michael Röckner in honor of his 60th birthday.

Abstract We give a brief introduction and overview of the topic of regularization and well-posedness by noise for ordinary and partial differential equations. The article is an attempt to outline in a concise fashion different directions of research in this field that have attracted attention in recent years. We close the article with a look on more recent developments in the field of nonlinear SPDE, focusing on stochastic scalar conservation laws and porous media equations. The article is tailored at master/PhD level, trying to allow a smooth introduction to the subject and pointing at a large list of references to allow further in-depth study.

Keywords Regularization and well-posedness by noise · Stochastic scalar conservation laws · Stochastic porous medium equation · Stochastic hamilton-jacobi equation

2000 Mathematics Subject Classification 60H15 · 35R60 · 35L65

1 Introduction

1.1 Finite Dimensional Case

In this section we will briefly recall some well-posedness by noise results for the case of ordinary differential equations. Since the available literature is vast and since we are mostly interested in the case of (S)PDE, we will restrict to pointing out some basic examples and a few selected results.

B. Gess (✉) Max-Planck Institute for Mathematics in the Sciences,
Inselstrasse 22, 04103 Leipzig, Germany
e-mail: bgess@mis.mpg.de

A. Eberle et al. (eds.), *Stochastic Partial Differential Equations and Related Fields*, Springer Proceedings in Mathematics & Statistics 229,
https://doi.org/10.1007/978-3-319-74929-7_3

The classical Cauchy–Lipschitz Theorem yields that ordinary differential equations

$$dX_t^x = b(t, X_t^x)dt, \quad X_0^x = x \in \mathbb{R}^d. \tag{1.1}$$

with $b : [0, T] \times \mathbb{R}^d \to \mathbb{R}^d$, have a unique global solution, provided that b is regular enough. For example, in a standard first course in analysis the well-posedness of solutions to (1.1) is usually proven assuming that b is locally Lipschitz continuous and of sublinear growth, that is, for some constant $C \geq 0$,

$$|b(t, x)| \leq C(1 + |x|) \quad \text{for all } t \geq 0, x \in \mathbb{R}^d. \tag{1.2}$$

There are several possibilities to relax these assumptions. For example, the growth condition (1.2) may be replaced by a one-sided growth condition

$$(b(t, x), x) \leq C(1 + |x|) \quad \text{for all } t \geq 0, x \in \mathbb{R}^d \tag{1.3}$$

and (local) Lipschitz continuity can be replaced, to some extent, by (local) one-sided Lipschitz continuity, that is, for some constant $C \geq 0$ (continuous function $C : \mathbb{R}^d \to \mathbb{R}_+$ resp.),

$$(b(t, x) - b(t, y), x - y) \leq C(y)|x - y|^2 \quad \text{for all } x, y \in \mathbb{R}^d. \tag{1.4}$$

Such type of extensions are of particular importance in the case of (S)PDE, leading to the notion of (locally) monotone operators (cf. e.g. [1, 2] and the references therein). In fact, the Cauchy–Lipschitz Theorem provides more information, namely the existence and uniqueness of a continuous flow of solutions $\phi : [0, T] \times [0, T] \times \mathbb{R}^d \to \mathbb{R}^d$ such that

$$\phi_{s,t} = \phi_{r,t} \circ \phi_{s,r}, \quad \phi_{s,s} = Id \quad \forall s \leq r \leq t, \ s, r, t \in [0, T] \tag{1.5}$$

and $\phi_{s,t}(x) = X_t^x$ is the unique solution to (1.1) started in x at time $s \in [0, T]$.

These classical results were extended in the seminal work [3] under less restrictive assumptions on the drift b, more precisely, assuming

$$\frac{|b|}{1 + |x|} \in L^1([0, T]; L^\infty(\mathbb{R}^d)) + L^1([0, T]; L^1(\mathbb{R}^d)), \tag{1.6}$$

and

$$b \in L^1_{loc}([0, T]; W^{1,1}_{loc}(\mathbb{R}^d)) \text{ and } \operatorname{div} b \in L^1([0, T]; L^\infty(\mathbb{R}^d)). \tag{1.7}$$

Under these assumptions it was shown in [3] that there is a unique map $\phi : [0, T] \times [0, T] \times \mathbb{R}^d \to \mathbb{R}^d$ satisfying (1.5) a.e. on $\mathbb{R}^d$ and for all $s \in [0, T]$,

$$\frac{d}{dt}\eta(\phi_{s,t}(x)) = D\eta(\phi_{s,t}(x)) \cdot b(t, \phi_{s,t}(x)), \quad \eta(\phi_{s,s}(x)) = \eta(x)$$

for all $(t, x) \in [0, T] \times \mathbb{R}^d$, η sufficiently smooth and integrable and

$$e^{-|A(t)-A(s)|}\lambda \leq (\phi_{s,t})_*\lambda \leq e^{-|A(t)-A(s)|}\lambda \tag{1.8}$$

for some constant $C_0 \geq 0$, where A is absolutely continuous, $A(0) = 0$ and nondecreasing, λ denotes the Lebesgue measure on $\mathbb{R}^d$ and $(\phi_{s,t})_*\lambda$ the push-forward of λ under $x \mapsto \phi_{s,t}(x)$. In fact, in the construction of solutions the function A is given by $A(t) := \int_0^t \|\text{div}b(r, \cdot)\|_{L^\infty(\mathbb{R}^d)}dr$. The proof of [3] relies on the interpretation of (1.1) as the system of characteristics associated to the linear transport equation

$$du + b(t, x) \cdot \nabla u dt = 0, \quad \text{on } [0, T] \times \mathbb{R}^d, \tag{1.9}$$

that is, on the relation,

$$u(t, x) = u_0(\phi_{0,t}^{-1}(x)),$$

In the sequel of this article this relation between finite dimensional ODE and linear transport equations will be exploited several times. The results of [3] have been generalized in [4] where (1.7) were relaxed to

$$b \in L^1_{loc}([0, T]; BV_{loc}(\mathbb{R}^d)) \text{ and div } b \in L^1([0, T]; L^\infty(\mathbb{R}^d)), \tag{1.10}$$

with several subsequent generalizations, cf. e.g. [5] and [6] for a nice survey.

On the other hand, relaxing the bounded divergence condition on b can lead to simple counter-examples, a standard one being $b(x) = \text{sgn}(x)\sqrt{|x|}$. Clearly, Lipschitz continuity of b fails at zero, causing non-uniqueness of solutions. Indeed, for each given $T \geq 0$,

$$X_t^0 := \begin{cases} 0 & \text{for } t \in [0, T] \\ \pm\frac{1}{2}(t-T)^2 & \text{for } t \geq T \end{cases} \tag{1.11}$$

defines a solution with initial condition $X_0 = 0$. Hence, (forward) uniqueness of solutions fails (branching), but backward uniqueness is satisfied. Another example is given by $b(x) = \sqrt{|x|}$. In this case, solutions are not unique for negative initial conditions x, indeed, for each $T > 0$, setting $c = \sqrt{-x}$,

$$X_t^x := \begin{cases} -\frac{1}{4}(t-2c)^2 & \text{for } t \in [0, 2c] \\ 0 & \text{for } t \in [2c, 2c+T] \\ \frac{1}{4}(t-T-2c)^2 & \text{for } t \geq 2c+T \end{cases} \tag{1.12}$$

defines a solution to (1.1) with $b(x) = \sqrt{|x|}$. In this example, we observe both non-uniqueness of solutions (branching) as well as collisions of solutions (coalescence). In this sense both forward and backward uniqueness of solutions are violated for (1.12). As a last example, consider $b(x) = -\text{sgn}(x)\sqrt{|x|}$. In this case, with $c = \sqrt{|x|}$, solutions are given by

$$X_t^x := \operatorname{sgn}(x) \begin{cases} \frac{1}{4}(t-2c)^2 & \text{for } t \in [0, 2c] \\ 0 & \text{for } t \geq 2c. \end{cases} \tag{1.13}$$

Hence, forward uniqueness is satisfied (no branching), while backward uniqueness is not. Several more counter-examples may be found in [3, 7]. In particular, in [3] it has been shown that the conditions (1.6), (1.10) are essentially sharp and uniqueness of solutions can fail if one of the conditions is "significantly" weakened.

This dramatically changes if the system (1.1) is perturbed by non-degenerate noise. The simplest case of well-posedness by noise can be observed in the case of finite dimensional stochastic differential equations perturbed by additive Brownian noise, that is, for $\sigma > 0$,

$$dX_t^x = b(t, X_t^x)dt + \sigma d\beta_t, \quad X_0^x = x \in \mathbb{R}^d, \tag{1.14}$$

where β denotes a standard Brownian motion in $\mathbb{R}^d$. Classical methods relying on the Girsanov transformation (cf. e.g. [8, 9]) can be used to prove the existence and weak uniqueness for weak solutions to (1.14) if b is only measurable and bounded. In [10] this was significantly strengthened by proving pathwise uniqueness for (1.14) and thus, via the Yamada–Watanabe theorem the existence of strong solutions assuming $b \in L^\infty([0, T] \times \mathbb{R}^d)$. A direct approach to the construction of strong solutions under the same assumption on b has been given in [11, 12], based on Malliavin calculus. A further extension was obtained in [13] where pathwise uniqueness was shown assuming that b satisfies the Krylov–Röckner condition, also sometimes called the strong Ladyzhenskaya-Prodi-Serrin (LPS) condition, that is, $b \in L^q([0, T]; L^p(\mathbb{R}^d))$ with $\frac{d}{p} + \frac{2}{q} < 1, p, q \geq 2$. In these works, however, no continuous dependence of the solutions X_t^x on the initial datum x was shown. This was resolved more recently in [14], again under the assumption that b satisfies the Krylov–Röckner condition. Under the same condition, in [15] the Malliavin differentiability of the solution X_t^x was shown. The natural follow-up question if the equivalence to the Lebesgue measure (1.8) is satisfied under these assumptions in some sense is positively answered in [16]. In addition, the important contribution [17] provides the existence of an associated stochastic flow ϕ and estimates on the derivative of the flow $D_x\phi$ for drifts $b \in L^\infty([0, T]; C_b^\alpha(\mathbb{R}^d))$, $\alpha \in (0, 1)$, with significant impact on the well-posedness of stochastic transport equations as we shall see below. In [18] these results have been partially extended by proving Sobolev differentiability of the stochastic flow ϕ in the sense that $\phi \in L^2(\Omega; W^{1,p}(\mathbb{R}^d; w))$, where $W^{1,p}(\mathbb{R}^d; w)$ are weighted Sobolev spaces with weight w satisfying an integrability condition, assuming only $b \in L^\infty([0, T] \times \mathbb{R}^d)$.

One approach to prove weak uniqueness of solutions to (1.14) is based on the analysis of the associated Fokker–Planck–Kolmogorov equation, satisfied by the evolution of the law, for which the regularizing effect of the noise becomes apparent. More precisely, at least informally, the law $u(t, x) := \text{Law}(X_t^x)$ satisfies the Fokker–Planck–Kolmogorov equation

$$\partial_t u = \frac{\sigma^2}{2}\Delta u + \operatorname{div}(bu), \quad \text{on } [0, T] \times \mathbb{R}^d, \tag{1.15}$$

where the strongly elliptic term $\frac{\sigma^2}{2}\Delta$ is due to the random perturbation in (1.14). The analysis of Fokker–Planck–Kolmogorov equations and their implication on the well-posedness of SDE has attracted much interest in recent years and many results can be found in the recent monograph [19] and the references therein. An extension of the DiPerna–Lions–Ambrosio theory for conservation equations to Fokker–Planck–Kolmogorov equations (1.15) and its consequences for weak/martingale solutions for SDE with multiplicative noise

$$dX_t^x = b(t, X_t^x)dt + \sigma(t, X_t^x)d\beta_t, \quad X_0^x = x \in \mathbb{R}^d. \tag{1.16}$$

has been put forward in [20–22]. A uniqueness result for (1.16) relying on a form of local one-sided Lipschitz condition, related to (1.4) can be found in [23]. The attempt of proving (weak) uniqueness of solutions for SPDE via the associated Fokker–Planck–Kolmogorov equations also seems to be a very promising approach in the theory of SPDE. For some references see Sect. 2 below.

Further extensions of the results recalled above, concerning the question of SDE driven by jump noise have been considered in [24–26]. Extensions to fractional Brownian motion can be found in [15, 27].

We have already encountered different notions of uniqueness of solutions to (1.14). Namely, weak and pathwise uniqueness. Roughly speaking, pathwise uniqueness means that any two strong solutions X_t^x, Y_t^x to (1.14) coincide $\mathbb{P}$-almost surely. Here, the corresponding $\mathbb{P}$-zero set on which these solutions possibly do not coincide is allowed to depend on the initial condition x. This leads to the notion of path-by-path uniqueness. Roughly speaking, we say that (1.14) satisfies path-by-path uniqueness if there is a set of full measure $\Omega_0 \subseteq C([0, T]; \mathbb{R}^d)$ with respect to the Wiener measure such that for each $\omega \in \Omega_0$ solutions to

$$X_t^x = X_0^x + \int_0^t b(r, X_r^x)dr + \omega_t, \quad X_0^x = x \in \mathbb{R}^d \tag{1.17}$$

are unique. Thus, in contrast to pathwise uniqueness, the exceptional zero set is not allowed to depend on the initial datum x in this case. Path-by-path uniqueness for (1.14) for drifts b being measurable and bounded was shown in [28] (cf. [29] for an extension to multiplicative noise). More recently, this was extended in [15] for drifts b satisfying the LPS condition, that is, $b \in L^q([0, T]; L^p(\mathbb{R}^d))$ with $\frac{d}{p} + \frac{2}{q} \le 1$, $p, q \ge 2$. See Sect. 2 for a recent extension of the result of [28] to the setting of SPDE.

In view of this result, a natural question to ask is which properties of a path $\omega \in C([0, T]; \mathbb{R}^d)$ lead to well-posedness of (1.17). In other words, to give a characterization of such paths in Ω_0. In this regard, the notion of an irregular path has been introduced in [27], where it is shown that assuming ω to be irregular in their sense implies the uniqueness of solutions to (1.17) under appropriate assumptions on b.

In conclusion, many details of the effect of well-posedness by noise for finite dimensional differential equations are quite well-understood by now. As we will see in the following, this changes drastically when we pass to the infinite dimensional case, that is, when considering the effect of well-posedness by noise for partial differential equations. At the same time, the uniqueness of solutions to nonlinear PDE arising in fluid dynamics is one of the key open questions in PDE theory. The hope to obtain uniqueness at least in the stochastically perturbed situation can be seen as one of the driving forces in the development of the field of SPDE.

1.2 Linear SPDE

In view of the success of perturbing ODE by additive noise, recalled in the last section, it is a natural idea to try to obtain analogous effects in the case of PDE. Indeed, in several instances, non-degenerate additive noise has been proven to produce regularizing or even well-posedness effects. Typically, these equations are of semilinear type with a leading non-degenerate, linear operator (cf. (2.1), (2.2) below). On the other hand, the phenomenon of turbulence in fluid dynamics is strongly related to vanishing viscosity, i.e. to hyperbolic systems such as the incompressible Euler equations

$$\begin{aligned} \partial_t u + (u \cdot \nabla)u &= 0 \\ \operatorname{div} u &= 0. \end{aligned}$$

In a more recent line of developments it has been uncovered that in the case of hyperbolic (linear) PDE, linear transport noise, rather than additive noise, induces both regularizing and well-posedness effects. In view of the (informal) relation to non-uniqueness and turbulence in fluid dynamics, these results have attracted a lot of interest.

More precisely, in a series of works Flandoli, Gubinelli, Priola and coworkers [7, 17, 30–33] have considered stochastic transport equations of the type, for $\sigma > 0$,

$$du + b(t,x) \cdot \nabla u dt + \sigma \nabla u \circ d\beta_t = 0, \quad \text{on } [0,T] \times \mathbb{R}^d, \tag{1.18}$$

where β is an $\mathbb{R}^d$-valued standard Brownian motion. The inclusion of noise of the above type corresponds to a random perturbation of the transport term b at the level of the characteristics, that is, the system of characteristics informally corresponding to (1.18) is given by

$$dX_t = b(t, X_t)dt + \sigma d\beta_t, \tag{1.19}$$

thus underlining the close relation to the results recalled in the last section. This viewpoint was adopted in [17] in order to prove that (essentially) bounded, weak solutions to (1.18) are unique as long as $b \in L^\infty([0,T]; C_b^\alpha(\mathbb{R}^d))$ and $\operatorname{div} b \in L^p([0,T] \times \mathbb{R}^d)$ for some $p > 2$. In fact, a good part of the proof in [17] consists in showing the existence of a $C^{1+\alpha'}$-flow of solutions $\phi_{0,t}(x,\omega)$ to (1.19), for $0 < \alpha' < \alpha$, and a bound on the corresponding Jacobian $D\phi_{0,t}(\omega, x)$ in $L^2([0,T]; W^{1,2}_{loc}(\mathbb{R}^d))$ a.s.. Via the representation formula

$$u(t,x) = u_0(\phi_{0,t}^{-1}(x)) \tag{1.20}$$

this immediately implies the existence of a bounded, weak solution to (1.18) for each $u_0 \in L^\infty(\mathbb{R}^d)$. Based on the estimate on the Jacobian also a commutator estimate in the spirit of [3] is shown leading to the uniqueness of weak solutions to (1.18). Note that the conditions imposed on b in [17] improve the ones from the deterministic case (cf. (1.7)). In particular, the model example (1.11) provides an example for which the deterministic transport equation (1.9) does not satisfy uniqueness of weak solutions, whereas the stochastically perturbed version (1.18) does.

Also the results obtained in [31] have a nice reinterpretation in terms of (1.18): Based on the representation formula (1.20) one can see that noise prevents the occurrence of shocks, that is, if $b \in L^q([0,T]; L^p(\mathbb{R}^d, \mathbb{R}^d))$, $\frac{d}{p} + \frac{2}{q} < 1$, $p, q \geq 2$ and $u_0 \in \bigcap_{p \geq 1} W^{1,p}(\mathbb{R}^d)$ then

$$\mathbb{P}\left(u(t,\cdot) \in \bigcap_{p\geq 1} W^{1,p}_{loc}(\mathbb{R}^d)\right) = 1 \quad \text{for all } t \geq 0.$$

As we have recalled above, the arguments introduced in [17] heavily rely on the representation of solutions via stochastic characteristics (1.20). This simple representation is lost for nonlinear PDE. In [30], an alternative argument was proposed under different assumptions on b: Let u be a bounded, weak solution to (1.18). Informally, using Itô's formula yields

$$d\eta(u) + b \cdot \nabla\eta(u)dt = -\nabla\eta(u) \circ d\beta_t, \tag{1.21}$$

for all $\eta \in C^1(\mathbb{R})$. We say that a weak solution to (1.18) is renormalized, if $\eta(u)$ satisfies (1.21) for all $\eta \in C^1(\mathbb{R})$. Hence, under appropriate assumptions on b we expect that weak solutions to (1.18) are renormalized. Indeed, this is justified in [30] under the (classical) assumption $b \in L^1([0,T]; BV_{loc}(\mathbb{R}^d))$. Rewriting the Stratonovich integral in the Itô sense and taking expectation then gives

$$d\mathbb{E}\eta(u) + b \cdot \nabla\mathbb{E}\eta(u)dt = \Delta\mathbb{E}\eta(u)dt.$$

Due to the non-degenerate, parabolic structure of this PDE this implies the uniqueness of bounded, weak solutions to (1.18), assuming only $b \in L^2([0,T]; L^\infty(\mathbb{R}^d)) \cap L^1([0,T]; BV_{loc}(\mathbb{R}^d))$ with $\int_0^T \int_{\mathbb{R}^d} \frac{|\operatorname{div} b|}{(1+|x|)^{N_0}} dxdt < \infty$ for some $N_0 \in \mathbb{N}$.

An extension of the above mentioned results to drifts b being only bounded and measurable has been given in [18]. The results on well-posedness of weak solutions by noise have been furthermore complemented in [18, 31, 33] by proving better regularity properties (even obtaining *classical* solutions in [33]) for solutions to the stochastically perturbed PDE than in the deterministic case, under various assumptions on b. In [15, 32, 34, 35] similar results have been established for the continuity equation that is (1.18) in divergence form (in the sense that noise prevents the concentration of mass).

The proof of well-posedness by noise for (1.18) opens the way to study selection principles based on vanishing noise. That is, one aims to let $\sigma \downarrow 0$ in (1.18) and to prove that the corresponding sequence of solutions u^σ to (1.18) converges to a limit u being a solution to the deterministic problem. In the case of the linear transport equation (1.18) such a selection principle has been studied in [36]. Concerning vanishing noise selection principles for SDE we refer to [37–40].

Based on these result one may hope for a similar effect of well-posedness by noise for PDE appearing in fluid dynamics. However, as pointed out in [7] the underlying reasons for non-uniqueness are quite different. In fact, the following negative results for the non-viscous Burgers equation may be found in [7]: Consider

$$du + \partial_x u^2 + \partial_x u \circ d\beta_t = \quad \text{on } [0, T] \times \mathbb{R}. \tag{1.22}$$

Then, setting $v(t, x) := u(t, x + \beta_t)$ we see that, informally, v is a solution to the deterministic Burgers equation

$$dv + \partial_x v^2 = 0 \quad \text{on } [0, T] \times \mathbb{R}. \tag{1.23}$$

In particular, shocks and non-uniqueness of weak solutions still appear in (1.22). Hence, no well-posedness by noise, nor regularization by noise seems to be present in this case. It will be one purpose of the following sections to analyze the validity of this conclusion and to propose different forms of noise that do lead to improvements by noise in the case of nonlinear PDE.

2 Nonlinear SPDE

As compared to the case of linear (stochastic) partial differential equations, much less is known concerning regularizing effects of noise and well-posedness by noise for nonlinear PDE. Historically, the first results in this direction were obtained in the case of viscous PDE perturbed by additive noise. For example, in [41] reaction diffusion equations perturbed by space time white noise of the type

$$du = \Delta u \, dt + f(u) \, dt + dW_t \quad \text{on } [0, 1] \tag{2.1}$$

with Neumann boundary conditions were considered and well-posedness was shown assuming only that f is measurable and satisfies a growth condition. Recently, this result was partially sharpened in [42] to path-by-path uniqueness for (2.1) posed on $\mathbb{R}$ and for f being measurable and bounded, thus extending Davie's result [28] from SDE to SPDE. A further extension of the results obtained in [41] was given in [43] where Malliavin differentiability of the solution to (2.1) was shown under the assumption of f being bounded and measurable.

As mentioned above, one of the key aims in understanding well-posedness by noise effects for nonlinear SPDE is the hope to be able to generate uniqueness of solutions in the case of nonlinear PDE arising in fluid dynamics; the most prominent example being the incompressible 3d Navier–Stokes equations. Despite its relevance and considerable effort, only partial, but highly interesting, results could be obtained in this regard. For example, in [44, 45] the 3d Navier–Stokes equations perturbed by sufficiently non-degenerate additive noise were considered

$$\begin{aligned} du + (u \cdot \nabla)u\, dt + \nabla p\, dt &= \Delta u\, dt + dW_t \\ \operatorname{div} u &= 0. \end{aligned} \tag{2.2}$$

The classical regularity result by Cafarelli-Kohn-Nirenberg [46] states, roughly speaking, that the set of singular points of solutions to the 3d Navier–Stokes equations is a Lebesgue zero set. In contrast, for the stochastically perturbed case (2.2) it was shown in [44] that for each fixed time $t \geq 0$, the set of singular points of the solution $u(t)$ to (2.2) is empty $\mathbb{P}$-a.s.. One should note, however, that the $\mathbb{P}$-zero set in this statement is allowed to depend on the time $t \geq 0$.

In [47] it was shown that noise can prevent the collapse of Vlasov–Poisson point charges. For the deterministic Vlasov–Poisson equation

$$\partial_t f + v\partial_x f + E(t,x)\partial_v f = 0$$

with

$$\rho(t,x) = \int_{\mathbb{R}} f(t,x,v)dv, \quad E(t,x) = \int_{\mathbb{R}} F(x-y)\rho(t,y)dy,$$

where $F \in C(\mathbb{R} \setminus \{0\})$, it is known that measure-valued solutions can develop singularities (cf. [48]). In contrast, in [47] it was shown that in the stochastically perturbed case

$$df + v\partial_x f dt + \Big(E(t,x) + \varepsilon \sum_{k=1}^{\infty} \sigma_k(x) \circ d\beta_t^k\Big)\partial_v f dt = 0,$$

where $\varepsilon > 0$, β^k are independent Brownian motions and σ_k satisfy a non-degeneracy condition, no singularities appear. One should also note the related work [49] proving the prevention of coalescence by noise in point vortex dynamics informally corresponding to stochastic 2D Euler equations. In addition, regularizing effects of noise for stochastic kinetic equations have been found in [50].

As we have recalled in the previous section, one way to prove the well-posedness by noise for SDE relies on considering the associated Fokker–Planck–Kolmogorov equation (1.15). In the case of (S)PDE the associated Fokker–Planck–Kolmogorov equation becomes a PDE on a space of functions with infinitely many independent variables. A considerable amount of work has been done in this direction in recent years. For reference let us refer to the books [51–53] and the recent works [54–59]. For example, in the recent work [59], using the theory of quasi-regular Dirichlet forms and maximal regularity results for Kolmogorov equations in infinite dimensions, the existence and uniqueness of strong solutions to SPDE

$$dX_t = AX_t dt - \nabla V(X_t)dt + B(X_t)dt + dW_t$$

on Hilbert spaces H was shown, where A is a self-adjoint, negative operator with trace class inverse, $V : H \to \bar{\mathbb{R}}$ is convex, proper, lower-semicontinuous satisfying appropriate bounds on its first and second derivative, $B : H \to H$ is measurable and bounded and W is a cylindrical Wiener process on H.

Several further examples of regularization and well-posedness by noise in non-linear situations can be found in [7].

2.1 Scalar Conservation Laws

In the following sections we will investigate regularizing and well-posedness effects of noise in the case of (stochastic) scalar conservation laws (SCL). The current section offers some introductory material and a short overview of some available results for deterministic and stochastic conservation laws. Since a full account of the theory is well beyond the scope of this introduction, we will restrict to pointing out some results that seem most relevant for the following sections.

Scalar conservation laws

$$\partial_t u + \operatorname{div} F(t, x, u) = 0, \quad \text{on } [0, T] \times \mathbb{R}^d \tag{2.3}$$

model the convective transport of a scalar quantity u induced by a possibly non-linear flux $F : [0, T] \times \mathbb{R}^d \times \mathbb{R} \to \mathbb{R}^d$. In the following we will consider either the Cauchy problem or periodic data for (2.3) unless specified otherwise. Typical examples are given by the inviscid Burgers equation and the Buckley-Leverett equation. It is well-known that the hyperbolic structure of these equations may cause the existence of singularities, such as shocks and mass concentration. On the other hand, distributional solutions to (2.3) can be shown to be non-unique in general. In order to select a physically meaningful distributional solution, motivated from the Second Law of thermodynamics (cf. [60, Sect. 4.5]) the notion of entropy solutions has been introduced in [61, 62], with their analogs in terms of mild solutions [63], kinetic solutions [64, 65] and dissipative solutions [66]. In the case of a homogeneous flux, that is $F(t, x, u) = F(u)$, roughly speaking, an entropy solution to (2.3) is a function u such that for all convex functions $\eta \in C^1(\mathbb{R})$, in the sense of distributions,

$$\operatorname{div}_{t,x}\begin{pmatrix}\eta(u)\\ \eta^F(u)\end{pmatrix} := \partial_t\eta(u) + \operatorname{div}_x \eta^F(u) \le 0,$$

where η^F is such that $(\eta^F)'(u) = F'(u)\eta'(u)$. In other words, for entropy solutions entropy-entropy flux pairs (η, η^F) are dissipated at a non-positive rate. Entropy solutions to (2.3) appear as limits of vanishing viscosity approximations

$$\partial_t u^\varepsilon + \operatorname{div} F(t,x,u^\varepsilon) = \varepsilon\Delta u^\varepsilon, \quad \text{on } [0,T]\times\mathbb{R}^d.$$

Noise may enter (nonlinear) SCL in several ways. A statistical analysis of Burgers' equation with white noise as initial condition was given in [67–70]. The first approach to (inviscid) Burgers equations with random forcing appeared in [71] where the stochastic Burgers equation driven by additive noise was considered. In the groundbreaking work [72] the existence and uniqueness of an invariant measure for the stochastic Burgers equation driven by additive Wiener noise

$$du + \frac{1}{2}\partial_x u^2 = dW_t, \tag{2.4}$$

has been shown. The existence of such an invariant measure is not obvious, since the dissipation mechanism needed in order to compensate the addition of energy via the stochastic term is not readily apparent in (2.4). Roughly speaking, the dissipation of energy is due to the loss of energy in the shocks. Generalizations to more general fluxes, boundary value problems and fractional Brownian motion may be found in [73–76].

Simultaneously, significant progress has been made in the case of scalar conservation laws perturbed by multiplicative noise. In [77] Lipschitz multiplicative perturbations to general scalar conservation laws with smooth flux F, i.e.

$$du + \partial_x F(u)dt = h(t,x,u)dt + g(u)dW_t \tag{2.5}$$

have been considered via an operator splitting approach. Feng and Nualart extended the theory of multidimensional stochastic SCL in [78] to multiplicative noise by adapting the (deterministic) notion of entropy solutions to the stochastic case, which led to the notion of *strong entropy solutions*. While this notion allowed an immediate adaptation of the classical Kruzkov uniqueness method to the stochastic case, the existence of strong entropy solutions could be proven only in one spatial dimension via the compensated compactness method. This obstacle was (partially) resolved in [79, 80] by adapting the notion of kinetic solutions to stochastic SCL at least for fluxes not depending on time nor on space ($F(t,x,u) \equiv F(u)$). An alternative proof of existence of strong entropy solutions was given more recently in [81] by proving uniform spatial and temporal regularity properties for the vanishing viscosity approximations in $L^1(\mathbb{R}^d)$. Notably, the methods employed in [81] may also be used for time and space dependent fluxes. Finally, the problem of existence of strong entropy solutions was resolved in [82] since the authors managed to modify

Kruzkov's uniqueness technique in such a way that uniqueness could be shown for a weaker notion of entropy solution (whose existence is simple). There are several extensions available [83–85]. It remains to mention [86], where well-posedness for (linear) transport equations with multiplicative noise has been shown.

All of the above mentioned works consider *semilinear* stochastic scalar conservation laws in the sense that the noise coefficients do not depend on the derivative(s) of the solution. In contrast, in the recent works [87–90] stochastic perturbations of the flux F are considered, which in general lead to SPDE of the type

$$du_t + \sum_{k=1}^{m} \partial_k F_k(t, x, u_t) \circ d\beta_t^k = 0. \tag{2.6}$$

One motivation of SPDE of this type, besides their occurrence in mean field games (cf. [88]), comes from the relation to stochastic Hamilton–Jacobi–Bellman (HJB) equations. More precisely, if we consider the stochastic HJB equation

$$dv_t + F(t, x, \partial_x v_t) + H(t, x, \partial_x v_t) \circ d\beta_t = 0$$

and set $u_t = \partial_x v_t$ then u_t informally satisfies

$$du_t + \partial_x F(t, x, u_t) + \partial_x H(t, x, u_t) \circ d\beta_t = 0.$$

In the case of homogeneous flux (i.e. $F(t, x, u) \equiv F(u)$) solutions to (2.6) can easily be defined by first passing to the corresponding kinetic formulation. Since this turns (2.6) into a (linear) kinetic equation, one may then define solutions via the corresponding (random) flow of characteristics. This approach was extended based on rough path estimates in [87], in order to treat spatially inhomogeneous fluxes, as they appear in (2.6).

2.2 Well-Posedness by Noise for Stochastic Inhomogeneous Scalar Conservation Laws

In Sect. 1.2 we have recalled that stochastic perturbations may lead to well-posedness in the case of transport equations with irregular drift. While part of the motivation of these results was the (informal) relation to questions of uniqueness of solutions to nonlinear PDE arising in fluid mechanics, the methods recalled in Sect. 1.2 highly depend on the linear structure of the transport equation. In addition, the example (1.22) seems to indicate that an analogous well-posedness by noise effect is not expected for nonlinear PDE. Accordingly, it was concluded in [17]: *The generalization to nonlinear transport equations, where b depends on u itself, would be a major next step for applications to fluid dynamics but it turns out to be a difficult problem* (cf. [17, p.6, l.11 ff.]).

In contrast to this, in the recent work [91] it was shown that a similar effect of well-posedness by noise holds for inhomogeneous scalar conservation laws. More precisely, as a special case, in [91] inhomogeneous Burgers equations of the type

$$\partial_t u + b(x) \cdot \nabla u^2 = 0 \quad \text{on } \mathbb{R}^d, \tag{2.7}$$

were considered. Distributional solutions to (2.7) are, in general, non-unique even if $d = 1$, $b \equiv 1$. However, in this case uniqueness can be restored by restricting to so-called entropy solutions. In the spatially inhomogeneous case (2.7) this ceases to be the case if b is not regular enough. For example, we may consider the model example

$$b(x) = \operatorname{sgn}(x)(\sqrt{|x|} \wedge K) \tag{2.8}$$

for some $K > 0$, $u_0(\cdot) = 1_{[0,1]}(\cdot)$, $d = 1$. For any given time $T > 0$ and choosing $K > T + 1$ for simplicity, we have at least two entropy solutions to (2.7) given by

$$u^1(t, x) := \begin{cases} 1 & \text{if } 0 \le x \le (t+1)^2 \\ 0 & \text{otherwise} \end{cases}$$

and

$$u^2(t, x) := \begin{cases} 1 & \text{if } -t^2 \le x \le (t+1)^2 \\ 0 & \text{otherwise.} \end{cases}$$

Hence, entropy solutions to (2.7) are not necessarily unique.

The study of scalar conservation laws with irregular flux has attracted much interest in recent years, see [92–96] among many more, and several selection criteria among entropy solutions have been introduced. One should note, however, that typically divergence form equations with discontinuous fluxes are studied in these works.

In [91] it is shown that for (2.7) noise can have a regularizing effect, in the sense that entropy solutions to

$$du + b(x) \cdot \nabla u^2 \, dt + \nabla u \circ d\beta_t = 0 \quad \text{on } \mathbb{R}^d, \tag{2.9}$$

are unique if we assume that $b \in (L^\infty \cap W^{1,1}_{loc})(\mathbb{R}^d)$ and $\operatorname{div} b \in (L^1 \cap L^p)(\mathbb{R}^d)$ for some $p > d$. In particular, these assumptions are satisfied by the model example (2.8). Hence, a well-posedness by noise effect rather similar to the one observed in the linear case in [17] also appears in the case of nonlinear conservation laws. The regularizing effect of the noise appears with regard to irregularities of the flux function rather than with regard to irregularities of the solution caused by the nonlinear nature of the PDE. In fact, also in the stochastically perturbed case shocks still appear and distributional solutions to (2.9) will still be non-unique.

The following result is a special case of the main result obtained in [91].

Theorem 1 *Assume that $b \in (L^\infty \cap W^{1,1}_{loc})(\mathbb{R}^d)$ and that $\operatorname{div} b \in (L^1 \cap L^p)(\mathbb{R}^d)$ for some $p > d$, $p \le \infty$. For every initial condition $u_0 \in (L^1 \cap L^\infty)(\mathbb{R}^d)$ there exists a unique entropy solution u to (2.9).*

The proof of this result relies on passing to a kinetic form of (2.9). In the deterministic case this goes back to [64]. Given a function $u \in L^2([0,T] \times \mathbb{R}^d)$ we introduce the kinetic function

$$\chi(t,x,\xi) = \chi(u(t,x),\xi) := \begin{cases} 1 & \text{for } 0 < \xi < u(t,x) \\ -1 & \text{for } u(t,x) < \xi < 0 \\ 0 & \text{otherwise.} \end{cases} \tag{2.10}$$

Informally, u is an entropy solution to (2.9) if and only if χ solves

$$\partial_t \chi + \xi b(x) \cdot \nabla \chi + \nabla \chi \circ d\beta_t = \partial_\xi m, \tag{2.11}$$

where m is a nonnegative random measure on $[0,T] \times \mathbb{R}^d \times \mathbb{R}$. An advantage of the kinetic form (2.11) is that, in contrast to (2.9), the kinetic equation (2.11) is a linear equation in χ.

The proof of the existence of entropy solutions to (2.9) relies on considering smooth approximations $(b^\varepsilon, u^\varepsilon)$ of (b, u_0). The existence and uniqueness of an entropy solution to

$$\begin{aligned} du^\varepsilon(t,x) + b^\varepsilon(x) \cdot \nabla (u^\varepsilon)^2(t,x)\,dt + \nabla u^\varepsilon(t,x) \circ d\beta_t &= 0 \\ u^\varepsilon(0,x) &= u_0^\varepsilon(x) \end{aligned}$$

can then be shown by a simple transformation method. In addition, one may obtain uniform bounds of the type, for all $p \in [1,\infty)$,

$$\operatorname{esssup}_{\omega\in\Omega} \sup_{t\in[0,T]} \|u^\varepsilon(t)\|_{L^p(\mathbb{R}^d)} \lesssim \|u_0\|^p_{L^p(\mathbb{R}^d)} + \|u_0\|^{p+1}_{L^\infty(\mathbb{R}^d)} \|\operatorname{div} b\|_{L^1([0,T]\times\mathbb{R}^d)}. \tag{2.12}$$

This allows to extract weakly and weakly* convergent subsequences $u^{\varepsilon_n} \rightharpoonup u, m^{\varepsilon_n} \rightharpoonup m$ as well as $\chi^{\varepsilon_n} = \chi(u^{\varepsilon_n}) \rightharpoonup f$ for some $f \in (L^1_{t,x,\xi} \cap L^\infty_{t,x,\xi})(\mathbb{R}^d)$. The difficulty now is that the limits u and f are not known to satisfy the nonlinear relation (2.10) anymore. This naturally leads to the definition of a generalized entropy solution to (2.11), which relies on weakening the nonlinear relation (2.10) which thereby becomes stable under weak limits. Roughly speaking, a function f is defined to be a generalized entropy solution to (2.9) if f solves (2.11) for some nonnegative measure m and

$$|f| = \operatorname{sgn}(\xi) f \le 1, \; \partial_\xi f = 2\delta_0 - \nu \tag{2.13}$$

for some nonnegative measure ν. The reader may easily verify that entropy solutions are generalized entropy solutions. In view of (2.12) the existence of generalized entropy solutions to (2.11) is obtained in [91] under the weaker assumptions

$b \in L^1_{loc}(\mathbb{R}^d)$, $\operatorname{div} b \in L^1(\mathbb{R}^d)$ and $u_0 \in (L^1 \cap L^\infty)(\mathbb{R}^d)$. One should note that for the proof of the existence of a generalized entropy solution, the presence of the stochastic perturbation in (2.11) was not used. The main difficulty thus becomes the proof that generalized entropy solutions to (2.12) are in fact entropy solutions and that entropy solutions are unique.

In order to show that a generalized entropy solution f is an entropy solution, one needs to show that there is a function u such that $f = \chi(u)$. In view of (2.13) it is enough to prove $|f| \in \{0, 1\}$ a.e.. To do so, in [91] the difference $|f| - f^2$ is estimated using (2.11). In a first step, one realizes that, since $b \in W^{1,1}_{loc}(\mathbb{R}^d)$, weak solutions to (2.11) are renormalized. Informally, this means that $|f| - f^2$ satisfies

$$\partial_t(|f| - f^2) + \xi b(x) \cdot \nabla(|f| - f^2) + \nabla(|f| - f^2) \circ d\beta_t = (\operatorname{sgn}(\xi) - 2f)\partial_\xi m.$$

Passing to the Itô form, integrating in ξ, taking expectation and using $\partial_\xi f = 2\delta_0 - \nu \leq 2\delta_0$ this implies

$$\int \varphi_t \mathbb{E}(|f_t| - f_t^2) \leq \int \varphi_0 \mathbb{E}(|f_0| - f_0^2) + \int_0^t \int \mathbb{E}(|f| - f^2)(\partial_t \varphi + \xi \operatorname{div}(b(x)\varphi) + \Delta\varphi) dx d\xi dr$$

for each nonnegative test function $\varphi \in C_c^\infty(\mathbb{R}_+ \times \mathbb{R}^d)$ not depending on the ξ-variable. Consequently, we would like to solve the parabolic, linear backward PDE

$$\partial_t \varphi + \operatorname{div}(\xi b(x)\varphi) + \Delta\varphi \leq C,$$

for some constant $C > 0$ and then use Gronwall's inequality. The difficulty here is that the test-function φ should not depend on ξ, despite the ξ-dependence of the PDE. The main idea is to split up the PDE in two parts, that is,

$$\begin{aligned}\partial_t \varphi + \operatorname{div}(\xi b(x)\varphi) + \Delta\varphi &= \partial_t \varphi + \xi(\operatorname{div} b)\,\varphi + \xi b(x) \cdot \nabla\varphi + \Delta\varphi \\ &\leq \partial_t \varphi + R|\operatorname{div} b|\,\varphi + R\|b\|_{L^\infty}\|\nabla\varphi\|_{L^\infty} + \Delta\varphi,\end{aligned}$$

where we used that, since $\|u_0\|_\infty < \infty$, we can restrict to $|\xi| \leq \|u_0\|_\infty =: R$. Hence, choosing φ to be a solution to

$$\partial_t \varphi + R|\operatorname{div} b|\,\varphi + \Delta\varphi = 0, \quad \varphi_t = \psi,$$

we obtain that

$$\int \varphi_t \mathbb{E}(|f_t| - f_t^2) \leq \int \varphi_0 \mathbb{E}(|f_0| - f_0^2) + \int \mathbb{E}(|f| - f^2)(R\|b\|_{L^\infty(\mathbb{R}^d)}\|\nabla\varphi\|_{L^\infty(\mathbb{R}^d)}) dt dx d\xi.$$

It then only remains to prove that $\|\varphi\|_{L^\infty(\mathbb{R}^d)} + \|\nabla\varphi\|_{L^\infty(\mathbb{R}^d)}$ is finite, which follows from heat kernel estimates.

2.3 Regularization by Noise for Stochastic Scalar Conservation Laws

Section 2.2 we have seen that linear multiplicative transport noise can lead to well-posedness of entropy solutions in the case of (nonlinear) scalar conservation laws with spatially irregular flux in a similar fashion as it does in the linear case. In other words, also in the nonlinear situation spatial inhomogeneities can be regularized by this type of noise. On the other hand, the example (1.22) shows that singularities that are due to the nonlinear structure of the equation, e.g. shocks, are not affected by linear transport noise. The aim of the following two sections is to present results obtained in [89, 97–99] where it is shown that, in contrast, a certain type of *nonlinear* noise can be used to regularize singularities caused by nonlinear effects in some PDE.

The example (1.22) shows that linear multiplicative noise does not lead to higher regularity for solutions to Burgers' equation. We recall that the kinetic form of the Burgers equation

$$\partial_t u + \partial_x u^2 = 0 \quad \text{on } [0, T] \times \mathbb{T} \tag{2.14}$$

introduced in the last section, reads

$$\partial_t \chi + \xi \partial_x \chi = \partial_\xi m, \tag{2.15}$$

for some nonnegative measure m, were $\mathbb{T}$ denotes the one-dimensional torus. The nonnegativity of the entropy defect measure m corresponds to the restriction to entropy solutions to (2.14). When we drop this restriction, that is, we consider weak solutions to (2.14) such that the kinetic function $\chi = \chi(u)$ satisfies (2.15) for some finite Radon measure m, not necessarily nonnegative, we are let to the class of quasi-solutions to (2.14). One reason to work with this class of solutions is that regularity estimates obtained from averaging techniques are essentially sharp for quasi-solutions. More precisely, following the arguments of [100, 101] one can see that quasi-solutions to (2.14) satisfy $u(t) \in W^{\lambda,1}$ for every $\lambda \in (0, \frac{1}{3})$. As shown in [102], however, there are quasi-solutions u such that $u(t) \notin W^{\lambda,1}$ for every $\lambda > \frac{1}{3}$.

In [98], among other results, the regularity of solutions to scalar conservation laws, including as a special case the stochastic Burgers equation

$$du + \partial_x u^2 \circ d\beta_t = 0 \ \text{ in } \ \mathbb{T} \times (0, \infty), \tag{2.16}$$

has been considered and it was shown that quasi-solutions to (2.16) satisfy $u(t) \in W^{\lambda,1}$ for every $\lambda \in (0, \frac{1}{2})$. In this sense, we see that the noise included in (2.16) has a regularizing effect. More precisely, the corresponding result in [98] reads

Theorem 2 *Let u be a pathwise quasi-solution to* (2.16) *with $u_0 \in L^2(\mathbb{T})$. Then, for all $\lambda \in (0, \frac{1}{2})$ and $T > 0$, there is a $C > 0$ such that*

$$\mathbb{E} \int_0^T \|u(t)\|_{W^{\lambda,1}} dt \leq C(1 + \|u_0\|_2^2 + \mathbb{E}|m|([0, T] \times \mathbb{T} \times \mathbb{R}))$$

and, if u is an entropy solution, for all $\delta > 0$,

$$\sup_{t\geq\delta} \mathbb{E}\|u(t)\|_{W^{\lambda,1}} < \infty.$$

While this shows an improvement of regularity of quasi-solutions by the inclusion of noise, one should note that if $u_0 \in BV(\mathbb{T})$ then each entropy solution to (2.14) satisfies $u(t) \in BV(\mathbb{T})$ for all $t \geq 0$. Hence, in this case no regularizing effect becomes apparent by the methods of [98]. This obstacle will be addressed in the following section.

2.4 *Open Interfaces and Porous Media Equations*

In order to outline the developments of [97] we shall concentrate on the model case of the porous medium equation

$$\partial_t w = \frac{1}{12}\partial_{xx} w^3 \quad \text{on } \mathbb{R}, \tag{2.17}$$

with initial condition w_0 being non-negative, smooth and compactly supported. Informally, one may rewrite this equation as

$$\partial_t w = \frac{1}{4} w^2 \partial_{xx} w + \frac{1}{2} w|\partial_x w|^2.$$

Concentrating at the leading order term $w^2\partial_{xx}w$ we see that the diffusivity coefficient w^2 is large for large values of w but decays to zero for $w \downarrow 0$. This leads to mass building up at the open interface

$$I(t) := \partial \operatorname{supp} w(t, \cdot)$$

and causes the possibility of singularities in terms of a blow-up of $\|\partial_x w(t)\|_{L^\infty}$ even if w_0 is smooth. Indeed, in the long-run solutions converge to the so-called Barenblatt solutions (cf. [103]) given by

$$w(t, x) = t^{-\frac{1}{4}}(C - \frac{1}{12}|x|^2 t^{-\frac{1}{2}})_+^{\frac{1}{2}},$$

where C depends on the L^1 norm of the initial condition. Clearly, at the open interface

$$I(t) = \{x \in \mathbb{R} : C - \frac{1}{12}|x|^2 t^{-\frac{1}{2}} = 0\} = \{\pm\sqrt{12C}t^{\frac{1}{4}}\},$$

the (informal) derivative

$$\partial_x w(t,x) = -\frac{1}{12} t^{-\frac{3}{4}} (C - \frac{1}{12}|x|^2 t^{-\frac{1}{2}})_+^{-\frac{1}{2}} 1_{C \geq \frac{1}{12}|x|^2 t^{-\frac{1}{2}}} x,$$

is unbounded. In [97] the possibility of regularizing this singularity by perturbation with nonlinear noise was investigated. The principle idea put forward in [97] is to regularize the singularity observed for (2.17) by the inclusion of noise of the type (2.16), that is, to consider, for $\sigma > 0$,

$$dv + \frac{\sigma}{2}\partial_x v^2 \circ d\beta_t = \frac{1}{12}\partial_{xx} v^3 \, dt \quad \text{on } \mathbb{R}. \tag{2.18}$$

In [104], the well-posedness and regularity of solutions to (2.18) was shown. As in [98] it remained an open question if this regularity is optimal. More importantly, the regularity estimates for solutions to (2.18) proven in [104] do not improve the regularity known in the deterministic case (2.17). These questions are addressed in [97].

Indeed, the results obtained in [97] prove that the solution v to (2.18) satisfies, if $\sigma > 1$, for all $t \geq 0$,

$$v(t) \in W^{1,\infty}(\mathbb{R}) \quad \mathbb{P}\text{-a.s.},$$

whereas (at least for some choice of initial conditions), if $\sigma \leq 1$, one has

$$\mathbb{P}\text{-a.s.} \quad \exists T > 0, \forall t \geq T, v(t) \notin W^{1,\infty}(\mathbb{R}).$$

More precisely, in [97] the following *sharp* bound is shown:

$$\|\partial_x v(t)\|_{L^\infty} \leq \frac{1}{L^+(t) \wedge L^-(t)}, \tag{2.19}$$

where L^+, L^- are the maximal continuous solutions to the reflected SDE on $(0,\infty)$ given by

$$\begin{aligned} dL^+ &= -\frac{1}{2L^+(t)} dt + \sigma d\beta_t, \quad L^+(0) = \frac{1}{\|(\partial_x v_0)_+\|_{L^\infty}} \\ dL^- &= -\frac{1}{2L^-(t)} dt - \sigma d\beta_t, \quad L^-(0) = \frac{1}{\|(\partial_x v_0)_-\|_{L^\infty}}. \end{aligned}$$

This demonstrates that, when the noise coefficient is large enough, the stochastic perturbation in (2.18) has a regularizing effect as compared to the non-perturbed situation (2.17) for which $\|\partial_x w\|_{L^\infty}$ may blow up in finite time. An interesting point about this result is that the observed regularizing effect depends on the strength of the noise σ, in contrast to the linear case (1.18). Moreover, it is shown in [97] that the estimate in (2.19) is optimal, in the sense that for a class of initial conditions equality in (2.19) holds.

We shall close the account of the results obtained in [97] by stating the main result in its general form. Consider SPDE of the type

$$du + \frac{1}{2}|Du|^2 \circ d\xi_t = F(x, u, Du, D^2u)\,dt \quad \text{on } \mathbb{R}^d, \tag{2.20}$$

for $F : \mathbb{R}^d \times \mathbb{R} \times \mathbb{R}^d \times \mathbb{R}^{d\times d} \to \mathbb{R}$ and $\xi : \mathbb{R}_+ \to \mathbb{R}$ being continuous functions. The map F is supposed to satisfy the typical conditions from viscosity theory allowing to prove the existence and uniqueness of viscosity solutions to (2.20) (cf. e.g. [105, 106]). In addition, and more importantly for our sake, F is assumed to satisfy the following: there exists a locally Lipschitz continuous function $V_F : (0, \infty) \to \mathbb{R}$, bounded from above on $[1, \infty)$ such that for all $g \in BUC(\mathbb{R}^d)$, $t \geq 0$, one has, in the sense of distributions,

$$D^2 g \leq \ell_0^{-1} Id \quad \Rightarrow \quad D^2(S_F(t, g)) \leq \frac{Id}{\varphi^{V_F}(t)(\ell_0)},$$

where $\varphi^{V_F}(t) : \mathbb{R}_+ \to \bar{\mathbb{R}}_+$ is the flow of solutions to the ODE

$$\dot{\ell}(t) = V(\ell), \quad \ell(0) = \frac{1}{\ell_0},$$

stopped when reaching the boundaries 0 or $+\infty$ and $S_F(t, g)$ is the viscosity solution to the deterministic PDE

$$du = F(x, u, Du, D^2u)\,dt \quad \text{on } \mathbb{R}^d.$$

This assumption yields a control on the rate of loss of semiconcavity for S_F. Note that φ^{V_F} may take the value 0 and thus no preservation of semiconcavity is assumed. In the following $BUC(\mathbb{R}^d) = UC_b(\mathbb{R}^d)$ will denote the space of bounded, uniformly continuous functions on $\mathbb{R}^d$.

Theorem 3 *Let $u_0 \in BUC(\mathbb{R}^d)$, $\xi \in C(\mathbb{R}_+)$, assume that F satisfies the above assumptions and let u be the solution to*

$$\begin{cases} du + \frac{1}{2}|Du|^2 \circ d\xi(t) = F(t, x, u, Du, D^2u)dt, \\ u(0, \cdot) = u_0. \end{cases} \tag{2.21}$$

Suppose that $D^2 u_0 \leq \frac{Id}{\ell_0}$ for some $\ell_0 \in [0, \infty)$, in the sense of distributions. Then, for each $t \geq 0$,

$$D^2 u(t, \cdot) \leq \frac{Id}{L(t)}, \tag{2.22}$$

in the sense of distributions, where L is the maximal continuous solution on $[0,\infty)$ *to*

$$\begin{aligned} dL(t) &= V_F(L(t))dt + d\xi(t) \text{ on } \{t \geq 0 : L(t) > 0\}, \quad L \geq 0, \\ L(0) &= \ell_0. \end{aligned} \tag{2.23}$$

The proof is based on a Trotter-Kato splitting scheme for (2.20). The estimate (2.22) is then proven for the corresponding approximating solutions u^n with respect to a discretization L^n of L, based on semiconcavity estimates for S_H, with $H(p) = \frac{1}{2}|p|^2$, where S_H denotes the solution to

$$\partial_t u + H(Du) = 0.$$

It is well-known that S_H and S_{-H} allow to obtain one-sided bounds (of the opposite sign) on the second derivative (cf e.g. [107]), and the fact that one can combine these two bounds to obtain $C^{1,1}$ bounds goes back to Lasry and Lions [108].

More precisely, setting $t_i^n = \frac{ti}{n}$ the approximations

$$u^n(t) := S_H(\xi_{t_{n-1}^n, t_n^n}) \circ S_F(\frac{t}{n}) \circ \cdots \circ S_H(\xi_{t_0^n, t_1^n}) \circ S_F(\frac{t}{n}) u^0$$

are considered, where $\xi_{s,t} := \xi_t - \xi_s$. By an extension of the continuity property of solutions to (2.20) with respect to the continuous driving path ξ to piecewise constant paths, it is shown in [97] that

$$u(t,\cdot) = \lim_{n\to\infty} u^n(t,\cdot).$$

Combining the above mentioned semiconcavity estimates for S_H with the assumption on S_F yields the following discrete analog of the bound (2.22)

$$D^2 u^n(t,\cdot) \leq \frac{Id}{L^n(t)},$$

where L^n is defined by induction

$$L^n(0) = \ell_0, \quad L^n(t_i^n) = \left(\varphi^{V_F}(\frac{t}{n})(L^n(t_{i-1}^n)) - \xi_{t_{i+1}^n, t_i^n}\right)_+ .$$

The remaining difficulty is to prove the convergence of L^n to L, treated in detail in [97].

References

1. Barbu, V.: Nonlinear differential equations of monotone types in Banach spaces. Springer monographs in mathematics. Springer, New York (2010)
2. Liu, W., Röckner, M.: Stochastic partial differential equations: an introduction. Universitext. Springer, Cham (2015)
3. DiPerna, R.J., Lions, P.-L.: Ordinary differential equations, transport theory and Sobolev spaces. Invent. Math. **98**(3), 511–547 (1989)
4. Ambrosio, L.: Transport equation and Cauchy problem for BV vector fields. Invent. Math. **158**(2), 227–260 (2004)
5. Crippa, G., De Lellis, C.: Estimates and regularity results for the diperna-lions flow. J. Reine Angew. Math. **616**, 15–46 (2008)
6. Ambrosio, L., Crippa, G.: Continuity equations and ODE flows with non-smooth velocity. Proc. R. Soc. Edinb. Sect. A **144**(6), 1191–1244 (2014)
7. Flandoli, F.: Random perturbation of PDEs and fluid dynamic models. In: Lecture notes in mathematics, vol. 2015. Springer, Heidelberg (2011). Lectures from the 40th Probability Summer School held in Saint-Flour (2010)
8. Karatzas, I., Shreve, S.E.: Brownian motion and stochastic calculus. In: Graduate texts in mathematics, vol. 113, 2nd edn. Springer, New York (1991)
9. Revuz, D., Yor, M.: Fundamental principles of mathematical sciences. In: Continuous martingales and Brownian motion, vol 293 of Grundlehren der Mathematischen Wissenschaften, 3rd edn. Springer, Berlin (1999)
10. Veretennikov, A.J.: Strong solutions and explicit formulas for solutions of stochastic integral equations. Mat. Sb. (N.S.), 111(153)(3):434–452, 480 (1980)
11. Menoukeu-Pamen, O., Meyer-Brandis, T., Nilssen, T., Proske, F., Zhang, T.: A variational approach to the construction and Malliavin differentiability of strong solutions of SDE's. Math. Ann. **357**(2), 761–799 (2013)
12. Meyer-Brandis, T., Proske, F.: Construction of strong solutions of SDE's via Malliavin calculus. J. Funct. Anal. **258**(11), 3922–3953 (2010)
13. Krylov, N.V., Röckner, M.: Strong solutions of stochastic equations with singular time dependent drift. Probab. Theory Relat. Fields **131**(2), 154–196 (2005)
14. Fedrizzi, E., Flandoli, F.: Pathwise uniqueness and continuous dependence of SDEs with non-regular drift. Stochastics **83**(3), 241–257 (2011)
15. Banos, D., Duedahl, S., Meyer-Brandis, T., Proske, F.: Construction of malliavin differentiable strong solutions of SDEs under an integrability condition on the drift without the yamada-watanabe principle, (2015). arXiv:1503.09019
16. Luo, D.: Quasi-invariance of the stochastic flow associated to Itô's SDE with singular time-dependent drift. J. Theor. Probab. **28**(4), 1743–1762 (2015)
17. Flandoli, F., Gubinelli, M., Priola, E.: Well-posedness of the transport equation by stochastic perturbation. Invent. Math. **180**(1), 1–53 (2010)
18. Mohammed, S.-E.A., Nilssen, T.K., Proske, F.N.: Sobolev differentiable stochastic flows for SDEs with singular coefficients: applications to the transport equation. Ann. Probab. **43**(3), 1535–1576 (2015)
19. Bogachev, V.I., Krylov, N.V., Röckner, M., Shaposhnikov, S.V.: Fokker-Planck-Kolmogorov equations. In: Mathematical surveys and monographs, vol. 207. American Mathematical Society, Providence, RI (2015)
20. Figalli, A.: Existence and uniqueness of martingale solutions for SDEs with rough or degenerate coefficients. J. Funct. Anal. **254**(1), 109–153 (2008)
21. Le Bris, C., Lions, P.-L.: Existence and uniqueness of solutions to Fokker–Planck type equations with irregular coefficients. Commun. Partial Differ. Equ. **33**(7–9), 1272–1317 (2008)
22. Trevisan, D.: Well-posedness of multidimensional diffusion processes with weakly differentiable coefficients. Electron. J. Probab. **21**(22), 41 (2016)
23. Röckner, M., Zhang, X.: Weak uniqueness of Fokker-Planck equations with degenerate and bounded coefficients. C. R. Math. Acad. Sci. Paris **348**(7–8), 435–438 (2010)

24. Haadem, S., Proske, F.: On the construction and Malliavin differentiability of solutions of Lévy noise driven SDE's with singular coefficients. J. Funct. Anal. **266**(8), 5321–5359 (2014)
25. Priola, E.: Pathwise uniqueness for singular SDEs driven by stable processes. Osaka J. Math. **49**(2), 421–447 (2012)
26. Priola, E. Stochastic flow for SDEs with jumps and irregular drift term. In: Stochastic analysis, volume 105 of Banach Center Publications, pages 193–210. Polish Academy of Science Institute of Mathematics, Warsaw (2015)
27. Catellier, R., Gubinelli, M.: Averaging along irregular curves and regularisation of ODEs. Stoch. Process. Appl. **126**(8), 2323–2366 (2016)
28. Davie, A.M. Uniqueness of solutions of stochastic differential equations. Int. Math. Res. Not. IMRN, (24):Art. ID rnm124, 26 (2007)
29. Davie, A.M. Individual path uniqueness of solutions of stochastic differential equations. In: Stochastic analysis 2010, pp. 213–225. Springer, Heidelberg (2011)
30. Attanasio, S., Flandoli, F.: Renormalized solutions for stochastic transport equations and the regularization by bilinear multiplication noise. Commun. Partial Differ. Equ. **36**(8), 1455–1474 (2011)
31. Fedrizzi, E., Flandoli, F.: Noise prevents singularities in linear transport equations. J. Funct. Anal. **264**(6), 1329–1354 (2013)
32. Flandoli, F., Gubinelli, M., Priola, E.: Does noise improve well-posedness of fluid dynamic equations? In: Stochastic partial differential equations and applications, vol. 25 of Quad. Mat., pages 139–155. Deptartment of Mathematics, Seconda University Napoli, Caserta, 2010
33. Flandoli, F., Gubinelli, M., Priola, E.: Remarks on the stochastic transport equation with Hölder drift. Rend. Semin. Mat. Univ. Politec. Torino **70**(1), 53–73 (2012)
34. Maurelli, M. Thesis (2011)
35. Neves, W., Olivera, C.: Wellposedness for stochastic continuity equations with Ladyzhenskaya-Prodi-Serrin condition. NoDEA Nonlinear Differ. Equ. Appl. **22**(5), 1247–1258 (2015)
36. Attanasio, S., Flandoli, F.: Zero-noise solutions of linear transport equations without uniqueness: an example. C. R. Math. Acad. Sci, Paris (2009)
37. Buckdahn, R., Ouknine, Y., Quincampoix, M.: On limiting values of stochastic differential equations with small noise intensity tending to zero. Bull. Sci. Math. **133**(3), 229–237 (2009)
38. Delarue, F., Flandoli, F.: The transition point in the zero noise limit for a 1D Peano example. Discret. Contin. Dyn. Syst. **34**(10), 4071–4083 (2014)
39. Pilipenko, A. Proske, F.: On a selection problem for small noise perturbation in multidimensional case, (2015). arXiv:1510.00966
40. Trevisan, D. Zero noise limits using local times. Electron. Commun. Probab. *18*(31), 7 (2013)
41. Gyöngy, I., Pardoux, É.: On the regularization effect of space-time white noise on quasi-linear parabolic partial differential equations. Probab. Theory Relat. Fields **97**(1–2), 211–229 (1993)
42. Butkovsky, O., Mytnik, L.: Regularization by noise and flows of solutions for a stochastic heat equation, 2016. arXiv:1610.02553
43. Nilssen, T.: Quasi-linear stochastic partial differential equations with irregular coefficients: Malliavin regularity of the solutions. Stoch. Partial Differ. Equ. Anal. Comput. **3**(3), 339–359 (2015)
44. Flandoli, F., Romito. M.: Probabilistic analysis of singularities for the 3D Navier-Stokes equations. In: Proceedings of EQUADIFF, 10 (Prague, 2001), vol. 127, pages 211–218 (2002)
45. Flandoli, F., Romito, M.: Markov selections for the 3D stochastic Navier–Stokes equations. Probab. Theory Relat. Fields **140**(3–4), 407–458 (2008)
46. Caffarelli, L., Kohn, R., Nirenberg, L.: Partial regularity of suitable weak solutions of the Navier–Stokes equations. Commun. Pure Appl. Math. **35**(6), 771–831 (1982)
47. Delarue, F., Flandoli, F., Vincenzi, D.: Noise prevents collapse of Vlasov–Poisson point charges. Commun. Pure Appl. Math. **67**(10), 1700–1736 (2014)
48. Majda, A.J., Bertozzi, A.L.: Vorticity and incompressible flow. In: Cambridge texts in applied mathematics, vol. 27. Cambridge University Press, Cambridge (2002)

49. Flandoli, F., Gubinelli, M., Priola, E.: Full well-posedness of point vortex dynamics corresponding to stochastic 2D Euler equations. Stoch. Process. Appl. **121**(7), 1445–1463 (2011)
50. Fedrizzi, E., Flandoli, F., Priola, E., Vovelle, J.: Regularity of stochastic kinetic equations, (2016). arXiv:1606.01088
51. Cerrai, S.: Second order PDE's in finite and infinite dimension: a probabilistic approach. Lecture notes in mathematics, vol. 1762. Springer, Berlin (2001)
52. Da Prato, G.: Kolmogorov equations for stochastic PDEs.In: Advanced Courses in Mathematics. CRM Barcelona. Birkhäuser Verlag, Basel (2004)
53. Da Prato, G., Zabczyk, J.: Stochastic equations in infinite dimensions. In: Encyclopedia of mathematics and its applications, vol. 44. Cambridge University Press, Cambridge (1992)
54. Da Prato, G., Flandoli, F.: Pathwise uniqueness for a class of SDE in Hilbert spaces and applications. J. Funct. Anal. **259**(1), 243–267 (2010)
55. Da Prato, G., Flandoli, F., Priola, E., Röckner, M.: Strong uniqueness for stochastic evolution equations in Hilbert spaces perturbed by a bounded measurable drift. Ann. Probab. **41**(5), 3306–3344 (2013)
56. Da Prato, G., Flandoli, F., Priola, E., Röckner, M.: Strong uniqueness for stochastic evolution equations with unbounded measurable drift term. J. Theor. Probab. **28**(4), 1571–1600 (2015)
57. Da Prato, G., Flandoli, F., Röckner, M.: Fokker-Planck equations for SPDE with non-trace-class noise. Commun. Math. Stat. **1**(3), 281–304 (2013)
58. Da Prato, G., Flandoli, F., Röckner, M.: Uniqueness for continuity equations in Hilbert spaces with weakly differentiable drift. Stoch. Partial Differ. Equ. Anal. Comput. **2**(2), 121–145 (2014)
59. Da Prato, G., Flandoli, F., Röckner, M., Veretennikov, A.Y.: Strong uniqueness for SDEs in Hilbert spaces with nonregular drift. Ann. Probab. **44**(3), 1985–2023 (2016)
60. Dafermos, C.M.: Fundamental principles of mathematical sciences. In: Hyperbolic conservation laws in continuum physics, vol 325 of Grundlehren der Mathematischen Wissenschaften, 3rd edn. Springer, Berlin (2010)
61. Kružkov, S.N. First order quasilinear equations with several independent variables. Mat. Sb. (N.S.), **81**(123), 228–255 (1970)
62. Lax, P.D.: Hyperbolic systems of conservation laws II. Commun. Pure Appl. Math. **10**, 537–566 (1957)
63. Crandall, M.G.: The semigroup approach to first order quasilinear equations in several space variables. Isr. J. Math. **12**, 108–132 (1972)
64. Lions, P.-L., Perthame, B., Tadmor, E.: A kinetic formulation of multidimensional scalar conservation laws and related equations. J. Am. Math. Soc. **7**(1), 169–191 (1994)
65. Perthame, B.: Kinetic formulation of conservation laws. In: Oxford lecture series in mathematics and its applications, vol. 21. Oxford University Press, Oxford (2002)
66. Perthame, B., Souganidis, P.E.: Dissipative and entropy solutions to non-isotropic degenerate parabolic balance laws. Arch. Ration. Mech. Anal. **170**(4), 359–370 (2003)
67. Avellaneda, M., Weinan, E.: Statistical properties of shocks in Burgers turbulence. Commun. Math. Phys. **172**(1), 13–38 (1995)
68. Burgers, J.: The non-linear diffusion equation: asymptotic solutions and statistical problems. Springer, Lecture series (1974)
69. Ryan, R.: Large-deviation analysis of Burgers turbulence with white-noise initial data. Commun. Pure Appl. Math. **51**(1), 47–75 (1998)
70. Sinaĭ, Y.G.: Statistics of shocks in solutions of inviscid Burgers equation. Commun. Math. Phys. **148**(3), 601–621 (1992)
71. Nakazawa, H.: Stochastic Burgers' equation in the inviscid limit. Adv. Appl. Math. **3**(1), 18–42 (1982)
72. Weinan, E., Khanin,K., Mazel, A., Sinai, Y.: Invariant measures for Burgers equation with stochastic forcing. Ann. Math. (2). **151**(3), 877–960 (2000)
73. Debussche, A., Vovelle, J.: Invariant measure of scalar first-order conservation laws with stochastic forcing. Probab. Theory Relat. Fields **163**(3–4), 575–611 (2015)

74. Kim, J.U.: On a stochastic scalar conservation law. Indiana Univ. Math. J. **52**(1), 227–256 (2003)
75. Saussereau, B., Stoica, I.L.: Scalar conservation laws with fractional stochastic forcing: existence, uniqueness and invariant measure. Stoch. Process. Appl. **122**(4), 1456–1486 (2012)
76. Vallet, G., Wittbold, P.: On a stochastic first-order hyperbolic equation in a bounded domain. Infin. Dimens. Anal. Quantum Probab. Relat. Top. **12**(4), 613–651 (2009)
77. Holden, H., Risebro, N.H.: Conservation laws with a random source. Appl. Math. Optim. **36**(2), 229–241 (1997)
78. Feng, J., Nualart, D.: Stochastic scalar conservation laws. J. Funct. Anal. **255**(2), 313–373 (2008)
79. Debussche, A., Vovelle, J.: Scalar conservation laws with stochastic forcing. J. Funct. Anal. **259**(4), 1014–1042 (2010)
80. Hofmanová, M.: Degenerate parabolic stochastic partial differential equations. Stoch. Process. Appl. **123**(12), 4294–4336 (2013)
81. Chen, G.-Q., Ding, Q., Karlsen, K.H.: On nonlinear stochastic balance laws. Arch. Ration. Mech. Anal. **204**(3), 707–743 (2012)
82. Bauzet, C., Vallet, G., Wittbold, P.: The cauchy problem for conservation laws with a multiplicative stochastic perturbation. J. Hyperbolic Differ. Equ. **9**(4), 661–709 (2013)
83. Bauzet, C., Vallet, G., Wittbold, P.: The Dirichlet problem for a conservation law with a multiplicative stochastic perturbation. J. Funct. Anal. **266**(4), 2503–2545 (2014)
84. Bauzet, C., Vallet, G., Wittbold, P.: A degenerate parabolic-hyperbolic Cauchy problem with a stochastic force. J. Hyperbolic Differ. Equ. **12**(3), 501–533 (2015)
85. Bauzet, C., Vallet, G., Wittbold, P., Zimmermann, A.: On a $p(t, x)$-Laplace evolution equation with a stochastic force. Stoch. Partial Differ. Equ. Anal. Comput. **1**(3), 552–570 (2013)
86. Kim, J.U.: On the Cauchy problem for the transport equation with random noise. J. Funct. Anal. **259**(12), 3328–3359 (2010)
87. Gess, B., Souganidis, P.E.: Scalar conservation laws with multiple rough fluxes. Commun. Math. Sci. **13**(6), 1569–1597 (2015)
88. Lions, P.-L., Perthame, B., Souganidis, P.E.: Scalar conservation laws with rough (stochastic) fluxes. Stoch. Partial Differ. Equ. Anal. Comput. **1**(4), 664–686 (2013)
89. Lions, P.-L., Perthame, B., Souganidis, P.E.: Stochastic averaging lemmas for kinetic equations. In: Séminaire Laurent Schwartz—Équations aux dérivées partielles et applications. Année 2011–2012, Sémin. Équ. Dériv. Partielles, pages Exp. No. XXVI, 17. École Polytech., Palaiseau (2013)
90. Lions, P.-L., Perthame, B., Souganidis, P.E.: Scalar conservation laws with rough (stochastic) fluxes: the spatially dependent case. Stoch. Partial Differ. Equ. Anal. Comput. **2**(4), 517–538 (2014)
91. Gess, B., Maurelli, M.: Well-posedness by noise for scalar conservation laws, (2017). arXiv:1701.05393
92. Andreianov, B., Karlsen, K.H., Risebro, N.H.: On vanishing viscosity approximation of conservation laws with discontinuous flux. Netw. Heterog. Media **5**(3), 617–633 (2010)
93. Andreianov, B., Karlsen, K.H., Risebro, N.H.: A theory of L^1-dissipative solvers for scalar conservation laws with discontinuous flux. Arch. Ration. Mech. Anal. **201**(1), 27–86 (2011)
94. Andreianov, B., Mitrović, D.: Entropy conditions for scalar conservation laws with discontinuous flux revisited. Ann. Inst. H. Poincaré Anal. Non Linéaire **32**(6), 1307–1335 (2015)
95. Crasta, G., De Cicco, V., De Philippis, G.: Kinetic formulation and uniqueness for scalar conservation laws with discontinuous flux. Commun. Partial Differ. Equ. **40**(4), 694–726 (2015)
96. Crasta, G., De Cicco, V., De Philippis, G., Ghiraldin, F.: Structure of solutions of multidimensional conservation laws with discontinuous flux and applications to uniqueness. Arch. Ration. Mech. Anal. **221**(2), 961–985 (2016)
97. Gassiat, P., Gess, B. Regularization by noise for stochastic Hamilton–Jacobi equations, (2016). arXiv:1609.07074

98. Gess, B., Souganidis, P.E.: Long-time behavior, invariant measures, and regularizing effects for stochastic scalar conservation laws. Commun. Pure Appl. Math. **70**(8), 1562–1597 (2017)
99. Gess, B., Souganidis, P.E.: Long-time behaviour, invariant measures and regularizing effects for stochastic scalar conservation laws - revised version, (2017) (preprint)
100. Golse, F., Perthame, B.: Optimal regularizing effect for scalar conservation laws. Rev. Mat. Iberoam. **29**(4), 1477–1504 (2013)
101. Jabin, P.-E., Perthame, B. Regularity in kinetic formulations via averaging lemmas. ESAIM Control Optim. Calc. Var. **8**, 761–774 (2002) (electronic). A tribute to J. L. Lions
102. De Lellis, C., Westdickenberg, M.: On the optimality of velocity averaging lemmas. Ann. Inst. H. Poincaré Anal. Non Linéaire **20**(6), 1075–1085 (2003)
103. Vázquez, J.L.: The porous medium equation: mathematical theory. Oxford mathematical monographs. The Clarendon Press Oxford University Press, Oxford (2007)
104. Gianazza, U., Schwarzacher, S.: Self-improving property of degenerate parabolic equations of porous medium-type, (2016). arXiv:1603.07241
105. Friz, P.K., Gassiat, P., Lions, P.-L., Souganidis, P.E.: Eikonal equations and pathwise solutions to fully non-linear SPDEs, (2016). arXiv:1602.04746
106. Lions, P.-L., Souganidis, P.E.: Stochastic viscosity solutions. (Book, in preparation)
107. Lions, P.-L.: Generalized solutions of Hamilton–Jacobi equations. In: Research notes in mathematics, vol. 69. Pitman (Advanced Publishing Program), Boston, Mass-London (1982)
108. Lasry, J.-M., Lions, P.-L: A remark on regularization in Hilbert spaces. Israel J. Math. **55**(3) (1986)

An Introduction to Singular SPDEs

Massimiliano Gubinelli and Nicolas Perkowski

Abstract We review recent results on the analysis of singular stochastic partial differential equations in the language of paracontrolled distributions.

Keywords Singular SPDEs · Analysis of PDEs · Paradifferential operators

MSC (2010) 60H15

1 Introduction

In recent years there has been much progress in the mathematical understanding of certain non-linear random PDEs which are not well posed in the classical analytic or probabilistic theory but which become amenable to rigorous analysis as soon as specific non-linear properties of the randomness are taken into account. In the related literature it has become common to refer to such equations as *singular SPDEs*, mainly to distinguish them from standard SPDEs. The difference is that singular SPDEs can be posed only in small subspaces of the standard function spaces (e.g. Hölder, Sobolev or even Besov spaces) and that the operations involved in such equations sometimes require *renormalisation*. Renormalisation in this context can be understood as the fact that only specific non-linearities can be formed meaningfully and that, in order to do so, subtractions of infinite quantities are often needed.

Financial support by the DFG via CRC 1060 (MG) and Research Unit FOR 2402 (NP) and is gratefully acknowledged.

M. Gubinelli (✉)
Hausdorff Center for Mathematics & Institute for Applied Mathematics, Universität Bonn, Bonn, Germany
e-mail: gubinelli@iam.uni-bonn.de

N. Perkowski
Institut für Mathematik, Humboldt–Universität zu Berlin, Berlin, Germany
e-mail: perkowsk@math.hu-berlin.de

A. Eberle et al. (eds.), *Stochastic Partial Differential Equations and Related Fields*, Springer Proceedings in Mathematics & Statistics 229,
https://doi.org/10.1007/978-3-319-74929-7_4

The aim of this short review is to present some of the ideas underlying singular SPDEs and their pathwise analysis and to make the connection with other parts of probability theory and mathematical physics: to renormalisation group theory and to scaling limits of interacting particle models. An important motivation to consider singular SPDEs is indeed that in some cases they appear in the description of universal large scale fluctuations for certain spatially extended probabilistic models. Universality here means that irrespective of most specific features of the model its large scale fluctuations are described by a generic equation that usually can be guessed by first principles and then hopefully confirmed by rigorous analysis. One of the main open problem in singular SPDEs is to enlarge the spectrum of models for which the universality can be rigorously proven.

Currently we dispose of four main approaches in order to study singular equations: *regularity structures*, *paracontrolled distributions*, *the renormalisation group approach*, and Otto's and Weber's rough path based approach. Apart from the renormalisation group approach, the other three techniques are all inspired by T. Lyons's *rough path* theory [30–32] and by the related notion of *controlled paths* [18, 25]. Regularity structures have been introduced by M. Hairer in his remarkable work [27] where they were in particular used to give, for the first time, a solution theory for the dynamic Φ^4_3 model. Regularity structures allow a detailed description of the *local* action of distributions on test functions in terms of a given *model* which usually is constructed from certain non-linear features of an underlying random process. Paracontrolled distributions have been introduced by the authors together with P. Imkeller [21], more or less at the same time as M. Hairer was developing his theory, as a tool to describe the "spectral" features of a function (or distribution) in terms of simpler objects, very similar to Hairer's models. Some time after, Kupiainen [29] observed that the renormalisation group approach can be also used to analyse singular equations by decomposing random fields in a multiscale fashion and by introducing recursive equations for each scale. The most recent approach is due to Otto and Weber [37]. In their approach a semigroup is used to provide a multiscale resolution of various singular objects and the scale parameter is handled in the spirit of the time parameter in rough path theory. In this review we will not address the connections of paracontrolled distributions with the other techniques.

We will illustrate the analysis of singular SPDEs on a series of models:

1. The 1d generalised Stochastic Burgers equation (gSBE)

$$\partial_t u(t,x) = \Delta u(t,x) + G(u(t,x))\partial_x u(t,x) + \xi(t,x), \qquad t \geqslant 0, x \in \mathbb{T}, \quad (1)$$

   where $u\colon \mathbb{R}_+ \times \mathbb{T} \to \mathbb{R}$ and $G : \mathbb{R} \to \mathbb{R}$ is a smooth function.
2. The 1d Kardar–Parisi–Zhang equation (KPZ)

$$\partial_t h(t,x) = \Delta h(t,x) + (\partial_x h(t,x))^2 - C + \xi(t,x), \qquad t \geqslant 0, x \in \mathbb{T} \quad (2)$$

   with $h\colon \mathbb{R}_+ \times \mathbb{T} \to \mathbb{R}$ and $C \in \mathbb{R}$ and the related conservative stochastic Burgers equation (CSBE)

$$\partial_t u(t,x) = \Delta u(t,x) + \partial_x(u(t,x)^2) + \partial_x \xi(t,x), \qquad t \geqslant 0, x \in \mathbb{T}, \tag{3}$$

where $u = \partial_x h \colon \mathbb{R}_+ \times \mathbb{T} \to \mathbb{R}$.

3. The dynamic Φ^4_d model or stochastic quantisation equation ($d = 2, 3$) (SQE)

$$\partial_t \varphi(t,x) = \Delta \varphi(t,x) - \varphi(t,x)^3 + C\varphi(t,x) + \xi(t,x), \qquad t \geqslant 0, x \in \mathbb{T}^d, \tag{4}$$

where $\varphi \colon \mathbb{R}_+ \times \mathbb{T}^d \to \mathbb{R}$ and $C \in \mathbb{R}$.

4. The generalised two–dimensional parabolic Anderson model (gPAM)

$$\partial_t u(t,x) = \Delta u(t,x) + G(u(t,x))\xi(x) - CG'(u(t,x))G(u(t,x)), \qquad t \geqslant 0, x \in \mathbb{T}^2, \tag{5}$$

where $u \colon \mathbb{R}_+ \times \mathbb{T}^2 \to \mathbb{R}$, $G : \mathbb{R} \to \mathbb{R}$ is a smooth function, and $C \in \mathbb{R}$.

In all these examples $\mathbb{T}^d = \mathbb{R}^d / \mathbb{Z}^d$ is the d-dimensional torus and ξ denotes a Gaussian white noise (space–time or space dependent only, according to the model). The specific choice of the dimensionality and/or space–time dependence of the noise is related to the degree of singularity of the equation and is motivated by the fact that in this review we will use the specific language of paracontrolled distributions which has the advantage of requiring very little background and is more directly related to standard PDE theory than the other approaches.

This choice has the drawback that we will not address the discussion of natural models like the generalised form of the KPZ equation, given by

$$\partial_t h(t,x) = \Delta h(t,x) + G(h(t,x))(\partial_x h(t,x))^2 + F(h(t,x))\xi(t,x), \qquad t \geqslant 0, x \in \mathbb{T},$$

which is within reach of regularity structure theory but still out of reach for paracontrolled distributions. The generalised KPZ equation is in a way the "ultimate" singular SPDE and its singularities are very challenging. Its analysis via regularity structures requires a great deal of work and a deeper understanding of the algebraic and analytic structures underlying the construction of an appropriate model and of the renormalisation [8, 14].

As we already remarked the main difficulty shared by all singular SPDEs is the presence of non-linear operations which are not well defined in classical function spaces. This difficulty blocks the analysis from the very beginning because it is not even possible to *formulate* the equation rigorously. All the equations we wrote in the introduction are classically ill-posed and the notation is just used in an informal and suggestive way. Indeed most of them have to be modified to take into account *renormalisation* which means that the constant C that appear in the equations is actually to be read as an infinite quantity and not a real number. A standard approach to make the analysis rigorous is to construct a series of approximate problems whose solutions converge to a well-defined limit. The characterisation of this limit and its independence of the details of the approximation procedure will constitute a rigorous definition of the solution to a singular SPDE. Taking as an example the 2d gPAM, we can (loosely) state a typical convergence result:

Theorem 1 *Let $G : \mathbb{R} \to \mathbb{R}$ be a smooth function, $\varepsilon > 0$, and let $u_\varepsilon : \mathbb{R}_+ \times \mathbb{T}^2 \to \mathbb{R}$ be the unique classical solution of the Cauchy problem*

$$\partial_t u_\varepsilon(t,x) = \Delta u_\varepsilon(t,x) + G(u_\varepsilon(t,x))\xi_\varepsilon(x) - G_2(u_\varepsilon(t,x))c_\varepsilon, \qquad t \geqslant 0, x \in \mathbb{T}^2,$$

where ξ_ε is a smooth approximation of the space white noise $\xi : \mathbb{T}^2 \to \mathbb{R}$ by convolution, $c_\varepsilon \in \mathbb{R}$, and $G_2(u) = G'(u)G(u)$. Then there exists a choice for c_ε such that locally in time the sequence u_ε converges, as $\varepsilon \to 0$, to a function u that does not depend on the exact form of the convolutional approximation of ξ_ε.

This formulation is quite minimal, in reality the analysis which has to be put forward to prove this kind of results give as a byproduct also much more detailed information about the function u. In particular it is possible to characterise u via a standard PDE for a different unknown and the assumptions on the approximations of the noise ξ become assumptions about the convergence of certain non-linear functionals of ξ_ε. The general scheme underlying the convergence is the following:

$$\begin{array}{ccccccc} \xi_\varepsilon & \overset{J}{\longleftrightarrow} & \Xi_\varepsilon & \overset{\Phi}{\longmapsto} & U_\varepsilon & \overset{\Pi}{\longmapsto} & u_\varepsilon \\ \downarrow & & \downarrow & & \downarrow & & \downarrow \\ \xi & \longmapsfrom & \Xi & \overset{\Phi}{\longmapsto} & U & \overset{\Pi}{\longmapsto} & u \end{array}$$

Here the vertical arrows represent limits for $\varepsilon \to 0$ (in appropriate function spaces). The upper row features $\Xi_\varepsilon = J(\xi_\varepsilon)$, an injective collection of non-linear quantities constructed from the approximate data ξ_ε (one could also consider the initial condition to be part of this data but we will refrain from doing so), and $U_\varepsilon = \Phi(\Xi_\varepsilon)$, an *enhanced* notion of solution, with *continuous* dependence on Ξ_ε, from which one can recover the classical solution u_ε through a continuous projection Π. The bottom row describes the situation after the limit $\varepsilon \to 0$ has been taken. The *enhanced data* Ξ still determines the limit noise ξ, however the reverse in not true and different approximation procedures for the same ξ can lead to different values of Ξ. But the remaining relations are preserved: from Ξ we can still recover an enhanced notion of solution U through the same continuous *solution map* $\Phi : \Xi \mapsto U$, and from U we obtain u by a projection in such a way that the convergence $u_\varepsilon \to u$ holds by the continuity of the solution map and of the projection. In the limit the situation is more complex than before passing to the limit. As mentioned before a different approximation procedure $\tilde{\xi}_\varepsilon \to \xi$ can lead to different enhanced data $\tilde{\Xi}$ in the limit, and as a consequence of the continuity of Φ to a different limiting solution $\tilde{u}$. This situation is depicted in the graph below.

$$
\begin{array}{ccccccc}
\xi_\varepsilon & \stackrel{J}{\longleftrightarrow} & \Xi_\varepsilon & \stackrel{\Phi}{\longmapsto} & U_\varepsilon & \stackrel{\Pi}{\longmapsto} & u_\varepsilon \\
\downarrow & & \downarrow & & \downarrow & & \downarrow \\
\xi & \longleftarrow & \Xi & \stackrel{\Phi}{\longmapsto} & U & \stackrel{\Pi}{\longmapsto} & u \\
= & & \neq & & \neq & & \neq \\
\xi & \longleftarrow & \tilde{\Xi} & \stackrel{\Phi}{\longmapsto} & \tilde{U} & \stackrel{\Pi}{\longmapsto} & \tilde{u} \\
\uparrow & & \uparrow & & \uparrow & & \uparrow \\
\tilde{\xi}_\varepsilon & \stackrel{J}{\longleftrightarrow} & \tilde{\Xi}_\varepsilon & \stackrel{\Phi}{\longmapsto} & \tilde{U}_\varepsilon & \stackrel{\Pi}{\longmapsto} & \tilde{u}_\varepsilon
\end{array}
$$

So the relation between the data ξ and the solution is not well defined unless some information on the non-linear features Ξ is provided as additional input.

Proving a convergence result therefore requires two conceptually different steps:

1. define and analyse the continuity of the map Φ, where an important task is to identify its domain and co–domain;
2. prove the convergence of the enhanced data $\Xi_\varepsilon \to \Xi$.

The convergence of the enhanced data is, in most of the applications, a purely probabilistic step which sometimes requires development of efficient tools but for which the main tools are classical and already present in the probabilistic literature for a long time. Some keywords in this context are Gaussian analysis, hypercontractivity, Besov embeddings, almost sure regularity of stochastic processes, martingale theory, Wick products, and chaos expansions. We would like to concentrate our discussion to the analytic part of the theory involving the construction of the enhanced spaces which constitute the domain and co-domain of the solution map Φ.

In Sects. 2 and 3 we present the basic ideas and analytic ingredients for paracontrolled distributions. Section 4 briefly discusses the need for renormalisation of the enhanced data and how this renormalisation translates in the equation. In Sect. 5 we sketch recent work of Bailleul and Bernicot on higher order expansions via paraproducts. Section 6 is dedicated to convergence results for singular SPDEs and we illustrate how to derive the Hairer-Quastel weak universality principle for the KPZ equation using paracontrolled distributions. In Sects. 7 and 8 we will see that paracontrolled distributions can not only be used to study singular SPDEs, but as noted by Cannizzaro and Chouk respectively Allez and Chouk they also allow us to construct certain singular operators.

Finally let us point out that there are many fascinating recent results that are based on paracontrolled distributions and that we have to omit here due to space constraints. To name just a few: non-explosion results for the dynamic Φ^4_3 model [34], the KPZ equation [24], and the multi-component KPZ equation [19], a formulation of paracontrolled distributions that allows to study equations on manifolds [3], convergence results for discrete dynamics [13, 24, 40, 41], a solution theory for quasilinear equations [6], nonlinear extensions of paraproducts [17], and a support theorem for gPAM [11] – not to mention all the exciting results that have been shown in the setting of regularity structures or the other approaches.

2 Paraproducts

In order to develop the paracontrolled analysis of the solution map we will introduce in this section Besov spaces and paraproducts. See [5] for details. Let $\mathscr{S}'(\mathbb{T}^d)$ denote the Schwartz space of distributions on $\mathbb{T}^d = \mathbb{R}^d/\mathbb{Z}^d$. Then any element $f \in \mathscr{S}'(\mathbb{T}^d)$ can be decomposed as

$$f = \sum_{i \geqslant -1} \Delta_i f,$$

where the sum is over $i = -1, 0, 1, 2, \dots$ and $\Delta_i f$ are smooth functions whose Fourier support is contained in a ball $\mathscr{B} \subseteq \mathbb{R}^d$ for $i = -1$ and in rescaled dyadic annuli $2^i \mathscr{A}$ for $i \geqslant 0$. The operators $\Delta_i : f \mapsto \Delta_i f$ are called Littlewood–Paley operators and can be constructed to enjoy nice analytic properties, for example they satisfy the Bernstein inequalities

$$\|D^\alpha \Delta_i f\|_{L^p(\mathbb{T}^d)} \lesssim 2^{i|\alpha| + i\left(\frac{d}{q} - \frac{d}{p}\right)} \|\Delta_i f\|_{L^q(\mathbb{T}^d)}, \qquad i \geqslant -1, \qquad p \geqslant q,$$

where α is a d-dimensional multiindex and D^α denotes the related mixed derivative of order $|\alpha|$. The (inhomogeneous) Besov space $B^\alpha_{p,q}$ is defined as the set of all distributions $f \in \mathscr{S}'(\mathbb{T}^d)$ such that the norm

$$\|f\|_{B^\alpha_{p,q}} = \|(\|2^{i\alpha} \Delta_i f\|_{L^p(\mathbb{T}^d)})_{i \geqslant -1}\|_{\ell^q(\mathbb{Z})}$$

is finite. In the analysis of singular SPDEs we will mainly use the Hölder–Besov spaces $\mathscr{C}^\alpha = B^\alpha_{\infty,\infty}$ to get rid of the integrability exponents. This turns out to be very convenient for non-linear estimates because $\|fg\|_{L^\infty} \leqslant \|f\|_{L^\infty} \|g\|_{L^\infty}$ and similarly for ℓ^∞ which is not true for the L^p respectively ℓ^p norms with $p < \infty$. To gain an intuitive understanding of the $\mathscr{C}^\alpha$ spaces it is useful to note that for $\alpha \in \mathbb{R}_+ \setminus \mathbb{Z}$ the space $\mathscr{C}^\alpha$ consists exactly of the $\lfloor \alpha \rfloor$ times continuously differentiable functions from $\mathbb{T}^d \to \mathbb{R}$ for which the classical increment-based $(\alpha - \lfloor \alpha \rfloor)$-Hölder norm of all partial derivatives of order $\lfloor \alpha \rfloor$ is finite. And roughly speaking for $\alpha \in (-1, 0)$ every $f \in \mathscr{C}^\alpha$ is the distributional derivative of some $F \in \mathscr{C}^{\alpha+1}$, and similarly distributions of lower regularity are higher order derivatives of Hölder continuous functions.

The Littlewood–Paley decomposition induces a natural decomposition of products of Besov functions in terms of paraproducts. Given f, g we have

$$fg = \sum_{i,j \geqslant -1} \Delta_i f \Delta_j g = f \prec g + f \circ g + f \succ g,$$

where the *paraproducts* $f \prec g$ and $f \succ g$ and the *resonant product* $f \circ g$ by are defined, respectively, by

$$f \prec g = g \succ f := \sum_{i \geqslant -1} (\Delta_{<i-1} f) \Delta_i g, \qquad f \circ g := \sum_{i,j:|i-j| \leqslant 1} \Delta_i f \Delta_j g,$$

and where we introduced the notation $\Delta_{<k} f = \sum_{\ell<k} \Delta_\ell f$. Paraproducts are continuous bilinear operations on the following function spaces

$$* \prec * : \mathscr{C}^\alpha \times \mathscr{C}^\beta \to \mathscr{C}^{\beta+\alpha}, \qquad \alpha \leqslant 0, \beta \in \mathbb{R},$$

$$* \prec * : L^\infty \times \mathscr{C}^\beta \to \mathscr{C}^\beta, \qquad \beta \in \mathbb{R},$$

while the resonant product is well defined only if $\alpha + \beta > 0$ and in that case it is a continuous bilinear operator

$$* \circ * : \mathscr{C}^\alpha \times \mathscr{C}^\beta \to \mathscr{C}^{\beta+\alpha}.$$

In particular we see that the usual product can be extended by continuity from smooth functions to Hölder–Besov distributions as a bilinear map

$$(f, g) \mapsto fg : \mathscr{C}^\alpha \times \mathscr{C}^\beta \to \mathscr{C}^{\min(\alpha,\beta)}$$

provided $\alpha + \beta > 0$.

Singular SPDEs are characterised by this condition not being satisfied in the non-linear terms. The products then become problematic and undefined for general inputs. To explain the difficulty let us note that given $\alpha, \beta \in \mathbb{R}$ with $\alpha + \beta < 0$ and given smooth functions (f, g) it is easy to construct sequences of functions (f_n, g_n) such that $(f_n, g_n) \to (f, g)$ in $\mathscr{C}^\alpha \times \mathscr{C}^\beta$ and such that the pointwise limit $\lim_n f_n g_n$ exists and is smooth but nonetheless

$$\lim_n f_n g_n \neq fg.$$

If we want a "robust" way to consider the product fg in the situation $\alpha + \beta < 0$, then we should take this *ambiguity* into account from the start and think about the product as describing a manifold of possibilities and not just a single deterministic operation on the inputs.

Part of the analysis of singular SPDEs can be understood as a classification of these ambiguities: we track the extent to which the possible outcomes of undefined operations can propagate into the solution theory of a given equation. We will come back to this point below when we discuss renormalisations.

Paraproducts and related paradifferential operators have been introduced in the seminal work of Bony on the propagation of singularities for non-linear hyperbolic equations [7, 33]. They provide a good approximation of non-linear operations, in this case the product, but can be used also to linearise other operations. For example, the following paralinearisation result is useful in order to deal with equations with non-polynomial coefficients: Given a smooth function $f : \mathbb{R} \to \mathbb{R}$ we have

$$u \in \mathscr{C}^{\alpha} \mapsto R_f(u) := f(u) - f'(u) \prec u \in \mathscr{C}^{2\alpha}, \qquad \alpha > 0, \tag{6}$$

which shows that the composition $f(u) \in \mathscr{C}^{\alpha}$ behaves like the paraproduct $(f'(u) \prec u) \in \mathscr{C}^{\alpha}$ modulo a smoother correction term.

One main tool in the paracontrolled analysis is the following commutator lemma which describes the interaction between the resonant product and the paraproduct. For a proof see [21].

Lemma 1 *Let $\alpha \in (0, 1)$, $\beta, \gamma \in \mathbb{R}$ be exponents such that $\beta + \gamma < 0$ and $\alpha + \beta + \gamma > 0$. Then there exists a continuous trilinear map $C : \mathscr{C}^{\alpha} \times \mathscr{C}^{\beta} \times \mathscr{C}^{\gamma} \to \mathscr{C}^{\alpha+\beta+\gamma}$ such that if f, g, h are smooth functions we have*

$$C(f, g, h) = (f \prec g) \circ h - f(g \circ h).$$

3 Paracontrolled Analysis

The construction of the solution maps proceeds via perturbative analysis with respect to a linearised approximation of the equation. We look for an expansion in terms of regularities of the various terms. In the following we ignore the initial conditions for the equations that we treat and when considering a solution w to $\mathscr{L} w = v$ we always silently assume that $w(0) = 0$. Once this is understood the case of general initial conditions does not add much conceptual (although some technical) difficulty. We also ignore the need for renormalisation at the moment and set the (infinite) constants C appearing in the equations equal to 0 in the following discussion. In general we omit many technical details, an introduction to paracontrolled distributions that provides more details can be found in the lecture notes [22].

Among the examples we treat, the simplest is the dynamic Φ^4_2 model (4), $\varphi : \mathbb{R}_+ \times \mathbb{T}^2 \to \mathbb{R}$,

$$\partial_t \varphi(t, x) = \Delta \varphi(t, x) - \varphi(t, x)^3 + \xi(t, x), \qquad t \geqslant 0, x \in \mathbb{T}^2.$$

This equation was first solved by Albeverio and Röckner [2] in 1991, a solution that is more related to the tools we present here was given in 2003 by Da Prato and Debussche [16], and the link between the two solution concepts was recently understood by Röckner, Zhu, and Zhu [39]. Da Prato and Debussche observed that we can decompose the solution as a sum of two terms which we postulate to be of increasing regularities: Setting $\varphi = X + \psi$, the equation takes the form

$$\mathscr{L} X + \mathscr{L} \psi = -X^3 - 3X^2 \psi - 3X \psi^2 - \psi^3 + \xi$$

where we use the notation $\mathscr{L} = \partial_t - \Delta$ and we choose X in order to cancel the most irregular term on the right hand side, namely the additive noise. So if $\mathscr{L}X = \xi$, a simple inhomogeneous linear equation that can be explicitly solved by convolving ξ against the heat kernel, we are left with an equation for ψ,

$$\mathscr{L}\psi = -X^3 - 3X^2\psi - 3X\psi^2 - \psi^3. \tag{7}$$

Using stochastic analysis it is possible to prove that

$$X \in C_T\mathscr{C}^{0-} := \bigcap_{\delta>0} C([0,T], \mathscr{C}^{-\delta}),$$

and therefore the right hand side of the equation for ψ is still not well defined because it features the products X^2 and X^3 of the distribution X of negative regularity. For the moment we *assume* that X^3, X^2 exist as elements of $C_T\mathscr{C}^{0-}$. They will be part of the non-linear features needed to define the solution map. If we also assume that

$$\psi \in C_T\mathscr{C}^{0+} := \bigcup_{\delta>0} C([0,T], \mathscr{C}^{\delta}),$$

then the right hand side of Eq. (7) is well defined and the estimates for the product give us $X^2\psi, X\psi^2 \in C_T\mathscr{C}^{0-}$. By standard estimates for the heat semigroup generated by the Laplace operator Δ the map that sends v to the solution w of $\mathscr{L}w = v$ is continuous from $C_T\mathscr{C}^{\alpha} = C([0,T], \mathscr{C}^{\alpha})$ equipped with $\|v\|_{C_T\mathscr{C}^{\alpha}} = \sup_{t\in[0,T]} \|v(t)\|_{\mathscr{C}^{\alpha}}$ to $C_T\mathscr{C}^{\alpha+2}$. So it follows that we can pose the Eq. (7) for ψ as a standard PDE with unique weak solution $\psi \in C_T\mathscr{C}^{2-}$ for some[1] $T > 0$. Here the change of variable has therefore been quite simple: The enhancement J, the solution map Φ and the projection Π take the form

$$J : \xi \mapsto (X, X^2, X^3),$$

$$\Phi : \Xi = (\zeta_1, \zeta_2, \zeta_3) \mapsto U = (\zeta_1, \psi), \qquad \Pi : U = (\zeta_1, \psi) \mapsto \varphi = \zeta_1 + \psi,$$

where ψ is a weak solution to

$$\mathscr{L}\psi = \zeta_3 + 3\zeta_2\psi + 3\zeta_1\psi^2 + \psi^3. \tag{8}$$

Note that ξ can be recovered from $(\zeta_1, \zeta_2, \zeta_3) = J(\xi)$ via $\xi = \mathscr{L}X$. Observe also that if ξ is a smooth function, the composition $\Pi \circ \Phi \circ J : \xi \mapsto \varphi$ gives back the classical weak solution to Eq. (4). We will come back again later to this equation to

---

[1] Possibly T is quite small because when setting up the Picard iteration we pick up a superlinear estimate so at this point we cannot exclude the possibility that the solution blows up in finite time. To prove existence for all times we have to make use of the sign of the nonlinearity $-\varphi^3$ in (4), see [35].

discuss the construction of the enhanced data Ξ which will require a renormalisation in the case of white noise. For details on the analysis that we sketched above see [16].

While the method we just discussed is simple and elegant, the other singular equations that we mentioned in the introduction apart from the dynamic Φ_2^4 model cannot be handled by a simple additive change of variables. Consider for example the generalised stochastic Burgers equation (1). Since the noise is additive as in the dynamic Φ_2^4 equation, we can proceed with the same decomposition of the solution. We let X be the solution to $\mathscr{L}X = \xi$ and write $u = X + v$, where v solves

$$\mathscr{L}v = G(X+v)\partial_x X + G(X+v)\partial_x v. \tag{9}$$

The analysis of the regularity of X now gives $X \in C_T\mathscr{C}^{1/2-}$ (it is better behaved than before because we are in one space-dimension and the white noise becomes more and more irregular in higher dimensions) and therefore the best we can hope for the right hand side of (9) is that it takes values in $C_T\mathscr{C}^{-1/2-}$ (the regularity of $\partial_x X$) which would put v in $C_T\mathscr{C}^{3/2-}$, that is v has two degrees of regularity more than the right hand side which follows from the regularising effect of the inversion of $\mathscr{L}$ that we discussed above. This would in turn mean that $G(X+v) \in C_T\mathscr{C}^{1/2-}$ provided G is at least C^1. In this setting the product $G(X+v)\partial_x v$ is well defined since $\partial_x v$ has regularity $C_T\mathscr{C}^{1/2-}$, but $G(X+v)\partial_x X$ is not since $\partial_x X$ has negative regularity $C_T\mathscr{C}^{-1/2-}$ and we barely fail to ensure that the sum of the regularities is positive. But no other simple additive subtraction is available and therefore we need to understand better the structure of the problematic product in order to determine sufficient conditions to control it. The paraproduct decomposition gives

$$G(X+v)\partial_x X = \underbrace{G(X+v)\prec\partial_x X}_{C_T\mathscr{C}^{-1/2-}} + \underbrace{G(X+v)\circ\partial_x X}_{!!} + \underbrace{G(X+v)\succ\partial_x X}_{C_T\mathscr{C}^{0-}},$$

where the underbraces denote the respective regularities of the two paraproducts and the difficulty is isolated in the resonant term. The paralinearisation result (6) applied to $G(X+v)$ shows that

$$G(X+v) = \underbrace{G'(X+v)\prec(X+v)}_{C_T\mathscr{C}^{1/2-}} + \underbrace{R_G(X+v)}_{C_T\mathscr{C}^{1-}},$$

and we can decompose the paraproduct on the right hand side into

$$G'(X+v)\prec(X+v) = \underbrace{G'(X+v)\prec X}_{C_T\mathscr{C}^{1/2-}} + \underbrace{G'(X+v)\prec v}_{C_T\mathscr{C}^{3/2-}},$$

which shows that the irregularity of $G(X+v)$ comes from the paraproduct $G'(X+v)\prec X$ and we can further isolate the difficulty in the resonant product:

$$G(X+v)\circ\partial_x X = \underbrace{(G'(X+v)\prec X)\circ\partial_x X}_{!!} + \underbrace{G'(X+v)\prec v}_{C_T\mathscr{C}^{1-}} + \underbrace{R_G(X+v)\circ\partial_x X}_{C_T\mathscr{C}^{1/2-}},$$

where the last two terms on the right hand side can be controlled by the estimates for the resonant product because here the sum of the regularities is strictly positive. To deal with the remaining ill-defined resonant product we apply the commutator estimate from Lemma 1 which gives us

$$(G'(X+v)\prec X)\circ\partial_x X = \underbrace{C(G'(X+v), X, \partial_x X)}_{C_T\mathscr{C}^{1/2-}} + G'(X+v)\underbrace{(X\circ\partial_x X)}_{!!},$$

where we continue to denote with the underbrace "!!" the term which is still problematic according to the deterministic a priori regularities. The reader can convince herself that all the other terms are indeed well defined. This analysis allowed us to *isolate* the singular nature of the product into some non-linear feature pertaining only to the data X. Much like the simpler algebraic analysis allowed by the Φ^4_2 model. Similarly we will now *assume* a prescribed natural regularity for $X\circ\partial_x X$, namely $X\circ\partial_x X \in C_T\mathscr{C}^{0-}$, and include this term in the enhanced data for this problem. Then the remaining product $G'(X+v)(X\circ\partial_x X)$ is well defined because $G'(X+v) \in C_T\mathscr{C}^{1/2-}$ and therefore the sum of the regularities of the factors is strictly positive. From here we can continue to solve the equation for v by a Picard iteration. The enhancement J, the solution map Φ and the projection Π now take the form

$$J : \xi \mapsto (X, X\circ\partial_x X)$$

$$\Phi : \Xi = (\zeta_1, \zeta_2) \mapsto U = (\zeta_1, v), \qquad \Pi : U = (\zeta_1, v) \mapsto u = \zeta_1 + v,$$

where $v \in C_T\mathscr{C}^{3/2-}$ is the solution to the PDE

$$\mathscr{L}v = G(\zeta_1+v)\prec\partial_x\zeta_1 + G'(\zeta_1+v)\zeta_2 + F(\zeta_1, v)$$

and where $F(\zeta_1, v)$ is a suitable continuous function taking values in $C_T\mathscr{C}^{0-}$. For details on this equation we refer to [21, 26].

A further level of complexity is exemplified by the gPAM (5). In this case it is not even possible to start the analysis by an additive change of variables. The two dimensional space white noise has regularity $\mathscr{C}^{-1-}$, so the best regularity we can hope for v is $v \in C_T\mathscr{C}^{1-}$ and then the non-linear term $G(u)\xi$ is not well defined. The paraproduct decomposition gives

$$G(u)\xi = \underbrace{G(u)\prec\xi}_{C_T\mathscr{C}^{-1-}} + \underbrace{G(u)\circ\xi}_{!!} + \underbrace{G(u)\succ\xi}_{C_T\mathscr{C}^{0-}}$$

and proceeding by paralinearisation and commutation we obtain the following decomposition of the resonant term

$$G(u)\circ\xi = (G'(u)\prec u)\circ\xi + R_G(u)\circ\xi \tag{10}$$

$$= \underbrace{C(G'(u), u, \xi)}_{C_T\mathscr{C}^{1-}} + G'(u)\underbrace{(u\circ\xi)}_{!!} + \underbrace{R_G(u)\circ\xi}_{C_T\mathscr{C}^{1-}}, \tag{11}$$

where we note that $u\circ\xi$ is still not well defined but if we assume it has its natural regularity $u\circ\xi \in C_T\mathscr{C}^{0-}$, then the product $G'(u)(u\circ\xi)$ poses no problem. This means that we can control the product $G(u)\xi$ once we have a control of the resonant term $u\circ\xi$. Contrary to the simpler analysis of the gSBE this term is still quite complex since involves the unknown u and cannot be "postulated" as we did with $X\circ\partial_x X$ before. However, our analysis shows that the right hand side of the equation can be decomposed in a series of terms of different regularities, where the worst is $G(u)\prec\xi$ (assuming for $u\circ\xi$ a better regularity). So the solution should satisfy

$$\mathscr{L}u = G(u)\prec\xi + \cdots,$$

where we neglected more regular terms. The idea is then to make a change of variables to remove this irregular term in the right hand side. A natural approach is to look for u with a similar form as the right hand side of the equation, namely a paraproduct plus a smoother remainder, $u = v\prec X + v^\sharp$, where $v \in C_T\mathscr{C}^{1-}$, $X \in C_T\mathscr{C}^{1-}$, $v^\sharp \in C_T\mathscr{C}^{2-}$ are functions to be dermined with $v^\sharp$ more regular than X and u. In order to perform this change of variables in the equation we need to modify the paraproduct $*\prec*$ a bit by introducing some time-smoothing and defining a modified paraproduct $*\prec\!\!\prec*$ in terms of which we make the *paracontrolled Ansatz*:

$$u = v\prec\!\!\prec X + v^\sharp. \tag{12}$$

This modification of the paraproduct is a small technical point which is not very relevant to the overall picture. The essential property of $*\prec\!\!\prec*$ is that the operator

$$(f, g) \mapsto H(f, g) = \mathscr{L}(f\prec\!\!\prec g) - f\prec\mathscr{L}g$$

maps[2] the space $C_T\mathscr{C}^{1-} \times C_T\mathscr{C}^{1-}$ to $C_T\mathscr{C}^{0-}$ despite the fact that both summands only live in $C_T\mathscr{C}^{-1-}$. For the usual paraproduct $*\prec*$ this is not possible because if we expand $\mathscr{L}(f\prec g) - f\prec\mathscr{L}g$ using Leibniz's rule we pick up the term $\partial_t f\prec g$ which cannot be controlled in terms of the $C_T\mathscr{C}^{1-}$ norm of f. The modified paraproduct overcomes this difficulty, but there exist other solutions: either use space-time parabolic Besov spaces and related paraproducts (for which the commutator of paraproduct and $\mathscr{L}$ can be controlled by standard estimates), or define a paraproduct which intertwines exactly with the heat kernel so $H(f, g) = 0$ (we will discuss this

[2]This is not exactly true, we also need some time regularity of f but for simplicity we omit this in the discussion.

last strategy in more detail in Sect. 5). In the following we will mostly neglect the difference between these two paraproducts and always consider $H(f, g)$ as a smoother remainder term. With this proviso we can expand both sides of the equation and get

$$\mathscr{L}u = v \prec \mathscr{L}X + H(v, X) + \mathscr{L}v^\sharp = G(u) \prec \xi + \tilde{F}(u, \xi),$$

where $\tilde{F}(u, \xi)$ denotes terms that *should be* in $C_T\mathscr{C}^{0-}$. Choosing $v = G(u)$ and $\mathscr{L}X = \xi$ we can get rid of the irregular term $G(u) \prec \xi$ on the right hand side and obtain an equation for $v^\sharp$ which sets it in the good space $C_T\mathscr{C}^{2-}$. Now that we have a better description of the solution u via the paracontrolled Ansatz we can go back to the analysis of the resonant term $u \circ \xi$ and observe that the commutator lemma gives

$$\underbrace{u \circ \xi}_{!!} = \underbrace{(G(u) \prec X) \circ \xi}_{!!} + \underbrace{v^\sharp \circ \xi}_{C_T\mathscr{C}^{1-}} = G(u)\, \underbrace{(X \circ \xi)}_{!!} + \underbrace{C(G(u), X, \xi)}_{C_T\mathscr{C}^{1-}} + \underbrace{v^\sharp \circ \xi}_{C_T\mathscr{C}^{1-}}$$

which again reduces the well-posedness of the right hand side to that of a polynomial non-linear feature constructed from ξ, in this case $X \circ \xi$. We will assume that $X \circ \xi$ is given as an element of $C_T\mathscr{C}^{0-}$ so in particular the product $G(u)(X \circ \xi)$ is then well defined and in $C_T\mathscr{C}^{0-}$. Resuming this analysis we can conclude that the enhancement J, the solution map Φ and the projection Π take here the form

$$\begin{aligned} &J : \xi \mapsto (X, X \circ \xi) \\ &\Phi : \Xi = (\zeta_1, \zeta_2) \mapsto U = (\zeta_1, u, v^\sharp), \qquad \Pi : U = (\zeta_1, u, v^\sharp) \mapsto u, \end{aligned} \tag{13}$$

where $(u, v^\sharp)$ is a solution to the system

$$\begin{cases} \mathscr{L}v^\sharp = G'(u)G(u)\zeta_2 + F(u, v^\sharp, \zeta_1) \\ u = G(u) \prec\!\!\prec \zeta_1 + v^\sharp \end{cases}$$

where $F(u, v^\sharp, \zeta_1)$ takes values in $C_T\mathscr{C}^{0-}$. The equation for $(u, v^\sharp)$, albeit not a standard PDE, can nonetheless be solved by usual fixpoint methods, at least locally in time.[3]

This last example allowed us to introduce the paracontrolled ansatz and the use of paraproducts to describe spaces of distributions with specific behaviour. All the other examples of singular SPDEs which we mentioned in the introduction can be analysed via a change of variables involving linear combinations of paraproducts and smooth remainder terms. We illustrate the final result of the analysis instead of going step by step as we did so far. For the dynamic Φ^4_3 model (4) we proceed as for Φ^4_2 and introduce further terms by writing

[3] We pick up a quadratic estimate from the paralinearisation result (6) because it is based on a second order Taylor expansion, and therefore we cannot exclude that the solutions blows up in finite time. But given an a priori bound on the L^∞ norm of u one can show that $(u, v^\sharp)$ stays bounded in $C_T\mathscr{C}^{1-} \times C_T\mathscr{C}^{2-}$, see [21], and such an a priori bound can for certain nonlinearities G be derived from the maximum principle, see [12].

$$\varphi = X + Y + \varphi^{Q}, \qquad \varphi^{Q} = \psi \prec\!\!\prec Q + \varphi^{\sharp},$$

where the functions $X \in C_T\mathscr{C}^{-1/2-}$, $Y \in C_T\mathscr{C}^{1/2-}$, $Q \in C_T\mathscr{C}^{1-}$, $\psi \in C_T\mathscr{C}^{1/2-}$, $\varphi^{\sharp} \in C_T\mathscr{C}^{3/2-}$ solve

$$\mathscr{L}X = \xi, \qquad \mathscr{L}Y = -X^3, \qquad \mathscr{L}Q = -X^2,$$

$$\mathscr{L}\varphi^{\sharp} = -3X^2 \circ Y - 3\psi(X^2 \circ Q) + F(\psi, \varphi^{\sharp}, X, Y, Q), \qquad \psi = 3(Y + \varphi^{Q}), \tag{14}$$

where F is a continuous function mapping into $C_T\mathscr{C}^{-1/2-}$. As before the main goal of this decomposition is to rewrite all the problematic resonant products in terms of simple expressions of the driving noise ξ. The enhancement J, the solution map Φ and the projection Π take here the form

$$\begin{aligned} J &: \xi \mapsto (X, Y, Q, X^2 \circ Y, X^2 \circ Q) \\ \Phi &: \Xi = (\zeta_1, \ldots, \zeta_5) \mapsto U = (\zeta_1, \zeta_2, \zeta_3, \psi, \varphi^{\sharp}), \\ \Pi &: U = (\zeta_1, \zeta_2, \zeta_3, \varphi, \psi, \varphi^{\sharp}) \mapsto \zeta_1 + \zeta_2 + \psi \prec\!\!\prec \zeta_3 + \varphi^{\sharp}, \end{aligned} \tag{15}$$

where the pair $(\psi, \varphi^{\sharp})$ solves the Eq. (14) above with the driving features $(X, Y, Q, X^2 \circ Y, X^2 \circ Q)$ replaced by generic functions $\Xi = (\zeta_1, \ldots, \zeta_5)$ with specific regularities, which in this case can be taken as

$$\Xi \in C_T\mathscr{C}^{-1/2-} \times C_T\mathscr{C}^{1/2-} \times C_T\mathscr{C}^{1-} \times C_T\mathscr{C}^{-1/2-} \times C_T\mathscr{C}^{-0-}.$$

The details can be found in [9], see also [34] for a proof that solutions exist for all times.

In the case of the KPZ equation (2) the decomposition is even more involved and the enhancement J, the solution map Φ and the projection Π take the form

$$\begin{aligned} J &: \xi \mapsto (Y, Y^{\vee}, Y^{\curlyvee}, Y^{\Psi}, Y^{\vee\vee}, \partial_x Y \circ \partial_x \mathscr{I} Y) \\ \Phi &: \Xi = (\zeta_1, \ldots, \zeta_6) \mapsto U = (\zeta_1, \zeta_2, \zeta_3, \psi, h^{\sharp}), \\ \Pi &: U = (\zeta_1, \zeta_2, \zeta_3, \psi, h^{\sharp}) \mapsto h = \zeta_1 + \zeta_2 + 2\zeta_3 + \psi \prec\!\!\prec \mathscr{I}\zeta_1 + h^{\sharp}, \end{aligned}$$

where $\mathscr{I} v(t) = \int_0^t P_{t-s} v(s) \mathrm{d}s$ and we recall that $(P_t)_{t \ge 0}$ is the heat semigroup, and where

$$\mathscr{L}Y = \xi, \quad \mathscr{L}Y^{\vee} = (\partial_x Y)^2, \quad \mathscr{L}Y^{\curlyvee} = \partial_x Y^{\vee} \partial_x Y, \quad \mathscr{L}Y^{\Psi} = \partial_x Y^{\curlyvee} \circ \partial_x Y,$$
$$\mathscr{L}Y^{\vee\vee} = (\partial_x Y^{\vee})^2, \quad \psi = 2(\psi \prec\!\!\prec \mathscr{I}\zeta_1 + h^{\sharp}) + 4Y^{\curlyvee}, \quad \mathscr{L}h^{\sharp} = F(\Xi, \psi, h^{\sharp})$$

for a continuous function F, see [24]. We provide more details for the CSBE equation (3)

$$\mathscr{L}u = \chi \partial_x u^2 + \partial_x \xi, \tag{16}$$

with a general constant χ in front of the nonlinearity because this will be needed in Sect. 6. The change of variables reads

$$u = X + \chi X^{\vee} + 2\chi^2 X^{\curlyvee} + u^{\mathcal{Q}}, \qquad u' = 2u^{\mathcal{Q}} + 4\chi^2 X^{\curlyvee}, \quad u^{\mathcal{Q}} = u' \prec\!\!\prec Q + u^{\sharp}, \tag{17}$$

where $u^{\sharp}$ solves

$$\begin{aligned}
\mathscr{L} u^{\sharp} &= \chi \partial_x u^2 - \chi \partial_x X^2 - 2\chi^2 \partial_x (X^{\vee} X) - \mathscr{L}(u' \prec\!\!\prec Q) \\
&= \chi \partial_x (\chi X^{\vee} + 2\chi^2 X^{\curlyvee} + u^{\mathcal{Q}})^2 + 2\chi \partial_x [X(2\chi^2 X^{\curlyvee} + u^{\mathcal{Q}})] - \mathscr{L}(u' \prec\!\!\prec Q) \\
&= \chi^3 \mathscr{L} X^{\vee\!\vee} + 2\chi \partial_x [u^{\mathcal{Q}} X - u^{\mathcal{Q}} \prec X] + 2\chi [\partial_x (u^{\mathcal{Q}} \prec X) - u^{\mathcal{Q}} \prec \partial_x X] \\
&\quad + 4\chi^3 \mathscr{L} X^{\Psi} + 4\chi^3 \partial_x [X^{\curlyvee} \succ X] + 4\chi^3 [\partial_x (X^{\curlyvee} \prec X) - X^{\curlyvee} \prec \partial_x X] \\
&\quad + \chi \partial_x [2\chi X^{\vee} (2\chi^2 X^{\curlyvee} + u^{\mathcal{Q}}) + (2\chi^2 X^{\curlyvee} + u^{\mathcal{Q}})^2] - [\mathscr{L}(u' \prec\!\!\prec Q) - u' \prec (\mathscr{L} Q)],
\end{aligned}$$

and the enhanced noise is defined by

$$\Xi = J(\xi) = (X, X, X^{\vee}, X^{\curlyvee}, X^{\Psi}, X^{\vee\!\vee}, Q, Q \circ X),$$

where $\mathscr{L} X = \partial_x \xi$, $\mathscr{L} Q = \partial_x X$, and

$$\mathscr{L} X^{\vee} = \partial_x X^2, \quad \mathscr{L} X^{\curlyvee} = \partial_x (X X^{\vee}), \quad \mathscr{L} X^{\Psi} = \partial_x (X^{\curlyvee} \circ X), \quad \mathscr{L} X^{\vee\!\vee} = \partial_x (X^{\vee})^2.$$

4 Ambiguities and Renormalisation

The previous analysis reduces the study of a singular SPDE to that of the enhancement J and of the enhanced solution map Φ. The enhanced version of the singular equation is a standard PDE for a new unknown together with some paradifferential relations. This factorisation "distillates" in the definition of the enhancement $\Xi = J(\xi)$ all the problematic products (or resonant products). One cannot expect to be able to analyse in full generality the map J without using specific properties of the driving function ξ. Two basic difficulties are:

1. the lack of continuity of the resonant products implies that these products are essentially undefined and the enhancement map J can be extended to irregular inputs ξ only within a specific approximation procedure (or not at all);
2. for most of the above examples, even within the more restricted context where we only try to extend J to a given stochastic process ξ through a specific approximation procedure, the enhancement map J fails to extend due to divergences: the only natural limit of the resonant products is infinite.

The first difficulty means that any resonant product which is formed without sufficient regularity should be considered inherently *ambiguous*, that is non-robust to different approximation procedures. A satisfying analysis of such ambiguities is still lacking in the paracontrolled setting and much more developed in the setting of regularity structures [8] where deep connection with algebra and renormalization procedures in Quantum Field Theory have been pointed out. For some details we suggest the reader to have a look at L. Zambotti's contribution in this volume. As an example we take here the case of the solution theory for the gPAM (13). In order to describe the effects of the ambiguity on the equation we construct an extended solution map Ψ_{ext} for smooth inputs η by translations $T_C J(\eta) = (\eta, (\mathscr{I}\eta)\circ\eta + C)$ of the enhanced gPAM noise $J(\eta)$ (here we denote as usual with $\mathscr{I}\eta$ a suitable solution to the parabolic problem $\mathscr{L}\mathscr{I}\eta = \eta$). Setting

$$\hat{u} = \Psi_{\text{ext}}(\xi, C) := \Pi \circ \Phi(T_C J(\eta)),$$

the reader can easily check that the function $\hat{u}$ satisfies a *modified* PDE which reads

$$\mathscr{L}\hat{u} = G(\hat{u})\eta + G'(\hat{u})G(\hat{u})C. \tag{18}$$

But given $\eta \in C^\infty(\mathbb{T}^2)$ and $C > 0$ one can find a sequence of smooth functions $(\eta_n) \subset C^\infty(\mathbb{T}^2)$ such that η_n converges to η in $\mathscr{C}^{-1-}$ but $J(\eta_n)$ converges to $T_C J(\eta)$. So the analysis of the gPAM model (5) together with the requirement of stability under perturbations which are only small in very weak topologies *generates* quite naturally a relaxed equation of the form (18). Similar considerations are applicable, at least in principle, to all the other singular SPDEs.

The problem of renormalisation is related to this ambiguity and to the robustness of the form of the equations under irregular perturbations. Here the problem is however that certain products are intrinsically impossible to define due to the presence of infinities and that some subtraction is required for them to have a finite limit. One of the simplest situations is still that of the 2d gPAM driven by space white noise. In this case the product $X\circ\xi$, understood as limit of smooth convolutional approximations is almost surely infinite. So if want the equations with mollified noise to have a well defined limit, then we need to start with a suitably modified equation of the form given by Eq. (18) and conceived in such a way that the additional term provides the necessary cancellations to remove the divergence in the resonant product. Denoting by ξ_ε the convolutional regularisation of the white noise ξ and by c_ε a family of constants we see that if $\hat{u}_\varepsilon$ is the solution to

$$\mathscr{L}\hat{u}_\varepsilon = G(\hat{u}_\varepsilon)\xi_\varepsilon - G'(\hat{u}_\varepsilon)G(\hat{u}_\varepsilon)c_\varepsilon,$$

then $\hat{u}_\varepsilon = \Psi_{\text{ext}}(\xi_\varepsilon, -c_\varepsilon) = \Pi \circ \Phi(T_{-c_\varepsilon} J(\eta)) = \Pi \circ \Phi((\xi_\varepsilon, \mathscr{I}\xi_\varepsilon\circ\xi_\varepsilon - c_\varepsilon))$. In other words $\hat{u}_\varepsilon$ is a continuous function of the quantities $\Xi_\varepsilon := (\xi_\varepsilon, \mathscr{I}\xi_\varepsilon\circ\xi_\varepsilon - c_\varepsilon)$. A similar result is also true for the right hand side of the equation, namely

$$G(\hat{u}_\varepsilon) \diamond \Xi_\varepsilon := G(\hat{u}_\varepsilon)\xi_\varepsilon - G'(\hat{u}_\varepsilon)G(\hat{u}_\varepsilon)c_\varepsilon.$$

By a probabilistic analysis it can be checked that a (non-unique) choice of $(c_\varepsilon)_\varepsilon$ exists for which $(\Xi_\varepsilon)_\varepsilon$ converges (in probability and in L^p with respect to the randomness) in the appropriate topology. Denoting by Ξ the limit we have that also the solution converges $\hat{u}_\varepsilon \to \hat{u}$ and satisfies the *renormalised* singular SPDE

$$\mathscr{L}\hat{u} = G(\hat{u}) \diamond \Xi,$$

where the right hand side now is a certain non-linear function of $\hat{u}$ and Ξ which can be identified with the limit $G(\hat{u}) \diamond \Xi = \lim_{\varepsilon\to 0} G(\hat{u}_\varepsilon) \diamond \Xi_\varepsilon$. In this particular case the limit is controlled via the paracontrolled Ansatz for $\hat{u}_\varepsilon$ and via the continuity results which follow from it and from the convergence of the enhanced noise Ξ_ε.

More complex renormalisations are necessary in other equations. For example, in the case of the Φ^4_3 model the ill-defined products are

$$X^2, X^3, X^2 \circ \mathscr{J}(X^3), X^2 \circ \mathscr{J}(X^2),$$

see Eq. (15), where $X = \mathscr{J}(\xi)$. Stochastic analysis shows that there exists a choice of constants $(c^1_\varepsilon, c^2_\varepsilon)_{\varepsilon>0}$ such that $c^1_\varepsilon, c^2_\varepsilon \to \infty$ as $\varepsilon \to 0$ for which the convergence in probability

$$X_\varepsilon^{\diamond 3} := X_\varepsilon^3 - 3c^1_\varepsilon X_\varepsilon \to X^{\diamond 3}, \qquad X_\varepsilon^{\diamond 2} := X_\varepsilon^2 - c^1_\varepsilon \to X^{\diamond 2},$$

$$X_\varepsilon^{\diamond 2} \diamond_\varepsilon \mathscr{J}(X_\varepsilon^{\diamond 3}) := X_\varepsilon^{\diamond 2} \circ \mathscr{J}(X_\varepsilon^{\diamond 3}) - 3c^2_\varepsilon X_\varepsilon \to X^{\diamond 2} \diamond \mathscr{J}(X^{\diamond 3}),$$

$$X_\varepsilon^{\diamond 2} \diamond_\varepsilon \mathscr{J}(X_\varepsilon^{\diamond 2}) := X_\varepsilon^{\diamond 2} \circ \mathscr{J}(X_\varepsilon^{\diamond 2}) - c^2_\varepsilon \to X^{\diamond 2} \diamond \mathscr{J}(X^{\diamond 2})$$

holds in the appropriate spaces, where $X_\varepsilon = \rho_\varepsilon * X$ is a convolutional regularisation of X with a smooth kernel. As a consequence of these convergence results and of the structure of the solution theory described by Eq. (15) the function $\varphi_\varepsilon = \Pi \circ \Phi(\Xi_\varepsilon)$, where

$$\Xi_\varepsilon = (X_\varepsilon, \mathscr{J}(X_\varepsilon^{\diamond 3}), \mathscr{J}(X_\varepsilon^{\diamond 2}), X_\varepsilon^{\diamond 2} \diamond_\varepsilon \mathscr{J}(X_\varepsilon^{\diamond 3}), X_\varepsilon^{\diamond 2} \diamond_\varepsilon \mathscr{J}(X_\varepsilon^{\diamond 2})),$$

solves the renormalised Φ^4_3 equation

$$\mathscr{L}\varphi_\varepsilon = -\varphi_\varepsilon^3 + 3(c^1_\varepsilon + c^2_\varepsilon)\varphi_\varepsilon + \xi_\varepsilon$$

and converges to $\varphi = \Pi \circ \Phi(\Xi)$ where $\Xi := \lim_\varepsilon \Xi_\varepsilon$ which again can be described in terms of a limiting renormalised singular SPDE. In this review we will not discuss details of the convergence results $\Xi_\varepsilon \to \Xi$. These are mostly handled with standard probabilistic techniques. A very comprehensive convergence theory for the stochastic terms exists in terms of regularity structures [8, 14] but most of the ideas can be employed also within the paracontrolled approach, see also [36].

5 Higher Order Expansions

The theory of paracontrolled distributions that we discussed so far is essentially a first order calculus. For example, in the parabolic Anderson model (gPAM with $G(u) = u$) we expand the solution in terms of a paraproduct $u \prec X$ plus a smoother remainder $u^\sharp$, where $\mathscr{L}X = \xi$ and $X \in C_T\mathscr{C}^\alpha$, $u^\sharp \in C_T\mathscr{C}^\alpha$ for $\alpha = 1-$. Then terms like the resonant product $u^\sharp \circ \xi$ pose no problem because $\xi \in \mathscr{C}^{\alpha-2}$ and $u^\sharp \in C_T\mathscr{C}^{2\alpha}$ and the sum of the regularities is $3\alpha - 2 > 0$. But what if ξ has worse regularity, so if $\alpha \leqslant 2/3$ and therefore $3\alpha - 2 \leqslant 0$? One relevant example is the parabolic Anderson model in $d = 3$, where the space white noise has regularity strictly less than $-3/2$ and therefore $\alpha < 1/2$. The idea, inspired by (controlled) rough paths, is then to find a higher order expansion of u which should be of the type

$$u = \sum_{k=1}^{n-1} \sum_{\tau \in \mathscr{I}_k} u^\tau \prec X^\tau + u^\sharp$$

for some $u^\sharp \in C_T\mathscr{C}^{n\alpha}$ and suitable $(X^\tau)_{\tau \in \bigcup_k \mathscr{I}_k}$, with $X^\tau \in C_T\mathscr{C}^{k\alpha}$ for all $\tau \in \mathscr{I}_k$. If $(n+1)\alpha - 2 > 0$, then at least the product $u^\sharp \circ \xi$ is well defined and this gives us some hope to construct a continuous map $(u, (u^\tau)_\tau, u^\sharp, \xi, (X^\tau)_\tau, \dots) \mapsto u\xi$. But what should the (X^τ) be and how do we define the map? This is not very obvious and for quite some time the extension of paracontrolled distributions to more irregular driving noises remained open. But recently Bailleul and Bernicot [4] have made significant progress in that direction.

To understand their results let us have a look at the parabolic Anderson model

$$\mathscr{L}u = u\xi$$

for $\xi \in \mathscr{C}^{\alpha-2}$ with $\alpha \in (1/2, 2/3)$, so slightly better than the white noise in $d = 3$ but worse than the white noise in $d = 2$. We guess the expansion $u = \sum_{k=1}^2 \sum_{\tau \in \mathscr{I}_k} u^\tau \prec X^\tau + u^\sharp$ with $u^\sharp \in C_T\mathscr{C}^{3\alpha}$ and (u^τ) and (X^τ) to be determined. Then

$$u\xi = u \prec \xi + u \succ \xi + \underbrace{\left(\sum_{k=1}^{2} \sum_{\tau \in \mathscr{I}_k} u^\tau \prec X^\tau \right) \circ \xi}_{!!} + u^\sharp \circ \xi,$$

where as before we single out the problematic resonant product with the underbrace "!!". If we assume that $u^\tau \in C_T\mathscr{C}^\alpha$ for all $\tau \in \mathscr{I}_1 \cup \mathscr{I}_2$ and $X^\tau \in C_T\mathscr{C}^{k\alpha}$ for all $\tau \in \mathscr{I}_k$, then for $\tau \in \mathscr{I}_2$ the resonant product $(u^\tau \prec X^\tau)\circ\xi$ can be controlled with the commutator estimate:

$$(u^\tau \prec X^\tau)\circ\xi = (\underbrace{C(u^\tau, X^\tau, \xi)}_{C_T\mathscr{C}^{4\alpha-2}} + u^\tau(X^\tau\circ\xi)),$$

where we used that $\alpha < 1$ and $4\alpha - 2 > 0$ to see that the commutator is bounded, and where we need to assume that $X^\tau\circ\xi \in C_T\mathscr{C}^{3\alpha-2}$ is extrinsically given. However, for $\tau \in \mathscr{I}_1$ the term $C(u^\tau, X^\tau, \xi)$ is still not well defined because the sum of the regularities is $3\alpha - 2 < 0$. To deal with the resonant product $(u^\tau \prec X^\tau)\circ\xi$ we therefore assume that u^τ is itself paracontrolled of order 2α for all $\tau \in \mathscr{I}_1$:

$$u^\tau = \sum_{\tau' \in \mathscr{I}_1} u^{\tau,\tau'} \prec X^{\tau'} + u^{\tau,\sharp}$$

with $u^{\tau,\tau'} \in C_T\mathscr{C}^\alpha$ and $u^{\tau,\sharp} \in C_T\mathscr{C}^{2\alpha}$. Then

$$(u^{\tau,\sharp} \prec X^\tau)\circ\xi = C(u^{\tau,\sharp}, X^\tau, \xi) + u^{\tau,\sharp}(X^\tau\circ\xi).$$

At this point we would like to gain 2α degrees of regularity from $u^{\tau,\sharp}$ in the commutator to see that it is in $C_T\mathscr{C}^{4\alpha-2}$, but this is not possible because $2\alpha > 1$ and the commutator estimate Lemma 1 allows us only to gain less than one derivative. However, the sum of the regularities of X^τ and ξ is $2\alpha - 2 > -1$, and therefore we can use that $u^{\tau,\sharp} \in C_T\mathscr{C}^{1-}$ to obtain

$$\begin{aligned}\|C(u^{\tau,\sharp}, X^\tau, \xi)\|_{C_T\mathscr{C}^{(1-)+2\alpha-2}} &\lesssim \|u^{\tau,\sharp}\|_{C_T\mathscr{C}^{1-}} \|X^\tau\|_{C_T\mathscr{C}^\alpha} \|\xi\|_{C_T\mathscr{C}^{\alpha-2}} \\ &\leqslant \|u^{\tau,\sharp}\|_{C_T\mathscr{C}^{2\alpha}} \|X^\tau\|_{C_T\mathscr{C}^\alpha} \|\xi\|_{C_T\mathscr{C}^{\alpha-2}}.\end{aligned}$$

Since we estimate the commutator in a space of positive regularity and $3\alpha - 2 < 0$, we still get $C(u^{\tau,\sharp}, X^\tau, \xi) \in C_T\mathscr{C}^{3\alpha-2}$.

Remark 1 This argument is particular to the not-so-singular problem studied here and breaks down if $\alpha < 1/2$. In that case it may be necessary to develop a version of the commutator estimate which allows to gain more than one derivative from $u^{\tau,\sharp}$. This can be achieved by subtracting not only $u^{\tau,\sharp}(X^\tau\circ\xi)$ from $(u^{\tau,\sharp} \prec X^\tau)\circ\xi$ but also further correction terms that involve modified Littlewood-Paley blocks and roughly speaking correspond to polynomial terms in regularity structures. Currently there is no reference where this is worked out.

Next, we note that the product $u^{\tau,\sharp}(X^\tau\circ\xi)$ is well defined if $X^\tau\circ\xi \in C_T\mathscr{C}^{2\alpha-2}$ is given because then the sum of the regularities of its factors is $4\alpha - 2 > 0$. It remains to understand the resonant product

$$((u^{\tau,\tau'} \prec X^{\tau'}) \prec X^{\tau}) \circ \xi = C(u^{\tau,\tau'} \prec X^{\tau'}, X^{\tau}, \xi) + (u^{\tau,\tau'} \prec X^{\tau'})(X^{\tau} \circ \xi)$$

where our commutator estimate really fails: After all $u^{\tau,\tau'} \prec X^{\tau'} \in C_T\mathscr{C}^{\alpha}$, $X^{\tau} \in C_T\mathscr{C}^{\alpha}$, and $\xi \in C_T\mathscr{C}^{\alpha-2}$ so the sum of the regularities is $3\alpha - 2 < 0$. But Bailleul and Bernicot realised that one can iterate the commutator estimate in the following way:

Lemma 2 ([4], formula (3.8)) *Let $\alpha, \beta, \gamma, \delta \in \mathbb{R}$ be exponents such that $\alpha + \beta + \gamma + \delta > 0$, then there exists a four-linear map $C^{(2)} : \mathscr{C}^{\alpha} \times \mathscr{C}^{\beta} \times \mathscr{C}^{\gamma} \times \mathscr{C}^{\delta} \to \mathscr{C}^{\alpha+\beta+\gamma+\delta}$ such that if f, g, h, ζ are smooth functions we have*

$$C^{(2)}(f, g, h, \zeta) = C(f \prec g, h, \zeta) - f C(g, h, \zeta).$$

Therefore, we can set

$$C(u^{\tau,\tau'} \prec X^{\tau'}, X^{\tau}, \xi) = C^{(2)}(u^{\tau,\tau'}, X^{\tau'}, X^{\tau}, \xi) + u^{\tau,\tau'} C(X^{\tau'}, X^{\tau}, \xi)$$

which is well defined provided that $C(X^{\tau'}, X^{\tau}, \xi)$ is given and has its natural regularity $C_T\mathscr{C}^{3\alpha-2}$. The only remaining product that we still need to control is then

$$\begin{aligned}(u^{\tau,\tau'} \prec X^{\tau'})(X^{\tau} \circ \xi) &= (u^{\tau,\tau'} \prec X^{\tau'}) \prec (X^{\tau} \circ \xi) + (u^{\tau,\tau'} \prec X^{\tau'}) \succ (X^{\tau} \circ \xi) \\ &\quad + C(u^{\tau,\tau'}, X^{\tau'}, X^{\tau} \circ \xi) + u^{\tau,\tau'}(X^{\tau'} \circ (X^{\tau} \circ \xi)),\end{aligned}$$

which is under control as long as $X^{\tau'} \circ (X^{\tau} \circ \xi) \in C_T\mathscr{C}^{3\alpha-2}$ is given. So let us put everything together:

$$\begin{aligned}u\xi &= \underbrace{u \succ \xi}_{2\alpha-2} + \underbrace{u \prec \xi}_{\alpha-2} + \underbrace{u^{\sharp} \circ \xi}_{4\alpha-2} + \sum_{\tau \in \mathscr{I}_2} (\underbrace{C(u^{\tau}, X^{\tau}, \xi)}_{4\alpha-2} + \underbrace{u^{\tau}(X^{\tau} \circ \xi)}_{3\alpha-2}) \\ &\quad + \sum_{\tau \in \mathscr{I}_1} (\underbrace{C(u^{\tau,\sharp}, X^{\tau}, \xi)}_{3\alpha-2} + \underbrace{u^{\tau,\sharp}(X^{\tau} \circ \xi)}_{2\alpha-2}) \\ &\quad + \sum_{\tau,\tau' \in \mathscr{I}_1} (\underbrace{C^{(2)}(u^{\tau,\tau'}, X^{\tau'}, X^{\tau}, \xi)}_{4\alpha-2} + \underbrace{u^{\tau,\tau'} C(X^{\tau'}, X^{\tau}, \xi)}_{3\alpha-2}) \\ &\quad + \sum_{\tau,\tau' \in \mathscr{I}_1} (\underbrace{C(u^{\tau,\tau'}, X^{\tau'}, X^{\tau}, \xi)}_{4\alpha-2} + \underbrace{u^{\tau,\tau'}(X^{\tau'} \circ (X^{\tau} \circ \xi))}_{3\alpha-2}) \\ &\quad + \sum_{\tau,\tau' \in \mathscr{I}_1} (\underbrace{(u^{\tau,\tau'} \prec X^{\tau'}) \prec (X^{\tau} \circ \xi)}_{2\alpha-2} + \underbrace{(u^{\tau,\tau'} \prec X^{\tau'}) \succ (X^{\tau} \circ \xi)}_{3\alpha-2}),\end{aligned}$$

where the underbrace indicates the regularity of each term. Therefore, the product $u\xi$ is under control if all of the following terms are extrinsically given

$$\{X^{\sigma} \circ \xi, X^{\tau} \circ \xi, C(X^{\tau'}, X^{\tau}, \xi), X^{\tau'} \circ (X^{\tau} \circ \xi) : \tau, \tau' \in \mathscr{I}_1, \sigma \in \mathscr{I}_2\} \tag{19}$$

and have their natural regularity.

But making sense of the product $u\xi$ is only the first step in the analysis of the equation. Next, we should check that for a paracontrolled u also the solution v to $\mathscr{L}v = u\xi$ is paracontrolled. Here a problem occurs: above we discussed that if we make the paracontrolled Ansatz

$$v = \sum_{k=1}^{2} \sum_{\tau \in \mathscr{I}_k} v^\tau \lll X^\tau + u^\sharp$$

(now with the modified paraproduct $\lll$ rather than $\prec$), then we can commute the operator $\mathscr{L}$ with $\lll$ in the sense that $\mathscr{L}(v^\tau \lll X^\tau) - v^\tau \prec \mathscr{L} X^\tau$ has higher regularity. However, the commutation can gain at most one degree of regularity from v^τ, and as we have just seen in Remark 1 this may not always be sufficient. So Bailleul and Bernicot introduce an "intertwined" paraproduct defined as

$$(f \otimes g)(t) = \int_0^t P_{t-s}(f \prec \mathscr{L} g)(s) \mathrm{d}s,$$

where $(P_t)_{t \geqslant 0}$ is the heat semigroup generated by Δ. Then by definition the exact relation $\mathscr{L}(f \otimes g) = f \prec \mathscr{L} g$ holds, without error term. So we make the modified paracontrolled Ansatz

$$v = \sum_{k=1}^{2} \sum_{\tau \in \mathscr{I}_k} v^\tau \otimes X^\tau + v^\sharp$$

with the same regularities as above. Then

$$\mathscr{L} v = \sum_{k=1}^{2} \sum_{\tau \in \mathscr{I}_k} v^\tau \prec \mathscr{L} X^\tau + \mathscr{L} v^\sharp,$$

and if we take $\tau_1 \in \mathscr{I}_1$ with $\mathscr{L} X^{\tau_1} = \xi$ and $v^{\tau_1} = u$, then $v^{\tau_1} \prec \mathscr{L} \xi^\tau$ cancels the worst regularity contribution $u \prec \xi \in C_T \mathscr{C}^\alpha$ to $u\xi$. The remaining X^τ and v^τ have to be chosen such that all contributions of regularity $2\alpha - 2$ cancel. The most tricky term to deal with is $u \succ \xi = \xi \prec u$ which we decompose as

$$\xi \prec u = \xi \prec \left(\sum_{k=1}^{2} \sum_{\tau \in \mathscr{I}_k} u^\tau \prec X^\tau + u^\sharp \right) = \sum_{\tau \in \mathscr{I}_1} \xi \prec (u^\tau \prec X^\tau) + \sum_{\tau \in \mathscr{I}_2} \underbrace{\xi \prec (u^\tau \prec X^\tau)}_{3\alpha - 2} + \underbrace{\xi \prec u^\sharp}_{4\alpha - 2}.$$

The first sum on the right hand side still has regularity $2\alpha - 2$, but Theorem 8 of [4] gives

$$T(\xi, u^\tau, X^\tau) = \xi \prec (u^\tau \prec X^\tau) - u^\tau \prec (\xi \prec X^\tau) \in C_T \mathscr{C}^{3\alpha - 2}.$$

Thus, we have

$$\xi \prec u = \sum_{\tau \in \mathscr{I}_1} \underbrace{T(\xi, u^\tau, X^\tau)}_{3\alpha-2} + \sum_{\tau \in \mathscr{I}_2} \underbrace{\xi \prec (u^\tau \prec X^\tau)}_{3\alpha-2} + \underbrace{\xi \prec u^\sharp}_{4\alpha-2} + \sum_{\tau \in \mathscr{I}_1} \underbrace{u^\tau \prec (\xi \prec X^\tau)}_{2\alpha-2},$$

and we need a contribution from $\mathscr{L}v$ to cancel the last term on the right hand side. The other terms in the product $u\xi$ which have regularity worse than $3\alpha - 2$ are $\sum_{\tau \in \mathscr{I}_1} u^{\tau,\sharp}(X^\tau \circ \xi)$ and $\sum_{\tau,\tau' \in \mathscr{I}_1} (u^{\tau,\tau'} \prec X^{\tau'}) \prec (X^\tau \circ \xi)$, and for fixed $\tau \in \mathscr{I}_1$ we have

$$u^{\tau,\sharp}(X^\tau \circ \xi) - u^{\tau,\sharp} \prec (X^\tau \circ \xi) \in C_T \mathscr{C}^{4\alpha-2},$$

so overall the contribution of regularity $2\alpha - 2$ is

$$\sum_{\tau \in \mathscr{I}_1} \left(u^{\tau,\sharp} \prec (X^\tau \circ \xi) + \sum_{\tau' \in \mathscr{I}_1} (u^{\tau,\tau'} \prec X^{\tau'}) \prec (X^\tau \circ \xi) \right) = \sum_{\tau \in \mathscr{I}_1} u^\tau \prec (X^\tau \circ \xi).$$

In conclusion, we obtain the decomposition

$$u\xi - \underbrace{u \prec \xi}_{\alpha-2} - \underbrace{\sum_{\tau \in \mathscr{I}_1} u^\tau \prec \{(\xi \prec X^\tau) + (X^\tau \circ \xi)\}}_{2\alpha-2} \in C_T \mathscr{C}^{3\alpha-2}.$$

We can rewrite $X^\tau \circ \xi + \xi \prec X^\tau = X^\tau \xi - X^\tau \prec \xi$, and therefore we should set $\mathscr{I}_1 = \{1\}$ with $\mathscr{L}X^1 = \xi$ and $\mathscr{I}_2 = \{2\}$ with $\mathscr{L}X^2 = X^1\xi - X^1 \prec \xi$. Then we get with the Ansatz $v^1 = u$ and $v^2 = u^1$ that

$$\mathscr{L}v = u \prec \xi + u^1 \prec (X^1 \xi - X^1 \prec \xi) + \mathscr{L}v^\sharp,$$

so if we set $\mathscr{L}v = u\xi$, then

$$\mathscr{L}v^\sharp = u\xi - u \prec \xi - u^1 \prec (X^1 \xi - X^1 \prec \xi) \in C_T \mathscr{C}^{3\alpha-2},$$

which proves that $v^\sharp \in C_T \mathscr{C}^{3\alpha}$ and therefore the paracontrolled Ansatz was justified. Moreover, since $v^1 = u$ we have with $v^{1,1} = u^1$ that $v^1 - v^{1,1} \prec X^1 = u - u^1 \prec X^1 \in C_T \mathscr{C}^{2\alpha}$ and therefore also v^1 is paracontrolled. The terms in (19) that we need to construct in order to make sense of all the products are

$$X^1 \circ \xi, X^2 \circ \xi, C(X^1, X^1, \xi), X^1 \circ (X^1 \circ \xi).$$

But of course now we were inconsistent: We started with $u = u^1 \prec X^1 + u^2 \prec X^2 + u^\sharp$ and ended up with a v that on the first level is paracontrolled in terms of the new intertwined paraproduct,

$$v - v^1 \circledcirc X^1 - v^2 \circledcirc X^2 \in C_T\mathscr{C}^{3\alpha},$$

but on the second level is paracontrolled in terms of the usual paraproduct, $v^1 - v^{1,1} \prec X^1 \in C_T\mathscr{C}^{2\alpha}$. To set up a Picard iteration the map that sends u to the solution v of $\mathscr{L}v = u\xi$ should map the space of paracontrolled distributions into itself, so we should assume that also u was paracontrolled in terms of $* \circledcirc *$ and also v^1 is paracontrolled in terms of $* \circledcirc *$. For that we need to understand the relation between the two paraproducts $* \prec *$ and $* \circledcirc *$ and also some commutator estimates involving $* \circledcirc *$. All this is worked out in [4], where also the nonlinear case of gPAM with ξ of regularity $\xi \in \mathscr{C}^{\alpha-2}$ with $\alpha < 2/3$ is treated. In that case we also need a higher order version of the paralinearisation result (6), but this is relatively easy to derive.

6 Weak Universality

One prominent application of the theory of singular SPDEs is the derivation of scaling limits of random fields described by local non-linear stochastic dynamics. As an example we will sketch the case of the CSBE equation which has been first analysed via regularity structures by Hairer and Quastel [28]. As a mesoscopic model of a weakly-asymmetric diffusion we will consider the solution of the following SPDE. Take a small $\varepsilon > 0$ and let $\mathbb{T}_\varepsilon = \mathbb{T}/\varepsilon$ and $v\colon \mathbb{R}_+ \times \mathbb{T}_\varepsilon \to \mathbb{R}$ be the solution to

$$\mathscr{L}v = \varepsilon^{1/2}\partial_x P(v) + \partial_x \eta$$

where η is a Gaussian noise on $\mathbb{R}_+ \times \mathbb{T}_\varepsilon$ with finite-range space-time correlations and $P\colon \mathbb{R} \to \mathbb{R}$ is a given smooth function. We assume that η is centred with covariance

$$\mathbb{E}[\eta(t,x)\eta(s,y)] = C_\varepsilon(t-s, x-y), \qquad t,s \in \mathbb{R}, x,y \in \mathbb{T}_\varepsilon,$$

where C_ε is (in the second variable) the $\mathbb{T}_\varepsilon$-periodised version of a function $C\colon \mathbb{R} \times \mathbb{R} \to \mathbb{R}$ with sufficient polynomial decay in both variables. The parabolic change of variables $v_\varepsilon(t,x) = \varepsilon^{-1/2} v(t/\varepsilon^2, x/\varepsilon)$ gives the equation

$$\mathscr{L}v_\varepsilon = \varepsilon^{-1}\partial_x P(\varepsilon^{1/2} v_\varepsilon) + \partial_x \xi_\varepsilon$$

where $\xi_\varepsilon = \varepsilon^{-3/2}\eta(\cdot/\varepsilon^2, \cdot/\varepsilon)$ is a noise which converges to a space-time white noise ξ. Indeed the rescaled fields $v_\varepsilon, \xi_\varepsilon$ live on the standard torus $\mathbb{T}$ and

$$\mathbb{E}[\xi_\varepsilon(t,x)\xi_\varepsilon(s,y)] = \varepsilon^{-3} C_\varepsilon((t-s)/\varepsilon^2, (x-y)/\varepsilon), \qquad t,s \in \mathbb{R}, x,y \in \mathbb{T},$$

where, as $\varepsilon \to 0$ we have $\varepsilon^{-3}C_\varepsilon((t-s)/\varepsilon^2, (x-y)/\varepsilon) \to \delta(t-s)\delta(x-y)$ weakly as a space-time distribution. The goal of the analysis is to show that as $\varepsilon \to 0$ the function v_ε converges to the solution of the CSBE equation (3) with a specific constant χ in front of the non-linearity. The constant will depend only on the shape of the function P. By going to a reference frame via a constant velocity a_ε change of variables we have that $v_\varepsilon = v_\varepsilon(t, x) = \varepsilon^{-1/2} v(t/\varepsilon^2, (x + a_\varepsilon t)/\varepsilon)$ is the solution to

$$\mathscr{L} v_\varepsilon = a_\varepsilon \partial_x v_\varepsilon + \varepsilon^{-1} \partial_x P(\varepsilon^{1/2} v_\varepsilon) + \partial_x \xi_\varepsilon. = \varepsilon^{-1} \partial_x \tilde P(\varepsilon^{1/2} v_\varepsilon) + \partial_x \xi_\varepsilon,$$

where $\tilde P(x) = P(x) + a_\varepsilon \varepsilon^{1/2} x$. We make an Ansatz of the form

$$v_\varepsilon = X_\varepsilon + \chi \tilde X^{\mathsf{V}}_\varepsilon + 2\chi^2 \tilde X^{\mathsf{Vr}}_\varepsilon + v^Q_\varepsilon,$$

where χ is a real number different from 0 and $\tilde X^{\mathsf{V}}_\varepsilon$ and $\tilde X^{\mathsf{Vr}}_\varepsilon$ are functions, all to be determined later. We then use a Taylor expansion to get

$$\begin{aligned}\frac{1}{\varepsilon} \tilde P(\varepsilon^{1/2} v_\varepsilon) &= \frac{1}{\varepsilon} \tilde P(\varepsilon^{1/2} X_\varepsilon) + \frac{1}{\varepsilon^{1/2}} \tilde P'(\varepsilon^{1/2} X_\varepsilon)(\chi \tilde X^{\mathsf{V}}_\varepsilon + 2\chi^2 \tilde X^{\mathsf{Vr}}_\varepsilon + v^Q_\varepsilon) \\ &\quad + \frac{1}{2} \tilde P''(\varepsilon^{1/2} X_\varepsilon)(\chi \tilde X^{\mathsf{V}}_\varepsilon + 2\chi^2 \tilde X^{\mathsf{Vr}}_\varepsilon + v^Q_\varepsilon)^2 + R_\varepsilon,\end{aligned}$$

where

$$\begin{aligned}R_\varepsilon = \varepsilon^{1/2} \frac{1}{2} \int_0^1 &\mathrm{d}\tau (1-\tau)^2 \tilde P^{(3)}(\varepsilon^{1/2} X_\varepsilon + \tau \varepsilon^{1/2}(\chi \tilde X^{\mathsf{V}}_\varepsilon + 2\chi^2 \tilde X^{\mathsf{Vr}}_\varepsilon + v^Q_\varepsilon)) \\ &\times (\chi \tilde X^{\mathsf{V}}_\varepsilon + 2\chi^2 \tilde X^{\mathsf{Vr}}_\varepsilon + v^Q_\varepsilon)^3.\end{aligned}$$

Let us assume that $\tilde P^{(3)}(x) = P^{(3)}(x)$ has polynomial growth of order M and that we have the following bounds

$$\|X_\varepsilon\|_{L^\infty} \lesssim \varepsilon^{-1/2-\kappa}, \quad \|\chi \tilde X^{\mathsf{V}}_\varepsilon + 2\chi^2 \tilde X^{\mathsf{Vr}}_\varepsilon + v^Q_\varepsilon\|_{L^\infty} \lesssim \varepsilon^{0-\kappa}$$

for some small $\kappa > 0$. Then

$$\|R_\varepsilon\|_{L^\infty} \lesssim \varepsilon^{1/2} \varepsilon^{-\kappa M} \varepsilon^{-3\kappa} \lesssim \varepsilon^{1/2-\kappa(M+3)}$$

so if κ is small enough this remainder goes to zero in L^∞. This justifies the Taylor expansion at least under the assumptions we made.

Now we set

$$\chi_\varepsilon = \tilde P''(\varepsilon^{1/2} X_\varepsilon), \quad \mathscr{L} X_\varepsilon = \partial_x \xi_\varepsilon,$$

$$2\chi \tilde X_\varepsilon = \tfrac{1}{\varepsilon^{1/2}} \tilde P'(\varepsilon^{1/2} X_\varepsilon), \quad \mathscr{L} \tilde Q_\varepsilon = \partial_x \tilde X_\varepsilon, \quad \chi \mathscr{L} \tilde X^{\mathsf{V}}_\varepsilon = \tfrac{1}{\varepsilon} \partial_x \tilde P(\varepsilon^{1/2} X_\varepsilon),$$

and

$$\mathscr{L} \tilde X^{\mathsf{Vr}}_\varepsilon = \partial_x (\tilde X^{\mathsf{V}}_\varepsilon \circ \tilde X_\varepsilon), \quad \mathscr{L} \tilde X^{\mathsf{Vrr}}_\varepsilon = \partial_x (\tilde X^{\mathsf{Vr}}_\varepsilon \circ \tilde X_\varepsilon), \quad \chi \mathscr{L} \tilde X^{\mathsf{VV}}_\varepsilon = \frac{1}{2} \partial_x [\chi_\varepsilon (\tilde X^{\mathsf{V}}_\varepsilon)^2].$$

Note that χ is a real number and χ_ε is a function and the two do not agree. The reason for the notation is that in the end χ_ε will converge to χ. With these definitions at hand we see that setting

$$v_\varepsilon = X_\varepsilon + \chi \tilde{X}^{\vee}_\varepsilon + 2\chi^2 \tilde{X}^{\curlyvee}_\varepsilon + v^Q_\varepsilon, \qquad v'_\varepsilon = 2\chi v^Q_\varepsilon + 4\chi^3 \tilde{X}^{\vee}_\varepsilon, \quad v^Q_\varepsilon = v'_\varepsilon \prec\!\!\prec \tilde{Q}_\varepsilon + v^\sharp_\varepsilon$$

we have

$$\begin{aligned}
\mathscr{L} v^\sharp_\varepsilon &= \frac{1}{\varepsilon}\partial_x[\tilde{P}(\varepsilon^{1/2} v_\varepsilon)] - \frac{1}{\varepsilon}\partial_x[\tilde{P}(\varepsilon^{1/2} X_\varepsilon)] - 2\chi^2 \partial_x[(\tilde{X}^{\vee}_\varepsilon \tilde{X}_\varepsilon)] - \mathscr{L}(v'_\varepsilon \prec\!\!\prec \tilde{Q}_\varepsilon) \\
&= \frac{1}{\varepsilon^{1/2}}\partial_x[\tilde{P}'(\varepsilon^{1/2} X_\varepsilon)(2\chi^2 \tilde{X}^{\curlyvee}_\varepsilon + v^Q_\varepsilon)] + \frac{1}{2}\partial_x[\tilde{P}''(\varepsilon^{1/2} X_\varepsilon)(\chi \tilde{X}^{\vee}_\varepsilon + 2\chi^2 \tilde{X}^{\curlyvee}_\varepsilon + v^Q_\varepsilon)^2] \\
&\quad + \partial_x R_\varepsilon - \mathscr{L}(v'_\varepsilon \prec\!\!\prec \tilde{Q}_\varepsilon) \\
&= \chi 2 \partial_x[\tilde{X}_\varepsilon(2\chi^2 \tilde{X}^{\curlyvee}_\varepsilon + v^Q_\varepsilon)] + \frac{1}{2}\partial_x[\chi_\varepsilon(\chi \tilde{X}^{\vee}_\varepsilon + 2\chi^2 \tilde{X}^{\curlyvee}_\varepsilon + v^Q_\varepsilon)^2] + \partial_x R_\varepsilon - \mathscr{L}(v'_\varepsilon \prec\!\!\prec \tilde{Q}_\varepsilon) \\
&= \chi^3 \mathscr{L} \tilde{X}^{\vee\!\vee}_\varepsilon + 2\chi \partial_x[v^Q_\varepsilon \tilde{X}_\varepsilon - v^Q_\varepsilon \prec \tilde{X}_\varepsilon] + 2\chi[\partial_x(v^Q_\varepsilon \prec \tilde{X}_\varepsilon) - v^Q_\varepsilon \prec \partial_x \tilde{X}_\varepsilon] \\
&\quad + 4\chi^3 \mathscr{L} \tilde{X}^{\curlyvee^{\!\vee}}_\varepsilon + 4\chi^3 \partial_x[\tilde{X}^{\curlyvee}_\varepsilon \succ \tilde{X}_\varepsilon] + 4\chi^3[\partial_x(\tilde{X}^{\curlyvee}_\varepsilon \prec \tilde{X}_\varepsilon) - \tilde{X}^{\curlyvee}_\varepsilon \prec \partial_x \tilde{X}_\varepsilon] \\
&\quad + \partial_x \chi_\varepsilon[2\chi \tilde{X}^{\vee}_\varepsilon(2\chi^2 \tilde{X}^{\curlyvee}_\varepsilon + v^Q_\varepsilon) + (2\chi^2 \tilde{X}^{\curlyvee}_\varepsilon + v^Q_\varepsilon)^2] - \left[\mathscr{L}(v'_\varepsilon \prec\!\!\prec \tilde{Q}) - v'_\varepsilon \prec (\mathscr{L}\tilde{Q})\right] \\
&\quad + \partial_x R_\varepsilon,
\end{aligned}$$

which has to be compared with the expansion that we used for the solution to the CSBE (17). The only structural difference is the presence of random fields χ_ε in the place of some of the χ and the additional source term $\partial_x R_\varepsilon$ which however goes to zero in a topology that is compatible with the required regularity of $v^\sharp_\varepsilon$. The role of the enhancement Ξ is taken by the (slightly modified) family of random fields

$$\tilde{\Xi}_\varepsilon = (\chi_\varepsilon, X_\varepsilon, \tilde{X}_\varepsilon, \tilde{X}^{\vee}_\varepsilon, \tilde{X}^{\curlyvee}_\varepsilon, \tilde{X}^{\curlyvee^{\!\vee}}_\varepsilon, \tilde{X}^{\vee\!\vee}_\varepsilon, \tilde{Q}_\varepsilon, \tilde{Q}_\varepsilon \circ \tilde{X}_\varepsilon).$$

If we prove that $\tilde{\Xi}_\varepsilon$ converges to an enhancement $\tilde{\Xi}$

$$\tilde{\Xi} = (\chi, X, X, X^{\vee}, X^{\curlyvee}, X^{\curlyvee^{\!\vee}}, X^{\vee\!\vee}, Q, Q \circ X),$$

in suitable topologies, where χ is a constant, we will have automatically that $(v_\varepsilon, v'_\varepsilon, v^\sharp_\varepsilon) \to (u, u', u^\sharp)$ in the appropriate topologies. It is not our aim here to fully develop this sketch of proof, especially because a non-trivial part consists in proving the convergence of the stochastic data for which some powerful machinery has been devised in the paper by Hairer and Quastel [28].

We would like to end by illustrating a specific phenomena related to convergence of non-linear functions of random distributions. Observe that the enhanced data for v^ε involves non-linear functions of the random field $\varepsilon^{1/2} X_\varepsilon$. A direct computation shows that for any given (t, x) the family of random variables $(\varepsilon^{1/2} X_\varepsilon(t, x))_\varepsilon$ converges to a Gaussian random variable with a finite variance. In general the covariance of the random field X_ε is given by

$$Q_\varepsilon(y-z,t-s) = \mathbb{E}[X_\varepsilon(t,y)X_\varepsilon(s,z)] = \varepsilon^{-1}(e^{\Delta(t-s)/\varepsilon^2}C)((y-z)/\varepsilon).$$

This quantity allows for various bounds:

$$|Q_\varepsilon(y-z,t-s)| \lesssim \varepsilon^{-1} \wedge (t-s)^{-1/2} \wedge (y-z)^{-1}.$$

In particular the random field X_ε converges only a random space–time distribution and not as a continuous stochastic process. It is then non-trivial to discuss the limit of the non-linear functional. Some Gaussian analysis can be used to show that if we have a sequence of polynomial functions $(F_\varepsilon : \mathbb{R} \to \mathbb{R})_\varepsilon$ such that $\mathbb{E}[F_\varepsilon(\varepsilon^{1/2}X_\varepsilon)] = 0$ then for all $p \geqslant 1$ and under some technical assumptions we have for every $\psi \in C_c^\infty(\mathbb{R}, \mathbb{R})$

$$\psi(t)F_\varepsilon(\varepsilon^{1/2}X_\varepsilon(t,x)) \to 0$$

almost surely along subsequences in $B^{-\kappa}_{\infty,\infty}(\mathbb{R} \times \mathbb{T})$ for some $\kappa > 0$. Moreover, if the component of $F_\varepsilon(\varepsilon^{1/2}X_\varepsilon)$ in the first chaos of the random field X_ε vanishes, then the above convergence can be improved at the cost of reducing the space-time regularity:

$$\varepsilon^{-1/2-\kappa}\psi(t)F_\varepsilon(\varepsilon^{1/2}X_\varepsilon(t,x)) \to 0$$

as a space–time distribution of parabolic regularity $-1/2-\kappa$, almost surely at least along subsequences. This implies that non-linear functions of $(\varepsilon^{1/2}X_\varepsilon(t,x))_\varepsilon$ have limits related to the first terms of the chaos expansion of the r.v. $F_\varepsilon(\varepsilon^{1/2}X_\varepsilon(t,x))$ and moreover that the whole shape of F_ε is involved in these limits and not only its behaviours near zero as could be naively deduced from the assumption that $\varepsilon^{1/2}X_\varepsilon(t,x)$ should behave as a small quantity. See also [23] where similar results are derived in a specific stationary setting based on the chaos expansion under the stationary measure of X_ε.

7 Anderson Hamiltonian

Paracontrolled distributions and related tools can not only be used to solve singular SPDEs, they also allow us to construct certain operators that are a priori ill-defined. Consider for example the parabolic Anderson model $\partial_t u = \Delta u + u\xi$, that is gPAM with $G(u) = u$. If we consider the *Anderson Hamiltonian*

$$\mathscr{H}u = (\Delta + \xi)u,$$

then formally the solution to the parabolic Anderson model is given by $u(t) = e^{t\mathscr{H}}u_0$, where $(e^{t\mathscr{H}})_{t\geqslant 0}$ is the semigroup generated by $\mathscr{H}$. So by understanding $\mathscr{H}$ we should also gain a better understanding of the parabolic equation. In particular it will be interesting to study the spectrum of $\mathscr{H}$ and the structure of its eigenfunctions in order to learn something about the long time behavior of $u(t)$.

We would like to see $\mathscr{H}$ as an unbounded operator on $L^2(\mathbb{T}^d)$ and not $\mathscr{C}^\alpha(\mathbb{T}^d)$, because L^2 is a Hilbert space and the spectral analysis of operators is much easier on Hilbert spaces than on Banach spaces. For $d = 1$ Fukushima and Nakao [20] constructed $\mathscr{H}$ already in 1977, but the case $d \in \{2, 3\}$ was only very recently understood by Allez and Chouk [1]. In the following we sketch their results for $d = 2$. Of course, there is no problem to make sense of $\mathscr{H}u$ if u is a smooth function. The problem is rather that $\mathscr{H}u$ should be in L^2, but for $u \in C^\infty$ the product ξu will not be better behaved than ξ because the multiplication with a smooth function does not increase the regularity. Since ξ is a distribution and not an L^2 function, C^∞ will not be contained in the domain of $\mathscr{H}$! On the other side if u is too irregular, the product ξu may not be defined. So the idea of Allez and Chouk is to define a domain of paracontrolled functions u for which $\mathscr{H}u$ takes values in L^2. For $\alpha \in \mathbb{R}$ we write $H^\alpha = B^\alpha_{2,2}$ for the L^2-Sobolev space with regularity α and let

$$\mathscr{D} = \{u \in H^{1-} : u^\sharp = u - u \prec X \in H^{2-}\},$$

where X is to be determined. Now we have to deal with Besov spaces other than $B^\alpha_{\infty,\infty}$, but it is easy to see that we have analogous estimates for the paraproduct and resonant term on H^α spaces, more precisely

$$\begin{aligned} * \prec * : H^\alpha \times \mathscr{C}^\beta &\to H^{\beta \wedge (\beta+\alpha)}, & \alpha, \beta \in \mathbb{R}, \\ * \circ * : H^\alpha \times \mathscr{C}^\beta &\to \mathscr{C}^{\beta+\alpha}, & \alpha + \beta > 0. \end{aligned}$$

The commutator estimate also extends to more general Besov spaces, see [38]: For $\alpha \in (0, 1)$ and $\beta, \gamma \in \mathbb{R}$ with $\beta + \gamma < 0$ and $\alpha + \beta + \gamma > 0$ we have

$$C \colon H^\alpha \times \mathscr{C}^\beta \times \mathscr{C}^\gamma \to H^{\alpha+\beta+\gamma}.$$

Recall that $\xi \in \mathscr{C}^{-1-}$, so for $X \in \mathscr{C}^{1-}$ and $u \in \mathscr{D}$ we obtain

$$\begin{aligned} \Delta u + u\xi = {} & \underbrace{(\Delta(u \prec X) - u \prec \Delta X)}_{H^{0-}} + \underbrace{\Delta u^\sharp}_{H^{0-}} + \underbrace{u \prec \Delta X}_{H^{-1-}} + \underbrace{u \prec \xi}_{H^{-1-}} + \underbrace{u \succ \xi}_{H^{0-}} \\ & + \underbrace{u^\sharp \circ \xi}_{H^{1-}} + \underbrace{C(u, X, \xi)}_{H^{1-}} + \underbrace{u(X \circ \xi)}_{H^{0-}}. \end{aligned} \tag{20}$$

Choosing X such that $\Delta X = -\xi$ we get $\mathscr{H}u \in H^{0-}$ for all $u \in \mathscr{D}$, and therefore $\mathscr{H}$ is an unbounded operator on H^{0-} with domain $\mathscr{D}$. Moreover, $\mathscr{D}$ is dense in H^{0-} and a Hilbert space when equipped with the norm $\|u\|_{\mathscr{D}} = \|u\|_{H^{1-}} + \|u^\sharp\|_{H^{2-}}$. But we are interested in the spectral theory for $\mathscr{H}$ and a generic $u \in \mathscr{D}$ can never be an eigenfunction because $\mathscr{H}u \in H^{0-} \not\subset \mathscr{D}$. To find the eigenfunctions of $\mathscr{H}$ we should identify a subspace of $\mathscr{D}$ on which $\mathscr{H}$ has better regularity. The idea of Allez and Chouk is to consider what they call *strongly paracontrolled distributions*: From the expansion (20) we see that if we want $\mathscr{H}u \in H^{1-}$, then all terms of regularity H^{0-} on the right hand side should cancel up to a remainder of regularity H^{1-}, so we should have

$$(\Delta(u \prec X) - u \prec \xi) + \Delta u^\sharp + u \succ \xi + u(X \circ \xi) \in H^{1-},$$

or in other words

$$-\Delta u^\sharp = (\Delta(u \prec X) - u \prec \xi) + u \succ \xi + u(X \circ \xi) + u^R,$$

where $u^R \in H^{1-}$, from where we get[4]

$$u^\sharp = (-\Delta)^{-1}((\Delta(u \prec X) - u \prec \xi) + u \succ \xi + u(X \circ \xi)) + u^{\sharp\sharp} =: \Phi(u) + u^{\sharp\sharp},$$

where $\Phi(u)$ is defined through the equation and $u^{\sharp\sharp} \in H^{3-}$. Thus, the space of strongly paracontrolled distributions is

$$\mathrm{dom}(\mathscr{H}) = \{u \in H^{1-} : u^{\sharp\sharp} = u - u \prec X - \Phi(u) \in H^{3-}\} \subset \mathscr{D},$$

and for $u \in \mathrm{dom}(\mathscr{H})$ we get

$$\mathscr{H}u = \Delta u + u\xi = (\Phi(u) + u^{\sharp\sharp}) \circ \xi + C(u, X, \xi) \in H^{1-} \subset L^2.$$

It is not at all trivial that $\mathrm{dom}(\mathscr{H})$ contains more functions than just 0, but in [1] it is even shown that $\mathrm{dom}(\mathscr{H})$ is dense in L^2. Moreover, it is shown that $\mathscr{H}$ is a symmetric operator:

$$\langle \mathscr{H}u, v\rangle_{L^2} = \langle u, \mathscr{H}v\rangle_{L^2}$$

for all $u, v \in \mathrm{dom}(\mathscr{H})$, and there exists a decreasing sequence $\lambda_1 \geqslant \lambda_2 \geqslant \dots$ of real eigenvalues with $\lim_{n\to\infty} \lambda_n = -\infty$ and an L^2-orthonormal basis of corresponding eigenvectors $(e_n)_{n\in\mathbb{N}}$ such that $\mathscr{H}e_n = \lambda_n e_n$ for all $n \in \mathbb{N}$ and such that

$$\mathscr{H}u = \sum_{n=1}^{\infty} \lambda_n \langle u, e_n\rangle_{L^2}, \qquad u \in \mathrm{dom}(\mathscr{H}).$$

There is just one thing that we omitted: as for the parabolic Anderson model it is of course necessary to renormalise the operator, because if $(\xi_\varepsilon)_{\varepsilon>0}$ is a convolution approximation of ξ and $-\Delta X_\varepsilon = \xi_\varepsilon$, then the term $X_\varepsilon \circ \xi_\varepsilon$ does not converge but only $X_\varepsilon \circ \xi_\varepsilon - c_\varepsilon$ converges for a suitable sequence of diverging constants $(c_\varepsilon)_{\varepsilon>0}$. Replacing $X \circ \xi$ by $X \circ \xi - \infty$ has the effect of changing $\mathscr{H}u = \Delta u + u\xi$ to

[4] Strictly speaking it is not possible to invert the Laplace operator and we have to shift it and consider $(1 - \Delta)^{-1}$ instead, but for simplicity we ignore this here.

$$\Delta u + u(\xi - \infty) = \Delta u + u \prec \xi + u \succ \xi + u^{\sharp} \circ \xi + C(u, X, \xi) + u(X \circ \xi - \infty).$$

From here we see that if ξ is the space white noise, the operator $\mathscr{H}$ cannot be continuously extended from $\mathrm{dom}(\mathscr{H})$ to the smooth functions because for $u \in C^\infty$ the product $u\xi$ does not create any divergences so $u(\xi - \infty)$ does not make any sense!

Allez and Chouk [1] then proceed to study how the largest eigenvalue λ_1 behaves in the white noise case, and they show that there exist constants $C_1, C_2 > 0$ such that

$$e^{-C_1 x} \leqslant \mathbb{P}(\lambda_1 \geqslant x) \leqslant e^{-C_2 x} \tag{21}$$

for $x \to \infty$. From here we learn at least heuristically that at large times the solution u to the parabolic Anderson model should not have any moments, because

$$u(t) = e^{t\mathscr{H}} u_0 = \sum_{n=1}^{\infty} e^{t\lambda_n} \langle u_0, e_n \rangle_{L^2},$$

and omitting the contribution from all eigenvalues except λ_1 we get

$$\mathbb{E}[|e^{t\lambda_1} \langle u_0, e_n \rangle_{L^2}|^p] = \mathbb{E}[e^{tp\lambda_1} |\langle u_0, e_n \rangle_{L^2}|^p].$$

By (21) we have $\mathbb{E}[e^{tp\lambda_1}] = \infty$ as soon as $tp > C_1$, so in that case we expect that also $\mathbb{E}[|u(t)|^p] = \infty$. But it remains an open problem how to make this intuitive argumentation rigorous.

8 Singular Martingale Problem

Similar ideas that we developed to study the Anderson Hamiltonian $\mathscr{H}u = (\Delta + \xi)u$ have also been used by Cannizzaro and Chouk [10], inspired by [15], to make sense of certain diffusions with distributional drift: Let[5] $\xi \in \mathscr{C}^{\alpha-1}(\mathbb{R})$ with $\alpha \in (1/3, 1/2)$ and consider the SDE $x \colon \mathbb{R}_+ \to \mathbb{R}$,

$$\mathrm{d}x_t = \xi(x_t)\mathrm{d}t + \sqrt{2}\mathrm{d}w_t,$$

where w is a Brownian motion. Of course, this equation does not make any sense because ξ is a distribution and cannot be evaluated in x_t. We still formally write down the martingale problem and call a continuous stochastic process x a solution if for all suitable u the process

[5]Besov spaces on $\mathbb{R}$ are defined exactly in the same way as on $\mathbb{T}$ and they have essentially the same properties.

$$M_t^u = u(x_t) - u(x_0) - \int_0^t \mathscr{G}u(x_s)\mathrm{d}s, \qquad t \geqslant 0,$$

is a continuous martingale, where $\mathscr{G}u = \xi\,\partial_x u + \Delta u$. But just as for the Anderson Hamiltonian the problem is that $\mathscr{G}u$ is only a distribution whenever u is a smooth function, and therefore we first need to identify a suitable domain of functions for which $\mathscr{G}u$ is continuous. We can do this by solving the equation

$$(\mathscr{G} - \lambda)u = \varphi, \tag{22}$$

for $\varphi \in C_b(\mathbb{R})$, the continuous and bounded functions on $\mathbb{R}$, and $\lambda > 0$. Then we get for the solution u that $\mathscr{G}u = \varphi + \lambda u$, so provided that u itself is a continuous function the martingale problem above makes sense. To solve (22) we make the paracontrolled Ansatz $u = u' \prec X + u^\sharp$ with $u' \in \mathscr{C}^\alpha(\mathbb{R})$, $(\Delta - \lambda)X = \partial_x \xi \in \mathscr{C}^\alpha(\mathbb{R})$, and $u^\sharp \in \mathscr{C}^{2\alpha}(\mathbb{R})$, and we reformulate the equation as

$$(\Delta - \lambda)u = \xi\,\partial_x u + \varphi.$$

From here it is not difficult to see that for λ large enough (depending only on ξ and $X \circ \xi$ but not on φ) there exists a unique paracontrolled solution u to the equation, and the space of paracontrolled functions u which solve $(\mathscr{G} - \lambda)u = \varphi$ for some $\varphi \in C_b(\mathbb{R})$ is a domain for $\mathscr{G}$. Moreover, there exists a unique (in law) solution x to the martingale problem defined above. For details see [10].

References

1. Allez, R., Chouk, K.: The Continuous Anderson Hamiltonian in Dimension Two (2015). arXiv:1511.02718
2. Albeverio, S., Röckner, M.: Stochastic differential equations in infinite dimensions: solutions via Dirichlet forms. Probab. Theory Related Fields **89**(3), 347–386 (1991)
3. Bailleul, I., Bernicot, F.: Heat semigroup and singular PDEs. J. Funct. Anal. **270**(9), 3344–3452 (2016)
4. Bailleul, I., Bernicot, F.: Higher Order Paracontrolled Calculus (2016). arXiv:1609.06966
5. Bahouri, H., Chemin, J.-Y., Danchin, R.: Fourier Analysis and Nonlinear Partial Differential Equations. Springer, Berlin (2011)
6. Bailleul, I., Debussche, A., Hofmanova, M.: Quasilinear Generalized Parabolic Anderson Model (2016). arXiv:1610.06726
7. Bony, J.-M.: Calcul symbolique et propagation des singularités pour les équations aux dérivées partielles non linéaires. Annales scientifiques de l'École Normale supérieure **14**, 209–246 (1981)
8. Brunded, Y.: Algebraic Renormalisation of Regularity Structures (2016). arXiv:1610.08468
9. Catellier, R., Chouk, K.: Paracontrolled Distributions and the 3-dimensional Stochastic Quantization Equation (2013). arXiv:1310.6869
10. Cannizzaro, G., Chouk, K.: Multidimensional SDEs with Singular Drift and Universal Construction of the Polymer Measure with White Noise Potential (2015). arXiv:1501.04751
11. Chouk, K., Friz, P.K.: Support Theorem for a Singular Semilinear Stochastic Partial Differential Equation (2014). arXiv:1409.4250

12. Cannizzaro, G., Friz, P.K,, Gassiat, P.: Malliavin calculus for regularity structures: the case of gPAM. J. Funct. Anal. **272**(1), 363–419 (2017)
13. Chouk, K., Gairing, J., Perkowski, N.: An Invariance Principle for the Two-dimensional Parabolic Anderson Model with Small Potential (2016). arXiv:1609.02471
14. Chandra, A., Hairer, M.: An Analytic BPHZ Theorem for Regularity Structures (2016). arXiv:1612.08138
15. Delarue, F., Diel, R.: Rough paths and 1d SDE with a time dependent distributional drift: application to polymers. Probab. Theory Related Fields **165**(1–2), 1–63 (2016)
16. Da Prato, G., Debussche, A.: Strong solutions to the stochastic quantization equations. Ann. Probab. **31**(4), 1900–1916 (2003)
17. Furlan, M., Gubinelli, M.: Paracontrolled Quasilinear SPDEs (2016). arXiv:1610.07886
18. Friz, P.K., Hairer, M.: A Course on Rough Paths: with an Introduction to Regularity Structures. Springer, Berlin (2014)
19. Funaki, T., Hoshino, M.: A coupled KPZ equation, its two types of approximations and existence of global solutions. J. Funct. Anal. **273**(3), 1165–1204 (2017)
20. Fukushima, M., Nakao, S.: On spectra of the Schrödinger operator with a white Gaussian noise potential. Zeitschrift für Wahrscheinlichkeitstheorie und Verwandte Gebiete **37**(3), 267–274 (1977)
21. Gubinelli, M., Imkeller, P., Perkowski, N.: Paracontrolled Distributions and Singular PDEs. Forum of Mathematics. Pi, 3:e6, 75 (2015)
22. Gubinelli, M., Perkowski, N.: Lectures on singular stochastic PDEs. Ensaios Matemáticos [Mathematical Surveys], vol. 29. Sociedade Brasileira de Matemática, Rio de Janeiro (2015)
23. Gubinelli, M., Perkowski, N.: The Hairer-Quastel universality result at stationarity. In: Stochastic Analysis on Large Scale Interacting Systems, RIMS Kôkyûroku Bessatsu, B59, pp. 101–115. Research Institute for Mathematical Sciences (RIMS), Kyoto (2016)
24. Gubinelli, M., Perkowski, N.: KPZ reloaded. Commun. Math. Phys. **349**(1), 165–269 (2017)
25. Gubinelli, M.: Controlling rough paths. J. Funct. Anal. **216**(1), 86–140 (2004)
26. Hairer, M.: Rough stochastic PDEs. Commun. Pure Appl. Math. **64**(11), 1547–1585 (2011)
27. Hairer, M.: A theory of regularity structures. Invent. Math. **198**(2), 269–504 (2014)
28. Hairer, M., Quastel, J.: A Class of Growth Models Rescaling to KPZ (2015). arXiv:1512.07845
29. Kupiainen, A.: Renormalization group and stochastic PDEs. Annales Henri Poincar. A J. Theor. Math. Phys. **17**(3), 497–535 (2016)
30. Lyons, T.J., Caruana, M.J., Lvy, T.: Differential Equations Driven by Rough Paths: Ecole d'Et de Probabilits de Saint-Flour XXXIV-2004, 1 edn. Springer, Berlin (2007)
31. Lyons, T., Qian, Z.: System Control and Rough Paths. Oxford University Press, Oxford (2002)
32. Lyons, T.: Differential equations driven by rough signals, pp. 215–310. Revista Matemtica, Iberoamericana (1998)
33. Meyer, Y.: Remarques sur un théorème de J.-M. Bony. In: Rendiconti del Circolo Matematico di Palermo. Serie II, pp. 1–20 (1981)
34. Mourrat, J.-C., Weber, H.: Global Well-posedness of the Dynamic ϕ^4_3 Model on the Torus (2016). arXiv:1601.01234
35. Mourrat, J.-C., Weber, H.: Global well-posedness of the dynamic Φ^4 model in the plane. Ann. Probab. **45**(4), 2398–2476 (2017)
36. Mourrat, J.-C., Weber, H., Xu, W.: Construction of ϕ^4_3 Diagrams for Pedestrians (2016). arXiv:1610.08897
37. Otto, F., Weber, H.: Quasilinear SPDEs via Rough Paths (2016). arXiv:1605.09744
38. Prömel, D.J., Trabs, M.: Rough differential equations driven by signals in Besov spaces. J. Differ. Equ. **260**(6), 5202–5249 (2016)
39. Röckner, M., Zhu, R., Zhu, X.: Restricted Markov uniqueness for the stochastic quantization of $P(\Phi)_2$ and its applications. J. Funct. Anal. **272**(10), 4263–4303 (2017)
40. Zhu, R., Zhu, X.: Approximating Three-dimensional Navier–Stokes Equations Driven by Space-time White Noise (2014). arXiv:1409.4864
41. Zhu, R., Zhu, X.: A Wong-Zakai Theorem for ϕ^4_3 Model (2015). arXiv:1504.04143

Fokker–Planck Equations in Hilbert Spaces

Giuseppe Da Prato

Abstract This paper includes in an unified way several results about existence and uniqueness of solutions of Fokker–Planck equations from (Bogachev et al., J Funct Anal, 256:1269–1298, 2009) [2], (Bogachev et al., J Evol Equ, 10(3):487–509, 2010) [3], (Bogachev et al., Partial Differ Equ, 36:925–939, 2011) [4] and (Bogachev et al., Bull London Math Soc 39:631–640, 2007) [1], using probabilistic methods. Several applications are provided including Burgers and $2D$-Navier–Stokes equations perturbed by noise. Some of these applications were also studied by a different analytic approach in (Bogachev et al., J Differ Equ, 259(8):3854–3873, 2015) [5], (Bogachev et al., Ann Sc Norm Super Pisa Cl Sci 14(3):983–1023, 2015) [6], (Da Prato et al., Commun Math Stat, 1(3):281–304, 2013) [11].

Keywords Fokker–Planck equations · Kolmogorov operators · Parabolic equations for measures · Stochastic PDEs

2000 Mathematics Subject Classification AMS 60H15 · 60J35 · 60J60 · 47D07

1 Introduction and Setting of the Problem

We are given a separable Hilbert space H with norm $|\cdot|$ and inner product $\langle\cdot,\cdot\rangle$. We denote by $\mathcal{B}(H)$ the set of all Borel subsets of H and by $\mathcal{P}(H)$ the set of all Borel probability measures on H. If $\psi : H \to [0,+\infty]$ is convex and lower semi-continuous we denote by $\mathcal{P}_\psi(H)$ the set of all elements $\zeta \in \mathcal{P}(H)$ such that $\int_H \psi(x)\zeta(dx) < \infty$. Clearly $\mathcal{P}_\psi(H)$ is a convex subset of $\mathcal{P}(H)$. We shall assume throughout the paper the following

Hypothesis 1 *(i) $A : D(A) \subset H \to H$ is self-adjoint and there exists $\omega > 0$ such that $\langle Ax, x\rangle \le -\omega|x|^2$, $\forall\, x \in D(A)$.*
(ii) There exists $\epsilon_0 \in (0, 1/2)$ such that $(-A)^{-1+2\epsilon_0}$ is of trace class.

G. Da Prato (✉)
Scuola Normale Superiore, Pisa, Italy
e-mail: giuseppe.daprato@sns.it

A. Eberle et al. (eds.), *Stochastic Partial Differential Equations and Related Fields*, Springer Proceedings in Mathematics & Statistics 229,
https://doi.org/10.1007/978-3-319-74929-7_5

(iii) $C : H \to H$ *is bounded, symmetric and nonnegative.*
(iv) $F : D(F) \subset [0, T] \times H \to H$ *is Borel, whereas* $D(F)$ *is a Borel subset of* $[0, T] \times H$.

We have taken A and C time independent and A self-adjoint for the sake of simplicity, but more general cases (including: A infinitesimal generator of a C_0 semigroup and time dependent and equations with multiplicative noise) could be considered without important changes in the proofs.

For any $\beta \in (0, 1)$ we shall use the notation $\|x\|_\beta := |(-A)^\beta x|,\ \forall\, x \in D((-A)^\beta)$. We recall that for any $\beta > 0$ the embedding $D((-A)^\beta) \subset H$ is compact and that there exists $k_\beta > 0$ such that

$$\|(-A)^\beta e^{tA}\| \le k_\beta t^{-\beta}, \quad \forall\, t > 0. \tag{1}$$

A *probability kernel* $[0, T] \to \mathcal{P}(H),\ t \mapsto \mu_t$, is a mapping $[0, T] \to \mathbb{R},\ t \mapsto \mu_t(I)$ such that is measurable for any $I \in \mathcal{B}(H)$. We identify a probability kernel $(\mu_t)_{t\in[0,T]}$ with the measure $\underline{\mu}(dx\, dt) = \mu_t(dx)dt$ on $([0, T] \times H, \mathcal{B}([0, T]) \times H))$ defined as

$$\underline{\mu}([a, b] \times I) = \int_a^b \mu_t(I)dt, \quad \forall\, a, b \in \mathbb{R},\ I \in \mathcal{B}(H).$$

Let us introduce the space of *exponential functions*; they will play the role of *test functions*. For any $h \in H$ we set $\varphi_h = e^{i\langle h,x\rangle},\ x \in H$. Moreover, we denote by $\mathcal{E}_A(H)$ the linear span of the real parts of functions φ_h such that $h \in D(A)$ and by $\mathcal{E}_A([0, T] \times H)$ the linear span of all functions of the form $u(t, x) = g(t)\varphi(x),\ (t, x) \in [0, T] \times H$, where $g \in C^1([0, T]),\ g(T) = 0$ and $\varphi \in \mathcal{E}_A(H)$. Finally, we define the *Kolmogorov* operator K_F, by setting

$$K_F u(t, x) = D_t u(t, x) + \frac{1}{2}\,\mathrm{Tr}\,[C D_x^2 u(t, x)] + \langle x, A D_x u(t, x)\rangle + \langle F(t, x), D_x u(t, x)\rangle, \tag{2}$$

for any $u \in \mathcal{E}_A([0, T] \times H)$ and $(t, x) \in [0, T] \times H$.

We consider the following problem. Given $\zeta \in \mathcal{P}(H)$, we want to find a probability kernel $\underline{\mu} = (\mu_t)_{t\in[0,T]}$ such that $\mu_0 = \zeta$ and for all $u \in \mathcal{E}_A([0, T] \times H)$, $K_F u$ is $\underline{\mu}$-integrable and it results

$$\int_{H_T} K_F u(t, x)\, \mu_t(dx)dt = -\int_H u(0, x)\, \zeta(dx), \tag{3}$$

where $H_T = [0, T] \times H$. $\underline{\mu}$ will be called a *solution* of the Fokker–Planck equation (3).

Let us describe the content of the paper. Section 2 is devoted to some preliminaries about the Ornstein–Uhlenbeck semigroup, following [2]. In Sect. 3 we present some existence result for the Fokker–Planck equation (3) under additional assumptions, Hypotheses 2 or 3. The first one is closely related to the paper [3], whereas the slightly different second one allows us to consider more general nonlinearities as Burgers and $2D$-Navier–Stokes equations. Finally, Sect. 4 is devoted to uniqueness. Here we start from a basic rank condition, introduced in [1] and distinguish two cases: (i) C is invertible and C^{-1} is bounded and (ii) C is of trace class. We note that an important consequence of uniqueness is that in this case the Chapman–Kolmogorov equation is well posed, see [3] for a thorough discussion of this fact.

We end this section with some notations used in what follows. By $C_b(H)$ we mean the space of all real continuous and bounded mappings $\varphi\colon H \to \mathbb{R}$ endowed with the sup norm $\|\cdot\|_0$. Moreover, $C_b^1(H)$ is the subspace of $C_b(H)$ of all continuously differentiable functions with bounded derivatives. Finally, $B_b(H)$ is the space of all real Borel mappings $\varphi\colon H \to \mathbb{R}$ endowed with the sup norm $\|\cdot\|_0$.

2 Preliminaries on the Ornstein–Uhlenbeck Semigroup

It is well known (see e.g. [8] or [13]) that under Hypothesis 1 the equation

$$\begin{cases} dZ = AZdt + \sqrt{C}\,dW(t), & t \geq 0 \\ X(s) = x, & s \leq t, \end{cases} \tag{4}$$

has a unique mild solution given by

$$Z(t,s,x) = e^{(t-s)A}x + W_A(t,s), \tag{5}$$

where the *stochastic convolution* $W_A(t,s)$ is defined as

$$W_A(t,s) = \int_s^t e^{(t-r)A}\sqrt{C}\,dW(r), \quad \forall\, 0 \leq s \leq t. \tag{6}$$

The following lemma will be used later

Lemma 1 *For all $\beta \in [0, \epsilon_0]$ and $0 \leq s \leq t \leq T$ we have*

$$\mathbb{E}|(-A)^{\beta}\, W_A(t,s)|^2 \leq \|C\|\,\mathrm{Tr}\,[(-A)^{-1+2\beta}] =: C(\beta) \tag{7}$$

Proof We have in fact

$$\begin{aligned} \mathbb{E}|(-A)^{\beta} W_A(t,s)|^2 &= \mathrm{Tr}\left[C\int_s^t (-A)^{2\beta}e^{2rA}dr\right] \\ &\leq \frac{1}{2}\,\|C\|\,\mathrm{Tr}\,[(-A)^{2\beta-1}(e^{2sA}-e^{2tA})] \leq \|C\|\,\mathrm{Tr}\,[(-A)^{2\beta-1}. \end{aligned}$$

□

Let us consider the transition evolution operator $R_{s,t}$, corresponding to Eq. (4), which acts, in particular, on $C_{b,1}(H)$, the space of all real continuous mappings φ in H such that $\sup_{x\in H} \frac{|\varphi(x)|}{1+|x|} < \infty$. For $0 \le s \le t \le T$ we have

$$R_{s,t}\varphi(x) := \int_H \varphi(y) N_{e^{(t-s)A}x, Q_{t-s}}(dy) = \mathbb{E}[\varphi(Z(t,s,x))], \quad \forall\, \varphi \in C_{b,1}(H). \tag{8}$$

Here $N_{e^{(t-s)A}x, Q_{t-s}}$ represents the Gaussian measure in H with mean $e^{(t-s)A}x$ and covariance

$$Q_{t-s} = \int_0^{t-s} e^{rA} C e^{rA} dr.$$

See e.g. [8]. Let us define a semigroup $S_\tau^{0,1}$, $\tau \ge 0$, in the space

$$C_T([0,T]; C_{b,1}(H)) = \{u \in C([0,T]; C_{b,1}(H)) : \ u(T,x) = 0, \forall\, x \in H\},$$

by setting

$$\begin{aligned}(S_\tau^{0,1}u)(t,x) &= (R_{t,t+\tau}u(t+\tau,\cdot)(x)\, \mathbb{1}_{[0,T-\tau]}(t) \\ &= \mathbb{E}[u(t+\tau, Z(t+\tau,t,x))]\, \mathbb{1}_{[0,T-\tau]}(t), \quad \forall u \in C_T([0,T]; C_{b,1}(H)).\end{aligned} \tag{9}$$

Note that $S_\tau^{0,1} = 0$, $\forall\, \tau \ge T$. Let denote by $\mathcal{K}_0^1$[1] the infinitesimal generator of $S_\tau^{0,1}$, defined through its resolvent following [7] (see (10) below). The resolvent set of $\mathcal{K}_0^1$ coincides with $\mathbb{R}$ and its resolvent is given, for all $f \in C_T([0,T]; C_{b,1}(H))$ and all $\lambda \in \mathbb{R}$, by

$$\begin{aligned}R(\lambda, \mathcal{K}_0^1) f(t,x) &= \int_0^\infty e^{-\lambda\tau} S_\tau^{0,1} f(t,x) d\tau \\ &= \int_t^T e^{-\lambda(r-t)} \mathbb{E}[f(Z(r,t,x))] dr, \quad \forall\, (t,x) \in [0,T] \times H.\end{aligned} \tag{10}$$

$\mathcal{K}_0^1$ can also be defined as follows, following [14]. We say that a function $u \in C_T([0,T]; C_{b,1}(H))$ belongs to the domain of $\mathcal{K}_0^1$ if there exists a function $f \in C_T([0,T]; C_{b,1}(H))$ such that

(i)
$$\lim_{\tau\to 0} \frac{1}{\tau} (S_\tau^{0,1}u(t,x) - u(t,x)) = f(t,x), \quad \forall\, (t,x) \in [0,T] \times H.$$

(ii) There exists $C_u > 0$ such that

$$\left| \frac{1}{\tau} (S_\tau^{0,1}u(t,x) - u(t,x)) \right| \le C_u, \quad \forall\, (t,x) \in [0,T] \times H, \ \forall\, \tau \in [0,1].$$

[1] The upper index 1 in the definitions of $S^{0,1}$ and $\mathcal{K}_0^1$ recalls the space $C_{b,1}(H)$.

We set $\mathcal{K}_0^1 u = f$ and call $\mathcal{K}_0^1$ the *infinitesimal generator* of S_τ on $C_T([0,T]; C_{b,1}(H))$. It is not difficult to check that $\mathcal{E}_A([0,T]\times H) \subset D(\mathcal{K}_0^1)$ and that for all $u \in \mathcal{E}_A([0,T]\times H)$ it results $\mathcal{K}_0^1 u = K_0 u$. So, the *abstract* operator $\mathcal{K}_0^1$ is an extension of the differential *Kolmogorov* operator K_0. If, in particular, $u(t,x) = e^{i\langle x,h\rangle} g(t)$, $t \in [0,T]$, $x \in H$, where $h \in D(A)$, $g \in C^1([0,T])$ and $g(T) = 0$, we have

$$K_0 u(t,x) = [g'(t) - (\tfrac{1}{2}\,|C^{1/2}h|^2 + i\langle x, Ah\rangle)g(t)]e^{i\langle x,h\rangle}.$$

So, $K_0 u$ has a linear growth in x; this is the reason for introducing the space $C_{b,1}(H)$ and for requiring that $h \in D(A)$.

One can show that functions from $C_T([0,T]; C_{b,1}(H))$ can be approximated point-wise by sequences (more precisely multi-sequences) of elements of $\mathcal{E}_A([0,T]\times H)$. To describe this approximation, we shall use the following notation

$$\lim_{n_1\to\infty}\lim_{n_2\to\infty}\lim_{n_3\to\infty} u_{n_1,n_2,n_3} = \lim_{\mathbf{n}\to\infty} u_{\mathbf{n}}.$$

The following three propositions were proven in [2], see also [9].

Proposition 2 *For all $u \in C_T([0,T]; C_{b,1}(H))$ there exists $(u_{\mathbf{n}}) \subset \mathcal{E}_A([0,T]\times H)$ such that*

(i) $\lim_{\mathbf{n}\to\infty} u_{\mathbf{n}}(t,x) = u(t,x), \quad \forall\,(t,x) \in [0,T]\times H.$

(ii) $|u_{\mathbf{n}}(t,x)| \le \|u\|_{C_T([0,T];C_{b,1}(H))}(1+|x|), \quad \forall\,(t,x) \in [0,T]\times H.$

Proposition 3 *For any $u \in D(\mathcal{K}_0^1)$ there exists $(u_{\mathbf{n}}) \subset \mathcal{E}_A([0,T]\times H)$ and $C > 0$ such that*

(i) $\lim_{\mathbf{n}\to\infty} u_{\mathbf{n}}(t,x) = u(t,x), \quad \forall\,(t,x) \in [0,T]\times H.$

(ii) $\lim_{\mathbf{n}\to\infty} K_0\, u_{\mathbf{n}}(t,x) = \mathcal{K}_0^1 u(t,x), \quad \forall\,(t,x) \in [0,T]\times H.$

(iii) $|u_{\mathbf{n}}(x)| + |K_0 u_{\mathbf{n}}(t,x)| \le C(1+|x|), \quad \forall\,(t,x) \in [0,T]\times H$

We say that $\mathcal{E}_A([0,T]\times H)$ is a *core* for $\mathcal{K}_0^1$.

Proposition 4 *Let $f, D_x f \in C_T([0,T]; C_{b,1}(H;H))$, $\lambda \in \mathbb{R}$ and let $u = R(\lambda, \mathcal{K}_0^1)f$. Then there exists $u_{\mathbf{n}} \subset \mathcal{E}_A([0,T]\times H)$ and $C > 0$ such that*

(i) $\lim_{\mathbf{n}\to\infty} u_{\mathbf{n}}(t,x) = u(t,x), \quad \forall\,(t,x) \in [0,T]\times H.$

(ii) $\lim_{\mathbf{n}\to\infty} K_0\, u_{\mathbf{n}}(t,x) = \mathcal{K}_0^1 u(t,x), \quad \forall\,(t,x) \in [0,T]\times H.$

(iii) $\lim_{\mathbf{n}\to\infty} D_x u_{\mathbf{n}}(t,x) = D_x u(t,x), \quad \forall\,(t,x) \in [0,T]\times H.$

(iv) $|u_{\mathbf{n}}(t,x)| + |K_0 u_{\mathbf{n}}(t,x)| + |D_x \varphi_{\mathbf{n}}(t,x)| \le C(1+|x|), \forall\,(t,x) \in [0,T]\times H.$

3 Existence

3.1 Basic Assumptions

We shall first introduce a suitable approximation F_α of F, $\alpha \in (0, 1]$, such that the problem

$$\begin{cases} dX_\alpha = (AX_\alpha + F_\alpha(t, X_\alpha))dt + \sqrt{C}\, dW(t), & t \in [0, T], \\ X_\alpha(s) = x, \quad 0 \le s \le t \le T \end{cases} \tag{11}$$

has a unique mild solution $X_\alpha(t, s, x)$, that is

$$X_\alpha(t, s, x) = e^{(t-s)A}x + \int_s^t e^{(t-r)A} F_\alpha(r, X_\alpha(r, s, x))dr + W_A(t, s). \tag{12}$$

More precisely, we shall assume that

Hypothesis 2 *For any* $\alpha \in (0, 1]$ *there exists a mapping* $F_\alpha : [0, T] \times H \to H$, $(t, x) \to F_\alpha(t, x)$ *with the following properties:*

(i) F_α *is continuous and bounded and Eq.* (11) *has a unique mild solution.*

(ii) There exists a convex and lower semicontinuous mapping $V : H \to [1, +\infty]$ *such that*

$$|F_\alpha(t, x)|^2 \le |F(t, x)|^2 \le V(x), \quad \forall\, (t, x) \in [0, T] \times H, \tag{13}$$

and

$$|F(t, x) - F_\alpha(t, x)| \le \alpha\, V(x), \quad \forall\, (t, x) \in [0, T] \times H, \ \alpha \in (0, 1]. \tag{14}$$

(iii) There exists $M > 0$ *such that*

$$P^\alpha_{s,t} V(x) \le MV(x), \quad \forall\, 0 \le s \le t \le T, \ \alpha \in (0, 1], \ x \in H, \tag{15}$$

where P^α *is the transition evolution operator,* $P^\alpha_{s,t}\, \varphi(x) = \mathbb{E}[\varphi(X_\alpha(t, s, x))]$, $0 \le s \le t \le T$, $\varphi \in B_b(H)$.

Now, fix $\zeta \in \mathcal{P}(H)$; then for any $t \in (0, T]$ and any $\alpha \in (0, 1]$ we define a probability measure μ_t^α in H such that (we take here $s = 0$ for simplicity)

$$\begin{aligned} \int_H \varphi(x)\mu_t^\alpha(dx) &= \int_H P^\alpha_{0,t}\, \varphi(x)\zeta(dx) \\ &= \int_H \mathbb{E}[\varphi(X_\alpha(t, 0, x))]\zeta(dx), \quad \forall\, \varphi \in \mathcal{E}_A(H) \end{aligned} \tag{16}$$

and set $\underline{\mu}^\alpha(dt\,dx) = \mu_t^\alpha(dx)dt$. Let us deduce some consequences of Hypothesis 2 assuming that $\int_H V(x)\,\zeta(dx) < \infty$. First from (15) and (16) we find

$$\int_H V(x)\,\mu_t^\alpha(dx) = \int_H P_{0,t}V(x)\,\zeta(dx) \le M\int_H V(x)\,\zeta(dx) \tag{17}$$

and so, from (13)

$$\int_H |F_\alpha(t,x)|^2\,\mu_t^\alpha(dx) \le \int_H |F(t,x)|^2\,\mu_t^\alpha(dx) \le M\int_H V(x)\,\zeta(dx). \tag{18}$$

Other useful estimates are provided by the following lemma.

Lemma 5 *Assume that Hypotheses 1 and 2 are fulfilled. Let $x \in H$, $0 < s \le t \le T$ and $\alpha \in (0,1]$; then $X_\alpha(t,s,x) \in D((-A)^{\epsilon_0})$, $\mathbb{P}$-a.s.*[2] *Moreover there exists $C > 0$ such that*

$$P_{s,t}^\alpha(|x|^2) = \mathbb{E}[|X_\alpha(t,s,x)|^2] \le C(1 + |x|^2 + V(x)) \tag{19}$$

and

$$P_{s,t}^\alpha(\|x\|_{\epsilon_0}) = \mathbb{E}[|(-A)^{\epsilon_0}X_\alpha(t,s,x)|] \le Ct-s)^{-\epsilon_0}(1 + |x|^2 + V(x)). \tag{20}$$

Proof Let us write $X_\alpha(t,s,x) = X_\alpha(t)$ for short. Then by (12) and Hölder's inequality we have

$$|X_\alpha(t)|^2 \le 3|x|^2 + 3T\int_s^t |F_\alpha(r, X_\alpha(r))|^2 dr + 3|W_A(t,s)|^2.$$

Now, taking expectation, recalling (7) and invoking (13) we obtain

$$\mathbb{E}|X_\alpha(t)|^2 \le 3|x|^2 + 3T\int_s^t \mathbb{E}[V_\alpha(X_\alpha(r)]dr + 3C(0).$$

Finally, taking into account (15), yields

$$\mathbb{E}|X_\alpha(t)|^2 \le 3|x|^2 + 3T^2MV(x) + 3C(0),$$

which implies (19) with a suitable $C > 0$. To prove (20) write

$$(-A)^{\epsilon_0}X_\alpha(t) = (-A)^{\epsilon_0}e^{(t-s)A}x + \int_s^t (-A)^{\epsilon_0}e^{(t-r)A}F_\alpha(r, X_\alpha(r))dr + (-A)^{\epsilon_0}W_A(t,s),$$

[2] ϵ_0 was defined in Hypothesis 1(ii).

from which, recalling (1) and using that $|F_\alpha(r, X_\alpha(r))| \le 1 + |F_\alpha(r, X_\alpha(r))|^2$, we have

$$|(-A)^{\epsilon_0} X_\alpha(t)| \le \kappa_{\epsilon_0} (t-s)^{-\epsilon_0}|x| + \kappa_{\epsilon_0} \int_s^t (t-r)^{-\epsilon_0}(1 + |F_\alpha(r, X_\alpha(r))|^2)dr \\ + |(-A)^{\epsilon_0} W_A(t,s)|.$$

Taking expectation and proceeding as before, we obtain

$$\begin{aligned}\mathbb{E}[|(-A)^{\epsilon_0} X_\alpha(t)|] &\le \kappa_{\epsilon_0} (t-s)^{-\epsilon_0}|x| + \kappa_{\epsilon_0} \int_s^t (t-r)^{-\epsilon_0}(1 + V(X_\alpha(r))dr \\ &\quad + (\mathbb{E}|(-A)^{\epsilon_0} W_A(t,s)|^2)^{1/2} \\ &\le \kappa_{\epsilon_0} (t-s)^{-\epsilon_0}|x|^2 + \kappa_{\epsilon_0} \frac{T^{1-\epsilon_0}}{1-\epsilon_0}(1 + MV_\alpha(x)) + C^{1/2}(\epsilon_0)\end{aligned}$$

which easily yields (20). □

In the following we set

$$\psi(x) := 1 + |x|^2 + V(x), \quad \forall\, x \in H. \tag{21}$$

ψ is convex and lower semicontinuous.

Corollary 6 *Assume that Hypotheses 1 and 2 are fulfilled and let $\zeta \in \mathcal{P}_\psi(H)$. Then we have*

$$\int_H |x|^2\, \mu_t^\alpha(dx) \le C \int_H \psi(x)\zeta(dx), \quad \forall\, \alpha \in (0,1] \tag{22}$$

and

$$\int_H \|x\|_{\epsilon_0}\, \mu_t^\alpha(dx) \le C t^{-\epsilon_0} \int_H \psi(x)\zeta(dx), \quad \forall\, \alpha \in (0,1]. \tag{23}$$

Proof The proof follows by integrating both sides of (19) and (20) (with $s = 0$) with respect to ζ over H. □

We notice now that by Itô's formula the measure $\underline{\mu}^\alpha(dt\,dx) = \mu_t^\alpha(dx)$ solves the approximating equation

$$\int_{H_T} K_{F_\alpha} u(t,x)\mu_t^\alpha(dx)\,dt = -\int_H u(0,x)\,\zeta(dx), \quad \forall\, u \in \mathcal{E}_A([0,T]\times H), \tag{24}$$

where K_{F_α} is the Kolmogorov operator

$$K_{F_\alpha} u = D_t u + \frac{1}{2}\,\mathrm{Tr}\,[CD_x^2 u] + \langle x, AD_x u\rangle + \langle F_\alpha, D_x u\rangle, \quad \forall u \in \mathcal{E}_A([0,T]\times H). \tag{25}$$

We are going to construct a sequence $(\mu_t^{\alpha_h})$ of measures weakly convergent to a measure μ_t for all $t \in [0, T]$ and finally, we shall pass to the limit as $h \to \infty$ in (24) (with α_h replacing α) and prove that $\underline{\mu}(dx\,dt) = \mu_t(dx)\,dt$ is a solution of the Fokker–Planck equation (3).

3.2 Tightness

Proposition 7 *Assume that $\zeta \in \mathcal{P}_\psi(H)$. Then $(\mu_t^\alpha)_{\alpha\in(0,1]}$ is tight for all $t \in (0, T]$ and we have*

$$\mu_t^\alpha(\|x\|_{\epsilon_0} \geq R) \leq \frac{Ct^{-\epsilon_0}}{R}\int_H \psi(x)\zeta(dx), \quad \forall\, R > 0. \tag{26}$$

Proof Thanks to (23), we have for all $R > 0$

$$\mu_t^\alpha(\|x\|_{\epsilon_0} \geq R) = \int_{\|x\|_{\epsilon_0}\geq R}\mu_t^\alpha(dx) \leq \frac{1}{R}\int_H \|x\|_{\epsilon_0}\,\mu_t^\alpha(dx) \leq \frac{Ct^{-\epsilon_0}}{R}\int_H \psi(x)\zeta(dx).$$

The conclusion follows from the arbitrariness of R because balls in the topology of $D((-A)^{\epsilon_0})$ are compact in H. □

Theorem 8 *Assume that Hypotheses 1 and 2 are fulfilled and let $\zeta \in \mathcal{P}_\psi(H)$.*[3] *Then there is a probability kernel $\underline{\mu}(dt\,dx) = \mu_t(dx)\,dt$ solving* (3) *and such that $\mu_t \in \mathcal{P}_\psi$ for all $t \in [0, T]$, where ψ is defined by* (21).

Proof Since for any $t \in (0, T]$, $(\mu_t^\alpha)_{\alpha\in(0,1]}$ is tight, there exists a sequence $\alpha_h(t) \to 0$ and $\mu_t \in \mathcal{P}(H)$ such that

$$\lim_{h\to\infty}\int_H \varphi\,d\mu_t^{\alpha_h(t)} = \int_H \varphi\,d\mu_t, \quad \forall\,\varphi \in C_b(H)\, t \in [0, T].$$

By a diagonal extraction argument we can find a sequence $(\alpha_h) \to 0$ and a family of measures $(\mu_t)_{t\in\mathbb{Q}}$ such that[4]

$$\lim_{h\to\infty}\int_H \varphi\,d\mu_t^{\alpha_h} = \int_H \varphi\,d\mu_t, \quad \forall\,\varphi \in C_b(H), \quad \forall\, t \in [0, T]\cap\mathbb{Q}. \tag{27}$$

Now we are going to extend the family $(\mu_t^{\alpha_h})_{t\in\mathbb{Q}}$ to the whole interval $(0, T]$ in such a way that the extension is a solution of (3). We shall proceed in three steps.

[3] ψ is defined in (21).

[4] $\mathbb{Q}$ denotes the set of rational numbers.

Step 1. For each $\varphi \in \mathcal{E}_A(H)$ there exists $C(\varphi) > 0$ such that for all $\alpha \in (0, 1]$ we have

$$\left|\int_H \varphi\, d\mu_t^\alpha - \int_H \varphi\, d\mu_s^\alpha\right| \le C(\varphi)(t-s)\int_H \psi\, d\zeta, \quad 0 \le s \le t \le T. \tag{28}$$

In fact by Itô's formula we have

$$\mathbb{E}[\varphi(X_\alpha(t,0,x))] - \mathbb{E}[\varphi(X_\alpha(s,0,x))] = \mathbb{E}\int_s^t (L_r^\alpha \varphi)(X_\alpha(r,0,x))dr, \tag{29}$$

where L_t^α, $t \in [0, T]$, is the Kolmogorov operator

$$L_t^\alpha \varphi = \tfrac{1}{2}\,\mathrm{Tr}\,[CD^2\varphi] + \langle x, AD\varphi\rangle + \langle F_\alpha(t,x), D\varphi\rangle, \quad \forall \varphi \in \mathcal{E}_A(H).$$

Now, taking into account (16), write

$$\begin{aligned}\int_H \varphi\, d\mu_t^\alpha - \int_H \varphi\, d\mu_s^\alpha &= \int_H (\mathbb{E}[\varphi(X_\alpha(t,0,x))] - \mathbb{E}[\varphi(X_\alpha(s,0,x))])\zeta(dx)\\ &= \mathbb{E}\int_H \int_s^t (L_t^\alpha \varphi)(X(r,0,x))\,\zeta(dx)\,dr.\end{aligned} \tag{30}$$

Let $C_1(\varphi) > 0$ be such that

$$\begin{aligned}|(L_t^\alpha \varphi)(x)| &\le C_1(\varphi)(1 + |x| + |F_\alpha(t,x)|)\\ &\le 3C_1(\varphi)(1 + |x|^2 + |F_\alpha(t,x)|^2), \quad \forall\, x \in H.\end{aligned}$$

Therefore from (30), taking into account (15), (13) and (22), we find

$$\begin{aligned}\left|\int_H \varphi\, d\mu_t^\alpha - \int_H \varphi\, d\mu_s^\alpha\right| &\le 3C_1(\varphi)\int_H\int_s^t (1 + \mathbb{E}|X_\alpha(t,0,x))|^2\\ &\quad + \mathbb{E}|F_\alpha(X_\alpha(t,0,x))|^2)\zeta(dx)dr\\ &\le 3C_1(\varphi)(t-s)\int_H (1 + C((1+|x|^2 + V(x))\\ &\quad + MV(x))\zeta(dx).\end{aligned}$$

So, (28) follows with a suitable constant $C(\varphi)$.

Step 2. Construction of μ_t for all $t \in (0, T]$.

Let $t_0 \in (0, T] \setminus \mathbb{Q}$ and let (t_j) be a sequence in $(0, T] \cap \mathbb{Q}$ convergent to t_0. The sequence (μ_{t_j}) is tight by Proposition 7. Let us choose a limit point of (μ_{t_j}) which we call μ_{t_0}, that is (for a subsequence which we still call (t_j))

$$\lim_{j\to\infty}\int_H \varphi\, d\mu_{t_j} = \int_H \varphi\, d\mu_{t_0}, \quad \forall\, \varphi \in C_b(H). \tag{31}$$

Claim. We have

$$\lim_{h\to\infty}\int_H \varphi\, d\mu_{t_0}^{\alpha_h} = \int_H \varphi\, d\mu_{t_0}, \quad \forall\, \varphi \in C_b(H). \tag{32}$$

To prove the claim it is enough to show that

$$\lim_{h\to\infty}\int_H \varphi\, d\mu_{t_0}^{\alpha_h} = \int_H \varphi\, d\mu_{t_0}, \quad \forall\, \varphi \in \mathcal{E}_A(H). \tag{33}$$

In fact, assume that we have proved (33). Since $(\mu_{t_0}^{\alpha_h})$ is tight by (26), there exists a subsequence $(\mu_{t_0}^{\alpha_{h_k}})$ weakly convergent to a probability measure ν, that is such that

$$\lim_{k\to\infty}\int_H \varphi\, d\mu_{t_0}^{\alpha_{h_k}} = \int_H \varphi\, d\nu, \quad \forall\, \varphi \in C_b(H).$$

On the other hand, by (33) it follows that

$$\lim_{k\to\infty}\int_H \varphi\, d\mu_{t_0}^{\alpha_{h_k}} = \int_H \varphi\, d\mu_{t_0}, \quad \forall\, \varphi \in \mathcal{E}_A(H)$$

so that

$$\int_H \varphi\, d\nu = \int_H \varphi\, d\mu_{t_0}, \quad \forall\, \varphi \in \mathcal{E}_A(H),$$

which implies $\nu = \mu_{t_0}$ because $\mathcal{E}_A(H)$ is dense in $L^1(H, \nu)$.

Now we can prove the Claim. Let $\varphi \in \mathcal{E}_A(H)$. Then we have for any $j \in \mathbb{N}$

$$\begin{aligned}\left|\int_H \varphi\, d\mu_{t_0} - \int_H \varphi\, d\mu_{t_0}^{\alpha_h}\right| &\le \left|\int_H \varphi\, d\mu_{t_0} - \int_H \varphi\, d\mu_{t_j}\right| \\ &\quad + \left|\int_H \varphi\, d\mu_{t_j} - \int_H \varphi\, d\mu_{t_j}^{\alpha_h}\right| + \left|\int_H \varphi\, d\mu_{t_j}^{\alpha_h} - \int_H \varphi\, d\mu_{t_0}^{\alpha_h}\right|.\end{aligned}$$

Taking into account (28), yields

$$\begin{aligned}\left|\int_H \varphi\, d\mu_{t_0} - \int_H \varphi\, d\mu_{t_0}^{\alpha_h}\right| &\le \left|\int_H \varphi\, d\mu_{t_0} - \int_H \varphi\, d\mu_{t_j}\right| \\ &\quad + \left|\int_H \varphi\, d\mu_{t_j} - \int_H \varphi\, d\mu_{t_j}^{\alpha_h}\right| \\ &\quad + C(\varphi)|t_j - t_0| \int_H \psi\, d\zeta =: J_1 + J_2 + J_3.\end{aligned} \tag{34}$$

Given $\delta > 0$ there is $j_\delta \in \mathbb{N}$ such that $J_1 + J_3 < \delta$. Therefore by (34) we have

$$\left|\int_H \varphi\, d\mu_{t_0} - \int_H \varphi\, d\mu_{t_0}^{\alpha_h}\right| \le \left|\int_H \varphi\, d\mu_{t_{j_\delta}} - \int_H \varphi\, d\mu_{t_{j_\delta}}^{\alpha_h}\right| + \delta.$$

Now the conclusion follows from (27) and the arbitrariness of δ.

Step 3. Conclusion

Let $(\mu_t)_{t\in[0,T]}$ be the family defined in Step 2. We are going to prove that for all $u \in \mathcal{E}_A([0,T]\times H)$

$$\lim_{h\to\infty} \int_{H_T} K_{F_{\alpha_h}} u(t,x)\, \mu_t^{\alpha_h}(dx)dt = \int_{H_T} K_F\, u(t,x)\, \mu_t(dx)dt. \tag{35}$$

This will imply

$$\int_{H_T} K_F u(t,x)\, \mu_t(dx)dt = -\int_H u(0,x)\zeta(dx), \tag{36}$$

so that $(\mu_t)_{t\in[0,T]}$ is a solution of (3). To prove (35) is enough to show that

$$\lim_{h\to\infty} \int_{H_T} \langle AD_x u(t,x), x\rangle\, \mu_t^{\alpha_h}(dx)dt = \int_{H_T} \langle AD_x u(t,x), x\rangle\, \mu_t(dx)dt \tag{37}$$

and

$$\lim_{h\to\infty} \int_{H_T} \langle D_x u(t,x), F_{\alpha_h}(t,x)\rangle\, \mu_t^{\alpha_h}(dx)dt = \int_{H_T} \langle D_x u(t,x), F(t,x)\rangle\, \mu_t(dx)dt. \tag{38}$$

Let us prove (37). Recalling that $D_x u(t,x) \in D(A)$, that Au is bounded and that $|x|\,|\langle AD_x u(t,x), x\rangle| \le \|ADu\|_0\, |x|^2$, we have for any $\epsilon > 0$, thanks to (19),

$$\begin{aligned}
&\left|\int_{H_T} \langle AD_x u(t,x), x\rangle\, \mu_t^{\alpha_h}(dx)dt - \int_{H_T} \langle AD_x u(t,x), x\rangle\, \mu_t(dx)dt\right| \le \epsilon \left|\int_{H_T} \frac{|x|\langle AD_x u(t,x), x\rangle}{1+\epsilon|x|}\, \mu_t^{\alpha_h}(dx)dt\right| \\
&+\left|\int_{H_T} \frac{\langle AD_x u(t,x), x\rangle}{1+\epsilon|x|}\, \mu_t^{\alpha_h}(dx)dt - \int_{H_T} \frac{\langle AD_x u(t,x), x\rangle}{1+\epsilon|x|}\, \mu_t(dx)dt\right| + \epsilon \left|\int_{H_T} \frac{|x|\langle AD_x u(t,x), x\rangle}{1+\epsilon|x|}\, \mu_t(dx)dt\right|. \\
&\le 2\epsilon C\|ADu\|_\infty \int_{H_T} \psi(x)\, \zeta(x)\, (dx) + \left|\int_{H_T} \frac{\langle AD_x u(t,x), x\rangle}{1+\epsilon|x|}\, \mu_t^{\alpha_h}(dx)dt - \int_{H_T} \frac{\langle AD_x u(t,x), x\rangle}{1+\epsilon|x|}\, \mu_t(dx)dt\right|
\end{aligned}$$

Now (37) follows letting first $h \to \infty$ and then $\epsilon \to 0$.

Next let us prove (38). Set $g_h(t,x) := \langle Du(t,x), F_{\alpha_h}(t,x)\rangle$ and $g(t,x) := \langle Du(t,x), F(t,x)\rangle$, so that (38) is equivalent to

$$\lim_{h\to\infty} \int_{H_T} g_h(t,x)\, \mu_t^{\alpha_h}(dx)dt = \int_{H_T} g(t,x)\, \mu_t(dx)dt. \tag{39}$$

Notice that by (14) we have

$$|g(t,x) - g_h(t,x)| \le \alpha_h \|Du\|_0 \,(1 + V(x)). \tag{40}$$

Now write

$$\begin{aligned}\left|\int_{H_T} g\, d\mu_t\, dt - \int_{H_T} g_h\, d\mu_t^{\alpha_h}\, dt\right| &\le \left|\int_{H_T} g\, d\mu_t\, dt - \int_{H_T} g\, d\mu_t^{\alpha_h}\right|\\ &\quad + \left|\int_{H_T} (g - g_h)\, d\mu_t^{\alpha_h}\, dt\right| =: I_h + J_h.\end{aligned}$$

As far as I_h is concerned, we have for any $j \in \mathbb{N}$,

$$\begin{aligned}I_h := \left|\int_{H_T} g\, d\mu_t\, dt - \int_{H_T} g\, d\mu_t^{\alpha_h}\, dt\right| &\le \left|\int_{H_T} (g - g_j)\, d\mu_t\, dt\right|\\ + \left|\int_{H_T} g_j\, d\mu_t\, dt - \int_{H} g_j\, d\mu_t^{\alpha_h}\, dt\right| &+ \left|\int_{H_T} (g - g_j)\, d\mu_t^{\alpha_h}\, dt\right|.\end{aligned}$$

Therefore, taking into account (40), yields

$$I_h \le 2\alpha_j M \|Du\|_0 \int_{H_T} (1 + V(x))\, d\mu_t\, dt + \left|\int_{H_T} g_j\, d\mu_t\, dt - \int_{H_T} g_j\, d\mu_t^{\alpha_h}\, dt\right|.$$

Now, given $\epsilon > 0$, choose j_0 such that $2\alpha_{j_0} \|Du\|_0 \int_{H_T} (1 + V(x))\, d\mu_t\, dt < \frac{\epsilon}{2}$. Then

$$I_h \le \frac{\epsilon}{2} + \left|\int_{H_T} g_{j_0}\, d\mu_t\, dt - \int_{H_T} g_{j_0}\, d\mu_t^{\alpha_h}\, dt\right|,$$

so that $\lim_{h\to\infty} I_h = 0$, for the arbitrariness of ϵ. Finally, as far as J_h is concerned, we have

$$J_h \le \alpha_h M \|Du\|_0 \int_H (1 + V(x))\, \mu_t^{\alpha_h}(dx)\, dt \to 0,$$

as $h \to \infty$. The proof is complete. □

Let us present now some examples.

Example 9 (Tr $C < \infty$) We assume here that Tr $C < \infty$, $F : [0,T] \times H \to H$ is continuous and there exist $k > 0$ and $N \in \mathbb{N}$ such that

$$\langle F(t,x), x\rangle \le k, \quad |F(t,x)| \le k(1 + |x|^{2N_1}), \quad \forall\, (t,x) \in [0,T] \times H. \tag{41}$$

Then we set

$$F_\alpha(t,x) = \frac{F(t,x)}{1 + \alpha |F(t,x)|}, \quad \forall\, (t,x) \in [0,T] \times H, \tag{42}$$

so that

$$|F(t,x) - F_\alpha(t,x)| \le \alpha |F(t,x)|^2, \quad \forall\, (t,x) \in [0,T] \times H, \tag{43}$$

Let X_α be the solution to (11). Then by Itô' formula and (41), for any $m \in \mathbb{N}$ there exists $a_m > 0$ such that

$$\mathbb{E}[|X_\alpha(t,s,x)|^{2m}] \le e^{-2m\omega(t-s)}|x|^{2m} + a_m, \quad \forall\, (t,x) \in [0,T] \times H,\ \alpha \in (0,1]. \tag{44}$$

Therefore Hypothesis 2 is fulfilled with $V(x) = C(1 + |x|^{4N}),\ x \in H$.

Example 10 ($C = I$) We assume here that $C = I$, $F : [0,T] \times H \to H$ is continuous and there exist $k > 0$, $N_1,\ N \in \mathbb{N}$ such that for all $(t,x) \in [0,T] \times H$ we have

$$\langle F(t,x+z), x\rangle \le k(1 + |z|^{2N_1}), \quad \forall\, z \in H \quad , \quad |F(t,x)| \le k(1 + |x|^{2N}). \tag{45}$$

Then we define F_α by (42). Let X_α be the solution to (11). Set $Y_\alpha(t,s,x) := X_\alpha(t,s,x) - W_A(t,s)$, so that

$$\frac{d}{dt} Y_\alpha(t,s,x) = AY_\alpha(t,s,x) + F_\alpha(t, Y_\alpha(t,s,x)). \tag{46}$$

Now, setting $Y(t) = Y_\alpha(t,s,x)$, it follows by a simple computation that for any $m \in \mathbb{N}$ we have

$$\begin{aligned}\frac{1}{2m}\frac{d}{dt}|Y(t)|^{2m} &= -|(-A)^{1/2}Y(t)|^2\,|Y(t)|^{2m-2} + |Y(t)|^{2m-2}\langle Y(t) + W_A(t,s), Y(t)\rangle\\ &\le k(1 + |W_A(t,s)|^{2N_2}).\end{aligned}$$

Recalling [8, Proposition 4.3] it follows that there is $C(T) > 0$ such that $\frac{d}{dt}\,\mathbb{E}|Y(t)|^{2m} \le C(T)$, which implies

$$\mathbb{E}|Y_\alpha(t,s,x)|^{2m} \le |x|^{2m} + C(T), \quad \forall\, x \in H. \tag{47}$$

Therefore Hypothesis 2 is fulfilled with $V(x) = C(1 + |x|^{4N})$.

Example 11 (*Reaction–diffusion equations*) Here we take $H = L^2(0,1)$, $Ax = D_\xi^2$ for all $x \in D(A) = H^2(0,1) \cap H_0^1(0,1)$ and $F(t,x)(\xi) = \sum_{i=0}^N a_i(t)(x(\xi))^i$, $t \in [0,T]$, $\xi \in [0,1]$, $x \in L^2(0,1)$, where $N \in \mathbb{N}$ is odd and greater than 1, $a_i \in C([0,T])$, $i = 1, \ldots, N$, and $a_N(t) < 0$ for all $t \in [0,T]$. Then Hypothesis 1 is fulfilled. Moreover, setting

$$F_\alpha(t,x) = \frac{F(t,x)}{1 + \alpha|F(t,x)|}, \quad (t,x) \in [0,T] \times [0,1], \tag{48}$$

we have

$$|F_\alpha(t,x)|_{L^2(0,1)} \le C\left(1+|x|^{2N}_{L^{2N}(0,1)}\right) \tag{49}$$

and

$$|F(t,x)-F_\alpha(t,x)|_{L^2(0,1)} \le C\alpha|F(t,x)|^2_{L^2(0,1)} \le C_1\left(1+|x|^{4N}_{L^{2N}(0,1)}\right). \tag{50}$$

Now Hypothesis 2(i)(ii) are fulfilled with $V(x) = C\left(1+|x|^{4N}_{L^{2N}(0,1)}\right)$. Finally, Hypothesis 2(iii) is fulfilled as well, see [8, Theorem 4.8] and [3, p. 505] where a more general example is also presented. Then Theorem 8 applies.

3.3 Other Assumptions

In this section we set $G(t,x) = (-A)^{-1/2}F(t,x)$, $J_\alpha = (1-\alpha A)^{-1}$, $\alpha \in (0,1]$.

Hypothesis 3 *We set $F_\alpha(t,x) = (-A)^{1/2}J_\alpha G_\alpha(t,x)$ and assume that:*
(i) G_α is continuous and bounded. and the mild equation

$$X_\alpha(t,s,x) = e^{(t-s)A}x + \int_s^t (-A)^{1/2}e^{(t-r)A}J_\alpha G_\alpha(r, X_\alpha(r,s,x))dr + W_A(t,s), \tag{51}$$

has a unique solution.

(ii) There exists a convex and lower semicontinuous mapping $V : H \to [1,+\infty]$ such that

$$|G_\alpha(t,x)|^2 \le |G(t,x)|^2 \le V(x), \quad \forall\,(t,x) \in [0,T]\times H, \tag{52}$$

and

$$|G(t,x) - G_\alpha(t,x)| \le \alpha\, V(x), \quad \forall\,(t,x) \in [0,T]\times H. \tag{53}$$

(iii) There exists $M \ge 0$ such that

$$P^\alpha_{0,t}|x|^2 \le M|x|^2, \quad P^\alpha_{0,t}V(x) \le MV(x), \quad \forall\, t\in[0,T],\ \alpha\in(0,1], \tag{54}$$

where P^α is the transition evolution operator $P^\alpha_{s,t}\varphi(x) = \mathbb{E}[\varphi(X_\alpha(t,s,x))]$, $0 \le s \le T$, $\varphi \in B_b(H)$.

(Note the difference between Hypothesis 2(iii) and Hypothesis 3(iii); the reason is that from (51) we are not able to estimate $\mathbb{E}|X_\alpha(t,s,x)|^2$ independently of α.)

Now fix $\zeta \in \mathcal{P}(H)$ and for any $t \in (0, T]$ and any $\alpha \in (0, 1]$ define the probability measure μ_t^α in H setting as in Sect. 3.1

$$\int_H \varphi(x)\mu_t^\alpha(dx) = \int_H P_{0,t}^\alpha \varphi(x)\zeta(dx) = \int_H \mathbb{E}[\varphi(X_\alpha(t, 0, x))]\zeta(dx), \quad \forall\, \varphi \in \mathcal{E}_A(H) \tag{55}$$

and $\underline{\mu}^\alpha(dt\,dx) = \mu_t^\alpha(dx)dt$.

Let us deduce some estimates from Hypothesis 3. First from (54) we have

$$\int_H |x|^2\,\mu_t^\alpha(dx) = \int_H P_{0,t}|x|^2\,\zeta(dx) \le M \int_H |x|^2\,\zeta(dx) \tag{56}$$

and

$$\int_H V(x)\,\mu_t^\alpha(dx) = \int_H P_{0,t}V(x)\,\zeta(dx) \le M \int_H V(x)\,\zeta(dx)$$

and so, from (52)

$$\int_H |F(t, x)|^2\,\mu_t^\alpha(dx) \le M \int_H V(x)\,\zeta(dx). \tag{57}$$

Lemma 12 *Assume that Hypotheses 1 and 3 are fulfilled and let $x \in H$, $0 < s \le t \le T$ and $\alpha \in (0, 1]$. Then $X_\alpha(t, s, x) \in D((-A)^{\epsilon_0})$, $\mathbb{P}$-a.s. and there exists $C > 0$ such that*

$$\mathbb{E}[|(-A)^{\epsilon_0} X_\alpha(t, s, x)|] \le Ct - s)^{-\epsilon_0}(1 + |x|^2 + V(x)). \tag{58}$$

Proof Writing $X_\alpha(t, s, x) = X_\alpha(t)$ we have

$$\begin{aligned}(-A)^{\epsilon_0} X_\alpha(t) = (-A)^{\epsilon_0} e^{(t-s)A}x &+ \int_s^t (-A)^{\epsilon_0+1/2} J_\alpha e^{(t-r)A} G_\alpha(r, X_\alpha(r))dr \\ &+ (-A)^{\epsilon_0} W_A(t, s).\end{aligned} \tag{59}$$

Therefore, since $\|J_\alpha\| \le 1$ and using that $|G_\alpha(r, X_\alpha(r))| \le 1 + |G_\alpha(r, X_\alpha(r))|^2$, we have

$$\begin{aligned}|(-A)^{\epsilon_0} X_\alpha(t)| \le \kappa_{\epsilon_0}(t - s)^{-\epsilon_0}\,|x| &+ \kappa_{(\epsilon_0+1/2)} \int_s^t (t - r)^{-\epsilon_0-1/2}(1 + |G_\alpha(r, X_\alpha(r))|^2)dr \\ &+ |(-A)^{\epsilon_0} W_A(t, s)|.\end{aligned}$$

Taking expectation, yields

$$\begin{aligned}\mathbb{E}|(-A)^{\epsilon_0} X_\alpha(t)| \le \kappa_{\epsilon_0}(t - s)^{-\epsilon_0}\,|x| &+ \kappa_{(\epsilon_0+1/2)} \int_s^t (t - r)^{-\epsilon_0-1/2}(1 + \mathbb{E}|G_\alpha(r, X_\alpha(r))|^2)dr \\ &+ (\mathbb{E}|(-A)^{\epsilon_0} W_A(t, s)|^2)^{1/2}.\end{aligned}$$

Consequently, taking into account (7) and (52) we find

$$\mathbb{E}|(-A)^{\epsilon_0} X_\alpha(t)| \le \kappa_{\epsilon_0}(t-s)^{-\epsilon_0}|x| + \kappa_{(\epsilon_0+1/2)}\int_s^t (t-r)^{-\epsilon_0-1/2}(1+MV(x))dr$$
$$+ (\mathbb{E}|(-A)^{\epsilon_0} W_A(t,s)|^2)^{1/2}$$

and (58) follows. □

In the following we shall set as before

$$\psi(x) := 1 + |x|^2 + V(x), \quad \forall\, x \in H. \tag{60}$$

Now, integrating (58) with respect to ζ we obtain

Corollary 13 *Assume that Hypotheses 1 and 3 are fulfilled. Then for all* $\zeta \in \mathcal{P}_\psi(H)$, $t \in (0, T]$, $x \in H$ *and* $\alpha \in (0, 1]$ *we have*

$$\int_H \|x\|_{\epsilon_0}\, \mu_t^\alpha(dx) \le Ct^{-\epsilon_0}\int_H \psi(x)\zeta(dx). \tag{61}$$

Now by Itô's formula we see that $\underline{\mu}^\alpha(dt\,dx) = \mu_t^\alpha(dx)dt$ *solves the approximating Fokker–Planck equation* (24) *where* K_{F_α} *is the Kolmogorov operator* (25)*, moreover the tightness of* μ_t^α *follows as in the proof of Proposition 7.*

Proposition 14 *Assume that* $\zeta \in \mathcal{P}_\psi(H)$. *Then* $(\mu_t^\alpha)_{\alpha\in(0,1]}$ *is tight for all* $t \in (0, T]$ *and we have*

$$\mu_t^\alpha(\|x\|_{\epsilon_0} \ge R) \le \frac{Ct^{-\epsilon_0}}{R}\int_H \psi(x)\zeta(dx), \quad \forall\, R > 0. \tag{62}$$

Theorem 15 *Assume that Hypotheses 1 and 3 are fulfilled and let* $\zeta \in \mathcal{P}_\psi(H)$*. Then there is a probability kernel* $\underline{\mu}(dt\,dx) = \mu_t(dx)\,dt$ *solving* (3) *and such that* $\mu_t \in \mathcal{P}_\psi$ *for all* $t \in [0, T]$, *where* ψ *is defined by* (60).

Proof We proceed as in the proof of Theorem 8. The Step 1 is similar: we construct a sequence $(\alpha_h) \to 0$ and a family of measures $(\mu_t)_{t\in\mathbb{Q}}$ such that

$$\lim_{h\to\infty}\int_H \varphi\, d\mu_t^{\alpha_h} = \int_H \varphi\, d\mu_t, \quad \forall\, \varphi \in C_b(H), \quad \forall\, t \in [0, T] \cap \mathbb{Q}. \tag{63}$$

Then we extend the family $(\mu_t^{\alpha_h})_{t\in\mathbb{Q}}$ to the whole interval $(0, T]$ in such a way that the extension is a solution of (3) proving that

Step 2. For each $\varphi \in \mathcal{E}_A(H)$ there exists $C(\varphi) > 0$ such that for all $\alpha \in (0, 1]$ we have

$$\left|\int_H \varphi\, d\mu_t^\alpha - \int_H \varphi\, d\mu_s^\alpha\right| \le C(\varphi)(t-s)\int_H \psi\, d\zeta, \quad 0 \le s \le t \le T. \tag{64}$$

In fact by Itô's formula we have

$$\mathbb{E}[\varphi(X_\alpha(t,0,x))] - \mathbb{E}[\varphi(X_\alpha(s,0,x))] = \mathbb{E}\int_s^t (L_r^\alpha \varphi)(X_\alpha(r,0,x))dr, \tag{65}$$

where L_t^α, $t \in [0,T]$, is the Kolmogorov operator

$$L_t^\alpha \varphi = \frac{1}{2}\operatorname{Tr}[CD^2\varphi] + \langle x, AD\varphi\rangle + \langle G_\alpha(t,x), (-A)^{1/2} J_\alpha D\varphi\rangle, \quad \forall \varphi \in \mathcal{E}_A(H).$$

Now we write

$$\int_H \varphi\, d\mu_t^\alpha - \int_H \varphi\, d\mu_s^\alpha = \mathbb{E}\int_H \int_s^t (L_t^\alpha \varphi)(X(r,0,x))\, d\zeta\, dr \tag{66}$$

and choose $C_1(\varphi) > 0$ such that

$$\begin{aligned}|(L_t^\alpha \varphi)(x)| &\le \tfrac{1}{2}\,\|\operatorname{Tr}[CD^2\varphi]\|_0 + |x|\,\|AD\varphi\|_0 + |G_\alpha(t,x)|\,\|(-A)^{1/2}J_\alpha D\varphi\|_0\\ &\le C_1(\varphi)(1+|x|+|G_\alpha(t,x)|) \le 3C_1(\varphi)(1+|x|^2+|G_\alpha(t,x)|^2), \quad \forall\, x \in H.\end{aligned}$$

Therefore (64) follows arguing as before as well as Step 2. We go now to the Step 3. We have still to show statements (37) and (38); the proof of (37) is exactly the same, whereas for the proof of (38) we have just to write

$$\langle D_x u(t,x), F_{\alpha_h}(t,x)\rangle = \langle (-A)^{1/2} J_{\alpha_h} D_x u(t,x), G_{\alpha_h}(t,x)\rangle$$

and notice that $(-A)^{1/2}D_x u$ is bounded. Then the proof ends as before. □

Example 16 (Burgers type equations) We take here $H = L^2(0,1)$, C of trace class and denote by A the realisation of the Laplace operator in $[0,1]$ with periodic boundary conditions: $Ax = D_\xi^2$, $D(A) = H_\#^2(0,1)$, where

$$H_\#^2(0,1) = \{x \in H^2(0,1) : x(0) = x(1),\ D_\xi x(0) = D_\xi x(1)\}.$$

Moreover we take $F(t,x)(\xi) = D_\xi(g(t,x(\xi))$, $x \in H_\#^1(0,1)$, where $g : [0,T]\times \mathbb{R} \to \mathbb{R}$ is continuous. Then Hypothesis 1 is fulfilled. Set $G(t,x)(\xi) = g(t,x(\xi))$, so that $F(t,x)(\xi) = (-A)^{1/2}G(t,x)$, $\forall\, x \in H_\#^1(0,1)$. Set moreover for $\alpha \in (0,1]$

$$F_\alpha(t,x)(\xi) = (-A)^{1/2} J_\alpha G_\alpha(t,x) = (-A)^{1/2} J_\alpha \left(\frac{G(t,x)}{1+\alpha|G(t,x)|}\right), \quad \forall\, x \in H_\#^1(0,1). \tag{67}$$

To check Hypothesis 3(i)(ii) we need suitable estimates of the solution $X_\alpha(t,s,x)$ of Eq. (11). For any $m \in \mathbb{N}$ set $\varphi_m(x) = \frac{1}{2m}\int_0^1 |x(\xi)|^{2m} d\xi$. By Itô's formula we have

$$d\varphi_m(X(t)) = (2m-1)\langle X(t)^{2m-1}, AX(t) + F_\alpha(t, X(t))dt + \sqrt{C}\, dW(t)\rangle$$
$$+ \frac{1}{2}(2m-1)\sum_{h=1}^{\infty} c_h \int_0^1 X(t)^{2m-2} e_h^2\, d\xi\, dt, \tag{68}$$

where (e_h) is an orthonormal basis in H and $Ce_h = c_h e_h,\ h \in \mathbb{N}$. Therefore

$$\frac{d}{dt}\mathbb{E}[\varphi_m(X(t))] \le (2m-1)\mathbb{E}\langle X(t)^{2m-1}, AX(t) + F_\alpha(t, X(t))\rangle$$
$$+ \frac{1}{2}(2m-1)\mathrm{Tr}\, C\, \mathbb{E}\int_0^1 X(t)^{2m-2}\, d\xi \tag{69}$$

In particular, we have $\frac{d}{dt}\mathbb{E}|X(t)|^2 \le -2\omega\mathbb{E}|X(t)|^2 + \mathrm{Tr}\, C$, so that

$$\mathbb{E}|X(t)|^2 \le e^{-2\omega t}|x|^2 + \frac{\mathrm{Tr}\, C}{2\omega} \tag{70}$$

Moreover by (69)

$$\frac{1}{4}\frac{d}{dt}\mathbb{E}[|X(t)|^4_{L^4(0,1)}] = 3\mathbb{E}\langle X(t)^3, AX(t) + D_\xi F(t, X(t))\rangle + \frac{3}{2}\mathrm{Tr}\, C\, \mathbb{E}[|X(t)|^2]. \tag{71}$$

But

$$\langle X(t)^3, AX(t)\rangle = -3\int_0^1 X(t)^2 (D_\xi X(t)^2 d\xi \le 0$$

and

$$\langle X(t)^3, D_\xi F(t, X(t))\rangle = -3\int_0^1 D_\xi X(t)\, X(t)^2 g(t, X(t)) d\xi = 0.$$

So,

$$\frac{1}{4}\frac{d}{dt}\mathbb{E}\left[|X(t)|^4_{L^4(0,1)}\right] \le \frac{3}{2}\mathrm{Tr}\, C\, \mathbb{E}[|X(t)|^2] \tag{72}$$

and taking into account (70) we have

$$\frac{d}{dt}\mathbb{E}\left[|X(t)|^4_{L^4(0,1)}\right] \le 6\,\mathrm{Tr}\, C\left(e^{-2\omega t}|x|^2 + \frac{\mathrm{Tr}\, C}{2\omega}\right).$$

It follows that

$$\mathbb{E}\left[|X(t)|^4_{L^4(0,1)}\right] \le |x|^4 + 6T\,\mathrm{Tr}\, C\left(|x|^2 + \frac{\mathrm{Tr}\, C}{2\omega}\right) \tag{73}$$

Iterating this procedure we find that there exists $C_m(T) > 0$ such that $\mathbb{E}[|X(t)|^{2m}_{L^{2m}(0,1)}] \le C_m(T)(1+|x|^{2m})$. Therefore Hypothesis 3(i)(ii)(iii) is fulfilled with $V(x) = C(1+|x|^4_{L^4(0,1)})$.

Example 17 (*Burgers equation perturbed by white noise*) We take here $H = L^2(0,1)$, $A = Ax = D^2_\xi$, $D(A) = H^2_\#(0,1)$, and $F(t,x)(\xi) = D_\xi(x^2(\xi))$, $\forall\, x \in H^1_\#(0,1)$.[5] Then Hypothesis 1 is fulfilled. Set $G(t,x)(\xi) = x^2(\xi)$, $\forall\, x \in H^1_\#(0,1)$, so that $F(t,x)(\xi) = (-A)^{1/2}G(t,x)$, $\forall\, x \in H^1_\#(0,1)$. Set moreover for $\alpha \in (0,1]$

$$F_\alpha(t,x)(\xi) = (-A)^{1/2} J_\alpha G_\alpha(t,x) = (-A)^{1/2} J_\alpha \left(\frac{x^2}{1+\alpha x^2}\right), \quad \forall\, x \in H^1_\#(0,1). \tag{74}$$

Then Hypothesis 3(i)(ii) is fulfilled with $V(x) = C(1+|x|^4_{L^4(0,1)})$. Finally, Hypothesis 3(iii) follows from [10, Proposition 2.2].[6]

Example 18 (*2D-Navier–Stokes equation perturbed by coloured noise*) Let us consider the space $L^2_\#$ of all real 2π-periodic functions in the real variables ξ_1 and ξ_2, which are measurable and square integrable on $[0,2\pi]\times[0,2\pi]$ endowed with the usual scalar product $\langle\cdot,\cdot\rangle$ and norm $|\cdot|$. We denote by $(L^2_\#)^2$ the space consisting of all pairs $x = \begin{pmatrix} x^1 \\ x^2 \end{pmatrix}$ of elements of $L^2_\#$ endowed with the inner product $\langle x, y\rangle = \int_{\mathcal{O}} [x^1(\xi)y^1(\xi) + x^2(\xi)y^2(\xi)]d\xi$, $x, y \in (L^2_\#)^2$. Moreover, for any $x \in (L^2_\#)^2$ we set $|x| = \langle x, x\rangle^{\frac{1}{2}}$. (We shall consider everywhere also the corresponding complexified spaces). Let $f_{h,k} = \begin{pmatrix} e_h \\ e_k \end{pmatrix}$, $h, k \in \mathbb{Z}^2$, be the complete orthonormal system of $(L^2_\#)^2$, where $e_k(\xi) = \frac{1}{2\pi}\, e^{i\langle k,\xi\rangle}$, $k = (k_1,k_2) \in \mathbb{Z}^2$, $\xi = (\xi_1,\xi_2)$, and $\langle k, \xi\rangle = k_1\xi_1 + k_2\xi_2$. Then we shall denote by H the closed subspace of $(L^2_\#)^2$ of all *divergence free vectors*, that is: $H =$ linear span $\{g_k : k \in \mathbb{Z}^2\}$, where

$$g_k = \frac{k^\perp}{|k|}\, e_k = \begin{pmatrix} -\frac{k_2}{|k|}\, e_k \\ \\ \frac{k_1}{|k|}\, e_k \end{pmatrix}, \qquad k^\perp = \begin{pmatrix} -k_2 \\ k_1 \end{pmatrix}, \quad k \in \mathbb{Z}^2.$$

Note in fact that div $g_k(\xi) = \frac{i}{|k|}\, e_k(-k_1k_2 + k_1k_2) = 0$, $k \in \mathbb{Z}^2$. Any element $x = \begin{pmatrix} x^1 \\ x^2 \end{pmatrix}$ of H can be developed as a Fourier series $x = \sum_{k\in\mathbb{Z}^2} x_k g_k$, where

[5] We take F independent of t for simplicity.

[6] The aforementioned paper concerns Dirichlet boundary conditions but all its results generalise easily to periodic ones.

$$
\begin{aligned}
x_k = \langle x, g_k\rangle_2 &= -\frac{k_2}{|k|}\,\langle x^1, e_k\rangle_2 + \frac{k_1}{|k|}\,\langle x^2, e_k\rangle_2 \\
&= \frac{1}{2\pi\,|k|}\,\int_{\mathcal{O}} [-k_2 x^1(\eta) + k_1 x^2(\eta)] e^{-i\langle k,\eta\rangle} d\eta.
\end{aligned} \tag{75}
$$

Moreover, we shall denote by $L^p_\#$, $p \geq 1$, the subspace of $(L^p_\#)^2$ of all divergence free vectors and by $|\cdot|_p$ the norm in $L^p_\#$. Let us define the *Stokes* operator $A\colon D(A) \to H$ setting $Ax = \mathcal{P}(\Delta_\xi x - x)$, $x \in D(A) = H^2_\#$, A is self-adjoint and $Ag_k = -(1+|k|^2)g_k$, $k \in \mathbb{Z}^2$. We also define the non linear operator F setting

$$
F(x) = \mathcal{P}(D_\xi x \cdot x), \quad x \in H^1_\#.
$$

For any $\sigma > 0$ we have

$$
H^{2\sigma}_\# := \left\{ x \in H : \ \|(-A)^\sigma x\|^2 = \sum_{k\in\mathbb{Z}^2} (1 + |k|^2)^{\sigma/2} |x_k|^2 \right\}.
$$

see [8, Chap. 6] for details. If $x \in H^1_\#$, taking into account that div $x = 0$, we have

$$
F(x) = \begin{pmatrix} x_1 D_{\xi_1} x_1 + x_2 D_{\xi_2} x_1 \\ x_1 D_{\xi_1} x_2 + x_2 D_{\xi_2} x_2 \end{pmatrix} = \begin{pmatrix} D_{\xi_1}(x_1^2) + D_{\xi_2}(x_1 x_2) \\ D_{\xi_1}(x_1 x_2) + D_{\xi_2}(x_2^2). \end{pmatrix} \tag{76}
$$

Set $G(x) = (-A)^{1/2} F(x)$, and moreover for $\alpha \in (0, 1]$

$$
F_\alpha(x) = \begin{pmatrix} D_{\xi_1}(\frac{x_1^2}{1+\alpha(x_1^2+x_2^2)}) + D_{\xi_2}(\frac{x_1 x_2}{1+\alpha(x_1^2+x_2^2)}) \\ D_{\xi_1}(\frac{x_1 x_2}{1+\alpha(x_1^2+x_2^2)}) + D_{\xi_2}(\frac{x_2^2}{1+\alpha(x_1^2+x_2^2)}) \end{pmatrix}
$$

and $G_\alpha(x) = (-A)^{-1/2} F_\alpha(x)$. Then for $h \in H^1_\#$ and $\alpha \in (0, 1]$ we have

$$
\begin{aligned}
\langle F_\alpha(x), h\rangle &= \left\langle \begin{pmatrix} D_{\xi_1}(\frac{x_1^2}{1+\alpha(x_1^2+x_2^2)}) + D_{\xi_2}(\frac{x_1 x_2}{1+\alpha(x_1^2+x_2^2)}) \\ D_{\xi_1}(\frac{x_1 x_2}{1+\alpha(x_1^2+x_2^2)}) + D_{\xi_2}(\frac{x_2^2}{1+\alpha(x_1^2+x_2^2)}) \end{pmatrix}, \begin{pmatrix} h_1 \\ h_2 \end{pmatrix} \right\rangle \\
&= -\langle \tfrac{x_1^2}{1+\alpha(x_1^2+x_2^2)}, D_{\xi_1} h_1\rangle - \langle \tfrac{x_1 x_2}{1+\alpha(x_1^2+x_2^2)}, D_{\xi_2} h_1 + D_{\xi_1} h_2\rangle - \langle \tfrac{x_2^2}{1+\alpha(x_1^2+x_2^2)}, D_{\xi_2} h_2\rangle
\end{aligned}
$$

Finally, assume that Tr $[CA^{-1]} < \infty$. then it is easy to see that, thanks to [8, Lemma 6.7], Hypothesis 3 is fulfilled with $G(x) = (1 + |x|^4_{L^4_\#})$.

4 Uniqueness

4.1 The Rank Condition

We start with the following crucial result, proved in [1].

Theorem 19 *Let $\psi : H \to [0, +\infty]$ be convex and lover semi-continuous and let $\zeta \in \mathcal{P}_\psi(H)$. Assume that for any solution $\underline{\mu} = (\mu_t)_{t\in[0,T]}$ of* (3) *such that $\mu_0 = \zeta$ the following statement holds*

$$K_F(\mathcal{E}_A([0, T] \times H)) \text{ is dense in } L^1([0, T] \times H, \underline{\mu}). \tag{77}$$

Then Eq. (3) *has at most one solution.*

The condition (77) is called the *rank condition.*

Proof Assume that $\underline{\mu}_1 = (\mu_{1,t})$ and $\underline{\mu}_2 = (\mu_{2,t})$ are solution of (3) such that $\mu_{1,0} = \mu_{2,0} = \zeta \in \mathcal{P}_\psi(H)$ and set

$$\underline{\lambda} := \frac{1}{2}\,(\underline{\mu}_1 + \underline{\mu}_2).$$

Then $\lambda_0 = \zeta$ and $\underline{\mu}_1 \ll \underline{\lambda}$, $\underline{\mu}_2 \ll \underline{\lambda}$. Denote by ρ_1 and ρ_2 the respective densities

$$\rho_1 := \frac{d\underline{\mu}_1}{d\underline{\lambda}}, \quad \rho_2 := \frac{d\underline{\mu}_2}{d\underline{\lambda}}.$$

We claim that

$$0 \le \rho_1(t, x) \le 2, \quad 0 \le \rho_2(t, x) \le 2, \quad \underline{\lambda}\text{-a.s.}, \quad \forall\, (t, x) \in [0, T] \times H. \tag{78}$$

Let us show for instance that $0 \le \rho_1(t, x) \le 2$. In fact, for any $A \in \mathcal{B}([0, T] \times H)$ we have $\underline{\mu}_2(A) = 2\underline{\lambda}(A) - \underline{\mu}_1(A)$, so that $\underline{\mu}_2(A) = \int_A (2 - \rho_1)d\underline{\lambda}$, which implies that $0 \le \rho_1(t, x) \le 2$, $\underline{\lambda}$-a.e. by the arbitrariness of A.

Now for any $u \in \mathcal{E}_A([0, T] \times H)$ we have (recall that $H_T = [0, T] \times H$))

$$\int_{H_T} K_F u\, (d\underline{\mu}_1 - d\underline{\mu}_2) = \int_{H_T} K_F u\, (\rho_1 - \rho_2) d\underline{\lambda} = 0.$$

Since $\rho_1 - \rho_2 \in L^\infty([0, T] \times H; \underline{\lambda})$, by the rank condition it follows that $\rho_1 = \rho_2$. □

A useful remark for dealing with the rank condition is that for any solution $\underline{\mu}$ to Eq. (3) such that $\mu_0 \in \mathcal{P}_\psi(H)$, K_F is *dissipative* in $L^1([0, T] \times H; \underline{\mu})$, as the next proposition shows.

Proposition 20 *Let $F : D(F) \subset [0, T] \times H \to H$ be Borel and $\zeta \in \mathcal{P}(H)$. Assume that $\underline{\mu}$ is a solution to* (3)*. Then K_F is dissipative in $L^p([0, T] \times H; \underline{\mu})$ for all $p \ge 1$.*

Proof First we show that for any $u \in \mathcal{E}_A([0, T] \times H)$ we have

$$\int_{H_T} K_F u(t, x)\, u(t, x)\, \mu_t(dx)\, dt = -\frac{1}{2} \int_{H_T} |C^{1/2} D_x u(t, x)|^2\, \mu_t(dx)\, dt - \frac{1}{2} \int_H u^2(0, x)\, \zeta(dx). \tag{79}$$

In fact if $u \in \mathcal{E}_A([0, T] \times H)$ we have $u^2 \in \mathcal{E}_A([0, T] \times H)$ as well and

$$D_t(u^2) = 2u D_t u, \quad D_x(u^2) = 2u D_x u, \quad D_x^2(u^2) = 2D_x u \otimes D_x u + 2u D_x^2 u.$$

It follows that

$$K_F(u^2(t, x)) = 2u(t, x)\, K_F u(t, x) + |C^{1/2} D_x u(t, x)|^2. \tag{80}$$

Integrating both sides of (80) with respect to $\underline{\mu}$ over $[0, T] \times H$, yields (79). Now, by (79) we have

$$\int_{H_T} K_F\, u(t, x)\, u(t, x)\, \mu_t(dx)\, dt \le 0, \quad \forall\, u \in \mathcal{E}_A([0, T] \times H),$$

which implies the dissipativity of K_F in $L^2([0, T] \times H, \underline{\mu})$. The case $p \neq 2$ follows from standard arguments about diffusions operators, see [12]. □

Remark 21 Assume that $\underline{\mu}$ is a solution to (3). By Proposition 20 it follows that K_F is closable in $L^1([0, T] \times H, \underline{\mu})$. We shall denote by $\overline{K_F}$ its closure. Clearly, if $\psi : H \to [0, +\infty]$ is convex lover semi-continuous and

$$\overline{K_F}\,(D(\overline{K_F})) \text{ is dense in } L^1([0, T] \times H, \underline{\mu}) \text{ for all } \underline{\mu} \text{ such that } \mu_0 \in \mathcal{P}_\psi(H),$$

then the rank condition (77) is fulfilled.

4.2 The Semigroup Associated to a Non Autonomous Problem

We assume here that Hypothesis 2 (resp. Hypothesis 3) is fulfilled. As usual in studying non autonomous systems, it is convenient to consider, besides (11) (resp. (51)), a problem in the unknowns $(H(\tau), y(\tau))$ (intended in the mild sense.)[7]

[7] We proceed here as in Sect. 2 (with $F_\alpha = 0$), but working in $C_b(H)$ rather than in $C_{1,b}(H)$.

$$\begin{cases} dH(\tau) = AH(\tau)d\tau + F_\alpha(y(\tau), H(\tau))d\tau + \sqrt{C}\, W(\tau), & \tau \geq 0, \\ dy(\tau) = d\tau, \quad \tau \geq 0 \\ H(0) = x, \quad y(0) = t. \end{cases} \tag{81}$$

(resp. a similar problem for (51)). To solve problem (81) we set $y(\tau, t) = t + \tau$, so that (81) reduces to

$$\begin{cases} dH(\tau) = AH(\tau)d\tau + F_\alpha(t + \tau, H(\tau))d\tau + \sqrt{C}\, dW(\tau) \\ \\ H(0) = x. \end{cases} \tag{82}$$

We denote by $H(\tau) = H(\tau, t, x)$ the mild solution of (81) which, however, is only defined for $\tau \in [0, T - t]$.

Let us define a semigroup $S_\tau^{F_\alpha}$, $\tau \geq 0$, in the space $C_T([0, T] \times H) = \{u \in C([0, T] \times H) : u(T, x) = 0, \forall\, x \in H\}$ by setting

$$(S_\tau^{F_\alpha} u)(t, x) = \mathbb{E}[u(t + \tau, H(\tau + t, x))]\, \mathbb{1}_{[0,T-\tau]}(t), \; u \in C_T([0, T] \times H).$$

Notice that $S_\tau^{F_\alpha} = 0$, $\forall\, \tau \geq T$. Since the law of $H(\tau, t, x)$ coincides, as easily seen, with that of $X_\alpha(t + \tau, t, x)$, for any $x \in H$ and any τ such that $t + \tau \leq T$, it results

$$\begin{aligned} (S_\tau^{F_\alpha} u)(t, x) &= (P_{t,t+\tau}^\alpha u(t + \tau, \cdot)(x)\, \mathbb{1}_{[0,T-\tau]}(t) \\ &= \mathbb{E}[u(t + \tau, X_\alpha(t + \tau, t, x))]\, \mathbb{1}_{[0,T-\tau]}(t), \quad \forall u \in C_T([0, T] \times H). \end{aligned} \tag{83}$$

Let denote by $\mathcal{K}_{F_\alpha}$ the infinitesimal generator of $S_\tau^{F_\alpha}$, defined through its resolvent (see (84) below). Then the resolvent set of $\mathcal{K}_{F_\alpha}$ coincides with $\mathbb{R}$ and its resolvent is given, for all $f \in C_T([0, T] \times H)$ and all $\lambda \in \mathbb{R}$, by

$$\begin{aligned} R(\lambda, \mathcal{K}_{F_\alpha}) f(t, x) &= \int_0^\infty e^{-\lambda \tau} S_\tau^{F_\alpha} f(t, x) d\tau \\ &= \int_t^T e^{-\lambda(r-t)} \mathbb{E}[f(X_\alpha(r, t, x))] dr, \quad (t, x) \in [0, T] \times H. \end{aligned} \tag{84}$$

$\mathcal{K}_{F_\alpha}$ can also be defined as follows (following [14]). We say that a function $u \in C_T([0, T] \times H)$ belongs to the domain of $\mathcal{K}_{F_\alpha}$ if there exists a function $f \in C_T([0, T] \times H)$ such that

(i)
$$\lim_{\tau \to 0} \frac{1}{\tau} (S_\tau^{F_\alpha} u(t, x) - u(t, x)) = f(t, x), \quad \forall\, (t, x) \in [0, T] \times H.$$

(ii) There exists $M_u > 0$ such that

$$\left| \frac{1}{\tau} (S_\tau^{F_\alpha} u(t, x) - u(t, x))) \right| \leq M_u, \quad \forall\, (t, x) \in [0, T] \times H, \; \tau \in [0, 1].$$

We set $\mathcal{K}_{F_\alpha} u = f$ and call $\mathcal{K}_{F_\alpha}$ the *infinitesimal generator* of S_τ on $C_T([0,T] \times [0,H])$. As easily seen, the abstract operator $\mathcal{K}_{F_\alpha}$ is an extension of the differential *Kolmogorov* operator K_{F_α} defined by (2) (with F_α replacing F).

4.3 The Case When C^{-1} is Bounded

Theorem 22 *Assume, besides Hypotheses 1 and 2 that there exists* $M_1 > 0$ *such that*

$$P_{0,t} V^2(x) \leq M_1 V^2(x), \quad \forall\, x \in H. \tag{85}$$

Let moreover $\zeta \in \mathcal{P}_{\psi+V^2}(H)$*, where* ψ *is defined by* (21)*. Then the Fokker–Planck equation* (3) *has a unique solution* $\underline{\mu} = (\mu_t)_{t\in[0,T]}$ *and* $\mu_t \in \mathcal{P}_{\psi+V^2}(H)$ *for any* $t \geq 0$.

Proof Let $f \in \mathcal{E}_A([0,T] \times H))$ and consider the approximating equation

$$\mathcal{K}_{F_\alpha} u_\alpha = f. \tag{86}$$

Thanks to (84) Eq. (86) has a unique solution u_α given by

$$u_\alpha(t,x) = -\int_0^{T-t} \mathbb{E}[f(t+\tau, X_\alpha(t+\tau,t,x))]d\tau, \quad (t,x) \in [0,T] \times H \tag{87}$$

and therefore

$$|u_\alpha(t,x)| \leq T\|f\|_0, \quad \forall\,(t,x) \in [0,T] \times H,\ \alpha \in (0,1]. \tag{88}$$

Step 1. $u_\alpha \in D(\mathcal{K}_0^1) \cap C_b^1(H)$ and it results

$$\mathcal{K}_{F_\alpha} u_\alpha = \mathcal{K}_0^1 u_\alpha + \langle F_\alpha(t,x), D_x u_\alpha \rangle = f. \tag{89}$$

Fix in fact $t \in [0,T]$ and $h > 0$ such that $t + h \leq T$. Then, recalling (12) we write

$$X_\alpha(t+h,t,x) = Z(t+h,t,x) + g_\alpha(t+h,t,x), \tag{90}$$

where

$$Z(t+h,t,x) = e^{hA}x + \int_t^{t+h} e^{(t+h-r)A}\sqrt{C}\; dW(r)$$

and

$$g_\alpha(t+h,t,x) = \int_t^{t+h} e^{(t+h-r)A} F_\alpha(r, X(r,t,x))dr.$$

Then by (9) and (83) we have

$$S_h^{0,1} u_\alpha(t,x) = \mathbb{E}[u_\alpha(t+h, Z(t+h,t,x))]$$

and

$$\begin{aligned} S_h^{F_\alpha} u_\alpha(t,x) &= \mathbb{E}[u_\alpha(t+h, X(t+h,t,x))] \\ &= \mathbb{E}[u_\alpha[t+h, Z(t+h,t,x) + g_\alpha(t+h,t,x))]. \end{aligned}$$

Consequently,

$$\begin{aligned} S_h^{F_\alpha} u_\alpha(t,x) = S_h^{0,1} u_\alpha(t,x) + \mathbb{E}\int_0^1 &\langle D_x u_\alpha(t+h, Z(t+h,t,x) \\ &+ \xi g_\alpha(t+h,t,x)), g_\alpha(t+h,t,x)\rangle\, d\xi, \end{aligned}$$

which implies

$$\begin{aligned} \frac{1}{h}(S_h^{F_\alpha} u_\alpha(t,x) - u_\alpha(t,x)) &= \frac{1}{h}(S_h^{0,1} u_\alpha(t,x) - u_\alpha(t,x)) \\ &+ \frac{1}{h}\mathbb{E}\int_0^1 \langle D_x u_\alpha(t+h, Z(t+h,t,x) + \xi g_\alpha(t+h,t,x)), g_\alpha(t+h,t,x)]\, d\xi. \end{aligned}$$

Now, letting $h \to 0$ and taking into account that $\lim_{h\to 0} \frac{1}{h}\, g(t+h,t,x) = F_\alpha(t,x)$, yields (89).

Notice that from Step 1 does not follow that $u_\alpha \in D(K_F) = \mathcal{E}_A([0,T]\times H)$, but we are going to show in the next step that u_α belongs to the domain of the closure $\overline{K_F}$ of K_F. (Recall Remark 21).

Step 2. $u_\alpha \in D(\overline{K_F})$ and we have

$$\overline{K_F} u_\alpha = \mathcal{K}_0^1 u_\alpha + \langle F(t,x), D_x u_\alpha\rangle = f + \langle F(t,x) - F_\alpha(t,x), D_x u_\alpha\rangle. \tag{91}$$

By Proposition 4 there is a multi-sequence $(u_{\alpha,\mathrm{j}})$ in $\mathcal{E}_A([0,T]\times H)$ such that for all $(t,x) \in [0,T]\times H$

(i) $\lim_{\mathrm{j}\to\infty} u_{\alpha,\mathrm{j}}(t,x) = u_\alpha(t,x)$,
(ii) $\lim_{\mathrm{j}\to\infty} K_0 u_{\alpha,\mathrm{j}}(t,x) = \mathcal{K}_0^1 u_\alpha(t,x)$,
(iii) $\lim_{\mathrm{j}\to\infty} D_x u_{\alpha,\mathrm{j}}(t,x) = D_x u_\alpha(t,x)$,
(iv) $|u_{\alpha,\mathrm{j}}(t,x)| + |K_0\, u_{\alpha,\mathrm{j}}(t,x)| + |D_x u_{\alpha,\mathrm{j}}(t,x)| \le C(1+|x|), \quad x \in H.$

Then we have

$$\begin{aligned} \lim_{\mathrm{j}\to\infty} K_F u_{\alpha,\mathrm{j}}(t,x) &= \lim_{\mathrm{j}\to\infty} (K_0 u_{\alpha,\mathrm{j}}(t,x) + \langle D_x u_{\alpha,\mathrm{j}}(t,x), F(t,x)\rangle) \\ &= \mathcal{K}_0^1 u_\alpha(t,x) + \langle D_x u_\alpha(t,x), F(t,x)\rangle, \quad \forall\, (t,x) \in [0,T]\times H. \end{aligned}$$

Since

$$|K_F\, u_{\alpha,\mathrm{j}}(t,x)| \le C(1+|x|) + C(1+|x|)|F(t,x)|, \quad \forall\, (t,x) \in [0,T]\times H, \tag{92}$$

we have by Hypothesis 2

$$\begin{aligned}\int_{H_T} |K_F\, u_{\alpha,\mathrm{j}}(t,x)|\, d\mu_t\, dt &\le \int_{H_T} [C(1+|x|) + C(1+|x|)|F(t,x)|]\, d\mu_t\, dt \\ &\le C' \int_{H_T} (1+|x|)V(x)\mu_t(dx)\, dt.\end{aligned}$$

By the dominated convergence theorem it follows that $u_\alpha \in D(\overline{K_F})$ and

$$\lim_{\mathrm{j}\to\infty} K_F u_{\alpha,\mathrm{j}} = \overline{K_F}\, u_\alpha = \mathcal{K}_0^1 u_\alpha + \langle F(t,x), D_x u_\alpha\rangle \quad \text{in } L^1([0,T]\times H; \underline{\mu}).$$

Therefore $u_\alpha \in D(\overline{K_F})$ and (91) is fulfilled .

Step 3. There is $C(T) > 0$ such that

$$\int_{H_T} |D_x u_\alpha|^2\, d\underline{\mu} \le C(T)\|f\|_0^2 \left(1 + \int_{H_T} |F - F_\alpha|^2\, d\underline{\mu}\right). \tag{93}$$

Multiplying both sides of Eq. (91) by u_α and integrating in $\underline{\mu}$ over H_T, yields

$$\int_{H_T} \overline{K_F} u_\alpha\, u_\alpha\, d\underline{\mu} = \int_{H_T} f\, u_\alpha\, d\underline{\mu} + \int_{H_T} \langle F - F_\alpha, D_x u_\alpha\rangle\, u_\alpha\, d\underline{\mu}.$$

Taking into account (79), we obtain

$$\begin{aligned}\frac{1}{2}\int_{H_T} |C^{1/2} D_x u_\alpha|^2\, d\underline{\mu} &\le -\int_{H_T} \overline{K_F} u_\alpha\, u_\alpha\, d\underline{\mu} \\ &= -\int_{H_T} f\, u_\alpha\, d\underline{\mu} + \int_{H_T} \langle F_\alpha - F, D_x u_\alpha\rangle\, u_\alpha\, d\underline{\mu}.\end{aligned}$$

Now, recalling (88), we obtain

$$\begin{aligned}\frac{1}{2}\int_{H_T} |C^{1/2} D_x u_\alpha|^2\, d\underline{\mu} &\le T^2\|f\|_0^2 + T\|f\|_0 \int_{H_T} |F_\alpha - F|\,|D_x u_\alpha|\, d\underline{\mu} \\ &\le T^2\|f\|_0^2 + T\|f\|_0 \|C^{-1/2}\| \int_{H_T} |F_\alpha - F|\,|C^{1/2} D_x u_\alpha|\, d\underline{\mu}.\end{aligned} \tag{94}$$

Now the conclusion follows by standard arguments.

Step 4.

$$\overline{K_F}(D(K_F)) \text{ is dense in } L^1([0,T]\times H, \underline{\mu}).$$

In view of (91) it is enough to show that

$$\lim_{\alpha\to 0}\int_{H_T} |\langle F_\alpha - F, D_x u_\alpha\rangle|\, d\underline{\mu} = 0. \tag{95}$$

We have in fact by Hölder's inequality, taking into account (85), (88) and (94) that,

$$\left(\int_{H_T} |\langle F_\alpha - F, D_x u_\alpha\rangle|\, d\underline{\mu}\right)^2 \le \int_{H_T} |F_\alpha - F|^2\, d\underline{\mu} \int_{H_T} |D_x u_\alpha|^2\, d\underline{\mu}$$

$$\le C(T)\|f\|_0^2\left(1+\int_{H_T}|F-F_\alpha|^2\, d\underline{\mu}\right)\int_{H_T}|F_\alpha - F|^2\, d\underline{\mu}$$

$$\le C(T)\|f\|_0^2\left(1+\alpha^2 M_1\int_{H_T} V^2(x)\, d\underline{\mu}\right)\alpha^2 M_1\int_{H_T} V^2(x)\, d\underline{\mu}\to 0$$

as $\alpha\to 0$. The proof is complete. □

Example 23 We continue here Example 10 using notations and assumptions there. As we have seen, Hypothesis 3 is fulfilled in this case. To apply Theorem 22 it remains to check that (85) is fulfilled. This clearly holds by (47), thus Theorem 22 applies.

Now we assume Hypotheses 1 and 3. Then, to repeat the proof of Theorem 22 we should find an estimate for $|(-A)^{1/2}D_x u_\alpha|^2$ rather than for $|D_x u_\alpha|^2$ which seems to be difficult. So, we shall assume (95) and, arguing as before, we can prove

Theorem 24 *Assume, besides Hypotheses 1 and 3 that*

$$\lim_{\alpha\to 0}\int_{H_T} |\langle F_\alpha - F, D_x u_\alpha\rangle|\, d\underline{\mu} = 0. \tag{96}$$

Let moreover $\zeta\in\mathcal{P}_\psi(H)$, where ψ is defined by (21). *Then the Fokker–Planck equation* (3) *has a unique solution $\underline{\mu} = (\mu_t)_{t\in[0,T]}$ and $\mu_t\in\mathcal{P}_\psi(H)$ for any $t\ge 0$.*

Example 25 (*Burgers equation*) We consider here the setting of Example 17. Then statement (96) follows from [10, Lemma 4.1] and so Theorem 24 applies.

4.4 The Case When Tr $C<\infty$

Trying to repeat the proof of Theorem 22, whereas Steps 1 and 2 can be repeated without anyproblems, there is a difficulty for the proof of step 3 which requires $C^{-1}\in L(H)$. The key point is again to prove the statement (95).

Theorem 26 *Assume, besides Hypotheses 1 and 3 that*

$$\lim_{\alpha\to 0}\int_{H_T} |\langle F_\alpha - F, D_x u_\alpha\rangle|\, d\underline{\mu} = 0. \tag{97}$$

Then there is a unique solution $\underline{\mu}$ *of the Fokker–Planck equation* (3).

Example 27 Consider the $2D$-Navier–Stokes equation from Example 18 and assume that $\mathrm{Tr}[(-A)C] < \infty$. Then (97) is fulfilled by [8, Eq. 5.4.8]. So, Theorem 26 applies.

References

1. Bogachev, V., Da Prato, G., Röckner, M., Stannat, W.: Uniqueness of solutions to weak parabolic equations for measures. Bull. London Math. Soc. **39**, 631–640 (2007)
2. Bogachev, V., Da Prato, G., Röckner, M.: Fokker-Planck equations and maximal dissipativity for Kolmogorov operators with time dependent singular drifts in Hilbert spaces. J. Funct. Anal. **256**, 1269–1298 (2009)
3. Bogachev, V., Da Prato, G., Röckner, M.: Existence and uniqueness of solutions for Fokker-Planck equations on Hilbert spaces. J. Evol. Equ. **10**(3), 487–509 (2010)
4. Bogachev, V., Da Prato, G., Röckner, M.: Uniqueness for solutions of Fokker-Planck equations on infinite dimensional spaces. Commun. Partial Differ. Equ. **36**, 925–939 (2011)
5. Bogachev, V., Da Prato, G., Röckner, M., Shaposhnikov, S.: On the uniqueness of solutions to continuity equations. J. Differ. Equ. **259**(8), 3854–3873 (2015)
6. Bogachev, V., Da Prato, G., Röckner, M., Shaposhnikov, S.: An analytic approach to infinite-dimensional continuity and Fokker-Planck-Kolmogorov equations. Ann. Sc. Norm. Super. Pisa Cl. Sci. **14**(3), 983–1023 (2015)
7. Cerrai, S.: A Hille-Yosida theorem for weakly continuous semigroups. Semigroup Forum **49**, 349–367 (1994)
8. Da Prato, G.: Kolmogorov Equations for Stochastic PDEs. Birkhäuser, Boston (2004)
9. Da Prato, G., Tubaro, L.: Some results about dissipativity of Kolmogorov operators. Czechoslov. Math. J. **51**(126), 685–699 (2001)
10. Da Prato, G., Debussche, A.: m-dissipativity of Kolmogorov operators corresponding to Burgers equations with space-time white noise. Potential Anal **26**, 31–55 (2007)
11. Da Prato, G., Flandoli, F., Röckner, M.: Fokker-Planck equations for SPDE with non-trace class noise. Commun. Math. Stat. **1**(3), 281–304 (2013)
12. Eberle, A.: Uniqueness and Non-uniqueness of Semigroups Generated by Singular Diffusion Operators. Lecture Notes in Mathematics, vol. 1718. Springer, Berlin (1999)
13. Liu, W., Röckner, M.: Stochastic Partial Differential Equations: An Introduction. Universitext. Springer, Berlin (2015)
14. Priola, E.: On a class of Markov type semigroups in spaces of uniformly continuous and bounded functions. Studia Math. **136**, 271–295 (1999)

Part II
Stochastic Partial Differential Equations and Regularity Structures

Stochastic and Deterministic Constrained Partial Differential Equations

Zdzisław Brzeźniak, Gaurav Dhariwal, Javed Hussain and Mauro Mariani

Abstract We present some results recently obtained by the authors for the so-called constrained stochastic and parabolic equations, including Navier–Stokes Equations.

Keywords Partial differential equations (deterministic and stochastic) · Constrained energy · Navier–Stokes equations · Stratonovich differential

Mathematics Subject Classification 60H15 · 35K05 · 35K55 · 35Q30 · 35Q60 58J65 · 60J25 · 76M35

1 Introduction

In the theory of partial differential equations one often studies equations with constraints on the values of the unknown function. Here primary examples are geometric heat and wave equations where it is required that the solution is a manifold-valued function. Such models have been extensively studied, one could mention Eells-Sampson [24], Struwe and Shatah [31–33] for the deterministic problems; Funaki [25], Carroll [20] and [8, 13–15] by the the first named authour et al., for the

Z. Brzeźniak (✉)
Department of Mathematics, University of York, Heslington, York YO10 5DD, UK
e-mail: zdzislaw.brzezniak@york.ac.uk

G. Dhariwal
Institute of Analysis and Scientific Computing, Vienna University of Technology, Wiedner Hauptstrasse 8–10, 1040 Vienna, Austria
e-mail: gaurav.dhariwal@tuwien.ac.at

J. Hussain
Sukkur IBA University, Sukkur Sindh, Pakistan
e-mail: javed.brohi@iba-suk.edu.pk

M. Mariani
Faculty of Mathematics, National Research University Higher School of Economics, 6 Usacheva St., 119048 Moscow, Russia
e-mail: mmariani@hse.ru

A. Eberle et al. (eds.), *Stochastic Partial Differential Equations and Related Fields*, Springer Proceedings in Mathematics & Statistics 229,
https://doi.org/10.1007/978-3-319-74929-7_6

stochastic problems. If the target manifold is a sphere, one can study a generalisation of the heat flow map, called the Landau-Lifshitz-Gilbert Equations [1, 2, 5, 7, 16]. Recently, different kind of constraints, the nonlocal ones, were investigated by Rybka [30], Caffarelli-Lin [18] and Caglioti et al. [19]. For instance, one imposes the constraint that the L^p norm of the solution remains constant. There are various motivations for these studies, one can refer to the above mentioned papers as well as to the recent papers [4, 9, 10] authored by different combinations of authors of the current paper. The aim of this paper is to explain a simple gradient-flow approach to such questions for deterministic equations. We also present, without proof, some results on the stochastic versions, which however would not be properly understood without the necessary deterministic background.

2 A Geometric Approach

It is well understood how to construct a stochastic or deterministic equation on hypersurfaces of an Euclidean space (or even general Hilbert space) from a given equation on an ambient space, provided the latter is given in terms of smooth vector fields. To be precise let us describe this procedure.

Suppose that $\big(\mathrm{H}, \langle\cdot,\cdot\rangle\big)$ is a Hilbert space and $\mathcal{M} = \varphi^{-1}(\{1\}) \subset \mathrm{H}$ is a hypersurface for some non-degenerate smooth function $\varphi\colon \mathrm{H} \to [0,\infty)$. Each element $m \in \mathcal{M}$ has a tangent space $T_m\mathcal{M}$ which can be identified with the closed subspace of H (of co-dimension 1) given by $\ker \mathrm{d_m}\varphi = \{\mathrm{x} \in \mathrm{H}\colon (\mathrm{d_m}\varphi)(\mathrm{x}) = 0\}$, where $\mathrm{d_m}\varphi \in \mathcal{L}(\mathrm{H},\mathbb{R})$ is the Fréchet derivative of φ at m. By the Riesz Lemma there exists a unique element in H, denoted by $(\mathrm{D}\varphi)(m)$, such that $(\mathrm{d_m}\varphi)(\mathrm{x}) = \langle(\mathrm{D}\varphi)(\mathrm{m}),\mathrm{x}\rangle$, $x \in \mathrm{H}$. Since $(\mathrm{D}\varphi)(m) \neq 0$, the orthogonal projection $\pi_m : \mathrm{H} \to T_m\mathcal{M}$ is given (with $|\cdot|$ being the norm on H) by the formula

$$\pi_m(x) = x - \langle x, \mathbf{n}(m)\rangle\mathbf{n}(m), \quad x \in \mathrm{H}\,, \tag{1}$$

where

$$\mathbf{n}(m) = \frac{\mathrm{D}\varphi(m)}{|\mathrm{D}\varphi(m)|}, \quad m \in \mathcal{M}\,.$$

Given a vector field $f\colon \mathrm{H} \to \mathrm{H}$ we can consider the "tangent projection" $\hat{f}$ of the restriction of f to $\mathcal{M}$ (which is a "tangent" vector field on $\mathcal{M}$) defined by

$$\hat{f}(m) := \pi_m(f(m)) \in T_m\mathcal{M}\,, \quad m \in \mathcal{M}\,. \tag{2}$$

The associated ODE

$$\frac{\mathrm{d}x(t)}{\mathrm{d}t} = f(x(t))\,, \quad t \geq 0\,, \tag{3}$$

takes the following well-known form on $\mathscr{M}$

$$\frac{\mathrm{d}x(t)}{\mathrm{d}t} = \hat{f}(x(t)), \quad t \geq 0. \tag{4}$$

Note that $\hat{f}$ has a smooth extension to an open neighbourhood of $\mathscr{M}$ and the ODE (4) is locally well-posed on that neighbourhood. One can then show that given $x_0 \in \mathscr{M}$, the local solution stays on $\mathscr{M}$ by either using a local diffeomorphism of some neighbourhood of x_0 in $\mathscr{M}$ (i.e. using de facto the Hilbert manifold structure of $\mathscr{M}$) or by showing that $\varphi(x(t)) = 0$ for t in the domain of the solution. If $\mathscr{M}$ is a compact set (that requires H to be finite dimensional) then we can easily deduce that each solution starting at $x_0 \in \mathscr{M}$ is a global one, i.e. defined on $[0, \infty)$. However, if $\mathscr{M}$ is not compact the solutions may not be global-in-time.

Similar argument can be made for stochastic differential equations, with one small but important difference. Suppose $f_0, f_1, \ldots, f_N$ is a finite collection of vector fields on H and $W = \big(W(t), t \geq 0\big)$ is an $\mathbb{R}^N$-valued Wiener process, we write $W(t) = (W_j(t))$ defined on some filtered probability space $(\Omega, \mathscr{F}, \mathbb{F}, \mathbb{P})$, $\mathbb{F} = (\mathscr{F}_t)_{t\geq 0}$, satisfying the so-called usual conditions.

In the whole ambient space we can study stochastic differential equations either in the Itô or Stratonovich form, the latter requiring more regularity assumptions on the vector fields $f_1, \ldots, f_N:$, i.e.

$$\mathrm{d}x = f_0(x)\,dt + \sum_{i=1}^{N} f_i(x)\,dW_i(t), \tag{5}$$

or

$$\begin{aligned} \mathrm{d}x &= f_0(x)\,dt + \sum_{i=1}^{N} f_i(x) \circ dW_i(t) \\ &= f_0(x)\,dt + \left[\frac{1}{2}\sum_{i=1}^{N} f_i'(x) f_i(x)\right] dt + \sum_{i=1}^{N} f_i(x)\,dW_i, \end{aligned} \tag{6}$$

where $f_i'(x) = \mathrm{d}_x f_i$, $x \in \mathrm{H}$. On the other hand, it turns out that the correct form of these equations on $\mathscr{M}$ is the Stratonovich one. This fact is related to the Wong-Zakai type theorems, see e.g. [3], and the rough paths theory proposed recently by Terry Lyons [28]. With the same notation as before one can consider an equation

$$\begin{aligned} \mathrm{d}x &= \hat{f}_0(x)\,dt + \sum_{j=1}^{N} \hat{f}_j(x) \circ dW_j \\ &= \hat{f}_0(x)\,dt + \sum_{j=1}^{N} \hat{f}_j(x)\,dW_j + \frac{1}{2}\sum_{j=1}^{N} \hat{f}_j'(x)\hat{f}_j(x)\,dt. \end{aligned} \tag{7}$$

The issues of local and global solutions to the above problem can be solved through a similar approach as the one used to answer the analogous questions in the deterministic case, see for instance [11] and references therein.

However, when the vector fields are not smooth or not everywhere defined or both, the situation changes. For instance, let us consider an unbounded, self-adjoint and non-negative operator A with domain of A on a Hilbert space H. The domain D(A) is a Hilbert space endowed with the "graph norm" : $\|x\|^2 = |x|^2 + |\mathrm{A}x|^2$, $x \in \mathrm{D(A)}$. Such an operator A induces a (only densely defined) vector field $f_0(x) = -\mathrm{A}x$. Theory of corresponding deterministic and stochastic problems related to Eqs. (5) and (6) is now well developed and understood, see e.g. a monograph [21] by Da Prato and Zabczyk. However, this is not the case for Eq. (7) with a vector field $\hat{f}_0$ defined by

$$\hat{f}_0(x) = f_0(x) - \langle f_0(x), \mathbf{n}(x)\rangle \mathbf{n}(x)\ , \quad x \in \mathscr{M} \cap \mathrm{D(A)}\ , \tag{8}$$

in the view of (1) and (2).

The aim of this paper is to present an intuitive introduction to such problems. In order to keep the paper concise, we will concentrate on the deterministic theory, which is essential to understand the corresponding stochastic counterpart. The main references for this are PhD thesis [26] by J Hussain and [23] by G Dhariwal, as well as [4] and [9].

3 Constrained "Heat" Equation

From now on we will consider a basic example of the hypersurface $\mathscr{M}$:

$$\mathscr{M} = \{x \in \mathrm{H}\colon |x| = 1\}\ , \tag{9}$$

i.e. $\mathscr{M}$ is a hypersurface of H with the function $\varphi(x) = \frac{1}{2}\left(|x|^2 - 1\right)$.

Note that in this case $\mathbf{n}(x) = x$, $x \in \mathrm{H}$. Therefore

$$\hat{f}(x) = -\mathrm{A}x + \langle \mathrm{A}x, x\rangle x = -\mathrm{A}x + |\mathrm{A}^{1/2}x|^2 x\ , \quad x \in \mathrm{D(A)}\ . \tag{10}$$

Hence, the Eq. (3) with $f(x) = -\mathrm{A}x$, projected onto the tangent space $T_x\mathscr{M}$ takes the following form

$$\frac{\mathrm{d}x}{\mathrm{d}t} = -\mathrm{A}x + |\mathrm{A}^{1/2}x|^2 x\ , \quad x(0) = x_0\ . \tag{11}$$

We cannot use the arguments from Sect. 2 in order to establish the existence and uniqueness of a local or global solution, as the right hand side of (11), i.e. $\hat{f}$ is neither a smooth vector field nor well-defined on H. In fact, as we will see later, we can only establish local existence of solutions for initial data x_0 from a small, yet dense subset of $\mathscr{M}$.

At this moment we should point out that the Eq. (11) is specially built to accommodate the constraint $|x| = 1$, i.e. specially built for the purpose of studying a tangent version of the Eq. (3) (with $f(x) = -\mathrm{A}x$) on the hypersurface $\mathscr{M} = \{x \in \mathrm{H}\colon |x| = 1\}$. Indeed, if A has an eigenvector $e_0 \in \mathrm{H}$ with an eigenvalue $\lambda_0 > 0$ then a solution to problem (11) with $x_0 = ae_0$ is of the form $x(t) = z(t)e_0$, where the $\mathbb{R}-$valued function z solves

$$\frac{\mathrm{d}z}{\mathrm{d}t} = \lambda_0(|z|^2 - 1)z \ , \qquad z(0) = a \ . \tag{12}$$

If $a > 1$, then the lifespan of z is finite and hence the solution to (11) is not global.

The previous example shows that even a linear problem of the form

$$\frac{\mathrm{d}u}{\mathrm{d}t} = -\mathrm{A}u \ , \tag{13}$$

leads to a non-linear problem of the form (11).

There are two natural generalisations. In order to describe the first one, we will have to look at (13) differently. We define the following "energy" on the space H:

$$\mathscr{E}_0(u) = \frac{1}{2}|\mathrm{A}^{1/2}u|^2, \quad u \in \mathrm{D}(\mathrm{A}^{1/2}) \ . \tag{14}$$

At a heuristic level, the gradient of $\mathscr{E}_0$ at u with respect to the H scalar product is equal to $\mathrm{A}u$, i.e. the Eq. (13) can be rewritten as

$$\frac{\mathrm{d}u}{\mathrm{d}t} = -\mathrm{grad}\, \mathscr{E}_0(u) \ , \tag{15}$$

where $\mathrm{grad}\, \mathscr{E}_0$ is meant to be w.r.t H. We can consider the restriction of $\mathscr{E}_0$ to $\mathscr{M}$:

$$\tilde{\mathscr{E}}_0(u) = \mathscr{E}_0(u) \ , \quad u \in \mathscr{M} \ ,$$

and denote by $\widetilde{\mathrm{grad}}\,\cdot$ the gradient of $\tilde{\mathscr{E}}_0$ with respect to the pseudo metric on $T_u\mathscr{M}$ induced by the scalar product on H. It is not difficult to see that

$$\widetilde{\mathrm{grad}}\, \tilde{\mathscr{E}}_0(u) = \pi_u\, (\mathrm{grad} \mathscr{E}(u)) \ . \tag{16}$$

Thus, we get the following representation of the problem (11)

$$\frac{du}{dt} = -\widetilde{\mathrm{grad}}\, \tilde{\mathscr{E}}_0(u) = -\pi_u\, (\mathrm{grad}\, \mathscr{E}_0(u)) \ . \tag{17}$$

The "gradient-flow" point of view has the advantage that it provides, at least at the heuristic level, a justification of the following equality

$$\tilde{\mathscr{E}}_0(u(t)) + \int_0^t \left|\widetilde{\text{grad}}\, \tilde{\mathscr{E}}_0(u(s))\right|^2 ds = \tilde{\mathscr{E}}_0(u_0)\,. \tag{18}$$

A similar identity holds for the unconstrained equation in the ambient space H:

$$\mathscr{E}_0(u(t)) + \int_0^t |\text{grad}\, \mathscr{E}_0(u(s))|^2\, ds = \mathscr{E}_0(u_0)\,. \tag{19}$$

Note that in the view of Eqs. (13) and (14), the above equation simply reads:

$$\frac{1}{2}|\mathrm{A}^{1/2}u(t)|^2 + \int_0^t |\mathrm{A}u(s)|^2 ds = \frac{1}{2}|\mathrm{A}^{1/2}u_0|^2\,. \tag{20}$$

Now we are ready to describe the first generalisation (promised earlier). We begin by considering a more general energy function

$$\mathscr{E}(u) = \mathscr{E}_0(u) + \Phi(u) = \frac{1}{2}|\mathrm{A}^{1/2}u|^2 + \Phi(u)\,, \quad u \in \mathrm{D}(\mathrm{A})\,, \tag{21}$$

where Φ is a non-negative function on D(A). In fact it is natural to assume that $\mathrm{D}(\mathrm{A}^\alpha) \subset \mathrm{D}(\Phi)$ for some appropriate $\alpha \in (0, 1)$ (for details see [10]). Then we can consider the gradient flows on H and $\mathscr{M}$:

$$\frac{du}{dt} = -\text{grad}\, \mathscr{E}(u) = -\mathrm{A}u - \text{grad}\, \Phi(u)\,, \tag{22}$$

and respectively

$$\frac{du}{dt} = -\widetilde{\text{grad}}\, \tilde{\mathscr{E}}(u) = -\mathrm{A}u + |\mathrm{A}^{1/2}u|^2 u - \pi_u\, (\text{grad}\, \Phi(u))\,. \tag{23}$$

As in the previous cases, the solutions of Eqs. (22) and (23) are expected to satisfy the following identities

$$\mathscr{E}(u(t)) + \int_0^t |\text{grad}\, \mathscr{E}(u(s))|^2\, ds = \mathscr{E}(u_0)\,, \quad t \geq 0\,, \tag{24}$$

$$\tilde{\mathscr{E}}(u(t)) + \int_0^t \left|\widetilde{\text{grad}}\, \tilde{\mathscr{E}}(u(s))\right|^2 ds = \tilde{\mathscr{E}}(u_0)\,, \quad t \geq 0\,. \tag{25}$$

One can guess that such equations could yield global existence of solutions once a local existence of a solution for a given $u_0 \in \mathscr{M}$ with finite energy (i.e. $\mathscr{E}(u_0) < \infty$) is established.

This procedure needs details, but in principle it is well established for evolution equations (deterministic and even stochastic) in Hilbert (or Banach) spaces [4, 26]. In the next section we will present some details based on works by the authors.

The second generalisation is to consider a map B in H, defined on $\mathrm{D}(\mathrm{A}^{\alpha})$ for a suitable $\alpha \in (0, 1)$ such that

$$\langle B(u), u \rangle = 0 , \quad u \in \mathrm{D}(\mathrm{A}^{\alpha}) . \tag{26}$$

An important example of such a map B is provided by the convection term in the Navier–Stokes Equations. Note that the orthogonality (26) proposed above implies that the "projected vector field" $\hat{B}$ is equal to B. Indeed,

$$\hat{B}(u) = B(u) - \langle B(u), u \rangle u = B(u) , \quad u \in \mathrm{D}(\mathrm{A}^{\alpha}) . \tag{27}$$

Thus the equation constrained to $\mathscr{M}$ corresponding to $\frac{du}{dt} = -\mathrm{A}u - B(u)$, is

$$\frac{du}{dt} = -\mathrm{A}u + |\mathrm{A}^{1/2}u|^2 u - B(u) . \tag{28}$$

We will see later that the problem (28) is locally well-posed for $u_0 \in \mathrm{D}(\mathrm{A}^{1/2}) \cap \mathscr{M}$. However, the existence of a global solution is an open question. But under the following additional assumption

$$\langle \mathrm{A}u, B(u) \rangle = 0 , \quad u \in \mathrm{D}(\mathrm{A}) , \tag{29}$$

one can prove the existence of a global solution to (28) for every $u_0 \in \mathrm{D}(\mathrm{A}^{1/2}) \cap \mathscr{M}$ [4]. This is an important case, as it covers the 2-d Navier–Stokes Equations with the periodic boundary conditions. In this case, Eq. (28) can be written as

$$\frac{du}{dt} = -\widetilde{\mathrm{grad}}\, \tilde{\mathscr{E}}(u(t)) - B(u(t)) , \tag{30}$$

where $\widetilde{\mathrm{grad}}\, \tilde{\mathscr{E}}(u) = \mathrm{A}u - |\mathrm{A}^{1/2}u|^2 u$. The assumptions on B imply that $B(u)$ is perpendicular to $\widetilde{\mathrm{grad}}\, \tilde{\mathscr{E}}(u)$. Thus, at a purely heuristic level we have

$$\begin{aligned} \frac{d}{dt}\tilde{\mathscr{E}}(u(t)) &= \left\langle \widetilde{\mathrm{grad}}\, \tilde{\mathscr{E}}(u(t)), \frac{du}{dt} \right\rangle = -\left|\widetilde{\mathrm{grad}}\, \tilde{\mathscr{E}}(u(t))\right|^2 - \left\langle \widetilde{\mathrm{grad}}\, \tilde{\mathscr{E}}(u(t)), B(u(t)) \right\rangle \\ &= -\left|\widetilde{\mathrm{grad}}\, \tilde{\mathscr{E}}(u(t))\right|^2 . \end{aligned} \tag{31}$$

So, we get the following equality

$$\tilde{\mathscr{E}}(u(t)) + \int_0^t \left|\widetilde{\mathrm{grad}}\, \tilde{\mathscr{E}}(u(s))\right|^2 ds = \tilde{\mathscr{E}}(u_0) , \quad t \geq 0 , \tag{32}$$

which will be used in order to prove non blow-up (thus global existence) of solutions.

4 Local Existence and Invariance

In this section along with the previously established assumptions on the Hilbert space H and the operator A, we also assume that F is a (non-linear) map defined on D(A) with values in H satisfying some assumptions listed below. We are interested in proving the local existence of the following problem

$$\frac{du}{dt} = -\mathrm{A}u + |\mathrm{A}^{1/2}u|^2 u + \hat{F}(u) , \quad u(0) = u_0 , \tag{33}$$

where $u_0 \in \mathcal{M}$, $\mathcal{M} = \{y \in \mathrm{H} : |y|_{\mathrm{H}} = 1\}$ is the unit sphere in H and

$$\hat{F}(u) = F(u) - \langle F(u), u\rangle u , \quad u \in \mathrm{D(A)} . \tag{34}$$

Let us consider the Hilbert spaces D(A) and V = D($\mathrm{A}^{1/2}$) endowed with the graph norms

$$|u|^2_{\mathrm{D(A)}} = |\mathrm{A}u|^2 + |u|^2 , \quad u \in \mathrm{D(A)} , \tag{35}$$

$$|u|^2_{\mathrm{V}} = |\mathrm{A}^{1/2}u|^2 + |u|^2 , \quad u \in \mathrm{D}(\mathrm{A}^{1/2}) . \tag{36}$$

We impose the following condition on F:

Assumption 1 There exists $\alpha \in [0, 1)$ and an increasing continuous function $R \colon [0, \infty] \to [0, \infty)$ such that $R(0) = 0$ and, for all $u, \mathrm{v} \in \mathrm{D(A)}$,

$$|F(u) - F(\mathrm{v})| \leq R(|u|_{\mathrm{V}} + |\mathrm{v}|_{\mathrm{V}})|u - \mathrm{v}|^{1-\alpha}_{\mathrm{V}} |u - \mathrm{v}|^{\alpha}_{\mathrm{D(A)}} , \tag{37}$$

$$|F(u)|^2 \leq R^2(|u|_{\mathrm{V}}) \left[|u|^{2(1-\alpha)}_{\mathrm{V}} |u|^{2\alpha}_{\mathrm{D(A)}} + 1 \right] . \tag{38}$$

Note that if $F(0) = 0$ then (37) implies (38). We have the following lemma, see [26].

Lemma 1 *If F satisfies Assumption 1 then the map Φ defined by*

$$\Phi : X_T \ni u \mapsto \left\{[0, T] \ni t \mapsto F(u(t))\right\} \in L^2(0, T; \mathrm{H}) , \tag{39}$$

where $X_T = L^2(0, T; \mathrm{D(A)}) \cap C([0, T]; \mathrm{V})$, satisfies, for all $u, \mathrm{v} \in X_T$,

$$|\Phi(u) - \Phi(\mathrm{v})|^2 \leq T^{1-\alpha} R^2 \left(|u|_{L^\infty(0,T;\mathrm{V})} + |\mathrm{v}|_{L^\infty(0,T;\mathrm{V})}\right) |u - \mathrm{v}|^{2\alpha}_{L^2(0,T;\mathrm{D(A)})} |u - \mathrm{v}|^{2(1-\alpha)}_{L^\infty(0,T;\mathrm{V})} , \tag{40}$$

$$|\Phi(u)|^2 \leq R^2 |u|_{L^\infty(0,T;\mathrm{V})} \left[|u|^{2(1-\alpha)}_{L^\infty(0,T;\mathrm{V})} |u|^{2\alpha}_{L^2(0,T;\mathrm{D(A)})} + 1 \right] . \tag{41}$$

Remark 1 If $u \in X_T = \mathscr{C}([0, T]; \mathrm{V}) \cap L^2(0, T; \mathrm{D(A)})$ then under Assumption 1, $F \in L^2(0, T; \mathrm{H})$.

The so-called maximal regularity property [22] of A[1] states that for every $u_0 \in \mathrm{V}$, and every $f \in L^2(0, T; \mathrm{H})$ there exists a unique solution $u = e^{-\cdot \mathrm{A}}u_0 + S * f$ of

$$\frac{du}{dt} = -\mathrm{A}u + f(t), \quad \text{on } (0, T)\ , \quad u(0) = u_0\ , \tag{42}$$

where $S = (e^{-t\mathrm{A}})_{t \geq 0}$, satisfying

$$\int_0^T |u(t)|^2_{\mathrm{D(A)}}\, dt + |u|^2_{C([0,T];\mathrm{V})} \leq C \left[|u_0|^2_{\mathrm{V}} + \int_0^T |f(s)|^2 ds \right], \tag{43}$$

for some $C = C(T)$ such that $\sup_{t \in [0,T]} C(t) < \infty$.

On combining the maximal regularity property of A with Lemma 1, we yield the following result:

Theorem 1 *If F satisfies Assumption 1 then for every $u_0 \in \mathrm{V}$ there exists $T > 0$, depending only on $|u_0|_{\mathrm{V}}$, and unique $u \in X_T$ which solves*

$$\frac{du}{dt} = -\mathrm{A}u(t) + F(u(t))\ , \quad u(0) = u_0\ . \tag{44}$$

The Eq. (44) although similar to Eq. (33), but is not identical, as it lacks the non-linearity which is probable to pose a problem while dealing with the existence and uniqueness of the solution. However, Theorem 1 suggests a way to proceed in order to prove the existence of solutions to (33).

We need to show that the functions

$$F_0 : \mathrm{D(A)} \ni \mathrm{u} \mapsto |\mathrm{A}^{1/2}\mathrm{u}|^2\mathrm{u} \in \mathrm{H}\ ,$$

and $\hat{F} - F$ satisfies Assumption 1 given that the function F satisfies the same assumption.

Lemma 2 *For all $u, \mathrm{v} \in \mathrm{V}$ we have*

$$|F_0(u) - F_0(\mathrm{v})| \leq |u - \mathrm{v}|_{\mathrm{V}} \left(|u|_{\mathrm{V}} + |\mathrm{v}|_{\mathrm{V}}\right)^2\ ,$$
$$|F_0(u)| \leq |u||u|^2_{\mathrm{V}} \leq |u|^3_{\mathrm{V}}.$$

Hence, we see that F_0 satisfies Assumption 1 with $\alpha = 0$. Concerning $\hat{F} - F_0$ the situation may be more difficult. However, since by (34)

$$F(u) - \hat{F}(u) = \langle F(u), u \rangle u\ , \tag{45}$$

we infer that $F_0 - \hat{F}$ satisfies (38) given that F satisfies it as well. In order to verify that $F_0 - \hat{F}$ satisfies (37) we introduce an auxiliary function

[1] At this stage we can figure out the assumptions on A so as to consider a more general operator than a self-adjoint operator (as considered in the current paper).

$$\psi : \mathrm{H} \times \mathrm{H} \ni (x, y) \mapsto \langle x, y \rangle y \in \mathrm{H}$$

which satisfies

$$|\psi(x_2, y_2) - \psi(x_1, y_1)| \leq |x_2 - x_1| \, |y_2|^2 + |y_2 - y_1| \, |x_1|(|y_1| + |y_2|) ,$$

and we easily deduce our claim from the identity $(F_0 - \hat{F})(u) = \psi(F_0(u), u)$. In this way using Theorem 1, we have proved the first part of the following result:

Theorem 2 *For every $\rho > 0$ there exists $T_0 = T_0(\rho)$ such that for each $u_0 \in \mathrm{V}$ with $|u_0|_{\mathrm{V}} \leq \rho$, there exists a unique $u \in X_{T_0}$ which solves* (33). *Moreover, if $u_0 \in \mathcal{M}$, then*

$$u(t) \in \mathcal{M}, \quad t \in [0, T_0] .$$

It remains to prove the second part of the above theorem. In some sense we follow the method described for ODEs in the previous section. Using Lions' Lemma [34, Theorem III.1.12] we have

$$\begin{aligned} \frac{1}{2}\frac{d}{dt}\left[|u(t)|^2 - 1\right] &= \langle u'(t), u(t)\rangle = \langle -\mathrm{A}u(t) + |\mathrm{A}^{1/2}u(t)|^2 u(t), u(t)\rangle + \langle \hat{F}(u(t)), u(t)\rangle \\ &= -\langle \mathrm{A}u(t), u(t)\rangle + |\mathrm{A}^{1/2}u(t)|^2 |u(t)|^2 - \langle F(u(t)), u(t)\rangle \left(|u(t)|^2 - 1\right) \\ &= \left[|\mathrm{A}^{1/2}u(t)|^2 - \langle F(u(t)), u(t)\rangle\right]\left[|u(t)|^2 - 1\right] . \end{aligned}$$

Since $|u(0)|^2 - 1 = 0$ and $\int_0^{T_0} \left[|\mathrm{A}^{1/2}u(t)|^2 - \langle F(u(t)), u(t)\rangle\right] dt < \infty$, by Gronwall Lemma we infer

$$|u(t)|^2 - 1 = 0, \quad t \in [0, T_0] .$$

5 Applications

In this section we will show that the results from Sect. 4 are applicable to two classical examples: the reaction diffusion equation [26] and Navier–Stokes Equations [4], both in two dimensional domains.

5.1 *Reaction Diffusion Equation*

Let $\mathcal{O} \subset \mathbb{R}^2$ be a bounded domain with sufficiently regular boundary. Put

$$\mathrm{D}(\mathrm{A}) = H^{2,2}(\mathcal{O}) \cap H_0^{1,2}(\mathcal{O}) , \quad \mathrm{A}u = -\Delta u. \tag{46}$$

Note that $D(A^{1/2}) = H_0^{1,2}(\mathcal{O})$.

Define, for some $p > 0$, the following two functions

$$\Phi(u) = \frac{1}{p+2} \int_{\mathcal{O}} |u(x)|^{p+2} dx, \qquad F(u) = -\text{grad}\, \Phi(u) = -|u|^p u \,. \tag{47}$$

Using the embedding $V \hookrightarrow L^q$, for each $q < \infty$, we can show

Lemma 3 *For every $u, v \in H_0^{1,2}(\mathcal{O})$ there exists $C > 0$ such that*

$$|F(u) - F(v)|_{L^2} \leq C|u - v|_{H_0^{1,2}} \left(|u|_{H_0^{1,2}} + |v|_{H_0^{1,2}}\right)^p \,. \tag{48}$$

We are interested in proving the existence of a global solution of the following reaction diffusion equation

$$\frac{du}{dt} - \Delta u = |\nabla u|_{L^2}^2 u + |u|^p u - |u|_{L^{p+2}}^{p+2} u \,, \qquad u(0) = u_0 \,, \tag{49}$$

where $u_0 \in H_0^{1,2}$.

By results from previous section and Lemma 3 we deduce the following theorem:

Theorem 3 *For every $u_0 \in H_0^{1,2}(\mathcal{O})$ such that $|u_0|_{L^2} = 1$, there exists T_0 depending on $|u_0|_{H_0^{1,2}}$ and a unique solution $u \in L^2(0, T_0; D(A)) \cap C([0, T]; H_0^{1,2})$ of (49) such that*

$$|u(t)|_{L^2} = 1, \quad t \in [0, T_0] \,. \tag{50}$$

Proof In order to prove the theorem it is essential to observe that for $u \in D(A)$, $\langle \Delta u, u \rangle_{L^2} = -\int_{\mathcal{O}} |\nabla u(x)|^2 dx = -|\nabla u|_{L^2}^2$ and

$$\begin{aligned} \hat{F}(u) = F(u) - \langle F(u), u \rangle_{L^2} u &= -|u|^p u + u \int_{\mathcal{O}} |u(x)|^{p+2} dx \\ &= -|u|^p u + |u|_{L^{p+2}}^{p+2} u \,. \end{aligned} \tag{51}$$

By Lemmata 2 and 3 the RHS of (49) satisfies Assumption 1 and thus we can deduce the proof of the theorem by Theorem 2.

5.2 *Navier–Stokes Equations on a Torus $\mathbb{T}^2$*

Let us denote by H the Hilbert space consisting of all measurable vector fields u on $\mathbb{T}^2$ such that

$$\text{div}\, u = 0 \,, \quad \int_{\mathbb{T}^2} |u(x)|^2 dx < \infty \text{ and } \int_{\mathbb{T}^2} u(x) dx = 0 \,. \tag{52}$$

We denote by $\pi : \mathbb{L}^2(\mathbb{T}^2) \to \mathrm{H}$ the orthogonal projection and put $\mathrm{D}(\mathrm{A}) = \mathrm{H} \cap H^{2,2}(\mathscr{O})$, $\mathrm{V} := \mathrm{D}(\mathrm{A}^{\frac{1}{2}}) = \mathrm{H} \cap H^{1,2}(\mathbb{T}^2)$ and, for $u, \mathrm{v} \in \mathrm{D}(\mathrm{A})$

$$\mathrm{A}u = -\pi \Delta u \ , \quad B(u, \mathrm{v}) = \pi(u \cdot \nabla \mathrm{v}) \quad \text{and} \quad B(u) = B(u, u) \ . \tag{53}$$

It is known (and can be verified easily) that B satisfies Assumption 1 with some $\alpha \in (\frac{1}{2}, 1)$ and properties (26) and (29) [4]. Thus, using the results from Sect. 2 essentially Theorem 2 we have:

Theorem 4 *For every $u_0 \in \mathrm{V}$ there exists a unique function $u \in C([0, \infty), \mathrm{V}) \cap L^2_{\mathrm{loc}}([0, \infty); \mathrm{D}(\mathrm{A}))$ such that $|u(t)|_{\mathrm{H}} = 1$ for all $t \geq 0$, which solves*

$$\frac{\mathrm{d}u}{\mathrm{d}t} = \mathrm{A}u + |\nabla u|^2_{L^2} u - B(u, u), \quad u(0) = u_0 \ . \tag{54}$$

6 Generalisation to Stochastic PDEs

In this section we are concerned with the following stochastic evolution equation

$$\begin{cases} du(t) + [Au(t) + B(u(t))]\, dt = |\nabla u(t)|^2_{L^2} u(t)\, dt \\ \qquad\qquad\qquad\qquad + \sum_{j=1}^m \pi\left((c_j \cdot \nabla) u(t)\right) \circ dW_j(t) \ , \\ u(0) = u_0 \ , \end{cases} \tag{55}$$

where $\circ$ denotes the Stratonovich differential and the operators A and B were defined earlier in Sect. 5.2. We assume that

Assumption 2 (A.1) The vector fields $c_1, \ldots, c_m$ are constant vector fields. (A.2) W_j, $j = 1, \ldots, m$, are $\mathbb{R}$−valued i.i.d standard Brownian Motions.

Now we formulate the main result of this section whose proof can be found in [9] along with other important results related to a'priori estimates, invariance of the manifold $\mathscr{M}$, tightness of measures and existence of a weak martingale solution.

Theorem 5 *Under Assumption 2, for every $u_0 \in \mathrm{V} \cap \mathscr{M}$ there exists a pathwise unique strong solution u of stochastic constrained Navier–Stokes equation* (55) *such that for each $T > 0$,*

$$\mathbb{E}\left[\int_0^T |u(t)|^2_{\mathrm{D(A)}}\, dt + \sup_{t \in [0,T]} |u(t)|^2_{\mathrm{V}}\right] < \infty \ . \tag{56}$$

The proof of this result is quite long and involved, yet can be summarised as follows. We first consider the Galerkin approximation of (55) with suitably chosen initial data. We prove that these approximated equations have global solutions satisfying suitable a'priori estimates (which hold as in the deterministic case due to properties (26) and (29)). Then we use suitable version of Prokhorov, Jakubowski-Skorokhod Theorems [15, 27] and martingale representation theorem [21] to construct a weak martingale solution. Finally, invoking the Yamada-Watanabe type theorem, see e.g. [21] or [29] we deduce the existence of a pathwise unique strong and regular solution as stated above.

Finally, let us point out that one should not mix up the constrained parabolic problems, stochastic or deterministic, with equations of Schrödinger type, where the constraint appears in a natural way, see for instance [12].

References

1. Alouges, F., Soyeur, A.: On global weak solutions for Landau-Lifshitz equations: existence and nonuniqueness. Nonlinear Anal. **18**, 1071–1084 (1992)
2. Banas, L., Brzeźniak, Z., Neklyudov, M., Prohl, A.: Stochastic Ferromagnetism: Analysis and Nmerics. Studies in mathematics. De Gruyter, Boston (2013)
3. Brzeźniak, Z., Carroll, A.: Approximations of the WongZakai type for stochastic differential equations in M-type 2 Banach spaces with applications to loop spaces. Sémi. de Probabilities XXXVII. Lect. Notes in Math **1832**, 25117289 (2003)
4. Brzeźniak, Z., Dhariwal, G., Mariani, M.: 2D constrained Navier-Stokes equations. J. Differ. Equ. **264**(4), 2833–2864 (2016)
5. Brzeźniak, Z., Goldys, B., Jegraj, T.: Large deviations and transitions between equilibria for stochastic Landau-Lifshitz equation. Arch. Ration. Mech. Anal. **226**(2), 497558 (2017)
6. Brzeźniak, Z., Goldys, B. and Ondrejat, M.: Stochastic Geometric Partial Differential Equations. New Trends in Stoc. Ana. and Related Topics. A Volume in Honour of Professor K. D. Elworthy **12**. World Scientific Publishing Co. 1–32 (2012)
7. Brzeźniak, Z., Li, L.: Weak solutions of the Stochastic LandauLifshitzGilbert Equations with nonzero anisotrophy energy. Appl. Math. Res. Express **2**, 334–375 (2016)
8. Brzeźniak, Z. and Carroll, A.: The stochastic geometric heat equation, in preparation
9. Brzeźniak, Z. and Dhariwal, G.: Stochastic constrained Navier-Stokes equations on T^2. https://arxiv.org/abs/1701.01385 Submitted (2017)
10. Brzeźniak, Z. and Hussain, J.: Global solutions of non-linear heat equation with solutions in a Hilbert manifold. In preparation
11. Brzeźniak, Z., Elworthy, K.D.: Stochastic differential equations on Banach manifolds. Methods Funct. Anal. Topol. **6**, 43–84 (2000)
12. Brzeźniak, Z., Millet, A.: On the stochastic Strichartz estimates and the stochastic nonlinear Schrödinger equation on a compact Riemannian manifold. Potential Anal. **41**, 269–315 (2014)
13. Brzeźniak, Z., Ondreját, M.: Stochastic geometric wave equations. Stoch. Anal. Lect. Ser. **68**, 157–188 (2010)
14. Brzeźniak, Z., Ondreját, M.: Weak solutions to stochastic wave equations with values in Riemannian manifolds. Commun. Part. Differ. Equ. **36**, 1624–1653 (2011)
15. Brzeźniak, Z., Ondreját, M.: Stochastic geometric wave equations with values in compact Riemanninan homogeneous spaces. Ann. Probab. **41**, 1938–1977 (2013)
16. Brzeźniak, Z., Goldys, B., Jegaraj, T.: Weak solutions of a stochastic Landau-Lifshitz-Gilbert Equation. Appl. Math. Res. Express **1**, 1–33 (2013)
17. Burq, N., Gérard, P., Tzvetkov, N.: The Cauchy problem for the nonlinear Schrödinger equation on a compact manifold. J. Nonlinear Math. Phys. **10**, 12–27 (2003)
18. Caffarelli, L., Lin, F.: Nonlocal Heat Flows preserving the L^2 Energy. Discret. Contin. Dyn. Syst. **23**, 49–64 (2009)

19. Caglioti, E., Pulvirenti, M., Rousset, F.: On a constrained 2D Navier- Stokes equation. Commun. Math. Phys. **290**, 651–677 (2009)
20. Carroll, A.: The stochastic nonlinear heat equation. Ph.D. thesis, University of Hull (1999)
21. Da Prato, G., Zabczyk, J.: Stochastic Equations in Infinite Dimensions, 2nd edn. Cambridge University Press, Cambridge (2014)
22. de Simon, L.: Unapplicazione della toria degli integrali singolari allo studio delle equazioni differenziali lineari astratte del primo ordine. Rend. Sem. Mat. Univ. Padova **34**, 205–223 (1964)
23. Dhariwal, G.: A study of constrained Navier-Stokes equations and related problems. Ph.D. thesis, University of York (2017)
24. Eells, J., Sampson, J.H.: Harmonic mappings of Riemannian manifolds. Am. J. Math. **86**, 109–160 (1964)
25. Funaki, T.: A stochastic partial differential equation with values in manifold. J. Funct. Anal. **109**, 257–258 (1992)
26. Hussain, J.: Analysis of some deterministic and stochastic evolution equations with solutions taking values in an infinite dimensional Hilbert manifold. Ph.D. thesis, University of York (2015)
27. Jakubowski, A.: The almost sure Skorokhod representation for subsequences in nonmetric spaces. Teor. Veroyatn. Primen. **42**, 209-216 (1997); Trans. Theory Probab. Appl. **42**, 167-174 (1998)
28. Lyons, T.J.: Differential equations driven by rough signals. Rev. Mat. Iberoamericana **14**(2), 21517310 (1998)
29. Ondreját, M.: Uniqueness for stochastic evolution equations in Banach spaces. Dissertationes Mathematicae **426**, 1–63 (2004)
30. Rybka, P.: Convergence of a heat flow on a Hilbert manifold. Proc. R. Soc. Edinb. **136A**, 851–862 (2006)
31. Shatah, J., Struwe, M.: Geometric Wave Equations. Courant lecture notes in mathematics 2. New York University, Courant Institute of Mathematical Sciences, New York (1998)
32. Shatah, J., Struwe, M.: Regularity Results for Nonlinear Wave Equations. Ann. Math. Second Ser. **138**, 503–518 (1993)
33. Shatah, J., Struwe, M.: The cauchy problem for wave maps. Int. Math. Res. Not. **11**, 555–571 (2002)
34. Temam, R.: Navier-Stokes Equations: Theory and Numerical Analysis. North-Holland Publishing Company, Amsterdam (1979)

SPDEs with Volterra Noise

Petr Čoupek, Bohdan Maslowski and Jana Šnupárková

Abstract Recent results on linear stochastic partial differential equations driven by Volterra processes with linear or bilinear noise are briefly reviewed and partially extended. In the linear case, existence and regularity properties of stochastic convolution integral are established and the results are applied to 1D linear parabolic PDEs with boundary noise of Volterra type. For the equations with bilinear noise, existence and large time behaviour of solutions are studied.

Keywords Volterra process · Rosenblatt process · Stochastic evolution equation Additive noise · Bilinear noise

2010 Mathematics Subject Classification 60H15 · 60G22

1 Introduction

Stochastic PDEs with non-Markovian noise have been recently studied in numerous papers. However, in most cases, the driving process is the fractional Brownian motion which is a very important, nevertheless specific stochastic process. A natural question may arise whether these standard results may be obtained for more general, Gaussian or non-Gaussian perturbations. In the present paper, we summarize some recent

P. Čoupek · B. Maslowski (✉)
Faculty of Mathematics and Physics, Charles University, Sokolovská 83,
186 75 Praha 8, Czech Republic
e-mail: maslow@karlin.mff.cuni.cz

P. Čoupek
e-mail: coupek@karlin.mff.cuni.cz

J. Šnupárková
Faculty of Chemical Engineering, University of Chemistry and Technology Prague,
Technická 3, 166 28 Praha 6, Czech Republic
e-mail: jana.snuparkova@vscht.cz

A. Eberle et al. (eds.), *Stochastic Partial Differential Equations and Related Fields*, Springer Proceedings in Mathematics & Statistics 229,
https://doi.org/10.1007/978-3-319-74929-7_7

results in this direction, notably [3, 12], for linear and bilinear noise, respectively. The result from [3] is partly extended and applied to the case of boundary noise. Similarly, the existence theorem from [12] is slightly modified to cover time-dependent diffusion coefficients (which is applied in a forthcoming paper [8] to stochastic LQ control problems).

In the rest of the section, regular Volterra processes are defined and several examples are given. Our main examples are (Liouville) multifractional Brownian motion in the Gaussian case and the Rosenblatt process in the non-Gaussian case. A Wiener-type integral is introduced following the lines of [1, 3]. In Sect. 2, the additive noise case is dealt with. Properties of stochastic convolution integral (existence, regularity) are studied and applied to the heat equation with Volterra noise on the boundary. Section 3 is devoted to equations with multiplicative noise (only scalar Gaussian noise is considered here). The integration is understood in the Skorokhod sense. Existence of solutions is shown for a general parabolic equation of higher order and some noise stabilization or destabilization properties are illustrated by simple examples.

Consider a function $K : [0, \infty) \times [0, \infty) \to [0, \infty)$ which is

- Volterra and vanishing, i.e.

(i) $K(0, 0) = 0$ and $K(t, r) = 0$ on $\{0 \leq t < r < \infty\}$,
(ii) $\lim_{t \to r+} K(t, r) = 0$ for all $r \geq 0$,

- square integrable and α-regular, i.e.

(iii) $K(t, \cdot) \in L^2(0, t)$ for all $t \geq 0$,
(iv) $K(\cdot, r) \in \mathscr{C}^1(r, T)$ for all $r \in [0, T)$, $T > 0$. Furthermore, there are $\alpha \in (0, \frac{1}{2})$ and $C_T > 0$ such that

$$\left| \frac{\partial K}{\partial u}(u, r) \right| \leq C_T (u - r)^{\alpha - 1} \left(\frac{u}{r} \right)^{\alpha}$$

on $\{0 < r < u < T\}$.

Such K is called an α-regular, vanishing Volterra kernel. Unless stated otherwise, we shall assume that every kernel K appearing in the sequel satisfies the hypotheses above and we shall only refer to it as a Volterra kernel.

Definition 1 We say that a centered, continuous stochastic process $b = (b_t, t \geq 0)$ is an (α-regular, vanishing) Volterra process, if $b_0 = 0$ and its covariance function takes the form

$$\mathbb{E}(b_s b_t) = \int_0^{s \wedge t} K(s, r) K(t, r) \mathrm{d}r, \quad s, t \geq 0, \tag{1}$$

for some Volterra kernel K.

Example 1 A standard example of a (Gaussian) α-regular Volterra process is the fractional Brownian motion (fBm) with Hurst parameter $H \in (\frac{1}{2}, 1)$, in which case $\alpha = H - \frac{1}{2}$ and the kernel is defined by

$$K^H(t,r) := c_H \int_r^t (u-r)^{H-\frac{3}{2}} \left(\frac{u}{r}\right)^{H-\frac{1}{2}} \mathrm{d}u$$

where c_H is the normalizing constant such that $\mathbb{E}(b_1^H)^2 = 1$ (cf. [2] or [5]).

Example 2 Consider a kernel $K : [0,1] \times [0,1] \to [0,\infty)$ defined as

$$K(t,r) := (t-r)^{H(t)-\frac{1}{2}} \mathbf{1}_{(0,t]}(r),$$

where $H : [0,1] \to \mathbb{R}$ is a function which satisfies

- $H \in \mathscr{C}^1(0,1)$,
- there is an $\varepsilon \in \left(0, \frac{1}{2}\right)$ such that $H(t) \in \left[\frac{1}{2}+\varepsilon, 1\right)$ for all $t \in [0,1]$,
- there is a constant $c_\varepsilon > 0$ such that

$$|H'(t)| \le c_\varepsilon \cdot \min_{u \in (0,t)} \left[\left(\frac{t}{u}\right)^{\varepsilon} \frac{1}{|\log(t-u)|\,(t-u)} \right].$$

  for all $t \in (0,1)$.

Then the stochastic process $X = (X_t, t \in [0,1])$ defined as $X_t := \int_0^t K(t,r)\,\mathrm{d}W_r$, where $W = (W_t, t \ge 0)$ is the standard (scalar) Wiener process, is an (ε-regular, vanishing) Gaussian Volterra process which is called the (Liouville) multifractional Brownian motion (cf. [3]).

Example 3 Let $H \in (\frac{1}{2}, 1)$. The Rosenblatt process (cf. [13] or [14]) is defined as

$$Z_t^H := \int_{\mathbb{R}} \int_{\mathbb{R}} \left(\int_0^t (u-r_1)_+^{-\frac{2-H}{2}} (u-r_1)_+^{-\frac{2-H}{2}} \mathrm{d}r \right) \mathrm{d}W_{r_1}\,\mathrm{d}W_{r_2}, \quad t \ge 0,$$

where z_+ denotes $z\mathbf{1}_{\{z \ge 0\}}$ and $W = (W_t, t \in \mathbb{R})$ is the standard two-sided Wiener process. Clearly, the Rosenblatt process is not Gaussian. Despite this fact, its second-order characteristics are the same as those of the fBm, in particular, the covariance function (cf. [14]). Hence, Z^H is again an α-regular Volterra process with $\alpha = H - \frac{1}{2}$ and with the same kernel as the fBm. Note that b^H may be also defined in a similar way as Z^H as both are examples of the so-called Hermite processes (cf. [14]).

In order to treat SPDEs driven by Volterra processes, a theory of stochastic integration must be developed. In general, we can neither appeal to the semimartingale property of Volterra processes (and use the Itô integral), nor to its Gaussianity.

Let $(V, \langle \cdot, \cdot \rangle_V)$ be a real separable Hilbert space and consider a linear space of V-valued deterministic step functions $\mathscr{E}$, i.e.

$$\mathscr{E} := \Bigg\{ \varphi : [0,T] \to V, \varphi = \sum_{i=1}^{n-1} \varphi_i \mathbf{1}_{[t_i, t_{i+1})} + \varphi_n \mathbf{1}_{[t_n, t_{n+1}]},$$

$$\varphi_i \in V, i \in \{1, \cdots, n\}, 0 = t_1 < t_2 <, \cdots, < t_{n+1} = T, n \in \mathbb{N} \Bigg\}.$$

Define an operator $\mathscr{K}_T^* : \mathscr{E} \to L^2([0,T]; V)$ by

$$(\mathscr{K}_T^* \varphi)(r) := \int_r^T \varphi(u) \frac{\partial K}{\partial u}(u, r)\,\mathrm{d}u, \quad \varphi \in \mathscr{E}. \tag{2}$$

For $\varphi \in \mathscr{E}$, the integral $\int_0^T \varphi \,\mathrm{d}b$ is defined as

$$i(\varphi) := \int_0^T \varphi \,\mathrm{d}b := \sum_{i=1}^{n-1} \varphi_i (b_{t_{i+1}} - b_{t_i}).$$

Now, using (1) and (2), we can show that

$$\|i(\varphi)\|_{L^2(\Omega;V)} = \|\mathscr{K}_T^* \varphi\|_{L^2([0,T];V)} \tag{3}$$

which is an Itô-type isometry for Volterra processes. Assuming, for simplicity, $\mathscr{K}^*$ is injective, the space of integrable functions $\mathscr{D}$ is obtained by completion of $\mathscr{E}$ with respect to the norm $\|\varphi\|_{\mathscr{D}} := \|\mathscr{K}_T^* \varphi\|_{L^2([0,T];V)}$ and the integral is the usual extension of i to $\mathscr{D}$ (cf. [3] for details). By Proposition 2.9 in [3] and its proof, we have that

$$L^{\frac{2}{1+2\alpha}}([0,T]; V) \hookrightarrow \mathscr{D}.$$

This is a standard result when the Volterra process under consideration is the fBm with $H = \alpha + \frac{1}{2} > \frac{1}{2}$ (cf. [6]). The above construction is standard for the fBm and it was generalized to Gaussian Volterra processes in [1]. For the non-Gaussian case see [3].

2 SPDEs with Additive Volterra Noise

We now turn our attention to linear SPDEs with additive Volterra noise. We model these equations as linear evolution equations in a separable (infinite-dimensional) Hilbert space.

Let U be a (real, separable) Hilbert space and $\{e_n\}$ its orthonormal basis. Let $\{b^{(n)}\}$ be a sequence of independent copies of a scalar Volterra process b. The U-cylindrical Volterra process $B = (B_t, t \in [0,T])$ is given by the (formal) series

$$B_t := \sum_n e_n b_t^{(n)}.$$

Let V be another (real, separable) Hilbert space and consider the equation

$$\begin{cases} \mathrm{d}X_t = AX_t \,\mathrm{d}t + \Phi \,\mathrm{d}B_t, \\ X_0 = x, \end{cases} \tag{4}$$

for $t \in [0, T]$ where $A : V \supset \mathrm{Dom}\,(A) \to V$ is an infinitesimal generator of a C_0-semigroup of bounded linear operators $(S(t), t \geq 0)$ acting on V; $x \in V$; and $\Phi \in \mathscr{L}(U, V)$. The solution to (4) is given in the mild form, i.e.

$$X_t = S(t)x + \int_0^t S(t-r)\Phi \,\mathrm{d}B_r, \quad t \in [0, T], \tag{5}$$

where the stochastic convolution integral on the RHS of (5) is defined by

$$\int_0^t S(t-r)\Phi \,\mathrm{d}B_r := \sum_n \int_0^t S(t-r)\Phi e_n \,\mathrm{d}b_r^{(n)}$$

provided that $\mathbf{1}_{[0,t]}(\cdot)S(t-\cdot)\Phi e_n \in \mathscr{D}$ for every n and that the whole series converges in $L^2(\Omega; V)$. We have the following result:

Proposition 1 (cf. [3], Propositions 4.1 and 4.2) *Let b be an α-regular kernel and assume that $S(t)\Phi \in \mathscr{L}_2(U, V)$ for every $t \in (0, T]$. Let there further be a $T_0 \in (0, T]$ such that*

$$\int_0^{T_0} \|S(u)\Phi\|_{\mathscr{L}_2(U,V)}^{\frac{2}{1+2\alpha}} \,\mathrm{d}r < \infty. \tag{6}$$

Then X is a well-defined, V-valued, mean-square right continuous process.

In particular, it follows from Proposition 1 that X admits a version with measurable sample paths. Henceforth, we shall consider only this version. The following proposition gives sufficient conditions for X to admit a version with continuous sample paths.

Proposition 2 (cf. [3], Proposition 4.5) *Let b be an α-regular Volterra process. Assume that $S(t)\Phi \in \mathscr{L}_2(U, V)$ for every $t \in (0, T]$. Further, let there be a $\beta > \frac{1}{2}$ and $T_0 \in (0, T]$ such that*

$$\int_0^{T_0} \left(r^{-\beta}\|S(u)\Phi\|_{\mathscr{L}_2(U,V)}\right)^{\frac{2}{1+2\alpha}} \,\mathrm{d}r < \infty. \tag{7}$$

Then X has a continuous version.

Remark 1 In the case of Gaussian Volterra processes, we may relax the condition on β. In particular, it would suffice that (7) is satisfied with some $\beta > 0$.

Sketch of proof (Proposition 2). We may assume that $\beta \in (\frac{1}{2}, \alpha + \frac{1}{2})$. The proof is based on the classical factorization method of Da Prato and Zabczyk (cf. e.g. [4]). First it is shown that

$$Y_s := \int_0^s (s-u)^{\beta-1} S(s-u)\Phi \,\mathrm{d}B_r \tag{8}$$

is a well-defined process such that its sample paths lie in $L^2([0,T];V)\mathbb{P}-$ a.s. Then one has to show that

$$X_t = \int_0^t (t-s)^{\beta-1} S(t-s) Y_s \,\mathrm{d}s, \quad \mathbb{P}-\text{a.s.}, \tag{9}$$

for every $t \in [0,T]$, and the claim of the proposition follows by Theorem 5.9 from [4] with $p=2$, $\alpha=\beta$, $r=0$ and $E_1=E_2=V$. If b is Gaussian, then $Y \in L^2$ implies $Y \in L^q$ for every $q \geq 2$. Thus, we may take any $p \geq 2$ in this situation (cf. Remark 1).

1. Y is well-defined and has paths in $L^2([0,T];V)$: In order to show that $Y = (Y_s, s \in [0,T])$ is well-defined, one has to follow the steps of the proof of Proposition 1 with (7) used instead of (6). Moreover, we have that

$$\sup_{s\in(0,T]} \mathbb{E}|Y_s|_V^2 \leq C \sup_{s\in(0,T]} \int_0^s \left(r^{-\beta}\|S(r)\Phi\|_{\mathscr{L}_2(U,V)}\right)^{\frac{2}{1+2\alpha}} \mathrm{d}r$$

for some finite constant $C>0$ which implies

$$\mathbb{E}\int_0^T |Y_s|^2 \,\mathrm{d}s \leq CT \int_0^T \left(r^{-\beta}\|S(r)\Phi\|_{\mathscr{L}_2(U,V)}\right)^{\frac{2}{1+2\alpha}} \mathrm{d}r < \infty$$

and hence, $Y \in L^2([0,T];V)\mathbb{P}-$ a.s.

2. The formula (9) holds: Recall the formula

$$\int_r^t (t-u)^{\beta-1}(u-r)^{-\beta}\,\mathrm{d}u = \frac{\pi}{\sin \pi\beta} =: \frac{1}{\Lambda}$$

for $r \in [0,t]$. Using this fact, we can write

$$\begin{aligned}
&\int_0^t S(t-r)\Phi \,\mathrm{d}B_r = \\
&\quad = \Lambda \sum_{n=1}^{\infty} \int_0^t \int_r^t (t-u)^{\beta-1}(u-r)^{-\beta} S(t-u)S(u-r)\Phi e_n \,\mathrm{d}u\,\mathrm{d}b_r^{(n)} \\
&\quad = \Lambda \sum_{n=1}^{\infty} \int_0^t (t-u)^{\beta-1} S(t-u)\left(\int_0^u (u-r)^{-\beta} S(u-r)\Phi e_n \,\mathrm{d}b_r^{(n)}\right)\mathrm{d}u
\end{aligned}$$

It remains to show that the stochastic Fubini-type theorem holds if $\beta \in (0, \alpha+\frac{1}{2}))$ (at least for this particular integrand, cf. [3], Lemma 4.4) and that the sum and the first integral can be interchanged.

For the purposes of the next proposition, let us remark that if S is analytic, then there exists a $\lambda \in \mathbb{R}$ such that $(\lambda I - A)$ is strictly positive. Hence, we may define

$$V_\delta := \mathrm{Dom}\left((\lambda I - A)^\delta\right)$$

for $\delta \geq 0$. The space V_δ equipped with the graph norm topology is a Hilbert space. Slightly altering the previous proof and using properties of the operator $(\lambda I - A)^\delta$, we obtain the following result.

Proposition 3 *If there is $\gamma \in [0, \alpha)$ and a finite constant $c > 0$ such that*

$$\|S(u)\Phi\|_{\mathscr{L}_2(U,V)} \leq cu^{-\gamma},$$

then X has a version which belongs to $\mathscr{C}^\nu([0, T]; V_\delta)$ for every $\delta, \nu \geq 0$ such that $\delta + \nu < \alpha - \gamma$.

Example 4 (1D Parabolic equations with boundary noise of Volterra type) Consider the deterministic parabolic equation of the second order

$$\partial_t u = \partial_{xx}^2 u$$

with the initial condition $u(0, \cdot) = x_0$ which is square-integrable on an open interval $(a, b) \subset \mathbb{R}$ and the Neumann boundary conditions

$$\partial_x u(\cdot, a) = -\eta_1, \quad \partial_x u(\cdot, b) = -\eta_2$$

on $[0, T]$ where $\eta = (\eta_1, \eta_2)$ is a 2-dimensional Volterra noise. We rewrite the problem as a stochastic evolution Eq. (4) by taking $U := \mathbb{R}^2$, $V := L^2(a, b)$ and $A := \partial_{xx}^2|_{\mathrm{Dom}(A)}$ with

$$\mathrm{Dom}(A) := \left\{ f \in V : f, f' \text{ is AC }, f'' \in V, \frac{\mathrm{d}f}{\mathrm{d}x}(a) = \frac{\mathrm{d}f}{\mathrm{d}x}(b) = 0 \right\}.$$

Then A generates an analytic semigroup on V. The noise process η is formally written as $\eta = \frac{\mathrm{d}}{\mathrm{d}t} B$ where $B = (B_t, t \in [0, T])$ is a U-cylindrical (i.e. 2-dimensional) Volterra process. Moreover, $\Phi := (\lambda I - A) N Q^{\frac{1}{2}}$ where $Q^{\frac{1}{2}} \in \mathbb{R}^{2\times 2}$ and N is the Neumann map, i.e. the operator given by $N : (g_1, g_2) \mapsto h$ where h satisfies the equation

$$\left(\frac{\mathrm{d}^2}{\mathrm{d}x^2} - \lambda I\right) h = 0,$$
$$h'(a) = g_1,$$
$$h'(b) = g_2.$$

The operator N belongs to $\mathscr{L}(U, V_\varepsilon)$ if $\varepsilon \in (0, \frac{3}{4})$ (cf. [10] and references therein). If $\alpha > \frac{1}{4}$, we can choose $\varepsilon \in (1 - \alpha, \frac{3}{4})$ so that $N \in \mathscr{L}(U, V_\varepsilon)$. By Proposition 3 with $\gamma := 1 - \varepsilon$, we have that the solution has a version which belongs to $\mathscr{C}^\nu([0, T]; V_\delta)$ for every $\delta, \nu \geq 0$ such that $\delta + \nu < \alpha + \varepsilon - 1$.

3 SPDEs with Multiplicative Gaussian Volterra Noise

Let $b = (b_t, t \geq 0)$ be a scalar **Gaussian** Volterra process (with an α-regular, vanishing Volterra kernel K) and consider the equation

$$\begin{cases} \mathrm{d}X_t &= AX_t \,\mathrm{d}t + \sigma(t)X_t \,\mathrm{d}b_t, \\ X_0 &= x_0, \end{cases} \tag{10}$$

in a real separable Hilbert space V where $A : V \supset \mathrm{Dom}\,(A) \to V$ is an infinitesimal generator of an analytic semigroup $(S(t), t \geq 0)$ on V, $\sigma : [0, \infty) \to \mathbb{R}$ is a deterministic continuous function and $x_0 \in V$. The solution $X = (X_t, t \geq 0)$ to Eq. (10) is considered in a weak form, i.e. for any $z \in \mathrm{Dom}\,(A^*)$ we have

$$\langle X_t, z\rangle_V = \langle x_0, z\rangle_V + \int_0^t \langle X_s, A^*z\rangle_V \,\mathrm{d}s + \int_0^t \sigma(s)\langle X_s, z\rangle_V \,\mathrm{d}b_s, \quad \mathbb{P} - \text{a.s.}$$

for $t \geq 0$. The stochastic integral is defined in the Skorokhod sense (cf. e.g. [1] for the particular case of Volterra processes or [11] for general Gaussian integrators) and we assume that both the integrals in the above expression are well defined.

Remark 2 In this (Gaussian) case, the process b admits the representation

$$b_t = \int_0^t K(t, r)\,\mathrm{d}W_r, \quad t \geq 0,$$

where $W = \{W_t, t \geq 0\}$ is the standard (scalar) Wiener process. Furthermore, if δ_W and δ_b are the Skorokhod integrals associated with W and b, respectively, then the operator $\mathcal{K}_T^*$, $T > 0$, provides the relation between δ_W and δ_b

$$\delta_b(u) =: \int_0^T u(r)\,\mathrm{d}b_r = \int_0^T (\mathcal{K}_T^* u)(r)\,\mathrm{d}W_r := \delta_W(u)$$

for any $u \in \mathrm{Dom}\,(\delta_b) = \mathrm{Dom}\,\big((\mathcal{K}_T^*)^{-1}(\delta_W)\big)$ (cf. [1] for further reference).

We impose the following assumption on the noise term:

(K) The function α_σ defined by

$$\alpha_\sigma(t) := \frac{\mathrm{d}}{\mathrm{d}t}\left(\int_0^t (\mathcal{K}_t^*\sigma)^2(r)\,\mathrm{d}r\right), \quad t \geq 0,$$

is continuous.

Remark 3 (i) α_σ is well-defined.
(ii) If σ is constant, it follows that

$$\alpha_\sigma(t) = \sigma^2 \frac{\mathrm{d}}{\mathrm{d}t}\left(\int_0^t K^2(t,r)\,\mathrm{d}r\right) = \sigma^2 \frac{\mathrm{d}}{\mathrm{d}t}\left(\mathbb{E}b_t^2\right).$$

(iii) If b is the fBm, then

$$\alpha_\sigma(t) = \sigma(t)\int_0^t \sigma(s)\phi_H(t-s)\,\mathrm{d}s$$

where $\phi_H(r) := H(2H-1)|r|^{2H-2}$ and, hence, the condition **(K)** is satisfied since σ is continuous.

Let $Z = (Z_t, t \geq 0)$ and $U = (U(t,s), 0 \leq s \leq t < \infty)$ be defined as

$$Z_t := \int_0^t \sigma(s)\,\mathrm{d}b_s, \qquad U(t,s) := S(t-s)\exp\left\{-\frac{1}{2}\int_s^t \alpha_\sigma(r)\,\mathrm{d}r\right\}.$$

Obviously, U is a strongly continuous evolution system on V generated by

$$\left(A - \frac{1}{2}\alpha_\sigma(t)I,\ t \geq 0\right).$$

We have the following existence result:

Proposition 4 *For $x_0 \in V$, the process $X = (X_t, t \geq 0)$ defined as*

$$X_t := \exp\{Z_t\}U(t,0)x_0, \quad t \geq 0,$$

is a weak solution to the system (10).

Proof The statement is a simplified version of a more general result that will appear in [8] (see also [7] for the particular case of fBm).

Assume at first that $x_0 \in \mathrm{Dom}\,(A)$ and take $z \in \mathrm{Dom}\,(A^*)$ arbitrary. The Itô formula for Gaussian Volterra processes ([9], Theorem 4.4; see also [1], Theorem 4) applied to the process $t \mapsto v(t, Z_t)$, where $v(t,y) := \langle U(t,0)x_0, \mathrm{e}^y z\rangle_V$, yields

$$\begin{aligned}
\langle X_t, z\rangle_V &= v(t, Z_t) \\
&= \langle x_0, z\rangle_V + \int_0^t \left(\frac{\partial v}{\partial t}(s, Z_s) + \frac{1}{2}\alpha_\sigma(s)\frac{\partial^2 v}{\partial y^2}(s, Z_s)\right)\mathrm{d}s + \int_0^t \sigma(s)\frac{\partial v}{\partial y}(s, Z_s)\,\mathrm{d}b_s \\
&= \langle x_0, z\rangle_V + \int_0^t \exp\{Z_s\}\Big(\big\langle AU(s,0)x_0, z\big\rangle_V - \frac{1}{2}\alpha_\sigma(s)\big\langle U(s,0)x_0, z\big\rangle_V\Big)\,\mathrm{d}s \\
&\quad + \frac{1}{2}\int_0^t \exp\{Z_s\}\alpha_\sigma(s)\big\langle U(s,0)x_0, z\big\rangle_V\,\mathrm{d}s \\
&\quad + \int_0^t \sigma(s)\exp\{Z_s\}\big\langle U(s,0)x_0, z\big\rangle_V\,\mathrm{d}b_s \\
&= \langle x_0, z\rangle_V + \int_0^t \langle X_s, A^*z\rangle_V\,\mathrm{d}s + \int_0^t \sigma(s)\langle X_s, z\rangle_V\,\mathrm{d}b_s.
\end{aligned}$$

Therefore, the process $(X_t, t \geq 0)$ is a weak (in fact, even strong) solution to (10).

In the general case $x_0 \in V$, consider an approximating sequence $x_n \in \mathrm{Dom}\,(A)$, $x_n \to x_0$ in V. For $X_t^n := \exp\{Z_s\}U(t,0)x_n$ we have that

$$\langle X_t^n, z\rangle_V = \langle x_n, z\rangle_V + \int_0^t \langle X_s^n, A^* z\rangle_V \,\mathrm{d}s + \int_0^t \sigma(s)\langle X_s^n, z\rangle_V \,\mathrm{d}b_s, \quad t \geq 0, \tag{11}$$

for $z \in \mathrm{Dom}\,(A^*)$, and $X^n \to X$ in $\mathscr{C}\big([0,T]; L^p(\Omega; V)\big)$ for each $p \geq 1$. Hence

$$\lim_{n\to\infty} \int_0^t \sigma(s)\langle X_s^n, z\rangle_V \,\mathrm{d}b_s = \langle X_t, z\rangle_V - \langle x_0, z\rangle_V - \int_0^t \langle X_s, A^* z\rangle_V \,\mathrm{d}s$$

in $L^2(\Omega)$. It is easy to see that

$$\sigma(\cdot)\langle X_\cdot^n, z\rangle_V \to \sigma(\cdot)\langle X_\cdot, z\rangle_V \quad \text{in} \quad L^2(\Omega; \mathscr{D}),$$

so the closedness of the stochastic integral as an operator $L^2(\Omega; \mathscr{D}) \to L^2(\Omega)$ allows to pass to the limit in (11), which completes the proof.

This result can be illustrated by the next example.

Example 5 Consider the stochastic parabolic equation of the second order

$$\frac{\partial u}{\partial t}(t,x) = \big(Lu(t,\,.)\big)(x) + \sigma(t)u(t,x)\frac{\mathrm{d}b}{\mathrm{d}t}, \quad (t,x) \in [0,\infty) \times \mathscr{O}, \tag{12}$$

with the initial condition $u(0,\cdot) = x_0$ on $\mathscr{O}$ and the Dirichlet boundary condition $u = 0$ on $[0,\infty) \times \partial\mathscr{O}$. Here, $\mathscr{O} \subset \mathbb{R}^d$ is a bounded domain with the boundary of class $\mathscr{C}^2$, σ satisfies **(K)**, and

$$(Lu(t,\,.))\,(x) := a_0(x)u(t,x) + \sum_{i=1}^{d} a_i(x)\frac{\partial u}{\partial x_i}(t,x) + \sum_{i,j=1}^{d} a_{ij}(x)\frac{\partial^2 u}{\partial x_i \partial x_j}(t,x)$$

is a strongly elliptic operator on $\mathscr{O}$. Suppose that the functions $a_0, a_i, a_{ij} \in \mathscr{C}^\infty(\bar{\mathscr{O}})$ for $i, j = 1, \ldots, d$.

Equation (12) can be written in the form of Eq. (10). Let $V := L^2(\mathscr{O})$ and

$$\big(Au(t,\,.)\big)(x) := \big(Lu(t,\,.)\big)(x),$$

with $\mathrm{Dom}\,(A) := H^2(\mathscr{O}) \cap H_0^1(\mathscr{O})$. Then A generates an analytic semigroup on V. Assuming that **(K)** is satisfied, it follows from Proposition 4 that Eq. (12) admits a weak solution X given by

$$X_t = \exp\Big\{\int_0^t \sigma(r)\,\mathrm{d}b_r - \frac{1}{2}\int_0^t \alpha_\sigma(r)\,\mathrm{d}r\Big\}S(t)x_0, \quad t \geq 0.$$

This formula can be used to study large-time behaviour of the solution. This task, however, turns out to be very difficult with a general σ and thus, it is assumed that σ is constant in the sequel. We obtain

$$\|X_t\|_V \leq M_A \|x_0\|_V \exp\left\{\sigma |b_t| - \frac{1}{2}\sigma^2 \mathbb{E} b_t^2 + \omega_A t\right\}, \quad t \geq 0,$$

where $M_A \geq 1$ and $\omega_A < 0$ are constants such that

$$\|S_A(t)\|_{\mathscr{L}(V)} \leq M_A \exp\{\omega_A t\}, \quad t \geq 0.$$

The limiting behaviour of the solution is then determined by the behaviour of the exponent $\sigma |b_t| - 1/2\sigma^2 \mathbb{E} b_t^2 + \omega_A t$. For example, in the case of fBm or Liouville fBm, the function α_σ takes the explicit form

$$\alpha_\sigma(t) = C_H t^{2H-1}, \quad t \geq 0,$$

with the constant being different for each of these processes, and we obtain the final estimate

$$\|X_t\|_V \leq M_A \|x_0\|_V \exp\left\{\sigma |b_t| - \frac{1}{2}\sigma^2 \frac{C_H}{2H} t^{2H} + \omega_A t\right\}, \quad t \geq 0.$$

Using the Law of Iterated Logarithm for Gaussian processes, we obtain that

$$\|X_t\|_V \to [t \to \infty]0, \quad \mathbb{P} - \text{a.s.}$$

which means that the noise stabilizes the solution pathwise. However, for $p > 0$, we have that

$$\begin{aligned}
\mathbb{E}\|X_t\|_V^p &= \|S(t)x_0\|_V^p \, \mathbb{E} \exp\left\{p\sigma b_t - \frac{p}{2}\sigma^2 \, \mathbb{E} b_t^2\right\} \\
&= \|S(t)x_0\|_V^p \, \exp\left\{\frac{p^2 - p}{2}\sigma^2 \, \mathbb{E} b_t^2\right\}.
\end{aligned}$$

Hence, for $p > 1$, the p-the moment of X_t may be destabilized by the noise, e.g. for the fBm, we have that

$$\mathbb{E}\|X_t\|_V^p = \exp\left\{\frac{p^2 - p}{2} \frac{C_H}{2H} t^{2H}\right\} \to \infty$$

(cf. [12] for more details).

Acknowledgements The authors are grateful to the anonymous referee for their valuable suggestions. The first author was supported by the Charles University grant GAUK No. 322715 and SVV 2016 No. 260334. The second author was supported by the Czech Science Foundation grant GAČR No. 15-08819S.

References

1. Alòs, E., Mazet, O., Nualart, D.: Stochastic calculus with respect to Gaussian processes. Ann. Probab. **29**(2), 766–801 (2001)
2. Alòs, E., Nualart, D.: Stochastic integration with respect to the fractional Brownian motion. Stoch. Stoch. Rep. **75**(3), 129–152 (2003)
3. Čoupek, P., Maslowski, B.: Stochastic evolution equations with Volterra noise. Stoch. Proc. Appl. **127**(3), 877–900 (2017)
4. Da Prato, G., Zabczyk, J.: Stochastic Equations in Infinite Dimensions, 2nd edn. Oxford University Press, Encyclopedia of Mathematics and its Applications (2014)
5. Decreusefond, L., Üstünel, A.S.: Stochastic analysis of the fractional Brownian motion. Potential Anal. **10**(2), 177–214 (1999)
6. Duncan, T.E., Maslowski, B., Pasik-Duncan, B.: Fractional Brownian motion and stochastic equations in Hilbert spaces. Stoch. Dyn. **2**(2), 225–250 (2002)
7. Duncan, T.E., Maslowski, B., Pasik-Duncan, B.: Stochastic equations in Hilbert spaces with a multiplicative fractional Gaussian noise. Stoch. Proc. Appl. **115**(8), 1357–1383 (2005)
8. Duncan, T.E., Maslowski, B., Pasik-Duncan, B.: Stochastic linear-quadratic control for bilinear evolution equations driven by Gauss-Volterra processes (2016).
9. Lebovits, J.: Stochastic calculus with respect to Gaussian processes: Part I (2017). URL https://arxiv.org/abs/1703.08393
10. Maslowski, B.: Stability of semilinear equations with boundary and pointwise noise. Ann. Scuola Norm. - Sci. **22**(1), 55–93 (1995)
11. Nualart, D.: The Malliavin Calculus and Related Topics. Springer, Probability and its applications (2006)
12. Šnupárková, J., Maslowski, B.: Stochastic affine evolution equations with multiplicative fractional noise (2016). URL https://arxiv.org/abs/1609.00582
13. Taqqu, M.S.: The Rosenblatt process. In: Davis, R.A., Lii, K.S., Politis, D.N. (eds.) Selected Works of Murray Rosenblatt, pp. 29–45. Springer, New York (2011)
14. Tudor, C.A.: Analysis of the Rosenblatt process. ESAIM Probab. Stat. **12**, 230–257 (2008)

Hitting Probabilities for Systems of Stochastic PDEs: An Overview

Robert C. Dalang

Abstract We consider a d-dimensional random field that solves a possibly non-linear system of stochastic partial differential equations, such as stochastic heat or wave equations. We present results, obtained in joint works with Davar Khoshnevisan and Eulalia Nualart, and with Marta Sanz-Solé, on upper and lower bounds on the probabilities that the random field visits a deterministic subset of $\mathbb{R}^d$, in terms, respectively, of Hausdorff measure and Newtonian capacity of the subset. These bounds determine the critical dimension above which points are polar, but do not, in general, determine whether points are polar in the critical dimension. For linear SPDEs, we discuss, based on joint work with Carl Mueller and Yimin Xiao, how the issue of polarity of points can be resolved in the critical dimension.

Keywords Hitting probabilities · Systems of stochastics PDEs · Malliavin calculus · Stochastic heat equation · Stochastic wave equation · Spatially homogeneous Gaussian noise · Capacity · Hausdorff measure

2010 Mathematics Subject Classification Primary: 60H15; Secondary: 60J45 · 60G15 · 60H07 · 60G60.

1 Introduction

A basic problem in probabilistic potential theory, according to [14, Chap. 10], is the following. Let $U = (U(x),\ x \in \mathbb{R}^k)$ be an $\mathbb{R}^d$-valued continuous stochastic process defined on a probability space $(\Omega, \mathscr{F}, P)$. Let $I \subset \mathbb{R}^k$ be a fixed compact set with

The research of the author is partly supported by the Swiss National Foundation for Scientific Research.

R. C. Dalang (✉)
Institut de Mathématiques, Ecole Polytechnique Fédérale de Lausanne,
Station 8, CH-1015, Lausanne, Switzerland
e-mail: robert.dalang@epfl.ch

A. Eberle et al. (eds.), *Stochastic Partial Differential Equations and Related Fields*, Springer Proceedings in Mathematics & Statistics 229,
https://doi.org/10.1007/978-3-319-74929-7_8

positive Lebesgue measure. The range of U over I is the random compact subset $U(I)$ of $\mathbb{R}^d$ consisting of all points visited by U: for $\omega \in \Omega$,

$$U(I)(\omega) = \{U(x)(\omega),\ x \in I\}.$$

Given a compact subset $A \subset \mathbb{R}^d$, is $P\{U(I) \cap A \neq \emptyset\}$ *positive?* If the answer is "no," then we say that A is *polar* for U, otherwise, A is *non-polar.* If $A = \{z\}$ consists of a single point $z \in \mathbb{R}^d$, then $\{U(I) \cap A \neq \emptyset\} = \{\exists x \in I : U(x) = z\}$, and we say that z is *polar* for U if the singleton $\{z\}$ is polar for U.

A related question concerns *hitting probabilities.* Namely, for a compact subset $A \subset \mathbb{R}^d$, what are bounds on

$$P\{U(I) \cap A \neq \emptyset\}\,?$$

In particular, one would like upper and lower bounds that are sufficient to determine whether or not points are polar for U.

In the case where the d components of U are i.i.d., there is typically a *critical value* $Q(k)$ such that:

- if $d > Q(k)$, then there is lots of room to move around in the value space and points are polar;
- if $1 \leq d < Q(k)$, then points are *non-polar*;
- at the critical value $d = Q(k)$, either situation may occur and it is usually more difficult to decide whether or not points are polar.

The oldest results in this direction concern standard Brownian motion and are due to Paul Lévy [18] (see also [15, Theorem 2.2]), where it is shown that points are polar for Brownian motion in all dimensions $d \geq 2$ (and they are non-polar in dimension $d = 1$). This kind of result was extended to classical Markov processes (see [3, 13, 22]). Extensions to multiparameter processes came later and a discussion of these extensions can be found in [14].

In this paper, we shall mainly discuss results on hitting probabilities for systems of stochastic partial differential equations (SPDEs), though we shall begin with the results on Gaussian random fields that motivated this study and that also serve as benchmarks for research on non-Gaussian random fields and solutions of nonlinear systems of SPDEs.

2 Benchmark Results for Gaussian Random Fields

In view of the many special properties of Gaussian random fields, it is natural to begin the study of hitting probabilities with such random fields. We start with a very particular random field, namely the Brownian sheet. Indeed, the results obtained by Khoshnevisan and Shi [16] were the starting point for much of the later research on this topic.

2.1 First Example: The Brownian Sheet

Let $(W(x),\ x \in \mathbb{R}_+^k)$ denote an k-parameter $\mathbb{R}^d$-valued *Brownian sheet*, that is, a centered continuous Gaussian random field

$$W(x) = (W_1(x), \ldots, W_d(x))$$

with covariance

$$E[W_i(x)W_j(y)] = \delta_{i,j} \prod_{\ell=1}^{k} \min(x_\ell, y_\ell), \qquad i, j \in \{1, \ldots, d\},$$

where $x = (x_1, \ldots, x_k)$ and $y = (y_1, \ldots, y_k)$, and $\delta_{i,j} = 1$ if $i = j$ and $\delta_{i,j} = 0$ otherwise. It is clear that the Brownian sheet is a multi-parameter extension of standard Brownian motion, which corresponds to the case $k = 1$.

In order to state the first theorem, we introduce some notation concerning potential theory. For all Borel sets $F \subset \mathbb{R}^d$, let $\mathscr{P}(F)$ denote the set of all probability measures with compact support in F. For all $\alpha \in \mathbb{R}$ and $\mu \in \mathscr{P}(\mathbb{R}^k)$, we let $I_\alpha(\mu)$ denote the *α-dimensional energy* of μ, that is,

$$I_\alpha(\mu) := \iint \mathrm{K}_\alpha(\|x - y\|)\, \mu(dx)\, \mu(dy),$$

where, for $r > 0$,

$$\mathrm{K}_\alpha(r) := \begin{cases} r^{-\alpha} & \text{if } \alpha > 0, \\ \max(\log(1/r), 1) & \text{if } \alpha = 0, \\ 1 & \text{if } \alpha < 0. \end{cases} \tag{1}$$

For all $\alpha \in \mathbb{R}$ and Borel sets $F \subset \mathbb{R}^k$, $\mathrm{Cap}_\alpha(F)$ denotes the *α-dimensional Bessel-Riesz capacity of F*, that is,

$$\mathrm{Cap}_\alpha(F) := \left[\inf_{\mu \in \mathscr{P}(F)} I_\alpha(\mu) \right]^{-1},$$

where, by definition, $1/\infty := 0$.

Finally, for $M > 0$, we let $B(0, M)$ denote the open ball in $\mathbb{R}^d$ of radius M centered at the origin.

Theorem 1 ([16]) *Fix $M > 0$. Let I be a* box, *that is, $I = [a_1, b_1] \times \cdots \times [a_k, b_k]$, where $0 < a_\ell < b_\ell < \infty$, $\ell = 1, \ldots, k$. There exists $0 < C < \infty$ (C depends only on $M, k, d, \min_{\ell=1,\ldots,k} a_\ell, \max_{\ell=1,\ldots,k} b_\ell$) such that for all compact sets $A \subset B(0, M)$ $(\subset \mathbb{R}^d)$,*

$$\frac{1}{C} Cap_{d-2k}(A) \le P\{W(I) \cap A \neq \emptyset\} \le C Cap_{d-2k}(A). \tag{2}$$

The statement and proof of this theorem can be found in [16, Theorem 1.1]. Notice that the constraint $A \subset B(0, M)$ is needed for the lower bound. Indeed, imagine translating the set A off to infinity. Then the probability that the random field W hits A would become less and less likely, simply because W has continuous sample paths and the parameter set I is compact.

Example 1 What does Theorem 1 tell us about polarity of points? In the case where $A = \{z\}$, it is not difficult to check that

$$\mathrm{Cap}_{d-2k}(\{z\}) = \begin{cases} 1 \text{ if } d < 2k, \\ 0 \text{ if } d \geq 2k. \end{cases}$$

Therefore, the upper bound in (2) tells us that *points are polar for W in all dimensions $d \geq 2k$*, and the lower bound in (2) tells us that *points are non-polar for W in dimensions $1 \leq d < 2k$*. In particular, $d = 2k$ is the critical dimension and points are polar in this critical dimension.

The result of Theorem 1 is essentially the optimal result to aim for. Indeed, the upper and lower bounds in (2) are identical up to a constant. As we will see, bounds as good as (2) are not available for wider classes of Gaussian random fields.

2.2 Anisotropic Gaussian Random Fields

We consider here a wider class of Gaussian random fields, studied in particular in [2, 27]. These random fields will typically have different behaviors in different directions, hence the name "anisotropic."

Let $(V(x),\ x \in \mathbb{R}^k)$ be a centered continuous Gaussian random field with values in $\mathbb{R}^d$. We write $V(x) = (V_1(x), \ldots, V_d(x))$ and we assume that the components $V_i = (V_i(x),\ x \in \mathbb{R}^k)$ are i.i.d. real-valued random fields. The canonical metric associated with these random fields is

$$\Delta(x, y) = \| V_1(x) - V_1(y) \|_{L^2} .$$

Let I be a box as in Theorem 1. Assume the following condition:

(C) There exists $0 < c < \infty$ and $H_1, \ldots, H_k \in]0, 1[$ such that for all $x \in I$,

$$c^{-1} \leq \Delta(0, x) \leq c,$$

$x \mapsto \Delta(0, x)$ is differentiable on I, and for all $x, y \in I$,

$$c^{-1} \sum_{j=1}^{k} |x_j - y_j|^{H_j} \leq \Delta(x, y) \leq c \sum_{j=1}^{k} |x_j - y_j|^{H_j}.$$

Given $\alpha \geq 0$, recall that the α-dimensional *Hausdorff measure* of F is defined by

$$\mathscr{H}_\alpha(F) = \lim_{\varepsilon \to 0^+} \inf \left\{ \sum_{i=1}^{\infty} (2r_i)^\alpha : F \subseteq \bigcup_{i=1}^{\infty} B(x_i\,, r_i),\ \sup_{i \geq 1} r_i \leq \varepsilon \right\},$$

where $B(x\,, r)$ denotes the open (Euclidean) ball of radius $r > 0$ centered at $x \in \mathbb{R}^d$. When $\alpha < 0$, we define $\mathscr{H}_\alpha(F)$ to be infinite.

Theorem 2 ([2]) *Fix $M > 0$. Set $Q = \sum_{j=1}^{k} \frac{1}{H_j}$. Then there is $0 < C < \infty$ such that for every compact set $A \subset B(0, M)$ $(\subset \mathbb{R}^d)$,*

$$C^{-1} Cap_{d-Q}(A) \leq P\{V(I) \cap A \neq \emptyset\} \leq C \mathscr{H}_{d-Q}(A). \tag{3}$$

Compared to Theorem 1, Theorem 2 applies to a wide class of Gaussian random fields, but there is a major difference: Hausdorff measure appears on the right-hand side of (3), instead of capacity, as in (2). To see why this is significant, we again consider the issue of polarity of points.

Example 2 Suppose that $A = \{z\}$. Then

$$\text{Cap}_{d-Q}(\{z\}) = \begin{cases} 1 \text{ if } d < Q, \\ 0 \text{ if } d = Q, \\ 0 \text{ if } d > Q, \end{cases} \qquad \mathscr{H}_{d-Q}(\{z\}) = \begin{cases} \infty \text{ if } d < Q, \\ 1 \ \text{ if } d = Q, \\ 0 \ \text{ if } d > Q. \end{cases}$$

In particular, points are polar for V when $d > Q$, and are non-polar when $q < Q$. Therefore, if Q is an integer, then $d = Q$ is the critical dimension for hitting points. However, if $d = Q$, then the statement of Theorem 2 reduces essentially to $0 \leq P\{\exists x \in I : V(x) = z\} \leq 1$, which is not particularly informative, and the issue of polarity of points in this critical dimension is not answered by Theorem 2.

2.3 Funaki's Random String

There is one further important result on Gaussian random fields, which brings us closer to SPDEs. This concerns the solution to a system of stochastic heat equations driven by space-time white noise, also known as Funaki's random string.

Let $(u(t, x),\ (t, x) \in \mathbb{R}_+ \times \mathbb{R})$ be an $\mathbb{R}^d$-valued random field such that

$$\frac{\partial}{\partial t} u(t, x) = \frac{\partial^2}{\partial x^2} u(t, x) + \dot{W}(t, x), \qquad x \in \mathbb{R},\ t > 0, \tag{4}$$

where $u(0, \cdot) : \mathbb{R} \to \mathbb{R}^d$ is a given deterministic and smooth function, and $\dot{W}(t, x)$ is an $\mathbb{R}^d$-valued space-time white noise (that is, each of the d components of $\dot{W}$ is a real-valued space-time white noise, and these components are independent random fields).

Theorem 3 ([19]) *The critical dimension for hitting points is $d = 6$ and points are polar in this dimension.*

The proof of Mueller and Tribe [19] uses a rather clever extension of the argument used by Paul Lévy for Brownian motion. Namely, since the solution of (4) is not stationary in time, they reduce the problem to the study of polarity of points for a "stationary pinned string," which solves the stochastic heat equation with a random initial condition and has stationary spatial increments, then they use a scaling property and a time reversal argument, as was used by Paul Lévy. We note that [19] also treats the issue of double points for the solution of (4).

The method of Mueller and Tribe is quite specific to the stochastic heat equation. For instance, it does not apply to the stochastic wave equation, nor even to the stochastic heat equation with deterministic non-constant coefficients: if $(t, x) \mapsto \sigma(t, x)$ is a smooth but non-constant function, then the method of [19] does not apply to the solution $(u(t, x))$ of the SPDE.

$$\frac{\partial}{\partial t}u(t, x) = \frac{\partial^2}{\partial x^2}u(t, x) + \sigma(t, x)\dot{W}(t, x), \qquad x \in \mathbb{R},\ t > 0.$$

3 Hitting Probabilities for Non-Gaussian Random Fields

The results presented in the previous section, concerning Gaussian random fields, tell us what inequalities it is reasonable to aim for in the case of non-Gaussian random fields. We now present the results that have been obtained in this direction.

3.1 Systems of Nonlinear Wave Equations in Spatial Dimension 1

In the paper [11], E. Nualart and the author considered the reduced stochastic wave equation in two parameters, which is obtained from the classical stochastic wave equation in one spatial dimension by a rotation of coordinates. We state however their results here for the classical stochastic wave equation.

Let $u = (u(t, x),\ (t, x) \in \mathbb{R}_+ \times \mathbb{R})$ be an $\mathbb{R}^d$-valued random field such that

$$\frac{\partial^2}{\partial t^2}u(t, x) = \frac{\partial^2}{\partial x^2}u(t, x) + \sigma(u(t, x))\dot{W}(t, x), \qquad x \in \mathbb{R},\ t > 0, \tag{5}$$

where $u(0, \cdot)$ and $\frac{\partial}{\partial t}u(0, \cdot)$ are given smooth functions from $\mathbb{R}$ into $\mathbb{R}^d$, $\dot{W}(t, x)$ is an $\mathbb{R}^d$-valued space-time white noise, and $v \mapsto \sigma(v)$ is a matrix-valued function $\sigma(v) = (\sigma_{i,j}(v),\ i, j = 1, \dots, d)$, where each $v \mapsto \sigma_{i,j}(v)$ infinitely differentiable with bounded partial derivatives. Assume also that σ is *strongly elliptic*, that is, there is $\rho > 0$ such that for all $v \in \mathbb{R}^d$ and $z \in \mathbb{R}^d$ with $\|z\| = 1$,

$$\|\sigma(v)z\|^2 = \sum_{i=1}^{d} \left(\sum_{j=1}^{d} \sigma_{i,j}(v) z_j \right)^2 \geq \rho^2.$$

The next theorem was obtained in [11, Theorem 5.1] for the reduced stochastic wave equation. In the case where the SPDE (5) also contains a nonlinear drift term, a slightly weaker result in given in [11, Corollary 5.3].

Theorem 4 *Let u be the solution of* (5). *Let $I = [t_0, t_1] \times [x_0, x_1]$ be a rectangle, where $0 < t_0 < t_1$ and $x_0 < x_1$ and let $M > 0$. Under the assumptions just stated, there exists a finite positive constant C such that, for all compacts sets $A \subset B(0, M)$ $(\subset \mathbb{R}^d)$,*

$$\frac{1}{C} \operatorname{Cap}_{d-4}(A) \leq P\{u(I) \cap A \neq \emptyset\} \leq C \operatorname{Cap}_{d-4}(A).$$

It follows from this theorem (and the properties of capacity mentioned in Example 1) that $d = 4$ is the critical dimension for hitting points and points are polar in this critical dimension.

The proof of Theorem 4 uses Malliavin calculus, and, for the upper bound, Cairoli's maximal inequality for multiparameter martingales (see [14, Chap. 7]). This last property, which was already used by Khoshnevisan and Shi [16] for Theorem 1, will not be available for the other SPDEs that we will consider in this paper.

3.2 *Other Non-linear Systems of SPDEs*

In this subsection, we consider a wide class of systems of SPDEs, that includes in particular systems of heat and wave equations.

Let L be a partial differential operator. For instance,

$$L = \frac{\partial}{\partial t} - \Delta \quad \text{or} \quad L = \frac{\partial^2}{\partial t^2} - \Delta$$

in the case of the heat (respectively wave) operator, where Δ is the Laplacian in the spatial variables,

Let $u = (u(t,x),\ (t,x) \in \mathbb{R}_+ \times \mathbb{R}^k)$, where

$$u(t,x) = (u^1(t,x), \dots, u^d(t,x)) \in \mathbb{R}^d$$

is the solution of

$$\begin{cases} Lu^1(t,x) = b^1(u(t,x)) + \sum_{j=1}^d \sigma_{1,j}(u(t,x)) \dot{W}_j(t,x), \\ \quad \vdots \\ Lu^d(t,x) = b^d(u(t,x)) + \sum_{j=1}^d \sigma_{d,j}(u(t,x)) \dot{W}_j(t,x), \end{cases} \tag{6}$$

where $t \in]0, T]$, $x \in \mathbb{R}^k$, with suitable initial conditions. The functions $b^i : \mathbb{R}^d \to \mathbb{R}$ and $\sigma_{i,j} : \mathbb{R}^d \to \mathbb{R}$, $i, j = 1, \ldots, d$, are assumed to be C^∞-functions with bounded derivatives of all positive orders. The matrix $\sigma = (\sigma_{i,j})$ is assumed to be strongly elliptic.

The choice of initial conditions should ensure that the system (6) is well-posed. For instance, in the case of the heat operator, the initial value $u(0, \cdot)$ is given, whereas for the wave operator, the initial velocity $\frac{\partial}{\partial t} u(0, \cdot)$ should also be given.

The noise process $\dot{W}(t, x) = (\dot{W}_1(t, x), \ldots, \dot{W}_d(t, x))$ is assumed to be Gaussian, white in time, and possibly correlated in space. We will only consider the spatially homogeneous case, with a spatial correlation given by a Riesz kernel. More precisely, we fix $k \geq 1$ and suppose $\beta \in]0, k \wedge 2[$ or $k = 1 = \beta$. We suppose that the covariance of the noise is

$$E(\dot{W}_\ell(t, x)\dot{W}_j(s, y)) = \delta(t - s)\|x - y\|^{-\beta}\delta_{\ell,j}, \tag{7}$$

unless $k = 1 = \beta$, in which $\dot{W}$ is an $\mathbb{R}^d$-valued space-time white noise.

For many choices of the operator L, the system (6) has a unique solution, and the optimal Hölder exponents for the solution can be determined. Often, the Hölder regularity is different in the time variable and in the spatial variable (or could even be different in each of the spatial variables). We assume here that we have, for all $p \geq 2$, the bounds

$$c(p)\, \Delta(t, x; s, y) \leq \|u(t, x) - u(s, y)\|_{L^p} \leq C(p)\, \Delta(t, x; s, y), \tag{8}$$

where

$$\Delta(t, x; s, y) = |t - s|^{H_1} + \|x - y\|^{H_2} \tag{9}$$

and $H_1, H_2 \in]0, 1]$. Define

$$Q = \frac{1}{H_1} + \frac{k}{H_2}. \tag{10}$$

The type of result that has been obtained in many cases, which we will summarize in Sect. 3.5, takes the following form.

Typical result 5 *Fix $\eta > 0$ and $M > 0$. Let $I = [t_0, t_1]$, with $0 < t_0 < t_1$, and let J be a box in $\mathbb{R}^k$. Then there are positive constants c_η and C_η such that, for all compact sets $A \subset B(0, M)$ $(\subset \mathbb{R}^d)$,*

$$c_\eta \operatorname{Cap}_{d-Q+\eta}(A) \leq P\{u(I \times J) \cap A \neq \emptyset\} \leq C_\eta \mathscr{H}_{d-Q-\eta}(A). \tag{11}$$

The bounds in (11) are similar to those in Theorem 2, with Hausdorff measure on the right-hand side. They are slightly less good, since there is an additional term $+\eta$ (resp. $-\eta$) in the dimension of the capacity (resp. the Hausdorff measure).

The inequalities in (11) imply that the critical dimension for hitting points is $d = Q$, with Q defined in (10) (assuming that Q is an integer). However, in the critical dimension $d = Q$, the issue of polarity of points is not answered.

The bounds in (11) and the methods used to obtain these bounds also lead to information about the Hausdorff dimensions of level sets of u and of the range of u, as well as to bounds on the probability that a level set of u meets a given subset of $\mathbb{R}_+ \times \mathbb{R}^k$: see for instance [6, Theorems 2.4 and 3.2] as well as [7, Corollary 1.5 and Theorem 1.6].

In the next two sections, we explain the main methods that have been developed to prove this kind of result. Then we will indicate specific PDE operators L for which this program has been carried out.

3.3 Proving the Upper Bound

Beginning with [6, Theorem 3.3], various sufficient conditions for obtaining the upper bound in (11) have been identified (see also [8, Sect. 2.3]). These apply in principle to any continuous random field $u = \{u(t,x),\ (t,x) \in \mathbb{R}_+ \times \mathbb{R}^k\}$.

Theorem 6 *Let $D \subset \mathbb{R}^d$. In addition to knowing the Hölder exponents H_1 and H_2 that appear in the upper bound in* (8)*, assume that for any $x \in \mathbb{R}^k$, $u(t,x)$ has a probability density function $p_{t,x}$, and*

$$\sup_{z \in D} \sup_{(t,x) \in I \times J} p_{t,x}(z) \leq C < \infty. \tag{12}$$

Then for any $\eta > 0$, for every Borel set $A \subset D$,

$$P\{u(I) \cap A \neq \emptyset\} \leq C \mathscr{H}_{d-Q-\eta}(A). \tag{13}$$

We note that the conclusion (13) only uses the upper bound in (8) (the lower bound is not needed). In the case of a Gaussian random field, condition (12) is usually easy to verify. In the non-Gaussian case, the existence of a probability density function and the uniform bound in (12) can often be obtained by using Malliavin calculus.

3.4 Proving the Lower Bound

Sufficient conditions for obtaining the lower bound in (11) have also been identified in [6]. These conditions were later weakened in [12, Remark 3.9]. Let $u = \{u(t,x),\ (t,x) \in \mathbb{R}_+ \times \mathbb{R}^k\}$ be an $\mathbb{R}^d$-valued continuous process.

Theorem 7 *Let $\Delta(s, y; t, x)$ be defined as in* (9). *Assume that:*

(a) *for all $(t, x) \in I \times J$, the probability density function of $u(t, x)$ exists, is continuous and strictly positive;*
(b) *for any $(t, x) \neq (s, y)$ in $I \times J$, the (two-point) probability density function $p_{s,y;t,x}$ of $(u(s, y), u(t, x))$ exists, and there are $c > 0$, $\gamma \geq \frac{1}{H_1} + \frac{k}{H_2}$ and $p > d(\gamma - \frac{1}{H_1} - \frac{k}{H_2})$ such that for all $z_1, z_2 \in [-N, N]^d$,*

$$p_{s,y;t,x}(z_1, z_2) \leq c[\Delta(s, y; t, x)]^{-\gamma} \left[\frac{(\Delta(s, y; t, x))^2}{\|z_1 - z_2\|^2} \wedge 1 \right]^{p/2d}. \qquad (14)$$

Then

$$P\{u(I \times J) \cap A \neq \emptyset\} \geq c \, \mathrm{Cap}_{\gamma - Q}(A),$$

where $Q = \frac{1}{H_1} + \frac{k}{H_2}$.

Usually, the best possible choice of γ is $\gamma = d$. Both properties (a) and (b) can be obtained by using Malliavin calculus. For instance, an early result on the positivity of densities was obtained by Kohatsu-Higa [17] (see also [20]). The upper bound in (b) is often difficult to obtain. In some cases [7], a stronger upper bound has been obtained.

Since Theorem 7 is not stated as such anywhere in the literature, we shall sketch its proof here. The main technical ingredient is the following anisotropic extension of [12, Theorem 3.8] (see also [12, Remark 3.9]). Recall the definition of the function $K_\alpha(r)$ in (1).

Lemma 8 *Suppose that condition (b) of Theorem 7 holds. For any I, K compact subsets of $[0, T]$ and $\mathbb{R}^k$, respectively, both with diameter ≤ 1, there exists a constant $C = C(H_1, H_2, \gamma, d, k, N)$ such that, for every $z_1, z_2 \in \mathbb{R}^d$ with $0 \leq \|z_1 - z_2\| \leq N$,*

$$\mathscr{I} := \int_{I \times K} dt dx \int_{I \times K} ds dy \, p_{s,y;t,x}(z_1, z_2) \leq C \, K_{\gamma - \frac{1}{H_1} - \frac{k}{H_2}}(\|z_1 - z_2\|).$$

Proof Define $\eta = \|z_1 - z_2\|$, and suppose that $\rho_0 > 0$ is such that $I \times K \subset \{(s, y) \in \mathbb{R}_+ \times \mathbb{R}^k : |s|^{H_1} + \|y\|^{H_2} \leq \rho_0\}$. Clearly, $\mathscr{I} = \mathscr{I}_1 + \mathscr{I}_2$, where

$$\mathscr{I}_1 = \int_{I \times K} dt dx \int_{I \times K} ds dy \, 1_{\{\Delta(s,y;t,x) \leq \frac{\rho_0 \eta}{N}\}} \, p_{s,y;t,x}(z_1, z_2),$$

$$\mathscr{I}_2 = \int_{I \times K} dt dx \int_{I \times K} ds dy \, 1_{\{\Delta(s,y;t,x) > \frac{\rho_0 \eta}{N}\}} \, p_{s,y;t,x}(z_1, z_2).$$

We bound each term separately. Observe that by (14) and since $p > 0$,

$$\begin{aligned}
\mathscr{I}_1 &\leq c \int_{I\times K} dtdx \int_{I\times K} dsdy\, 1_{\{\Delta(s,y;t,x)\leq \frac{\rho_0\eta}{N}\}} [\Delta(s, y; t, x)]^{-\gamma} \left[\frac{(\Delta(s, y; t, x))^2}{\|z_1 - z_2\|^2}\right]^{p/2d} \\
&= c \int_{I\times K} dtdx \int_{I\times K} dsdy\, 1_{\{\Delta(s,y;t,x)\leq \frac{\rho_0\eta}{N}\}} [\Delta(s, y; t, x)]^{-\gamma+\frac{p}{d}}\, \eta^{-\frac{p}{d}} \\
&\leq \tilde{c}\, \eta^{-\frac{p}{d}} \int_0^{\rho_0^{H_1^{-1}}} dr \int_0^{\rho_0^{H_2^{-1}}} du\, u^{k-1}\, 1_{\{r^{H_1}+u^{H_2}\leq \frac{\rho_0\eta}{N}\}} (r^{H_1} + u^{H_2})^{-\gamma+\frac{p}{d}}.
\end{aligned}$$

Use the change of variables $w = u^{H_2/H_1}$ to see that

$$\mathscr{I}_1 \leq \tilde{c}'\, \eta^{-\frac{p}{d}} \int_0^{\rho_0^{H_1^{-1}}} dr \int_0^{\rho_0^{H_1^{-1}}} dw\, w^{\frac{H_1}{H_2}-1}\, w^{(k-1)\frac{H_1}{H_2}}\, 1_{\{r^{H_1}+w^{H_1}\leq \frac{\rho\eta}{N}\}} (r^{H_1} + w^{H_1})^{-\gamma+\frac{p}{d}}.$$

Pass to polar coordinates in the variables (r, w) to see that

$$\mathscr{I}_1 \leq \tilde{c}''\, \eta^{-\frac{p}{d}} \int_0^{2(\frac{\rho\eta}{N})^{H_1^{-1}}} d\rho\, \rho\, \rho^{k\frac{H_1}{H_2}-1}\, \rho^{-\gamma H_1+\frac{pH_1}{d}}.$$

Since $k\frac{H_1}{H_2} - \gamma H_1 + \frac{pH_1}{d} > -1$ because $p > d(\gamma - \frac{1}{H_1} - \frac{k}{H_2})$ by hypothesis, the integral is finite and we obtain

$$\mathscr{I}_1 \leq \hat{c}''\, \eta^{-\frac{p}{d}}\, \eta^{\frac{k}{H_2}-\gamma+\frac{p}{d}+\frac{1}{H_1}} = \hat{c}''\eta^{-(\gamma-\frac{1}{H_1}-\frac{k}{H_2})}. \tag{15}$$

Now observe that by (14) and since $p > 0$,

$$\begin{aligned}
\mathscr{I}_2 &\leq c \int_{I\times K} dtdx \int_{I\times K} dsdy\, 1_{\{\Delta(s,y;t,x)> \frac{\rho_0\eta}{N}\}} [\Delta(s, y; t, x)]^{-\gamma} \\
&\leq \tilde{c} \int_0^{\rho_0^{H_1^{-1}}} dr \int_0^{\rho_0^{H_2^{-1}}} du\, 1_{\{r^{H_1}+u^{H_2}> \frac{\rho_0\eta}{N}\}}\, u^{k-1}(r^{H_1} + u^{H_2})^{-\gamma} \\
&\leq \tilde{c}\, (\mathscr{I}_{2,1} + \mathscr{I}_{2,2} + \mathscr{I}_{2,3}),
\end{aligned}$$

where

$$\begin{aligned}
\mathscr{I}_{2,1} &= \int_0^{(\frac{\rho_0\eta}{2N})^{H_2^{-1}}} du \int_{(\frac{\rho_0\eta}{2N})^{H_1^{-1}}}^{\rho_0^{H_1^{-1}}} dr\, u^{k-1}(r^{H_1} + u^{H_2})^{-\gamma}, \\
\mathscr{I}_{2,2} &= \int_{(\frac{\rho_0\eta}{2N})^{H_2^{-1}}}^{\rho_0^{H_2^{-1}}} du \int_0^{(\frac{\rho_0\eta}{2N})^{H_1^{-1}}} dr\, u^{k-1}(r^{H_1} + u^{H_2})^{-\gamma}, \\
\mathscr{I}_{2,3} &= \int_{(\frac{\rho_0\eta}{2N})^{H_1^{-1}}}^{\rho_0^{H_1^{-1}}} dr \int_{(\frac{\rho_0\eta}{2N})^{H_2^{-1}}}^{\rho_0^{H_2^{-1}}} du\, u^{k-1}(r^{H_1} + u^{H_2})^{-\gamma}.
\end{aligned}$$

Clearly, since $\gamma > H_1^{-1}$,

$$\begin{aligned}\mathscr{I}_{2,1} &\leq c \int_0^{(\frac{\rho_0\eta}{2N})^{H_2^{-1}}} du\, u^{k-1} \int_{(\frac{\rho_0\eta}{2N})^{H_1^{-1}}}^{\rho_0^{H_1^{-1}}} dr\, r^{-\gamma H_1} \leq c'\, \eta^{kH_2^{-1}}\, \eta^{-\gamma+H_1^{-1}} \\ &= c'\, \eta^{-(\gamma-\frac{1}{H_1}-\frac{k}{H_2})}. \end{aligned} \tag{16}$$

Similarly, since $\gamma > kH_2^{-1}$,

$$\begin{aligned}\mathscr{I}_{2,2} &\leq c \int_0^{(\frac{\rho_0\eta}{2N})^{H_1^{-1}}} dr \int_{(\frac{\rho_0\eta}{2N})^{H_2^{-1}}}^{\rho_0^{H_2^{-1}}} du\, u^{k-1-\gamma H_2} \leq c'\eta^{H_1^{-1}}\, \eta^{kH_2^{-1}-\gamma} \\ &\leq c'\eta^{-(\gamma-\frac{1}{H_1}-\frac{k}{H_2})}. \end{aligned} \tag{17}$$

Finally,

$$\mathscr{I}_{2,3} \leq c \int_{(\frac{\rho_0\eta}{2N})^{H_1^{-1}}}^{\rho_0^{H_1^{-1}}} dr \int_{(\frac{\rho_0\eta}{2N})^{H_2^{-1}}}^{\rho_0^{H_2^{-1}}} du\, \frac{u^{k-1}}{r^{\gamma H_1} + u^{\gamma H_2}}.$$

Use the change of variables $w = u^{H_2/H_1}$ to see that

$$\mathscr{I}_{2,3} \leq \tilde{c} \int_{(\frac{\rho_0\eta}{2N})^{H_1^{-1}}}^{\rho_0^{H_1^{-1}}} dr \int_{(\frac{\rho_0\eta}{2N})^{H_1^{-1}}}^{\rho_0^{H_1^{-1}}} dw\, \frac{w^{\frac{H_1}{H_2}-1}\, w^{(k-1)\frac{H_1}{H_2}}}{r^{\gamma H_1} + w^{\gamma H_1}}.$$

We bound the integrand above by $cw^{k\frac{H_1}{H_2}-1}(r+w)^{-\gamma H_1}$, then pass to polar coordinates in the variables (r, w) to see that

$$\mathscr{I}_{2,3} \leq c' \int_{(\frac{\rho_0\eta}{2N})^{H_1^{-1}}}^{2\rho_0^{H_1^{-1}}} d\rho\, \rho\, \rho^{k\frac{H_1}{H_2}-1}\, \rho^{-\gamma H_1}. \tag{18}$$

If $\gamma > \frac{1}{H_1} + \frac{k}{H_2}$, then we replace $2\rho^{H_1^{-1}}$ by $+\infty$ in the upper bound, to get

$$\mathscr{I}_{2,3} \leq \tilde{c}'\, \eta^{-(\gamma-\frac{1}{H_1}-\frac{k}{H_2})}. \tag{19}$$

Putting together (15)–(17) and (19) proves the lemma when $\gamma > \frac{1}{H_1} + \frac{k}{H_2}$.

If $\gamma = \frac{1}{H_1} + \frac{k}{H_2}$, then from (18),

$$\mathscr{I}_{2,3} \leq c \int_{(\frac{\rho_0\eta}{2N})^{H_1^{-1}}}^{2\rho_0^{H_1^{-1}}} d\rho\, \rho^{-1} = c\left[\log(2\rho_0^{H_1^{-1}}) + H_1^{-1}\log\left(\frac{2N}{\rho_0\,\eta}\right)\right], \tag{20}$$

and this is bounded above by $cK_0(\eta)$. Putting together (15)–(17) and (20) proves the lemma when $\gamma = \frac{1}{H_1} + \frac{k}{H_2}$. □

With Lemma 8, the proof of Theorem 7 follows as in [12, Sect. 3.2].

3.5 Results for Systems of Non-linear Equations

We now list the specific operators for which the methods outlined in Sect. 3.2–3.4 have been completely carried out.

Heat equation, $k = 1$, space-time white noise [7]. In this case,

$$Lu^i(t,x) = \frac{\partial u^i}{\partial t}(t,x) - \frac{\partial^2 u^i}{\partial x^2}(t,x)$$

and $\dot{W}_j(t,x)$ is space-time white noise. For any fixed $\eta > 0$, the bound (14) has been proved with $\gamma = d + \eta$. It is well-known that $H_1 = \frac{1}{4}$, $H_2 = \frac{1}{2}$, so the bounds on hitting probabilities are

$$c_\eta \operatorname{Cap}_{d+\eta-6}(A) \le P\{u(I \times J) \cap A \neq \emptyset\} \le C_\eta \mathscr{H}_{d-\eta-6}(A)$$

(see [7, Theorem 1.2]). These bounds are consistent with the result of Theorem 3.

Heat equation, $k \ge 1$, spatially homogeneous noise [8]. In this case,

$$Lu^i(t,x) = \frac{\partial u^i}{\partial t}(t,x) - \Delta u^i(t,x)$$

and $\dot{W}(t,x)$ is spatially homogeneous noise with covariance given in (7).

It is shown in [23, Theorem 2.1] that for any $H_1 < \frac{2-\beta}{4}$ and $H_2 < \frac{2-\beta}{2}$, the upper bound of (8) holds. For any $\eta > 0$, the inequality (14) has been obtained for $\gamma = d + \eta$. Setting $Q = \frac{4+6k}{2-\beta}$, the bounds on hitting probabilities are

$$c_\eta \operatorname{Cap}_{d+\eta-Q}(A) \le P\{u(I \times J) \cap A \neq \emptyset\} \le C_\eta \mathscr{H}_{d-\eta-Q}(A)$$

(see [8, Theorem 1.2]).

Wave equation, $k \in \{1, 2, 3\}$, spatially homogeneous noise [12]. In this case,

$$Lu^i(t,x) = \frac{\partial^2 u^i}{\partial t^2}(t,x) - \Delta u^i(t,x)$$

and $\dot{W}_j(t,x)$ is spatially homogeneous noise with covariance given in (7).

It is shown in [12, Proposition 2.2] that the upper bound of (8) holds with

$$H_1 = H_2 = \frac{2-\beta}{2}.$$

For any $\eta > 0$, the inequality (14) has been obtained for $\gamma = d + \eta + \frac{4d^2}{2-\beta}$ $(\eta > 0)$. Defining Q as in (10), the bounds on hitting probabilities are

$$c_\eta \operatorname{Cap}_{d+\eta-Q+\frac{4d^2}{2-\beta}}(A) \leq P\{u(I \times J) \cap A \neq \emptyset\} \leq C_\eta \mathscr{H}_{d-\eta-Q}(A) \qquad (21)$$

(see [12, Theorems 2.1 and 4.3]). It is expected that the optimal value of γ should be no more than $\gamma = d + \eta$, and this would considerably improve the lower bound in (21).

4 Polarity of Points in Critical Dimensions

As we have mentioned, the results in Sects. 3.2–3.5 do not answer the question of polarity of points in critical dimensions. Building on works of Talagrand [25, 26] for fractional Brownian motion, we have obtained results in this direction for a wide class of Gaussian random fields.

Let $v = (v(t,x),\ t \geq 0,\ x \in \mathbb{R}^k)$ be an $\mathbb{R}^d$-valued Gaussian random field with i.i.d. components. Fix $I \times J$ as in (11). Suppose that there are $C > 0$ and $H_1, H_2 \in\,]0,1[$ such that for all $(t,x), (s,y) \in I \times J$,

$$\|v(t,x) - v(s,y)\|_{L^2} \leq C\, \Delta(t,x;s,y) := |t-s|^{H_1} + \|x-y\|^{H_2}. \qquad (22)$$

We consider the following assumption on v.

Assumption 9 There is a random field $(V(A,t,x),\ A \in \mathscr{B}(\mathbb{R}_+),\ t \geq 0,\ x \in \mathbb{R}^k)$ and $\varepsilon_0 > 0$ such that:

(a) for fixed $(t,x) \in (I \times J)^{(\varepsilon_0)}$ (this denotes an ε_0-enlargement of $I \times J$), $A \mapsto V(A,t,x)$ is an $\mathbb{R}^d$-valued Gaussian white noise with i.i.d. components;
(b) when $A \cap B = \emptyset$, $V(A,\cdot,\cdot)$ and $V(B,\cdot,\cdot)$ are independent;
(c) there are constants $c_0 \in \mathbb{R}_+$, $a_0 \in \mathbb{R}$ and $\gamma_1 > 0$, $\gamma_2 > 0$ such that for all $a_0 \leq a \leq b \leq +\infty$, $(t,x), (s,y) \in (I \times J)^{(\varepsilon_0)}$:

$$\begin{aligned}\|v(t,x) - v(s,y) - (V([a,b[,t,x) - V([a,b[,s,y))\|_{L^2} \\ \leq c\left[a^{\gamma_1}|t-s| + a^{\gamma_2}\|x-y\| + b^{-1}\right]\end{aligned} \qquad (23)$$

and

$$\|V([0,a_0],t,x) - V([0,a_0],s,y)\|_{L^2} \leq c_0(|t-s| + \|x-y\|). \qquad (24)$$

(d) There is a constant $\tilde{c} > 0$ such that for all $(t, x) \in (I \times J)^{(\varepsilon_0)}$, and $i = 1, \dots, d$, we have $\|v_i(t, x)\|_{L^2} \geq \tilde{c}$;

(e) There is $\rho > 0$ with the following property. For $(t, x) \in I \times J$, there are $(t', x') \in (I \times J)^{(\varepsilon_0)}$, $\delta_j \in]\alpha_j, 1]$, $j = 1, 2$, and $C > 0$ such that for all $i = 1, \dots, d$, $(s, y), (\bar{s}, \bar{y}) \in (I \times J)^{(\varepsilon_0)}$ with $\Delta(s, y; t, x) \leq 2\rho$ and $\Delta(\bar{s}, \bar{y}; t, x) \leq 2\rho$,

$$\left|E[(v_i(s, y) - v_i(\bar{s}, \bar{y}))v_i(t', x')]\right| \leq C(|s - \bar{s}|^{\delta_1} + \|y - \bar{y}\|^{\delta_2}).$$

Remark 1 (1) If there exist exponents γ_j such that (23) holds, then it can be checked that a possible choice for the Hölder exponents in (22) is that they satisfy

$$\gamma_j = \tfrac{1}{H_j} - 1 \tag{25}$$

(see [10, Proposition 2.2]).

(2) Condition (c) states that if $|t - s| \sim 2^{-n/H_1}$ and $\|x - y\| \sim 2^{-n/H_2}$, then the increment $\|v(t, x) - v(s, y)\|$ is well-approximated by the increment $\|V([2^n, 2^{n+1}[, t, x) - V([2^n, 2^{n+1}[, s, y)\|$. If we are considering several increments over boxes of different sizes, then this approximation is useful because V has lots of independence built into it.

(3) Property (d) is a non-degeneracy assumption, while property (e) states that covariances of v_i are smoother than sample paths of v (since $\delta_j > \alpha_j$).

Theorem 10 ([10, Theorem 2.6] *Under Assumption 9, if $d = Q := \frac{1}{H_1} + \frac{k}{H_2}$, then for all $z \in \mathbb{R}^d$,*

$$P\{\exists (t, x) \in \mathbb{R}_+^* \times \mathbb{R}^k : v(t, x) = z\} = 0,$$

that is, points are polar for v.

It turns out that Theorem 10 is widely applicable, since Assumption 9 is satisfied by the solutions to many systems of linear SPDEs. In addition, in the case where $\Delta(t, x; s, y)$ not only satisfies the upper bound (22) but also the lower bound (8), we know from Sect. 3.5 that Q defined in Theorem 10 is the critical dimension for the SPDEs considered there. So the next proposition establishes polarity of points in the critical dimension for several types of SPDEs.

Proposition 11 *[10, Sects. 7–9] Let $v = (v(t, x))$ be the solution of a system of linear SPDEs. Assumption 9 is satisfied in the following cases:*

(a) systems of linear heat *equations in spatial dimension $k = 1$, driven by space-time white noise, with possibly non-constant coefficients;*

(b) systems of linear wave *equations in spatial dimension $k = 1$, driven by space-time white noise;*

(c) systems of linear heat *equations in spatial dimension $k \geq 1$, driven by spatially homogeneous Gaussian noise with covariance given in (7);*

(d) systems of linear wave *equations in spatial dimension $k \geq 1$, driven by spatially homogeneous Gaussian noise with covariance given in* (7).

In particular, in each of these four cases, points are polar for v in the critical dimension.

Via Case (a) of Proposition 11, one recovers the results of Theorem 3. As as example of how one uses Theorem 10 to establish the claims of Proposition 11, we consider the case of *wave* equations in spatial dimension 1 driven by space-time white noise $\hat{W}$. Let $v = (v(t,x),\ t \in \mathbb{R}_+,\ x \in \mathbb{R})$ solve

$$\begin{cases} \frac{\partial^2}{\partial t^2} v_j(t,x) = \frac{\partial^2}{\partial x^2} v_j(t,x) + \dot{\hat{W}}_j(t,x), & j = 1,\dots,d, \\ v(0,x) = 0, \quad \frac{\partial}{\partial t} v(0,x) = 0, & x \in \mathbb{R}. \end{cases} \tag{26}$$

Corollary 12 ([10, Theorem 9.1]) *Suppose $d = 4$ (critical dimension). Then points are polar for v.*

The main point, in order to apply Theorem 10, is to determine the random field V of Assumption 9. Letting $G(t,x)$ denote the fundamental solution of the wave equation, it is well-known that the solution of (26) has the representation

$$v(t,x) = \int_0^t \int_{\mathbb{R}} G(t-s, x-y)\, \hat{W}(ds,dy). \tag{27}$$

Taking space-time Fourier transforms of G and $\hat{W}$ leads to the representation

$$v(t,x) = \int_{\mathbb{R}} \int_{\mathbb{R}} \frac{e^{-i\xi\cdot x - i\tau t}}{2|\xi|} \left[\frac{1 - e^{it(\tau+|\xi|)}}{\tau+|\xi|} - \frac{1 - e^{it(\tau-|\xi|)}}{\tau-|\xi|} \right] W(d\tau, d\xi), \tag{28}$$

where W is another space-time white noise. Note that this representation already appears in [1]. We call it a *harmonizable* representation of v.

With the representation (28), it is shown in [10, Lemmas 9.3 and 9.6] that Assumption 9 is satisfied by setting

$$V(A,t,x) := \iint_{\{\max(|\tau|^{\frac{1}{2}}, |\xi|^{\frac{1}{2}}) \in A\}} \frac{e^{-i\xi\cdot x - i\tau t}}{2|\xi|} \left[\frac{1 - e^{it(\tau+|\xi|)}}{\tau+|\xi|} - \frac{1 - e^{it(\tau-|\xi|)}}{\tau-|\xi|} \right] W(d\tau, d\xi),$$

where W is again a space-time white noise.

For condition (c) of Assumption 9, the formula (25) applied to the Hölder exponent $\frac{1}{2}$ gives $\gamma := (\frac{1}{2})^{-1} - 1 = 1$, so it is necessary to check that

$$\begin{aligned} &\|v(t,x) - v(s,y) - (V([a,b[,t,x) - V([a,b[,s,y))\|_{L^2} \\ &\quad \leq c_0 \left[a^1 |t-s| + a^1 |x-y| + b^{-1} \right]. \end{aligned}$$

Proving this inequality requires estimating some double integrals.

It turns out that for the other examples mentioned in Proposition 11, the same Fourier transform method applied to the standard representation (27) applies, and then the candidate process V is easily obtained as above from the harmonizable representation of v.

In future work, we plan to extend the above results to the solution of systems of nonlinear SPDEs, and to study the issue of existence/nonexistence of multiple point in critical dimensions, extending the recent result of [9].

Finally, we also mention the papers [4, 5, 21, 24], which are also concerned with hitting probabilities for SPDEs.

References

1. Balan, R.: Linear SPDEs driven by stationary random distributions. J. Fourier Anal. Appl. **18**(6), 1113–1145 (2012)
2. Biermé, H., Lacaux, C., Xiao, Y.: Hitting probabilities and the Hausdorff dimension of the inverse images of anisotropic Gaussian random fields. Bull. Lond. Math. Soc. **41**, 253–273 (2009)
3. Blumenthal, R.M., Getoor, R.K.: Markov Processes and Potential Theory. Pure and Applied Mathematics, vol. 29. Academic Press, New York (1968)
4. Chen, Z.L., Zhou, Q.: Hitting probabilities and the Hausdorff dimension of the inverse images of a class of anisotropic random fields. Acta Math. Sin. **31**, 1895–1922 (2015)
5. Clarke de la Cerda, J., Tudor, C.A.: Hitting times for the stochastic wave equation with fractional colored noise. Rev. Mat. Iberoam. **30**, 685–709 (2014)
6. Dalang, R.C., Khoshnevisan, D., Nualart, E.: Hitting probabilities for systems of non-linear stochastic heat equations with additive noise. Latin Am. J. Probab. Stat. (ALEA) **3**, 231–271 (2007)
7. Dalang, R.C., Khoshnevisan, D., Nualart, E.: Hitting probabilities for systems for non-linear stochastic heat equations with multiplicative noise. Probab. Theory Related Fields **144**, 371–427 (2009)
8. Dalang, R.C., Khoshnevisan, D., Nualart, E.: Hitting probabilities for systems of non-linear stochastic heat equations in spatial dimension $k \geq 1$. J. SPDE's Anal. Comput. **1–1**, 94–151 (2013)
9. Dalang, R.C., Mueller, C.: Multiple points of the Brownian sheet in critical dimensions. Ann. Probab. **43**(4), 1577–1593 (2015)
10. Dalang, R.C., Mueller, C., Xiao, Y.: Polarity, of points for Gaussian random fields. Ann. Probab. **45**(6B), 4700–4751 (2017)
11. Dalang, R.C., Nualart, E.: Potential theory for hyperbolic SPDEs. Ann. Probab. **32**, 2099–2148 (2004)
12. Dalang, R.C., Sanz-Solé, M.: Hitting probabilities for non-linear systems of stochastic waves. Mem. Am. Math. Soc. **237**(1120), 1–75 (2015)
13. Doob, J.L.: Classical potential theory and its probabilistic counterpart. Grundlehren der Mathematischen Wissenschaften, vol. 262. Springer, New York (1984)
14. Khoshnevisan, D.: Multiparameter Processes. An Introduction to Random Fields. Springer Monographs in Mathematics. Springer, New York (2002)
15. Khoshnevisan, D.: Intersections of Brownian motions. Exp. Math. **21**(3), 97–114 (2003)
16. Khoshnevisan, D., Shi, Z.: Brownian sheet and capacity. Ann. Probab. **27**(3), 1135–1159 (1999)
17. Kohatsu-Higa, A.: Lower bound estimates for densities of uniformly elliptic random variables on Wiener space. Probab. Theory Related Fields **126**, 421–457 (2003)

18. Lévy, P.: Processus Stochastiques et Mouvement Brownien. Gauthier-Villars, Paris (1948)
19. Mueller, C., Tribe, R.: Hitting properties of a random string. Electron. J. Probab. **7**(10) (2002)
20. Nualart, E.: On the density of systems of non-linear spatially homogeneous SPDEs. Stochastics **85**(1), 48–70 (2013)
21. Nualart, E., Viens, F.: The fractional stochastic heat equation on the circle: time regularity and potential theory. Stoch. Process. Appl. **119**(5), 1505–1540 (2009)
22. Port, S.C., Stone, C.J.: Brownian Motion and Classical Potential Theory. Academic Press, New York (1978)
23. Sanz-Solé, M., Sarrà, M.: Hölder continuity for the stochastic heat equation with spatially correlated noise, Seminar on Stochastic Analysis, Random Fields ans Applications, III (Ascona, 1999). Progr. Prob. **52**, 259–268 (2002)
24. Sanz-Solé, M., Viles, N.: Systems of stochastic Poisson equations: hitting probabilities. Stochastic Process. Appl. (2016). (to appear). arXiv:1612.04567
25. Talagrand, M.: Hausdorff measure of trajectories of multiparameter fractional Brownian motion. Ann. Probab. **23**(2), 767–775 (1995)
26. Talagrand, M.: Multiple points of trajectories of multiparameter fractional Brownian motion. Probab. Theory Related Fields **112–4**, 545–563 (1998)
27. Xiao, Y.: Sample path properties of anisotropic Gaussian random fields. In: Khoshnevisan, D., Rassoul-Agha, F. (eds.) A Minicourse on Stochastic Partial Differential Equations. Lecture Notes in Mathematics, vol. 1962, pp. 145–212. Springer, New York (2009)

Curvature Motion Perturbed by a Direction-Dependent Colored Noise

Clément Denis, Tadahisa Funaki and Satoshi Yokoyama

Abstract The aim of this paper is twofold. First we give a brief overview of several results on the deterministic and stochastic motions by mean curvature and their derivation under the so-called sharp interface limit. Then, we study the motions by mean curvature perturbed by a direction-dependent Gaussian colored noise described by $V = \kappa + \dot{W}(t, \mathbf{n})$. This part is a generalization of (Funaki, Acta Math Sin (Engl Ser), 15:407–438, 1999) [10] where the noise is independent from space. We derive a uniform moment estimate on solutions of approximating equations and prove a Wong–Zakai type convergence theorem (in law) for the SPDEs for the curvature of a convex curve in two-dimensional space before the time the curve exhibits a singularity.

Keywords Stochastic partial differential equation · Motion by mean curvature · Wong–Zakai theorem · Colored noise

MSC 2010 60H15 · 35K93

C. Denis
Département de mathématiques, Ecole Normale Supérieure (ENS) de Paris -Saclay,
61 avenue du Président Wilson, Cachan, France
e-mail: clement.denis@ens-cachan.fr

T. Funaki · S. Yokoyama
Graduate School of Mathematical Sciences, University of Tokyo, Komaba,
Tokyo 153 -8914, Japan
e-mail: funaki@ms.u-tokyo.ac.jp

S. Yokoyama
e-mail: satoshi2@ms.u-tokyo.ac.jp

T. Funaki (✉) · S. Yokoyama
Department of Mathematics, School of Fundamental Science and Engineering,
Waseda University, 3-4-1 Okubo, Shinjuku-ku, Tokyo 169-8555, Japan
e-mail: t-funaki@waseda.jp

S. Yokoyama
e-mail: satoshiyokoyama@aoni.waseda.jp

A. Eberle et al. (eds.), *Stochastic Partial Differential Equations and Related Fields*, Springer Proceedings in Mathematics & Statistics 229,
https://doi.org/10.1007/978-3-319-74929-7_9

1 Introduction

One of the authors presented the result of [12] at the conference held in honor of Michael Röckner. The present paper is linked to the talk. We first give a brief overview of related results and then study the motion by mean curvature (MMC) perturbed by a direction-dependent noise, which is a generalization of the result of [10]. Our main result is a Wong–Zakai type theorem for the SPDE (11), which describes the MMC with a direction-dependent noise, or its cut-off version (12). We consider a random PDE (25) by replacing the noise in (12) with smooth one and show the convergence of the solution in law sense; see Theorem 2. Such approximation appears in a study of the sharp interface limit for the stochastic Allen–Cahn equation, cf. [10].

1.1 A Brief Overview

The MMC is the time evolution of $(d-1)$-dimensional hypersurface Γ_t embedded in $\mathbb{R}^d$, defined by

$$V = \kappa, \tag{1}$$

where V is the inward normal velocity of Γ_t and κ is the mean curvature of Γ_t multiplied by $d-1$. For example, if Γ_t is a sphere, then under the evolution (1) it shrinks to a single point within a finite time.

There are a large number of references for the MMC in particular in nonlinear PDEs, whereas those for the stochastic MMC are rather limited; see Sect. 4.3 of [11] for a brief survey of this subject.

The evolution (1) can be described by nonlinear PDEs, e.g., for signed distance functions $d(t,x) = \text{signed dist}(x, \Gamma_t)$, $x \in \mathbb{R}^d$ (cf. [3]); for height functions $h(t,x)$ when Γ_t is represented as a graph $\{y = h(t,x); x \in \mathbb{R}^{d-1}\}$; a formulation due to geometric measure theory (Brakke [4]); the level set formulation and viscosity solutions (cf. [14]).

The derivation of the MMC from the Allen–Cahn equation [2], which is a reaction-diffusion equation with bistable reaction term satisfying the balance condition:

$$\frac{\partial u^\epsilon}{\partial t} = \Delta u^\epsilon + \frac{1}{\epsilon^2} f(u^\epsilon), \tag{2}$$

in the limit as $\epsilon \downarrow 0$ called the sharp interface limit is well-studied from several different view points.

A variant of the MMC (1) is the volume preserving mean curvature flow. The time evolution of hypersurface Γ_t is governed by

$$V = \kappa - \text{av}_{\Gamma_t}(\kappa), \tag{3}$$

where $\text{av}_{\Gamma_t}(\kappa) = \frac{1}{|\Gamma_t|}\int_{\Gamma_t} \kappa$ is the average of κ over Γ_t (cf. [3]).

We now consider the stochastic perturbation of the MMC (1) of the form:

$$V = \kappa + \dot{W}(t, x), \quad x \in \Gamma_t, \tag{4}$$

where $\dot{W}(t, x)$ is a certain space-time dependent noise in general.

Funaki [10] studied (4) when $\dot{W}(t, x) = \dot{W}(t)$, the white noise only in time, by rewriting it as an SPDE for the curvature κ of Γ_t under the Gauss map; see Sect. 1.2.1. This approach has a limitation only for a convex setting in two-dimensional space. Lions and Souganidis [22, 23] give a meaning to (4) by introducing the notion of stochastic viscosity solutions for the level set function when $\dot{W}(t, x) = \alpha \dot{W}(t)$ with a constant α. The signed distance functions' approach was taken by Dirr, Luckhaus and Novaga [6] and by Weber [26]. This approach is also limited to the noise $\dot{W} = \dot{W}(t)$ dependent only on time. Von Renesse and others [7, 17] studied (4) when Γ_t is represented as a graph.

The derivation of the stochastic MMC (4) in the sharp interface limit for the stochastic Allen–Cahn equation:

$$\frac{\partial u^\epsilon}{\partial t} = \Delta u^\epsilon + \frac{1}{\epsilon^2} f(u^\epsilon) + \frac{1}{\epsilon} \dot{W}^\epsilon(t, x) \tag{5}$$

was studied by a physics paper by Kawasaki and Ohta [18]. Then, Funaki [9] studied the problem in one-dimension for $\dot{W}^\epsilon(t, x) = \epsilon^\gamma a(x) \dot{W}(t, x)$, the space-time white noise multiplied by a small scaling factor ϵ^γ, $\gamma > 19/2$, with a spatial cut-off $a(x)$ and Funaki [10], Weber [26] in higher dimensions for $\dot{W}^\epsilon(t, x) = \dot{W}^\epsilon(t)$, only time dependent mild (smooth) noise converging to the white noise $\dot{W}(t)$ in time as $\epsilon \downarrow 0$. Röger and Weber [25] discussed the problem for a multiplicative transport type noise having spatial dependence.

Generation of interfaces, which means the creation of interfaces in a short time starting from rather general initial values, was studied in a stochastic setting by Alfaro et al. [1] and Lee [20, 21].

Funaki and Yokoyama [12] studied the sharp interface limit for the stochastic mass conserving Allen–Cahn equation (only with time-dependent mild noise $\dot{W}^\epsilon(t)$ converging extremely slowly to $\dot{W}(t)$) and derived a mixture of (3) and (4) given by

$$V = \kappa - \text{av}_{\Gamma_t}(\kappa) + \frac{c}{|\Gamma_t|} \circ \dot{W}(t), \tag{6}$$

in two-dimensional space at least if Γ_t is convex. Funaki et al. [13] studied the evolution (6) with a simpler additive noise $c\dot{W}(t)$ instead of the multiplicative noise $\frac{c}{|\Gamma_t|} \circ \dot{W}(t)$.

Several results are known on the stochastic Cahn–Hilliard equation, which is a fourth order SPDE having a mass-conservation law.

1.2 The MMC Perturbed by a Direction-Dependent Noise

1.2.1 An SPDE for the Curvature

We consider the stochastic evolution (4) with $\dot{W}(t,x)$ replaced by a direction-dependent Gaussian colored noise $\dot{W}(t,\mathbf{n})$ in two-dimensional space under the convex setting. More precisely, we consider the motion of a closed convex curve Γ_t in a domain $D \subset \mathbb{R}^2$ which is governed by

$$V = \kappa + \circ \dot{W}(t,\mathbf{n}(t,x)), \quad x \in \Gamma_t, \tag{7}$$

where $\mathbf{n}(t,x)$ is the inward normal vector at $x \in \Gamma_t$. This type of equation with more general $\dot{W}(t,\mathbf{n},x)$ instead of $\dot{W}(t,\mathbf{n})$ is proposed by [18]; see also [17]. In our case, the noise $\dot{W}(t,\mathbf{n})$ is a formal time derivative of a direction-dependent Brownian motion $W(t,\mathbf{n})$ which is defined by

$$W(t,\mathbf{n}) = \sum_{i=1}^{\infty} \psi_i(\mathbf{n}) w_i(t), \quad t > 0, \ \mathbf{n} = (\cos\theta, \sin\theta), \quad \theta \in S \simeq [0, 2\pi), \tag{8}$$

where $(w_i(t))_{i\in\mathbb{N}}$ is a family of independent one-dimensional Brownian motions and $(\psi_i)_{i\in\mathbb{N}}$ is a family of functions of $\mathbf{n}$ or equivalently those of θ; see Sect. 1.2.2 for details. The sign $\circ$ in (7) means the product in Stratonovich sense; see below.

For a given function $q = q(t,\mathbf{n})$, $\mathbf{n} \in S$, let us determine the motion of curves Γ_t in D by

$$V = \kappa + \dot{q}(t,\mathbf{n}). \tag{9}$$

If Γ_t is convex, it is known that (9) can be rewritten under the Gauss map $\theta \in S \mapsto X(\theta) \in \Gamma_t$ into the PDE for the curvature $\kappa = \kappa(t,\theta)$ of Γ_t at $X(\theta)$:

$$\frac{\partial \kappa}{\partial t} = \kappa^2 \left(\frac{\partial^2 \kappa}{\partial \theta^2} + \kappa + \frac{\partial^2 \tilde{q}}{\partial \theta^2} + \tilde{q} \right), \tag{10}$$

where $\tilde{q}(t,\theta) = \dot{q}\big(t, (\cos\theta, \sin\theta)\big)$; see [15, 16]. Once κ is determined, the motion of curves $\Gamma_t = \big(X(t,\theta) \in \mathbb{R}^2; \theta \in S\big)$ is given by:

$$\begin{aligned} X(t,\theta) = {} & \left(\int_0^\theta \frac{\sin y}{\kappa(t,y)} dy \, , \, - \int_0^\theta \frac{\cos y}{\kappa(t,y)} dy \right) \\ & + \left(\int_0^t (\kappa + \tilde{q})(s,0) ds, \int_0^t \frac{\partial}{\partial \theta} (\kappa + \tilde{q})(s,0) ds \right) + X(0,0). \end{aligned}$$

Thus one can expect that (7) is described by the SPDE for κ:

$$\frac{\partial \kappa}{\partial t} = \kappa^2 \left(\frac{\partial^2 \kappa}{\partial \theta^2} + \kappa + \circ \dot{w}(t, \theta) \right), \tag{11}$$

where $w(t, \theta) = \partial_\theta^2 W(t, \theta) + W(t, \theta) = \sum_{i=1}^\infty \varphi_i(\theta) w_i(t)$ with $\varphi_i(\theta) = \partial_\theta^2 \psi_i(\theta) + \psi_i(\theta)$ and $W(t, \theta) = W\big(t, (\cos\theta, \sin\theta)\big)$. The sign $\circ$ means the product in Stratonovich sense. The random motion of curves $\Gamma_t = (X(t, \theta) \in \mathbb{R}^2; \theta \in S)$ is recovered from $\kappa(t, \theta)$ by

$$X(t, \theta) = \left(\int_0^\theta \frac{\sin y}{\kappa(t, y)} dy \, , \, - \int_0^\theta \frac{\cos y}{\kappa(t, y)} dy \right)$$
$$+ \left(\int_0^t (\kappa + \circ \dot{w})(s, 0) ds, \int_0^t \frac{\partial}{\partial \theta} (\kappa + \circ \dot{w})(s, 0) ds \right) + X(0, 0).$$

In fact, $\int_0^t \circ \dot{w}(s, 0) ds = \sum_{i=1}^\infty \varphi_i(0) w_i(t)$ and $\int_0^t \circ \partial_\theta \dot{w}(s, 0) ds = \sum_{i=1}^\infty \partial_\theta \varphi_i(0)$ $w_i(t)$.

1.2.2 Assumptions for Our Models

We introduce a cut-off to the SPDE (11) similarly to [10] to avoid the singularity of the curve. Namely, denoting $\dot{w}$ in (11) by $\dot{W}^Q$ with the covariance operator Q specified later, we consider for every $N \in \mathbb{N}$,

$$\frac{\partial \kappa}{\partial t} = a(\kappa) \frac{\partial^2 \kappa}{\partial \theta^2} + b(\kappa) + h(\kappa) \circ \dot{W}^Q, \quad t > 0, \ \theta \in S, \tag{12}$$

where $a \equiv a_N, b \equiv b_N, h \equiv h_N \in C_b^\infty(S)$ satisfying $a(\kappa) \geq a_0 > 0$ (strong ellipticity) and $a(\kappa) = \kappa^2$, $b(\kappa) = \kappa^3$ and $h(\kappa) = \kappa^2$ for $N^{-1} \leq \kappa \leq N$, where $C_b^\infty(S)$ is the family of smooth functions on S having bounded derivatives of all orders. Let Q be a positive linear self-adjoint operator defined on $L^2(S)$ which is of trace class having the eigenvalues $\{\lambda_i^2\}_{i \in \mathbb{N}}$ and the corresponding normalized eigenfunctions $\{\alpha_i(\theta)\}_{i \in \mathbb{N}}$. Without loss of generality, we can assume that $\lambda_1 \geq \lambda_2 \geq \cdots > 0$. Note that $\mathrm{tr} Q = \sum_{i=1}^\infty \lambda_i^2 < \infty$ holds. We assume that the noise term $\dot{W}^Q$ has the form:

$$\dot{W}^Q(t, \theta) = \sum_{i=1}^\infty \lambda_i \alpha_i(\theta) \dot{w}_i(t), \tag{13}$$

where $(w_i(t))_{i \in \mathbb{N}}$ are independent one-dimensional Brownian motions. In other words, for the original noise $\dot{w}(t, \theta)$ in (11), we assume $\sum_{i=1}^\infty \|\varphi_i\|_{L^2(S)}^2 < \infty$.

We also assume that

$$\sum_{i=1}^{\infty} \lambda_i M_{i,n} < \infty, \tag{14}$$

holds for every $n \in \mathbb{N}$, where

$$M_{i,n} = \max\left[\sup_{0\le l\le n+1}\left(\left\|\alpha_i^{(l)}\right\|_{L^\infty}, \left\|\alpha_i^{(l)}\right\|_{L^\infty}^2\right), 1\right], \quad i \in \mathbb{N},\ n \in \mathbb{N}, \tag{15}$$

and $\alpha^{(l)} := \partial_\theta^l \alpha$.

Note that the SPDE (12) is equivalent to the following SPDE in Itô's form:

$$\frac{\partial \kappa}{\partial t} = a(\kappa)\frac{\partial^2 \kappa}{\partial \theta^2} + \tilde{b}(\kappa) + h(\kappa)\dot{W}^Q, \quad t > 0,\ \theta \in S, \tag{16}$$

where

$$\tilde{b}(\kappa) = b(\kappa) + \frac{1}{2}h(\kappa)h'(\kappa)\mathrm{tr}Q. \tag{17}$$

For later use, we set

$$Q(x, y) = \sum_{i=0}^{\infty} \lambda_i^2 \alpha_i(x)\alpha_i(y). \tag{18}$$

To introduce the approximation of the noise $\dot{W}^Q(t,\theta)$, we consider a family of stochastic processes $(\xi_i)_{i\in\mathbb{N}}$ defined on a probability space $(\Omega, \mathcal{F}, P)$ such that

1. $\xi_i = \big(\xi_i(t); t \ge 0\big)$, $i \in \mathbb{N}$, are stationary and strongly mixing, that is, for some $p > 3/2$,

$$\int_0^\infty \rho(t)^{1/p} dt < +\infty, \tag{19}$$

   holds, where

$$\rho(t) = \sup_{s\ge 0} \sup_{\substack{A\in\mathcal{F}_{s+t,\infty},\\ B\in\mathcal{F}_{0,s}}} \left|\frac{P(A\cap B) - P(A)P(B)}{P(B)}\right|, \quad t \ge 0, \tag{20}$$

   and $\mathcal{F}_{s,t} = \sigma\{\xi_i(r); i \in \mathbb{N},\ s \le r \le t\}$, $\mathcal{F}_{s,\infty} = \sigma\{\xi_i(r); i \in \mathbb{N},\ s \le r < \infty\}$.
2. ξ_i is C^1 in t a.s. and there exists a constant $\bar{C} > 0$ being non-random and independent of i such that $|\xi_i(t)|$, $|\dot{\xi}_i(t)| \le \bar{C}$ holds for every $i \in \mathbb{N}$.
3. $E[\xi_i(t)] = 0$ and $\sigma_i^2 = 1$ for every $i \in \mathbb{N}$, and $\{\xi_i\}_{i\in\mathbb{N}}$ are independent, where

$$\sigma_i^2 := 2\int_0^\infty E[\xi_i(0)\xi_i(t)]dt. \tag{21}$$

Let us define ξ_i^ϵ, $i \in \mathbb{N}, 0 < \epsilon < 1$, by

$$\xi_i^\epsilon(t) = \epsilon^{-\gamma}\xi_i(\epsilon^{-2\gamma}t), \quad t > 0, \tag{22}$$

for some $0 < \gamma < 1/3$. Then $\int_0^t \xi_i^\epsilon(s)ds$ converges to a one-dimensional Brownian motion $w_i(t)$ in law as $\epsilon \downarrow 0$. For $i \in \mathbb{N}$, $t > 0$ and $0 < \epsilon < 1$, we clearly have

$$||\xi_i^\epsilon(t)||_{L^\infty(\Omega)} \leq \bar{C}\epsilon^{-\gamma}. \tag{23}$$

Let us define $Z_\epsilon^Q = Z_\epsilon^Q(t,\theta)$ by

$$Z_\epsilon^Q(t,\theta) = \int_0^t \sum_{i=1}^\infty \lambda_i \alpha_i(\theta)\xi_i^\epsilon(s)ds. \tag{24}$$

Then, we consider a random PDE for $\kappa = \kappa^\epsilon(t,\theta)$:

$$\frac{\partial\kappa}{\partial t} = a(\kappa)\frac{\partial^2\kappa}{\partial\theta^2} + b(\kappa) + h(\kappa)\dot{Z}_\epsilon^Q, \quad t > 0, \ \theta \in S. \tag{25}$$

Since the coefficients of (25) are good enough for each fixed ω, the existence and uniqueness of the solutions of (25) for every $0 < \epsilon < 1$ follow by the standard theory of quasilinear parabolic PDEs.

2 Main Result

Let $\kappa_0 \in C^2(S)$ be a given function satisfying $\kappa_0(\theta) > 0$ for every $\theta \in S$. We assume $h(\kappa) = h_N(\kappa) > 0$ for every $\kappa \in \mathbb{R}$ and $N \in \mathbb{N}$ in the SPDE (12). One can always choose such cut-off function h_N. Then, we have the existence and uniqueness of the solutions of the SPDE (12).

Theorem 1 *For every $N \in \mathbb{N}$, there exists a unique solution κ of the SPDE* (12) *with the initial value κ_0.*

Indeed, set

$$A(\kappa) := \int_0^\kappa \frac{dy}{a(y)}, \quad \kappa \in \mathbb{R},$$

then A is a continuous and increasing function of κ. Let us denote by $\kappa = f(u)$ the inverse function of $u = A(\kappa)$. By a simple computation and noting $f'(u) = a(f(u)) > 0$, we see that the SPDE (12) is rewritten into the following SPDE of divergence form for $u = u(t,\theta) := A(\kappa(t,\theta))$, $t > 0$, $\theta \in S$:

$$\frac{\partial u}{\partial t} = \frac{\partial}{\partial \theta}\left(a(f(u))\frac{\partial u}{\partial \theta}\right) + \frac{b(f(u))}{a(f(u))} + \frac{h(f(u))}{a(f(u))} \circ \dot{W}^Q, \tag{26}$$

for which the result of Debussche, de Moor and Hofmanova [5] is applicable and Theorem 1 follows. Note that the SPDE (26) is easily rewritten into Itô's form. Otto and Weber [24] discussed the SPDE like (12) in rough path sense.

Our main result is a Wong–Zakai type theorem for the SPDE (12).

Theorem 2 *For every $n \in \mathbb{N}$ and $T > 0$, the joint distribution of $(\kappa^\epsilon, Z_\epsilon^Q)$ determined by* (25) *converges weakly to that of (κ, W^Q), i.e., the joint law of the solution κ of the SPDE* (12) *and its noise W^Q, on the space $\bar{\mathcal{X}} \equiv C\left([0,T], (C^n(S))^2\right)$ as $\epsilon \downarrow 0$.*

Note that this theorem implies the existence of the solution of the SPDE (12) and therefore (16) in law sense. In addition, one can easily show the pathwise uniqueness for the solution of (16) in a similar way to Lemma 5.2 in [10]. Therefore, the existence of the strong solution of (16) or (12) follows by the argument of Yamada–Watanabe's type. This is another way to show Theorem 1 in our setting.

To show Theorem 2, we need to show that the family of the joint distributions of $(\kappa^\epsilon, Z_\epsilon^Q)$ is tight in $\bar{\mathcal{X}}$ and the solution of the martingale problem corresponding to $(\kappa^\epsilon, Z_\epsilon^Q)$ converges to that of the limit process (κ, W^Q) as $\epsilon \downarrow 0$. To prove the tightness, it suffices to show the following two propositions.

Proposition 1 *For every $n \in \mathbb{N}$ and $T > 0$, let P^ϵ be the probability law in $C([0,T], C^n(S))$ of the solution $\kappa^\epsilon = \kappa^\epsilon(t,\theta)$ of* (25) *starting at κ_0. Then $\{P^\epsilon\}_{0<\epsilon<1}$ is tight.*

Proposition 2 *For every $n \in \mathbb{N}$ and $T > 0$, let Q^ϵ be the probability law in $C([0,T], C^n(S))$ of $Z_\epsilon^Q = Z_\epsilon^Q(t,\theta)$. Then $\{Q^\epsilon\}_{0<\epsilon<1}$ is tight.*

To prove Proposition 1, we need to show

Proposition 3 *For every $p \geq 1$ and $T > 0$, we have*

$$\sup_{0<\epsilon<1} E\left[\sup_{0\leq t\leq T} \|\kappa^{(n)}(t)\|_{L^p(S)}^p\right] < \infty, \tag{27}$$

where $\kappa = \kappa^\epsilon$ and $\kappa^{(n)} = \partial_\theta^n \kappa$.

The proof of Proposition 3 is given in Sect. 3. Once we have the uniform moment estimates (27), the conclusion of Proposition 1 follows by using a criterion due to Holley and Stroock (see [8]), if one can prove the weak tightness of $\{\kappa^\epsilon\}_{0<\epsilon<1}$:

Proposition 4 *For every $\phi \in C^\infty(S)$ and $p \geq 1$, there exists $C > 0$ independent of ϵ such that*

$$E[|\langle \kappa^\epsilon(t_2) - \kappa^\epsilon(t_1), \phi\rangle|^p] \leq C(t_2 - t_1)^{\frac{p}{2}}, \tag{28}$$

for every $0 \leq t_1 < t_2 \leq T$.

Proposition 4 can be obtained by a similar argument to Lemma 4.1 (cited as Lemma 9 in Sect. 3.3), Proposition 4.1 in [10], noting $\sigma_i^2 = 1$ for σ_i^2 defined in (21) and (23).

To prove Proposition 2, we need the following.

Proposition 5 *For every $p \geq 1$, $T > 0$ and $i \in \mathbb{N}$, there exists $C > 0$ independent of ϵ and i such that*

$$E[|\eta_i^\epsilon(t_2) - \eta_i^\epsilon(t_1)|^p] \leq C(t_2 - t_1)^{\frac{p}{2}}, \tag{29}$$

holds for every $0 \leq t_1 < t_2 \leq T$, where $\eta_i^\epsilon(t) = \int_0^t \xi_i^\epsilon(s)ds$.

Proposition 5 is shown similarly to and actually much simpler than Proposition 4 as pointed out in the proof of Theorem 5.1 in [10]. Note that, compared to the situation discussed in [10], $i \in \mathbb{N}$ moves in our case, but the mixing rate ρ, the variance $\sigma_i^2 = 1$ and the bound $\bar{C}$ in (23) are all uniform in i, so that the constant $C > 0$ in Proposition 5 can be taken independently of i.

For the proof of Proposition 2, recall that $Z_\epsilon^Q(t,\theta) = \sum_{i=1}^\infty \lambda_i \alpha_i(\theta)\eta_i^\epsilon(t)$. Since $\eta_i^\epsilon(t)$ converges to a Brownian motion in law as $\epsilon \downarrow 0$ for each $i \in \mathbb{N}$, the finite sum in $i \leq N$ with a fixed N in Z_ϵ^Q defines a tight family in $C([0,T], C^n(S))$. Therefore, to complete the proof of Proposition 2, it suffices to show that, for every $\delta_1, \delta_2 > 0$ and $0 \leq l \leq n$, there exists $N \in \mathbb{N}$ such that

$$\sup_{0<\epsilon<1} P\left(\sup_{0\leq t\leq T} \left\| \sum_{i\geq N} \lambda_i \alpha_i^{(l)}(\theta)\eta_i^\epsilon(t) \right\|_{L^\infty(S)} \geq \delta_1 \right) \leq \delta_2. \tag{30}$$

However, by Chebyshev's inequality, the above probability is bounded by

$$\frac{1}{\delta_1} \sum_{i\geq N} \lambda_i \|\alpha_i^{(l)}\|_{L^\infty(S)} E\left[\sup_{0\leq t\leq T} |\eta_i^\epsilon(t)| \right].$$

Here, the estimate given by Proposition 5 combined with $\eta_i^\epsilon(0) = 0$ implies

$$\sup_{i\in\mathbb{N}, 0<\epsilon<1} E\left[\sup_{0\leq t\leq T} |\eta_i^\epsilon(t)| \right] < \infty$$

by the celebrated Kolmogorov's criterion (see, e.g., Kunita [19], p. 31, Theorem 1.4.1), and this shows (30) for a sufficiently large N with the help of the condition (14).

Theorem 2 is shown in Sect. 4 based on the martingale method and the tightness result followed from Propositions 1 and 2.

3 Uniform Moment Estimates on $\kappa^{(n)}$

In this section, we show Proposition 3. In what follows, we assume that p is a positive even integer. We begin by computing the time derivative of $\|\kappa^{(n)}\|_{L^p}^p$ for κ determined by (25), that is, we have

$$\frac{d}{dt}\|\kappa^{(n)}\|_{L^p}^p = p\left(\Phi_1^n(\kappa) + \Phi_2^n(\kappa) + \sum_{i=1}^{\infty}\lambda_i \Phi_3^{n,i}(\kappa)\xi_i^\epsilon(t)\right), \tag{31}$$

for every $n \in \mathbb{N}$, where

$$\Phi_1^n(\kappa) = \int_S (\kappa^{(n)})^{p-1}[a(\kappa)\kappa'']^{(n)}d\theta,$$

$$\Phi_2^n(\kappa) = \int_S (\kappa^{(n)})^{p-1}[b(\kappa)]^{(n)}d\theta,$$

$$\Phi_3^{n,i}(\kappa) = \int_S (\kappa^{(n)})^{p-1}[\alpha_i(\theta)h(\kappa)]^{(n)}d\theta.$$

We also compute the time derivative of $\Phi_3^{n,i}$:

$$\frac{d}{dt}\Phi_3^{n,i}(\kappa) = \Psi_1^{n,i}(\kappa) + \sum_{j=1}^{\infty}\lambda_j \Psi_2^{n,i,j}(\kappa)\xi_j^\epsilon(t), \tag{32}$$

where

$$\Psi_1^{n,i}(\kappa) = (p-1)\int_S [a(\kappa)\kappa'' + b(\kappa)]^{(n)}(\kappa^{(n)})^{p-2}[\alpha_i(\theta)h(\kappa)]^{(n)}d\theta + \int_S (\kappa^{(n)})^{p-1}\left(\alpha_i(\theta)h'(x)[a(\kappa)\kappa'' + b(\kappa)]\right)^{(n)}d\theta,$$

$$\Psi_2^{n,i,j}(\kappa) = (p-1)\int_S [\alpha_i h(\kappa)]^{(n)}[\alpha_j h(\kappa)]^{(n)}(\kappa^{(n)})^{p-2}d\theta + \int_S (\kappa^{(n)})^{p-1}[\alpha_i\alpha_j h'h(\kappa)]^{(n)}d\theta.$$

We denote by $\mathcal{P}_n$, $n \in \mathbb{N}$, the family of polynomials of the forms $P(y_1,\dots,y_{n-1},\theta;\kappa) = \sum_\alpha g_{1,\alpha}(\kappa)g_{2,\alpha}(\theta)y^\alpha$, $y_i \in \mathbb{R}$, $1 \le i \le n-1$, $\kappa \in \mathbb{R}$, $\theta \in S$ with $g_{1,\alpha} \in C_b^\infty(\mathbb{R})$, $g_{2,\alpha} \in C_b^\infty(S)$, $\alpha = (\alpha_1,\dots,\alpha_{n-1}) \in \mathbb{Z}_0^n$, $y^\alpha = y_1^{\alpha_1}\dots y_{n-1}^{\alpha_{n-1}}$ and the sum $\sum_\alpha$ is finite.

3.1 Bounds for Φ_1^n, Φ_2^n, and $\Phi_3^{n,i}$ for $n \in \mathbb{N}$

Lemma 1 *For every $n \geq 1$, there exist $c, C > 0$ and $P_k = P_k(\kappa^{(1)}, \dots, \kappa^{(n-1)}; \theta, \kappa) \in \mathcal{P}_n$, $k = 1, 2, 3, 4$, such that*

$$\Phi_1^n(\kappa) \leq -c\psi_p^n(\kappa) + C\left(\|\kappa^{(n)}\|_{L^p}^p + \sum_{k=1}^{3} \|P_k\|_{L^p}^p \right), \tag{33}$$

$$\left|\Phi_2^n(\kappa)\right| \leq C\left(\|\kappa^{(n)}\|_{L^p}^p + \|P_4\|_{L^p}^p \right), \tag{34}$$

where

$$\psi_p^n(\kappa) = \int_S (\kappa^{(n)})^{p-2}(\kappa^{(n+1)})^2 d\theta.$$

Proof The proofs of (33) and (34) are the same as those of Lemmas 6.1 and 6.2 in [10], respectively. □

Lemma 2 *For every $n \geq 1$, there exist $C > 0$ and $P_{5,l} = P_{5,l}(\kappa^{(1)}, \dots, \kappa^{(n-1)}; \theta, \kappa) \in \mathcal{P}_n$, $l = 0, \dots, n$, such that for every $i \in \mathbb{N}$,*

$$\left|\Phi_3^{n,i}(\kappa)\right| \leq C \sum_{l=0}^{n} ||\alpha_i^{(n-l)}||_{L^\infty} \left(\|\kappa^{(n)}\|_{L^p}^p + \|P_{5,l}\|_{L^p}^p \right). \tag{35}$$

Proof This lemma is shown by a similar method to Lemma 6.2 in [10]. We need only to see the estimate for $[\alpha_i(\theta)h(\kappa)]^{(n)}$ and that for α_i. So we omit the details. □

By following the method used in Lemma 6.3 in [10], we obtain the following lemma.

Lemma 3 *For every $n \geq 1$ and $\delta > 0$, there exist $C > 0$ and $P_{6,i} = P_{6,i}(\kappa^{(1)}, \dots, \kappa^{(n-1)}; \theta, \kappa) \in \mathcal{P}_n$, $i = 1, \dots, n-1$, such that*

$$\left|\Phi_3^{n,i}(\kappa)\right| \leq C\epsilon^\gamma(p-1) \sum_{l=0}^{n-1} \left\|\alpha_i^{(n-1-l)}\right\|_{L^\infty} \left(\delta\psi_p^{(n)}(\kappa) + \frac{\|\kappa^{(n)}\|_{L^p}^p}{\delta} + \frac{\|P_{6,l}\|_{L^p}^p}{\delta\epsilon^{p\gamma}} \right).$$

Furthermore, for $n = 1$ we have the following estimate:

$$\left|\Phi_3^{1,i}\right| \leq (p-1)C\, \|\alpha_i\|_{L^\infty} \left[\delta\epsilon^\gamma \psi_p^{(n)}(\kappa) + \frac{1}{\delta\epsilon^\gamma} \|\kappa'\|_{L^{p-2}}^{p-2} \right].$$

3.2 Estimates on $\Psi_1^{n,i}$ and $\Psi_2^{n,i,j}$

Lemma 4 *For $n \geq 1$, there exist $C > 0$ and $P_{7,l,m}, P_{8,l,m}, P_{9,l} \in \mathcal{P}_n$, $0 \leq l, m \leq n$, such that for every $i, j \in \mathbb{N}$,*

$$\left|\Psi_2^{n,i,j}(\kappa)\right| \leq C \sum_{0\leq l,m\leq n} \left\|\alpha_i^{(n-l)}\right\|_{L^\infty} \left\|\alpha_j^{(n-m)}\right\|_{L^\infty} \left(\|\kappa^{(n)}\|_{L^p}^p + \|P_{7,l,m}\|_{L^p}^p + \|P_{8,l,m}\|_{L^p}^p\right)$$
$$+ C \sum_{0\leq l\leq n} \left\|(\alpha_i\alpha_j)^{(n-l)}\right\|_{L^\infty} \left(\|\kappa^{(n)}\|_{L^p}^p + \|P_{9,l}\|_{L^p}^p\right). \tag{36}$$

Proof Let us decompose $\Psi_2^{n,i,j}(\kappa)$ into the sum $I_1 + I_2$, where

$$I_1 = (p-1)\int_S [\alpha_i h(\kappa)]^{(n)} [\alpha_j h(\kappa)]^{(n)} (\kappa^{(n)})^{p-2} d\theta,$$
$$I_2 = \int_S (\kappa^{(n)})^{p-1} [\alpha_i \alpha_j h' h(\kappa)]^{(n)} d\theta.$$

Using the estimate for $\alpha_i^{(l)}(\theta)$, $l = 0, \ldots, n-1$, the assertion is easily obtained. □

Bound for $\Psi_1^{n,i}(\kappa)$: First we consider the case $n = 1$.

Lemma 5 *There exists $C > 0$ such that for every $\delta > 0$,*

$$\left|\Psi_1^{1,i}(\kappa)\right| \leq C \sum_{l=0}^{2} \left\|\alpha_i^{(l)}\right\|_{L^\infty} \left[1 + \|\kappa^{(n)}\|_{L^p}^p + \left(1 + \delta\epsilon^{-\gamma}\right)\psi_p^1(\kappa) + \epsilon^\gamma \delta^{-1} \|\kappa'\|_{L^{p+2}}^{p+2}\right].$$

Proof This is obtained by a similar argument to Lemma 6.5 in [10] by noting the estimate for the derivatives of $\alpha_i(\theta)$. □

We assume $n \geq 2$ for the following lemmas. First we use integration by parts to write

$$\Psi_1^{n,i}(\kappa) = -(p-1)(p-2)\Psi_a^{n,i}(\kappa) - (p-1)\Psi_b^{n,i}(\kappa) - (p-1)\Psi_c^{n,i}(\kappa),$$

where

$$\Psi_a^{n,i}(\kappa) = \int_S (\kappa^{(n)})^{p-3} \kappa^{(n+1)} [\alpha_i(\theta) h(\kappa)]^{(n)} \left[a(\kappa)\kappa'' + b(\kappa)\right]^{(n-1)} d\theta,$$
$$\Psi_b^{n,i}(\kappa) = \int_S (\kappa^{(n)})^{p-2} [\alpha_i(\theta) h(\kappa)]^{(n+1)} \left[a(\kappa)\kappa'' + b(\kappa)\right]^{(n-1)} d\theta,$$
$$\Psi_c^{n,i}(\kappa) = \int_S (\kappa^{(n)})^{p-2} \kappa^{(n+1)} \left[\alpha_i(\theta) h'[\, a(\kappa)\kappa'' + b(\kappa)\,]\right]^{(n-1)} d\theta.$$

Lemma 6 (Estimate on $\Psi_b^{n,i}(\kappa)$) *For $n \geq 2$, there exist $C > 0$ and $P_i \in \mathcal{P}_n$, $i = 10, 11, 12, 13$, such that*

$$\left|\Psi_b^{n,i}(\kappa)\right| \leq C(M_{i,n}+1)\left[\psi_p^n(\kappa) + \|\kappa^{(n)}\|_{L^p}^p + \sum_{k=10}^{13} \|P_k\|_{L^p}^p\right]. \tag{37}$$

Proof This is obtained by a similar argument to Lemma 6.6 in [10] by noting the estimate for the derivatives of $\alpha_i(\theta)$. □

Lemma 7 (Estimate on $\Psi_c^{n,i}(\kappa)$) *For $n \geq 2$, there exist $C > 0$ and $P_{k,l} \in \mathcal{P}_n$, $14 \leq k \leq 16, 0 \leq l \leq n-1$, such that*

$$\left|\Psi_c^{n,i}(\kappa)\right| \leq C\sum_{l=0}^{n-1} \left\|\alpha_i^{(n-1-l)}\right\|_{L^\infty} \left(\psi_p^n(\kappa) + \|\kappa^{(n)}\|_{L^p}^p + \sum_{k=14}^{16} \|P_{k,l}\|_{L^p}^p\right). \tag{38}$$

Proof This is obtained by similar argument to Lemma 6.7 in [10]. □

Bound for $\Psi_a^{n,i}(\kappa)$: The bound for $\Psi_a^{n,i}(\kappa)$ is obtained by a similar argument to Lemma 6.9 in [10].

Lemma 8 *For every $n \geq 2$ and $0 < \bar{\beta} < 1$, there exist $C, N, q > 0$ and $P_{k,l}(\kappa^{(1)}, \ldots, \kappa^{(n-1)}; \theta, \kappa)$, $0 \leq l \leq n$, $17 \leq k \leq 20$ such that*

$$\begin{aligned}|\Psi_a^{n,i}(\kappa)| \leq\ & C\sum_{l=0}^{n} \left\|\alpha_i^{(n-l)}\right\|_{L^\infty} \left[\psi_p^n(\kappa) + \|\kappa^{(n)}\|_{L^p}^p + \sum_{k=17}^{20} \|P_{k,l}\|_{L^p}^p\right] \\ & + C\sum_{l=0}^{n} \left\|\alpha_i^{(n-l)}\right\|_{L^\infty} \left(1 + \sum_{k=1}^{n-1} \left\|\kappa^{(k)}\right\|_{L^q}^N + \left\|\kappa^{(n)}\right\|_{L^q}^{\bar{\beta}}\right) \psi_2^n(\kappa).\end{aligned} \tag{39}$$

3.3 Estimates Based on the Mixing Property

This section gives the mixing lemma and an estimate on the conditional expectation of $\int_s^t \Phi_3^{n,i}(\kappa_r)\xi_i^\epsilon(r)dr$.

Lemma 9 (Lemma 4.1 in [10]) *We assume that $0 \leq r \leq s \leq t < \infty$.*

(1) *For $\mathcal{F}_{t,\infty}$-measurable random variables Z satisfying $E[Z] = 0$,*

$$|E[Z|\mathcal{F}_{0,s}]| \leq 2\|Z\|_{L^\infty(\Omega)}\rho(t-s) \quad a.s. \tag{40}$$

(2) *For $\mathcal{F}_{0,r}$-measurable X, $\mathcal{F}_{s,s}$-measurable Y and $\mathcal{F}_{t,t}$-measurable Z satisfying $E[Z] = 0$, we have*

$$|E[X(YZ - E[YZ])]| \leq 8\|X\|_{L^\infty(\Omega)}\|Y\|_{L^\infty(\Omega)}\|Z\|_{L^\infty(\Omega)} [\rho(t-s)\rho(s-r)]^{1/2}, \quad (41)$$

$$|E[X(YZ - E[YZ])]| \leq 2^{4-\frac{2}{p}}\|X\|_{L^p(\Omega)}\|Y\|_{L^\infty(\Omega)}\|Z\|_{L^\infty(\Omega)} [\rho(t-s)\rho(s-r)]^{\frac{p-1}{2p}}, \quad (42)$$

for every $p \geq 1$.

Lemma 10 *For* $0 \leq s \leq t \leq T$, *for all* $i, n \in \mathbb{N}$, *we have*

$$\left| E\left[\int_s^t \Phi_3^{n,i}(\kappa_r)\xi_i^\epsilon(r)dr \middle| \mathcal{F}_{0,s}^\epsilon \right] \right|$$

$$\leq C\epsilon^\gamma \left|\Phi_3^{n,i}(\kappa_s)\right| + C\int_s^t E\left[\epsilon^\gamma \left|\Psi_1^{n,i}(\kappa_{r'})\right| + \sum_{j=1}^\infty \lambda_j \left|\Psi_2^{n,i,j}(\kappa_{r'})\right| \middle| \mathcal{F}_{0,s}^\epsilon \right] dr', \quad (43)$$

for some constant $C > 0$ *independent of* i, *where* $\mathcal{F}_{0,s}^\epsilon = \sigma\{\xi_i^\epsilon(r); i \in \mathbb{N},\ 0 \leq r \leq s\}$.

Proof The assertion is obtained by applying the above Lemma 9 (1) and Lemma 6.10 in [10] and (23) for (32). □

3.4 Proof of Proposition 3

Since the proof is similar to that in Sect. 6.7 in [10], we only touch the different points. As a whole, we need to carefully check the estimate for the infinite sum of the right hand side of (31) by using the definition of $M_{i,n}$ and (14). The rest of the proof is not too much different. First we obtain the energy estimate for κ^ϵ as follows:

$$E\left[\left\|\kappa_t^{(n)}\right\|_{L^p}^p \middle| \mathcal{F}_{0,s}^\epsilon \right] + c'' \int_s^t E\left[\psi_p^n(\kappa_r) \middle| \mathcal{F}_{0,s}^\epsilon \right] dr$$

$$\leq (1 + C\epsilon^\gamma) \left\|\kappa_s^{(n)}\right\|_{L^p}^p + C\epsilon^\gamma \sum_{l=0}^n \left\|P_{5,l}\right\|_{L^p}^p$$

$$+ C\int_s^t E\left[\left\|\kappa_r^{(n)}\right\|_{L^p}^p \middle| \mathcal{F}_{0,s}^\epsilon \right] dr + C\int_s^t E\left[\mathcal{P}(\kappa_r) \middle| \mathcal{F}_{0,s}^\epsilon \right] dr$$

$$+ C\epsilon^\gamma(p-2)\int_s^t E\left[\left(1 + \epsilon^{-\delta} + \sum_{i=1}^{n-1} \|\kappa_r^{(i)}\|_{L^q}^N\right) \psi_2^{(n)}(\kappa_r) \middle| \mathcal{F}_{0,s}^\epsilon \right] dr,$$

for every $0 < \epsilon < \epsilon_0$ (with ϵ_0 small enough to satisfy $C\epsilon_0^\gamma \ll c$), $\delta > 0$ and for some $N = N_\delta$, $q = q_\delta > 1$, where $\mathcal{P}(\kappa) = \sum_{1 \leq k \leq 20, k \neq 5,6} \sum_{0 \leq l,m \leq n} \left\|P_{k,l,m}(\kappa)\right\|_{L^p}^p$, with $P_{k,l,m}$ being understood as each of $P_k(\kappa)$, $P_{k,l}(\kappa)$, $P_{k,l,m}(\kappa)$ and 0. Indeed, we first use (31) by integrating and taking the conditional expectation. We then apply (33) for $\Phi_1^{(n)}$ and (34) for $\Phi_2^{(n)}$. To estimate $\Phi_3^{n,i}$, we use (43), (35) for the first term (that is, containing $\Phi_3^{(n)}(\kappa_s)$) and then (36) for $\Psi_2^{n,i,j}(\kappa)$. Finally we use (37) to (39) for $\Psi_1^{n,i}$. Let us focus on the infinite sum in the right hand side of (31). By seeing (43) and the estimate (36) for $\Psi_2^{n,i,j}(\kappa)$, the infinite sum appearing from such terms converges due to the condition $\sum_{i=1}^{\infty} \lambda_i M_{i,n} < \infty$. The rest of the proof is carried out similarly to Theorem 6.1 in [10].

4 Martingale Method

4.1 Martingale Problem Corresponding to the Limit Process (κ, W^Q)

Let $\mathcal{D}$ be the class of all (tame) functions $\Psi = \Psi(\kappa)$ on the space $C^n(S)$ having the form:

$$\Psi(\kappa) = \psi(\langle \kappa, \phi_1 \rangle, \dots, \langle \kappa, \phi_m \rangle), \tag{44}$$

for some $m \in \mathbb{N}$, $\psi \in C_0^\infty(\mathbb{R}^m)$ and $(\phi_j)_{1 \le j \le m} \subset C^\infty(S)$, where $\langle \kappa, \phi \rangle = \int_S \kappa(\theta) \phi(\theta) d\theta$. We denote for $\Psi \in \mathcal{D}$:

$$D\Psi(\kappa; \theta) = \sum_{j=1}^{m} \frac{\partial \psi}{\partial x_j}(\langle \kappa, \phi_1 \rangle, \dots, \langle \kappa, \phi_m \rangle)\phi_j(\theta), \tag{45}$$

$$D^2\Psi(\kappa; \theta_1, \theta_2) = \sum_{i,j=1}^{m} \frac{\partial^2 \psi}{\partial x_j \partial x_i}(\langle \kappa, \phi_1 \rangle, \dots \langle \kappa, \phi_m \rangle)\phi_j(\theta_1)\phi_i(\theta_2). \tag{46}$$

We introduce the following operator $\mathcal{L}$ on $\mathcal{D}$:

$$\mathcal{L}\Psi(\kappa) = \langle D\Psi(\kappa; \cdot), a(\kappa)\kappa'' + \tilde{b}(\kappa) \rangle + \frac{1}{2}\langle D^2\Psi(\kappa; \cdot, \cdot), Q(\cdot, \cdot) h \otimes h \rangle, \quad \kappa \in C^2(S), \tag{47}$$

where $(h \otimes h) \equiv (h(\kappa) \otimes h(\kappa)) = h(\kappa(\theta_1))h(\kappa(\theta_2))$ and $\langle D^2\Psi(\kappa; \cdot, \cdot), Q(\cdot, \cdot)h \otimes h \rangle$ means

$$\int_S \int_S D^2\Psi(\kappa; \theta_1, \theta_2) Q(\theta_1, \theta_2) h(\kappa(\theta_1)) h(\kappa(\theta_2)) d\theta_1 d\theta_2,$$

which is shortened to $\langle D^2\Psi, Qh \otimes h \rangle$. Let $\bar{\mathcal{D}}$ be the class of all functions $\bar{\Psi}$ on $C^n(S) \times C^n(S)$ having the form $\bar{\Psi}(\kappa, \eta) = \Psi_1(\kappa)\varphi(\eta)$ for $\kappa, \eta \in C^n(S)$ with Ψ_1 and $\varphi \in \mathcal{D}$. We introduce the operator $\bar{\mathcal{L}}$ on $\bar{\mathcal{D}}$:

$$\bar{\mathcal{L}}\bar{\Psi}(\kappa, \eta) = (\mathcal{L}\Psi_1)\varphi + \langle D\Psi_1 \otimes D\varphi, h(\kappa) Q(\cdot, \cdot) \rangle + \frac{1}{2}\Psi_1 \langle D^2\varphi, Q \rangle, \tag{48}$$

where

$$\langle D\Psi_1 \otimes D\varphi, h(\kappa)Q(\cdot,\cdot)\rangle = \int_S \int_S D\Psi_1(\kappa;\theta_1)D\varphi(\eta;\theta_2)Q(\theta_1,\theta_2)h(\kappa(\theta_1))d\theta_1 d\theta_2.$$

Applying Itô's formula to $\bar{\Psi}$, it is shown that (κ, W^Q) is a solution for the $(\bar{\mathcal{L}}, \bar{\mathcal{D}})$-martingale problem. We denote by $\mathcal{P}(\bar{\mathcal{X}})$ a family of probability measures on $\bar{\mathcal{X}}$.

Proposition 6 *The $(\bar{\mathcal{L}}, \bar{\mathcal{D}})$ martingale problem is well-posed.*

To show Proposition 6, we prepare the following lemma:

Lemma 11 *Let $\bar{P} \in \mathcal{P}(\bar{\mathcal{X}})$ be a solution of the $(\bar{\mathcal{L}}, \bar{\mathcal{D}})$-martingale problem. For $\omega = (\kappa, W^Q) \in \bar{\mathcal{X}}$, let us set*

$$\mu_t(\theta) \equiv \mu(t,\theta,\omega) = \kappa(t,\theta) - \kappa(0,\theta) - \int_0^t \left(a\left(\kappa(s,\theta)\right) \frac{\partial^2 \kappa}{\partial \theta^2}(s,\theta) + \tilde{b}\left(\kappa(s,\theta)\right) \right) ds.$$

Then the processes $\{\mu_t(\theta)\ ;\ \theta \in S\}$ and $\{W^Q(t,\theta)\ ;\ \theta \in S\}$ are local martingales under $\bar{P}$ and the quadratic variational processes are given by

$$[\mu_t(\theta_1), \mu_t(\theta_2)] = \int_0^t h(\kappa(s,\theta_1))h(\kappa(s,\theta_2))Q(\theta_1,\theta_2)ds \tag{49}$$

$$[\mu_t(\theta_1), W^Q(t,\theta_2)] = \int_0^t h(\kappa(s,\theta_1))Q(\theta_1,\theta_2)ds \tag{50}$$

$$[W^Q(t,\theta_1), W^Q(t,\theta_2)] = Q(\theta_1,\theta_2)t \tag{51}$$

Proof Let us set $\bar{\Psi} = \langle \kappa, \phi \rangle$, then $\bar{\mathcal{L}}\bar{\Psi} = \langle a(\kappa)\kappa'' + \tilde{b}(\kappa), \phi \rangle$ and $\langle \mu_t, \phi \rangle$ is a local martingale for every $\phi \in C^\infty(S)$. Note that $\psi(x) = x \notin C_0^\infty(\mathbb{R})$, however, we apply for the cut-off argument. Letting $\phi \to \delta_\theta$, we see that $\mu_t(\theta)$ is a local martingale. Similarly, choosing $\bar{\Psi} = \langle W^Q, \phi \rangle$, it is a local martingale. Note that for $\bar{\Psi}^i \in \mathcal{D}$, $i = 1, 2$,

$$[\mu_t(\bar{\Psi}^1), \mu_t(\bar{\Psi}^2)] = \int_0^t \bar{\Gamma}(\bar{\Psi}^1, \bar{\Psi}^2)(\kappa(s), W^Q(s))ds,$$

holds, where $\mu_t(\bar{\Psi})$ is the martingale part of $\bar{\Psi}(\kappa(t), W^Q(t))$ and

$$\bar{\Gamma}(\bar{\Psi}^1, \bar{\Psi}^2) = \bar{\mathcal{L}}(\bar{\Psi}^1\bar{\Psi}^2) - \bar{\Psi}^1\bar{\mathcal{L}}\bar{\Psi}^2 - \bar{\Psi}^2\bar{\mathcal{L}}\bar{\Psi}^1.$$

Let us take $\bar{\Psi}^i = \langle \kappa, \phi_i \rangle$ for $\phi_i \in C^\infty(S)$, $i = 1.2$, then

$$\bar{\Gamma}(\bar{\Psi}^1, \bar{\Psi}^2) = \langle \phi_1 \otimes \phi_2, h \otimes hQ \rangle,$$

which proves (49) by letting $\phi_i \to \delta_{\theta_i}$, $i = 1, 2$. Similarly, (50) is shown by taking $\bar{\Psi}^1 = \langle \kappa, \phi_1 \rangle$, $\bar{\Psi}^2 = \langle W^Q, \phi_2 \rangle$ and then letting ϕ_i tend to δ_{θ_i}, $i = 1, 2$. As for (51), it is easily shown by taking $\bar{\Psi}^i = \langle W^Q, \phi_i \rangle$. □

Proof of Proposition 6. Recalling the condition $h > 0$, let

$$\tilde{w}_t(\theta) = \int_0^t \frac{1}{h(\kappa(s,\theta))} d\mu_s(\theta), \quad \theta \in S. \tag{52}$$

From Lemma 11, we see $[\tilde{w}_t(\theta_1), \tilde{w}_t(\theta_2)] = Q(\theta_1, \theta_2)t$ for every $\theta_1, \theta_2 \in S$. Thus $\bar{P}\left\{\tilde{w}_t(\theta) = W^Q(t,\theta),\ \theta \in S\right\} = 1$. In particular, κ_t satisfies (16) with $\tilde{w}_t$ defined above in (52) under the probability measure $\bar{P}$, where $(\kappa, \tilde{w}) \in \bar{\mathcal{X}}$ is the coordinate function. Hence, the pathwise uniqueness for the SPDE (16) proves the conclusion of Proposition 6 by the argument of Yamada–Watanabe's type. □

4.2 *Proof of Theorem 2*

We first prove the following proposition.

Proposition 7 *For every* $0 \le s \le t \le T$, $\bar{\Psi} \in \bar{\mathcal{D}}$ *and* $\mathcal{F}_s$*-measurable* $\Phi \in C_b(\bar{\mathcal{X}})$,

$$\lim_{\epsilon \downarrow 0} E\Big[\Big(\bar{\Psi}(\kappa^\epsilon(t), Z_\epsilon^Q(t)) - \bar{\Psi}(\kappa^\epsilon(s), Z_\epsilon^Q(s)) - \int_s^t \bar{\mathcal{L}}\bar{\Psi}(\kappa^\epsilon(r), Z_\epsilon^Q(r))dr\Big)\Phi\Big] = 0. \tag{53}$$

Proof A simple computation shows

$$\frac{d}{dt}\bar{\Psi}(\kappa^\epsilon(t), Z_\epsilon^Q(t)) = ((\mathcal{L} - \mathcal{L}_0)\Psi_1)\varphi + \sum_{i=0}^\infty \lambda_i \langle D\Psi_1, h\alpha_i\rangle \varphi \xi_i^\epsilon(t) + \Psi_1 \sum_{i=0}^\infty \lambda_i \langle D\varphi, \alpha_i\rangle \xi_i^\epsilon(t),$$

where $\mathcal{L}_0$ denotes the operator $\mathcal{L}$ with $a = b = 0$. Then the proof is completed if one can prove

$$\lim_{\epsilon \downarrow 0} E\Big[\Big(\int_s^t \Big(\sum_{i=0}^\infty \lambda_i \langle D\Psi_1, h\alpha_i\rangle \xi_i^\epsilon(r) + \mathcal{L}_0\Psi_1\Big)\varphi + \frac{1}{2}\langle D\Psi_1 \otimes D\varphi, h(\kappa)Q(\cdot,\cdot)\rangle dr\Big)\Phi\Big] = 0, \tag{54}$$

and

$$\lim_{\epsilon\downarrow 0} E\Big[\int_s^t \Psi_1 \sum_{i=0}^{\infty} \lambda_i \langle D\varphi, \alpha_i\rangle \xi_i^\epsilon(r) + \Big(\frac{1}{2}\langle D\Psi_1 \otimes D\varphi, h(\kappa)Q(\cdot,\cdot)\rangle\Big)$$

$$+ \frac{1}{2}\Psi_1\langle D^2\varphi, Q\rangle dr \Phi\Big] = 0. \tag{55}$$

Let us prove (54) first. Noting that for all $i \geq 1$,

$$\frac{d}{dt}\langle D\Psi_1, \alpha_i h\rangle = \langle D^2\Psi_1, \alpha_i h \otimes (a\kappa'' + b)\rangle + \langle D\Psi_1, \alpha_i h'(a\kappa'' + b)\rangle$$

$$+ \sum_{j=0}^{\infty} \lambda_j \langle D\Psi_1,\ \alpha_i\alpha_j h'h\rangle \xi_j^\epsilon(t) + \sum_{j=0}^{\infty} \lambda_j \langle D^2\Psi_1, \alpha_i h \otimes \alpha_j h\rangle \xi_j^\epsilon(t), \tag{56}$$

we get

$$\frac{d}{dt}[\langle D\Psi_1, \alpha_i h\rangle\varphi] = \tilde{\Psi}_i^{(1)} + \sum_{k=-1}^{0}\sum_{j=0}^{\infty} \lambda_j \tilde{\Psi}_{i,j,k}^{(2)} \xi_j^\epsilon(t), \tag{57}$$

where

$$\begin{aligned}
\tilde{\Psi}_i^{(1)} &= \big[\langle D^2\Psi_1, \alpha_i h \otimes (a\kappa'' + b)\rangle + \langle D\Psi_1, \alpha_i h'(a\kappa'' + b)\rangle\big]\varphi,\\
\tilde{\Psi}_{i,j,-1}^{(2)} &= \big[\langle D\Psi_1, \alpha_i\alpha_j h'h\rangle + \langle D^2\Psi_1, \alpha_i h \otimes \alpha_j h\rangle\big]\varphi,\\
\tilde{\Psi}_{i,j,0}^{(2)} &= \langle D\Psi_1, \alpha_i h\rangle\langle D\varphi, \alpha_j\rangle.
\end{aligned}$$

Since

$$\begin{aligned}
&\int_s^t \big(\langle D\Psi_1, h\alpha_i\rangle\varphi\big)(r)\xi_i^\epsilon(r)dr\\
&= \big(\langle D\Psi_1, \alpha_i h\rangle\varphi\big)(s)\int_s^t \xi_i^\epsilon(r)dr + \int_s^t \xi_i^\epsilon(r)dr \int_s^r \tilde{\Psi}_i^{(1)}(r')dr'\\
&\quad + \sum_{k=-1}^{0}\sum_{j=0}^{\infty}\lambda_j \int_s^t \xi_i^\epsilon(r)dr \int_s^r \tilde{\Psi}_{i,j,k}^{(2)}(r')\xi_j^\epsilon(r')dr'\\
&\equiv I_i^{(0)} + I_i^{(1)} + \sum_{k=-1}^{1}\sum_{j=0}^{\infty}\lambda_j I_{i,j,k}^{(2)},
\end{aligned}$$

using (14) and the dominated convergence theorem, (54) can be shown if we prove

$$\lim_{\epsilon \downarrow 0} E\left[\sum_{i=1}^{\infty} \lambda_i I_i^{(0)} \Phi\right] = 0, \tag{58}$$

$$\lim_{\epsilon \downarrow 0} E\left[\sum_{i=1}^{\infty} \lambda_i I_i^{(1)} \Phi\right] = 0, \tag{59}$$

$$\lim_{\epsilon \downarrow 0} E\Big[\Big(\sum_{i=1}^{\infty}\sum_{j=1}^{\infty} \lambda_i \lambda_j I_{i,j,-1}^{(2)} - \mathcal{L}_0 \Psi_1 \varphi\Big)\Phi\Big] = 0, \tag{60}$$

$$\lim_{\epsilon \downarrow 0} E\Big[\Big(\sum_{i=1}^{\infty}\sum_{j=1}^{\infty} \lambda_i \lambda_j I_{i,j,0}^{(2)} - \frac{1}{2}\langle D\Psi_1 \otimes D\varphi, Q(\cdot,\cdot)h(\kappa)\rangle\Big)\Phi\Big] = 0. \tag{61}$$

Recall that

$$E[\xi_i^\epsilon(t)] = 0 \quad and \quad |\xi_i^\epsilon(t)| \leq \bar{C}\epsilon^{-\gamma}, \tag{62}$$

for every i.

Proof of (58) Note that Φ, φ and $D\Psi_1$ are bounded and $\mathcal{F}_{0,s}^\epsilon$-measurable. From (14) and (62), it follows that the expectation of (58) is equal to

$$\sum_{i=1}^{\infty} E\left[\lambda_i I_i^{(0)} \Phi\right], \tag{63}$$

for each $\epsilon \in (0, 1)$, which is also equal to

$$\sum_{i=1}^{\infty} E\left[\lambda_i \left(\langle D\Psi_1, \alpha_i h\rangle \varphi\right)(s) E\Big[\int_s^t \xi_i^\epsilon(r)dr \Big| \mathcal{F}_{0,s}^\epsilon\Big]\Phi\right]. \tag{64}$$

From (62) and (40), we get

$$\left|E\left[\int_s^t \xi_i^\epsilon(r)dr \middle| \mathcal{F}_{0,s}^\epsilon\right]\right| \leq \bar{C}\Big(\int_0^\infty \rho(r)dr\Big)\epsilon^\gamma. \tag{65}$$

Thus, (64), (65) and using (14) again, we have

$$\left|\sum_{i=1}^{\infty} E[\lambda_i I_i^{(0)} \Phi]\right| \leq C\epsilon^\gamma, \tag{66}$$

for some constant $C > 0$. □

Proof of (59) From (14) and (62), the expectation of (59) is equal to

$$\sum_{i=1}^{\infty} E\left[\lambda_i I_i^{(1)} \Phi\right]. \tag{67}$$

Then we have

$$\begin{aligned}\sum_{i=1}^{\infty} \left|E\left[\lambda_i I_i^1 \Phi\right]\right| &\leq \sum_{i=1}^{\infty} \lambda_i \left|E\left[\Phi \int_s^t \tilde{\Psi}_i^{(1)}(r')dr' \int_{r'}^t E\left[\xi_i^\epsilon(r)\middle|\mathcal{F}_{0,r'}^\epsilon\right]dr\right]\right| \\ &\leq C\sum_{i=1}^{\infty} \lambda_i \|\Phi\|_{L^\infty(\Omega)} \int_s^t E\left[|\tilde{\Psi}_i^{(1)}(r')|\right]dr'\,\bar{C}\epsilon^\gamma, \end{aligned}\tag{68}$$

where the second inequality comes from (65) with r' instead of s. From boundedness of $D^2\Psi_1$, $D\Psi_1$, a, b, h, h' and φ, we have

$$|\tilde{\Psi}_i^{(1)}(r')| \leq C\|\alpha_i\|_{L^\infty(S)} \left(\left\|\kappa^{(2)}(r')\right\|_{L^1(S)} + 1\right),$$

for some constant $C > 0$. Using Proposition 3 and (14), it follows that (67) is bounded above by $C\epsilon^\gamma$ for some $C > 0$. □

Proof of (60) *and* (61) The expectations in (60) and (61) are equal to

$$E\Big[\sum_{i=1}^{\infty}\sum_{j=1}^{\infty} \lambda_i\lambda_j \int_s^t \tilde{\Psi}_{i,j,k}^{(2)}(r')\left(\xi_j^\epsilon(r')\int_{r'}^t \xi_i^\epsilon(r)dr - \frac{1}{2}\delta_{i,j}\right)dr'\Phi\Big], \tag{69}$$

with $k = -1$ and 0, respectively, for each $\epsilon \in (0, 1)$. These are further equal to

$$\sum_{i=1}^{\infty}\sum_{j=1}^{\infty} E\Big[\lambda_i\lambda_j \int_s^t \tilde{\Psi}_{i,j,k}^{(2)}(r')\left(\xi_j^\epsilon(r')\int_{r'}^t \xi_i^\epsilon(r)dr - \frac{1}{2}\delta_{ij}\right)dr'\Phi\Big], \quad k = -1, 0, \tag{70}$$

by interchanging the sum in i, j and the expectation by noting

$$\sup_{0<\epsilon<1} \|\tilde{\Psi}_{i,j,k}^{(2,\epsilon)}(r')\|_{L^\infty(\Omega)} \leq C\|\alpha_i\|_{L^\infty}\|\alpha_j\|_{L^\infty} \tag{71}$$

and using (62) and (14). Thus our goal is to show that (70) tends to 0 as $\epsilon \downarrow 0$. However, since if $r' < t$, we have

$$\lim_{\epsilon\downarrow 0} \int_{r'}^t E[\xi_j^\epsilon(r')\xi_i^\epsilon(r)]dr = \frac{1}{2}\delta_{ij}, \tag{72}$$

and

$$\sup_{0<\epsilon<1}\left|\int_{r'}^{t} E[\xi_j^\epsilon(r')\xi_i^\epsilon(r)]dr\right| < \infty, \tag{73}$$

by noting (14), (19) and (40), it suffices to show that

$$\sum_{i=1}^{\infty}\sum_{j=1}^{\infty} E\left[\lambda_i\lambda_j \int_s^t \tilde{\Psi}_{i,j,k}^{(2)}(r')\int_{r'}^{t}\left(\xi_j^\epsilon(r')\xi_i^\epsilon(r) - E[\xi_j^\epsilon(r')\xi_i^\epsilon(r)]\right)drdr'\Phi\right], \quad k=-1,0, \tag{74}$$

tends to 0 as $\epsilon \downarrow 0$ instead of (70). To show this, we prepare a lemma.

Lemma 12 *Assume that* $\tilde{\Psi}_{i,j,k}^{(2)}(r') \equiv \tilde{\Psi}_{i,j,k}^{(2,\epsilon)}(r')$, $i, j = 1, 2, \ldots, k = -1, 0$ *have the form*

$$\tilde{\Psi}_{i,j,k}^{(2,\epsilon)}(r') = \tilde{\Psi}_{i,j,k}^{(2,\epsilon)}(s) + \int_s^{r'} \partial\tilde{\Psi}_{i,j,k}^{(2,\epsilon)}(r'')dr'', \quad r' \in [s,t], \tag{75}$$

with $\mathcal{F}_{0,s}^\epsilon$*-measurable* $\tilde{\Psi}_{i,j,k}^{(2,\epsilon)}(s)$ *and* $\mathcal{F}_{0,r''}^\epsilon$*-measurable* $\partial\tilde{\Psi}_{i,j,k}^{(2,\epsilon)}(r'')$*. Furthermore, assume that* $\tilde{\Psi}_{i,j,k}^{(2,\epsilon)}$ *satisfies* (71) *and*

$$\sup_{0<\epsilon<1,\, r''\in[s,t]} \epsilon^\gamma \|\partial\tilde{\Psi}_{i,j,k}^{(2,\epsilon)}(r'')\|_{L^q(\Omega)} \le C\|\alpha_i\|_{L^\infty}\|\alpha_j\|_{L^\infty} \tag{76}$$

for every $q \ge 1$ *and some constant* $C > 0$*. Then* (74) *converges to* 0 *as* $\epsilon \downarrow 0$*.*

Proof Let us denote by $I_{i,j}^\epsilon$ the expectation in (74). Then, $I_{i,j}^\epsilon$ can be decomposed into the sum

$$I_{i,j}^\epsilon = I_{1,i,j}^\epsilon(s) + \int_s^t I_{2,i,j}^\epsilon(r'')dr'',$$

where

$$I_{1,i,j}^\epsilon(s) = E\left[\Phi\tilde{\Psi}_{i,j,k}^{(2,\epsilon)}(s)\lambda_i\lambda_j\int_s^t\int_{r'}^t \left(\xi_j^\epsilon(r')\xi_i^\epsilon(r) - E[\xi_j^\epsilon(r')\xi_i^\epsilon(r)]\right)drdr'\right], \tag{77}$$

$$I_{2,i,j}^\epsilon(r'') = E\left[\Phi\partial\tilde{\Psi}_{i,j,k}^{(2,\epsilon)}(r'')\lambda_i\lambda_j\int_{r''}^t\int_{r'}^t \left(\xi_j^\epsilon(r')\xi_i^\epsilon(r) - E[\xi_j^\epsilon(r')\xi_i^\epsilon(r)]\right)drdr'\right]. \tag{78}$$

Then, using (42), we have

$$|I_{1,i,j}^\epsilon(r'')| \le C\lambda_i\lambda_j\|\Phi\|_{L^\infty(\Omega)}$$
$$\times\, \epsilon^{-2\gamma}\int_{r''}^t\int_{r'}^t \|\tilde{\Psi}_{i,j,k}^{(2)}(r')\|_{L^q(\Omega)}\left[\rho\left(\frac{r-r'}{\epsilon^{2\gamma}}\right)\rho\left(\frac{r'-r''}{\epsilon^{2\gamma}}\right)\right]^{\frac{q-1}{2q}} drdr', \tag{79}$$

for every $q \geq 1$. Since $\rho(t)$ is decreasing in t and satisfies the condition (19), we get

$$\lim_{\epsilon\downarrow 0} \epsilon^{-3\gamma} \sup_{r''\in[s,t]} \int_{r''}^{t}\int_{r'}^{t} \left[\rho\left(\frac{r-r'}{\epsilon^{2\gamma}}\right)\rho\left(\frac{r'-r''}{\epsilon^{2\gamma}}\right)\right]^{\frac{q-1}{2q}} drdr' = 0, \tag{80}$$

for some $q \geq 1$ (choose q such that $\frac{3}{4} < \frac{p(q-1)}{2q} < 1$). From (14), (79), (80) and (71), for arbitrary small $\delta > 0$, we can choose a suitable number $N = N(\delta) \geq 1$ independent of $\epsilon \in (0, 1)$ such that $\sum_{i,j\geq N} |I^{\epsilon}_{i,j}| < \delta$ holds. On the other hand, using (42), we have

$$|I^{\epsilon}_{2,i,j}(r'')| \leq C\lambda_i\lambda_j \|\Phi\|_{L^\infty(\Omega)} \|\partial\tilde{\Psi}^{(2,\epsilon)}_{i,j,k}(r'')\|_{L^q(\Omega)}$$
$$\times \epsilon^{-2\gamma} \int_{r''}^{t}\int_{r'}^{t} \left[\rho\left(\frac{r-r'}{\epsilon^{2\gamma}}\right)\rho\left(\frac{r'-r''}{\epsilon^{2\gamma}}\right)\right]^{\frac{q-1}{2q}} drdr', \tag{81}$$

for every $q \geq 1$. From (14), (76), (81) and (80), we get

$$\lim_{\epsilon\downarrow 0} \sum_{i=1}^{\infty}\sum_{j=1}^{\infty} \int_{s}^{t} I^{\epsilon}_{2,i,j}(r'')dr'' = 0, \tag{82}$$

using (14). Thus the proof of Lemma 12 is completed. □

We now return to the proof of (60) and (61). It is reduced to showing that (74) tends to 0 as $\epsilon \downarrow 0$. For this, it only remains to check that $\tilde{\Psi}^{(2)}_{i,j,k}$ satisfies the assumptions of Lemma 12. Since $|\tilde{\Psi}^{(2)}_{i,j,k}| \leq C\|\alpha_i\|_{L^\infty}\|\alpha_j\|_{L^\infty}$, we only need to show

$$||\frac{d}{dt}\tilde{\Psi}^{(2)}_{i,j,k}||_{L^q(\Omega)} \leq C\lambda_i\lambda_j\epsilon^{-\gamma}||\alpha_i||_{L^\infty}||\alpha_j||_{L^\infty}, \quad t \in [0, T]. \tag{83}$$

However, from (14), (56), Proposition 3 and (62), and boundedness of $D\Psi_1$, $D\varphi$, Ψ_1, φ, a, b, h and h', we get:

$$\begin{aligned}
&\|\frac{d\langle D\Psi_1, \alpha_i h\rangle}{dt}\langle D\varphi, \alpha_j\rangle\|_{L^q(\Omega)} \\
\leq& C\left\|\alpha_i^{(0)}\right\|_{L^\infty}\left\|\alpha_j^{(0)}\right\|_{L^\infty}\left(1 + ||\kappa''||_{L^q(\Omega)}\sum_{l=1}^{\infty}\lambda_l\left\|\alpha_l^{(0)}\right\|_{L^\infty}\bar{C}\epsilon^{-\gamma}\right) \\
\leq& \tilde{C}\epsilon^{-\gamma},
\end{aligned}$$

for some constant $C, \tilde{C} > 0$. By a similar argument, the estimate of the other terms of $\frac{d}{dt}\tilde{\Psi}^{(2)}_{i,j,k}$ can be easily obtained, thus the assumptions of Lemma 12 hold. This completes the proof of (60) and (61). □

The proof of (55) is similar, so that we omit the details. We now give the proof of the main theorem, Theorem 2.

Proof The tightness of the family of joint distributions $\bar{P}^\epsilon$, $0 < \epsilon < 1$ of $(\kappa^\epsilon, Z^Q_\epsilon)$ on the space $\bar{\mathcal{X}}$ follows from Propositions 1 and 2 as we mentioned in Sect. 2. From Proposition 7, every limit $\bar{P}$ of $\bar{P}^\epsilon$ as $\epsilon \downarrow 0$ solves the $(\bar{\mathcal{L}}, \bar{\mathcal{X}})$-martingale problem. However, $\bar{P}$ is uniquely determined from Proposition 6, therefore the proof of Theorem 2 is completed. □

Acknowledgements We thank the referee who suggested the proof of Theorem 1, in particular, the SPDE (26) and the reference [5]. We acknowledge the support from the training course at ENS Cachan, under which C. Denis could visit Tokyo and stayed for four months in 2016.

References

1. Alfaro, M., Antonopoulou, D., Karali, G., Matano, H.: Generation and propagation of fine transition layers for the stochastic Allen-Cahn equation (preprint, 2016)
2. Allen, S.M., Cahn, J.W.: A macroscopic theory for antiphase boundary motion and its application to antiphase domain coarsening. Acta. Metall. **27**, 1085–1095 (1979)
3. Bellettini, G.: Lecture Notes on Mean Curvature Flow, Barriers and Singular Perturbations. Lecture Notes, vol. 12. Scuola Normale Superiore di Pisa (2013)
4. Brakke, K.A.: The Motion of a Surface by its Mean Curvature. Mathematical Notes, vol. 20. Princeton University Press, Princeton (1978)
5. Debussche, A., de Moor, S., Hofmanova, M.: A regularity result for quasilinear stochastic partial differential equations of parabolic type. SIAM J. Math. Anal. **47**, 1590–1614 (2015)
6. Dirr, N., Luckhaus, S., Novaga, M.: A stochastic selection principle in case of fattening for curvature flow. Calc. Var. Partial Differ. Equ. **13**, 405–425 (2001)
7. Es-Sarhir, A., von Renesse, M.: Ergodicity of stochastic curve shortening flow in the plane. SIAM J. Math. Anal. **44**, 224–244 (2012)
8. Funaki, T.: A stochastic partial differential equation with values in a manifold. J. Func. Anal. **109**, 257–288 (1992)
9. Funaki, T.: The scaling limit for a stochastic PDE and the separation of phases. Probab. Theory Relat. Fields **102**, 221–288 (1995)
10. Funaki, T.: Singular limit for stochastic reaction-diffusion equation and generation of random interfaces. Acta Math. Sin. (Engl. Ser.) **15**, 407–438 (1999)
11. Funaki, T.: Lectures on Random Interfaces. SpringerBriefs in Probability and Mathematical Statistics. Springer, Berlin (2016)
12. Funaki, T., Yokoyama, S.: Sharp interface limit for stochastically perturbed mass conserving Allen-Cahn equation. arXiv:1610.01263
13. Funaki, T., Nakada, S., Yokoyama, S.: A stochastically perturbed volume preserving mean curvature flow (preprint, 2017)
14. Giga, Y.: Surface Evolution Equations. A Level Set Approach. Monographs in Mathematics, vol. 99. Birkhäuser, Basel (2006)
15. Giga, Y., Mizoguchi, N.: Existence of periodic solutions for equations of evolving curves. SIAM J. Math. Anal. **27**, 5–39 (1996)
16. Gurtin, M.E.: Thermomechanics of Evolving Phase Boundaries in the Plane. Clarendon Press, Oxford (1993)
17. Hofmanová, M., Röger, M., von Renesse, M.: Weak solutions for a stochastic mean curvature flow of two-dimensional graphs. Probab. Theory Relat. Fields **168**, 373–408 (2017)

18. Kawasaki, K., Ohta, T.: Kinetic drumhead model of interface I. Prog. Theoret. Phys. **67**, 147–163 (1982)
19. Kunita, H.: Stochastic Flows and Stochastic Differential Equations. Cambridge University Press, Cambridge (1990)
20. Lee, K.: Generation and motion of interfaces in one-dimensional stochastic Allen-Cahn equation. J. Theor. Probab., published online (2016)
21. Lee, K.: Generation of interfaces for multi-dimensional stochastic Allen-Cahn equation with a noise smooth in space, arXiv:1604.06535
22. Lions, P.L., Souganidis, P.E.: Fully nonlinear stochastic partial differential equations. C. R. Acad. Sci. Paris Ser. I Math. **326**, 1085–1092 (1998)
23. Lions, P.L., Souganidis, P.E.: Fully nonlinear stochastic partial differential equations: nonsmooth equations and applications. C. R. Acad. Sci. Paris Ser. I Math. **327**, 735–741 (1998)
24. Otto, F., Weber, H.: Quasilinear SPDEs via rough paths. arXiv:1605.09744
25. Röger, M., Weber, H.: Tightness for a stochastic Allen-Cahn equation. Stoch. Partial Differ. Equ. Anal. Comput. **1**, 175–203 (2013)
26. Weber, H.: On the short time asymptotic of the stochastic Allen-Cahn equation. Ann. Inst. H. Poincaré Probab. Statist. **46**, 965–975 (2010)

Poisson Stochastic Process and Basic Schauder and Sobolev Estimates in the Theory of Parabolic Equations (Short Version)

N. V. Krylov and E. Priola

Abstract We show among other things how knowing Schauder or Sobolev-space estimates for the one-dimensional heat equation allows one to derive their multidimensional analogs for equations with coefficients depending only on time variable with the *same* constants as in the case of the one-dimensional heat equation. The method is quite general and is based on using the Poisson stochastic process. It also applies to equations involving non-local operators. It looks like no other method is available at this time and it is a very challenging problem to find a purely analytic approach to proving such results. We only give examples of applications of our results. Their proofs will appear elsewhere.

Keywords Schauder estimates · Sobolev-space estimates · Multidimensional parabolic equations · Poisson process

2010 Mathematics Subject Classification 35K10 · 35K15

1 General Setting. Main Results

Let W be a set consisting of real-valued (Borel) measurable functions $u = u_t = u_t(x)$ on $[0, T] \times \mathbb{R}^d$, where $T \in (0, \infty)$ is fixed.

Let $\mathscr{G}$ be a commutative group of affine volume-preserving transformations of $\mathbb{R}^d$. If $g, h \in \mathscr{G}$ by gh we mean the composition of the two transformations.

As usual, if $f(x)$ is a function on $\mathbb{R}^d$ and $g \in \mathscr{G}$, we define $(gf)(x) = f(gx)$, where gx is the image of x under mapping g.

N. V. Krylov (✉)
127 Vincent Hall, University of Minnesota, Minneapolis, MN 55455, USA
e-mail: nkrylov@umn.edu

E. Priola
Dipartimento di Matematica, Universita di Torino,
Via Carlo Alberto 10, 10123 Torino, Italy
e-mail: enrico.priola@unito.it

A. Eberle et al. (eds.), *Stochastic Partial Differential Equations and Related Fields*, Springer Proceedings in Mathematics & Statistics 229,
https://doi.org/10.1007/978-3-319-74929-7_10

By $B((0,T),\mathcal{G})$ we denote the set of bounded measurable $\mathcal{G}$-valued functions on $(0,T)$, $B(\mathbb{R}^d)$ is the set of Borel bounded functions on $\mathbb{R}^d$, $\mathcal{B}([0,T]\times\mathbb{R}^d)$ is the Borel σ-field in $[0,T]\times\mathbb{R}^d$.

Fix a constant $K\in[0,\infty)$.

Assumption 1.1 (i) For any $u\in W$ we have

$$\sup_{\substack{t\in[0,T],\\ x\in\mathbb{R}^d}} |u_t(x)|\le K.$$

(ii) (Convexity of W.) If $(\Omega,\mathcal{F},P)$ is a probability space and $u(\omega)=u_t(\omega,x)$ is an $\mathcal{F}\times\mathcal{B}([0,T]\times\mathbb{R}^d)$-measurable function such that $u(\omega)\in W$ for any ω, then the function $E[u_t(x)]$ belongs to W.

(iii) ("Shift" invariance of W.) For $u\in W$ and any bounded measurable $\mathcal{G}$-valued function g_t given on $[0,T]$, the function $u_t(g_t x)$ is in W.

Example 1 Fix a constant $K_0\in(0,\infty)$ and let W be the set of Borel functions on $[0,T]\times\mathbb{R}$ satisfying, for each $t\in[0,T]$,

$$0\le u_t(x)\le 1,\quad \int_{\mathbb{R}} u_t^2(x)\,dx\le K_0. \tag{1}$$

Then Assumption 1.1 is satisfied if $\mathcal{G}$ is the group of translations of $\mathbb{R}$.

Example 2 Denote points x in $\mathbb{R}^d$, $d\ge 2$, as (x^1,x'), where $x'=(x^2,\dots,x^d)$. Take zero as the initial data and consider the problem of solving

$$\partial_t u_t(x)=D_{11}u_t(x)+f_t(x),\quad t\in(0,T),x\in\mathbb{R}^d, \tag{2}$$

$$f\in B_c((0,T),C_0^\infty(\mathbb{R}^d))\quad\text{is fixed},$$

where for a real-valued function $f(t,x)$, $t\in(0,T)$, $x\in\mathbb{R}^d$, write

$$f\in B_c((0,T),C_0^\infty(\mathbb{R}^d))$$

if f is a Borel bounded function, such that $f(t,\cdot)\in C_0^\infty(\mathbb{R}^d)$ for any t, for any $n=0,1,\dots$, the $C^n(\mathbb{R}^d)$-norms of $f(t,\cdot)$ are bounded on $(0,T)$, and the supports of $f(t,\cdot)$ belong to the same ball.

Fix $p>1$. One knows (see, for instance, [4–6]) that there exists a unique solution u such that, for any x',

$$\sup_{\substack{t\in[0,T],\\ x^1\in\mathbb{R}}}|u_t(x)|\le T\sup_{\substack{t\in(0,T),\\ x^1\in\mathbb{R}}}|f_t(x)|,\quad \int_0^T\int_{\mathbb{R}}|u_t(x)|^p\,dtdx^1\le N_p(T)\int_0^T\int_{\mathbb{R}}|f_t(x)|^p\,dtdx^1,$$

$$\int_0^T\int_{\mathbb{R}}|D_{11}u_t(x)|^p\,dtdx^1\le N_p\int_0^T\int_{\mathbb{R}}|f_t(x)|^p\,dtdx^1.$$

Hence

$$\sup_{\substack{t\in[0,T],\\ x\in\mathbb{R}^d}} |u_t(x)| \le A, \quad \int_0^T \int_{\mathbb{R}^d} |u_t(x)|^p \, dtdx \le B, \quad \int_0^T \int_{\mathbb{R}^d} |D_{11}u_t(x)|^p \, dtdx \le C, \tag{3}$$

where

$$A = T \sup_{\substack{t\in(0,T),\\ x\in\mathbb{R}^d}} |f_t(x)|, \quad B = N_p(T) \int_0^T \int_{\mathbb{R}^d} |f_t(x)|^p \, dtdx, \quad C = N_p \int_0^T \int_{\mathbb{R}^d} |f_t(x)|^p \, dtdx.$$

Take W as the set of functions on $[0, T] \times \mathbb{R}^d$, satisfying (3). Also let $\mathcal{G}$ be the group of all shifts of $\mathbb{R}^d$. Then Assumption 1.1 is satisfied.

Next, let $L := \{L_t, t \in (0, T)\}$, be a family of linear operators $L_t : C_0^\infty(\mathbb{R}^d) \to B(\mathbb{R}^d)$ and take and fix

$$f \in B_c((0, T), C_0^\infty(\mathbb{R}^d)), \quad u_0 \in B(\mathbb{R}^d). \tag{4}$$

Assumption 1.2 The couple (L, f) is W-regular in the following sense.

(i) ($\mathcal{G}$ and L commute.) For any $t \in (0, T)$ and $g \in \mathcal{G}$, we have $gL_t = L_t g$.

(ii) For any $\zeta \in C_0^\infty(\mathbb{R}^d)$, $L_t\zeta(x) := (L_t\zeta)(x)$ is measurable with respect to (t, x) and

$$\int_{[0,T]\times\mathbb{R}^d} |L_t\zeta(x)| dtdx < \infty.$$

(iii) There is a mapping $B((0, T), \mathcal{G}) \to W$ sending $h \in B((0, T), \mathcal{G})$ into $u[h] \in W$ such that $u = u[h]$ has initial condition u_0 and satisfies

$$\partial_t u_t(x) = L_t^* u_t(x) + (h_t f_t)(x), \quad t \in [0, T], \ x \in \mathbb{R}^d \tag{5}$$

(iv) For any $h', h'' \in B((0, T), \mathcal{G})$ and $(t, x) \in [0, T] \times \mathbb{R}^d$, we have

$$|u_t[h'](x) - u_t[h''](x)| \le K \int_0^t \sup_{y\in\mathbb{R}^d} |f_r(h'_r y) - f_r(h''_r y)| \, dr. \tag{6}$$

Incidentally, Assumption 1.2(iv) implies that we are dealing with evolution equations in the sense that if $S \in (0, T)$, $h', h'' \in B((0, T), \mathcal{G})$ and $h'_t = h''_t$ for all $t \le S$, then $u_t[h'](x) = u_t[h''](x)$ for all $t \le S$ and $x \in \mathbb{R}^d$. It also provides some kind of continuous dependence of solutions with respect to the free term.

We say that $u \in W$ satisfies (5) with initial condition u_0 if, for any $\zeta \in C_0^\infty(\mathbb{R}^d)$ and $t \in [0, T]$,

$$(u_t, \zeta) := \int_{\mathbb{R}^d} u_t(x)\zeta(x)dx = (u_0, \zeta) + \int_0^t (u_s, L_s\zeta) \, ds + \int_0^t (h_s f_s, \zeta) \, ds. \tag{7}$$

Obviously, in Example 1 the couple $(0, 0)$ is W-regular if u_0 satisfies (1) for $t = 0$, and, as is easy to see, in Example 2 the couple (D_{11}, f) is also W-regular.

Theorem 1 *For any $g^{(1)}, \dots, g^{(n)} \in B((0, T), \mathcal{G})$ and $\lambda_1, \dots, \lambda_n \geq 0$, the couple, consisting of the family of operators $\hat{L}_t$, such that*

$$\hat{L}_t^* = L_t^* + \sum_{i=1}^{n} \lambda_i (g_t^{(i)} - 1), \tag{8}$$

where 1 stands for the operation of multiplying by one, and f, is W-regular.

Example 3 In Example 1 take $h > 0$ and let $g_t^{(1)}$ be the shift operator $g_t^{(1)} : x \to x + h$ and $g_t^{(2)}$ be the shift operator $g_t^{(1)} : x \to x - h$. Take $u_0(x)$ satisfying (1) for $t = 0$. Then Theorem 1 says that there is a solution $u_t(x)$ of class W of the equation

$$\partial_t u_t(x) = \frac{1}{h^2} \big[u_t(x + h) - 2u_t(x) + u_t(x - h \big]$$

with initial value u_0. In particular, (1) holds for this solution. Furthermore, passing to the limit as $h \to 0$ in the above equation (always understood as (7)) presents no difficulty and we obtain that there is a solution $u_t(x)$ of class W of the heat equation

$$\partial_t u_t(x) = D^2 u_t(x)$$

with initial value u_0. In particular, by adapting K_0, we see that for this solution and any $t \in [0, T]$

$$\int_{\mathbb{R}} u_t^2(x)\, dx \leq \int_{\mathbb{R}} u_0^2(x)\, dx.$$

Of course, this is a trivial fact from the point of view of the PDE theory. Yet what is remarkable is that it is obtained by quite elementary means (see Sect. 2).

Example 4 In Example 2 take $h > 0$, $\mathbb{R}^d$-valued bounded measurable functions $l^{(i)}(t), i = \pm 1, \dots, \pm n$ and let g_t^i be shift operators $x \to x + h l^{(i)}(t)$. Take $\lambda_i = h^{-2}$. Assume

$$l^{(-i)}(t) = -l^{(i)}(t).$$

Then Theorem 1 says that there is a solution $u_t(x)$ of class W of the equation

$$\partial_t u_t(x) = D_{11} u_t(x) + \sum_{i=1}^{n} \frac{1}{h^2} \big[u_t(x + h l_t^{(i)}) - 2u_t(x) + u_t(x - h l_t^{(i)}) \big] + f_t(x)$$

with zero initial value. In particular,

$$\sup_{\substack{t \in [0,T], \\ x \in \mathbb{R}^d}} |u_t(x)| \leq A, \quad \int_0^T \int_{\mathbb{R}^d} |u_t(x)|^p\, dt dx \leq B, \quad \int_0^T \int_{\mathbb{R}^d} |D_{11} u_t(x)|^p\, dt dx \leq C.$$

Assumption 1.3 For any sequence $u^k \in W$ and a bounded function $u = u_t(x)$, $(t, x) \in [0, T] \times \mathbb{R}^d$, such that

$$\int_{\mathbb{R}^d} u_t^k(x)\zeta(x)\,dx \to \int_{\mathbb{R}^d} u_t(x)\zeta(x)\,dx$$

for any $t \in [0, T]$ and $\zeta \in C_0^\infty(\mathbb{R}^d)$, there exists $w \in W$ such that $w_t = u_t$ (a.e.) on $\mathbb{R}^d$ for any $t \in [0, T]$.

The main implication of Assumption 1.3 is the following.

Theorem 2 *Suppose that Assumptions 1.1(i) and 1.3 are satisfied. Let* $\{L_t^k, t \in (0, T)\}$, $k = 0, 1, \dots,$ *be a sequence of families of linear operators mapping* $C_0^\infty(\mathbb{R}^d)$ *into* $B(\mathbb{R}^d)$ *subject to the following conditions:*

(a) For each k, Assumption 1.2(ii) is satisfied with L_t^k *in place of* L_t*;*

(b) For any $\zeta \in C_0^\infty(\mathbb{R}^d)$*, we have*

$$\lim_{k\to\infty} \int_{(0,T)\times\mathbb{R}^d} |(L_t^k - L_t^0)\zeta(x)|\,dtdx = 0;$$

(c) For each $k = 1, 2, \dots,$ *there exists* $u^k \in W$ *such that for any* $\zeta \in C_0^\infty(\mathbb{R}^d)$ *and* $t \in [0, T]$,

$$(u_t^k, \zeta) = (u_0, \zeta) + \int_0^t (u_s^k, L_s^k\zeta)\,dsk^2 + \int_0^t (f_s, \zeta)\,ds. \tag{9}$$

Then there exists $u^0 \in W$ *for which* (9) *holds with 0 in place of k for any* $\zeta \in C_0^\infty(\mathbb{R}^d)$ *and* $t \in [0, T]$.

Example 5 In Example 4 Theorem 2 says that there exists a solution of class W of the equation

$$\partial_t u_t(x) = D_{11}u_t(x) + a_t^{ij} D_{ij}u_t(x) + f_t(x)$$

with zero initial value, where

$$a_t^{ij} = \sum_{k=1}^n l_t^{(k)i} l_t^{(k)j}.$$

In particular,

$$\sup_{\substack{t\in[0,T],\\ x\in\mathbb{R}^d}} |u_t(x)| \le A, \quad \int_0^T \int_{\mathbb{R}^d} |u_t(x)|^p\,dtdx \le B, \quad \int_0^T \int_{\mathbb{R}^d} |D_{11}u_t(x)|^p\,dtdx \le C.$$

In particular, there exists a solution of class W of the equation

$$\partial_t u_t(x) = D_{11}u_t(x) + \cdots + D_{dd}u_t(x) + f_t(x) = \Delta u_t(x) + f_t(x)$$

with zero initial value, satisfying the same estimates. Since the Laplacian is rotation invariant, for any unit $l \in \mathbb{R}^d$

$$\int_0^T \int_{\mathbb{R}^d} |D_l^2 u_t(x)|^p \, dtdx \le N_p \int_0^T \int_{\mathbb{R}^d} |f_t(x)|^p \, dtdx,$$

where D_l^2 is the symbol of the second order derivative in the direction of l.

We see that in such estimates the constant is independent of the dimension.

2 Main Idea

We explain how the main idea works in the simplest situation of Example 3. For $t \in [0, T]$, $x \in \mathbb{R}$ consider the equation

$$\partial_t u_t(x) = 0$$

with initial condition $u_0(x)$ of class $C_0^\infty(\mathbb{R})$ and such that $0 \le u_0 \le 1$. The solution $u_t(x) = u_0(x)$ satisfies

$$0 \le u_t(x) \le 1, \quad \int_{\mathbb{R}} u_t^2(x)\, dx \le \int_{\mathbb{R}} u_0^2(x)\, dx. \tag{10}$$

Let π_t be a Poisson process with parameter $\lambda > 0$. Set

$$v_t(x) = u_t(x + \pi_t), \quad w_t = E v_t(x)$$

Then

$$v_t(x) = u_0(x) + \int_{(0,t]} \big[v_{s-}(x+1) - v_{s-}(x)\big]\, d\pi_s,$$

$$w_t(x) = u_0(x) + \lambda \int_0^t \big[w_s(x+1) - w_s(x)\big]\, ds.$$

In other words,

$$\partial_t w_t(x) = \lambda\big[w_t(x+1) - w_t(x)\big].$$

Now let

$$\hat{v}_t(x) = w_t(x - \pi_t), \quad \hat{u}_t(x) = E\hat{v}_t(x).$$

Then

$$d\hat{v}_t(x) = \lambda[\hat{v}_t(x+1) - \hat{v}_t(x)]\,dt + [\hat{v}_{t-}(x-1) - \hat{v}_{t-}(x)]\,d\pi_t,$$

$$\partial\hat{u}_t(x) = \lambda[\hat{u}_t(x+1) - 2\hat{u}_t(x) + \hat{u}(x-1)].$$

Obviously (10) holds true for w_t and $\hat{u}_t$. By replacing π_t with $h\pi_t$, taking $\lambda = h^{-2}$ and letting $h \downarrow 0$ we get that there exists a solution of the heat equation

$$\partial_t u_t(x) = D^2 u_t(x)$$

with initial data u_0, satisfying (10):

$$0 \le u_t(x) \le 1, \quad \int_{\mathbb{R}} u_t^2(x)\,dx \le \int_{\mathbb{R}} u_0^2(x)\,dx.$$

3 Generalizations

Let $\mathfrak{N}$ be a subset of the space of affine transformations of $\mathbb{R}^d$ and suppose that the above $\mathcal{G}$ is given as

$$\mathcal{G} = \{e^{t\nu} : t \in \mathbb{R}, \nu \in \mathfrak{N}\}, \tag{11}$$

where by $e^{t\nu}$ we mean a transformation $g(t)$ defined as a unique solution of the equation

$$g(t) = 1 + \int_0^t \nu g(s)\,ds. \tag{12}$$

Of course, we keep the assumption that $\mathcal{G}$ is a commutative group of volume-preserving transformations.

With any $\nu \in \mathfrak{N}$ we associate an operator M_ν acting on smooth functions $\phi : \mathbb{R}^d \to \mathbb{R}$ by the formula

$$M_\nu\phi(x) = \frac{d^2}{(d\varepsilon)^2}\phi(e^{\varepsilon\nu}x)\big|_{\varepsilon=0} = (\nu x)^i(\nu x)^j D_{ij}\phi(x) + (\nu^2 x - \nu 0)^i D_i\phi(x).$$

Example 6 Let l be a unit vector in $\mathbb{R}^d$ and define a transformation $\nu = \nu_l$ by $\nu_l x \equiv l$ on $\mathbb{R}^d$. Then (12) becomes

$$g(t)x = x + \int_0^t \nu g(s)x\,ds = x + \int_0^t l\,ds = x + tl.$$

Observe that in this example, for smooth ϕ, we have $M_\nu\phi(x) = D_l^2\phi(x)$. Thus, if $\mathfrak{N} = \{\nu_l : l \in \mathbb{R}^d, |l| = 1\}$, then $\mathcal{G}$ is the set of shifts of $\mathbb{R}^d$ and $\mathcal{G}$ is a commutative group.

Example 7 Let $\nu x = Qx$, where Q is a skew-symmetric $d \times d$-matrix. Then $g_t x = e^{t\nu}x = (\exp[tQ])x$, where $\exp[tQ]$ is an orthogonal matrix. In this example, for smooth ϕ,

$$M_\nu\phi(x) = (Qx)^i(Qx)^j D_{ij}\phi(x) + (Q^2x)^i D_i\phi(x).$$

Theorem 3 *Suppose that W, $\mathcal{G}$, L, u_0, and f satisfy Assumptions 1.1 and 1.2 with $\mathcal{G}$ from* (11) *and suppose that W also satisfies Assumption 1.3. Then, for any $\mu^{(1)}, \dots, \mu^{(n)} \in B((0,T), \mathfrak{M})$ Eq.* (5) *with*

$$L_t^* + \sum_{i=1}^n M_{\mu_t^{(i)}}$$

in place of L_t^* and initial condition u_0 has a solution in W.

Example 8 Let $d = 2$, $\alpha \in (0,1)$, and $L_t = \Delta$. We know that for any

$$f \in B_c((0,T), C_0^\infty(\mathbb{R}^2)) \tag{13}$$

the equation

$$u_t(x) = \int_0^t [\Delta u_s(x) + f_s(x)]\,ds, \quad t \le T, x \in \mathbb{R}^2, \tag{14}$$

has a unique continuous solution such that

$$\sup_{\substack{t\in[0,T]\\ x\in\mathbb{R}^2}} |u_t(x)| + \sup_{t\in[0,T]} \int_{\mathbb{R}^2} |u_t(x)|\,dx \le N_0\Big[\int_0^T \int_{\mathbb{R}^2} |f_t(x)|\,dxdt + \sup_{\substack{t\in(0,T)\\ x\in\mathbb{R}^2}} |f_t(x)|\Big], \tag{15}$$

$$\sup_{t\in[0,T]} [D_l^2 u_t]_{C^\alpha(\mathbb{R}^2)} \le N_\alpha \sup_{t\in[0,T]} [f_t]_{C^\alpha(\mathbb{R}^2)}, \ \forall l : |l| = 1, \tag{16}$$

where N_0 and N_α are some constants.

According to Theorem 3, if (13) holds, the equation

$$u_t(x) = \int_0^t [\Delta u_s(x) + Mu_s(x) + f_s(x)]\,ds,$$

where

$$M\phi(x) = (x^2)^2 D_{11}\phi(x) - 2x^1x^2 D_{12}\phi(x) + (x^1)^2 D_{22}\phi(x)$$
$$-x^1 D_1\phi(x) - x^2 D_2\phi(x),$$

has a continuous solution, which satisfies estimates (15) and (16) (with the same N_0 and N_α).

It seems to us that this is a nontrivial and maybe unexpected new result.

Here we let $\mathcal{G} = \{e^{\lambda Q}; \lambda \in \mathbb{R}\}$ be the group of all rotations of $\mathbb{R}^2$ about the origin, where

$$Q = \begin{pmatrix} 0 & 1 \\ -1 & 0 \end{pmatrix}.$$

4 Extensions to Non-local Operators

Assumption 4.1 We are given a family $\{\nu_t(A), t \in (0, T)\}$ of measures on Borel subsets of $\mathbb{R}^d$ such that

(i) $\nu_t(\{0\}) = 0$ for any $t \in (0, T)$,

(ii) $\nu_t(A)$ is a (Borel) measurable function of $t \in (0, T)$,

(iii) we have

$$\int_{\mathbb{R}^d} (1 \wedge |x|^2)\, \nu_t(dx) < \infty \quad \forall t \in (0, T), \quad \int_{(0,T)\times\mathbb{R}^d} (1 \wedge |x|^2)\, \nu_t(dx)dt < \infty.$$

Assumption 4.2 We are given $W, \mathcal{G}, K, L, u_0, f$ as in Theorem 3 with $\mathcal{G}$ being the group of translations.

Introduce

$$L_t^0 = L_t + J_{\nu_t}, \tag{17}$$

where, for $\phi \in C_0^\infty(\mathbb{R}^d)$ and measure ν,

$$J_\nu \phi(x) = \int_{\mathbb{R}^d} \Big[\phi(x + y) - \phi(x) - y \cdot D\phi(x)\, I_{\{|y|\le 1\}}(y)\Big]\nu(dy).$$

As a side observation recall that if ν is a measure on $\mathbb{R}^d \setminus \{0\}$ such that

$$\int_{\mathbb{R}^d} (1 \wedge |x|^2)\, \nu(dx) < \infty,$$

the operator J_ν is known in probability theory as the generator of a unique in law Lévy process associated to ν (this process is without Gaussian part; see [3, 8]).

One knows that, owing to Assumption 4.1, $J_{\nu_t}\phi(x)$ is well defined for any $\phi \in C_0^\infty(\mathbb{R}^d)$. Standard measure theoretic arguments show that $J_{\nu_t}\phi(x)$ is a measurable function of (t, x) for any $\phi \in C_0^\infty(\mathbb{R}^d)$.

Theorem 4 *Under the above assumptions for any $h \in B((0, T), \mathcal{G})$ there exists $u \in W$ such that (7) holds for any $\zeta \in C_0^\infty(\mathbb{R}^d)$ and $t \in [0, T]$ with L_s^0 in place of L_s.*

5 The Origin of the Main Idea

The origin of our ideas lies in the theory of stochastic partial differential equations (SPDEs) and can be found in the proof of Theorem 2.1 of [2]. This idea can be implemented quite formally without using the theory of SPDEs, see, for instance, [1, 7], where still one needs to be familiar with the Itô stochastic integral with respect to the Wiener process albeit of nonrandom functions.

In case $d = 1$ Theorem 2.1 of [2] deals, in particular, with extending the third estimate in (3) to solutions of Eq. (2) with $a_t D_{11}u$ in place of $D_{11}u$, where $a_t \geq 1$. To do this one denotes $\sigma_t = (2a_t - 2)^{1/2}$, takes a one-dimensional Wiener process $w_t = w_t(\omega)$, sets

$$B_t = \int_0^t \sigma_s \, dw_s,$$

and for each ω, solves (2) with $f(t, x - B_t)$ in place of $f_t(x)$ with zero initial data. Call the solution $\phi_t(x)$ and set $\psi_t(x) = \phi_t(x + B_t)$. It turns out (Itô's formula) that

$$\psi_t(x) = \int_0^t [a_t D_{11}\psi_s(x) + f_s(x)] \, ds + \int_0^t \sigma_s D\psi_s(x) \, dw_s,$$

where the last term is an Itô stochastic integral with mean zero. By taking mathematical expectations of both sides, one sees that $E\psi_t(x)$ is a solution of Eq. (2) with $a_t D_{11}u$ in place of $D_{11}u$. In addition, obviously, the third estimate in (3) holds with the same C with $E\psi_t(x)$ in place of $u_t(x)$.

Acknowledgements The article is based on the talk given by the first author at the international conference "Stochastic Partial Differential Equations and Related Topics" October 10–14, 2016, Bielefeld University. The opportunity to give this talk is greatly appreciated.

References

1. Krylov, N.V.: A parabolic Littlewood–Paley inequality with applications to parabolic equations. Topol. Methods Nonlinear Anal.; J. Juliusz Schauder Cent. **4**(2), 355–364 (1994)
2. Krylov, N.V.: On L_p-theory of stochastic partial differential equations in the whole space. SIAM J. Math. Anal. **27**(2), 313–340 (1996)
3. Krylov, N.V.: Introduction to the Theory of Random Processes. American Mathematical Society, Providence (2002)
4. Krylov, N.V.: Lectures on Elliptic and Parabolic Equations in Sobolev Spaces. American Mathematical Society, Providence (2008)
5. Ladyzhenskaya, O.A., Solonnikov, V.A., Ural'tseva, N.N.: Linear and quasi-linear parabolic equations, Nauka, Moscow (1967), in Russian; English translation: American Mathematical Society, Providence, RI (1968)
6. Lieberman, G.M.: Second Order Parabolic Differential Equations. World Scientific Publishing Co., Inc., River Edge (1996)

7. Priola, E.: L^p-parabolic regularity and non-degenerate Ornstein-Uhlenbeck type operators. In: Citti, G., et al. (eds.) Geometric Methods in PDEs. Springer INdAM Series, vol. 13, pp. 121–139. Springer, Berlin (2015)
8. Sato, K.I.: Lévy Processes and Infinite Divisible Distributions. Cambridge University Press, Cambridge (1999)

Dynamics of SPDEs Driven by a Small Fractional Brownian Motion with Hurst Parameter Larger than 1/2

L. H. Duc, M. J. Garrido-Atienza and B. Schmalfuß

Abstract We consider mild solutions of an SPDE driven by a time dependent perturbation which is Hölder continuous with a Hölder exponent larger than 1/2. In particular, such a perturbation is given by a fractional Brownian motion with Hurst parameter larger than 1/2. The coefficient in front of this noise is an operator with bounded first and second derivatives. We formulate conditions such that the equation has a unique pathwise solution. Further we investigate the globally exponential stability of the trivial solution.

Keywords Stability, stochastic partial differential equations · Fractional Brownian motion

Mathematical Subject Classification 2010 Primary: 35B35 · Secondary: 34A34, 34F05

Dedicated to Michael Röckner on occasion of his Sixtieth Birthday.

L. H. Duc
Max-Planck-Institut Für Mathematik in den Naturwissenschaften,
Inselstr. 22, 04103 Leipzig, Germany
e-mail: duc.luu@mis.mpg.de

L. H. Duc
Institute of Mathematics, Vietnam Academy of Science and Technology, 18 Hoang Quoc Viet Road, Cau Giay disk, 10307 Hanoi, Vietnam

M. J. Garrido-Atienza
Dpto. Ecuaciones Diferenciales y Análisis Numérico, Universidad de Sevilla,
Avda. Reina Mercedes, s/n, 41012 Sevilla, Spain
e-mail: mgarrido@us.es

B. Schmalfuß (✉)
Institut für Mathematik, Friedrich Schiller Universität Jena,
Ernst Abbe Platz 2, 77043 Jena, Germany
e-mail: bjoern.schmalfuss@uni-jena.de

A. Eberle et al. (eds.), *Stochastic Partial Differential Equations and Related Fields*, Springer Proceedings in Mathematics & Statistics 229,
https://doi.org/10.1007/978-3-319-74929-7_11

1 Introduction

Stability of ordinary and partial differential equations motivated by many applications is an interesting object of mathematical research. The foundation of this theory can be found in the monograph by Amann [1]. However, just four decades ago the stability of stochastic differential equations (SDEs), namely Ito equations, became an object of interests. In the fundamental monograph by Khasmiskii [14] different kinds of stability were studied, like stability in probability, moment stability and exponential almost sure stability. All these methods are based on the fact that an Ito equation with sufficiently smooth coefficients generates a Markov semigroup or a Markov evolution family. Estimates of the generators of theses objects by appropriate Lyapunov functions lead to conclusions about the stability of these stochastic equations. Further stability investigations of more general stochastic differential equations are presented by Mao [17]. Moreover, there also exist conditions ensuring stability of parabolic stochastic differential equations, see [15].

During the last two decades, stochastic differential equations driven by other noises than the Brownian motions came in the center of interests. One class of noises is the fractional Brownian motion (fBm) which is a centered Gauß process with a special covariance function, determined by a parameter $H \in (0, 1)$ known as the *Hurst* parameter. For $H \neq 1/2$ this stochastic process does not have the *nice* properties of the Brownian motion, despite the fact that the latter is nothing but an fBm with parameter $H = 1/2$. Especially this process is not a semi-martingale or a Markov process.

To define stochastic integrals with respect to this noise new techniques like Young integrals or rough path theory were necessary. From these integrals one then can formulate ordinary and partial stochastic differential equations driven by an fBm, see [2, 9, 18] for the case of stochastic parabolic equations with multiplicative nontrivial noise.

Yet, there is not so much work done regarding the asymptotic behavior of stochastic differential equations driven by an fBm. The paper [21] addresses stability for delay equations driven by fBm, but with additive noise only. In [20], the measure concentration property of the fBm is used to derive longtime behavior of an SDE solution driven by an fBm. In [6] it is studied the stability in the mean, asymptotic stability, global stability, and Mittag-Leffler stability for fractional systems with a function as initial condition and an additive noise, by using comparison results. In [19] the fixed point theory is used to investigate the stability for stochastic functional partial differential equations driven by additive fBm with $H > 1/2$. The paper [5] addresses the pth moment and almost surely exponential stabilities with the exponential rate function t^{2H} of stochastic parabolic equation driven by multiplicative fBm with $H > 1/2$ with Markovian switching. There are also several papers, see [10–13], where the existence of adapted stationary solutions to dissipative finite-dimensional SDEs driven by fBm (with additive noise with any Hurst parameter and multiplicative noise with $H > 1/3$) and their speed of convergence to the stationary state is investigated, but the analysis is built on suitable extensions of Markovian notions as strong Feller property. As we will explain below, our approach differs from the one in those papers.

The purpose of this article is to formulate conditions ensuring almost sure exponential stability of stochastic parabolic equations driven by a fractional Brownian motion with parameter $H \in (1/2, 1)$. In particular, the set of paths where we have this stability is invariant with respect to the Wiener shift and has measure one. To solve this problem we have to assume that the linear part of such an equation generates an analytic semigroup which is exponentially stable. Under some smallness conditions on the noise and boundedness of the first and second derivatives of the coefficient in front of the noise, we will obtain exponential stability of the trivial solution of our equation. For a more detailed description of the results given in this paper, we refer to the forthcoming article [4].

2 Preliminaires

Let $(\Omega, \mathscr{F}, \mathbb{P})$ a probability space. We introduce a mapping

$$\theta : \mathbb{R} \times \Omega \to \Omega$$

which is $\mathscr{B}(\mathbb{R}) \otimes \mathscr{F}, \mathscr{F}$ measurable and satisfies the flow property

$$\theta_t \theta_\tau := \theta_t \circ \theta_\tau = \theta_{t+\tau}, \quad \theta_0 = \mathrm{id}_\Omega.$$

The quadruple $(\Omega, \mathscr{F}, \mathbb{P}, \theta)$ is called a metric dynamical system. We assume that the flow θ is $\mathbb{P}$ ergodic, which means that every $A \in \mathscr{F}$ satisfying $\theta_t A = A$ for all $t \in \mathbb{R}$ is such that either $\mathbb{P}(A) = 0$ or $\mathbb{P}(A) = 1$.

For $t \in \mathbb{R}$ and $H \in (0, 1)$, let $B^H(t)$ be a fractional Brownian motion in $\mathbb{R}$ with Hurst parameter H, defined by a centered Gauß process with covariance

$$R(s, t) = \frac{1}{2}(|t|^{2H} + |s|^{2H} - |t - s|^{2H}).$$

It is not hard to lift this fBm into a separable Hilbert space V with a covariance Q of trace class. In particular, it is possible to find a canonical version of this process $(C_0(\mathbb{R}, V), \mathscr{B}(C_0(\mathbb{R}, V)), \mathbb{P}_H)$ such that $\mathbb{P}_H$ is ergodic with respect to the so-called Wiener shift $\theta_t\omega(\cdot) = \omega(\cdot + t) - \omega(t)$, see [7]. Furthermore, Kolmogorov's theorem about the existence of a Hölder continuous version ensures the existence of a measurable subset Ω of $C_0(\mathbb{R}, V)$ consisting of continuous functions which are β'-Hölder continuous in any interval $[-n, n]$ with $\beta' < H$. In what follows, we denote the trace σ–algebra $\mathscr{B}(C_0(\mathbb{R}, V)) \cap \Omega$ by $\mathscr{F}$ and $\mathbb{P}(\cdot) := \mathbb{P}_H(\cdot \cap \Omega)$ on $\mathscr{F}$. Then θ is $\mathbb{P}$ ergodic, see [7].

3 Evolution Equations Driven by an fBm with $H \in (1/2, 1)$

On the separable Hilbert space $(V, (\cdot,\cdot), \|\cdot\|)$, we consider the evolution equation

$$u(t) = S(t)u_0 + \int_0^t S(t-r)F(u(r))dr + \int_0^t S(t-r)G(u(r))d\omega \qquad (1)$$

where ω is assumed to be the canonical continuous Q-fractional Brownian motion with values in V introduced in Sect. 2. The stochastic integral is interpreted as a Hilbert-space valued Young integral. Regarding the definition of this integral, let us mention that there are several techniques to define it, but we will consider the one based on fractional derivatives, whose definition and main properties can be found in Chen et al. [2]. One interesting property of such a stochastic integral is that it can be defined in a pathwise sense provided that the integrand satisfies a special Hölder condition, which is in contrast to the well-known Ito–integral with integrator given by a Brownian motion.

In particular, the following shift property for the integral is satisfied, see [2]:

$$\int_\tau^{\tau+\cdot} S(\cdot+\tau-r)G(u(r))d\omega(r) = \int_0^\cdot S(\cdot-r)G(u(r+\tau))d\theta_\tau\omega(r). \qquad (2)$$

To formulate an existence theorem we need to define the following function spaces and corresponding norms. Let $L(V)$ be the space of continuous linear operators on V and $L_2(V)$ be the space of Hilbert–Schmidt operators on V. Let $C_b^i(V,V)$, $C_b^i(V, L_2(V))$ be the spaces of i-times continuously differentiable functions where all the derivatives up to the i derivative are bounded. In addition, for $\beta \in (0,1)$ we consider the following modified Hölder norm

$$\|u\|_{\beta,\beta} = \|u\|_{\beta,\beta,T_1,T_2} = \|u\|_{\infty,T_1,T_2} + |||u|||_{\beta,\beta,T_1,T_2},$$

where

$$\|u\|_{\infty,T_1,T_2} = \sup_{T_1\le r\le T_2}\|u(r)\|, \quad |||u|||_{\beta,\beta,T_1,T_2} = \sup_{T_1<s<t\le T_2}(s-T_1)^\beta \frac{\|u(t)-u(s)\|}{(t-s)^\beta},$$

see Lunardi [16]. This norm defines the Banach space $C_\beta^\beta([T_1,T_2],V)$.

Let A be the generator of the analytic semigroup S on V. $-A$ is also assumed to be positive definite and symmetric with a compact inverse. Then the eigenvectors of $-A$ generate a complete orthonormal system in V related to the point spectrum $(\lambda_i)_{i\in\mathbb{N}}$ which is positive and of finite multiplicity and with limit point ∞. We can consider the fractional powers of $-A$ and introduce the spaces $V_\delta := D((-A)^\delta)$ with norm $\|\cdot\|_{V_\delta} = \|(-A)^\delta\cdot\|$ for $\delta \ge 0$ such that $V = V_0$, see [2]. Note that, thanks to the analyticity of the semigroup, there exists a constant $c_S > 0$ such that

$$\|S(t)\|_{L(V_\zeta,V_\gamma)} = \|(-A)^\gamma S(t)\|_{L(V_\zeta,V)} \le c_S t^{\zeta-\gamma} e^{-\lambda t} \quad \text{for } \gamma \ge \zeta \ge 0, \quad t \in [0,T]$$
$$\|S(t) - \mathrm{id}\|_{L(V_\sigma,V_\eta)} \le c_S t^{\sigma-\eta}, \quad \text{for } \eta \ge 0, \quad \sigma \in [\eta, 1+\eta], \tag{3}$$

where $0 < \lambda < \lambda_1$ and c_S depends on the appearing parameters. We emphasize that the reason to consider the modified Hölder continuous space $C_\beta^\beta([0,T],V)$ is that S has a finite $C_\beta^\beta([0,T],V)$–norm but not a finite $C^\beta([0,T],V)$–norm, where $C^\beta([0,T],V)$ denotes as usual the space of β–Hölder–continuous functions.

The following theorem is proven in [2]:

Theorem 1 *Let $T > 0$, ω the canonical Q–fractional Brownian motion with $H \in (1/2, 1)$, $1/2 < \beta < H$ and $u_0 \in V$. Under the assumptions that A generates the semigroup introduced above, $F \in C_b^1(V,V)$ and $G \in C_b^2(V, L_2(V))$, there exists a unique solution $u \in C_\beta^\beta([0,T],V)$ of (1).*

This proof is based on the Banach fixed point theorem, for which we need to estimate the mappings

$$t \mapsto \int_0^t S(t-r)F(u(r))dr \in C_\beta^\beta([0,T];V),$$
$$t \mapsto \int_0^t S(t-r)G(u(r))d\omega \in C_\beta^\beta([0,T];V).$$

If we denote

$$c_F = \|F(0)\|,\ c_G = \|G(0)\|_{L_2(V)},\ \|DF\|_{L(V)} \le c_{DF},\ \|DG\|_{L(V,L_2(V))} \le c_{DG}, \tag{4}$$

then we obtain

$$\left\| \int_0^\cdot S(\cdot - r)F(u(r))dr \right\|_{\beta,\beta,0,T} \le T c_S (c_F + c_{DF}\|u\|_{\infty,0,T})$$
$$\left\| \int_0^\cdot S(\cdot - r)G(u(r))d\omega \right\|_{\beta,\beta,0,T} \le T^{\beta'} c_{\alpha,\beta,\beta'} c_S (c_G + c_{DG}\|u\|_{\beta,\beta,0,T}) \,|||\omega|||_{\beta',0,T}\,, \tag{5}$$

where

$$|||\omega|||_{\beta',0,T} = \sup_{0\le s<t\le T} \frac{\|\omega(t)-\omega(s)\|}{(t-s)^{\beta'}}.$$

Note that to estimate the stochastic integral we apply techniques to estimate the fractional derivatives $D_{0+}^\alpha G(u)[r]$ and $D_{T-}^{1-\alpha}(\omega - \omega(T))[r]$, see [2]. Above we have taken $1/2 < \beta < \beta' < H$ and α is a constant such that $1 - \beta' < \alpha < \beta$. The constant $c_{\alpha,\beta,\beta'}$ is related to the estimates of the fractional derivatives, while the constant c_S is related to the estimates of the $C_\beta^\beta([0,T],V)$–norm of S. In fact, c_S is a constant depending on S that can change from line to line.

4 Exponential Stability of SPDEs Driven by fBm with $H \in (1/2, 1)$

From now on assume that $F(0) = 0$, $G(0) = 0$ (hence c_F and c_G in (4) are both null) such that $u \equiv 0$ solves (1) uniquely. In this section we will prove exponential stability to the trivial solution of the solution u to (1). In particular we will show that there exists a positive constant ρ such that for every bounded set B of V there exists a constant $C_B(\omega)$ such that for every solution u of (1) with initial condition $u_0 \in B$ we have

$$\|u(t)\| \leq C_B(\omega)e^{-\rho t} \quad \text{for all } t \geq 0. \tag{6}$$

The first idea could be to estimate $\|u\|_{\beta,\beta,n,n+1}$ and show the exponential decay for $n \to \infty$. To obtain this decay we should use a particular discrete Gronwall lemma, but this is not possible since the sequence of the seminorms $|||\omega|||_{\beta',n,n+1}$ is unbounded with probability one. Therefore, the strategy will consist of defining a sequence $(T_n(\omega))_{n\in\mathbb{Z}}$ such that the corresponding sequence

$$(|||\omega|||_{\beta',T_n(\omega),T_{n+1}(\omega)})_{n\in\mathbb{Z}}$$

is bounded by a sufficiently small constant. In particular, the sequence $(T_n(\omega))_{n\in\mathbb{Z}}$ does not have cluster points, see Theorem 2 below.

For some positive given constant μ we define

$$\begin{aligned} T(\omega) &= \inf\{\tau > 0 : |||\omega|||_{\beta',0,\tau}\, \tau^{\beta'} + \tau > \mu\}, \\ \hat{T}(\omega) &= \sup\{\tau < 0 : |||\omega|||_{\beta',\tau,0}\, (-\tau)^{\beta'} - \tau > \mu\}. \end{aligned} \tag{7}$$

In particular, we have

$$T(\omega) = -\hat{T}(\theta_{T(\omega)}\omega), \quad \hat{T}(\omega) = -T(\theta_{\hat{T}(\omega)}\omega).$$

Iterating these two first stopping times, we obtain

$$T_i(\omega) = \begin{cases} 0 & : \quad i = 0, \\ T_{i-1}(\omega) + T(\theta_{T_{i-1}(\omega)}\omega) : & i \in \mathbb{N}, \\ T_{i+1}(\omega) + \hat{T}(\theta_{T_{i+1}(\omega)}\omega) : & i \in -\mathbb{N}. \end{cases} \tag{8}$$

From this definition, by induction it is not difficult to derive that for any $i \in \mathbb{Z}$

$$T_{i+1}(\omega) - T_i(\omega) = T_1(\theta_{T_i(\omega)}\omega). \tag{9}$$

Denote by $N(\omega)$ the maximal index such that such that $T_0(\omega),\ T_1(\omega), \cdots, T_{N(\omega)-1}(\omega)$ are bounded by μ.

Theorem 2 *Let $\mu > 0$ be the constant appearing in the definition of (7) and define the constant $d \in (0,1)$ by*

$$\mathbb{E}_H N(\omega) \leq \mu \mathbb{E}_H \left(\frac{\sup_{r \in [-\mu,0]} |||\theta_r \omega|||_{\beta'',0,\mu} + \mu^{1-\beta''}}{\mu} \right)^{\frac{1}{\beta''}} =: \frac{1}{d}, \tag{10}$$

where $1/2 < \beta < \beta' < \beta'' < H$. Then, if in addition ω has a covariance with small trace, we have

$$\liminf_{k \to \pm\infty} \frac{|T_k(\omega)|}{|k|} \geq \mu d$$

on a θ-invariant set of full measure.

For the proof of this technical Lemma we refer to [4].

Remark 1 We would like to stress that, according to the definition of d given by (10), if the driven input signal ω is such that $\operatorname{tr} Q$ tends to zero, then d is close to one, and this is exactly what we need to ensure exponential stability of the trivial solution, as it is proven in the next theorem.

We now formulate the main result of this article:

Theorem 3 *Under the assumptions of Theorem 1 if in addition* $\operatorname{tr} Q$ *is assumed to be sufficiently small and*

$$\lambda > c_S(c_{DF} + c_{\alpha,\beta,\beta'} c_{DG}), \tag{11}$$

where the constants have been introduced in (3), (4) and (5), then the trivial solution to (1) is exponentially stable on a θ-invariant set of full measure with exponential rate fulfilling (21) below.

Proof Given $t \in [T_n, T_{n+1}]$, thanks to the additivity of the integrals and (2), we can consider the following splitting of the solution:

$$\begin{aligned}
u(t) &= S(t)u_0 + \sum_{i=0}^{n-1} \int_{T_i}^{T_{i+1}} S(t-r)F(u(r))dr + \sum_{i=0}^{n-1} \int_{T_i}^{T_{i+1}} S(t-r)G(u(r))d\omega(r) \\
&\quad + \int_{T_n}^{t} S(t-r)F(u(r))dr + \int_{T_n}^{t} S(t-r)G(u(r))d\omega(r) \\
&= S(t)u_0 \\
&\quad + \sum_{i=0}^{n-1} S(t-T_{i+1}) \int_0^{T_{i+1}-T_i} S(T_{i+1}-T_i-r)F(u(r+T_i))dr \\
&\quad + \sum_{i=0}^{n-1} S(t-T_{i+1}) \int_0^{T_{i+1}-T_i} S(T_{i+1}-T_i-r)G(u(r+T_i))d\theta_{T_i}\omega(r) \\
&\quad + \int_0^{t-T_n} S(t-T_n-r)F(u(r+T_n))dr
\end{aligned}$$

$$+\int_0^{t-T_n} S(t-T_n-r)G(u(r+T_n))d\theta_{T_n}\omega(r).$$

We want to estimate the $\|\cdot\|_{\beta,\beta,T_n,T_{n+1}}$ of the left hand side. For the first term on the right hand side, according to (3),

$$\begin{aligned}\|S(\cdot)u_0\|_{\beta,\beta,T_n,T_{n+1}} &= \sup_{t\in[T_n,T_{n+1}]}\|S(t)u_0\| + \sup_{T_n<s<t\le T_{n+1}}(s-T_n)^\beta\frac{\|S(t)u_0-S(s)u_0\|}{(t-s)^\beta}\\ &\le c_S\left(e^{-\lambda T_n(\omega)}\|u_0\| + \sup_{T_n<s<t\le T_{n+1}}(s-T_n)^\beta\frac{s^{-\beta}e^{-\lambda T_n}(t-s)^\beta}{(t-s)^\beta}\|u_0\|\right)\\ &\le c_S e^{-\lambda T_n(\omega)}\|u_0\|\end{aligned}$$

since $s^{-\beta}\le(s-T_n)^{-\beta}$. Let us abbreviate $u(\cdot+T_i)$ by $u^i(\cdot)$, that is,

$$u(\tau+T_i)=u^i(\tau),\;\;\tau\in[0,T_{i+1}(\omega)-T_i(\omega)],$$

where this interval equals to $[0,T_1(\theta_{T_i(\omega)}\omega)]$ due to (9). In particular

$$u(t)=u^n(t-T_n),\;\;t-T_n\in[0,T_1(\theta_{T_n(\omega)}\omega)].$$

Therefore, $\|u\|_{\beta,\beta,T_n(\omega),T_{n+1}(\omega)}=\|u^n\|_{\beta,\beta,0,T_1(\theta_{T_n(\omega)}\omega)}$.

Taking in (4) the constants $c_F=0$ and $c_G=0$, on account of (5) for $n=0,\;1,\cdots$ we have:

$$\begin{aligned}\|u^n\|_{\beta,\beta,0,T_1(\theta_{T_n(\omega)}\omega)} &\le c_S e^{-\lambda T_n(\omega)}\|u_0\|\\ &\quad + c_S\sum_{i=0}^{n-1}e^{-\lambda(T_n(\omega)-T_{i+1}(\omega))}T_1(\theta_{T_i(\omega)}\omega)c_{DF}\|u^i\|_{\beta,\beta,0,T_1(\theta_{T_i(\omega)}\omega)}\\ &\quad + c_Sc_{\alpha,\beta,\beta'}\sum_{i=0}^{n-1}e^{-\lambda(T_n(\omega)-T_{i+1}(\omega))}\left|\!\left|\!\left|\theta_{T_i(\omega)}\omega\right|\!\right|\!\right|_{\beta',0,T_1(\theta_{T_i(\omega)}\omega)}T_1(\theta_{T_i(\omega)}\omega)^{\beta'}c_{DG}\\ &\qquad\times\|u^i\|_{\beta,\beta,0,T_1(\theta_{T_i(\omega)}\omega)}\\ &\quad + c_ST_1(\theta_{T_n(\omega)}\omega)c_{DF}\|u^n\|_{\beta,\beta,0,T_1(\theta_{T_n(\omega)}\omega)}\\ &\quad + c_Sc_{\alpha,\beta,\beta'}\left|\!\left|\!\left|\theta_{T_n(\omega)}\omega\right|\!\right|\!\right|_{\beta',0,T_1(\theta_{T_n(\omega)})}T_1(\theta_{T_n(\omega)}\omega)^{\beta'}c_{DG}\|u^n\|_{\beta,\beta,0,T_1(\theta_{T_n(\omega)}\omega)}\\ &\le c_Se^{-\lambda T_n(\omega)}\|u_0\|\\ &\quad + \mu c_S(c_{DF}+c_{\alpha,\beta,\beta'}c_{DG})\sum_{i=0}^{n-1}e^{-\lambda(T_n(\omega)-T_{i+1}(\omega))}\|u^i\|_{\beta,\beta,0,T_1(\theta_{T_i(\omega)}\omega)}\\ &\quad + \mu c_S(c_{DF}+c_{\alpha,\beta,\beta'}c_{DG})\|u^n\|_{\beta,\beta,0,T_1(\theta_{T_n(\omega)}\omega)}\end{aligned}\qquad(12)$$

where this last inequality follows from (7).

First of all, the constant $\mu \in (0, 1]$ taken on (7) is chosen small enough such that

$$\mu c_S(c_{DF} + c_{\alpha,\beta,\beta'} c_{DG}) < 1. \tag{13}$$

Then, from (12) we derive

$$\begin{aligned} \|u^n\|_{\beta,\beta,0,T_1(\theta_{T_n(\omega)}\omega)} &\le \frac{c_S}{1-\mu c_S(c_{DF}+c_{\alpha,\beta,\beta'}c_{DG})} e^{-\lambda T_n(\omega)}\|u_0\| \\ &+ \frac{\mu c_S(c_{DF}+c_{\alpha,\beta,\beta'}c_{DG})}{1-\mu c_S(c_{DF}+c_{\alpha,\beta,\beta'}c_{DG})} \sum_{i=0}^{n-1} e^{-\lambda(T_n(\omega)-T_{i+1}(\omega))}\|u^i\|_{\beta,\beta,0,T_1(\theta_{T_i(\omega)}\omega)}. \end{aligned} \tag{14}$$

Denoting $y^n = e^{\lambda T_n}\|u^n\|_{\beta,\beta,0,T_1(\theta_{T_n(\omega)}\omega)}$, (14) can be rewritten as

$$y^n \le \frac{c_S}{1-\mu c_S(c_{DF}+c_{\alpha,\beta,\beta'}c_{DG})}\|u_0\| + \frac{\mu c_S(c_{DF}+c_{\alpha,\beta,\beta'}c_{DG})}{1-\mu c_S(c_{DF}+c_{\alpha,\beta,\beta'}c_{DG})} \sum_{i=0}^{n-1} e^{\lambda T_1(\theta_{T_i(\omega)}\omega))} y^i.$$

Applying a discrete version of Gronwall's lemma, see [3, 8], we obtain

$$\begin{aligned} y^n &\le \frac{c_S}{1-\mu c_S(c_{DF}+c_{\alpha,\beta,\beta'}c_{DG})}\|u_0\| \prod_{i=0}^{n-1}\left(1+\frac{\mu c_S(c_{DF}+c_{\alpha,\beta,\beta'}c_{DG})}{1-\mu c_S(c_{DF}+c_{\alpha,\beta,\beta'}c_{DG})} e^{\lambda T_1(\theta_{T_i(\omega)}\omega))}\right) \\ &\le \frac{c_S}{1-\mu c_S(c_{DF}+c_{\alpha,\beta,\beta'}c_{DG})}\|u_0\|\left(1+\frac{\mu c_S(c_{DF}+c_{\alpha,\beta,\beta'}c_{DG})}{1-\mu c_S(c_{DF}+c_{\alpha,\beta,\beta'}c_{DG})} e^{\lambda\mu}\right)^n, \end{aligned}$$

where the last inequality holds true due to the definition of the first stopping time, see (7) and (8). Hence

$$\begin{aligned} \|u^n\|_{\beta,\beta,0,T_1(\theta_{T_n(\omega)}\omega)} &\le \frac{c_S e^{-\lambda T_n(\omega)}}{1-\mu c_S(c_{DF}+c_{\alpha,\beta,\beta'}c_{DG})}\|u_0\| \\ &\times \left(1+\frac{\mu c_S(c_{DF}+c_{\alpha,\beta,\beta'}c_{DG})}{1-\mu c_S(c_{DF}+c_{\alpha,\beta,\beta'}c_{DG})} e^{\lambda\mu}\right)^n. \end{aligned} \tag{15}$$

We also choose a constant $\hat{d} \in (0, 1)$ and a small enough $\delta > 0$ such that

$$\lambda\hat{d} > (1+\delta)c_S(c_{DF} + c_{\alpha,\beta,\beta'}c_{DG}), \tag{16}$$

which is possible thanks to (11). In fact, we choose a small $\mu \in (0, 1]$ so that we also have for the above $\delta > 0$ that

$$\frac{e^{\lambda\mu}}{1-\mu c_S(c_{DF}+c_{\alpha,\beta,\beta'}c_{DG})} \le 1+\delta. \tag{17}$$

Hence

$$1+\frac{\mu c_S(c_{DF}+c_{\alpha,\beta,\beta'}c_{DG})}{1-\mu c_S(c_{DF}+c_{\alpha,\beta,\beta'}c_{DG})}e^{\lambda\mu} \leq e^{\frac{\mu c_S(c_{DF}+c_{\alpha,\beta,\beta'}c_{DG})}{1-\mu c_S(c_{DF}+c_{\alpha,\beta,\beta'}c_{DG})}e^{\lambda\mu}} \leq e^{(1+\delta)\mu c_S(c_{DF}+c_{\alpha,\beta,\beta'}c_{DG})}. \tag{18}$$

Furthermore, for the given $\hat{d}$ and μ we choose $\operatorname{tr} Q$ small enough such that

$$\hat{d} \leq d(\operatorname{tr} Q, \mu) = d < 1. \tag{19}$$

From Theorem 2 we know that for any $\varepsilon > 0$ there exists $n_0 = n_0(\varepsilon, \omega)$ such that

$$T_n \geq (\mu d - \varepsilon)n, \tag{20}$$

for $n \geq n_0$. Then coming back to (15), on account of (20) we have that

$$\begin{aligned}
\|u\|_{\beta,\beta,T_n(\omega),T_{n+1}(\omega)} &= \|u^n\|_{\beta,\beta,0,T_1(\theta_{T_n(\omega)}\omega)} \\
&\leq \frac{c_S}{1-\mu c_S(c_{DF}+c_{\alpha,\beta,\beta'}c_{DG})}\|u_0\|e^{-\lambda T_n(\omega)+(1+\delta)\mu c_S(c_{DF}+c_{\alpha,\beta,\beta'}c_{DG})n} \\
&\leq \frac{c_S}{1-\mu c_S(c_{DF}+c_{\alpha,\beta,\beta'}c_{DG})}\|u_0\|e^{-\left(\lambda-\frac{\mu(1+\delta)c_S(c_{DF}+c_{\alpha,\beta,\beta'}c_{DG})}{\mu d-\varepsilon}\right)T_n(\omega)}.
\end{aligned}$$

Given any $t > T_{n_0(\varepsilon,\omega)}$ there exists $n(t) > n_0(\varepsilon, \omega)$ such that $t \in [T_{n(t)}(\omega), T_{n(t)+1}(\omega))$, that is, $t - T_{n(t)}(\omega) \in [0, T_1(\theta_{T_{n(t)}(\omega)}\omega))$, and thus $t - T_{n(t)}(\omega) \leq \mu$. Hence

$$\|u(t)\| \leq \frac{c_S}{1-\mu c_S(c_{DF}+c_{\alpha,\beta,\beta'}c_{DG})}\|u_0\|e^{\lambda\mu}e^{-\left(\frac{\lambda(d-\frac{\varepsilon}{\mu})-(1+\delta)c_S(c_{DF}+c_{\alpha,\beta,\beta'}c_{DG})}{d-\frac{\varepsilon}{\mu}}\right)t}.$$

Consider $\bar{\rho}$ given by

$$\bar{\rho} := \lambda - \frac{c_S(c_{DF}+c_{\alpha,\beta,\beta'}c_{DG})}{d(\operatorname{tr} Q, \mu)}$$

which is positive by the assumptions on μ and $\operatorname{tr} Q$, see (16) and (19). Note that by choosing ε small enough, for δ satisfying (16) and (17), we still have

$$\rho_\varepsilon = \frac{\lambda(d-\frac{\varepsilon}{\mu})-(1+\delta)c_S(c_{DF}+c_{\alpha,\beta,\beta'}c_{DG})}{d-\frac{\varepsilon}{\mu}} > 0 \tag{21}$$

and hence (6) holds true when $t > T_{n_0(\varepsilon,\omega)}$ and $\rho = \rho_\varepsilon$.

On the other hand, for $t \leq T_{n_0}$ there exists $k \leq n_0 - 1$ such that $t \in [T_k(\omega), T_{k+1}(\omega)]$. Hence

$$\|u(t)\| \leq \|u\|_{\beta,\beta,T_k,T_{k+1}} \leq \|u^k\|_{\beta,\beta,0,T_1(\theta_{T_k(\omega)}\omega)},$$

thus, from (15) and (18),

$$
\begin{aligned}
\|u(t)\| &\leq \sup_{r\in[0,T_{n_0(\omega)}]}\|u(r)\| \leq \sup_{0\leq k\leq n_0-1}\|u^k\|_{\beta,\beta,0,T_1(\theta_{T_k(\omega)}\omega)}\\
&\leq \frac{c_S}{1-\mu c_S(c_{DF}+c_{\alpha,\beta,\beta'}c_{DG})}\|u_0\|\sup_{0\leq k\leq n_0-1} e^{-\lambda T_k}e^{(1+\delta)\mu c_S(c_{DF}+c_{\alpha,\beta,\beta'}c_{DG})k}\\
&\leq \frac{c_S}{1-\mu c_S(c_{DF}+c_{\alpha,\beta,\beta'}c_{DG})}\|u_0\|\sup_{0\leq k\leq n_0-1}\left(e^{-\lambda T_k}e^{(1+\delta)\mu c_S(c_{DF}+c_{\alpha,\beta,\beta'}c_{DG})k}\right)e^{\rho_\varepsilon T_{n_0}}e^{-\rho_\varepsilon t}.
\end{aligned}
$$

Assume that $u_0 \in B$, a bounded set of V. Then there exists $R > 0$ such that $\|u_0\| \leq R$, hence for all $t \geq 0$

$$
\|u(t)\| \leq \sup_{t\leq T_{n_0}(\omega)}\|u(t)\| + \mathbf{1}_{(T_{n_0}(\omega),\infty)}(t)\|u(t)\| \leq C_{B,\varepsilon}(\omega)e^{-\rho_\varepsilon t}
$$

namely, the solution u is exponentially stable, where $C(\omega) = C_{B,\varepsilon}(\omega)$ is given by

$$
C_{B,\varepsilon}(\omega) = \frac{c_S R}{1-\mu c_S(c_{DF}+c_{\alpha,\beta,\beta'}c_{DG})}K_\varepsilon(\omega),
$$

with

$$
K_\varepsilon(\omega) = e^{\lambda\mu} + \sup_{0\leq k\leq n_0-1}\left(e^{-\lambda T_k}e^{(1+\delta)\mu c_S(c_{DF}+c_{\alpha,\beta,\beta'}c_{DG})k}\right)e^{\rho_\varepsilon T_{n_0}}.
$$

References

1. Amann, H.: Ordinary Differential Equations: An Introduction to Nonlinear Analysis. de Gruyter Studies in Mathematics, vol. 13. Walter de Gruyter and Co., Berlin (1990). Translated from the German by Gerhard Metzen
2. Chen, Y., Gao, H., Garrido-Atienza, M.J., Schmalfuß, B.: Pathwise solutions of SPDEs driven by Hölder-continuous integrators with exponent larger than 1/2 and random dynamical systems. Discret. Contin. Dyn. Syst. **34**(1), 79–98 (2014)
3. Dragomir, S.S.: Some Gronwall Type Inequalities and Applications. Nova Science Publishers Inc, Hauppauge (2003)
4. Duc, L.H., Garrido-Atienza, M.J., Neuenkirch, A., Schmalfuß, B.: Exponential stability of stochastic evolution equations driven by small fractional Brownian motion with Hurst parameter in (1/2, 1). J. Differ. Equ. **264**(2), 1119–1145 (2018)
5. Fan, X., Yuan, C.: Lyapunov exponents of PDEs driven by fractional noise with Markovian switching. Stat. Probab. Lett. **110**, 39–50 (2016)
6. Fiel, A., León, J.A., Márquez-Carreras, D.: Stability for a class of semilinear fractional stochastic integral equations. Adv. Differ. Equ. **2016**(166), 20 (2016)
7. Garrido-Atienza, M.J., Schmalfuß, B.: Ergodicity of the infinite dimensional fractional Brownian motion. J. Dyn. Differ. Equ. **23**(3), 671–681 (2011)
8. Garrido-Atienza, M.J., Neuenkirch, A., Schmalfuß, B.: Asymptotical stability of differential equations driven by Holder continuous paths. J. Dyn. Differ. Equ. (2017). https://doi.org/10.1007/s10884-017-9574-6
9. Gubinelli, M., Tindel, S.: Rough evolution equations. Ann. Probab. **38**(1), 1–75 (2010)

10. Hairer, M.: Ergodicity of stochastic differential equations driven by fractional Brownian motion. Ann. Probab. **33**(2), 703–758 (2005)
11. Hairer, M., Ohashi, A.: Ergodic theory for SDEs with extrinsic memory. Ann. Probab. **35**(5), 1950–1977 (2007)
12. Hairer, M., Pillai, N.S.: Ergodicity of hypoelliptic SDEs driven by fractional Brownian motion. Ann. Inst. Henri Poincaré Probab. Stat. **47**(2), 601–628 (2011)
13. Hairer, M., Pillai, N.S.: Regularity of laws and ergodicity of hypoelliptic SDEs driven by rough paths. Ann. Probab. **41**(4), 2544–2598 (2013)
14. Khasminskii, R.: Stochastic Stability of Differential Equations. Stochastic Modelling and Applied Probability, vol. 66, 2nd edn. Springer, Heidelberg (2012). With contributions by Milstein, G.N., Nevelson, M.B
15. Liu, K., Truman, A.: A note on almost sure exponential stability for stochastic partial functional differential equations. Stat. Probab. Lett. **50**(3), 273–278 (2000)
16. Lunardi, A.: Analytic Semigroups and Optimal Regularity in Parabolic Problems. Progress in Nonlinear Differential Equations and their Applications, vol. 16. Birkhäuser, Basel (1995)
17. Mao, X.: Exponential Stability of Stochastic Differential Equations. Monographs and Textbooks in Pure and Applied Mathematics, vol. 182. Marcel Dekker Inc, New York (1994)
18. Maslowski, B., Nualart, D.: Evolution equations driven by a fractional Brownian motion. J. Funct. Anal. **202**(1), 277–305 (2003)
19. Ruan, D., Luo, J.: Fixed points and exponential stability of stochastic functional partial differential equations driven by fractional Brownian motion. Publ. Math. Debrecen **86**(3–4), 285–293 (2015)
20. Saussereau, B.: Transportation inequalities for stochastic differential equations driven by a fractional Brownian motion. Bernoulli **18**(1), 1–23 (2012)
21. Tan, L.: Exponential stability of fractional stochastic differential equations with distributed delay. Adv. Differ. Equ. **2014**(321), 8 (2014)

On the Well-Posedness of SPDEs with Singular Drift in Divergence Form

Carlo Marinelli and Luca Scarpa

Abstract We prove existence and uniqueness of strong solutions for a class of second-order stochastic PDEs with multiplicative Wiener noise and drift of the form $\operatorname{div}\gamma(\nabla\cdot)$, where γ is a maximal monotone graph in $\mathbb{R}^n \times \mathbb{R}^n$ obtained as the subdifferential of a convex function satisfying very mild assumptions on its behavior at infinity. The well-posedness result complements the corresponding one in our recent work arXiv:1612.08260 where, under the additional assumption that γ is single-valued, a solution with better integrability and regularity properties is constructed. The proof given here, however, is self-contained.

Keywords Stochastic evolution equations · Singular drift · Divergence form · Multiplicative noise · Monotone operators

2010 Mathematics Subject Classification Primary: 60H15 · 47H06 · Secondary: 46N30

1 Introduction and Main Result

Let us consider the stochastic partial differential equation

$$du(t) - \operatorname{div}\gamma(\nabla u(t))\,dt \ni B(t, u(t))\,dW(t), \qquad u(0) = u_0, \tag{1}$$

posed on $L^2(D)$, with D a bounded domain of $\mathbb{R}^n$ with smooth boundary. The following assumptions will be in force: (a) γ is the subdifferential of a lower semicontinuous convex function $k : \mathbb{R}^n \to \mathbb{R}_+$ with $k(0) = 0$ and such that

C. Marinelli · L. Scarpa (✉)
Department of Mathematics, University College London, Gower Street, London WC1E 6BT, UK
e-mail: luca.scarpa.15@ucl.ac.uk

A. Eberle et al. (eds.), *Stochastic Partial Differential Equations and Related Fields*, Springer Proceedings in Mathematics & Statistics 229,
https://doi.org/10.1007/978-3-319-74929-7_12

$$\lim_{|x|\to\infty} \frac{k(x)}{|x|} = +\infty, \qquad \limsup_{|x|\to\infty} \frac{k(-x)}{k(x)} < +\infty$$

(in particular, γ is a maximal monotone graph in $\mathbb{R}^n \times \mathbb{R}^n$ whose domain coincides with $\mathbb{R}^n$); (b) W is a cylindrical Wiener process on a separable Hilbert space H, supported by a filtered probability space $(\Omega, \mathscr{F}, (\mathscr{F}_t)_{t\in[0,T]}, \mathbb{P})$ satisfying the "usual conditions"; (c) B is a map from $\Omega \times [0, T] \times L^2(D)$ to $\mathscr{L}^2(H, L^2(D))$, the space of Hilbert-Schmidt operators from H to $L^2(D)$, that is Lipschitz-continuous and has linear growth with respect to its third argument, uniformly with respect to the other two, and is such that $B(\cdot, \cdot, a)$ is measurable and adapted for all $a \in L^2(D)$.

Under the additional assumption that γ is a (single-valued) continuous function, we proved in [7] that (1) admits a strong solution u, which is unique within a set of processes satisfying mild integrability conditions. The solution of [7] is constructed pathwise, i.e. for each $\omega \in \Omega$, so that, as is natural to expect, measurability problems arise with respect to the usual σ-algebras on $\Omega \times [0, T]$ used in the theory of stochastic processes. Precisely because of such an issue we needed to assume γ to be single-valued.

The purpose of this note is to provide an alternative approach to establish the well-posedness of (1) that, avoiding pathwise constructions, is simpler than that of [7] and does not need any extra assumption on γ. The price to pay is that the solution we obtain here is less regular than that of [7]. We also refer to [8] for an alternative approach as well as to [10] for a related result obtained by analogous methods.

Let us define the concept of solution to (1) we shall be working with.

Definition 1 Let u_0 be an $L^2(D)$-valued $\mathscr{F}_0$-measurable random variable. A *strong solution* to Eq. (1) is a couple (u, η) satisfying the following properties:

1. u is a measurable and adapted $L^2(D)$-valued process such that

$$u \in L^1(0, T; W_0^{1,1}(D)) \quad \text{and} \quad B(\cdot, u) \in L^2(0, T; \mathscr{L}^2(U, L^2(D))) \quad \mathbb{P}\text{-a.s.};$$

2. η is a measurable and adapted $L^1(D)^n$-valued process such that

$$\eta \in L^1(0, T; L^1(D)^n) \quad \mathbb{P}\text{-a.s.}, \qquad \eta \in \gamma(\nabla u) \quad \text{a.e. in } \Omega \times (0, T) \times D;$$

3. one has, as an equality in $L^2(D)$,

$$u(t) - \int_0^t \operatorname{div}\eta(s)\, ds = u_0 + \int_0^t B(s, u(s))\, dW(s) \quad \mathbb{P}\text{-a.s.} \quad \forall t \in [0, T]. \tag{2}$$

Note that (2) has to be intended in the sense of distributions. In particular, since $\eta \in L^1(D)^n$, the integrand in the second term of (2) does not, in general, take values in $L^2(D)$. However, the conditions on B imply that the stochastic integral in (2) is an $L^2(D)$-valued local martingale, hence the term involving the divergence of η turns out to be $L^2(D)$-valued by comparison.

We can now state our main result. Here and in the following $k^* : \mathbb{R}^n \to \mathbb{R}_+$ is the convex conjugate of k, defined as $k^*(y) := \sup_{x\in\mathbb{R}^n}\big(x \cdot y - k(x)\big)$.

Theorem 1 *Let* $u_0 \in L^2(\Omega, \mathscr{F}_0; L^2(D))$. *Then Eq.* (1) *admits a unique strong solution* (u, η) *such that*

$$\sup_{t\le T} \mathbb{E}\|u(t)\|^2_{L^2(D)} + \mathbb{E}\int_0^T \|u(t)\|_{W_0^{1,1}(D)}\,dt < \infty,$$
$$\mathbb{E}\int_0^T \|\eta(t)\|_{L^1(D)^n}\,dt < \infty,$$
$$\mathbb{E}\int_0^T \big(\|k(\nabla u(t))\|_{L^1(D)} + \|k^*(\eta(t))\|_{L^1(D)}\big)\,dt < \infty.$$

Moreover, the solution map $u_0 \mapsto u$ *is Lipschitz-continuous from* $L^2(\Omega; L^2(D))$ *to* $L^\infty(0, T; L^2(\Omega; L^2(D)))$, *and* u *is weakly continuous as a function on* $[0, T]$ *with values in* $L^2(\Omega; L^2(D))$.

Under the extra assumption of γ being single-valued, the solution obtained in [7] is more regular in the sense that $\mathbb{E}\sup_{t\le T}\|u(t)\|^2_{L^2(D)}$ is finite, the solution map is Lipschitz-continuous from $L^2(\Omega; L^2(D))$ to $L^2(\Omega; L^\infty(0, T; L^2(D)))$, and $u(\omega, \cdot)$ is weakly continuous as a function on $[0, T]$ with values in $L^2(D)$ for $\mathbb{P}$-a.a. $\omega \in \Omega$.

2 Well-Posedness of an Auxiliary Equation

The goal of this section is to prove well-posedness of a version of (1) with additive noise. Namely, we consider the initial value problem

$$du(t) - \operatorname{div}\gamma(\nabla u(t))\,dt \ni G(t)\,dW(t), \qquad u(0) = u_0, \tag{3}$$

where $G \in L^2(\Omega \times [0, T]; \mathscr{L}^2(H, L^2(D)))$ is a measurable and adapted process.

Proposition 1 *Equation* (3) *admits a unique strong solution* (u, η) *satisfying the same integrability and weak continuity conditions of Theorem 1.*

We introduce the regularized equation

$$du_\lambda(t) - \operatorname{div}\gamma_\lambda(\nabla u_\lambda(t))\,dt - \lambda\Delta u_\lambda(t)\,dt = G(t)\,dW(t), \qquad u_\lambda(0) = u_0,$$

where $\gamma_\lambda : \mathbb{R}^n \to \mathbb{R}^n$, $\gamma_\lambda := \frac{1}{\lambda}(I - (I + \lambda\gamma)^{-1})$, for any $\lambda > 0$, is the Yosida approximation of γ, and $\Delta : H_0^1(D) \to H^{-1}(D)$ is the (variational) Dirichlet Laplacian. Since γ_λ is monotone and Lipschitz-continuous, the classical variational approach (see [4, 9] as well as [5]) yields the existence of a unique predictable process u_λ with values in $H_0^1(D)$ such that

$$\mathbb{E}\|u_\lambda\|^2_{C([0,T];L^2(D))} + \mathbb{E}\int_0^T \|u_\lambda(t)\|^2_{H^1_0(D)}\,dt < \infty$$

and

$$u_\lambda(t) - \int_0^t \operatorname{div}\gamma_\lambda(\nabla u_\lambda(s))\,ds - \lambda\int_0^t \Delta u_\lambda(s)\,ds = u_0 + \int_0^t G(s)\,dW(s) \qquad (4)$$

$\mathbb{P}$-a.s. in $H^{-1}(D)$ for all $t \in [0,T]$.

We are now going to prove a priori estimates and weak compactness in suitable topologies for u_λ and related processes. These will allow us to pass to the limit as $\lambda \to 0$ in (4).

For notational parsimony, we shall often write, for any $p \geq 0$, L^p_ω, L^p_t, and L^p_x in place of $L^p(\Omega)$, $L^p(0,T)$, and $L^p(D)$, respectively, and C_t to denote $C([0,T])$. Other similar abbreviations are self-explanatory. The $L^2(D)$-norm will be denoted simply by $\|\cdot\|$. If a function $f : D \to \mathbb{R}^n$ is such that each component f^j, $j = 1,\dots,n$, belongs to $L^p(D)$, we shall just write $f \in L^p(D)$ rather than $f \in L^p(D)^n$. The notation $a \lesssim b$ means that $a \leq Nb$ for a positive constant N.

Lemma 1 *There exists a constant N such that*

$$\begin{aligned}\|u_\lambda\|_{L^2_\omega C_t L^2_x} + \lambda^{1/2}\|\nabla u_\lambda\|_{L^2_{t,\omega,x}} + \|\gamma_\lambda(\nabla u_\lambda)\cdot\nabla u_\lambda\|_{L^1_{t,\omega,x}},\\ < N\big(\|u_0\|_{L^2_{\omega,x}} + \|G\|_{L^2_{t,\omega}\mathscr{L}^2(H,L^2_x)}\big).\end{aligned}$$

Proof Itô's formula for the square of the norm in L^2_x yields

$$\begin{aligned}\|u_\lambda(t)\|^2 + 2\int_0^t\int_D \gamma(\nabla u_\lambda(s))\cdot\nabla u_\lambda(s)\,dx\,ds + 2\lambda\int_0^t \|\nabla u_\lambda(s)\|^2\,ds\\ = \|u_0\|^2 + 2\int_0^t u_\lambda(s)G(s)\,dW(s) + \int_0^t \|G(s)\|^2_{\mathscr{L}^2(H,L^2_x)}\,ds,\end{aligned}$$

hence, taking the supremum in time and expectation,

$$\begin{aligned}\mathbb{E}\|u_\lambda\|^2_{C_t L^2_x} + \mathbb{E}\int_0^T\int_D \gamma_\lambda(\nabla u_\lambda(s))\cdot\nabla u_\lambda(s)\,dx\,ds + \lambda\mathbb{E}\|\nabla u_\lambda\|^2_{L^2_{t,x}}\\ \lesssim \mathbb{E}\|u_0\|^2 + \mathbb{E}\|G\|^2_{L^2_t\mathscr{L}^2(H,L^2_x)} + \mathbb{E}\sup_{t\in[0,T]}\left|\int_0^t u_\lambda(s)G(s)\,dW(s)\right|,\end{aligned}$$

where, by Davis' inequality (see, e.g., [6]), the ideal property of Hilbert–Schmidt operators (see, e.g., [1, p. V.52]), and the elementary inequality $ab \leq \varepsilon a^2 + b^2/\varepsilon$ $\forall a,b \geq 0$, $\varepsilon > 0$,

$$\mathbb{E} \sup_{t\in[0,T]} \left| \int_0^t u_\lambda(s)G(s)\,dW(s) \right| \lesssim \mathbb{E}\biggl(\int_0^T \|u_\lambda(s)G(s)\|^2_{\mathscr{L}^2(H,\mathbb{R})}\,ds \biggr)^{1/2}$$
$$\leq \varepsilon \mathbb{E}\|u_\lambda\|^2_{C_t L^2_x} + N(\varepsilon)\mathbb{E}\int_0^T \|G(s)\|^2_{\mathscr{L}^2(H,L^2_x)}\,ds$$

for any $\varepsilon > 0$. To conclude it suffices to choose ε small enough. □

Lemma 2 *The families* (∇u_λ) *and* $(\gamma_\lambda(\nabla u_\lambda))$ *are relatively weakly compact in* $L^1(\Omega \times (0, T) \times D)$.

Proof Recall that, for any $y, r \in \mathbb{R}^n$, ones has $k(y) + k^*(r) = r \cdot y$ if and only if $r \in \partial k(y) = \gamma(y)$. Therefore, since

$$\gamma_\lambda(x) \in \partial k\bigl((I + \lambda\gamma)^{-1}x\bigr) = \gamma\bigl((I + \lambda\gamma)^{-1}x\bigr) \qquad \forall x \in \mathbb{R}^n,$$

we deduce by the definition of γ_λ that

$$\begin{aligned} k\bigl((I + \lambda\gamma)^{-1}x\bigr) + k^*\bigl(\gamma_\lambda(x)\bigr) &= \gamma_\lambda(x) \cdot (I + \lambda\gamma)^{-1}x \\ &= \gamma_\lambda(x) \cdot x - \lambda\,|\gamma_\lambda(x)|^2 \leq \gamma_\lambda(x) \cdot x \qquad \forall x \in \mathbb{R}^n\,. \end{aligned} \tag{5}$$

(See, e.g., [3] for all necessary facts from convex analysis used in this note.) Hence, taking Lemma 1 into account, there exists a constant $N > 0$, independent of λ, such that

$$\mathbb{E}\int_0^T\!\!\int_D k^*\bigl(\gamma_\lambda(\nabla u_\lambda)\bigr) \leq \mathbb{E}\int_0^T\!\!\int_D \gamma_\lambda(\nabla u_\lambda) \cdot \nabla u_\lambda < N.$$

The assumptions on k imply that its convex conjugate k^* is also convex, lower semicontinuous and such that $\lim_{|y|\to\infty} k^*(y)/|y| = +\infty$. Therefore a simple modification of the criterion by de la Vallée Poussin implies that $(\gamma_\lambda(\nabla u_\lambda))$ is uniformly integrable on $\Omega \times (0, T) \times D$, hence that it is relatively weakly compact in $L^1_{t,\omega,x}$ by the Dunford–Pettis theorem. A completely analogous argument shows that

$$\mathbb{E}\int_0^T\!\!\int_D k\bigl((I + \lambda\gamma)^{-1}\nabla u_\lambda\bigr) \leq \mathbb{E}\int_0^T\!\!\int_D \gamma_\lambda(\nabla u_\lambda) \cdot \nabla u_\lambda < N,$$

hence that $(I + \lambda\gamma)^{-1}\nabla u_\lambda$ is relatively weakly compact in $L^1_{t,\omega,x}$. Moreover, since $\nabla u_\lambda = (I + \lambda\gamma)^{-1}\nabla u_\lambda + \lambda\gamma_\lambda(\nabla u_\lambda)$, it also follows that (∇u_λ) is relatively weakly compact in $L^1_{t,\omega,x}$. □

Thanks to Lemmata 1 and 2, there exists a subsequence of λ, denoted by the same symbol, and processes $u \in L^\infty_t L^2_{\omega,x} \cap L^1_{t,\omega} W^{1,1}_0$ and $\eta \in L^1_{t,\omega,x}$ such that

$$\begin{aligned}
u_\lambda &\longrightarrow u && \text{weakly* in } L^\infty_t L^2_{\omega,x},\\
u_\lambda &\longrightarrow u && \text{weakly in } L^1_{t,\omega} W^{1,1}_0,\\
\gamma_\lambda(\nabla u_\lambda) &\longrightarrow \eta && \text{weakly in } L^1_{t,\omega,x},\\
\lambda u_\lambda &\longrightarrow 0 && \text{weakly in } L^2_{t,\omega} H^1_0.
\end{aligned}$$

as $\lambda \to 0$. Let $t \in [0, T]$ be arbitrary but fixed. The fourth convergence above implies

$$\lambda \int_0^t \Delta u_\lambda(s)\,ds \longrightarrow 0 \qquad \text{in } L^2_\omega H^{-1},$$

while the third yields, for any $\varphi \in L^\infty_\omega W^{1,\infty}$,

$$\mathbb{E}\int_0^t\int_D \gamma_\lambda(\nabla u_\lambda(s)) \cdot \nabla\varphi\,dx\,ds \longrightarrow \mathbb{E}\int_0^t\int_D \eta(s)\cdot\nabla\varphi\,dx\,ds,$$

hence $\mathbb{E}\int_0^t \langle \operatorname{div}\gamma_\lambda(\nabla u_\lambda(s)), \varphi\rangle\,ds \longrightarrow \mathbb{E}\int_0^t \langle \operatorname{div}\eta(s), \varphi\rangle\,ds$. Therefore, recalling (4), by difference we deduce that

$$\mathbb{E}\langle u_\lambda(t), \varphi\rangle \longrightarrow \mathbb{E}\langle u(t), \varphi\rangle.$$

Consequently, since $u_\lambda(t)$ is bounded in $L^2_\omega L^2_x$, we also have that $u_\lambda(t) \to u(t)$ weakly in $L^2_\omega L^2_x$. Taking the limit as $\lambda \to 0$ in (4) thus yields

$$u(t) - \int_0^t \operatorname{div}\eta(s)\,ds = u_0 + \int_0^t G(s)\,dW(s) \qquad \text{in } L^1_\omega V_0',$$

where V_0' is the (topological) dual of a separable Hilbert space V_0 embedded continuously and densely in H^1_0, and continuously in $W^{1,\infty}$. The identity immediately implies that $u \in C_t L^1_\omega V_0'$. Since $u \in L^\infty_t L^2_\omega L^2_x$, it follows by a result of Strauss (see [11, Theorem 2.1]) that u is a weakly continuous function on $[0, T]$ with values in $L^2_\omega L^2_x$.

By Mazur's lemma there exist sequences of convex combinations of $\gamma_\lambda(\nabla u_\lambda)$ that converge η in (the norm topology of) $L^1_{t,\omega,x}$, thus also, passing to a subsequence, $\mathbb{P}\otimes dt$-almost everywhere in L^1_x. Similarly, since $u_\lambda \to u$ weakly* in $L^\infty_t L^2_{\omega,x}$ implies that $u_\lambda \to u$ weakly in $L^2_{t,\omega,x}$, there exist sequences of convex combinations of u_λ that converge to u $\mathbb{P}\otimes dt$-almost everywhere in L^2_x. Since convex combinations of (u_λ) and of $(\gamma_\lambda(\nabla u_\lambda))$ are (at least) predictable and adapted, respectively, it follows that u is predictable and η is measurable and adapted. Moreover, thanks to the weak lower semicontinuity of convex integrals, one has

$$\mathbb{E}\int_0^T\int_D \big(k(\nabla u) + k^*(\eta)\big) < \infty.$$

In order to show that $\eta \in \gamma(\nabla u)$ for a.a. (ω, t, x), we need the following "energy identity".

Lemma 3 *Assume that*

$$y(t) + \alpha \int_0^t y(s)\, ds - \int_0^t \operatorname{div}\zeta(s)\, ds = y_0 + \int_0^t C(s)\, dW(s)$$

in L^2_x $\mathbb{P}$-a.s. for all $t \in [0, T]$, where $\alpha \in \mathbb{R}$, $y_0 \in L^2_{\omega,x}$ is $\mathscr{F}_0$-measurable, and

$$y \in L^\infty_t L^2_{\omega,x} \cap L^1_{t,\omega} W^{1,1}_0, \qquad \zeta \in L^1_{t,\omega,x}, \qquad C \in L^2_{t,\omega}\mathscr{L}^2(H, L^2_x)$$

are measurable and adapted processes such that $k(c\nabla y) + k^*(c\zeta) \in L^1_{t,\omega,x}$ for a constant $c > 0$. Then

$$\begin{aligned}\mathbb{E}\|y(t)\|^2 &+ 2\alpha\mathbb{E}\int_0^t \|y(s)\|^2\, ds + 2\mathbb{E}\int_0^t\int_D \zeta \cdot \nabla y\, dx\, ds\\ &= \mathbb{E}\|y_0\|^2 + \mathbb{E}\int_0^t \|C(s)\|^2_{\mathscr{L}^2(H,L^2_x)}\, ds \qquad \forall t \in [0, T].\end{aligned}$$

Proof Let $m \in \mathbb{N}$ be such that such that $(I - \delta\Delta)^{-m}$ maps L^1_x into $H^1_0 \cap W^{1,\infty}$, and use the notation $h^\delta := (I - \delta\Delta)^{-m} h$ for any h taking values in L^1_x. One has

$$y^\delta(t) + \alpha \int_0^t y^\delta(s)\, ds - \int_0^t \operatorname{div}\zeta^\delta(s)\, ds = y^\delta_0 + \int_0^t C^\delta(s)\, dW(s) \tag{6}$$

$\mathbb{P}$-a.s. for all $t \in [0, T]$, as an equality in L^2_x, for which Itô's formula yields

$$\begin{aligned}\|y^\delta(t)\|^2 &+ 2\alpha \int_0^t \|y^\delta(s)\|^2\, ds + 2\int_0^t\int_D \zeta^\delta \cdot \nabla y^\delta\, dx\, ds\\ &= \|y^\delta_0\|^2 + \int_0^t \|C^\delta(s)\|^2_{\mathscr{L}^2(H,L^2_x)}\, ds + \int_0^t y^\delta(s)C^\delta(s)dW(s).\end{aligned}$$

It is evident from (6) that y^δ is a continuous L^2_x-valued process, hence the stochastic integral $(y^\delta C^\delta) \cdot W$ on the right-hand side of the above identity is a continuous local martingale. Let (T_n) be a localizing sequence, and multiply the previous identity by $1_{[0,T_n]}$, to obtain, thanks to $\mathbb{E}(y^\delta C^\delta) \cdot W(\cdot \wedge T_n) = 0$,

$$\begin{aligned}\mathbb{E}\|y^\delta(t \wedge T_n)\|^2 &+ 2\alpha\mathbb{E}\int_0^{t\wedge T_n} \|y^\delta(s)\|^2\, ds + 2\mathbb{E}\int_0^{t\wedge T_n}\int_D \zeta^\delta \cdot \nabla y^\delta\, dx\, ds\\ &= \mathbb{E}\|y^\delta_0\|^2 + \mathbb{E}\int_0^{t\wedge T_n} \|C^\delta(s)\|^2_{\mathscr{L}^2(H,L^2_x)}\, ds.\end{aligned}$$

Letting n tend to ∞, the dominated convergence theorem yields

$$\begin{aligned}\mathbb{E}\|y^\delta(t)\|^2 + 2\alpha\mathbb{E}\int_0^t \|y^\delta(s)\|^2\,ds + 2\mathbb{E}\int_0^t\int_D \zeta^\delta\cdot\nabla y^\delta\,dx\,ds\\ = \mathbb{E}\|y_0^\delta\|^2 + \mathbb{E}\int_0^t \|C^\delta(s)\|^2_{\mathscr{L}^2(H,L^2_x)}\,ds\end{aligned}$$

for all $t\in[0,T]$. We are now going to pass to the limit as $\delta\to 0$: the first and second terms on the left-hand side and the first on the right-hand side clearly converge to $\mathbb{E}\|y(t)\|^2$, $2\alpha\mathbb{E}\int_0^t\|y(s)\|^2\,ds$ and $\mathbb{E}\|y_0\|^2$, respectively. Properties of Hilbert–Schmidt operators and the dominated convergence theorem also yield

$$\lim_{\delta\to 0}\mathbb{E}\int_0^t \|C^\delta(s)\|^2_{\mathscr{L}^2(H,L^2_x)}\,ds = \mathbb{E}\int_0^t \|C(s)\|^2_{\mathscr{L}^2(H,L^2_x)}\,ds$$

for all $t\in[0,T]$. To conclude it then suffices to show that $\nabla y^\delta\cdot\zeta^\delta\to\nabla y\cdot\zeta$ in $L^1_{t,\omega,x}$. Since $\nabla y^\delta\to\nabla y$ and $\zeta^\delta\to\zeta$ in measure in $\Omega\times(0,t)\times D$, Vitali's theorem implies strong convergence in $L^1_{t,\omega,x}$ if the sequence $(\nabla y^\delta\cdot\zeta^\delta)$ is uniformly integrable in $\Omega\times(0,t)\times D$. In turn, the latter is certainly true if $(|\nabla y^\delta\cdot\zeta^\delta|)$ is dominated by a sequence that converges strongly in $L^1_{t,\omega,x}$. Indeed, using the assumptions on the behavior of k at infinity as well as the generalized Jensen inequality for sub-Markovian operators (see [2]), one has

$$\pm c^2\zeta^\delta\cdot\nabla y^\delta \lesssim 1 + k(c\nabla y^\delta) + k^*(c\zeta^\delta) \le 1 + (I-\delta\Delta)^{-m}\left(k(c\nabla y) + k^*(c\zeta)\right),$$

where the sequence on the right-hand side converges in $L^1_{t,\omega,x}$ as $\delta\to 0$ because, by assumption, $k(c\nabla y)+k^*(c\zeta)\in L^1_{t,\omega,x}$. $\square$

Itô's formula yields

$$\begin{aligned}\mathbb{E}\|u_\lambda(t)\|^2 + 2\mathbb{E}\int_0^t\int_D \gamma_\lambda(\nabla u_\lambda)\cdot\nabla u_\lambda + 2\lambda\mathbb{E}\int_0^t\|\nabla u_\lambda\|^2\\ = \mathbb{E}\|u_0\|^2 + \mathbb{E}\int_0^t \|G(s)\|^2_{\mathscr{L}^2(H,L^2_x)}\,ds\end{aligned}$$

and, by Lemma 3,

$$\mathbb{E}\|u(t)\|^2 + 2\mathbb{E}\int_0^t\int_D \eta\cdot\nabla u = \mathbb{E}\|u_0\|^2 + \mathbb{E}\int_0^t \|G(s)\|^2_{\mathscr{L}^2(H,L^2_x)}\,ds.$$

One then has

$$
\begin{aligned}
2 \limsup_{\lambda\to 0} \mathbb{E} \int_0^T \int_D \gamma_\lambda(\nabla u_\lambda(s)) \cdot \nabla u_\lambda(s)\, dx\, ds \\
&\leq \mathbb{E}\|u_0\|^2 + \mathbb{E}\int_0^T \|G(s)\|^2_{\mathscr{L}^2(H,L^2_x)}\, ds - \liminf_{\lambda\to 0} \mathbb{E}\|u_\lambda(T)\|^2 \\
&\leq \mathbb{E}\|u_0\|^2 + \mathbb{E}\int_0^T \|G(s)\|^2_{\mathscr{L}^2(H,L^2_x)}\, ds - \mathbb{E}\|u(T)\|^2 \\
&= \mathbb{E}\int_0^T \int_D \eta(s) \cdot \nabla u(s)\, dx\, ds.
\end{aligned}
$$

Since $\nabla u_\lambda \to \nabla u$ and $\gamma_\lambda(\nabla u_\lambda) \to \eta$ weakly in $L^1_{t,\omega,x}$, this implies that $\eta \in \gamma(\nabla u)$ a.e. in $\Omega \times (0, T) \times D$. We have thus proved the existence and weak continuity statements of Proposition 1.

In order to show that the solution is unique, we are going to prove that *any* solution depends continuously on (u_0, G). Let (u_i, η_i), $i = 1, 2$, satisfy

$$
u_i(t) - \int_0^t \operatorname{div}\eta_i(s)\, ds = u_0 + \int_0^t G_i(s)\, ds
$$

in the sense of Definition 1, as well as the integrability conditions (on u and η) of Theorem 1. Setting $y := u_1 - u_2$, $y_0 := u_{01} - u_{02}$, $\zeta := \eta_1 - \eta_2$, and $F := G_1 - G_2$, one has

$$
y(t) - \int_0^t \operatorname{div}\zeta(s)\, ds = y_0 + \int_0^t F(s)\, dW(s)
$$

$\mathbb{P}$-a.s. in $L^2(D)$ for all $t \in [0, T]$. For any process h, let us use the notation $h^\alpha(t) := e^{-\alpha t} h(t)$. For any $\alpha > 0$, the integration-by-parts formula yields

$$
y^\alpha(t) + \int_0^t (-\operatorname{div}\zeta^\alpha(s) + \alpha y^\alpha(s))\, ds = y_0 + \int_0^t F^\alpha(s)\, dW(s),
$$

hence also, thanks to Lemma 3,

$$
\begin{aligned}
\mathbb{E}\|y^\alpha(t)\|^2 + 2\alpha\mathbb{E}\int_0^t \|y^\alpha(s)\|^2\, ds + 2\mathbb{E}\int_0^t \int_D \zeta^\alpha(s) \cdot \nabla y^\alpha(s)\, dx\, ds \\
&\leq \mathbb{E}\|y_0\|^2 + \mathbb{E}\int_0^t \|F^\alpha(s)\|^2_{\mathscr{L}^2(H,L^2_x)}\, ds,
\end{aligned}
$$

where $\zeta^\alpha \cdot \nabla y^\alpha \geq 0$ by monotonicity. Therefore, taking the L^∞_t norm,

$$
\|y^\alpha\|_{L^\infty_t L^2_{\omega,x}} + \sqrt{\alpha}\|y^\alpha\|_{L^2_{t,\omega,x}} \lesssim \|y_0\|_{L^2_{\omega,x}} + \|F^\alpha\|_{L^2_{t,\omega}\mathscr{L}^2(H,L^2_x)},
$$

that is, using the notation $L^p_t(\alpha) := L^p([0,T], e^{-\alpha t}dt)$ for any $p \geq 0$,

$$\begin{aligned}\|u_1 - u_2\|_{L^\infty_t(\alpha)L^2_{\omega,x}} &+ \sqrt{\alpha}\|u_1 - u_2\|_{L^2_t(\alpha)L^2_{\omega,x}} \\ &\lesssim \|u_{01} - u_{02}\|_{L^2_{\omega,x}} + \|G_1 - G_2\|_{L^2_t(\alpha)L^2_\omega \mathscr{L}^2(H,L^2_x)}.\end{aligned} \tag{7}$$

Taking $\alpha = 0$ and $G_1 = G_2$ immediately yields the uniqueness of solutions (as well as Lipschitz-continuous dependence on the initial datum). The proof of Proposition 1 is thus complete.

3 Proof of Theorem 1

For any $v \in L^2_{t,\omega,x}$ measurable and adapted, and any $\mathscr{F}_0$-measurable random variable $u_0 \in L^2_{\omega,x}$, the process $B(\cdot, v)$ is measurable, adapted, and belongs to $L^2_{t,\omega}\mathscr{L}^2(H, L^2_x)$, hence the equation

$$du(t) - \operatorname{div}\gamma(\nabla u(t))\,dt \ni B(t, v(t))\,dW(t), \qquad u(0) = u_0,$$

is well-posed in the sense of Proposition 1. Moreover, for any v_1, v_2 and u_{01}, u_{02} satisfying the same hypotheses on v and u_0, respectively, (7) yields

$$\begin{aligned}\|u_1 - u_2\|_{L^\infty_t(\alpha)L^2_{\omega,x}} &+ \sqrt{\alpha}\|u_1 - u_2\|_{L^2_t(\alpha)L^2_{\omega,x}} \\ &\lesssim \|u_{01} - u_{02}\|_{L^2_{\omega,x}} + \|B(\cdot, v_1) - B(\cdot, v_2)\|_{L^2_t(\alpha)L^2_\omega \mathscr{L}^2(H,L^2_x)}.\end{aligned}$$

It hence follows by the Lipschitz-continuity of B that

$$\|u_1 - u_2\|_{L^2_t(\alpha)L^2_{\omega,x}} \lesssim \frac{1}{\sqrt{\alpha}}\Big(\|u_{01} - u_{02}\|_{L^2_{\omega,x}} + \|v_1 - v_2\|_{L^2_t(\alpha)L^2_{\omega,x}}\Big), \tag{8}$$

where the implicit constant does not depend on α. In particular, denoting by Γ the map $(u_0, v) \mapsto u$, one has that $v \mapsto \Gamma(u_0, v)$ is a strict contraction of $L^2_t(\alpha)L^2_{\omega,x}$ for α large enough. Therefore, by equivalence of norms, $v \mapsto \Gamma(u_0, v)$ admits a unique fixed point in $L^2_{t,\omega,x}$, which solves (1) and satisfies all integrability conditions. Such solution is unique as any solution is a fixed point of $v \mapsto \Gamma(u_0, v)$.

Let us show that the solution map $u_0 \mapsto u$ is Lipschitz-continuous: (8) yields, choosing α large enough,

$$\|u_1 - u_2\|_{L^2_t(\alpha)L^2_{\omega,x}} \leq N_1\|u_{01} - u_{02}\|_{L^2_{\omega,x}} + N_2\|u_1 - u_2\|_{L^2_t(\alpha)L^2_{\omega,x}}$$

with constants $N_1 > 0$ and $0 < N_2 < 1$, hence, by equivalence of norms,

$$\|u_1 - u_2\|_{L^2_t L^2_{\omega,x}} \lesssim \|u_{01} - u_{02}\|_{L^2_{\omega,x}}.$$

This in turn implies, in view of (7) (with $\alpha = 0$) and the Lipschitz-continuity of B,

$$\begin{aligned}\|u_1 - u_2\|_{L_t^\infty L^2_{\omega,x}} &\lesssim \|u_{01} - u_{02}\|_{L^2_{\omega,x}} + \|B(\cdot, u_1) - B(\cdot, u_2)\|_{L^2_{t,\omega}\mathscr{L}^2(H,L^2_x)} \\ &\lesssim \|u_{01} - u_{02}\|_{L^2_{\omega,x}} + \|u_1 - u_2\|_{L^2_t L^2_{\omega,x}} \lesssim \|u_{01} - u_{02}\|_{L^2_{\omega,x}},\end{aligned}$$

which completes the proof.

Remark A priori estimates entirely analogous to those of Lemma 1, as well as weak compactness results exactly as in Lemma 2, can be proved for the regularized equation obtained by replacing γ with $\gamma_\lambda + \lambda\nabla$ directly in (1). It is however not immediately clear how to pass to the limit as $\lambda \to 0$ in the stochastic integrals appearing in such regularized equations with multiplicative noise, i.e. to show that $B(u_\lambda) \cdot W$ converges to $B(u) \cdot W$ in a suitable sense.

Acknowledgements The authors are partially supported by The Royal Society through its International Exchange Scheme. Parts of this chapter were written while the first-named author was visiting the Interdisziplinäres Zentrum für Komplexe Systeme at the University of Bonn, hosted by Prof. S. Albeverio.

References

1. Bourbaki, N.: Espaces vectoriels topologiques. Chapitres 1 à 5, new edn. Masson, Paris (1981). MR 633754
2. Haase, M.: Convexity inequalities for positive operators. Positivity **11**(1), 57–68 (2007). MR 2297322 (2008d:39034)
3. Hiriart-Urruty, J.-B., Lemaréchal, C.: Fundamentals of Convex Analysis. Springer, Berlin (2001). MR 1865628 (2002i:90002)
4. Krylov, N.V., Rozovskiĭ, B.L.: Current problems in mathematics. Stochastic Evolution Equations, vol. 14 (Russian), pp. 71–147, 256. Akad. Nauk SSSR, Vsesoyuz. Inst. Nauchn. i Tekhn. Informatsii, Moscow (1979). MR MR570795 (81m:60116)
5. Liu, W., Röckner, M.: Stochastic Partial Differential Equations: An Introduction. Springer, Cham (2015). MR 3410409
6. Marinelli, C., Röckner, M.: On the maximal inequalities of Burkholder, Davis and Gundy. Expo. Math. **34**(1), 1–26 (2016). MR 3463679
7. Marinelli, C., Scarpa, L.: Strong solutions to SPDEs with monotone drift in divergence form. arXiv:1612.08260
8. Marinelli, C., Scarpa, L.: A note on doubly nonlinear SPDEs with singular drift in divergence form. arXiv:1712.05595
9. Pardoux, E.: Equations aux derivées partielles stochastiques nonlinéaires monotones, Ph.D. thesis, Université Paris XI (1975)
10. Scarpa, L.: Well-posedness for a class of doubly nonlinear stochastic PDEs of divergence type. J. Differ. Eqn. **263**(4), 2113–2156 (2017)
11. Strauss, W.A.: On continuity of functions with values in various Banach spaces. Pacific J. Math. **19**, 543–551 (1966). MR 0205121 (34 #4956)

Lower Bounds for Weak Approximation Errors for Spatial Spectral Galerkin Approximations of Stochastic Wave Equations

Ladislas Jacobe de Naurois, Arnulf Jentzen and Timo Welti

Abstract Although for a number of semilinear stochastic wave equations existence and uniqueness results for corresponding solution processes are known from the literature, these solution processes are typically not explicitly known and numerical approximation methods are needed in order for mathematical modelling with stochastic wave equations to become relevant for real world applications. Therefore, the numerical analysis of convergence rates for such numerical approximation processes is required. A recent article by the authors proves upper bounds for weak errors for spatial spectral Galerkin approximations of a class of semilinear stochastic wave equations. The findings there are complemented by the main result of this work, that provides lower bounds for weak errors which show that in the general framework considered the established upper bounds can essentially not be improved.

Keywords Stochastic wave equations · Weak convergence · Lower bounds · Essentially sharp convergence rates · Spectral Galerkin approximations

Mathematics Subject Classification classes 60H15 · 65C30

1 Introduction

In this work we consider numerical approximation processes of solution processes of stochastic wave equations and examine corresponding weak convergence properties. As opposed to strong convergence, weak convergence even in the case of stochastic evolution equations with regular nonlinearities is still only poorly understood (see, e.g., [3, 6–8, 12] for several weak convergence results for stochastic wave equations and, e.g., the references in Sect. 1 in [4] for further results on weak convergence in the

L. Jacobe de Naurois · A. Jentzen (✉) · T. Welti (✉)
ETH Zürich, Zürich, Switzerland
e-mail: arnulf.jentzen@sam.math.ethz.ch

T. Welti
e-mail: timo.welti@sam.math.ethz.ch

A. Eberle et al. (eds.), *Stochastic Partial Differential Equations and Related Fields*, Springer Proceedings in Mathematics & Statistics 229,
https://doi.org/10.1007/978-3-319-74929-7_13

literature). Therefore, equations available to current numerical analysis are limited to model problems, such as the ones considered in the present article, that cannot take into account the full complexity of models for evolutionary processes under influence of randomness appearing in real world applications (see, e.g., the references in Sect. 1 in [4]). The recent article [4] by the authors provides upper bounds for weak errors for spatial spectral Galerkin approximations of a class of semilinear stochastic wave equations, including equations driven by multiplicative noise and, in particular, the hyperbolic Anderson model. The main result of this article, Theorem 1.1 below, in turn shows that the weak convergence rates for stochastic wave equations established in Theorem 1.1 in [4] can in the general setting there *essentially not be improved*. Theorem 1.1 is obtained by proving lower bounds for weak errors in the case of concrete examples of stochastic wave equations with additive noise and without drift nonlinearity (cf. Corollary 2.10 and (1.4) below). We argue similarly to the reasoning in Sect. 7 in Conus et al. [1] and Sect. 9 in Jentzen and Kurniawan [5]. First results on lower bounds for strong errors for two examples of stochastic heat equations were achieved in Davie and Gaines [2]. Furthermore, lower bounds for strong errors for examples and whole classes of stochastic heat equations have been established in Müller-Gronbach et al. [10] (see also the references therein) and in Müller-Gronbach and Ritter [9], respectively. Results on lower bounds for weak errors in the case of a few specific examples of stochastic heat equations can be found in Conus et al. [1] and in Jentzen and Kurniawan [5].

Theorem 1.1 *For all real numbers $\eta, T \in (0,\infty)$, every $\mathbb{R}$-Hilbert space $(H, \langle\cdot,\cdot\rangle_H, \|\cdot\|_H)$, every orthonormal basis $(e_n)_{n\in\mathbb{N}=\{1,2,3,\dots\}}\colon \mathbb{N}\to H$ of H, every probability space $(\Omega,\mathcal{F},\mathbb{P})$ with a normal filtration $(\mathbb{F}_t)_{t\in[0,T]}$, and every id_H-cylindrical $(\Omega,\mathcal{F},\mathbb{P},(\mathbb{F}_t)_{t\in[0,T]})$-Wiener process $(W_t)_{t\in[0,T]}$ there exist a strictly increasing sequence $(\lambda_n)_{n\in\mathbb{N}}\colon \mathbb{N}\to(0,\infty)$, a linear operator $A\colon D(A)\subseteq H\to H$ with $D(A)=\{v\in H\colon \sum_{n=1}^{\infty}|\lambda_n\langle e_n,v\rangle_H|^2<\infty\}$ and $\forall\, v\in D(A)\colon Av=\sum_{n=1}^{\infty}-\lambda_n\langle e_n,v\rangle_H e_n$, a family of interpolation spaces $(H_r,\langle\cdot,\cdot\rangle_{H_r},\|\cdot\|_{H_r})$, $r\in\mathbb{R}$, associated to $-A$ (cf., e.g., [11, Section 3.7]), a family of $\mathbb{R}$-Hilbert spaces $(\mathbf{H}_r,\langle\cdot,\cdot\rangle_{\mathbf{H}_r},\|\cdot\|_{\mathbf{H}_r})$, $r\in\mathbb{R}$, with $\forall\, r\in\mathbb{R}\colon \quad (\mathbf{H}_r,\langle\cdot,\cdot\rangle_{\mathbf{H}_r},\|\cdot\|_{\mathbf{H}_r}) = \big(H_{r/2}\times H_{r/2-1/2},\langle\cdot,\cdot\rangle_{H_{r/2}\times H_{r/2-1/2}},\|\cdot\|_{H_{r/2}\times H_{r/2-1/2}}\big)$, families of functions $P_N\colon \bigcup_{r\in\mathbb{R}}H_r\to\bigcup_{r\in\mathbb{R}}H_r$, $N\in\mathbb{N}\cup\{\infty\}$, and $\mathbf{P}_N\colon \bigcup_{r\in\mathbb{R}}\mathbf{H}_r\to\bigcup_{r\in\mathbb{R}}\mathbf{H}_r$, $N\in\mathbb{N}\cup\{\infty\}$, with $\forall\, N\in\mathbb{N}\cup\{\infty\}, r\in\mathbb{R}, u\in H_r, (v,w)\in\mathbf{H}_r\colon \big(P_N(u)=\sum_{n=1}^{N}\langle(\lambda_n)^{-r}e_n,u\rangle_{H_r}(\lambda_n)^{-r}e_n$ and $\mathbf{P}_N(v,w)=(P_N(v),P_N(w))\big)$, a linear operator $\mathbf{A}\colon D(\mathbf{A})\subseteq\mathbf{H}_0\to\mathbf{H}_0$ with $D(\mathbf{A})=\mathbf{H}_1$ and $\forall\,(v,w)\in\mathbf{H}_1\colon \mathbf{A}(v,w)=(w,Av)$, real numbers $\gamma,c\in(0,\infty)$, a vector $\xi\in\mathbf{H}_\gamma$, and functions $\varphi\in C^2_b(\mathbf{H}_0,\mathbb{R})$, $\mathbf{F}\in C^2_b(\mathbf{H}_0,\mathbf{H}_0)$, $\mathbf{B}\in C^2_b(\mathbf{H}_0,\mathrm{HS}(H,\mathbf{H}_0))$, and $(C_\varepsilon)_{\varepsilon\in(0,\infty)}\colon(0,\infty)\to[0,\infty)$ with $\forall\,\beta\in(\gamma/2,\gamma]\colon (-A)^{-\beta/2}\in\mathrm{HS}(H)$, $\mathbf{F}(\mathbf{H}_0)\subseteq\mathbf{H}_\gamma$, $(\mathbf{H}_0\ni v\mapsto\mathbf{F}(v)\in\mathbf{H}_\gamma)\in C^2_b(\mathbf{H}_0,\mathbf{H}_\gamma)$, $\forall\, v\in\mathbf{H}_0, u\in H\colon \mathbf{B}(v)u\in\mathbf{H}_\gamma$, $\forall\, v\in\mathbf{H}_0\colon (H\ni u\mapsto\mathbf{B}(v)u\in\mathbf{H}_\gamma)\in L(H,\mathbf{H}_\gamma)$, and $(\mathbf{H}_0\ni v\mapsto(H\ni u\mapsto\mathbf{B}(v)u\in\mathbf{H}_\gamma)\in L(H,\mathbf{H}_\gamma))\in C^2_b(\mathbf{H}_0,L(H,\mathbf{H}_\gamma))$ such that*

(i) it holds that there exist up to modifications unique $(\mathbb{F}_t)_{t\in[0,T]}$-predictable stochastic processes $\mathbf{X}^N\colon[0,T]\times\Omega\to\mathbf{P}_N(\mathbf{H}_0)$, $N\in\mathbb{N}\cup\{\infty\}$, which satisfy for all $N\in\mathbb{N}\cup\{\infty\}$, $t\in[0,T]$ that $\sup_{s\in[0,T]}\mathbb{E}\big[\|\mathbf{X}^N_s\|^2_{\mathbf{H}_0}\big]<\infty$ and $\mathbb{P}$-a.s. that

$$\mathbf{X}_t^N = e^{t\mathbf{A}}\mathbf{P}_N\xi + \int_0^t e^{(t-s)\mathbf{A}}\mathbf{P}_N\mathbf{F}(\mathbf{X}_s^N)\,ds + \int_0^t e^{(t-s)\mathbf{A}}\mathbf{P}_N\mathbf{B}(\mathbf{X}_s^N)\,dW_s \quad (1.1)$$

(ii) and it holds for all $\varepsilon \in (0, \infty)$, $N \in \mathbb{N}$ *that*

$$c \cdot (\lambda_N)^{-\eta} \le \left|\mathbb{E}\big[\varphi(\mathbf{X}_T^\infty)\big] - \mathbb{E}\big[\varphi(\mathbf{X}_T^N)\big]\right| \le C_\varepsilon \cdot (\lambda_N)^{\varepsilon-\eta}. \quad (1.2)$$

Here and below we denote for every non-trivial $\mathbb{R}$-Hilbert space $(V, \langle\cdot,\cdot\rangle_V, \|\cdot\|_V)$ and every $\mathbb{R}$-Hilbert space $(W, \langle\cdot,\cdot\rangle_W, \|\cdot\|_W)$ by $C_b^2(V, W)$ the set of all globally bounded twice continuously Fréchet differentiable functions from V to W with globally bounded derivatives. In the following we provide a few further comments regarding the statement and the proof of Theorem 1.1. The initial value ξ and the functions $\mathbf{F}$ and $\mathbf{B}$ in the setting of Theorem 1.1 can be chosen in such a way that there exist appropriate $\xi_0 \in H$, $\xi_1 \in H_{-1/2}$ and appropriate functions $F\colon \mathbf{H}_0 \to H_{-1/2}$, $B\colon \mathbf{H}_0 \to \mathrm{HS}(H, H_{-1/2})$ such that $\xi = (\xi_0, \xi_1)$, $\mathbf{F} = (0, F)$, and $\mathbf{B} = (0, B)$. In this case, for every $N \in \mathbb{N} \cup \{\infty\}$ the first component process $X^N\colon [0, T] \times \Omega \to P_N(H)$ of $\mathbf{X}^N$ is, roughly speaking, a mild solution of the stochastic wave-type evolution equation

$$\ddot{X}_t = AX_t + P_N F(X_t, \dot{X}_t) + P_N B(X_t, \dot{X}_t)\dot{W}_t \quad (1.3)$$

with $X_0 = P_N\xi_0$, $\dot{X}_0 = P_N\xi_1$ for $t \in [0, T]$. Theorem 1.1 is a direct consequence of Theorem 1.1 in [4] (with $\gamma = 2\eta$, $\beta = \min\{\eta + \varepsilon, 2\eta\}$, $\rho = 0$ in the notation of Theorem 1.1 in [4]) and Corollary 2.10 below (with $p = 1/\eta$, $\delta = 1/2 - \eta$ in the notation of Corollary 2.10 below). In the case $\eta = 1/2$, the lower bound in (1.2) is obtained, for example, for the stochastic wave equations

$$\ddot{X}_t(x) = \tfrac{\partial^2}{\partial x^2} X_t(x) + P_N \dot{W}_t(x) \quad (1.4)$$

with $X_0(x) = \dot{X}_0(x) = 0$ and $X_t(0) = X_t(1) = 0$ for $x \in (0, 1)$, $t \in [0, T]$, $N \in \mathbb{N} \cup \{\infty\}$, corresponding to the choices $H = L^2((0, 1); \mathbb{R})$, $\forall n \in \mathbb{N}$: $e_n = \sqrt{2}\sin(n\pi(\cdot)) \in H, \forall n \in \mathbb{N}\colon \lambda_n = \pi^2 n^2, \xi = 0, \mathbf{F} = 0, \mathbf{B} = (\mathbf{H}_0 \ni (v, w) \mapsto (H \ni u \mapsto (0, u) \in \mathbf{H}_0) \in \mathrm{HS}(H, \mathbf{H}_0))$ in the setting of Theorem 1.1 (cf. Corollary 2.11 below). Inequality (1.2) reveals that the weak convergence rates in Theorem 1.1 in [4] are essentially sharp. More details and further lower bounds for weak approximation errors for stochastic wave equations can be found in Corollary 2.8 and Corollary 2.10 below.

2 Lower Bounds for Weak Errors

2.1 Setting

Let $(H, \langle\cdot,\cdot\rangle_H, \|\cdot\|_H)$ be a separable $\mathbb{R}$-Hilbert space, for every set A let $\mathcal{P}(A)$ be the power set of A, let $T \in (0, \infty)$, let $(\Omega, \mathcal{F}, \mathbb{P})$ be a probability space with a normal filtration $(\mathbb{F}_t)_{t\in[0,T]}$, let $(W_t)_{t\in[0,T]}$ be an id_H-cylindrical $(\Omega, \mathcal{F}, \mathbb{P}, (\mathbb{F}_t)_{t\in[0,T]})$-Wiener process, let $\mathbb{H} \subseteq H$ be a non-empty orthonormal basis of H, let $\lambda\colon \mathbb{H} \to \mathbb{R}$ be a function with $\sup_{h\in\mathbb{H}} \lambda_h < 0$, let $A\colon D(A) \subseteq H \to H$ be the linear operator which satisfies $D(A) = \{v \in H\colon \sum_{h\in\mathbb{H}} |\lambda_h \langle h, v\rangle_H|^2 < \infty\}$ and $\forall\, v \in D(A)\colon Av = \sum_{h\in\mathbb{H}} \lambda_h \langle h, v\rangle_H h$, let $(H_r, \langle\cdot,\cdot\rangle_{H_r}, \|\cdot\|_{H_r})$, $r \in \mathbb{R}$, be a family of interpolation spaces associated to $-A$, let $(\mathbf{H}_r, \langle\cdot,\cdot\rangle_{\mathbf{H}_r}, \|\cdot\|_{\mathbf{H}_r})$, $r \in \mathbb{R}$, be the family of $\mathbb{R}$-Hilbert spaces which satisfies for all $r \in \mathbb{R}$ that $(\mathbf{H}_r, \langle\cdot,\cdot\rangle_{\mathbf{H}_r}, \|\cdot\|_{\mathbf{H}_r}) = \big(H_{r/2} \times H_{r/2-1/2}, \langle\cdot,\cdot\rangle_{H_{r/2}\times H_{r/2-1/2}}, \|\cdot\|_{H_{r/2}\times H_{r/2-1/2}}\big)$, let $P_I\colon \bigcup_{r\in\mathbb{R}} H_r \to \bigcup_{r\in\mathbb{R}} H_r$, $I \in \mathcal{P}(\mathbb{H})$, and $\mathbf{P}_I\colon \bigcup_{r\in\mathbb{R}} \mathbf{H}_r \to \bigcup_{r\in\mathbb{R}} \mathbf{H}_r$, $I \in \mathcal{P}(\mathbb{H})$, be the functions which satisfy for all $I \in \mathcal{P}(\mathbb{H})$, $r \in \mathbb{R}$, $u \in H_r$, $(v, w) \in \mathbf{H}_r$ that $P_I(u) = \sum_{h\in I} \langle |\lambda_h|^{-r} h, u\rangle_{H_r} |\lambda_h|^{-r} h$ and $\mathbf{P}_I(v, w) = \big(P_I(v), P_I(w)\big)$, let $\mathbf{A}\colon D(\mathbf{A}) \subseteq \mathbf{H}_0 \to \mathbf{H}_0$ be the linear operator which satisfies $D(\mathbf{A}) = \mathbf{H}_1$ and $\forall\, (v, w) \in \mathbf{H}_1\colon \mathbf{A}(v, w) = (w, Av)$, let $\mu\colon \mathbb{H} \to \mathbb{R}$ be a function which satisfies $\sum_{h\in\mathbb{H}} \frac{|\mu_h|^2}{|\lambda_h|} < \infty$, let $\mathbf{B} \in \mathrm{HS}(H, \mathbf{H}_0)$ be the linear operator which satisfies for all $v \in H$ that $\mathbf{B}v = \big(0, \sum_{h\in\mathbb{H}} \mu_h \langle h, v\rangle_H h\big)$, and let $\mathbf{X}^I = (X^{I,1}, X^{I,2})\colon \Omega \to \mathbf{P}_I(\mathbf{H}_0)$, $I \in \mathcal{P}(\mathbb{H})$, be random variables which satisfy for all $I \in \mathcal{P}(\mathbb{H})$ that it holds $\mathbb{P}$-a.s. that $\mathbf{X}^I = \int_0^T e^{(T-s)\mathbf{A}} \mathbf{P}_I \mathbf{B}\, dW_s$.

2.2 Lower Bounds for the Squared Norm

Lemma 2.1 *Assume the setting in Sect. 2.1. Then for all $I \in \mathcal{P}(\mathbb{H})$ it holds $\mathbb{P}$-a.s. that*

$$\mathbf{X}^I = \mathbf{P}_I \mathbf{X}^{\mathbb{H}} = \begin{pmatrix} X^{I,1} \\ X^{I,2} \end{pmatrix} = \begin{pmatrix} \sum_{h\in I}\Big(\frac{\mu_h}{|\lambda_h|^{1/2}} \int_0^T \sin\big(|\lambda_h|^{1/2}(T-s)\big)\, d\langle h, W_s\rangle_H\Big) h \\ \sum_{h\in I}\Big(\frac{\mu_h}{|\lambda_h|^{1/2}} \int_0^T \cos\big(|\lambda_h|^{1/2}(T-s)\big)\, d\langle h, W_s\rangle_H\Big) |\lambda_h|^{1/2} h \end{pmatrix}. \tag{2.1}$$

Proof of Lemma 2.1. Lemma 2.5 in [4] proves that it holds $\mathbb{P}$-a.s. that

$$\begin{aligned} \mathbf{X}^{\mathbb{H}} &= \int_0^T e^{(T-s)\mathbf{A}} \mathbf{B}\, dW_s = \sum_{h\in\mathbb{H}} \int_0^T e^{(T-s)\mathbf{A}} \mathbf{B} h\, d\langle h, W_s\rangle_H \\ &= \sum_{h\in\mathbb{H}} \begin{pmatrix} \frac{\mu_h}{|\lambda_h|^{1/2}} \int_0^T \sin\big(|\lambda_h|^{1/2}(T-s)\big) h\, d\langle h, W_s\rangle_H \\ \mu_h \int_0^T \cos\big(|\lambda_h|^{1/2}(T-s)\big) h\, d\langle h, W_s\rangle_H \end{pmatrix} \\ &= \begin{pmatrix} \sum_{h\in\mathbb{H}}\Big(\frac{\mu_h}{|\lambda_h|^{1/2}} \int_0^T \sin\big(|\lambda_h|^{1/2}(T-s)\big)\, d\langle h, W_s\rangle_H\Big) h \\ \sum_{h\in\mathbb{H}}\Big(\frac{\mu_h}{|\lambda_h|^{1/2}} \int_0^T \cos\big(|\lambda_h|^{1/2}(T-s)\big)\, d\langle h, W_s\rangle_H\Big) |\lambda_h|^{1/2} h \end{pmatrix}. \end{aligned} \tag{2.2}$$

Furthermore, Lemma 2.7 in [4] shows for all $I \in \mathcal{P}(\mathbb{H})$ that it holds $\mathbb{P}$-a.s. that

$$\mathbf{P}_I \mathbf{X}^{\mathbb{H}} = \int_0^T \mathbf{P}_I e^{(T-s)\mathbf{A}} \mathbf{B}\, dW_s = \int_0^T e^{(T-s)\mathbf{A}} \mathbf{P}_I \mathbf{B}\, dW_s = \mathbf{X}^I. \tag{2.3}$$

This and (2.2) complete the proof of Lemma 2.1. □

Lemma 2.2 *Assume the setting in Sect. 2.1 and let $I \in \mathcal{P}(\mathbb{H})$. Then*

(i) it holds that $\langle h, X^{I,1}\rangle_{H_0}$, $h \in \mathbb{H}$, is a family of independent centred Gaussian random variables,

(ii) it holds that $\big\langle |\lambda_h|^{1/2}h, X^{I,2}\big\rangle_{H_{-1/2}}$, $h \in \mathbb{H}$, is a family of independent centred Gaussian random variables, and

(iii) it holds for all $h \in \mathbb{H}$ that

$$\operatorname{Var}\big(\langle h, X^{I,1}\rangle_{H_0}\big) = \mathbb{1}_I(h)\frac{|\mu_h|^2}{|\lambda_h|}\frac{1}{2}\left(T - \frac{\sin(2|\lambda_h|^{1/2}T)}{2|\lambda_h|^{1/2}}\right), \tag{2.4}$$

$$\operatorname{Var}\Big(\big\langle |\lambda_h|^{1/2}h, X^{I,2}\big\rangle_{H_{-1/2}}\Big) = \mathbb{1}_I(h)\frac{|\mu_h|^2}{|\lambda_h|}\frac{1}{2}\left(T + \frac{\sin(2|\lambda_h|^{1/2}T)}{2|\lambda_h|^{1/2}}\right), \tag{2.5}$$

$$\operatorname{Cov}\Big(\langle h, X^{I,1}\rangle_{H_0}, \big\langle |\lambda_h|^{1/2}h, X^{I,2}\big\rangle_{H_{-1/2}}\Big) = \mathbb{1}_I(h)\frac{|\mu_h|^2}{|\lambda_h|}\left(\frac{1-\cos(2|\lambda_h|^{1/2}T)}{4|\lambda_h|^{1/2}}\right). \tag{2.6}$$

Proof of Lemma 2.2. Observe that Lemma 2.1 implies (i) and (ii). It thus remains to prove (iii). Lemma 2.1 assures for all $h \in \mathbb{H}$ that it holds $\mathbb{P}$-a.s. that

$$\langle h, X^{I,1}\rangle_{H_0} = \mathbb{1}_I(h)\frac{\mu_h}{|\lambda_h|^{1/2}}\int_0^T \sin\big(|\lambda_h|^{1/2}(T-s)\big)\,\mathrm{d}\langle h, W_s\rangle_H, \tag{2.7}$$

$$\big\langle |\lambda_h|^{1/2}h, X^{I,2}\big\rangle_{H_{-1/2}} = \mathbb{1}_I(h)\frac{\mu_h}{|\lambda_h|^{1/2}}\int_0^T \cos\big(|\lambda_h|^{1/2}(T-s)\big)\,\mathrm{d}\langle h, W_s\rangle_H. \tag{2.8}$$

Itô's isometry hence shows for all $h \in \mathbb{H}$ that

$$\begin{aligned}\operatorname{Var}\big(\langle h, X^{I,1}\rangle_{H_0}\big) &= \mathbb{E}\big[|\langle h, X^{I,1}\rangle_{H_0}|^2\big] \\ &= \mathbb{1}_I(h)\frac{|\mu_h|^2}{|\lambda_h|}\int_0^T \big|\sin\big(|\lambda_h|^{1/2}(T-s)\big)\big|^2\,\mathrm{d}s \\ &= \mathbb{1}_I(h)\frac{|\mu_h|^2}{|\lambda_h|}\frac{1}{2}\left(T - \frac{\sin(2|\lambda_h|^{1/2}T)}{2|\lambda_h|^{1/2}}\right),\end{aligned} \tag{2.9}$$

$$\begin{aligned}\operatorname{Var}\Big(\big\langle |\lambda_h|^{1/2}h, X^{I,2}\big\rangle_{H_{-1/2}}\Big) &= \mathbb{E}\Big[\Big|\big\langle |\lambda_h|^{1/2}h, X^{I,2}\big\rangle_{H_{-1/2}}\Big|^2\Big] \\ &= \mathbb{1}_I(h)\frac{|\mu_h|^2}{|\lambda_h|}\int_0^T \big|\cos\big(|\lambda_h|^{1/2}(T-s)\big)\big|^2\,\mathrm{d}s \\ &= \mathbb{1}_I(h)\frac{|\mu_h|^2}{|\lambda_h|}\frac{1}{2}\left(T + \frac{\sin(2|\lambda_h|^{1/2}T)}{2|\lambda_h|^{1/2}}\right).\end{aligned} \tag{2.10}$$

Furthermore, observe that it holds for all $h \in \mathbb{H}$ that

$$
\begin{aligned}
&\operatorname{Cov}\Big(\langle h, X^{I,1}\rangle_{H_0}, \big\langle |\lambda_h|^{1/2} h, X^{I,2}\big\rangle_{H_{-1/2}}\Big) = \mathbb{E}\Big[\langle h, X^{I,1}\rangle_{H_0} \big\langle |\lambda_h|^{1/2} h, X^{I,2}\big\rangle_{H_{-1/2}}\Big] \\
&= \mathbb{1}_I(h) \frac{|\mu_h|^2}{|\lambda_h|} \int_0^T \sin\big(|\lambda_h|^{1/2}(T-s)\big) \cos\big(|\lambda_h|^{1/2}(T-s)\big)\, ds \\
&= \mathbb{1}_I(h) \frac{|\mu_h|^2}{|\lambda_h|} \left(\frac{\big|\sin\big(|\lambda_h|^{1/2} T\big)\big|^2}{2|\lambda_h|^{1/2}} \right) \\
&= \mathbb{1}_I(h) \frac{|\mu_h|^2}{|\lambda_h|} \left(\frac{1 - \cos\big(2|\lambda_h|^{1/2} T\big)}{4|\lambda_h|^{1/2}} \right).
\end{aligned} \tag{2.11}
$$

The proof of Lemma 2.2 is thus completed. □

Lemma 2.3 *Assume the setting in Sect. 2.1 and let $I \in \mathcal{P}(\mathbb{H})$. Then it holds for all $i \in \{1, 2\}$ that*

$$
\mathbb{E}\big[\|\mathbf{X}^I\|_{\mathbf{H}_0}^2\big] = T \sum_{h \in I} \frac{|\mu_h|^2}{|\lambda_h|} < \infty, \tag{2.12}
$$

$$
\mathbb{E}\Big[\|X^{I,i}\|_{H_{1/2-i/2}}^2\Big] = \frac{1}{2} \sum_{h \in I} \frac{|\mu_h|^2}{|\lambda_h|} \left(T + \frac{\sin\big(2|\lambda_h|^{1/2} T\big)}{(-1)^i 2|\lambda_h|^{1/2}} \right) < \infty. \tag{2.13}
$$

Proof of Lemma 2.3. Itô's isometry and Lemma 2.6 in [4] imply that

$$
\begin{aligned}
\mathbb{E}\big[\|\mathbf{X}^I\|_{\mathbf{H}_0}^2\big] &= \mathbb{E}\left[\left\| \int_0^T e^{(T-s)\mathbf{A}} \mathbf{P}_I \mathbf{B}\, dW_s \right\|_{\mathbf{H}_0}^2 \right] \\
&= T \|\mathbf{P}_I \mathbf{B}\|_{\mathrm{HS}(H, \mathbf{H}_0)}^2 = T \sum_{h \in I} \frac{|\mu_h|^2}{|\lambda_h|} < \infty.
\end{aligned} \tag{2.14}
$$

In addition, Lemma 2.2 shows for all $i \in \{1, 2\}$ that

$$
\begin{aligned}
\mathbb{E}\Big[\|X^{I,i}\|_{H_{1/2-i/2}}^2\Big] &= \sum_{h \in \mathbb{H}} \mathbb{E}\Big[\big| \big\langle |\lambda_h|^{i/2-1/2} h, X^{I,i} \big\rangle_{H_{1/2-i/2}} \big|^2 \Big] \\
&= \frac{1}{2} \sum_{h \in I} \frac{|\mu_h|^2}{|\lambda_h|} \left(T + \frac{\sin\big(2|\lambda_h|^{1/2} T\big)}{(-1)^i 2|\lambda_h|^{1/2}} \right) < \infty.
\end{aligned} \tag{2.15}
$$

The proof of Lemma 2.3 is thus completed. □

Corollary 2.4 *Assume the setting in Sect. 2.1 and let $I \in \mathcal{P}(\mathbb{H})$. Then it holds for all $(v, w) \in \mathbf{P}_I(\mathbf{H}_0)$ that*

$$\begin{aligned}
\operatorname{CovOp}(\mathbf{X}^I)\begin{pmatrix} v \\ w \end{pmatrix} =& \frac{1}{2}\sum_{h\in I}\frac{|\mu_h|^2}{|\lambda_h|}\Bigg[\left(T - \frac{\sin(2|\lambda_h|^{1/2}T)}{2|\lambda_h|^{1/2}}\right)\langle h, v\rangle_{H_0}\begin{pmatrix} h \\ 0 \end{pmatrix} \\
&+ \left(\frac{1-\cos(2|\lambda_h|^{1/2}T)}{2|\lambda_h|^{1/2}}\right)\big\langle|\lambda_h|^{1/2}h, w\big\rangle_{H_{-1/2}}\begin{pmatrix} h \\ 0 \end{pmatrix} \\
&+ \left(\frac{1-\cos(2|\lambda_h|^{1/2}T)}{2|\lambda_h|^{1/2}}\right)\langle h, v\rangle_{H_0}\begin{pmatrix} 0 \\ |\lambda_h|^{1/2}h \end{pmatrix} \\
&+ \left(T + \frac{\sin(2|\lambda_h|^{1/2}T)}{2|\lambda_h|^{1/2}}\right)\big\langle|\lambda_h|^{1/2}h, w\big\rangle_{H_{-1/2}}\begin{pmatrix} 0 \\ |\lambda_h|^{1/2}h \end{pmatrix}\Bigg] \in \mathbf{P}_I(\mathbf{H}_0).
\end{aligned} \tag{2.16}$$

Proof of Corollary 2.4. Lemma 2.1, and Lemma 2.2 prove for all $x_1 = (v_1, w_1)$, $x_2 = (v_2, w_2) \in \mathbf{P}_I(\mathbf{H}_0)$ that

$$\begin{aligned}
&\langle x_1, \operatorname{CovOp}(\mathbf{X}^I)x_2\rangle_{\mathbf{H}_0} = \operatorname{Cov}\big(\langle x_1, \mathbf{X}^I\rangle_{\mathbf{H}_0}, \langle x_2, \mathbf{X}^I\rangle_{\mathbf{H}_0}\big) = \mathbb{E}\big[\langle x_1, \mathbf{X}^I\rangle_{\mathbf{H}_0}\langle x_2, \mathbf{X}^I\rangle_{\mathbf{H}_0}\big] \\
&= \mathbb{E}\big[\big(\langle v_1, X^{I,1}\rangle_{H_0} + \langle w_1, X^{I,2}\rangle_{H_{-1/2}}\big)\big(\langle v_2, X^{I,1}\rangle_{H_0} + \langle w_2, X^{I,2}\rangle_{H_{-1/2}}\big)\big] \\
&= \sum_{h\in\mathbb{H}}\operatorname{Var}\big(\langle h, X^{I,1}\rangle_{H_0}\big)\langle h, v_1\rangle_{H_0}\langle h, v_2\rangle_{H_0} \\
&\quad+ \sum_{h\in\mathbb{H}}\operatorname{Cov}\Big(\langle h, X^{I,1}\rangle_{H_0}, \big\langle|\lambda_h|^{1/2}h, X^{I,2}\big\rangle_{H_{-1/2}}\Big)\langle h, v_1\rangle_{H_0}\big\langle|\lambda_h|^{1/2}h, w_2\big\rangle_{H_{-1/2}} \\
&\quad+ \sum_{h\in\mathbb{H}}\operatorname{Cov}\Big(\langle h, X^{I,1}\rangle_{H_0}, \big\langle|\lambda_h|^{1/2}h, X^{I,2}\big\rangle_{H_{-1/2}}\Big)\langle h, v_2\rangle_{H_0}\big\langle|\lambda_h|^{1/2}h, w_1\big\rangle_{H_{-1/2}} \\
&\quad+ \sum_{h\in\mathbb{H}}\operatorname{Var}\Big(\big\langle|\lambda_h|^{1/2}h, X^{I,2}\big\rangle_{H_{-1/2}}\Big)\big\langle|\lambda_h|^{1/2}h, w_1\big\rangle_{H_{-1/2}}\big\langle|\lambda_h|^{1/2}h, w_2\big\rangle_{H_{-1/2}} \\
&= \Bigg\langle v_1, \sum_{h\in\mathbb{H}}\Big[\operatorname{Var}\big(\langle h, X^{I,1}\rangle_{H_0}\big)\langle h, v_2\rangle_{H_0} \\
&\qquad\qquad+ \operatorname{Cov}\Big(\langle h, X^{I,1}\rangle_{H_0}, \big\langle|\lambda_h|^{1/2}h, X^{I,2}\big\rangle_{H_{-1/2}}\Big)\big\langle|\lambda_h|^{1/2}h, w_2\big\rangle_{H_{-1/2}}\Big]h\Bigg\rangle_{H_0} \\
&\quad+ \Bigg\langle w_1, \sum_{h\in\mathbb{H}}\Big[\operatorname{Cov}\Big(\langle h, X^{I,1}\rangle_{H_0}, \big\langle|\lambda_h|^{1/2}h, X^{I,2}\big\rangle_{H_{-1/2}}\Big)\langle h, v_2\rangle_{H_0} \\
&\qquad\qquad+ \operatorname{Var}\Big(\big\langle|\lambda_h|^{1/2}h, X^{I,2}\big\rangle_{H_{-1/2}}\Big)\big\langle|\lambda_h|^{1/2}h, w_2\big\rangle_{H_{-1/2}}\Big]|\lambda_h|^{1/2}h\Bigg\rangle_{H_{-1/2}} \\
&= \Bigg\langle x_1, \sum_{h\in\mathbb{H}}\Big[\operatorname{Var}\big(\langle h, X^{I,1}\rangle_{H_0}\big)\langle h, v_2\rangle_{H_0} \\
&\qquad\qquad+ \operatorname{Cov}\Big(\langle h, X^{I,1}\rangle_{H_0}, \big\langle|\lambda_h|^{1/2}h, X^{I,2}\big\rangle_{H_{-1/2}}\Big)\big\langle|\lambda_h|^{1/2}h, w_2\big\rangle_{H_{-1/2}}\Big]\begin{pmatrix} h \\ 0 \end{pmatrix}\Bigg\rangle_{\mathbf{H}_0} \\
&\quad+ \Bigg\langle x_1, \sum_{h\in\mathbb{H}}\Big[\operatorname{Cov}\Big(\langle h, X^{I,1}\rangle_{H_0}, \big\langle|\lambda_h|^{1/2}h, X^{I,2}\big\rangle_{H_{-1/2}}\Big)\langle h, v_2\rangle_{H_0}
\end{aligned} \tag{2.17}$$

$$+ \operatorname{Var}\Big(\big\langle|\lambda_h|^{1/2}h, X^{I,2}\big\rangle_{H_{-1/2}}\Big)\big\langle|\lambda_h|^{1/2}h, w_2\big\rangle_{H_{-1/2}}\Big] \begin{pmatrix} 0 \\ |\lambda_h|^{1/2}h \end{pmatrix}\Big\rangle_{\mathbf{H}_0}.$$

This and again Lemma 2.2 complete the proof of Corollary 2.4. □

Proposition 2.5 *Assume the setting in Sect. 2.1. Then it holds for all $I \in \mathcal{P}(\mathbb{H})$ that*

$$\mathbb{E}\big[\|\mathbf{X}^{\mathbb{H}}\|^2_{\mathbf{H}_0}\big] - \mathbb{E}\big[\|\mathbf{X}^{I}\|^2_{\mathbf{H}_0}\big] = \mathbb{E}\big[\|\mathbf{X}^{\mathbb{H}\setminus I}\|^2_{\mathbf{H}_0}\big] \geq T \inf_{h\in\mathbb{H}} |\mu_h|^2 \sum_{h\in\mathbb{H}\setminus I} \frac{1}{|\lambda_h|}. \tag{2.18}$$

Proof of Proposition 2.5. Orthogonality and Lemma 2.1 imply for all $I \in \mathcal{P}(\mathbb{H})$ that

$$\begin{aligned} \mathbb{E}\big[\|\mathbf{X}^{I}\|^2_{\mathbf{H}_0}\big] + \mathbb{E}\big[\|\mathbf{X}^{\mathbb{H}\setminus I}\|^2_{\mathbf{H}_0}\big] &= \mathbb{E}\big[\|\mathbf{P}_I\mathbf{X}^{\mathbb{H}}\|^2_{\mathbf{H}_0}\big] + \mathbb{E}\big[\|\mathbf{P}_{\mathbb{H}\setminus I}\mathbf{X}^{\mathbb{H}}\|^2_{\mathbf{H}_0}\big] \\ &= \mathbb{E}\big[\|(\mathbf{P}_I + \mathbf{P}_{\mathbb{H}\setminus I})\mathbf{X}^{\mathbb{H}}\|^2_{\mathbf{H}_0}\big] = \mathbb{E}\big[\|\mathbf{X}^{\mathbb{H}}\|^2_{\mathbf{H}_0}\big]. \end{aligned} \tag{2.19}$$

This and Lemma 2.3 show for all $I \in \mathcal{P}(\mathbb{H})$ that

$$\begin{aligned} \mathbb{E}\big[\|\mathbf{X}^{\mathbb{H}}\|^2_{\mathbf{H}_0}\big] - \mathbb{E}\big[\|\mathbf{X}^{I}\|^2_{\mathbf{H}_0}\big] &= \mathbb{E}\big[\|\mathbf{X}^{\mathbb{H}\setminus I}\|^2_{\mathbf{H}_0}\big] \\ &= T \sum_{h\in\mathbb{H}\setminus I} \frac{|\mu_h|^2}{|\lambda_h|} \geq T \inf_{h\in\mathbb{H}} |\mu_h|^2 \sum_{h\in\mathbb{H}\setminus I} \frac{1}{|\lambda_h|}. \end{aligned} \tag{2.20}$$

The proof of Proposition 2.5 is thus completed. □

In Corollary 2.7 and Corollary 2.8 below lower bounds on the weak approximation error with the squared norm as test function are presented. Our proofs of Corollary 2.7 and Corollary 2.8 use the following elementary and well-known lemma (cf., e.g., Proposition 7.4 in Conus et al. [1]).

Lemma 2.6 *Let $p \in (0, \infty)$, $\delta \in (-\infty, 1/2 - 1/(2p))$. Then it holds for all $N \in \mathbb{N}$ that*

$$\sum_{n=N+1}^{\infty} n^{p(2\delta-1)} \geq \frac{N^{p(2\delta-1)+1}}{[p(1-2\delta)-1]2^{p(1-2\delta)-1}}. \tag{2.21}$$

Proof of Lemma 2.6. Observe that the assumption that $\delta \in (-\infty, 1/2 - 1/(2p))$ ensures that $p(2\delta - 1) \in (-\infty, -1)$. This implies for all $N \in \mathbb{N}$ that

$$\begin{aligned} \sum_{n=N+1}^{\infty} n^{p(2\delta-1)} &= \sum_{n=N+1}^{\infty} \int_n^{n+1} n^{p(2\delta-1)}\,dx \geq \sum_{n=N+1}^{\infty} \int_n^{n+1} x^{p(2\delta-1)}\,dx \\ &= \int_{N+1}^{\infty} x^{p(2\delta-1)}\,dx = -\frac{(N+1)^{p(2\delta-1)+1}}{p(2\delta-1)+1} \\ &\geq \frac{N^{p(2\delta-1)+1}}{[p(1-2\delta)-1]2^{p(1-2\delta)-1}}. \end{aligned} \tag{2.22}$$

This completes the proof of Lemma 2.6. □

Corollary 2.7 *Assume the setting in Sect. 2.1, let $c \in (0, \infty)$, $p \in (1, \infty)$, let $e \colon \mathbb{N} \to \mathbb{H}$ be a bijection which satisfies for all $n \in \mathbb{N}$ that $\lambda_{e_n} = -cn^p$, and let $I_N \in \mathcal{P}(\mathbb{H})$, $N \in \mathbb{N}$, be the sets which satisfy for all $N \in \mathbb{N}$ that $I_N = \{e_1, e_2, \ldots, e_N\} \subseteq \mathbb{H}$. Then it holds for all $N \in \mathbb{N}$ that*

$$\mathbb{E}\big[\|\mathbf{X}^{\mathbb{H}}\|_{\mathbf{H}_0}^2\big] - \mathbb{E}\big[\|\mathbf{X}^{I_N}\|_{\mathbf{H}_0}^2\big] \geq \frac{T \inf_{h \in \mathbb{H}} |\mu_h|^2 N^{1-p}}{c(p-1)2^{p-1}}. \tag{2.23}$$

Proof of Corollary 2.7. Proposition 2.5 and Lemma 2.6 prove for all $N \in \mathbb{N}$ that

$$\begin{aligned}\mathbb{E}\big[\|\mathbf{X}^{\mathbb{H}}\|_{\mathbf{H}_0}^2\big] - \mathbb{E}\big[\|\mathbf{X}^{I_N}\|_{\mathbf{H}_0}^2\big] &\geq T \inf_{h \in \mathbb{H}} |\mu_h|^2 \sum_{h \in \mathbb{H} \setminus I_N} \frac{1}{|\lambda_h|} = c^{-1} T \inf_{h \in \mathbb{H}} |\mu_h|^2 \sum_{n=N+1}^{\infty} \frac{1}{n^p} \\ &\geq \frac{T \inf_{h \in \mathbb{H}} |\mu_h|^2 N^{1-p}}{c(p-1)2^{p-1}}.\end{aligned} \tag{2.24}$$

The proof of Corollary 2.7 is thus completed. □

Corollary 2.8 *Assume the setting in Sect. 2.1, let $c, p \in (0, \infty)$, $\delta \in (-\infty, 1/2 - 1/(2p))$, let $e \colon \mathbb{N} \to \mathbb{H}$ be a bijection which satisfies for all $n \in \mathbb{N}$ that $\lambda_{e_n} = -cn^p$, let $I_N \in \mathcal{P}(\mathbb{H})$, $N \in \mathbb{N}$, be the sets which satisfy for all $N \in \mathbb{N}$ that $I_N = \{e_1, e_2, \ldots, e_N\} \subseteq \mathbb{H}$, and assume for all $h \in \mathbb{H}$ that $|\mu_h| = |\lambda_h|^\delta$. Then it holds for all $N \in \mathbb{N}$ that*

$$\mathbb{E}\big[\|\mathbf{X}^{\mathbb{H}}\|_{\mathbf{H}_0}^2\big] - \mathbb{E}\big[\|\mathbf{X}^{I_N}\|_{\mathbf{H}_0}^2\big] \geq \frac{T c^{2\delta-1} N^{p(2\delta-1)+1}}{[p(1-2\delta)-1]2^{p(1-2\delta)-1}}. \tag{2.25}$$

Proof of Corollary 2.8. Proposition 2.5, Lemma 2.3, and Lemma 2.6 show for all $N \in \mathbb{N}$ that

$$\begin{aligned}\mathbb{E}\big[\|\mathbf{X}^{\mathbb{H}}\|_{\mathbf{H}_0}^2\big] - \mathbb{E}\big[\|\mathbf{X}^{I_N}\|_{\mathbf{H}_0}^2\big] &= T \sum_{h \in \mathbb{H} \setminus I_N} \frac{|\mu_h|^2}{|\lambda_h|} = T \sum_{h \in \mathbb{H} \setminus I_N} |\lambda_h|^{2\delta-1} \\ &= T c^{2\delta-1} \sum_{n=N+1}^{\infty} n^{p(2\delta-1)} \geq \frac{T c^{2\delta-1} N^{p(2\delta-1)+1}}{[p(1-2\delta)-1]2^{p(1-2\delta)-1}}.\end{aligned} \tag{2.26}$$

This completes the proof of Corollary 2.8. □

2.3 Lower Bounds for the Weak Error of a Particular Regular Test Function

The next result, Proposition 2.9 below, follows directly from Lemma 2.2, and Lemma 2.3 above and Lemma 9.5 in Jentzen and Kurniawan [5].

Proposition 2.9 *Assume the setting in Sect. 2.1 and let $\varphi_i \colon \mathbf{H}_0 \to \mathbb{R}$, $i \in \{1, 2\}$, be the functions which satisfy for all $i \in \{1, 2\}$, $(v_1, v_2) \in \mathbf{H}_0$ that $\varphi_i(v_1, v_2) = \exp(-\|v_i\|^2_{H_{1/2-i/2}})$. Then it holds for all $i \in \{1, 2\}$, $I \in \mathcal{P}(\mathbb{H})$ that $\varphi_i \in C^2_b(\mathbf{H}_0, \mathbb{R})$ and*

$$\mathbb{E}[\varphi_i(\mathbf{X}^I)] - \mathbb{E}[\varphi_i(\mathbf{X}^{\mathbb{H}})] \geq \frac{\mathbb{E}\Big[\|X^{\mathbb{H},i}\|^2_{H_{1/2-i/2}}\Big] - \mathbb{E}\Big[\|X^{I,i}\|^2_{H_{1/2-i/2}}\Big]}{\exp\Big(6\,\mathbb{E}\Big[\|X^{\mathbb{H},i}\|^2_{H_{1/2-i/2}}\Big]\Big)}. \tag{2.27}$$

Corollary 2.10 *Assume the setting in Sect. 2.1, let $c, p \in (0, \infty)$, $\delta \in (-\infty, 1/2 - 1/(2p))$, let $e \colon \mathbb{N} \to \mathbb{H}$ be a bijection which satisfies for all $n \in \mathbb{N}$ that $\lambda_{e_n} = -cn^p$, let $I_N \in \mathcal{P}(\mathbb{H})$, $N \in \mathbb{N}$, be the sets which satisfy for all $N \in \mathbb{N}$ that $I_N = \{e_1, e_2, \ldots, e_N\} \subseteq \mathbb{H}$, assume for all $h \in \mathbb{H}$ that $|\mu_h| = |\lambda_h|^\delta$, and let $\varphi_i \colon \mathbf{H}_0 \to \mathbb{R}$, $i \in \{1, 2\}$, be the functions which satisfy for all $i \in \{1, 2\}$, $(v_1, v_2) \in \mathbf{H}_0$ that $\varphi_i(v_1, v_2) = \exp(-\|v_i\|^2_{H_{1/2-i/2}})$. Then it holds for all $i \in \{1, 2\}$, $N \in \mathbb{N}$ that $\varphi_i \in C^2_b(\mathbf{H}_0, \mathbb{R})$ and*

$$\begin{aligned}&\mathbb{E}[\varphi_i(\mathbf{X}^{I_N})] - \mathbb{E}[\varphi_i(\mathbf{X}^{\mathbb{H}})] \\ &\geq \left[1 + \inf_{x \in [2c^{1/2}T, \infty)} \frac{\sin(x)}{(-1)^i x}\right] \frac{Tc^{2\delta-1} 2^{p(2\delta-1)} N^{p(2\delta-1)+1}}{[p(1-2\delta)-1] \exp\left(\frac{6p(2\delta-1)Tc^{2\delta-1}}{p(2\delta-1)+1}\right)} > 0.\end{aligned} \tag{2.28}$$

Proof of Corollary 2.10. Lemma 2.3, and Lemma 2.6, and the fact that $\forall x \in (0, \infty)\colon \left|\frac{\sin(x)}{x}\right| < 1$ prove for all $i \in \{1, 2\}$, $N \in \mathbb{N}$ that

$$\begin{aligned}\mathbb{E}\Big[\|X^{\mathbb{H},i}\|^2_{H_{1/2-i/2}}\Big] - \mathbb{E}\Big[\|X^{I_N,i}\|^2_{H_{1/2-i/2}}\Big] &= \frac{1}{2} \sum_{h \in \mathbb{H} \setminus I_N} \frac{|\mu_h|^2}{|\lambda_h|} \left(T + \frac{\sin(2|\lambda_h|^{1/2}T)}{(-1)^i 2|\lambda_h|^{1/2}}\right) \\ &\geq \left(1 + \inf_{h \in \mathbb{H}} \frac{\sin(2|\lambda_h|^{1/2}T)}{(-1)^i 2|\lambda_h|^{1/2}T}\right) \frac{T}{2} \sum_{h \in \mathbb{H} \setminus I_N} |\lambda_h|^{2\delta-1} \\ &\geq \left(1 + \inf_{x \in [2c^{1/2}T, \infty)} \frac{\sin(x)}{(-1)^i x}\right) \frac{Tc^{2\delta-1}}{2} \sum_{n=N+1}^{\infty} n^{p(2\delta-1)} \\ &\geq \left(1 + \inf_{x \in [2c^{1/2}T, \infty)} \frac{\sin(x)}{(-1)^i x}\right) \frac{Tc^{2\delta-1} 2^{p(2\delta-1)} N^{p(2\delta-1)+1}}{[p(1-2\delta)-1]} > 0.\end{aligned} \tag{2.29}$$

Furthermore, note that the fact that $p(2\delta - 1) \in (-\infty, -1)$ ensures that

$$\begin{aligned}\sum_{n=1}^{\infty} n^{p(2\delta-1)} &\leq 1 + \sum_{n=1}^{\infty} \int_n^{n+1} x^{p(2\delta-1)}\, \mathrm{d}x = 1 + \int_1^{\infty} x^{p(2\delta-1)}\, \mathrm{d}x \\ &= 1 - \frac{1}{p(2\delta-1)+1} = \frac{p(2\delta-1)}{p(2\delta-1)+1}.\end{aligned} \tag{2.30}$$

Lemma 2.3 hence implies for all $i \in \{1, 2\}$ that

$$\begin{aligned}\exp\Big(-6\,\mathbb{E}\Big[\|X^{\mathbb{H},i}\|^2_{H_{1/2-i/2}}\Big]\Big) &\geq \exp\big(-6\,\mathbb{E}\big[\|\mathbf{X}^{\mathbb{H}}\|^2_{\mathbf{H}_0}\big]\big) = \exp\Big(-6Tc^{2\delta-1}\sum_{n=1}^{\infty} n^{p(2\delta-1)}\Big) \\ &\geq \exp\Big(-\frac{6p(2\delta-1)Tc^{2\delta-1}}{p(2\delta-1)+1}\Big) > 0.\end{aligned} \tag{2.31}$$

Combining this and (2.29) with Proposition 2.9 concludes the proof of Corollary 2.10. □

Roughly speaking, Corollary 2.11 below specifies Corollary 2.10 to the case where the linear operator $A\colon D(A) \subseteq H \to H$ in the setting in Sect. 2.1 is the Laplacian with Dirichlet boundary conditions on $H = L^2((0, 1); \mathbb{R})$. Corollary 2.11 is an immediate consequence of Corollary 2.10.

Corollary 2.11 *Assume the setting in Sect. 2.1, let $\delta \in (-\infty, 1/4)$, let $e\colon \mathbb{N} \to \mathbb{H}$ be a bijection which satisfies for all $n \in \mathbb{N}$ that $\lambda_{e_n} = -\pi^2 n^2$, let $I_N \in \mathcal{P}(\mathbb{H})$, $N \in \mathbb{N}$, be the sets which satisfy for all $N \in \mathbb{N}$ that $I_N = \{e_1, e_2, \ldots, e_N\} \subseteq \mathbb{H}$, assume for all $h \in \mathbb{H}$ that $|\mu_h| = |\lambda_h|^\delta$, and let $\varphi_i\colon \mathbf{H}_0 \to \mathbb{R}$, $i \in \{1, 2\}$, be the functions which satisfy for all $i \in \{1, 2\}$, $(v_1, v_2) \in \mathbf{H}_0$ that $\varphi_i(v_1, v_2) = \exp(-\|v_i\|^2_{H_{1/2-i/2}})$. Then it holds for all $i \in \{1, 2\}$, $N \in \mathbb{N}$ that $\varphi_i \in C^2_b(\mathbf{H}_0, \mathbb{R})$ and*

$$\begin{aligned}&\mathbb{E}[\varphi_i(\mathbf{X}^{I_N})] - \mathbb{E}[\varphi_i(\mathbf{X}^{\mathbb{H}})] \\ &\geq \Big[1 + \inf_{x\in[2\pi T,\infty)} \frac{\sin(x)}{(-1)^i x}\Big] \frac{T(4\pi^2)^{2\delta-1} N^{4\delta-1}}{[1-4\delta]\exp\big(\frac{12(2\delta-1)T\pi^{4\delta-2}}{4\delta-1}\big)} > 0.\end{aligned} \tag{2.32}$$

Acknowledgements This project has been partially supported through the ETH Research Grant ETH-47 15-2 "Mild stochastic calculus and numerical approximations for nonlinear stochastic evolution equations with Lévy noise".

References

1. Conus, D., Jentzen, A., Kurniawan, R.: Weak convergence rates of spectral Galerkin approximations for SPDEs with nonlinear diffusion coefficients. ArXiv e-prints (Aug 2014), 59 pages. arXiv: 1408.1108 [math.PR]. Accepted in Ann. Appl. Probab.

2. Davie, A.M., Gaines, J.G.: Convergence of numerical schemes for the solution of parabolic stochastic partial differential equations. Math. Comput. **70**, 233, 121–134 (2001). ISSN 0025-5718. https://doi.org/10.1090/S0025-5718-00-01224-2
3. Hausenblas, E.: Weak approximation of the stochastic wave equation. J. Comput. Appl. Math. **235**, 1, 33–58 (2010). ISSN 0377-0427. https://doi.org/10.1016/j.cam.2010.03.026
4. Jacobe de Naurois, L., Jentzen, A., Welti, T.: Weak convergence rates for spatial spectral Galerkin approximations of semilinear stochastic wave equations with multiplicative noise. ArXiv e-prints (Aug 2015), 27 pages. arXiv: 1508.05168 [math.PR]. Accepted in Appl. Math. Optim.
5. Jentzen, A., Kurniawan, R.: Weak convergence rates for Euler-type approximations of semilinear stochastic evolution equations with nonlinear diffusion coefficients. ArXiv e- prints (Jan 2015), 51 pages. arXiv: 1501.03539 [math.PR]
6. Kovács, M., Larsson, S., Lindgren, F.: Weak convergence of finite element approximations of linear stochastic evolution equations with additive noise. BIT 52, 1, 85–108 (2012). ISSN 0006-3835. https://doi.org/10.1007/s10543-011-0344-2
7. Kovács, M., Larsson, S., Lindgren, F.: Weak convergence of finite element approximations of linear stochastic evolution equations with additive noise II. Fully discrete schemes. BIT 53, 2, 497–525 (2013). ISSN 0006-3835
8. Kovács, M., Lindner, F., Schilling, R.L.: Weak convergence of finite element approximations of linear stochastic evolution equations with additive Lévy noise. SIAM/ASA J. Uncertain. Quantif. **3**, 1, 1159–1199 (2015). ISSN 2166-2525. https://doi.org/10.1137/15M1009792
9. Müller-Gronbach, T., Ritter, K.: Lower bounds and nonuniform time discretization for approximation of stochastic heat equations. Found. Comput. Math. **7**, 2, 135–181 (2007). ISSN 1615-3375. https://doi.org/10.1007/s10208-005-0166-6
10. Müller-Gronbach, T., Ritter, K., Wagner, T.: Optimal pointwise approximation of infinite-dimensional Ornstein-Uhlenbeck processes. Stoch. Dyn. **8**, 3, 519–541 (2008). ISSN 0219-4937. https://doi.org/10.1142/S0219493708002433
11. Sell, G.R., You, Y.: Dynamics of evolutionary equations, vol. 143. Applied Mathematical Sciences. Springer, New York, 2002, xiv+670. ISBN 0-387-98347-3. https://doi.org/10.1007/978-1-4757-5037-9
12. Wang, X.: An exponential integrator scheme for time discretization of nonlinear stochastic wave equation. J. Sci. Comput. **64**, 1, 234–263 (2015). ISSN 0885-7474. https://doi.org/10.1007/s10915-014-9931-0

Estimates for Nonlinear Stochastic Partial Differential Equations with Gradient Noise via Dirichlet Forms

Jonas M. Tölle

Abstract We present a priori estimates for nonlinear Stratonovich stochastic partial differential equations on the d-dimensional torus with p-Laplace-type drift with sublinear non-homogeneous nonlinearities and Gaussian gradient Stratonovich noise with C^1-vector field coefficients. Assuming a commutator bound, the results are obtained by using resolvent and Dirichlet form methods and an approximative Itô-formula.

Keywords Nonlinear stochastic partial differential equations · Dirichlet forms · A Priori estimate · Stochastic p-laplace evolution equation · Gradient Stratonovich noise · Commutator bound · Bakry-Émery curvature-dimension condition

2010 Mathematics Subject Classification 35K55 · 35K92 · 60H15

1 Introduction

We would like to study a priori estimates for the following (formal) Stratonovich stochastic partial differential equation (Stratonovich SPDE) for a nonlinear divergence form drift operator of p-Laplace type, $t \in (0, T]$,

$$dX_t = \operatorname{div}(\phi(\nabla X_t))\,dt + \sum_{i=1}^{N} \langle b_i, \nabla X_t \rangle \circ d\beta_t^i, \quad X_0 = x \in L^2(\mathbb{T}^d). \tag{1}$$

Here, $\phi : \mathbb{R}^d \to 2^{\mathbb{R}^d}$ is some (possibly multi-valued) monotone nonlinearity of at most linear growth, $b_i : \mathbb{T}^d \to \mathbb{R}^d$, $1 \le i \le N$ are C^1-vector fields on the

---

Dedicated to Michael Röckner on the occasion of his 60th birthday.

J. M. Tölle (✉)
Institut für Mathematik, Universität Augsburg, 86135 Augsburg, Germany
e-mail: jonas.toelle@math.uni-augsburg.de

A. Eberle et al. (eds.), *Stochastic Partial Differential Equations and Related Fields*, Springer Proceedings in Mathematics & Statistics 229,
https://doi.org/10.1007/978-3-319-74929-7_14

d-dimensional torus, such that

$$\operatorname{div} b_i = 0, \quad \forall 1 \le i \le N. \tag{2}$$

We set

$$\mathbf{b} := \begin{pmatrix} b_1 \\ \vdots \\ b_N \end{pmatrix} : \mathbb{T}^d \to \mathbb{R}^{N\times d}.$$

Furthermore, $\beta = (\beta^1, \dots, \beta^N)$ denotes an N-dimensional Wiener process on a filtered probability space $(\Omega, \mathscr{F}, (\mathscr{F}_t)_{t\ge 0}, \mathbb{P})$ that satisfies the usual conditions. In order to simplify the presentation, we consider periodic boundary conditions, so that the natural choice for the spatial domain is the the d-dimensional torus $\mathbb{T}^d$. The reader may notice that both expressions on the r.h.s. of (1) are merely formal. In fact, we shall study viscosity-type approximative estimates for the following Itô-SPDE

$$dX + \partial\Phi(X_t)\,dt = \frac{1}{2}L^{\mathbf{b}}X_t\,dt + \sum_{i=1}^{N}\langle b_i, \nabla X_t\rangle\,d\beta_t^i, \quad X_0 = x \in L^2(\mathbb{T}^d), \tag{3}$$

where $\partial\Phi$ denotes the subdifferential[1] of Φ, which, in turn, is the lower semi-continuous (l.s.c.) envelope of the map

$$u \mapsto \begin{cases} \int_{\mathbb{T}^d} \varphi(\nabla u)\,d\xi, & u \in H^1(\mathbb{T}^d), \\ +\infty, & u \in L^2(\mathbb{T}^d) \setminus H^1(\mathbb{T}^d), \end{cases}$$

where $\varphi : \mathbb{R}^d \to [0,\infty)$ is some convex potential of the nonlinearity ϕ, that is, $\langle \eta, \zeta - \xi\rangle \le \varphi(\zeta) - \varphi(\xi)$ for every $\eta \in \phi(\xi)$ and all $\xi, \zeta \in \mathbb{R}^d$. Regarding our assumptions on φ, see condition (N) below. In (3), $L^{\mathbf{b}}$ denotes the Dirichlet operator associated to the Dirichlet form

$$\mathscr{E}^{\mathbf{b}}(u, v) := \int_{\mathbb{T}^d} \langle \mathbf{b}\nabla u, \mathbf{b}\nabla v\rangle\,d\xi, \quad u, v \in H^1(\mathbb{T}^d),$$

where $\mathbf{b}z$ denotes the application of matrix-multiplication for $z \in \mathbb{R}^d$. For smooth functions $u \in C^\infty(\mathbb{T}^d)$, by assumption (2), we have that $L^{\mathbf{b}}u = \operatorname{div}(\mathbf{b}^*\mathbf{b}\nabla u)$, where $\mathbf{b}^*$ denotes the matrix-adjoint of $\mathbf{b}$.

Equation (1) has previously been studied in [5, 6, 10, 19]. One of the problems occurring in the proof of well-posedness for (1) is the characterization of limit solutions (see Definition 12 below) for initial data in $L^2(\mathbb{T}^d)$. For the case of gradient Stratonovich noise, even for initial data in $H^1(\mathbb{T}^d)$, the limit solutions may fail to

[1]The *subdifferential* $\partial\Phi : H \to 2^H$ of a convex, l.s.c. map $\Phi : H \to [0, +\infty]$ is defined by $y \in \partial\Phi(x)$, $x, y \in H$, whenever $\langle y, z - x\rangle \le \Phi(z) - \Phi(x)$ for every $z \in H$. We shall assume that $\Phi(0) = 0$, see condition (N) below.

satisfy an Itô equation. One possible resolution of this issue is the implementation of so called stochastic variational inequalities (SVI) via a viscosity method, that is, essentially a perturbation argument, where Eq. (3) is perturbed with the term $\varepsilon \Delta X_t\,dt$, $\varepsilon > 0$. In order to benefit from the additional regularity of the perturbed equation and in order to be able to pass to the limit $\varepsilon \to 0$, one needs to know suitable a priori estimates which explicitly depend on the viscosity parameter ε. This has been used to prove uniqueness of time-continuous SVI solutions e.g. in [11, 13]. In this work, we shall concentrate on the a priori estimate for viscosity approximations as in the main result Theorem 15 below. Its application to SVI can then be carried out by well-known arguments in a future work. See also the discussion in the concluding Sect. 5. Alternative approaches to SPDE of the type (3) can be found in [8, 19].

1.1 Notation

Denote by $\mathbb{T}^d := \mathbb{R}^d/\mathbb{Z}^d$ the standard flat torus of dimension $d \geq 1$, equipped with the d-dimensional Lebesgue measure $d\xi$ on the Borel σ-algebra $\mathscr{B}(\mathbb{T}^d)$.

Denote $H := L^2(\mathbb{T}^d)$, $S := H^1(\mathbb{T}^d)$, $S^* := H^{-1}(\mathbb{T}^d)$. Note that the embedding $S \hookrightarrow H$ is dense and compact. Let $(\mathscr{E}, D(\mathscr{E}))$ be the *Dirichlet form* of the *Laplace (-Beltrami) operator* $L := \Delta$ on H, that is, $D(\mathscr{E}) := S$ and

$$\mathscr{E}(u, v) := \int_{\mathbb{T}^d} \langle \nabla u, \nabla v \rangle \, d\xi, \quad u, v \in D(\mathscr{E}).$$

Let $(G_\alpha)_{\alpha>0}$ be the *resolvent*, i.e., for $\alpha > 0$, $G_\alpha u := (\alpha - L)^{-1}u$, $u \in H$. For convenience, we shall also introduce the alternative resolvent $J_\delta u := (1 - \delta L)^{-1}u$, where $u \in H$ and $\delta > 0$. Clearly, $J_\delta = \frac{1}{\delta} G_{1/\delta}$ for every $\delta > 0$. The resolvent J_δ, when considered both as a map from H to H or as a map from S to S, is a contraction. Denote the negative of the *Yosida-approximation* by $L^{(\delta)}u := LJ_\delta u = \frac{1}{\delta}(J_\delta - 1)u$, $u \in H$. $L^{(\delta)} : H \to H$ is a negative definite bounded operator with operator norm equal to $\frac{1}{\delta}$. Furthermore, for $\alpha > 0$, $u, v \in D(\mathscr{E})$, let $\mathscr{E}_\alpha(u, v) := \mathscr{E}(u, v) + \alpha(u, v)_H$ be a family of mutually equivalent inner products for S. For $\beta > 0$, define also *approximate forms* $\mathscr{E}^{(\beta)}(u, v) := \beta(u - \beta G_\beta u, v)_H$, $u, v \in H$, see e.g. [18, Chapter I, p. 20]. Set also $\mathscr{E}^{(\beta)}_\alpha(u, v) := \mathscr{E}^{(\beta)}(u, v) + \alpha(u, v)_H$, $\alpha, \beta > 0$, $u, v \in H$.

2 Preliminaries

Let us assume that there exists a constant $C > 0$ and a map $\psi : [0, \infty) \to [0, \infty)$ such that

(N) $\varphi(\zeta) = \psi(|\zeta|)$, for all $\zeta \in \mathbb{R}^d$ and ψ is convex, continuous, and satisfies $\psi(0) = 0$, $\lim_{r\to\infty} \psi(r) = +\infty$, $\psi(r) \leq C(1 + |r|^2)$ for every $r \geq 0$.

Example 1 Condition (N) is e.g. satisfied for the following choices of φ: for $\zeta \in \mathbb{R}^d$, let $\varphi(\zeta) := \frac{1}{p}|\zeta|^p$, $p \in [1,2]$ or let $\varphi(\zeta) := (1+|\zeta|)\log(1+|\zeta|) - |\zeta|$.

Let us further assume that there exists a constant $\kappa > 0$ such that

(E) Assume that $N = d$, $\mathbf{b} \in C^1(\mathbb{T}^d; \mathbb{R}^{N\times d})$, (2) holds, and $|\mathbf{b}(\xi)\zeta|^2 \geq \kappa|\zeta|^2$ for all $\zeta \in \mathbb{R}^d$ and all $\xi \in \mathbb{T}^d$.

Remark 2 Note that (E) implies that $D(\mathscr{E}^{\mathbf{b}}) = D(\mathscr{E}) = S$ and that $-L^{\mathbf{b}}$ is uniformly elliptic.

Lemma 3 *Assume that $N = d$, $\mathbf{b} \in C^1(\mathbb{T}^d; \mathbb{R}^{N\times d})$ and that* (2) *holds. Then condition (E) is equivalent to $b_1(\xi), \dots, b_N(\xi)$ being linearly independent in $\mathbb{R}^d$ for each $\xi \in \mathbb{T}^d$.*

Proof First note that $\mathbf{b}^*\mathbf{b}$ is precisely the Gram matrix of the vectors $b_1, \dots, b_N$. Consider the statement of $b_1, \dots, b_N$ being pointwise linearly independent. It is well-known that this is equivalent to the Gram matrix being positive definite, see e.g. [16, Chapter 10], which in turn is equivalent to (E) with $\kappa := \min_{\xi\in\mathbb{T}^d} \kappa(\xi)$, $\kappa(\xi)$ being the smallest eigenvalue of $\mathbf{b}^*(\xi)\mathbf{b}(\xi)$ (all eigenvalues are strictly positive and real). By continuous dependence of the eigenvalues of $\mathbf{b}^*\mathbf{b}$ on $\xi \in \mathbb{T}^d$, and since $\mathbb{T}^d$ is compact, we argue by contradiction to see that $\kappa > 0$. □

(R) Assume that there exists a constant $c \in \mathbb{R}$, such that for every $\beta > 0$, we have that

$$\beta \int_{\mathbb{T}^d} \langle \beta G_\beta \mathbf{b}\nabla f - \beta \mathbf{b}\nabla G_\beta f, \mathbf{b}\nabla f\rangle \, d\xi \geq c(\mathscr{E}^{\mathbf{b}})^{(\beta)}(f,f), \quad \forall f \in S.$$

As a matter of fact, in (R), it is necessary that $c \leq 0$. See Remark 6 below.

Example 4 By [23, Theorem 3.1, Proposition 3.2], (R) holds with $c = 0$, if L and the first order operator $S \ni u \mapsto \mathbf{b}\nabla u \in L^2(\mathbb{T}^d; \mathbb{R}^N)$ commute componentwise on some core of L which consists of smooth functions. This is related to the notion of so-called *Killing vector fields*, see [10] for examples.

Remark 5 If one lets $\beta \to \infty$ in (R) for $f \in C^\infty(\mathbb{T}^d)$, one obtains,

$$\int_{\mathbb{T}^d} [\langle \Delta \mathbf{b}\nabla f, \mathbf{b}\nabla f\rangle - \langle \mathbf{b}\nabla f, \mathbf{b}\nabla \Delta f\rangle] \, d\xi \geq c \int_{\mathbb{T}^d} |\mathbf{b}\nabla f|^2 \, d\xi, \quad \forall f \in C^\infty(\mathbb{T}^d), \tag{4}$$

which resembles an integrated version of the *Bakry-Émery curvature-dimension condition*, see (16) below. In fact, letting $g \in C^\infty(\mathbb{T}^d)$, $g \geq 0$, we easily see that (4) follows from the weak Γ_2-condition

$$\frac{1}{2}\int_{\mathbb{T}^d} \Delta g|\mathbf{b}\nabla f|^2 \, d\xi - \int_{\mathbb{T}^d} g\langle \mathbf{b}\nabla f, \mathbf{b}\nabla \Delta f\rangle \, d\xi \geq c \int_{\mathbb{T}^d} g|\mathbf{b}\nabla f|^2 \, d\xi, \quad \forall f, g \in C^\infty(\mathbb{T}^d),\ g \geq 0, \tag{5}$$

from [14, Theorem 4.6] for the Dirichlet form $\mathscr{E}^{\mathbf{b}}$. However, inequality (5) and the curvature-dimension condition (16) below are equivalent by results from [3, 17].

Remark 6 Let $\mathbf{b}$ satisfy (E). Observe that there exists a non-constant function $f \in S$ such that $\Delta f = -\lambda_1 f$, where $\lambda_1 > 0$ is the smallest non-zero eigenvalue of $-\Delta$. For such an eigenfunction f, we get by the Poincaré inequality (see e.g. [25, Example 1.1.2]) that

$$\int_{\mathbb{T}^d} [\langle \Delta \mathbf{b}\nabla f, \mathbf{b}\nabla f\rangle - \langle \mathbf{b}\nabla f, \mathbf{b}\nabla \Delta f\rangle]\, d\xi = -\int_{\mathbb{T}^d} |\nabla \mathbf{b}\nabla f|^2\, d\xi + \lambda_1 \int_{\mathbb{T}^d} |\mathbf{b}\nabla f|^2\, d\xi$$
$$\leq \lambda_1 \sum_{i=1}^{N} \left(\int_{\mathbb{T}^d} \langle b_i, \nabla f\rangle\, d\xi \right)^2 = 0, \tag{6}$$

where, in the last step, we have used integration by parts and (2). Hence, for general $\mathbf{b}$ satisfying (E), the estimate (6), combined with Remark 5, shows that $c \leq 0$ is necessary for (R) to hold.

Example 7 Let $N = d = 2$. Denote $g_t := g(t) := \sin(2\pi t)$, $h_t := h(t) := \cos(2\pi t)$, $t \in [0, 1]$. Consider $f(t, s) := g_t + h_s$, $\xi = (t, s) \in \mathbb{T}^2$. Clearly, $f \in C^\infty(\mathbb{T}^2)$, $\nabla f = 2\pi(h_t, -g_s)$, and $\Delta f = -4\pi^2 f$. For a constant $\gamma > \frac{1+\sqrt{3}}{\sqrt{2}}$, let

$$\mathbf{b}(\xi) = \begin{pmatrix} b_1(t,s) \\ b_2(t,s) \end{pmatrix} := \begin{pmatrix} g_t g_s + \gamma & h_t h_s \\ g_t h_s & -h_t g_s + \gamma \end{pmatrix}.$$

By a simple calculation and Lemma 3, we see that $\mathbf{b}$ satisfies (E) for such γ. We have that

$$\mathbf{b}\nabla f = 2\pi h_t (g_t g_s - g_s h_s, g_t h_s + g_s^2) + \gamma \nabla f,$$
$$|\mathbf{b}\nabla f|^2 = 4\pi^2 h_t^2 (g_t^2 g_s^2 + g_s^2 h_s^2 + g_t^2 h_s^2 + g_s^4) + 2\gamma \langle \mathbf{b}\nabla f - \gamma\nabla f, \nabla f\rangle + \gamma^2 |\nabla f|^2,$$
$$\langle \mathbf{b}\nabla f, \mathbf{b}\nabla \Delta f\rangle = -4\pi^2 |\mathbf{b}\nabla f|^2,$$
$$\nabla(\mathbf{b}\nabla f) = 4\pi^2 \left[\begin{pmatrix} g_s(h_{2t} + g_t h_s) \\ h_t(g_t h_s + h_{2s}) \end{pmatrix} \otimes \begin{pmatrix} h_{2t} h_s - g_t g_s^2 \\ h_t(g_{2s} - g_t g_s) \end{pmatrix} + \gamma \begin{pmatrix} -g_t & 0 \\ 0 & -h_s \end{pmatrix} \right].$$

Observe that $\mathbf{b}\nabla f - \gamma\nabla f$ and $\gamma\nabla f$ are orthogonal in $L^2(\mathbb{T}^2; \mathbb{R}^2)$. Also, $\nabla(\mathbf{b}\nabla f) - \gamma\, \mathrm{Hess}(f)$ and $\gamma\, \mathrm{Hess}(f)$ are orthogonal in $L^2(\mathbb{T}^2; \mathbb{R}^{2\times 2})$. Combining these identities, we obtain

$$-\int_{\mathbb{T}^2} |\nabla \mathbf{b}\nabla f|^2\, d\xi - \int_{\mathbb{T}^2} \langle \mathbf{b}\nabla f, \mathbf{b}\nabla \Delta f\rangle\, d\xi = -4\pi^2 \left(\left(\frac{11}{2} + 4\gamma^2 \right)\pi^2 - \left(\frac{3}{2} + 4\gamma^2 \right)\pi^2 \right) = -16\pi^4,$$

and

$$\int_{\mathbb{T}^2} |\mathbf{b}\nabla f|^2\, d\xi = \left(\frac{3}{2} + 4\gamma^2 \right)\pi^2.$$

As a consequence, by Remark 5, if (E) and (R) hold for $\mathbf{b}$ as above, it is necessary that $c \leq -\frac{32}{3+8\gamma^2}\pi^2$.

2.1 Commutator Estimates

We shall prove some approximative commutator bounds needed later.

Lemma 8 *Assume (E) to hold. Let $u \in S$. Then, for every $\beta > 0$,*

$$\mathscr{E}^{(\beta)}(u, J_\delta L^{\mathbf{b}} J_\delta u) \leq \frac{1}{\delta} \|\mathbf{b}\|_\infty^2 \|u\|_S^2.$$

Proof Let $\beta > 0$ and $u \in S$. Set $y_\delta := J_\delta u \in D(L)$. Then, noting that $L^{(\delta)}$ and $\sqrt{-L} G_\beta$ commute in H,

$$\begin{aligned}
\mathscr{E}^{(\beta)}(u, J_\delta L^{\mathbf{b}} J_\delta u) =& \beta(y_\delta - \beta G_\beta y_\delta, L^{\mathbf{b}} y_\delta)_H = \beta \mathscr{E}^{\mathbf{b}}(\beta G_\beta y_\delta - y_\delta, y_\delta) \\
\leq& \beta \mathscr{E}^{\mathbf{b}}(\beta G_\beta y_\delta - y_\delta, \beta G_\beta y_\delta - y_\delta)^{1/2} \mathscr{E}^{\mathbf{b}}(y_\delta, y_\delta)^{1/2} \\
\leq& \beta \|\mathbf{b}\|_\infty^2 \|\sqrt{-L}(\beta G_\beta y_\delta - y_\delta)\|_H \|y_\delta\|_S \\
\leq& \beta \|\mathbf{b}\|_\infty^2 \|\sqrt{-L} G_\beta L y_\delta\|_H \|u\|_S \\
=& \beta \|\mathbf{b}\|_\infty^2 \|L^{(\delta)} \sqrt{-L} G_\beta u\|_H \|u\|_S \\
\leq& \frac{1}{\delta} \|\mathbf{b}\|_\infty^2 \|\beta G_\beta u\|_S \|u\|_S \leq \frac{1}{\delta} \|\mathbf{b}\|_\infty^2 \|u\|_S^2,
\end{aligned}$$

where we have used that for $x \in D(L)$, $\beta > 0$, it holds that $\beta G_\beta u - u = G_\beta L u$. □

Lemma 9 *Assume (E) and (R) to hold. Let $u \in H$. Then there exists $c \leq 0$ such that for every $\delta > 0$ and every $\beta > 0$,*

$$\sum_{i=1}^N \mathscr{E}_1^{(\beta)}(\langle b_i, \nabla J_\delta u\rangle, \langle b_i, \nabla J_\delta u\rangle) + \mathscr{E}_1^{(\beta)}(J_\delta L^{\mathbf{b}} J_\delta u, u) \leq -c(\mathscr{E}^{\mathbf{b}})^{(\beta)}(J_\delta u, J_\delta u).$$

Proof Let $u \in H$, let $\delta, \beta > 0$. Set $y_\delta := J_\delta u$. Then by (R) and Remark 6, there exists $c \leq 0$, such that

$$\begin{aligned}
&\sum_{i=1}^N \mathscr{E}_1^{(\beta)}(\langle b_i, \nabla J_\delta u\rangle, \langle b_i, \nabla J_\delta u\rangle) + \mathscr{E}_1^{(\beta)}(J_\delta L^{\mathbf{b}} J_\delta u, u) \\
=& \beta \left[\mathscr{E}^{\mathbf{b}}(\beta G_\beta y_\delta, y_\delta) - \sum_{i=1}^N (\beta G_\beta(\langle b_i, \nabla y_\delta\rangle), \langle b_i, \nabla y_\delta\rangle)_H \right] \\
=& \beta \int_{\mathbb{T}^d} \left\langle \beta \mathbf{b} \nabla G_\beta y_\delta - \beta G_\beta \mathbf{b} \nabla y_\delta, \mathbf{b} \nabla y_\delta \right\rangle d\xi \\
\leq& -c(\mathscr{E}^{\mathbf{b}})^{(\beta)}(y_\delta, y_\delta).
\end{aligned}$$

□

Corollary 10 *Assume (E) and (R) to hold. Let $u \in S$. Then the estimate of Lemma 8 improves to*

$$\mathscr{E}^{(\beta)}(u, J_\delta L^{\mathbf{b}} J_\delta u) \le -c\|\mathbf{b}\|_\infty^2 \|u\|_S^2 \quad \forall \delta, \beta > 0,$$

where $c \le 0$ *does neither depend on* δ, *nor on* β.

3 An Approximating Equation

Let us recall the following conditions from [12], simplified with regard to the time-dependence of the drift coefficients, which is not needed here. Suppose that $A : S \to 2^{S^*}$ satisfies the following conditions: There is a constant $C > 0$ such that

(A1) The map $x \mapsto A(x)$ is maximal monotone with non-empty values.

(A2) For all $x \in S$, for all $y \in A(x)$:

$$\|y\|_{S^*} \le C\|x\|_S.$$

(A3) For all $x \in S$, for all $y \in A(x)$, and for all $\mu > 0$:

$$2_{S^*}\langle y, L^{(\mu)}x\rangle_S \le C\|x\|_S^2,$$

such that C is independent of μ.

Let U be a separable Hilbert space. Denote the *space of Hilbert-Schmidt operators* from U to H by $L_2(U, H)$. Suppose that $B : [0, T] \times \Omega \times S \to L_2(U, H)$ is *progressively measurable*[2] and that there exist constants $C_1, C_2, C_3 > 0$ such that

(B1) There is $h \in L^1([0, T] \times \Omega)$ such that

$$\|B_t(x)\|_{L_2(U,H)}^2 \le C_1\|x\|_S^2 + h_t$$

for all $t \in [0, T]$, $x \in S$ and $\omega \in \Omega$.

(B2)

$$\|B_t(x) - B_t(y)\|_{L_2(U,H)}^2 \le C_2\|x - y\|_H^2$$

for all $t \in [0, T]$, $x, y \in S$ and $\omega \in \Omega$.

(B3) There is $\tilde{h} \in L^1([0, T] \times \Omega)$ such that

$$\|B_t(x)\|_{L_2(U,S)}^2 \le C_3\|x\|_S^2 + \tilde{h}_t$$

for all $t \in [0, T]$, $x \in S$ and $\omega \in \Omega$.

Denote by $(W_t)_{t\ge 0}$ a cylindrical Wiener process in U for the stochastic basis $(\Omega, \mathscr{F}, (\mathscr{F}_t)_{t\ge 0}, \mathbb{P})$.

---

[2]That is, for every $t \in [0, T]$ the map $B : [0, t] \times \Omega \times S \to L_2(U, H)$ is $\mathscr{B}([0, t]) \otimes \mathscr{F}_t \otimes \mathscr{B}(S)$-measurable.

Definition 11 We say that a continuous $(\mathscr{F}_t)_{t\geq 0}$-adapted stochastic process $X : [0,T]\times\Omega\to H$ is a *solution* to

$$dX_t + A(X_t)\,dt \ni B_t(X_t)\,dW_t, \quad X_0 = x, \tag{7}$$

if $X\in L^2(\Omega; C([0,T];H))\cap L^2([0,T]\times\Omega; S)$ and solves the following integral equation in S^*

$$X_t = x - \int_0^t \eta_s\,ds + \int_0^t B_s(X_s)\,dW_s,$$

$\mathbb{P}$-a.s. for all $t\in[0,T]$, where $\eta\in A(X)$, $dt\otimes\mathbb{P}$-a.s.

Definition 12 An $(\mathscr{F}_t)_{t\in[0,T]}$-adapted stochastic process $X\in L^2(\Omega; C([0,T];H))$ is called a *limit solution* to (7) with starting point $x\in H$ if for all approximations $x^m\in S$, $m\in\mathbb{N}$ with $\|x^m - x\|_H\to 0$ as $m\to\infty$ and all B^m satisfying (B1)–(B3) and such that $B^m(y)\to B(y)$ strongly in $L^2([0,T]\times\Omega; L_2(U,H))$ for every $y\in S$, we have that $X^m\to X$ strongly in $L^2(\Omega; C([0,T];H))$ as $m\to\infty$.

Roughly rephrased, [12, Theorems 4.4 and 4.6] state that (A1)–(A3), (B1)–(B3) guarantee the existence and uniqueness of solutions to (7) (that are uniformly S-bounded in quadratic mean), whereas (A1)–(A3), (B1)–(B2) guarantee the existence of limit solutions to (7).

Let $\lambda,\delta,\varepsilon>0$. We write $\phi^\lambda = \partial\varphi^\lambda$, $\lambda>0$, meaning that ϕ^λ is the Yosida-approximation of ϕ, whereas φ^λ denotes the Moreau-Yosida approximation of φ, see [2, p. 266]. Let us verify (A1)–(A3), (B1)–(B2) for $x\in S$,

$$A(x) := A^{\lambda,\delta,\varepsilon}(x) := -J_\delta\operatorname{div}(\phi^\lambda(\nabla J_\delta x)) - \varepsilon J_\delta L^{(\delta)}x - \frac{1}{2}J_\delta L^{\mathbf{b}}J_\delta x, \tag{8}$$

and for $U := \mathbb{R}^N$, $\zeta = (\zeta^1,\dots,\zeta^N)\in U$, $x\in S$, $t\in[0,T]$,

$$B_t(x)\zeta := B^\delta(x)\zeta := \sum_{i=1}^N \langle b_i, \nabla J_\delta x\rangle\zeta^i. \tag{9}$$

Proposition 13 *Suppose that conditions (N) and (E) hold. Then A and B as defined in* (8), *and* (9) *respectively, satisfy conditions (A1)–(A3), and (B1)–(B2) respectively, for* $U := \mathbb{R}^N$ *and all* $\lambda,\delta,\varepsilon>0$.

Proof Fix $\lambda,\delta,\varepsilon>0$.

(A1): The map

$$x\mapsto -J_\delta\operatorname{div}(\phi^\lambda(\nabla J_\delta x))$$

is the maximal monotone subdifferential of the l.s.c. map

$$u \mapsto \int_{\mathbb{T}^d} \varphi^\lambda(\nabla J_\delta u)\, d\xi.$$

In this regard, see also [24, Proposition II.7.8] for the chain rule of a convex functional and a bounded linear operator. The map

$$x \mapsto -\varepsilon J_\delta L^{(\delta)} x - \frac{1}{2} J_\delta L^{\mathbf{b}} J_\delta x$$

is positive definite due to condition (E). As this map is also bounded linear, the conditions of Browder's theorem on the maximality of the sum of monotone operators are satisfied, see [22, Theorem, pp. 75–76], and hence (A1) holds.

(A2): As the linear part of the map A is bounded in H, (A2) easily follows together with arguments as in [12, Proposition 7.1].

(A3): The claim for the nonlinear part follows as in [12, Example 7.11]. For the cases of Dirichlet and Neumann boundary conditions, respectively, compare also to [9, 10]. Let $x \in S$. Regarding the linear part, clearly, since $-J_\delta L^{(\delta)}$ is positive definite and symmetric and commutes with $L^{(\mu)}$, we get for $y_\delta := \sqrt{-L^{(\delta)} J_\delta x}$ that

$$-2\varepsilon (J_\delta L^{(\delta)} x, L^{(\mu)} x)_H = 2\varepsilon (L^{(\mu)} y_\delta, y_\delta) \le 0 \quad \forall \mu > 0.$$

for all $\mu > 0$. Furthermore, by Lemma 8, for every $\mu > 0$,

$$-(J_\delta L^{\mathbf{b}} J_\delta x, L^{(\mu)} x)_H = \mathscr{E}^{(1/\mu)}(x, J_\delta L^{\mathbf{b}} J_\delta x) \le \frac{1}{\delta} \|\mathbf{b}\|_\infty^2 \|x\|_S^2.$$

(B1): By a straightforward calculation, we get for $x \in S$,

$$\|B^\delta(x)\|_{L_2(U,H)}^2 = \mathscr{E}^{\mathbf{b}}(J_\delta x, J_\delta x) \le \|\mathbf{b}\|_\infty^2 \|x\|_S^2.$$

(B2): By linearity, for $x, y \in S$,

$$\begin{aligned}
&\|B^\delta(x) - B^\delta(y)\|_{L_2(U,H)}^2 = \|B^\delta(x-y)\|_{L_2(U,H)}^2 \\
=&\mathscr{E}^{\mathbf{b}}(J_\delta(x-y), J_\delta(x-y)) \le \|\mathbf{b}\|_\infty^2 \mathscr{E}^{(1/\delta)}(x-y, x-y) \\
\le&\left(\frac{2}{\delta} + 1\right) \|\mathbf{b}\|_\infty^2 \|x-y\|_H^2,
\end{aligned}$$

compare also with [18, Chapter I, Lemma 2.11]. □

4 A Priori Estimates

Consider the SPDE

$$dX_t + A^{\lambda,\delta,\varepsilon}(X_t)\,dt = B^\delta(X_t)\,d\beta_t, \quad X_0 = x_n, \tag{10}$$

where $x_n \in S$, and where, from now on, whenever it seems convenient, we suppress the indices in the notation as in $X = X^{n,\lambda,\delta,\varepsilon}$. The existence and uniqueness of limit solutions to (10) follows from [12, Theorem 4.6].

For fixed $\delta > 0$ and for every $m \in \mathbb{N}$, let $y \mapsto B^{\delta,m}(y)$ be progressively measurable maps on S such that

(i) each $B^{\delta,m}$ satisfies (B1)–(B2) with constants C_1 and C_2 not depending on m,
(ii) each $B^{\delta,m}$ satisfies (B3) (with constants $C_3 = C_3(m)$ typically depending on m),
(iii) $\|B^{\delta,m}(y) - B^\delta(y)\|_{L_2(U,H)} \to 0$ for every $y \in S$ as $m \to \infty$.

The existence of such a sequence of maps can be proved e.g. by introducing an approximation step that employs standard mollifiers. Consider the sequence of Itô-processes

$$X_t^m = x_n - \int_0^t A^{\lambda,\delta,\varepsilon}(X_s^m)\,ds + \int_0^t B^{\delta,m}(X_s^m)\,d\beta_s, \tag{11}$$

where $x_n \in S, m \in \mathbb{N}$. The existence of such processes is guaranteed by [12, Theorem 4.4].

Proposition 14 *Suppose that conditions (N) and (E) hold. Let $\lambda, \delta, \varepsilon > 0$. Let $x_n \in S$, and let $X = X^{n,\lambda,\delta,\varepsilon}$ be a limit solution to* (10). *Then, we have that*

$$\operatorname*{ess\,sup}_{t\in[0,T]} \left[\mathbb{E}\|X_t\|_H^2\right] + 2\varepsilon\mathbb{E}\int_0^T \mathscr{E}(J_\delta X_s, J_\delta X_s)\,ds \le \|x_n\|_H^2. \tag{12}$$

Proof Let $x_n \in S, m \in \mathbb{N}$. We may apply the Itô formula [20, Theorem 4.2.5] for the Gelfand triple $S \subset H \subset S^*$ and the process (11). We get for $t \in [0, T]$, after taking the expected value,

$$\begin{aligned}
&\mathbb{E}\|X_t^m\|_H^2 - \|x_n\|_H^2\\
\le&\mathbb{E}\int_0^t \left[-2(A^{\lambda,\delta,\varepsilon}(X_s^m), X_s^m)_H + \|B^{\delta,m}(X_s^m)\|_{L_2(U,H)}^2\right] ds\\
\le&\mathbb{E}\int_0^t \left[-2(A^{\lambda,\delta,\varepsilon}(X_s^m), X_s^m)_H + \|B^{\delta,m}(X_s^m) - B^{\delta,m}(X_s)\|_{L_2(U,H)}^2 + \|B^{\delta,m}(X_s)\|_{L_2(U,H)}^2\right] ds\\
\le&\mathbb{E}\int_0^t \left[-2(A^{\lambda,\delta,\varepsilon}(X_s^m), X_s^m)_H + C\|X_s^m - X_s\|_H^2 + \|B^{\delta,m}(X_s)\|_{L_2(U,H)}^2\right] ds,
\end{aligned}$$

where X is as in (10) and C does not depend on m. By [12, Theorem 4.6], $X^m \to X$ in $L^2(\Omega; C([0, T]; H))$. Note that $A^{\lambda,\delta,\varepsilon} : H \to H$ is monotone, continuous and bounded and thus by Minty's trick (see e.g. [20, Remark 4.1.1]), $A^{\lambda,\delta,\varepsilon}(X^m) \to A^{\lambda,\delta,\varepsilon}(X)$ weakly in H. Hence by (i), (iii) above and Lebesgue's dominated convergence theorem, we converge to the inequality

$$\mathbb{E}\|X_t\|_H^2 - \|x_n\|_H^2 \le \mathbb{E}\int_0^t \left[-2(A^{\lambda,\delta,\varepsilon}(X_s), X_s)_H + \|B^\delta(X_s)\|_{L_2(U,H)}^2\right] ds.$$

Note that, as above, for $y \in H$: $\|B^{\delta}(y)\|^2_{L_2(U,H)} = \mathscr{E}^{\mathbf{b}}(J_\delta y, J_\delta y)$. On the other hand, $(J_\delta L^{\mathbf{b}} J_\delta y, y)_H = -\mathscr{E}^{\mathbf{b}}(J_\delta y, J_\delta y)$, for all $y \in H$. Hence, employing also the monotonicity of the nonlinear part of $A^{\lambda,\delta,\varepsilon}$, we get that

$$\begin{aligned}\mathbb{E}\|X_t\|^2_H - \|x_n\|^2_H \leq & 2\varepsilon\mathbb{E}\int_0^t (J_\delta L^{(\delta)} X_s, X_s)_H\, ds \\ \leq & -2\varepsilon\mathbb{E}\int_0^t \mathscr{E}(J_\delta X_s, J_\delta X_s)\, ds.\end{aligned}$$

□

Let $\beta > 0$. Define renormed spaces $H_\beta := L^2(\mathscr{X})$ with norm $\|u\|^2_\beta := \mathscr{E}_1^{(\beta)}(u,u)$. Obviously, $\|\cdot\|_H \leq \|\cdot\|_\beta \leq \sqrt{2\beta+1}\|\cdot\|_H$.

Theorem 15 *Suppose that conditions (N), (E) and (R) hold. Let $x_n \in S$, and let $X = X^{n,\lambda,\delta,\varepsilon}$ be a limit solution to* (10). *Then, we have for $\lambda, \delta, \varepsilon > 0$, that*

$$\operatorname*{ess\,sup}_{t\in[0,T]} \left[\mathbb{E}\|X_t\|^2_S\right] + 2\varepsilon \int_0^T \left\|\sqrt{J_\delta L^{(\delta)}} X_t\right\|^2_S dt \leq e^{-c\|\mathbf{b}\|^2_\infty T}\|x_n\|^2_S, \tag{13}$$

where $c \leq 0$ is as in condition (R).

Proof We shall apply the Itô formula [20, Theorem 4.2.5] for the Gelfand triple $S \subset H_\beta \subset S^*$ and the process (11). As in the proof of Proposition 14, we get for $t \in [0,T]$, after taking the expected value,

$$\mathbb{E}\|X_t^m\|^2_\beta - \|x_n\|^2_\beta \leq \mathbb{E}\int_0^t \left[-2(A^{\lambda,\delta,\varepsilon}(X_s^m), X_s^m)_\beta + C\|X_s^m - X_s\|^2_\beta + \|B^{\delta,m}(X_s)\|^2_{L_2(U,H_\beta)}\right] ds,$$

where X is as in (10) and C does not depend on m. Now, we argue essentially as in the proof of Proposition 14 and obtain that

$$\mathbb{E}\|X_t\|^2_\beta - \|x_n\|^2_\beta \leq \mathbb{E}\int_0^t \left[-2(A^{\lambda,\delta,\varepsilon}(X_s), X_s)_\beta + \|B^{\delta}(X_s)\|^2_{L_2(U,H_\beta)}\right] ds.$$

Note that $(J_\delta \operatorname{div}(\phi^\lambda(\nabla J_\delta y)), y)_\beta \leq 0$, $y \in H$, as in the proof of Proposition 13 (A3), compare also with [10, Step I in the proof of Theorem 4.1, p. 1780]. An application of Lemma 9 shows that

$$\mathbb{E}\|X_t\|^2_\beta - \|x_n\|^2_\beta \leq 2\varepsilon\mathbb{E}\int_0^t (J_\delta L^{(\delta)} X_s, X_s)_\beta\, ds - c\|\mathbf{b}\|^2_\infty \int_0^t \|J_\delta X_s\|^2_\beta\, ds.$$

The claim follows, after letting $\beta \to \infty$ (where we use Fatou's lemma and the Mosco convergence[3] of $\|\cdot\|^2_\beta \to \|\cdot\|^2_S$),[4] by resolvent contraction in S and Gronwall's lemma. □

[3]See [2] for the notion of *Mosco convergence*.

[4]Cf. [18, Chapter I, Lemma 2.12], which Michael Röckner, in his lectures, usually referred to as the "Wunderlemma".

5 Concluding Remarks

5.1 Applications to Stochastic Variational Inequalities (SVI)

Estimate (13) turns out useful for the purpose of studying existence and uniqueness of time-continuous SVI solutions to SPDE (1). This program for SVI has been e.g. implemented in [11, 13] for other noise types and in [10] for the Neumann p-Laplace with gradient noise for a subclass of vector fields $\mathbf{b}$. SVI for singular drift terms as the total variation flow first appeared in [7]. Let us mention also the works [9, 19], where the situation of Dirichlet boundary conditions is investigated.

In order to pass to the limit in (10), we have to assume that the nonlinear potential ψ in condition (N) further satisfies

$$\psi(2r) \leq K\psi(r) \quad \forall r \geq 0, \tag{14}$$

for some constant $K > 0$ independent of r. This condition is known as the (global) Δ_2-condition in the theory of Orlicz spaces, see e.g. [21]. The doubling property (14) implies that the Moreau-Yosida approximation φ^λ and φ become uniformly comparable up to a controllable error term. Then, for each $\varepsilon > 0$, we will be able to pass to the limit $\delta \to 0$ and $\lambda \to 0$ in (13).

All in all, the approximative a priori estimate can be used to prove the existence of time-continuous SVI solutions, and with some extra effort, also the uniqueness of those. This shall be investigated in a forthcoming work.

5.2 Approach via Curvature-Dimension Conditions

Set $\mathscr{X} := \mathbb{T}^d$. Note that $(\mathscr{X}, (\mathbf{b}^*\mathbf{b})^{-1})$ is a Riemannian manifold with metric $\mathbf{g} := (\mathbf{b}^*\mathbf{b})^{-1}$, which is *quasi-isometric*[5] to $\mathscr{X}$ with the flat metric. Let us write $\mathscr{X}^{\mathbf{g}}$ if we want to emphasize the choice of the metric. We denote the volume measure on $\mathscr{X}^{\mathbf{g}}$ by $d\nu := \sqrt{\det \mathbf{g}}\, d\xi$. Note that $(\mathscr{X}, \mathbf{g})$, equipped with the Lebesgue measure, is thus a *weighted manifold* with density $\rho := \sqrt{\det \mathbf{g}^{-1}}$ w.r.t. $d\nu$. Above, in order to verify (A3), we have used the following *gradient estimate*, see [12, Example 7.11].

(P) Let $(P_t^{\mathbf{b}})_{t\geq 0}$ be the heat semigroup associated to $\mathscr{E}^{\mathbf{b}}$. Assume that for any $f \in C^1(\mathscr{X})$, there exists a constant $K_{\mathbf{b}} \geq 0$, such that

$$|\nabla P_t^{\mathbf{b}} f| \leq e^{-2K_{\mathbf{b}}t} P_t^{\mathbf{b}} |\nabla f| \quad \forall t \geq 0. \tag{15}$$

Note that by [15, Exercise 3.12], $L^{\mathbf{b}} = \Delta^{\mathbf{g}}$ on the weighted manifold $(\mathscr{X}, \mathbf{g}, d\xi)$, where $\Delta^{\mathbf{g}}$ denotes the weighted Laplace-Beltrami operator of $\mathscr{X}^{\mathbf{g}}$. By e.g. [1, 4], (15)

[5] See [15, p. 93].

is equivalent to the existence of a constant $K_{\mathbf{b}} \geq 0$ such that the following pointwise *Bakry-Émery-curvature-dimension condition* $BE(K_{\mathbf{b}}, \infty)$ holds

$$L^{\mathbf{b}}|\mathbf{b}\nabla f|^2 - 2\langle \mathbf{b}\nabla f, \mathbf{b}\nabla L^{\mathbf{b}} f\rangle \geq \frac{K_{\mathbf{b}}}{2}|\mathbf{b}\nabla f|^2, \quad \forall f \in C^\infty(\mathscr{X}), \tag{16}$$

see [1, 4, 25, 26] for the terminology and further results on equivalent *curvature-dimension conditions* as well as *Ricci curvature bounds* in weighted Riemannian manifolds.

Assuming (P) would lead to an alternative approach for the a priori estimate (13), that is, the Laplace operator is to be replaced by $L^{\mathbf{b}}$, and the regularizing resolvents (J_δ) are to be replaced by $J_\delta^{\mathbf{b}} = (1 - \delta L^{\mathbf{b}})^{-1}$, $\delta > 0$. Of course, in (10), we would choose an "$\varepsilon L^{\mathbf{b}}$"-viscosity term, $\varepsilon > 0$. In this situation, estimate (13) would yield an integrated $D(L^{\mathbf{b}})$ bound rather than an $L^2(\Omega \times [0, T]; H^2(\mathscr{X}))$-bound. The investigation of this method is topic of a forthcoming work.

References

1. Ambrosio, L., Gigli, N., Savaré, G.: Bakry-Émery curvature-dimension condition and Riemannian Ricci curvature bounds. Ann. Probab. **43**(1), 339–404 (2015)
2. Attouch, H.: Variational Convergence for Functions and Operators. Pitman, Boston-London-Melbourne (1984)
3. Bakry, D.: On Sobolev and logarithmic Sobolev inequalities for Markov semigroups. In: New Trends in Stochastic Analysis (Charingworth, 1994), pp. 43–75. World Scientific Publishing, River Edge, N.J. (1997)
4. Bakry, D., Gentil, I., Ledoux, M.: Analysis and Geometry of Markov Diffusion Operators. Springer International Publishing, Cham (2014)
5. Barbu, V., Brzeźniak, Z., Hausenblas, E., Tubaro, L.: Existence and convergence results for infinite dimensional nonlinear stochastic equations with multiplicative noise. Stochastic Process. Appl. **123**(3), 934–951 (2013)
6. Barbu, V., Brzeźniak, Z., Tubaro, L.: Stochastic nonlinear parabolic equations with Stratonovich gradient noise. Appl. Math. Optim. **188**(3), 1–17 (2017)
7. Barbu, V., Da Prato, G., Röckner, M.: Stochastic nonlinear diffusion equations with singular diffusivity. SIAM J. Math. Anal. **41**(3), 1106–1120 (2009)
8. Barbu, V., Röckner, M.: An operatorial approach to stochastic partial differential equations driven by linear multiplicative noise. J. Eur. Math. Soc., 1789–1815 (2015)
9. Barbu, V., Rökner, M.: Stochastic variational inequalities and applications to the total variation flow perturbed by linear multiplicative noise. Arch. Ration. Mech. Anal. **209**(3), 797–834 (2013)
10. Ciotir, I., Tölle, J.M.: Nonlinear stochastic partial differential equations with singular diffusivity and gradient Stratonovich noise. J. Funct. Anal. **271**(7), 1764–1792 (2016)
11. Gess, B., Röckner, M.: Stochastic variational inequalities and regularity for degenerate stochastic partial differential equations. Trans. Am. Math. Soc. **369**(5), 3017–3045 (2017)
12. Gess, B., Tölle, J.M.: Multi-valued, singular stochastic evolution inclusions. J. Math. Pures Appl. **101**(6), 789–827 (2014)
13. Gess, B., Tölle, J.M.: Stability of solutions to stochastic partial differential equations. J. Differ. Equ. **260**(6), 4973–5025 (2016)
14. Gigli, N., Kuwada, K., Ohta, S.I.: Heat flow on Alexandrov spaces. Commun. Pure Appl. Math. **66**(3), 307–331 (2013)

15. Grigor'yan, A.: Heat kernel and analysis on manifolds. In: AMS/IP Studies in Advanced Mathematics, vol. 47. American Mathematical Society, Providence, RI; International Press, Boston, MA (2009)
16. Lax, P.D.: Linear Algebra and Its Applications. Wiley (2007)
17. Ledoux, M.: The geometry of Markov diffusion generators. Annales de la faculté des sciences de Toulouse Mathématiques **9**(2), 305–366 (2000)
18. Ma, Z.M., Röckner, M.: Introduction to the Theory of (Non-symmetric) Dirichlet forms. Universitext. Springer, Berlin-Heidelberg-New York (1992)
19. Munteanu, I., Röckner, M.: The Total Variation Flow Perturbed by Gradient Linear Multiplicative Noise. To appear in IDAQP pp. 1–22 (2016)
20. Prévôt, C., Röckner, M.: A Concise Course on Stochastic Partial Differential Equations. Lecture Notes in Mathematics, vol. 1905. Springer, Berlin (2007)
21. Rao, M.M., Ren, Z.D.: Applications of Orlicz spaces. Pure and Applied Mathematics. Marcel Dekker Inc, New York-Basel (2002)
22. Rockafellar, R.T.: On the maximality of sums of nonlinear monotone operators. Trans. Am. Math. Soc. **149**(1), 75–88 (1970)
23. Shigekawa, I.: Semigroup domination on a Riemannian manifold with boundary. Acta Appl. Math. **63**(1–3), 385–410 (2000)
24. Showalter, R.E.: Monotone operators in Banach space and nonlinear partial differential equations. Amer. Math. Soc, Mathematical surveys and monographs (1997)
25. Wang, F.Y.: Functional Inequalities. Markov Semigroups and Spectral Theory. Science Press, Beijing-New York (2005)
26. Wang, F.Y.: Equivalent semigroup properties for the curvature-dimension condition. Bull. Sci. Math. **135**(6–7), 803–815 (2011)

Random Data Cauchy Problem for Some Dispersive Equations

Wei Yan and Jinqiao Duan

Abstract Dispersive equations with Hamiltonian structures possess interesting dynamical properties. This subject has attracted a lot of attention recently. We present a brief overview of recent works on the random data Cauchy problems for a Schrödinger-type equation, a wave equation, and a KdV-type equation.

Keywords Random initial data · Dispersive partial differential equations · Noise Stochastic dynamics · Stochastic partial differential equations

Mathematics Subject Classifications (MSC) 35Q41 · 35Q53 · 60H15

1 Random Data Cauchy Problem for a Schrödinger-Type Equation

Partial differential equations in mathematical physics are often under the influence of uncertainties [1]. These fluctuations may be in the form of random initial data, random boundary conditions, random parameters, or random forcing. In the present article, we review some recent progress in the study of dispersive equations, including a Schrödinger-type equation, a wave equation and a KdV-type equation, with random initial conditions.

In 1988, Lebowitz et al. [2] introduced the Gibbs measure to study the statistical mechanics of the nonlinear Schrödinger equation on the torus

W. Yan
College of Mathematics and Information Science, Henan Normal University, Xinxiang, China
e-mail: yanwei19821115@sina.cn

J. Duan (✉)
Department of Applied Mathematics, Illinois Institute of Technology, Chicago, USA
e-mail: duan@iit.edu

A. Eberle et al. (eds.), *Stochastic Partial Differential Equations and Related Fields*, Springer Proceedings in Mathematics & Statistics 229,
https://doi.org/10.1007/978-3-319-74929-7_15

$$iu_t + u_{xx} + |u|^{p-2}u = 0, \tag{1}$$

with random initial data. Then, Bourgain [3, 4] investigated the periodic Schrödinger equation and invariant measures, using the Fourier restriction norm method and Gibbs measure to extend the local solution to global solution. In a sense, the Gibbs measure plays a similar role as the conservation law. Using the high-low frequency technique introduced in [5] in the probabilistic setting, Colliander and Oh [6, 7] studied the following one-dimensional cubic nonlinear Schrödinger equation

$$iu_t + u_{xx} + |u|^2u = 0, \tag{2}$$

with the aid of a suitable randomization. Moreover, Nahmod et al. [8] constructed a weighted invariant Wiener measure associated with the periodic derivative nonlinear Schrödinger equation

$$iu_t - u_{xx} = \lambda(|u|^2u)_x = 0 \tag{3}$$

in one dimension and established the global well-posedness for random initial data living in its support. In particular, the Wiener measure is shown to be invariant almost surely for data in a Fourier-Lebesgue space $FL^{s,r}(T)$ with $s \geq \frac{1}{2}$, $2 < r < 4$, $(s-1)r < -1$ and scaling like $H^{\frac{1}{2}-\varepsilon}(T)$, for $\varepsilon > 0$. By treating a one-dimensional nonlinear Schrödinger equation with a harmonic potential, Burq et. al [9] established the almost sure global existence and scattering for the defocusing nonlinear Schrödinger equation

$$iu_t + (\Delta - |x|^2)u = \pm|u|^{p-1}u \tag{4}$$

on the whole real line **R**. Deng [10], Poiret [11, 12] and Poiret et al. [13] further studied the defocusing nonlinear Schrödinger equation with harmonic potential, which is supercritical with respect to the initial data. Bourgain and Bulut [14] studied the almost sure global well-posedness for the radial nonlinear Schrödinger equation on the unit ball with random data. Bényi et. al [15, 16] used the Wiener randomization to establish the almost sure well-posedness in the supercritical case. Motivated by the work of Bourgain [17], Federico and de Suzzoni [18] then discussed the invariant measure for the Schrödinger equation on the real line. Hirayama and Okamoto [19] investigated the random data Cauchy problem for the nonlinear Schrödinger equation

$$iu_t + \Delta u = \pm\partial(\bar{u}^m) \tag{5}$$

on R^d, $d \geq 1$, where ∂ is a first order derivative with respect to the spatial variable, and proved the almost sure local in time well-posedness, small data global in time well-posedness and scattering property in $H^s(R^d)$ with

$$s > \max\left(\frac{d-1}{d}s_c, \frac{s_c}{2}, s_c - \frac{d}{2(d+1)}\right),$$

$d+m \geq 5$ with $s_c = \frac{d}{2} - \frac{1}{m-1}$. Moreover, Hirayama and Okamoto [20] examined the random data Cauchy problem for the fourth order Schrödinger equation with derivative nonlinearity

$$iu_t + \Delta^2 u = \partial(|u|^2 u) \tag{6}$$

and proved almost sure local in time well-posedness, small data global in time well-posedness and scattering property in $H^s(R^d)$ with

$$s > \max\left(\frac{d-5}{2}, \frac{d-5}{6}\right),$$

whose lower bound is smaller than the scale critical regularity $s_c = \frac{d-3}{2}$. Recently, motivated by [16], we studied the random data Cauchy problem for the cubic fourth-order Schrödinger equation in the supercritical regime [21]and proved that the almost sure local in time well-posedness, small data global in time well-posedness hold in $H^s(R^d)$ with $s \in \left(\frac{d-1}{d+3}\frac{d-4}{2}, \frac{d-4}{2}\right)$ $(5 \leq d \leq 7, d \in N)$, where $\frac{d-4}{2}$ is the scaling critical regularity index. We also established the small data global in time well-posedness in $H^s(R^d)$ with $s \in \left(\max\left(\frac{d-6}{2}, \frac{d-1}{4}\right), \frac{d-4}{2}\right)$ $(d \geq 8, d \in N)$, where $\frac{d-4}{2}$ is the scaling critical regularity index.

2 Random Data Cauchy Problem for a KdV-Type Equation

Now we consider the random data Cauchy problem for a KdV-type equation. Oh [22] investigated the invariance of the Gibbs measure for the Schrödinger–Benjamin–Ono system

$$iu_t + u_{xx} = \alpha vu, \tag{7}$$

$$v_t + \gamma H v_{xx} = \beta(|v|^2)_x \tag{8}$$

on the one dimensional torus. Oh [23] studied the periodic KdV equation with suitable randomization of the initial data. By constructing the Gibbs measure, Oh [24] further established the almost sure global well posedness for the coupled KdV systems

$$u_t + a_{11}u_{xxx} + a_{12}v_{xxx} + b_1 uu_x + b_2 uv_x + b_3 u_x v + b_4 vv_x = 0, \tag{9}$$

$$v_t + a_{21}u_{xxx} + a_{22}v_{xxx} + b_5 uu_x + b_6 uv_x + b_7 u_x v + b_8 vv_x = 0. \tag{10}$$

Deng [25] studied the invariance of the Gibbs measure for the Benjamin–Ono equation

$$u_t + Hu_{xx} = uu_x. \tag{11}$$

Richard [26] established the almost sure local well-posedness and global well-posedness and studied the invariance of the Gibbs measure for the periodic quartic gKdV

$$v_t + v_{xxx} = \frac{1}{4}P(v^3)v_x, \tag{12}$$

where $P(u) = u - \frac{1}{2\pi}\int_0^{2\pi} udx$. Kenig et al. [27] proved that the Cauchy problem for a generalized KdV equation is locally well-posed in $H^s(R)$ with $s > \frac{3}{14}$

$$u_t + u_{xxx} + \frac{1}{8}\partial_x(u^8) = 0 \tag{13}$$

on the real line. Birnir et al. [28] proved that the problem ((13) is ill-posed in $H^s(R)$ with $s < \frac{3}{14}$. We have recently established [29] the almost sure well-posedness of (13) in $H^s(R)$ with $s > \frac{17}{112}$ with random data.

3 Random Data Cauchy Problem for a Wave Equation

In this section, we consider random data Cauchy problem for a wave equation. Mckean and Vaninsky [30] studied the statistical mechanics of nonlinear wave equation

$$u_{tt} - \Delta u + |u|^{p-1}u = 0. \tag{14}$$

After a suitable randomization, Burq and Tzvetkov [31] investigated the invariant measure for a three dimensional nonlinear wave equation

$$u_{tt} - \Delta u + |u|^{\alpha}u = 0, \alpha < 2 \tag{15}$$

in the subcritical case. More precisely, Burq and Tzvetkov [32] studied the random data Cauchy problem for (14) in $H^{\sigma}(\boldsymbol{\Theta})$, where $\max(0, \frac{\alpha-1}{\alpha}) < \sigma < \frac{1}{2}$ and $\boldsymbol{\Theta}$ is the unit ball of R^3.Obviously, if $u(x,t)$ is the solution to (15), then $u_\lambda := \lambda^{-\frac{2}{\alpha}}u\left(\frac{x}{\lambda}, \frac{t}{\lambda}\right)$ is the solution to (15). By a direct computation, we have that $\|u_\lambda\|_{\dot{H}^{s_{critical}}} = \|u\|_{\dot{H}^{s_{critical}}}$, where $s_{critical} = \frac{3}{2} - \frac{\alpha}{2}$. We say that the Cauchy problem for (15) is subcritical, critical, supercritical if $\sigma > s_{critical}, \sigma = S_{critical}, \sigma < S_{critical}$, respectively. Furthermore, Burq and Tzvetkov [32] investigated the local random data Cauchy problem for the supercritical wave equation

$$u_{tt} - \Delta u + u^3 = 0 \tag{16}$$

on a three dimensional compact manifold in the supercritical regime. Burq and Tzvetkov [33] also investigated the global random data Cauchy problem for a supercritical wave equation

$$u_{tt} - \Delta u + |u|^{\alpha} u = 0, 2 < \alpha < 3 \tag{17}$$

on a three dimensional ball in the supercritical case. Moreover, de Suzzoni studied the invariant measure [34] for the cubic wave equation on the unit ball of R^3, and the large data low regularity scattering property [35, 36] for the wave equation on the Euclidean space. Bourgain and Bulut [37] studied the invariant Gibbs measure evolution for the radial nonlinear wave equation on a three dimensional ball. Burq and Tzvetkov [38] established the probabilistic well-posedness for the cubic wave equation on three dimensional torus. In fact, Burq et al. [39] further studied the global infinite energy solutions for the cubic wave equation on the d dimensional torus, $d \geq 4$. Xu [40] studied the invariant Gibbs measures for a three dimensional wave equation in infinite Volume. Recently, we examined the random data Cauchy problem for the cubic wave equation [41] on a three dimensional compact manifold with boundary and the quintic wave equation on two dimensional compact manifold in the supercritical regime. We have obtained three results. First, we constructed the local strong solution to the cubic nonlinear wave equation with random data for a large set of initial data in $H^s(M)$ with $s \geq \frac{5}{14}$, where M is a three dimensional compact manifold with boundary. Second, we constructed the local strong solution to the quintic nonlinear wave equation with random data for a large set of initial data in $H^s(M)$ with $s \geq \frac{1}{6}$, where M is a two dimensional compact boundaryless manifold. Finally, we constructed the local strong solution to the quintic nonlinear wave equation with random data for a large set of initial data in $H^s(M)$ with $s \geq \frac{23}{90}$, where M is a two dimensional compact manifold with boundary.

Acknowledgements Part of this work was done while Wei Yan was visiting Department of Applied Mathematics, Illinois Institute of Technology, Chicago, USA.

References

1. Duan, J., Wang, W.: Effective Dynamics of Stochastic Partial Differential Equations. Elsevier, New York (2014)
2. Lebowitz, J., Rose, H., Speer, E.: Statistical mechanics of the nonlinear Schrödinger equation. J. Stat. Phys. **50**, 657–687 (1988)
3. Bourgain, J.: Periodic Schrödinger equation and invariant measures. Commun. Math. Phys. **166**, 1–26 (1994)
4. Bourgain, J.: Invariant measures for the 2 D-defocusing nonlinear Schrödinger equation. Comm. Math. Phys. **176**, 421–445 (1996)
5. Bourgain, J.: Refinements of Strichartz' inequality and applications to 2D-NLS with critical nonlinearity. Int. Math. Res. Not. **5**, 253–283 (1998)
6. Colliander, J., Oh, T.: Almost sure well-posedness of the cubic nonlinear Schrödinger equation below $L^2(T)$. Duke Math. J. **161**, 367–414 (2012)
7. Oh, T., Sulem, C.: On the one-dimensional cubic nonlinear Schrödinger equation below L^2. Kyoto J. Math. **52**, 99–115 (2012)
8. Nahmod, A.R., Oh, T., Rey-Bellet, L., Staffilani, G.: Invariant weighted Wiener measures and almost sure global well-posedness for the periodic derivative NLS. J. Eur. Math. Soc. **14**, 1275–1330 (2012)

9. Burq, N., Thomann, L., Tzvetkov, N.: Long time dynamics for the one dimensional nonlinear Schrödinger equation. Annales de l'institut Fourier **63**, 2137–2198 (2013)
10. Deng, Y.: Two-dimensional nonlinear Schrödinger equation with random radial data. Anal. PDE. **5**, 913–960 (2012)
11. Poiret, A.: Solutions globales pour des équation de Schrödinger sur-critiques en toutes dimensions [Global solutions for supercritical Schrödinger equations in all dimensions]. arXiv:1207.3519
12. Poiret, A.: Solutions globales pour l'équation de Schrödinger cubique en dimension 3[Global solutions for the cubic Schrödinger equation in dimension 3]. arXiv: 1207.1578
13. Poiret, A., Robert, D., Thomann, L.: Probabilistic global well-posedness for the supercritical nonlinear harmonic oscillator. Anal. PDE. **7**, 997–1026 (2014)
14. Bourgain, J., Bulut, A.: Almost sure global well-posedness for the radial nonlinear Schrödinger equation on the unit ball II: the 3d case. J. Eur. Math. Soc. **16**, 1289–1325 (2014)
15. Bényi, A., Oh, T., Pocovnicu, O.: Wiener randomization on unbounded domains and an application to almost sure well-posedness of NLS. Excursions in Harmonic Analysis
16. Bényi, A., Oh, T., Pocovnicu, O.: On the probabilistic Cauchy theory of the cubic nonlinear Schrödinger equation on R^d, $d \geq 3$. Tran. Amer. Math. Soc. **2**, 1–50 (2015)
17. Bourgain, J.: Invariant measures for NLS in infinite volume. Commun. Math. Phy. **210**, 605–620 (2000)
18. Federico, C., de Suzzoni, A.S.: Invariant measure for the Schrödinger equation on the real line. J. Funct. Anal. **269**, 271–324 (2015)
19. Hirayama, H., Okamoto, M.: Random data Cauchy theory for the fourth order nonlinear Schrödinger equation with cubic nonlinearity. arXiv:1505.06497
20. Hirayama, H. Okamoto, M.: Random data Cauchy problem for the nonlinear Schrödinger equation with derivative nonlinearity. arXiv:1508.02161
21. Yan, W., Duan, J., Li Y.S.: Random data Cauchy problem for the fourth-order Schrödinger equation. Submitted (2017)
22. Oh, T.: Invariance of the Gibbs measure for the Schrödinger-Benjamin-Ono system. SIAM J. Math. Anal. **41**, 2207–2225 (2009)
23. Oh, T.: Remarks on nonlinear smoothing under randomization for the periodic KdV and the cubic Szegö equation. Funkcial. Ekvac. **54**, 335–365 (2011)
24. Oh, T.: Invariant Gibbs measures and a.s. global well posedness for coupled KdV systems. Diff. Int. Equ. **22**, 637–668 (2009)
25. Deng, Y.: Invariance of the Gibbs measure for the Benjamin-Ono equation. J. Eur. Math. Soc. **17**, 1107–1198 (2015)
26. Richard, G.: Invariance of the Gibbs measure for the periodic quartic gKdV. Ann. I. H. Poincaré-AN. **33**, 699–766 (2016)
27. Kenig, C.E., Ponce, G., Vega, L.: Well-posedness and scattering results for the generalized Korteweg-de Vries equation via the contraction principle. Commun. pure. Appl. Math. **46**, 527–620 (1993)
28. Birnir, B., Kenig, C.E., Ponce, G., Svanstedt, N., Vega, L.: On the ill-posedness of the IVP for generalized Korteweg-de Vries and nonlinear Schrödinger equations. J. London Math. Soc. **53**, 551–559 (1996)
29. Yan, W., Duan, J., Huang, J. H.: Random data Cauchy problem for a generalized KdV equation in the supercritical case. Submitted to J. Diff. Eqns. (2016)
30. Mckean, H.P., Vaninsky, K.L.: Trends and Perspective in Applied Mathematics. Statistical mechanics of nonlinear wave equations, pp. 239–264. Springer, New York (1994)
31. Burq, N., Tzvetkov, N.: Invariant measure for a three dimensional nonlinear wave equation. Int. Math. Res. Not. **22**(Art. ID rnm108), 26 (2007)
32. Burq, N., Tzvetkov, N.: Random data Cauchy theory for supercritical wave equations. I. Local Theory Invent. Math. **173**, 449–475 (2008)
33. Burq, N., Tzvetkov, N.: Random data Cauchy theory for supercritical wave equations, II. Glob. Exist. Result Invent. Math. **173**, 477–496 (2008)

34. de Suzzoni, A.S.: Invariant measure for the cubic wave equation on the unit ball of R^3.. Dyn. Partial Differ. Equ. **8**, 127–147 (2011)
35. de Suzzoni, A.S.: Large data low regularity scattering results for the wave equation on the Euclidean space. Commun. Partial Differ. Equ. **38**, 1–49 (2013)
36. de Suzzoni, A.S.: Convergence of the chioce of a particular basis of $L^2(S^3)$ for the cubic wave equation on the sphere and the Euclidean space. Commu. Pure Appl. Anal. **13**, 991–1015 (2014)
37. Bourgain, J., Bulut, A.: Invariant Gibbs measure evolution for the radial nonlinear wave equation on the 3d ball. J. Funct. Anal. **266**, 2319–2340 (2014)
38. Burq, N., Tzvetkov, N.: Probabilistic well-posedness for the cubic wave equation. J. Eur. Math. Soc. **16**, 1–30 (2014)
39. Burq, N., Thomann, L., Tzvetkov, N.: Global infinite energy solutions for the cubic wave equation. Bull. Soc. Math. Math. France **143**, 301–313 (2015)
40. Xu, S.: Invariant Gibbs measures for 3D NLW in infinite volume. arXiv:1405.3856
41. Duan, J., Huang, J. H., Li, Y. S., Yan, W.: Random data Cauchy problem for the supercritical wave equation on compact manifold. Submitted (2017)

SPDEs and Renormalisation

Lorenzo Zambotti

Abstract We review the main ideas of renormalisation of stochastic partial differential equations, as they appear in the theory of regularity structures. We discuss in an informal way noise regularisation, the transformation of canonical-to-renormalised models, the space of models and the role of the continuity of the solution map.

Keywords Stochastic partial differential equations · Renormalisation

Mathematics Subject Classification (2010) 60H15 · 82C28

In the foundational paper [7], Martin Hairer introduced a theory of regularity structures and applied it to solving two important equations, whose well-posedness was an open problem. The first part of [7] is truly a *theory*, in the sense that it can be applied in the same way to a large class of problems: the second part however, which deals with two concrete applications, contains more and more *ad hoc* arguments, which should be adapted if applied in different contexts. Worse, for many other interesting equations the approach proposed in [7] becomes impractical, since the combinatorial complexity can be huge.

Fortunately, the situation has changed recently. The trio of papers

- Martin Hairer (2014),
  A theory of regularity structures, Inventiones Math., [7]
- Yvain Bruned, M.H., L.Z. (2016),
  Algebraic renormalisation of regularity structures, arXiv, [2]
- Ajay Chandra, M.H. (2016),
  An analytic BPHZ theorem for regularity structures, arXiv, [3]

now gives a *completely automatic black box* for obtaining local existence and uniqueness theorems for a wide class of SPDEs. Prominent examples are the following:

L. Zambotti (✉)
LPMA, UPMC, 4 Place Jussieu, 75005 Paris, France
e-mail: lorenzo.zambotti@upmc.fr

A. Eberle et al. (eds.), *Stochastic Partial Differential Equations and Related Fields*, Springer Proceedings in Mathematics & Statistics 229,
https://doi.org/10.1007/978-3-319-74929-7_16

$$\text{(KPZ)} \qquad \partial_t u = \Delta u + (\partial_x u)^2 + \xi, \quad x \in \mathbb{R},$$

$$\text{(PAM)} \qquad \partial_t u = \Delta u + u\,\xi, \quad x \in \mathbb{R}^2,$$

$$(\Phi^4_3) \qquad \partial_t u = \Delta u - u^3 + \xi, \quad x \in \mathbb{R}^3,$$

for ξ a space time white noise (only depending in space for PAM).

Each of these equations contains a product between a distribution (generalised function) and another distribution or, in the case of PAM, a function without a sufficient regularity. In order to put the problem in a more general context, note that if $T \in \mathcal{S}'(\mathbb{R}^d)$ is a tempered distribution and $\psi \in \mathcal{S}(\mathbb{R}^d)$, then we can define canonically the product $\psi T = T\psi \in \mathcal{S}'(\mathbb{R}^d)$ by

$$\psi T(\varphi) = T\psi(\varphi) := T(\psi\varphi), \qquad \varphi \in \mathcal{S}(\mathbb{R}^d).$$

The products appearing in the above equations can not be defined in this way since ψ is not expected to belong to $\mathcal{S}(\mathbb{R}^d)$.

Therefore, even with polynomial non-linearities, we do not know how to properly multiply such (random) distributions. The randomness here is in fact not crucial, since the problems are the same if we consider ξ as a deterministic distribution with sufficiently low regularity.

In classical stochastic calculus we have a similar problem in the definition of *stochastic integrals*. If $(B_t)_{t\geq 0}$ is a Brownian motion and $(A_t)_{t\geq 0}$ a smooth process, then we can define canonically by means of a simple integration by parts

$$\int_0^t A_s \,\mathrm{d}B_s := A_t B_t - \int_0^t B_s \dot{A}_s \,\mathrm{d}s, \qquad t \geq 0.$$

However we have several possible extensions of this definition to a larger class of non-smooth A (Itô, Stratonovich...). This is related to the fact that, for A a generic process such that the Itô integral is well defined, the map $B \to \int_0^\bullet A_s \,\mathrm{d}B_s$ is *measurable but not continuous*.

This is the starting point of the Rough Paths theory, initiated by Terry Lyons [9] and later expanded by, among others, Massimiliano Gubinelli, whose ideas on controlled [4] and branched [5] rough paths have much inspired Martin Hairer in the elaboration of Regularity Structures.

Lack of continuity of stochastic integrals with respect to trajectories of the underlying BM has relevant consequences when one looks for approximations by noise-regularisation, as can be already seen in the case of SDEs. Let us consider for example the ODE in $\mathbb{R}^d$

$$\frac{\mathrm{d}}{\mathrm{d}t} x_\varepsilon = b(x_\varepsilon) + f(x_\varepsilon)\,\frac{\mathrm{d}B_\varepsilon}{\mathrm{d}t} \tag{1}$$

where B_ε is a smooth approximation of a Brownian motion B and b, f are smooth coefficients. Then the Wong–Zakai Theorem [11, 12] states that, as $\varepsilon \to 0$, we have $x_\varepsilon \to x$, solution to the Stratonovich SDE

$$\mathrm{d}x = b(x)\,\mathrm{d}t + f(x) \circ \mathrm{d}B.$$

In order to obtain the Itô SDE in the limit, one has to define rather

$$\frac{\mathrm{d}}{\mathrm{d}t}\hat{x}_\varepsilon = b(\hat{x}_\varepsilon) - \frac{1}{2}Df(\hat{x}_\varepsilon)f(\hat{x}_\varepsilon) + f(\hat{x}_\varepsilon)\,\frac{\mathrm{d}B_\varepsilon}{\mathrm{d}t} \tag{2}$$

and in this case, as $\varepsilon \to 0$, we have $\hat{x}_\varepsilon \to \hat{x}$, solution to

$$\mathrm{d}\hat{x} = b(\hat{x})\,\mathrm{d}t + f(\hat{x})\,\mathrm{d}B. \tag{3}$$

Note that this is already an example of a "renormalised" equation: although there is no diverging quantity, we have to modify the regularised equation (1) if we want to obtain in the limit the Itô SDE (3). In other words, if we want to approximate (3), the correct choice is the less intuitive (2), rather than (1).

The deep reason for that, if any, is that the "product" $f(\hat{x})\,\mathrm{d}B$ is less intuitive than we would like to think, and it has to be properly interpreted if we want to approximate it with smoother distributions.

Let us go back now to SPDEs. Let $\xi_\varepsilon = \rho_\varepsilon * \xi$ be a regularisation of ξ, with $(\rho_\varepsilon)_{\varepsilon>0}$ is a family of mollifiers, and let u_ε solve

$$\partial_t u_\varepsilon = \Delta u_\varepsilon + F(u_\varepsilon, \nabla u_\varepsilon, \xi_\varepsilon)$$

where F is a non-linear function from a certain class, which includes the three equations above. The question is of course: what happens as $\varepsilon \to 0$? In order to control this limit, we need a topology such that

1. the map $\xi_\varepsilon \mapsto u_\varepsilon$ is continuous
2. $\xi_\varepsilon \to \xi$ as $\varepsilon \to 0$.

The first point requires the topology to be "strong", while the second one requires it to be "weak". In fact, no simple solution seems to be available if the regularity of ξ is sufficiently low. The analytic part of the theory of Regularity Structures (RS) gives a framework to solve this problem, by constructing, for a given equation,

- a space of Models $\mathscr{M}$ endowed with a metric d
- a canonical lift of every smooth ξ_ε to a model $\mathbf{X}^\varepsilon \in \mathscr{M}$
- a continuous function $\Phi : \mathscr{M} \to \mathscr{S}'(\mathbb{R}^d)$ such that $u_\varepsilon = \Phi(\mathbf{X}^\varepsilon)$ solves the regularised equation
$$\partial_t u_\varepsilon = \Delta u_\varepsilon + F(u_\varepsilon, \nabla u_\varepsilon, \xi_\varepsilon).$$

Roughly speaking, the model $\mathbf{X}^\varepsilon \in \mathscr{M}$ contains *a finite number of explicit products* which are relevant to the given equation. By simplifying a lot, we can say that convergence in $(\mathcal{M}, \mathrm{d})$ means convergence of all these objects as distributions. For instance $\mathbf{X}^\varepsilon \in \mathcal{M}$ can contain

$$\xi_\varepsilon(G * \xi_\varepsilon)$$

(with G the heat kernel). These products can be *ill-defined* in the limit $\varepsilon \to 0$, e.g.

$$\mathbb{E}[\xi_\varepsilon(G * \xi_\varepsilon)] = \rho_\varepsilon * G * \rho_\varepsilon(0) \to G(0) = +\infty.$$

Therefore in general $\mathbf{X}^\varepsilon$ does *not* converge in $(\mathcal{M}, \mathrm{d})$ as $\varepsilon \to 0$.

The RS theory identifies a class of equations, called *subcritical*, for which it is enough to *modify a finite number of products* in order to obtain a convergent lift $\hat{\mathbf{X}}^\varepsilon \in \mathcal{M}$ of ξ_ε. For instance

$$\xi_\varepsilon(G * \xi_\varepsilon) \to \xi_\varepsilon(G * \xi_\varepsilon) - \mathbb{E}[\xi_\varepsilon(G * \xi_\varepsilon)]. \tag{4}$$

The model $\hat{\mathbf{X}}^\varepsilon \in \mathcal{M}$ contains all these modified *(renormalised)* products. Then we define the *renormalised solution* by $\hat{u}_\varepsilon := \Phi(\hat{\mathbf{X}}^\varepsilon)$.

One can summarize the procedure into three steps:

- *Analytic step*: Construction of the space of models $(\mathcal{M}, \mathrm{d})$ and continuity of the solution map $\Phi : \mathcal{M} \to \mathcal{S}'(\mathbb{R}^d)$, [7]
- *Algebraic step*: Renormalisation of the canonical model $\mathbf{X}^\varepsilon \to \hat{\mathbf{X}}^\varepsilon \in \mathcal{M}$, [2]
- *Probabilistic step*: Convergence a.s. of the renormalised model $\hat{\mathbf{X}}^\varepsilon$ to $\hat{\mathbf{X}}$ in $(\mathcal{M}, \mathrm{d})$, [3].

The final result is a *renormalised solution* $\hat{u} := \Phi(\hat{\mathbf{X}})$, which is also the unique solution of a fixed point problem. This works for very general noises, far beyond the Gaussian case (Fig. 1).

The Wong–Zakai result for SPDEs is much more subtle than for SDEs [8]: if

$$\partial_t u_\varepsilon = \partial_x^2 u_\varepsilon + H(u_\varepsilon) + F(u_\varepsilon)\,\xi_\varepsilon, \qquad x \in \mathbb{R}, \tag{5}$$

then $u_\varepsilon = \Phi(\mathbf{X}^\varepsilon)$ does *not* converge in general and, in order to obtain in the limit the correct Itô SPDE, it is necessary to renormalise the equation and study

$$\partial_t \hat{u}_\varepsilon = \partial_x^2 \hat{u}_\varepsilon + \bar{H}(\hat{u}_\varepsilon) - C_\varepsilon\, F'(\hat{u}_\varepsilon)\, F(\hat{u}_\varepsilon) + F(\hat{u}_\varepsilon)\,\xi_\varepsilon \tag{6}$$

where $C_\varepsilon = \mathbb{E}[\xi_\varepsilon(G * \xi_\varepsilon)] = \varepsilon^{-1} c_0$,

$$\bar{H} := H + c_1 F' F^3 + c_2 F'' F' F^2,$$

and $\{c_0, c_1, c_2\}$ are constants depending on the mollifier ρ used in the noise-regularisation. It turns out that $\hat{u}_\varepsilon = \Phi(\hat{\mathbf{X}}^\varepsilon)$, and the limit $\hat{u} := \Phi(\hat{\mathbf{X}})$ solves the Itô SPDE

$$\mathrm{d}\hat{u} = (\partial_x^2 \hat{u} + H(\hat{u}))\,\mathrm{d}t + F(\hat{u})\,\mathrm{d}W_t, \tag{7}$$

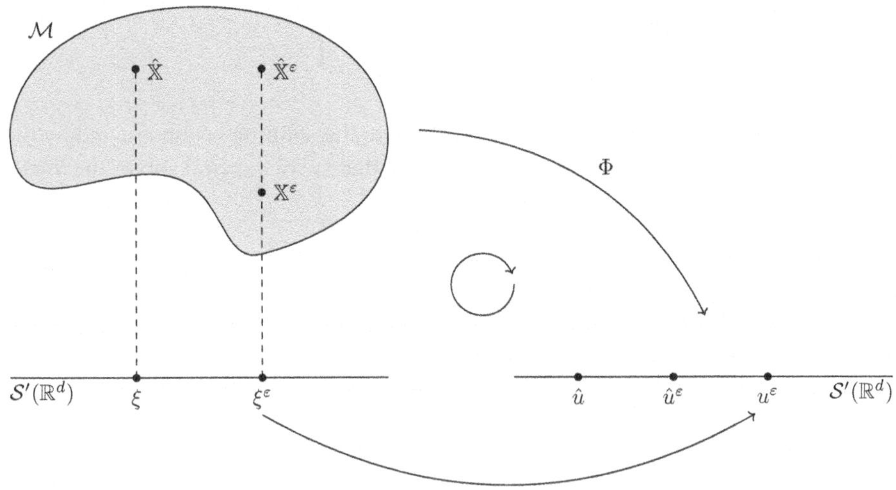

Fig. 1 In this figure we show the factorisation of the map $\xi_\varepsilon \mapsto u_\varepsilon$ into $\xi_\varepsilon \mapsto \mathbf{X}^\varepsilon \mapsto \Phi(\mathbf{X}^\varepsilon) = u_\varepsilon$. We also see that in the space of models $\mathscr{M}$ we have several possible lifts of $\xi_\varepsilon \in \mathscr{S}'(\mathbb{R}^d)$, e.g. the canonical model $\mathbf{X}^\varepsilon$ and the renormalised model $\hat{\mathbf{X}}^\varepsilon$; it is the latter that converges to a model $\hat{\mathbf{X}}$, thus providing a lift of ξ. Note that $\hat{u}_\varepsilon = \Phi(\hat{\mathbf{X}}_\varepsilon)$ and $\hat{u} = \Phi(\hat{\mathbf{X}})$

where W is the martingale measure associated with ξ, in the spirit of [10].

Therefore, the Wong–Zakai result for SPDEs is that the correct approximation to (7) is not (5) but (6). The Stratonovich integral with respect to W is not well-defined, since the covariation between $F(\hat{u})$ and W is infinite, which explains the diverging constant C_ε in (6). The renormalisation procedure should not be seen as a "modification of the equation" but as a *choice of the correct equation*.

It is very interesting to note that there is *nothing singular* in the SPDE (7). The convergence of $\hat{u}_\varepsilon$ to $\hat{u}$ is however far from simple and requires the full power of the RS theory [8].

Another important point is the following: the final aim is to renormalise the *unknown* solution $u_\varepsilon = \Phi(\mathbf{X}^\varepsilon)$. One of the main ideas of the RS theory is that, for this, it is enough to renormalise the *finitely many explicit* products defining the model $\mathbf{X}^\varepsilon$, and this can be done following [2]. Then [3] shows that the renormalised model $\hat{\mathbf{X}}^\varepsilon$ converges to $\hat{\mathbf{X}}$ in $(\mathscr{M}, \mathrm{d})$. Continuity of the solution map $\Phi : \mathscr{M} \to \mathscr{S}'(\mathbb{R}^d)$ yields convergence of the renormalised solution $\hat{u}_\varepsilon = \Phi(\hat{\mathbf{X}}^\varepsilon)$ to $\hat{u} = \Phi(\hat{\mathbf{X}})$ [7].

For instance, let us consider the regularised version of KPZ:

$$\partial_t u_\varepsilon = \partial_x^2 u_\varepsilon + (\partial_x u_\varepsilon)^2 + \xi_\varepsilon. \tag{8}$$

The renormalised version is

$$\partial_t \hat{u}_\varepsilon = \partial_x^2 \hat{u}_\varepsilon + (\partial_x \hat{u}_\varepsilon)^2 - C_\varepsilon + \xi_\varepsilon \tag{9}$$

with

$$C_\varepsilon = \mathbb{E}\left[(\partial_x G * \xi_\varepsilon)^2\right] \sim \frac{1}{\varepsilon}.$$

The first mathematical paper on KPZ is [1], where the solution is constructed by the Hopf–Cole transform, namely the simple remark that $z_\varepsilon := \exp(\hat{u}_\varepsilon)$ solves the *linear* equation

$$\partial_t z_\varepsilon = \partial_x^2 z_\varepsilon + z_\varepsilon \, \xi_\varepsilon. \tag{10}$$

It was not until [6] that a direct approach to (8) and (9) was obtained. The reason why mathematicians have been at loss with this equation for so long, is of course that it is not clear at all how one should handle the term $(\partial_x \hat{u}_\varepsilon)^2 - C_\varepsilon$ as $\varepsilon \to 0$. Well, now we know that it is enough to handle the convergence as a distribution of the explicit function

$$(\partial_x G * \xi_\varepsilon)^2 - \mathbb{E}[(\partial_x G * \xi_\varepsilon)^2] \tag{11}$$

which is the renormalised version of $(\partial_x G * \xi_\varepsilon)^2$, plus a few other terms which are less evident from the equation, and then the continuity of the solution map Φ does the rest of the job.

The Hopf–Cole transform shows that, as discussed above for (5), (6) and (9) is not merely the renormalisation of (8), it is rather *the correct equation* approximating KPZ. Indeed, $z_\varepsilon = \exp(\hat{u}_\varepsilon)$ converges to the solution z of the linear Itô SPDE

$$\mathrm{d}z = \partial_x^2 z \; \mathrm{d}t + z \, \mathrm{d}W_t$$

while $\exp(u_\varepsilon)$ does not converge.

Finally, let us notice that the two examples of renormalised products we have discussed in (4)–(11) are simply given by substraction of a constant. This might give the impression that the renormalisation procedure reduces to a mere *centering*. Although this is partly true, it should be emphasised that the transformation from $\mathbf{X}^\varepsilon$ to $\hat{\mathbf{X}}^\varepsilon$ is described in [2] by the action on $\mathscr{M}$ of a *renormalisation group* which is in general *non-linear and non-commutative*.

References

1. Bertini, L., Lorenzo, G.: Stochastic Burgers and KPZ equations from particle systems. Commun. Math. Phys. **183**(3), 571–607 (1997)
2. Bruned, Y., Hairer, M., Zambotti, L.: Algebraic renormalisation of regularity structures (2016). arXiv:1610.08468
3. Chandra, A., Hairer, M.: An analytic BPHZ theorem for regulariy structures (2016). arXiv:1612.08138
4. Gubinelli, M.: Controlling rough paths. J. Funct. Anal. **216**(1), 86–140 (2004)
5. Gubinelli, M.: Ramification of rough paths. J. Differ. Equ. **248**(4), 693–721 (2010)
6. Hairer, M.: Solving the KPZ equation. Ann. Math. **178**(2), 559–664 (2013)
7. Hairer, M.: A theory of regularity structures. Invent. Math. **198**(2), 269–504 (2014)

8. Hairer, M., Pardoux, E.: A Wong-Zakai theorem for stochastic PDEs. J. Math. Soc. Jpn. **67**(4), 1551–1604 (2015)
9. Lyons, T.J.: Differential equations driven by rough signals. Rev. Mat. Iberoamericana **14**(2), 215–310 (1998)
10. Walsh, J.B.: An introduction to stochastic partial differential equations. In: École d'été de probabilités de Saint-Flour, XIV–1984, Lecture Notes in Mathematics, vol. 1180, pp. 265–439. Springer, Berlin (1986)
11. Wong, E., Zakai, M.: On the convergence of ordinary integrals to stochastic integrals. Ann. Math. Stat. **36**, 1560–1564 (1965)
12. Wong, E., Zakai, M.: On the relation between ordinary and stochastic differential equations. Int. J. Eng. Sci. **3**, 213–229 (1965)

Recent Progress on Stochastic Nonlinear Schrödinger Equations

Deng Zhang

Abstract This note presents the recent results for stochastic nonlinear Schrödinger equations with linear multiplicative noise, including the global well-posedness, the noise effect on blow-up and the optimal bilinear control problem. An important role in the proof is played by the rescaling approach.

Keywords (Stochastic) Nonlinear Schrödinger equation · Maximal monotonicity · Noise effect · Optimal bilinear control · Strichartz estimates

2010 Mathematics Subject Classification 60H15 · 60H30 · 35Q55 · 47H05 · 81Q93

1 Introduction

We are concerned with the stochastic nonlinear Schrödinger equation with linear multiplicative noise

$$\begin{aligned} &i dX(t) = \Delta X(t)dt + F(X(t))dt - i\mu X(t)dt + iX(t)dW(t), \quad t \in (0,T), \qquad (1)\\ &X(0) = x. \end{aligned}$$

Here, X is a complex-valued function on $[0,T] \times \mathbb{R}^d$, $F(X)$ represents the nonlinearity, which is mainly of the pure power type $F(X) = \lambda |X|^{\alpha-1}X$ or of the logarithmic type $F(X) = \lambda X \log |X|^2$, where $\alpha \in (1,\infty)$, and $\lambda = -1$ (resp. 1) corresponding to the defocusing (resp. focusing) case. The term W is the colored Wiener process

$$W(t,\xi) = \sum_{j=1}^{N} \mu_j e_j(\xi)\beta_j(t), \ t \in (0,T), \ \xi \in \mathbb{R}^d, \qquad (2)$$

D. Zhang (✉)
School of Mathematical Sciences, Shanghai Jiao Tong University, Shanghai 200240, China
e-mail: dzhang@sjtu.edu.cn

A. Eberle et al. (eds.), *Stochastic Partial Differential Equations and Related Fields*, Springer Proceedings in Mathematics & Statistics 229,
https://doi.org/10.1007/978-3-319-74929-7_17

where $N \in \mathbb{N} \cup \{+\infty\}$, $\mu_j \in \mathbb{C}$, e_j are real-valued functions, and $\beta_j(t)$ are independent real Brownian motions on a probability space $(\Omega, \mathscr{F}, \mathbb{P})$ with natural filtration $(\mathscr{F}_t)_{t\geq 0}$, $1 \leq j \leq N$. For simplicity we assume $N < \infty$, but the arguments apply as well to the case where $N = \infty$. Moreover,

$$\mu(x) = \frac{1}{2}\sum_{j=1}^{N} |\mu_j|^2 e_j^2(\xi). \tag{3}$$

The significance of the stochastic nonlinear Schrödinger equation (1) is well-known. It describes the propagation of nonlinear dispersive waves in nonhomogeneous or random media. It is also closely related to the theory of measurements continuous in time in quantum mechanics [8]. One of the main feature is that $|X(t)|_2^2$ is on longer independent of time, but a positive continuous martingale (see (8) below). This key fact allows to define the "physical" probability law of events occurring in time $[0, T]$ and in particular implies the mean norm square conservation of X. It should be mentioned that, in the case where $Re\mu_j = 0$, $1 \leq j \leq N$, $|X(t)|_2^2$ is pathwisely conserved, and so this case is called *the conservative case*. In this case, the measurement provides no information on the quantum system. Thus, below we are concerned with the general case where $\mu_j \in \mathbb{C}$, $1 \leq j \leq N$, including both the conservative and non-conservative cases. For more physical interpretations, we refer to [8–10, 23] and references therein.

For (1) with the nonlinearity of the pure power type, earlier results are mainly proved in the conservative case. See e.g. [15, 17] for the global well-posedness, [16, 18] for noise effects on blow-up. See also [13] for the stochastic nonlinear Schrödinger equation on compact Riemannian manifolds.

Recently, based on the rescaling approach, it is proved in [5, 6] that the stochastic nonlinear Schrödinger equation (1) can be transformed equivalently to a random Schrödinger equation with time dependent lower order perturbations (see (5) below). This fact relates closely these two equations and allows to obtain the global well-posedness in the full subcritical range of the exponents of the nonlinearity, including also the non-conservative case. In addition, it gives as well the pathwise continuous dependence of solutions on initial data, which is quite a rare property for solutions to stochastic partial differential equations. See also [19] for the conservative case with one dimension Brownian motion.

Furthermore, the rescaling approach reveals the stabilization effect of the noise in the non-conservative case (see [7]). It also fits well with the theory of maximal monotone operators, which is first developed in [3] for a large class of stochastic partial differential equations with monotone nonlinearities under coercivity condition. See also [2] for the stochastic logarithmic Schrödinger equation. Recently, the rescaling approach has been also applied in [1] to the optimal bilinear control problem of the quantum mechanical system with final observation governed by (1).

This note presents these recent results for stochastic nonlinear Schrödinger equations. The remainder of this note is structured as follows. Section 2 is concerned with the rescaling approach. Section 3 is devoted to the stochastic nonlinear Schrödinger equation with the nonlinearity of pure power type. Finally, Sect. 4 contains the global well-posedness of the stochastic logarithmic Schrödinger equation.

2 Rescaling Approach

We introduce the rescaling transformation

$$y = e^{-W} X, \tag{4}$$

which can be seen as a Doss-Sussman type transformation generalized to infinite dimensions. By (4), the stochastic equation (1) can be reduced to an equivalent random Schrödinger equation with time dependent lower order perturbations below.

$$\begin{aligned} \partial_t y(t) &= A(t)y(t) + e^{-W(t)}F(t), \\ y(0) &= y_0, \end{aligned} \tag{5}$$

where $A(t) := -i(\Delta + b(t)\cdot\nabla + c(t))$, $b(t,x) = 2\nabla W(t,x)$, $c(t,x) = \sum_{j=1}^{d} (\partial_j W(t,x))^2 + \Delta W(t,x) - i\widehat{\mu}(x)$, and $\widehat{\mu} = \frac{1}{2}\sum_{j=1}^{N}(\mu_j^2 + |\mu_j|^2)e_j^2$.

This point of view allows to avoid the analysis of the classical stochastic convolution and to analyze the reduced equation pathwisely by deterministic tools, which in turn usually gives sharper pathwise results for the original stochastic equation. The equivalence of solutions to (1) and (5) is formulated in Theorem 1 below.

Definition 1 Let $x \in L^2$ (resp. H^1) and $0 < T < \infty$. A strong L^2(resp. H^1)-solution to (1) on $[0,T]$ is an L^2(resp. H^1)-valued continuous $(\mathscr{F}_t)_{t\geq 0}$-adapted process $X = X(t)$ such that $F(X) \in L^1(0,T;H^{-1})$ and $\mathbb{P}$-a.s.,

$$X(t) = x - \int_0^t (i\Delta X(s) + iF(X(s)) + \mu X(s))ds + \int_0^t X(s)dW(s),\ t\in[0,T], \tag{6}$$

as an Itô equation in H^{-2} (resp. H^{-1}). We say that uniqueness holds for (1), if for any two strong solutions X_i, $i = 1, 2$, it holds that $X_1 = X_2$ on $[0,T]$, $\mathbb{P}$-a.s.

The strong L^2(resp. H^1)-solution y to (5) is defined similarly as above.

Theorem 1 ([5, Lemma 6.1], [6, Lemma 2.4])

(i) *Let y be a strong L^2(resp. H^1)-solution to* (5). *Then, $X := e^W y$ is a strong L^2(resp. H^1)-solution to* (1).

(ii) *Suppose that X is a strong L^2(resp. H^1)-solution to* (1). *Then, $y := e^{-W}X$ is a strong L^2(resp. H^1)-solution to* (5).

Remark 1 Theorem 1 can be generalized to a general local version involving stopping times, see [6, Lemma 2.4] and [27, Theorems 1.1.4, 2.1.3].

3 SNLS with Pure Power Nonlinearity

In this section we are mainly concerned with (1) with the nonlinearity of pure power type $F(X) = \lambda |X(t)|^{\alpha-1}X(t)$, namely,

$$idX(t) = \Delta X(t)dt + \lambda |X(t)|^{\alpha-1}X(t)dt - i\mu X(t)dt + iX(t)dW(t), \quad (7)$$
$$X(0) = x,$$

where W and μ are as in (2) and (3) respectively, $\lambda = \pm 1$ and $\alpha \in (1, \infty)$.

3.1 Global Well-Posedness

Let us start with the L^2 case. (p, q) is called a Strichartz pair, if $\frac{2}{q} = \frac{d}{2} - \frac{d}{p}$, $(p, q) \in [2, \infty] \times [2, \infty]$, and $(p, q, d) \neq (\infty, 2, 2)$. We assume that

(H0) $e_j \in C_b^\infty(\mathbb{R}^d)$, $1 \leq j \leq N$, such that

$$\lim_{|\xi| \to \infty} \zeta(\xi)(|\nabla e_j(\xi)| + |\Delta e_j(\xi)|) = 0,$$

where

$$\zeta(\xi) = \begin{cases} 1 + |\xi|^2, & \text{if } d \neq 2, \\ (1 + |\xi|^2)(\ln(1 + |\xi|^2))^2, & \text{if } d = 2. \end{cases}$$

Theorem 2 ([5, Theorem 2.2]) *Assume (H0), $\lambda = \pm 1$, $\alpha \in (1, 1 + \frac{4}{d})$, $d \geq 1$. Then, for each $x \in L^2$ and $0 < T < \infty$, there exists a unique strong L^2-solution X to (7) in the sense of Definition 1, which satisfies*

$$X \in L^2(\Omega; C([0, T]; L^2)),$$
$$X \in L^q(0, T; L^{\alpha+1}), \ \mathbb{P} - a.s.,$$

where $q = \frac{4(\alpha+1)}{d(\alpha-1)}$.

Moreover, for $\mathbb{P} - a.e.\ \omega \in \Omega$, the map $x \to X(\cdot, x, \omega)$ is continuous from L^2 to $C([0, T]; L^2) \cap L^q(0, T; L^{\alpha+1})$, and $t \to |X(t)|_2^2$ is a continuous martingale with the representation

$$|X(t)|_2^2 = |x|_2^2 + 2 \sum_{k=1}^N \int_0^t \int_{\mathbb{R}^d} Re(\mu_k) e_k |X(s)|^2 d\xi d\beta_k(s), \ t \in [0, T]. \quad (8)$$

Next, in the H^1 case, we assume that

(H1) $e_j \in C_b^\infty(\mathbb{R}^d)$, $1 \leq j \leq N$, such that for any multi-index γ with $1 \leq |\gamma| \leq 3$,

$$\lim_{|\xi|\to\infty} \zeta(\xi)(|\partial^\gamma e_j(\xi)|) = 0,$$

where ζ is defined as in Assumption ($H0$).

Theorem 3 ([6, Theorem 1.2]) *Assume* ($H1$) *and let* $\alpha \in (1, 1 + \frac{4}{(d-2)_+})$ *in the defocusing case* $\lambda = -1$, $\alpha \in (1, 1 + \frac{4}{d})$ *in the focusing case* $\lambda = 1$. *Then for each* $x \in H^1$ *and* $0 < T < \infty$, *there exists a unique strong* H^1*-solution* X *to (7) in the sense of Definition 1, such that*

$$X \in L^2(\Omega; C([0, T]; H^1)) \cap L^{\alpha+1}(\Omega; C([0, T]; L^{\alpha+1})), \tag{9}$$

and

$$X \in L^\gamma(0, T; W^{1,\rho}), \ \ \mathbb{P} - a.s., \tag{10}$$

where (ρ, γ) *is any Strichartz pair.*

Furthermore, for $\mathbb{P} - a.e.\ \omega$, *the map* $x \to X(\cdot, x, \omega)$ *is continuous from* H^1 *to* $C([0, T]; H^1) \cap L^\gamma(0, T; W^{1,\rho})$.

Remark 2 In [5, 6], an additional decay condition $\lim_{|\xi|\to\infty} \zeta(\xi)|e_j(\xi)| = 0$, $1 \le j \le N$, is assumed, which actually can be removed. See [7, Remark 2.1].

Remark 3 The local existence, uniqueness and blow-up alternative in the (super)critical case is also obtained in [5, 6]. See [5, Proposition 5.1] for the mass-critical case ($\alpha = 1 + \frac{4}{d}$), see also [6, Theorem 2.1] for the energy-critical case ($\lambda = \pm 1$, $\alpha = 1 + \frac{4}{d-2}$ with $d \ge 3$) and for the focusing mass-(super)critical case ($\lambda = 1$, $1 + \frac{4}{d} \le \alpha < 1 + \frac{4}{(d-2)_+}$, $d \ge 1$).

Similar results hold as well in the case of infinitely many noises, i.e., $N = \infty$, under appropriate summability assumption of μ_j, $1 \le j \le N$. We refer to [27, Remarks 1.3.6, 2.3.13].

By virtue of Theorem 1, we reduce the proof of Theorems 2 and 3 to that of the random equation (5), to which we can apply the Strichartz estimates in [25] for Schrödinger equations with lower order perturbations. Another important role is played by Itô's formulae for L^p- and H^1-norms, based on the ideas in [4, 24] respectively.

3.2 Noise Effect on Blow-Up

The noise effect on blow-up was first studied in the conservative case in [18], which showed that the linear multiplicative noise can accelerate blow-up immediately with positive probability in the focusing mass-supercritical case. See also [16] for the case of additive noise.

Here, we are interested in the noise effect on blow-up in the non-conservative case, namely, $Re\mu_j \neq 0$ for some $1 \leq j \leq N$. Without loss of generality, we assume that $Re\mu_1 \neq 0$. Interestingly, the linear multiplicative noise in the non-conservative case has the different effect to stabilize the system. We assume that

(H2) $e_j = f_j + c_j$, $1 \leq j \leq N$, where c_j are real constants and f_j are real-valued functions satisfying Assumption $(H1)$.

(The constants c_j, $1 \leq j \leq N$, can be viewed as the strength of the noise.)

Theorem 4 ([7, Theorem 1.2]) *Consider* (7) *in the non-conservative focusing mass-(super)critical case, i.e.,* $\lambda = 1$, $1 + \frac{4}{d} \leq \alpha < 1 + \frac{4}{(d-2)_+}$, $d \geq 1$. *Assume* $(H2)$ *with* f_j, $1 \leq j \leq N$, *and* c_k, $2 \leq k \leq N$ *being fixed. If* f_j, $1 \leq j \leq N$, *are also constants, then for any* $x \in H^1$,

$$\mathbb{P}(X(t) \textit{ does not blow up on } [0, \infty)) \to 1, \quad \textit{as } c_1 \to \infty.$$

Similar result in the spatially dependent case is claimed in [7], we find a gap in the proof for this case and we are trying to fill the gap.

This stabilization effect is actually revealed by the reduced random equation (5), where the damped term $\widehat{\mu}$ of positive real part appears in the non-conservative case, but vanishes in the conservative case. This fact indicates the different effects of the noise in the conservative and non-conservative cases.

3.3 Optimal Bilinear Control

Control problems of nonlinear Schrödinger equations arise naturally in quantum theory. Consider the controlled stochastic nonlinear Schrödinger equation

$$\begin{aligned} idX =& \Delta X dt + \lambda |X|^{\alpha-1} X dt - i\mu X dt + V_0 X dt + \sum_{j=1}^{m} u_j V_j X dt + iX dW \\ X(0) =& x, \end{aligned} \tag{11}$$

where λ, α, W and μ are as in (7), $V_j \in W^{1,\infty}(\mathbb{R}^d)$ are real valued functions, $0 \leq j \leq m$. We work in the conservative case, i.e., $Re\mu_j = 0$, $1 \leq j \leq N$.

The real valued input control $u = (u_1, \ldots, u_m)$ is in the admissible set

$$\mathscr{U}_{ad} = \{u \in L^2_{ad}(0, T; \mathbb{R}^m);\ u \in U, \ \ a.e. \ on \ (0, T) \times \Omega.\}, \tag{12}$$

where U is a compact convex subset of $\mathbb{R}^m$, and $L^2_{ad}(0, T; \mathbb{R}^m)$ is the space of all $(\mathscr{F}_t)_{t \geq 0}$-adapted $\mathbb{R}^m$-valued processes u such that $u \in L^2((0, T) \times \Omega; \mathbb{R}^m)$. In most situations, u represents an external applied force due to the interaction of the quantum system with an electric field or a laser pulse applied to a quantum system.

The objective functional $\Phi : L^2_{ad}(0, T; \mathbb{R}^m) \to \mathbb{R}$ is defined by

$$\Phi(u) = \mathbb{E}|X^u(T) - \mathbb{X}_T|_2^2 + \gamma_1 \mathbb{E}\int_0^T |X^u(t) - \mathbb{X}_1(t)|_2^2 dt + \gamma_2 \mathbb{E}\int_0^T |u(t)|_m^2 dt \\ + \gamma_3 \mathbb{E}\int_0^T |u'(t)|_m^2 dt, \tag{13}$$

Here, u' is the time derivative in the sense of distributions ($\Phi(u) = \infty$ if there is no such derivative), $|\cdot|_m$ means the Euclidean norm, $\gamma_j \geq 0$, $1 \leq j \leq 3$, $\mathbb{X}_T \in L^2(\Omega, \mathscr{F}_T, \mathbb{P}; L^2)$ and $\mathbb{X}_1 \in L^2_{ad}(0, T; L^2(\Omega; L^2))$ are given, where $L^2_{ad}(0, T; L^2(\Omega; L^2))$ denotes the space of L^2-valued $(\mathscr{F}_t)_{t\geq 0}$-adapted processes $\mathbb{X}$ such that $\mathbb{X} \in L^2((0, T) \times \Omega \times \mathbb{R}^d)$.

The optimal control problem is formulated below.

$$\text{(P)} \quad \text{Min}\{\Phi(u);\ u \in \mathscr{U}_{ad},\ X^u\ satisfies\ (11)\}. \tag{14}$$

One of the main difficulties to study (14) lies in the fact that even if a space $\mathscr{Y}$ is compactly imbedded into another space $\mathscr{X}$, one generally does not have the compact imbedding from $L^p(\Omega; \mathscr{Y})$ to $L^p(\Omega; \mathscr{X})$, $1 \leq p \leq \infty$. Nevertheless, we consider the existence for relaxed versions of Problem (P).

Definition 2 Let $\mathscr{Y} := L^2(\mathbb{R}^d) \times L^2((0, T) \times \mathbb{R}^d) \times C([0, T]; \mathbb{R}^N) \times L^2(0, T; \mathbb{R}^m) \times L^2((0, T) \times \mathbb{R}^d)$ and $(\Omega^*, \mathscr{F}^*, (\mathscr{F}_t^*)_{t\geq 0})$ be a new filtered probability space, carrying $(\mathbb{X}_T^*, \mathbb{X}_1^*, \beta^*, u^*, X^*)$ in $\mathscr{Y}$. Define $L^2_{ad^*}(0, T; L^2(\Omega; L^2))$, $\mathscr{U}_{ad^*}$ and $\Phi^*(u^*)$ similarly as above on this new filtered probability space.

The system $(\Omega^*, \mathscr{F}^*, \mathbb{P}^*, (\mathscr{F}_t^*)_{t\geq 0}, \beta^*, u^*, X^*)$ is said to be *admissible*, if $\mathbb{X}_T^* \in L^2(\Omega, \mathscr{F}_T^*, \mathbb{P}^*; L^2)$, $\mathbb{X}_1^* \in L^2_{ad^*}(0, T; L^2(\Omega; L^2))$, $\beta^* = (\beta_1^*, \ldots, \beta_N^*)$ is an $(\mathscr{F}_t^*)_{t\geq 0}$-adapted $\mathbb{R}^N$-valued Wiener process, the joint distributions of $(\mathbb{X}_T^*, \mathbb{X}_1^*, \beta^*)$ and $(\mathbb{X}_T, \mathbb{X}_1, \beta)$ coincide, $u^* \in \mathscr{U}_{ad^*}$, and X^* is an $(\mathscr{F}_t^*)_{t\geq 0}$-adapted L^2-valued process satisfying equation (11) corresponding to (β^*, u^*).

The admissible system $(\Omega^*, \mathscr{F}^*, \mathbb{P}^*, (\mathscr{F}_t^*)_{t\geq 0}, \beta^*, u^*, X^*)$ is said to be a *relaxed solution* to the optimal control problem (P), if

$$\Phi^*(u^*) \leq \inf\{\Phi(u);\ u \in \mathscr{U}_{ad},\ X^u\ satisfies\ (11)\}. \tag{15}$$

In the difficult irregular case where $\gamma_3 = 0$, in order to construct a relaxed solution with equality in (15), we introduce the dual backward stochastic equation

$$dY = -i\Delta Y\,dt - \lambda i h_1(X^u) Y dt + \lambda i h_2(X^u)\overline{Y} dt + \mu Y dt - i V_0 Y dt - i u \cdot V Y dt \\ + \gamma_1 (X^u - \mathbb{X}_1) dt - \sum_{k=1}^N \overline{\mu_k} e_k Z_k dt + \sum_{k=1}^N Z_k d\beta_k(t), \tag{16}$$

$$Y(T) = -(X^u(T) - \mathbb{X}_T),$$

where $h_1(X^u) := \frac{\alpha+1}{2}|X^u|^{\alpha-1}$, $h_2(X^u) := \frac{\alpha-1}{2}|X^u|^{\alpha-3}(X^u)^2$. Moreover, assume that

(H3) $2 \leq \alpha < 1 + \frac{4}{d}$, $1 \leq d \leq 3$, and e_k are constants, $1 \leq k \leq N$.

Theorem 5 ([1, Theorem 2.6]) *Consider Φ with $\gamma_3 = 0$. Assume Assumption $(H3)$, and $\mathbb{X}_T \in L^{2+\nu}(\Omega; H^1)$, $\mathbb{X}_1 \in L^{2+\nu}(\Omega; L^2(0,T; H^1))$ for some small $\nu \in (0,1)$.*

Then, for each $x \in H^1$, $0 < T < \infty$, there exists a relaxed solution $(\Omega^*, \mathscr{F}^*, \mathbb{P}^*, (\mathscr{F}_t^*)_{t\geq 0}, \beta^*, u^*, X^*)$ to Problem (P) in the sense of Definition 2, such that

$$\Phi^*(u^*) = \inf\{\Phi(u);\ u \in \mathscr{U}_{ad},\ X^u\ satisfies\ (11)\}. \tag{17}$$

Moreover, we have (the stochastic maximum principle)

$$u^*(t) = P_U\left(\frac{1}{\gamma_2}\operatorname{Im}\int_{\mathbb{R}^d} V(\xi)X^*(t,\xi)\overline{Y^*}(t,\xi)d\xi\right),\ \forall t \in [0,T],\ \mathbb{P}^* - a.s., \tag{18}$$

where P_U is the projection on U, (Y^*, Z^*) is the solution to the backward stochastic equation (16) *with $\mathbb{X}_T, \mathbb{X}_1, \beta, u, X^u$ replaced by $\mathbb{X}_T^*, \mathbb{X}_1^*, \beta^*, u^*, X^*$ respectively.*

Remark 4 We also prove the existence of relaxed solutions in the easier regular case where $\gamma_3 > 0$. See [1, Theorem 2.5].

The proof of Theorem 5 is based on the Skorohod representation theorem, the Ekeland variational principle and the subdifferential developed by Rockafellar in [26]. A great effort is dedicated to the analysis of the linearized backward dual stochastic equation (16). Again, the rescaling approach allows to obtain delicate estimates of the solutions to (16).

4 SNLS with Logarithmic Nonlinearity

This section is concerned with the stochastic nonlinear Schrödinger equation (1) with the logarithmic nonlinearity $F(X) = \lambda X \log|X|^2$, namely,

$$\begin{aligned} &idX = \Delta X dt + \lambda X \log|X|^2 dt - i\mu X dt + iXdW,\ t \in (0,T), \\ &X(0) = x, \end{aligned} \tag{19}$$

where λ, μ, W are as in (1).

Actually, the logarithmic nonlinearity is the unique nonlinearity for which the separability hypothesis of noninteracting subsystems of the Schrödinger theory holds (see [11]). It also possesses many attractive features, including the additivity of the energy for noninteracting subsystems, the validity of the lower energy bound and Planck's relation for all stationary states. We refer to [11, 12, 22, 28] and references therein.

Since the logarithmic nonlinearity is not locally Lipschitz, the fixed point argument used in the proof of Theorems 2 and 3 is not applicable here. In view of the quasi-monotone feature of the logarithmic nonlinearity, we solve (19) by using the theory of maximal monotone operators, which fits well with the rescaling approach. A key ingredient is to introduce a convenient state space for (19) (see [14]). Set $U := H^1 \cap V$, where V is the Orlicz space $V = \{u \in L^1_{loc} : N(|u|) \in L^1\}$ equipped with the Luxembourg norm $\|u\|_V = \inf\{k > 0 : \int N(k^{-1}|u(\xi)|)d\xi \le 1\}$, and

$$N(x) = \begin{cases} -x^2 \log x^2, & \text{if} 0 \le x \le e^{-3}; \\ 3x^2 + 4e^{-3}x - e^{-6}, & \text{if} e^{-3} \le x. \end{cases}$$

Definition 3 A continuous L^2-valued $(\mathscr{F}_t)$-adapted process X is said to be a solution to (19) if for any $p \ge 3$, $X \in L^p(\Omega \times (0,T); U)$, $X \log |X|^2 \in L^{p'}(\Omega \times (0,T); U')$, U' is the dual space of U, and it satisfies $\mathbb{P}$-a.s.

$$X(t) = x - \int_0^t (i\Delta X(s)ds + \mu X(s) + \lambda i X(s) \log |X(s)|^2)ds + \int_0^t X(s)dW(s)$$

for all $t \in [0, T]$, where the stochastic term is taken in Itô's sense.

As in [21] (see also [20]) we assume that

(H4) $e_j \in \mathscr{B}^\infty(\mathbb{R}^d)$ such that for each $1 \le k \le d$, $1 \le m \le N$,

$$|\partial_k e_m(\xi)| \le \lambda(|\xi|), \quad \xi \in \mathbb{R}^d,$$

where $\mathscr{B}^\infty = \{f \in C^\infty(\mathbb{R}^d), \partial^\alpha f \in L^\infty, \ for \ all \ \alpha\}$, and $\lambda(\cdot)$ is a positive non-increasing function in $C([0,\infty]) \cap L^1([0,\infty))$.

Theorem 6 ([2, Theorem 1.2]) *Under Hypothesis (H4), for any initial datum $x \in U$ and $0 < T < \infty$, there exists a unique solution X to* (19) *in the sense of Definition 3.*

Moreover, for any $p \ge 2$,

$$\mathbb{E}\|X(t)\|^p_{L^\infty(0,T;U)} < \infty, \quad \mathbb{E}\|X(t) \log |X(t)|^2\|^p_{L^\infty(0,T;U')} < \infty,$$

and

$$\mathbb{E}\|e^{W(t)} \frac{d}{dt}(e^{-W(t)} X(t))\|^p_{L^\infty(0,T;U')} < \infty.$$

Again, we prove Theorem 6 from the reduced random equation via the rescaling transformation. An important role in the proof is played by the maximal monotonicity of the logarithmic nonlinearity in an appropriate rescaled space.

Acknowledgements The author is grateful to Professor Röckner for insightful discussions and constant encouragement. This work is supported by NSFC (No. 11501362). Financial support through SFB1283 at Bielefeld University is also gratefully acknowledged.

References

1. Barbu, V., Röckner, M., Zhang, D.: Optimal bilinear control of nonlinear stochastic Schrödinger equations driven by linear multiplicative noise, to appear in Ann. Probab
2. Barbu, V., Röckner, M., Zhang, D.: The stochastic logarithmic Schrödinger equation. J. Math. Pures Appl. **9**, 107 (2017) (no. 2, 123–149)
3. Barbu, V., Röckner, M.: An operatorial approach to stochastic partial differential equaitons driven by linear multiplicative noise. J. Eur. Math. Soc. **17**(7), 1789–1815 (2015)
4. Barbu, V., Da Prato, G., Röckner, M.: Stochastic porous media equations and self-organized criticality. Commun. Math. Phys. **285**(3), 901–923 (2009)
5. Barbu, V., Röckner, M., Zhang, D.: The stochastic nonlinear Schrödinger equation with multiplicative noise: the rescaling aproach. J. Nonlinear Sci. **24**, 383–409 (2014)
6. Barbu, V., Röckner, M., Zhang, D.: Stochastic nonlinear Schrödinger equations. Nonlinear Anal. **136**, 168–194 (2016)
7. Barbu, V., Röckner, M., Zhang, D.: Stochastic nonlinear Schrödinger equation: no blow-up in the non-conservative case. J. Differ. Equs. **263**(11), 7919–7940 (2017)
8. Barchielli, A., Gregoratti, M.: Quantum Trajectories and Measurements in Continuous Case. The Diffusive Case. Lecture Notes Physics, vol. 782. Springer, Berlin (2009)
9. Barchielli, A., Holevo, A.S.: Constructing quantum measurement processes via classical stochastic calculus. Stoch. Process. Appl. **58**(2), 293–317 (1995)
10. Barchielli, A., Paganoni, A.M., Zucca, F.: On stochastic differential equations and semigroups of probability operators in quantum probability. Stoch. Process. Appl. **73**(1), 69–86 (1998)
11. Bialynicki-Birula, I., Mycielski, J.: Nonlinear wave mechanics. Ann. Phys. **100**, 62–93 (1976)
12. Bialynicki-Birula, I., Mycielski, J.: Gaussons: solitons of the logarithmic Schröodinger equation. Special issue on solitons in physics. Phys. Scr. **20**(3–4), 539–544 (1979)
13. Brzeźniak, Z., Millet, A.: On the stochastic Strichartz estimates and the stochastic nonlinear Schrödinger equation on a compact Riemannian manifold. Potential Anal. **41**(2), 269–315 (2014)
14. Cazenave, T.: Stable solutions of the logarithmic Schrödinger equation. Nonlinear Anal. **7**(10), 1127–1140 (1983)
15. de Bouard, A., Debussche, A.: A stochastic nonlinear Schrodinger equation with multiplicative noise. Commun. Math. Phys. **205**, 161–181 (1999)
16. de Bouard, A., Debussche, A.: On the effect of a noise on the solutions of the focusing supercritical nonlinear Schrödinger equation. Probab. Theory Relat. Fields **123**(1), 76–96 (2002)
17. de Bouard, A., Debussche, A.: The stochastic nonlinear Schrödinger equation in H^1. Stoch. Anal. Appl. **21**, 97–126 (2003)
18. de Bouard, A., Debussche, A.: Blow up for the stochastic nonlinear Schrödinger equation with multiplicative noise. Ann. Probab. **33**, 1078–1110 (2005)
19. de Bouard, A., Fukuizumi, R.: Representation formula for stochastic Schrödinger evolution equations and applications. Nonlinearity **25**(11), 2993–3022 (2012)
20. Doi, S.: On the Cauchy problem for Schrödinger type equation and the regularity of solutions. J. Math. Kyoto Univ. **34**(2), 319–328 (1994)
21. Doi, S.: Remarks on the Cauchy problem for Schrödinger-type equations. Commun. PDE **21**, 163–178 (1996)
22. Guerrero, P., López, J.L., Nieto, J.: Global H^1 solvability of the 3D logarithmic Schrödinger equation. Nonlinear Anal.: Real World Appl. **11**, 79–87 (2010)

23. Holevo, A.S.: On dissipative stochastic equations in a Hilbert space. Probab. Theory Relat. Fields **104**(4), 483–500 (1996)
24. Krylov, N.V.: Itô's formula for the L_p-norm of stochastic W_p^1-valued processes. Probab. Theory Relat. Fields **147**(3–4), 583–605 (2010)
25. Marzuola, J., Metcalfe, J., Tataru, D.: Strichartz estimates and local smoothing estimates for asymptotically flat Schrödinger equations. J. Funct. Anal. **255**(6), 1479–1553 (2008)
26. Rockafellar, R.T.: Directionally Lipschitzian functions and subdifferential calculus. Proc. Lond. Math. Soc. **3**, 39 (1979) (no. 2, 331–355)
27. Zhang, D.: Stochastic nonlinear Schrödinger equation (Ph.D. thesis), Universität Bielefeld, 2014
28. Zloshchastiev, K.G.: Logarithmic nonlinearity in theories of quantum gravity: origin of time and observational consequences. Gravit. Cosmol. **16**(4), 288–297 (2010)

Part III
Stochastic Analysis Including Geometric Aspects

Generalized Solutions to Nonlinear Fokker–Planck Equations with Linear Drift

Viorel Barbu

Dedicated to Michael Röckner at his 60th birthday.

Abstract Existence and long-time behaviour of solutions to nonlinear Fokker–Planck equations (NFPEs) with linear drift are studied.

Keywords Fokker–Planck equation · Entropy · Accretive · Mild solution · Lyapunov function

MSC (2010) 35K55 · 35Q84 · 47H07

1 The Problem

Here, we shall consider the equation

$$\begin{aligned} &u_t(t,x) + \mathrm{div}_x(D(x)u(t,x)) - \Delta_x\beta(u(t,x)) = 0 \text{ in } (0,\infty)\times\mathbb{R}^d,\\ &u(0,x) = u_0(x), \quad x\in\mathbb{R}^d,\ d\geq 1, \end{aligned} \tag{1.1}$$

where

(i) $D \in L^\infty(\mathbb{R}^d;\mathbb{R}^d)$, div $D\in L^\infty(\mathbb{R}^d)$.

β is continuous, monotonically nondecreasing, $\beta(0)=0$,

$|\beta(r)| \leq C_1|r|^m + C_2$, $\forall r\in\mathbb{R}$, where $1\leq m<\infty$.

Equation (1.1) describes the evolution of a probability density $u = P$ associated to the Markovian stochastic processes with drift coefficients $(D_i)_{i=1}^d = D$ and diffusion

V. Barbu (✉)
Octav Mayer Institute of Mathematics of the Romanian Academy, Iasi, Romania
e-mail: vb41@uaic.ro

A. Eberle et al. (eds.), *Stochastic Partial Differential Equations and Related Fields*, Springer Proceedings in Mathematics & Statistics 229,
https://doi.org/10.1007/978-3-319-74929-7_18

$\sigma_{ij} = \delta_{ij}$. Moreover, it is related to the anomalous diffusion which describes particle transport in irregular media. In the special case $D \equiv 0$, (1.1) reduces to the nonlinear porous media equation in $\mathbb{R}^d$.

In $1D$, Eq. (1.1) can be derived from the entropy functional

$$S[u] = \int_{\mathbb{R}} \Phi[u(x)]dx, \tag{1.2}$$

where

$$\Phi \in C^\infty(0, \infty), \ \lim_{r\to 0} \Phi'(r) = \infty \text{ and } \Phi''(r) < 0 \text{ for } r > 0. \tag{1.3}$$

The corresponding Fokker–Planck equation is

$$P_t + \left(H(x)P - \frac{1}{\alpha} (\Phi(P) - P\Phi'(P))_x \right)_x = 0. \tag{1.4}$$

Here the drift function H is the gradient of a potential V (i.e., $H = -\frac{dV}{dx}$) and the constant α represents the strength of fluctuations [5]. A similar approach applies to higher dimensions.

In the special case of the Boltzmann–Gibbs entropy

$$S[u] = -\int u(x) \log u(x) dx,$$

Equation (1.4) reduces to

$$P_t + P_x - \frac{1}{\alpha} P_{xx} = 0, \tag{1.5}$$

while, for the entropy functional

$$S[u] = \frac{1}{p-1} \int (|u|^p - u) dx, \ p > 1, \tag{1.6}$$

Equation (1.4) with $H \equiv 1$ reads as the Plastino and Plastino model [7]

$$P_t + P_x - \frac{1}{\alpha} ((P)^p)_{xx} = 0.$$

Assumption (i) agrees with the key entropy condition (1.3). Indeed, if $\Phi \in C^1(0, \infty) \cap C[0, \infty)$ is a solution to the equation

$$\Phi(r) - r\Phi'(r) = \beta(r), \ \forall r > 0; \ \Phi'(0) = \infty, \tag{1.7}$$

such that

$$\Phi''(r) < 0, \ \Phi'(r) \geq 0, \ \forall r \in \mathbb{R},$$

where β satisfies (i), the NFPE reduces to (1.1) such that (1.3) holds. We note that, in particular, assumption (i) is satisfied for $\beta(u) = \frac{1}{\alpha} \ln(1+u)$, that is, for the Fokker–Planck equation of classical bosons (see [4, 5])

$$P_t + (DP)_x + \frac{1}{\alpha}(\ln(1+P))_{xx} = 0.$$

In [1], Eq. (1.1), was studied the existence of an entropy solution for the Fokker–Planck equation

$$\begin{aligned} &u_t + \operatorname{div}(D(x,u)u) - \Delta\beta(u) = 0, \text{ in } (0,T)\times\mathbb{R}^d, \\ &u(0,x) = u_0(x), \end{aligned} \tag{1.8}$$

where $D(x,u) \equiv b(u)$, with b continuous. In this work, we shall confine to the case of linear drift $D(x,u) \equiv D(x)$.

2 The Existence and Uniqueness of a Generalized Solution

To (1.1) we associate the operator $A : L^1(\mathbb{R}^d) \to L^1(\mathbb{R}^d)$ defined as the closure $\overline{A}_1$ in $L^1(\mathbb{R}^d) \times L^1(\mathbb{R}^d)$ of the operator

$$\begin{aligned} A_1 u &= -\Delta\beta(u) + \operatorname{div}(D(x)u),\ \forall u \in D(A_1), \\ D(A_1) &= \{u \in L^1(\mathbb{R}^d) \cap L^\infty(\mathbb{R}^d),\ \beta(u) \in H^1(\mathbb{R}^d),\ A_1 u \in L^1(\mathbb{R}^d)\}. \end{aligned} \tag{2.1}$$

We have also

Lemma 2.1 *The operator A_1 is accretive in $L^1(\mathbb{R}^d)$ and*

$$L^1(\mathbb{R}^d) \cap L^\infty(\mathbb{R}^d) \subset R(I + \lambda A_1),\ \forall \lambda > 0. \tag{2.2}$$

$$(I + \lambda A_1)^{-1} f \geq 0 \textit{ in } \mathbb{R}^d \textit{ if } f \geq 0 \textit{ in } \mathbb{R}^d \tag{2.3}$$

$$\int_{\mathbb{R}^d} (I+\lambda A_1)^{-1} f\,dx = \int_{\mathbb{R}^d} f\,dx \textit{ in } \mathbb{R}^d. \tag{2.4}$$

Proof The accretivity of A_1 follows by multiplying the equation

$$u - \bar{u} + \lambda(A_1 u - A_1 \bar{u}) = f - \bar{f},\ u, \bar{u} \in D(A_1),$$

in the duality pair ${}_{H^{-1}(\mathbb{R}^d)}\langle\cdot,\cdot\rangle_{H^1(\mathbb{R}^d)}$ with $\mathcal{X}_\varepsilon(u-\bar{u})$ and integrate over $\mathbb{R}^d$, where $\mathcal{X}_\varepsilon$ is a smooth approximation of the sign function for $\varepsilon \to 0$.

More precisely, χ_ε is defined by

$$\chi_\varepsilon(r) = \begin{cases} -1 & \text{for } r < -\varepsilon, \\ \frac{1}{\varepsilon} r & \text{for } |r| < \varepsilon, \\ 1 & \text{for } r > \varepsilon. \end{cases}$$

To prove (2.2), we fix $f \in L^1(\mathbb{R}^d) \cap L^\infty(\mathbb{R}^d) \subset L^2(\mathbb{R}^d)$ and consider the equation $u + \lambda A_1 u = f$, that is,

$$u - \lambda\Delta\beta(u) + \lambda\,\mathrm{div}(Du) = f \text{ in } \mathcal{D}'(\mathbb{R}^d),\ \lambda > 0. \tag{2.5}$$

(Here, $\mathcal{D}'(\mathbb{R}^d)$ is the space of distributions on $\mathbb{R}^d$.)

We set $\beta_\varepsilon = \frac{1}{\varepsilon}\,\beta(I + \varepsilon\beta)^{-1})$ and approximate (2.5) by

$$u - \lambda\Delta(\beta_\varepsilon(u) + \varepsilon u) + \lambda\,\mathrm{div}(Du) = f \text{ in } \mathbb{R}^d, \tag{2.6}$$

Equivalently,

$$(\varepsilon + \beta_\varepsilon)^{-1}(v) - \lambda\Delta v + \lambda\,\mathrm{div}(D(\varepsilon + \beta_\varepsilon)^{-1}(v)) = f \text{ in } \mathbb{R}^d. \tag{2.7}$$

The operator $v \to \beta_\varepsilon^{-1}(v) - \lambda\Delta v$ is coercive and maximal monotone in $H^1(\mathbb{R}^d) \times H^{-1}(\mathbb{R}^d)$ and so, for each $w \in L^2(\mathbb{R}^d)$,

$$(\varepsilon + \beta_\varepsilon)^{-1}(v) - \lambda\Delta v = -\lambda\,\mathrm{div}(D(\varepsilon + \beta_\varepsilon)^{-1}(w)) + f \quad \text{in } \mathbb{R}^d$$

has a unique solution $v = F(w) \in H^1(\mathbb{R}^d)$.

By the contraction principle, for $\lambda > \frac{1}{2L}\,\|D\|_\infty$, Eq. (2.7) has a unique solution $v_\varepsilon \in H^1(\mathbb{R}^d)$. This extends to all $\lambda > 0$.

We have by (2.6)

$$\begin{aligned} &|u_\varepsilon|_p \le |f|_p,\ \forall \varepsilon > 0,\ p \in [1, \infty), \\ &|\nabla\beta_\varepsilon(u_\varepsilon)|_2^2 + \varepsilon|\nabla u_\varepsilon|_2^2 \le C,\ \forall \varepsilon > 0. \end{aligned}$$

(Here, $|\cdot|_p$, $1 \le p \le \infty$, is the norm of $L^p(\mathbb{R}^d)$.)

On a subsequence $\varepsilon \to 0$, we have

$$\begin{aligned} u_\varepsilon &\to u && \text{weakly in } L^p,\ p \in (1, \infty), \\ \beta_\varepsilon(u_\varepsilon) &\to \eta && \text{weakly in } H^1, \\ \Delta(\beta_\varepsilon(u_\varepsilon) + \varepsilon u_\varepsilon) &\to \Delta\eta && \text{weakly in } H^{-1}, \\ \mathrm{div}(Du_\varepsilon) &\to \mathrm{div}(Du) && \text{in } H^{-1}. \end{aligned}$$

$$\limsup_{\varepsilon\to 0} \int_{\mathbb{R}^d} \beta_\varepsilon(u_\varepsilon)u_\varepsilon\,dx \le -\lambda\int_{\mathbb{R}^d} |\nabla\eta|^2 dx - \int_{\mathbb{R}^d} fu\,dx = \int_{\mathbb{R}^d} \eta u\,dx,$$

$$u - \lambda\Delta\eta + \lambda\,\mathrm{div}(Du) = f \text{ in } \mathcal{D}'(\mathbb{R}^d).$$

Hence $\eta = \beta(u)$, a.e. in $\mathbb{R}^d$ and $u + \lambda A_1 u = f$, as claimed. ∎

Proposition 2.2 *Under assumption* (i), *the operator* $A = \overline{A}_1$ *is m-accretive in* $L^1(\mathbb{R}^d)$. *Moreover, one has for all* $\lambda > 0$

$$(I + \lambda A)^{-1} = (I + \lambda A_1)^{-1} \text{ on } L^\infty(\mathbb{R}^d) \cap L^1(\mathbb{R}^d) \tag{2.8}$$

$$(I + \lambda A)^{-1} f \geq 0 \text{ if } f \geq 0 \text{ in } \mathbb{R}^d \tag{2.9}$$

$$\int_{\mathbb{R}^d} (I + \lambda A)^{-1} f\, dx = \int_{\mathbb{R}^d} f\, dx, \quad \forall \lambda > 0. \tag{2.10}$$

It follows also that $\overline{D(A)} = L^1(\mathbb{R}^d)$.

By Proposition 2.2, the finite difference scheme

$$\begin{aligned} u_h(t) + hAu_h(t) &= u_h(t-h), \ h > 0, \ t \geq 0, \\ u_h(t) &= u_0, \quad \text{for } t \leq 0, \end{aligned} \tag{2.11}$$

has a unique solution u_h and, by then, by the Crandall and Liggett exponential formula (see [3]),

$$u_h(t) \to u(t) \text{ strongly in } L^1(\mathbb{R}^d), \tag{2.12}$$

uniformly on compact intervals.

The function u is called *the mild solution to Eq.* (1.1).

Now, we can formulate the main existence result.

Theorem 2.3 *Under assumptions* (i), *for each* $u_0 \in L^1(\mathbb{R}^d) \cap L^\infty(\mathbb{R}^d)$, *Eq.* (1.1) *has a unique mild solution* $u \in C([0, T]; L^1(\mathbb{R}^d))$, $\forall T > 0$. *Moreover,* $S(t)u_0 = u(t)$ *is a continuous semigroup of contractions in* $L^1(\mathbb{R}^d)$,

$$|S(t)u_0|_p \leq |u_0|_p, \quad \forall u_0 \in L^p(\mathbb{R}^d), \ 1 \leq p \leq \infty \tag{2.13}$$

$$u(t,x) \geq 0, \ \text{ a.e. } x \in \mathbb{R}^d \ \text{ if } u_0(x) \geq 0, \ \text{ a.e. } x \in \mathbb{R}^d, \tag{2.14}$$

$$\int_{\mathbb{R}^d} u(t,x)dx = \int_{\mathbb{R}^d} u_0(x)dx, \ \forall t \geq 0. \tag{2.15}$$

Definition 2.4 The function $u \in C([0, T]; L^1(\mathbb{R}^d))$ is said to be a generalized solution to (1.1) if

$$\begin{aligned} &\frac{\partial u}{\partial t} + \mathrm{div}_x(D(x)u) - \Delta_x \beta(u) = 0 \ \text{ in } \mathcal{D}'((0,T) \times \mathbb{R}^d), \\ &u(0,x) = u_0(x) \qquad\qquad\qquad\qquad\qquad \text{in } \mathbb{R}^d. \end{aligned} \tag{2.16}$$

By (2.11)–(2.12), we see that

Theorem 2.5 *Under assumption* (i)*, for each* $u_0 \in L^1(\mathbb{R}^d) \cap L^\infty(\mathbb{R}^d)$*, the mild solution* u *to* (2.1) *is a generalized solution. Moreover, there is at most one generalized solution* $u \in C([0,T]; L^1(\mathbb{R}^d) \cap L^\infty((0,T) \times \mathbb{R}^d)$.

The uniqueness part of Theorem 2.5 follows as in [1] and it will be omitted.

3 Long-Time Dynamical Behavior

Let $S(t)$ be the semigroup of contractions generated on $\overline{D(A)}$ under assumption (i).

Definition 3.1 A real valued function $\psi : L^1(\mathbb{R}^d) \to \mathbb{R}^+$ is a Lyapunov function for $S(t)$ if $\psi(S(t)u_0) \le \psi(u_0)$, $\forall t \ge 0$, $\forall u_0 \in D(\psi) \cap \overline{D(A)}$, where $D(\psi) = \{u_0 \in L^1(\mathbb{R}^d);\ \psi(u_0) < \infty\}$.

As we shall see later on, the free energy of the system (the so-called H-functional) is the best candidate for the Lyapunov functions.

Again, by (2.11) and (2.12), we see that

Proposition 3.2 *Assume that* $\psi : L^1(\mathbb{R}^d) \to \mathbb{R}^+$ *is lower semicontinuous and, for all* $\lambda > 0$,

$$\psi((I + \lambda A)^{-1}u_0) \le \psi(u_0),\ \forall u_0 \in D(\psi) \cap \overline{D(A)}. \tag{3.1}$$

Then, ψ *is a Lyapunov function for the semigroup* $S(t)$.

Here, $D(\psi) = \{u;\ \psi(u) < \infty\}$.

We shall look at Lyapunov functions of the form

$$\psi(u) = \int_{\mathbb{R}^d} j(u(x))dx,\quad \forall u \in L^1(\mathbb{R}^d), \tag{3.2}$$

where $j : \mathbb{R} \to \mathbb{R}^+$ is convex, lower semicontinuous and $j(0) = 0$. Then, as well known, $\psi : L^1(\mathbb{R}^d) \to \mathbb{R}$ is convex, lower semicontinuous and

$$D(\psi) = \{u \in L^1(\mathbb{R}^d);\ j(u) \in L^1(\mathbb{R}^d)\}.$$

We have (see [1]).

Theorem 3.3 *Under hypothesis* (i)*, the function* ψ *is a Lyapunov function for* $S(t)$.

Remark 3.4 In particular, it follows by Proposition 3.2 that the operator A is m-completely accretive in sense of [2].

4 The H-Theorem

The so-called H-theorem amounts to saying that, for $t \to \infty$, $S(t)u_0 \to v$, where v is a stationary solution to Eq. (1.1), that is, $Av = 0$ (see [5, 6]).

Since, as easily seen by (2.11), for each ℓ and $u_0 \in \overline{D(A)}$,

$$\int_{\mathbb{R}^d} |(S(t)u_0)(x+\ell) - (S(t)u_0)(x)|dx \le \int_{\mathbb{R}^d} |u_0(x+\ell) - u_0(x)|dx,$$

by the Kolmogorov compactness theorem, the trajectory $\{S(t)u_0;\ t \ge 0\}$ is compact in $L^1_{loc}(R^d)$ and in every $L^p_{loc}(R^d)$, $1 \le p < \infty$, if $u_0 \in L^1(R^d) \cap L^\infty(R^d)$.

Hence the ω-limit set

$$\omega(u_0) = \left\{ v = \lim_{t_n \to \infty} S(t_n)u_0 \text{ in } L^1_{loc}(\mathbb{R}^d) \right\}$$

is nonempty and, if $u_0 \in L^1(R^d) \cap L^\infty(R^d)$, we have also by (2.13) that

$$\omega(u_0) = \left\{ v = \underset{t_n \to \infty}{\text{weak limit}}\ S(t_n)u_0 \text{ in each } L^p(\mathbb{R}^d) \right\}.$$

By weak lower-semicontinuity of ψ, we have

$$\psi(v) \le \lim_{t \to \infty} \psi(S(t)u_0), \quad \forall v \in \omega(u_0).$$

If ψ is continuous on L^1 and $\{S(t)u_0,\ t \ge 0\}$ is compact, then we have

$$\psi(S(t)v) \equiv \psi(v),\ \forall t \ge 0,\ v \in \omega(u_0), \tag{4.1}$$

which is a weak form of the H-theorem.

To give a specific example, we assume that

$$D = -\nabla V,\ V \in C^1(\mathbb{R}^d),\ V \ge 0, \tag{4.2}$$

$$\beta \in C^1(\mathbb{R}),\ \beta(0) = 0,\ \beta'(r) > 0,\ \forall r > 0, \tag{4.3}$$

$$\inf\{\|D(x)\|;\ x \in \mathbb{R}^d\} > 0. \tag{4.4}$$

Then, as easily seen by (4.2), (4.3), the energy functional

$$E(u) = \int_{\mathbb{R}^d} (V(x)u(x) - \Phi(u(x)))dx,$$

where Φ is given by (1.7), and so

$$\Phi''(r) = -\frac{\beta'(r)}{r} \quad \forall r > 0.$$

We note that E is convex and is a Lyapunov functional for the semigroup $S(t)$. Indeed, taking into account that $\partial E(u) = V - \Phi'(u)$, we get that

$$\begin{aligned}\langle \partial E(u), A_1 u\rangle &= \int_{\mathbb{R}^d} (V - \Phi'(u))(-\Delta\beta(u) + \mathrm{div}(Du))dx \\ &= \int_{\mathbb{R}^d} \left(\frac{(\beta'(u))^2}{u}\,|\nabla u|^2 + u|\nabla V|^2\right) dx \geq 0, \\ &\qquad\qquad \forall u \in D(A_1),\ u \geq 0,\end{aligned} \tag{4.5}$$

and, by density, this implies that

$$E(S(t)u_0) \leq \mathbb{E}(u_0),\ \forall t \geq 0. \tag{4.6}$$

Moreover, by (4.4), (4.6) and (2.11), we see that

$$\begin{aligned}&E(u_h(ih)) - E((u_h(i-1)h)) \\ &\quad \leq -h \int_{\mathbb{R}^d} \left(\frac{(\beta'(u_h(ih)))^2}{u_h(ih)}\right) |\nabla u_h(ih)|^2 + u_h(ih)|\nabla V|^2 dx \\ &\quad \leq -\rho \int_{\mathbb{R}^d} u_h(ih)dx,\ \forall i = 1, 2, \ldots, h > 0,\end{aligned}$$

where $\rho > 0$. This yields

$$\frac{1}{t-s}\,(E(S(t)u_0) - E(S(s)u_0)) \leq \rho|S(t)u_0|_1,\ \forall t > s > 0,$$

and, therefore,

$$E(S(t)u_0) \leq \exp(-\rho t)|u_0|_1,\ \forall t \geq 0.$$

Hence, if $u_\infty \in \omega(u_0)$, we have $E(u_\infty) = 0$. Assume further that

$$\inf_{x\in\mathbb{R}^d} V(x) > \sup_{r>0} \frac{\Phi(r)}{r}. \tag{4.7}$$

Then the latter implies that $u_\infty = 0$. We have, therefore,

Theorem 4.1 *Under assumptions* (4.2)–(4.6), *we have*

$$\lim_{t\to\infty} S(t)u_0 = 0 \textit{ in } L^1(\mathbb{R}^d) \textit{ for each } u_0 \in L^1(\mathbb{R}^d).$$

5 Final Remarks

For $D = -\nabla V$, we associate with Eq. (1.1) the free energy functional

$$F(u) = \int_{\mathbb{R}^d} \Phi(u)dx + \int_{\mathbb{R}^d} V u\, dx, \tag{5.1}$$

where Φ satisfies (1.7) and $V \in C^\infty(\mathbb{R}^d)$ is such that

$$V(x) \geq 0, \ |\nabla V(x)| \leq C(V(x) + 1), \ \forall x \in \mathbb{R}^d.$$

On the set

$$K = \left\{u : \mathbb{R}^d \to [0, \infty) \text{ measurable}, \int_{\mathbb{R}^d} u(x)dx = 1, \int_{\mathbb{R}^d} |x|^2 u(x)dx < \infty\right\},$$

consider the iterative scheme

$$u_{k+1} = \arg\min_{u \in K} \left\{\frac{1}{2h} d^2(u, u_k) + F(u)\right\}, \tag{5.2}$$

where d is the Wasserstein distance (see, e.g., [6]).

Consider the sequence $u^h : [0, T] \to \mathbb{R}^d$ of the step functions

$$u^h(t) = u_k \text{ for } t \in [kh, (k + 1)h], \ k = 0, 1. \tag{5.3}$$

Problem *Does the sequence* $\{u^h\}$ *strongly converge to* $S(t)u_0$ *for* $h \to 0$?

The answer is positive (see [6]) if $\Phi(u) = u \log u$ (the case of Gibbs–Boltzmann entropy), that is, for the Fokker–Planck equation

$$u_t + \text{div}(Du) - \Delta u = 0 \ \text{ in } (0, T) \times \mathbb{R}^d$$

and one might suspect that it is true in this case for other functions Φ satisfying (1.7).

References

1. Barbu, V.: Generalized solutions to nonlinear Fokker-Plank equations. J. Differ. Equ. **261**, 2446–2471 (2016)
2. Benilan, Ph, Crandall, M.G.: Completely accretive operators. Semigroup Theory and Evolution Equations. Lecture Notes in Pure and Applied Mathematics, pp. 41–75 (1989)
3. Crandall, M.G., Liggett, T.M.: Generation of semi-groups of nonlinear transformations on general Banach spaces. Amer. J. Math. **93**, 265–298 (1971)
4. Frank, T.D.: Nonlinear Fokker-Planck Equations. Springer, Berlin (2005)

5. Frank, T.D., Daffertshofer, A.: h-Theorem for nonlinear Fokker-Planck equations related to generalized thermostatistics. Phys. A **295**, 455–474 (2001)
6. Jordan, R., Kinderlehrer, D., Otto, F.: The variational formulation of the Fokker-Planck equation. SIAM J. Math. Anal. **29**, 1–17 (1998)
7. Plastino, A.R., Plastino, A.: Phys. A **222**, 347 (1995)

Examples of Renormalized SDEs

Y. Bruned, I. Chevyrev and P. K. Friz

Abstract We demonstrate two examples of stochastic processes whose lifts to geometric rough paths require a renormalisation procedure to obtain convergence in rough path topologies. Our first example involves a physical Brownian motion subject to a magnetic force which dominates over the friction forces in the small mass limit. Our second example involves a lead-lag process of discretised fractional Brownian motion with Hurst parameter $H \in (1/4, 1/2)$, in which the stochastic area captures the quadratic variation of the process. In both examples, a renormalisation of the second iterated integral is needed to ensure convergence of the processes, and we comment on how this procedure mimics negative renormalisation arising in the study of singular SPDEs and regularity structures.

Keywords Renormalization · Rough paths · Gaussian analysis

Subject classification (MSC 2010) Primary 60H99 · Secondary 60H10

1 Introduction

In recent years, the theory of regularity structures [1] has been proposed to give meaning to a wide class of singular SPDEs. A central feature of the theory is the notion

---

Y. Bruned
Department of Mathematics, Imperial College London, London, England
e-mail: y.bruned@imperial.ac.uk

I. Chevyrev
Mathematical Institute, University of Oxford, Oxford, England
e-mail: chevyrev@maths.ox.ac.uk

P. K. Friz (✉)
Institut für Mathematik, Technische Universität Berlin, Berlin, Germany
e-mail: friz@math.tu-berlin.de

P. K. Friz
Weierstraß–Institut für Angewandte Analysis und Stochastik, Berlin, Germany

A. Eberle et al. (eds.), *Stochastic Partial Differential Equations and Related Fields*, Springer Proceedings in Mathematics & Statistics 229,
https://doi.org/10.1007/978-3-319-74929-7_19

of renormalisation, specifically "negative renormalisation" [2], which is required to obtain convergence of random models to a meaningful limit. It is well-known that this procedure is inherit to the problem since naive approximations of such equations typically fail to converge (with a number of notable exceptions, including a special variant of gKPZ [3]). The same feature thus naturally appears in other solution theories which have been proposed to solve such equations, including the theories of paracontrolled distributions [4] and the Wilsonian renormalisation group [5].

Viewing regularity structures as a multidimensional generalisation of the theory of rough paths [6], it is natural to ask how renormalisation manifests itself in the latter. As solution theories to singular S(P)DEs, both share the common feature that one must give meaning to often analytically ill-posed higher order terms (iterated integrals) of distributions, which is typically done through some stochastic means. However, a key difference in applications of rough paths to SDEs and rough SPDEs [7] is that one can usually give meaning to such terms as limits of iterated integrals of mollifications of the irregular noise without the need of renormalisation.

The purpose of this note is to demonstrate situations in rough paths theory which fall outside this usual setting and for which renormalisation is a necessary feature. Specifically, we construct two stochastic processes whose lifts to geometric rough paths fail to converge without a renormalisation procedure akin to the one encountered in the theory of regularity structures.

The first is a physical Brownian motion subject to a magnetic force which dominates over the friction forces in the small mass limit. This example builds on the work [8], where a similar situation was considering with a constant magnetic field.

The second is a lead-lag process of a discretized path, which we take to be fractional Brownian motion with Hurst parameter $H \in (1/4, 1/2]$. The stochastic area of this lead-lag process captures the quadratic variation of the discretized path, and thus, as one can expect, the second iterated integral fails to converge as the mesh of the discretisation goes to zero (unless $H = 1/2$). This example is motivated from a similar Hoff process considered for semi-martinagles in [9].

In both examples we demonstrate an explicit renormalisation procedure of the second iterated integral under which the processes converge in rough path topologies. These (diverging) counter-terms serve precisely the same re-centring role encountered in regularity structures (for a direct comparison, consider the renormalisation of PAM [1, 4, 10] where only one diverging term needs to be considered). In turn, rough differential equations driven by the renormalised and unrenormalised rough paths are related to one another by the addition of diverging terms, which again mimics the situation encountered in singular SPDEs. We refer to the recent article [11] for a much more detailed study of this relation.

2 Magnetic Field Blow-Up

Consider a physical Brownian motion in a magnetic field with dynamics given by

$$m\ddot{x} = -A\dot{x} + B\dot{x} + \xi, \quad x(t) \in \mathbb{R}^d,$$

where A is a symmetric matrix with strictly positive spectrum (representing friction), B is an anti-symmetric matrix (representing the Lorentz force due to a magnetic field), and ξ is an $\mathbb{R}^d$-valued white noise in time. We shall consider the situation that A is constant whereas B is a function of the mass m.

We rewrite these dynamics as

$$\begin{aligned} dX_t &= \frac{1}{m} P_t dt, \quad X_0 = 0, \\ dP_t &= -\frac{1}{m} M P_t dt + dW_t, \quad P_0 = 0, \end{aligned}$$

where $M = A - B$, and we have chosen the starting point as zero simply for convenience. We furthermore introduce the parameter $\varepsilon^2 = m$ and write X_t^ε, P_t^ε, and $M^\varepsilon = A - B^\varepsilon$ to denote the dependence on ε.

We are interested in the convergence of the processes P^ε and $M^\varepsilon X^\varepsilon$ in rough path topologies. Let $G^2(\mathbb{R}^d)$ and $\mathfrak{g}^2(\mathbb{R}^d)$ denote the step-2 free nilpotent Lie group and Lie algebra respectively. Let us also write $\mathfrak{g}^2(\mathbb{R}^d) = \mathbb{R}^d \oplus \mathfrak{g}^{(2)}(\mathbb{R}^d)$ for the decomposition of $\mathfrak{g}^2(\mathbb{R}^d)$ into the first and second levels, where we identify $\mathfrak{g}^{(2)}(\mathbb{R}^d)$ with the space of anti-symmetric $d \times d$ matrices.

For every $\varepsilon > 0$, define the matrix

$$C^\varepsilon = \int_0^\infty e^{-M^\varepsilon s} e^{-(M^\varepsilon)^* s} ds,$$

and the element

$$v^\varepsilon = -\frac{1}{2}(M^\varepsilon C^\varepsilon - C^\varepsilon (M^\varepsilon)^*) \in \mathfrak{g}^{(2)}(\mathbb{R}^d).$$

For any $v \in \mathfrak{g}^{(2)}(\mathbb{R}^d)$, $p \in [1, 3)$, and p-rough path $(Z_{s,t}, \mathbb{Z}_{s,t}) \in G^2(\mathbb{R}^d)$ (where we ignore zeroth component 1), we define the translated rough path $T_v(Z_{s,t}, \mathbb{Z}_{s,t})$ by

$$T_v(Z_{s,t}, \mathbb{Z}_{s,t}) = (Z_{s,t}, \mathbb{Z}_{s,t} + (t - s)v^\varepsilon). \tag{1}$$

Consider the $G^2(\mathbb{R}^d)$-valued processes

$$\begin{aligned} (P_{s,t}^\varepsilon, \mathbb{P}_{s,t}^\varepsilon) &= \left(P_{s,t}^\varepsilon, \int_s^t P_{s,r}^\varepsilon \otimes \circ d P_r^\varepsilon \right), \\ (Z_{s,t}^\varepsilon, \mathbb{Z}_{s,t}^\varepsilon) &= \left(M^\varepsilon X_{s,t}^\varepsilon, \int_s^t M^\varepsilon X_{s,r} \otimes d(M^\varepsilon X^\varepsilon)_r \right), \end{aligned}$$

and the canonical lift of the Brownian motion W

$$(W_{s,t}, \mathbb{W}_{s,t}) = \left(W_{s,t}, \int_s^t W_{s,r} \otimes \circ d W_r \right),$$

where the integrals in the definition of $\mathbb{P}_{s,t}^\varepsilon$ and $\mathbb{W}_{s,t}$ are in the Stratonovich sense.

The following proposition establishes the convergence of the "renormalised" paths $T_{\nu^\varepsilon}(P^\varepsilon_{s,t}, \mathbb{P}^\varepsilon_{s,t})$ and $T_{\nu^\varepsilon}(Z^\varepsilon_{s,t}, \mathbb{Z}^\varepsilon_{s,t})$.

Theorem 1 *Suppose that*

$$\lim_{\varepsilon\to 0} |M^\varepsilon|\varepsilon^\kappa = 0 \textit{ for some } \kappa \in [0, 1]. \tag{2}$$

Then for any $\alpha \in [0, 1/2 - \kappa/4)$ and $q < \infty$, it holds that $T_{\nu^\varepsilon}(P^\varepsilon, \mathbb{P}^\varepsilon) \to (0, 0)$ and $T_{\nu^\varepsilon}(Z^\varepsilon, \mathbb{Z}^\varepsilon) \to (W, \mathbb{W})$ in L^q and α-Hölder topology as $\varepsilon \to 0$. More precisely, as $\varepsilon \to 0$, in L^q

$$\sup_{s,t\in[0,T]} \frac{|P^\varepsilon_{s,t}|}{|t-s|^\alpha} + \sup_{s,t\in[0,T]} \frac{|\mathbb{P}^\varepsilon_{s,t} + (t-s)\nu^\varepsilon|}{|t-s|^{2\alpha}} \to 0.$$

and

$$\sup_{s,t\in[0,T]} \frac{|Z^\varepsilon_{s,t} - W_{s,t}|}{|t-s|^\alpha} + \sup_{s,t\in[0,T]} \frac{|\mathbb{Z}^\varepsilon_{s,t} + (t-s)\nu^\varepsilon - \mathbb{W}_{s,t}|}{|t-s|^{2\alpha}} \to 0.$$

The rest of the section is devoted to the proof of Theorem 1 which builds on the proof of [8] Theorem 1.

We set $Y^\varepsilon = P^\varepsilon/\varepsilon$ and obtain that

$$\begin{aligned} dY^\varepsilon_t &= -\varepsilon^2 M^\varepsilon Y^\varepsilon_t dt + \varepsilon^{-1} dW_t \\ dX^\varepsilon_t &= \varepsilon^{-1} Y^\varepsilon_t dt. \end{aligned}$$

For fixed ε, we introduce the Brownian motion $\tilde{W}^\varepsilon_\cdot = \varepsilon W_{\varepsilon^{-2}\cdot}$ and consider

$$d\tilde{Y}^\varepsilon_t = -M^\varepsilon \tilde{Y}^\varepsilon_t dt + d\tilde{W}^\varepsilon_t.$$

Observe that we have the pathwise equalities

$$Y^\varepsilon_\cdot = \tilde{Y}^\varepsilon_{\varepsilon^{-2}\cdot}, \tag{3}$$

and since $Y^\varepsilon_0 = 0$, we have

$$\tilde{Y}^\varepsilon_t = \int_0^t e^{-M^\varepsilon(t-s)} d\tilde{W}^\varepsilon_s. \tag{4}$$

The dependence of M^ε on ε is, by construction, only though B^ε, the anti-symmetric part of M^ε. In particular, since the symmetric part A stays constant and has strictly positive spectrum, it follows that for some $\lambda > 0$, $\mathrm{Re}(\sigma(M^\varepsilon)) \subset (\lambda, \infty)$ for all $\varepsilon > 0$. In particular,

$$\sup_{\tau>0}\sup_{\varepsilon>0}\frac{|e^{-\tau M^\varepsilon}|}{e^{-\lambda\tau}}<\infty. \tag{5}$$

We see then that

$$\sup_{\varepsilon>0}|C^\varepsilon|<\infty$$

and

$$\sup_{\varepsilon>0}\sup_{0\leq t<\infty}\mathbb{E}\left[|\tilde{Y}_t^\varepsilon|^2\right]<\infty. \tag{6}$$

Lemma 2 *There exists $C_1>0$ such that for all $\varepsilon\in(0,1]$ and $s,t\in[0,T]$*

$$\mathbb{E}\left[\left|Y_{s,t}^\varepsilon\right|^2\right]^{1/2}\leq C_1\min\{\varepsilon^{-1}|t-s|^{1/2},1\}$$

and

$$\mathbb{E}\left[\left|\int_s^t Y_r^\varepsilon\otimes Y_r^\varepsilon dr-(t-s)C^\varepsilon\right|^2\right]^{1/2}\leq C_1\min\{\varepsilon|t-s|^{1/2},|t-s|\}.$$

Proof The first inequality is clear from (3), (4) and (6). For the second, from (4), we see that for every $r>0$, $\tilde{Y}_r^\varepsilon$ has distribution $\mathcal{N}(0,C_r^\varepsilon)$ where

$$C_r^\varepsilon=\int_0^r e^{-M^\varepsilon(r-u)}e^{-(M^\varepsilon)^*(r-u)}du=\int_0^r e^{-M^\varepsilon u}e^{-(M^\varepsilon)^*u}du.$$

Hence $Y_r^\varepsilon=\tilde{Y}_{\varepsilon^{-2}r}$ has distribution $\mathcal{N}(0,C_{\varepsilon^{-2}r}^\varepsilon)$. Thus

$$\mathbb{E}\left[\int_s^t Y_r^\varepsilon\otimes Y_r^\varepsilon dr\right]=\int_s^t C_{\varepsilon^{-2}r}^\varepsilon dr=\int_s^t\int_0^{\varepsilon^{-2}r}e^{-M^\varepsilon u}e^{-(M^\varepsilon)^*u}dudr=:\mu_{s,t}^\varepsilon.$$

Observe that from (5)

$$\begin{aligned}|\mu_{s,t}^\varepsilon-(t-s)C^\varepsilon|&\leq\int_s^t\int_{\varepsilon^{-2}r}^\infty|e^{-M^\varepsilon u}e^{-(M^\varepsilon)^*u}|dudr\\&\leq C_2\int_s^t\int_{\varepsilon^{-2}r}^\infty e^{-2\lambda u}dudr\\&\leq C_3\int_s^t e^{-2\lambda\varepsilon^{-2}r}dr\\&\leq C_4\min\{\varepsilon^2,|t-s|\}\\&\leq C_4\min\{\varepsilon|t-s|^{1/2},|t-s|\}.\end{aligned}$$

We now claim that

$$\mathbb{E}\left[\left|\int_s^t Y_r^\varepsilon \otimes Y_r^\varepsilon dr - \mu_{s,t}^\varepsilon\right|^2\right] \leq C_5 \min\{\varepsilon^2|t-s|, |t-s|^2\},$$

from which the conclusion follows. Indeed, by Fubini and Wick's formula

$$\begin{aligned}
\mathbb{E}\left[\left(\int_s^t Y_r^{\varepsilon,i} Y_r^{\varepsilon,j} dr\right)^2\right] &= \int_{[s,t]^2} \mathbb{E}\left[Y_r^{\varepsilon,i} Y_r^{\varepsilon,j} Y_u^{\varepsilon,i} Y_u^{\varepsilon,j}\right] drdu \\
&= \int_{[s,t]^2} \mathbb{E}\left[Y_r^{\varepsilon,i} Y_r^{\varepsilon,j}\right] \mathbb{E}\left[Y_u^{\varepsilon,i} Y_u^{\varepsilon,j}\right] drdu \\
&+ \int_{[s,t]^2} \mathbb{E}\left[Y_r^{\varepsilon,i} Y_u^{\varepsilon,i}\right] \mathbb{E}\left[Y_r^{\varepsilon,j} Y_u^{\varepsilon,j}\right] drdu \\
&+ \int_{[s,t]^2} \mathbb{E}\left[Y_r^{\varepsilon,i} Y_u^{\varepsilon,j}\right] \mathbb{E}\left[Y_r^{\varepsilon,j} Y_u^{\varepsilon,i}\right] drdu \\
&\leq (\mu_{i,j}^\varepsilon)_{s,t}^2 + 4 \int_{[s,t]^2} \left|\mathbb{E}\left[Y_u^\varepsilon \otimes Y_r^\varepsilon\right]\right|^2 \mathbf{1}\{r \leq u\} drdu.
\end{aligned}$$

Observe that for $r \leq u$

$$\mathbb{E}\left[Y_u^\varepsilon \mid Y_r^\varepsilon\right] = e^{-\varepsilon^{-2} M^\varepsilon (u-r)} Y_r^\varepsilon.$$

and so

$$|\mathbb{E}\left[Y_u^\varepsilon \otimes Y_r^\varepsilon\right]|^2 \mathbf{1}\{r \leq u\} \leq C_6 e^{-\varepsilon^{-2} 2\lambda(u-r)} |C_{\varepsilon^{-2} r}^\varepsilon| \leq C_7 e^{-\varepsilon^{-2} 2\lambda(u-r)}.$$

Thus

$$\begin{aligned}
\mathbb{E}\left[\left(\int_s^t Y_r^{\varepsilon,i} Y_r^{\varepsilon,j} dr - (\mu_{i,j}^\varepsilon)_{s,t}\right)^2\right] &= \mathbb{E}\left[\left(\int_s^t Y_r^{\varepsilon,i} Y_r^{\varepsilon,j} dr\right)^2\right] - (\mu_{i,j}^\varepsilon)_{s,t}^2 \\
&\leq C_8 \int_s^t \int_r^t e^{-\varepsilon^{-2} 2\lambda(u-r)} dudr \\
&\leq C_9 \min\{\varepsilon^2 |t-s|, |t-s|^2\}
\end{aligned}$$

as claimed.

Lemma 3 *There exists $C_{10} > 0$ such that for all $\varepsilon \in (0, 1]$ and $s, t \in [0, T]$*

$$||P_{s,t}^\varepsilon||_{L^2} \leq C_{10} \min\{\varepsilon, |t-s|^{1/2}\}$$

and

$$||\mathbb{P}_{s,t}^\varepsilon + (t-s)\nu^\varepsilon||_{L^2} \leq C_{10} |M^\varepsilon| \min\{\varepsilon |t-s|^{1/2}, |t-s|\}$$

Proof The first inequality is immediate from Lemma 2. For the second, we have

$$\begin{aligned}\mathbb{P}^\varepsilon_{s,t} &= \varepsilon^2 \int_s^t Y^\varepsilon_{s,r} \otimes \circ dY^\varepsilon_r \\ &= -\int_s^t Y^\varepsilon_{s,r} \otimes M^\varepsilon Y^\varepsilon_r dr + \varepsilon \int_s^t Y^\varepsilon_{s,r} \otimes dW_r + \frac{1}{2}(t-s)I.\end{aligned}$$

Since $Y^\varepsilon_{s,r} \otimes M^\varepsilon Y^\varepsilon_r = (Y^\varepsilon_{s,r} \otimes Y^\varepsilon_r)(M^\varepsilon)^*$ and we can directly verify that $v^\varepsilon = C^\varepsilon(M^\varepsilon)^* - \frac{1}{2}I$, we have

$$\mathbb{P}^\varepsilon_{s,t} + (t-s)v^\varepsilon = -\left(\int_s^t Y^\varepsilon_{s,r} \otimes Y^\varepsilon_r dr - (t-s)C^\varepsilon\right)(M^\varepsilon)^* + \varepsilon \int_s^t Y^\varepsilon_{s,r} \otimes dW_r.$$

From Lemma 2, we see that

$$\left|\left|\varepsilon \int_s^t Y^\varepsilon_{s,r} \otimes dW_r\right|\right|_{L^2} \le C_{11} \min\{\varepsilon|t-s|^{1/2}, |t-s|\}.$$

Furthermore, by Fubini and Wick's formula, we can readily show

$$\left|\left|\int_s^t Y^\varepsilon_s \otimes Y^\varepsilon_r dr\right|\right|_{L^2} \le C_{12} \min\{\varepsilon|t-s|^{1/2}, |t-s|\}.$$

It now follows from Lemma 2 that

$$||\mathbb{P}^\varepsilon_{s,t} + (t-s)v^\varepsilon||_{L^2} \le C_{13}|M^\varepsilon| \min\{\varepsilon|t-s|^{1/2}, |t-s|\}.$$

Lemma 4 *There exists $C_{14} > 0$ such that for all $\varepsilon \in (0,1]$ and $s,t \in [0,T]$*

$$\mathbb{E}\left[|Z^\varepsilon_{s,t} - W_{s,t}|^2\right]^{1/2} \le C_{14} \min\{\varepsilon, |t-s|^{1/2}\}$$

and

$$\mathbb{E}\left[\left|\mathbb{Z}^\varepsilon_{s,t} + (t-s)v^\varepsilon - \mathbb{W}_{s,t}\right|^2\right]^{1/2} \le C_{14}|M^\varepsilon| \min\{\varepsilon|t-s|^{1/2}, |t-s|\}.$$

Proof The first inequality follows from $Z^\varepsilon_{s,t} = W_{s,t} - \varepsilon Y^\varepsilon_{s,t}$ and Lemma 2. For the second, we have

$$\begin{aligned}\int_s^t Z^\varepsilon_{s,r} \otimes dZ^\varepsilon_r &= \int_s^t Z^\varepsilon_{s,r} \otimes dW_r - \varepsilon\left(\int_s^t Z^\varepsilon_r \otimes dY^\varepsilon_r - Z^\varepsilon_s \otimes Y^\varepsilon_{s,t}\right) \\ &= \int_s^t Z^\varepsilon_{s,r} \otimes dW_r - \varepsilon\left(Z^\varepsilon_t \otimes Y^\varepsilon_t - \int_s^t dZ_r \otimes Y^\varepsilon_r - Z^\varepsilon_s \otimes Y^\varepsilon_t\right) \\ &= \int_s^t Z^\varepsilon_{s,r} \otimes dW_r - \varepsilon Z^\varepsilon_{s,t} \otimes Y^\varepsilon_t + \int_s^t M^\varepsilon Y^\varepsilon_r \otimes Y^\varepsilon_r dr.\end{aligned}$$

We see that

$$\left\| \int_s^t Z^\varepsilon_{s,r} \otimes dW_r - \int_s^t W_{s,r} \otimes dW_r \right\|_{L^2}^2 \leq C_{15} \min\{\varepsilon^2|t-s|, |t-s|^2\}.$$

Furthermore, by Fubini and Wick's formula, we can readily show

$$||\varepsilon Z^\varepsilon_{s,t} \otimes Y^\varepsilon_t||_{L^2}^2 = \left\| \int_s^t M^\varepsilon Y^\varepsilon_r \otimes Y^\varepsilon_t dr \right\|_{L^2}^2 \leq C_{16}|M^\varepsilon| \min\{\varepsilon^2|t-s|, |t-s|^2\}.$$

Finally, by Lemma 2

$$\left\| \int_s^t M^\varepsilon Y^\varepsilon_r \otimes Y^\varepsilon_r dr - (t-s)M^\varepsilon C^\varepsilon| \right\|_{L^2} \leq C_{17}|M^\varepsilon| \min\{\varepsilon|t-s|^{1/2}, |t-s|\}.$$

It follows that

$$\left\| \mathbb{Z}^\varepsilon_{s,r} - \mathbb{W}_{s,t} - (t-s)(M^\varepsilon C^\varepsilon - \frac{1}{2}I) \right\|_{L^2} \leq C_{18}|M^\varepsilon| \min\{\varepsilon|t-s|^{1/2}, |t-s|\}.$$

We can directly verify $v^\varepsilon = -M^\varepsilon C^\varepsilon + \frac{1}{2}I$, from which the conclusion follows.

Proof (*Proof of Theorem* 1) Observe that condition (2) implies that

$$\lim_{\varepsilon \to 0} |M^\varepsilon|\varepsilon = 0.$$

From Lemmas 3 and 4, along with Gaussian chaos, we thus obtain the pointwise convergence as $\varepsilon \to 0$ for any $q < \infty$ and $s, t \in [0, T]$ in L^q

$$|P^\varepsilon_{s,t}| + |\mathbb{P}^\varepsilon_{s,t} + (t-s)v^\varepsilon|^{1/2} \to 0$$

and

$$|Z^\varepsilon_{s,t} - W_{s,t}| + |\mathbb{Z}^\varepsilon_{s,t} + (t-s)v^\varepsilon - \mathbb{W}_{s,t}|^{1/2} \to 0.$$

Furthermore, since $\min\{\varepsilon|t-s|^{1/2}, |t-s|\} \leq \varepsilon^\kappa|t-s|^{1-\kappa/2}$ for all $\kappa \in [0, 1]$, condition (2), Lemmas 3 and 4, and Gaussian chaos imply that for any $q < \infty$ there exists $C_q > 0$ such that for all $s, t \in [0, T]$ and

$$\sup_{\varepsilon \in (0,1]} \mathbb{E}\left[|P^\varepsilon_{s,t}|^q\right] \leq C_q|t-s|^{q/2},$$

$$\sup_{\varepsilon \in (0,1]} \mathbb{E}\left[|Z^\varepsilon_{s,t} - W_{s,t}|^q\right] \leq C_q|t-s|^{q/2}$$

and

$$\sup_{\varepsilon \in (0,1]} \mathbb{E}\left[|\mathbb{P}^{\varepsilon}_{s,t} + (t-s)v^{\varepsilon}|^q\right] \leq C_q|t-s|^{q(1-\kappa/2)},$$

$$\sup_{\varepsilon \in (0,1]} \mathbb{E}\left[|\mathbb{Z}^{\varepsilon}_{s,t} + (t-s)v^{\varepsilon} - \mathbb{W}_{s,t}|^q\right] \leq C_q|t-s|^{q(1-\kappa/2)}.$$

Applying [7, Theorem A.13] completes the proof.

3 Rough Lead-Lag Process

Consider a path $X : [0,1] \mapsto \mathbb{R}^d$. Let $n \geq 1$ be an integer and write for brevity $X^n_i = X_{i/n}$. Consider the piecewise linear path $\tilde{X}^n : [0,1] \mapsto \mathbb{R}^{2d}$ defined by

$$\tilde{X}^n_{2i/2n} = (X^n_i, X^n_i),$$
$$\tilde{X}^n_{(2i+1)/2n} = (X^n_i, X^n_{i+1}),$$

and linear on the intervals $\left[\frac{2i}{2n}, \frac{2i+1}{2n}\right]$ and $\left[\frac{2i+1}{2n}, \frac{2i+2}{2n}\right]$ for all $i = 0, \ldots, n-1$. Note that this is a variant of the Hoff process considered in [9].

Denote by $\tilde{\mathbf{X}}^n_{s,t} = \exp(\tilde{X}^n_{s,t} + \mathbb{A}^n_{s,t})$ the level-2 lift of $\tilde{X}^n$, where $\mathbb{A}^n_{s,t}$ is the $(2d) \times (2d)$ anti-symmetric Lévy area matrix given by

$$\mathbb{A}^n_{s,t} = \frac{1}{2}\left(\int_s^t \tilde{X}^n_{s,r} \otimes d\tilde{X}^n_r - \int_s^t \tilde{X}^n_{s,r} \otimes d\tilde{X}^n_r\right).$$

Let $H \in (0,1)$ and consider a fractional Brownian motion B^H with covariance $R(s,t) = \frac{1}{2}(t^{2H} + s^{2H} - |t-s|^{2H})$. Let $X : [0,1] \mapsto \mathbb{R}^d$ be d independent copies of B^H.

Recall the definition of T_v from (1). We are interested in the convergence in rough path topologies of $T_{\tilde{v}^n}(\tilde{\mathbf{X}}^n)$ where $\tilde{v}^n \in \mathfrak{g}^{(2)}(\mathbb{R}^{2d})$ is appropriately chosen. Define the (diagonal) $d \times d$ matrix

$$v^n = \frac{1}{2}\mathbb{E}\left[\sum_{k=0}^{n-1}(X^n_{k+1} - X^n_k) \otimes (X^n_{k+1} - X^n_k)\right] = \frac{n^{1-2H}}{2}I,$$

and the anti-symmetric $(2d) \times (2d)$ matrix

$$\tilde{v}^n = \begin{pmatrix} 0 & -v^n \\ v^n & 0 \end{pmatrix} \in \mathfrak{g}^{(2)}(\mathbb{R}^{2d}).$$

Finally, consider the path $\tilde{X} = (X, X) : [0, 1] \mapsto \mathbb{R}^{2d}$, its canonically defined Lévy area $\mathbb{A}$ (which exists for $1/4 < H \leq 1$), and its level-2 lift $\tilde{\mathbf{X}} = \exp(\tilde{X} + \mathbb{A})$. The following is the main result of this subsection.

Theorem 5 *Suppose $1/4 < H \leq 1/2$. Then for all $\alpha \in [0, H)$ and $q < \infty$, it holds that $T_{\tilde{\nu}^n}(\tilde{\mathbf{X}}^n) \to \tilde{\mathbf{X}}$ in L^q and α-Hölder topology. More precisely, as $n \to \infty$, in L^q*

$$\sup_{s,t\in[0,T]} \frac{|\tilde{X}^n_{s,t} - \tilde{X}_{s,t}|}{|t-s|^\alpha} + \sup_{s,t\in[0,T]} \frac{|\mathbb{A}^n_{s,t} + (t-s)\tilde{\nu}^n - \mathbb{A}_{s,t}|}{|t-s|^{2\alpha}} \to 0.$$

The rest of the section is devoted to the proof of Theorem 5. We first state two lemmas which are purely deterministic.

Let $Y^n : [0, 1] \mapsto \mathbb{R}^d$ be the piecewise linear interpolation of X over the partition $\left(0, \frac{1}{n}, \ldots, \frac{n-1}{n}, 1\right)$, let $\tilde{Y}^n = (Y^n, Y^n) : [0, 1] \mapsto \mathbb{R}^{2d}$, and let $\mathbb{Y}^n$ be the Lévy area of $\tilde{Y}^n$.

Lemma 6 *Let $s \in [\frac{m}{n}, \frac{m+1}{n}]$ and $t \in [\frac{k}{n}, \frac{k+1}{n}]$ with $s < t$, and define*

$$\begin{aligned}
\Delta_1 &= n\left(\frac{m+1}{n} \wedge t - s\right)|X^n_{m+1} - X^n_m|,\\
\Delta_2 &= |X^n_k - X^n_{m+1}| \text{ if } k > m, \quad 0 \text{ if } k = m\\
\Delta_3 &= n\left(t - \frac{k}{n} \vee s\right)|X^n_{k+1} - X^n_k|.
\end{aligned}$$

There exists a constant $C_1 > 0$ such that for all $n \geq 1$ and $0 \leq s < t \leq 1$, it holds that

$$|\tilde{X}^n_{s,t} - \tilde{Y}^n_{s,t}| \leq C_1 (\Delta_1 + \Delta_3).$$

and, if $k > m$, then $|\mathbb{A}^n_{s,t} - \mathbb{A}^n_{\frac{m+1}{n},\frac{k}{n}}|$ and $|\mathbb{Y}^n_{s,t} - \mathbb{Y}^n_{\frac{m+1}{n},\frac{k}{n}}|$ are bounded above by

$$C_1 \left(\Delta_1^2 + (\Delta_1 + \Delta_2 + \Delta_3)\Delta_3 + \Delta_1\Delta_2\right).$$

Proof Direct calculation and triangle inequality.

The second part of the above lemma essentially allows us to work over the partition $\left(0, \frac{1}{n}, \ldots, \frac{n-1}{n}, 1\right)$, on which computations are easier.

Lemma 7 *Suppose $0 \leq m \leq k \leq n$.*

(1) For all pairs $1 \leq i, j \leq d$ and $d + 1 \leq i, j \leq 2d$

$$\left(\mathbb{A}^n_{\frac{m}{n},\frac{k}{n}}\right)^{i,j} = \left(\mathbb{Y}^n_{\frac{m}{n},\frac{k}{n}}\right)^{i,j}.$$

(2) For all $1 \le i \le d < j \le 2d$

$$\left(\mathbb{A}^n_{\frac{m}{n},\frac{k}{n}}\right)^{i,j} = \left(\mathbb{Y}^n_{\frac{m}{n},\frac{k}{n}}\right)^{i,j} - \frac{1}{2}\sum_{r=m}^{k-1}(X^{n,i}_{r+1} - X^{n,i}_r)(X^{n,j}_{r+1} - X^{n,j}_r)$$

Proof Denote $\tilde{X}^n = (M^n, N^n)$, so that M^n is the lag component, and N^n is the lead. The first equality is clear since M^n and N^n are simply reparametrisations of Y^n over the interval $[\frac{m}{n}, \frac{k}{n}]$. For the second, observe that

$$\int_{m/n}^{k/n} M^{n,i}_{m/n,r} dN^{n,j}_r = \sum_{r=m}^{k-1}(X^{n,i}_r - X^{n,i}_m)(X^{n,j}_{r+1} - X^{n,j}_r)$$

and

$$\int_{m/n}^{k/n} N^{n,j}_{m/n,r} dM^{n,i}_r = \sum_{r=m}^{k-1}(X^{n,j}_{r+1} - X^{n,j}_m)(X^{n,i}_{r+1} - X^{n,i}_r).$$

Remark now that the signature of Y^n over $[\frac{m}{n}, \frac{k}{n}]$ is

$$e^{X_{m+1}-X_m} \dots e^{X_k - X_{k-1}},$$

so that a calculation with the CBH formula gives

$$\left(\mathbb{Y}^n_{m/n,k/n}\right)^{i,j} = \frac{1}{2}\sum_{r=m}^{k-1}(X^{n,i}_r - X^{n,i}_m)(X^{n,j}_{r+1} - X^{n,j}_r) - (X^{n,j}_r - X^{n,j}_m)(X^{n,i}_{r+1} - X^{n,i}_r).$$

Using the fact that

$$\left(\mathbb{A}^n_{m/n,k/n}\right)^{i,j} = \frac{1}{2}\left(\int_{m/n}^{k/n} M^{n,i}_{m/n,r} dN^{n,j}_r - \int_{m/n}^{k/n} N^{n,j}_{m/n,r} dM^{n,i}_r\right),$$

the conclusion readily follows.

We now return to the specific case that $X : [0, 1] \mapsto \mathbb{R}^d$ is given by d-independent copies of a fractional Brownian motion with Hurst parameter $H \in (0, 1)$. In particular, this implies that for all $s, t, a, b \in [0, 1]$

$$\mathbb{E}\left[(X^i_t - X^i_s)(X^j_b - X^j_a)\right] = \delta_{i,j}\frac{1}{2}(|t-a|^{2H} + |s-b|^{2H} - |t-b|^{2H} - |s-a|^{2H}). \tag{7}$$

Consider the $(2d) \times (2d)$ anti-symmetric matrix

$$\mathbb{P}^n_{s,t} = \mathbb{A}^n_{s,t} - \mathbb{Y}^n_{s,t} + (t-s)\tilde{v}^n.$$

Lemma 8 *There exists $C_2 > 0$ such that for all $H \le 1/2$, $n \ge 1$, and $0 \le m \le k \le n$*

$$\left|\left|\mathbb{P}^n_{m/n,k/n}\right|\right|_{L^2} \le C_2 \frac{(k-m)^{1/2}}{n^{2H}}.$$

Proof Denote $K = k - m$. By part (2) of Lemma 7, we have

$$\left|\mathbb{P}^n_{m/n,k/n}\right| \le \sum_{i,j=1}^{d} \left| \sum_{r=m}^{k-1} (X^{n,i}_{r+1} - X^{n,i}_r)(X^{n,j}_{r+1} - X^{n,j}_r) - \frac{K}{n} v^n_{i,j} \right|.$$

Observe moreover that

$$\frac{K}{n} v^n_{i,j} = \mathbb{E}\left[\sum_{r=m}^{k-1} (X^{n,i}_{r+1} - X^{n,i}_r)(X^{n,j}_{r+1} - X^{n,j}_r) \right] = \delta_{i,j} K n^{-2H},$$

and that for all $r, \ell \in \{0, \ldots, n-1\}$

$$\mathbb{E}\left[(X^{n,i}_{r+1} - X^{n,i}_r)(X^{n,i}_{\ell+1} - X^{n,i}_\ell)\right] = \frac{n^{-2H}}{2}(|r-\ell+1|^{2H} + |r-\ell-1)|^{2H} - 2|r-\ell|^{2H}).$$

Then for all $i \ne j$, by independence of the components of X,

$$\begin{aligned}
&\mathbb{E}\left[\left(\sum_{r=m}^{k-1} (X^{n,i}_{r+1} - X^{n,i}_r)(X^{n,j}_{r+1} - X^{n,j}_r)\right)^2\right] \\
&= \mathbb{E}\left[\sum_{r=m}^{k-1}\sum_{\ell=m}^{k-1} X^i_{r,r+1} X^i_{\ell,\ell+1} X^j_{r,r+1} X^j_{\ell,\ell+1}\right] \\
&= \sum_{r=m}^{k-1}\sum_{\ell=m}^{k-1} \mathbb{E}\left[(X^{n,i}_{r+1} - X^{n,i}_r)(X^{n,i}_{\ell+1} - X^{n,i}_\ell)\right]^2 \\
&= \sum_{r=m}^{k-1}\sum_{\ell=m}^{k-1} \frac{n^{-4H}}{4}\left(|r-\ell+1|^{2H} + |r-\ell-1|^{2H} - 2|r-\ell|^{2H}\right)^2 \\
&= \frac{n^{-4H}}{4} \sum_{x=-K+1}^{K-1} (K - |x|)(|x+1|^{2H} + |x-1|^{2H} - 2x^{2H})^2 \\
&=: \psi(n, K).
\end{aligned}$$

Likewise for $i = j$, by Wick's formula,

$$\begin{aligned}\mathbb{E}\left[\left(\sum_{r=m}^{k-1}(X^{n,i}_{r+1} - X^{n,i}_r)^2\right)^2\right] &= \sum_{r=m}^{k-1}\sum_{\ell=m}^{k-1}\mathbb{E}\left[(X^{n,i}_{r+1} - X^{n,i}_r)^2\right]\mathbb{E}\left[(X^{n,i}_{\ell+1} - X^{n,i}_\ell)^2\right] \\ &\quad + 2\mathbb{E}\left[(X^{n,i}_{r+1} - X^{n,i}_r)(X^{n,i}_{\ell+1} - X^{n,i}_\ell)\right]^2 \\ &= K^2 n^{-4H} + 2\psi(n, K) \\ &= \left(\frac{K}{n}v^n_{i,i}\right)^2 + 2\psi(n, K).\end{aligned}$$

It hence follows that

$$||\mathbb{P}^n_{m/n,k/n}||^2_{L^2} \leq C_3\psi(n, K).$$

The conclusion now follows since one can readily show that there exists $C_4 > 0$ such that for all $H \leq 1/2$, $n \geq 1$, and $0 \leq K \leq n$

$$\psi(n, K) \leq C_4 K n^{-4H}$$

(in fact the inequality holds for all $H < 3/4$, though with a constant in general depending on H).

Lemma 9 *There exists $C_5 > 0$ such that for all $H \in \left(\frac{1}{4}, \frac{1}{2}\right]$, $n \geq 1$ and $0 \leq s < t \leq 1$*

$$||\mathbb{P}^n_{s,t}||_{L^2} \leq C_5|t - s|^{2H}.$$

Proof Suppose $s \in [\frac{m}{n}, \frac{m+1}{n}]$ and $t \in [\frac{k}{n}, \frac{k+1}{n}]$. If $m = k$, then $\mathbb{A}^n_{s,t} = \mathbb{Y}^n_{s,t} = 0$ and $|t - s| < n^{-1}$, so that

$$|\mathbb{P}^n_{s,t}| = (t - s)\,|\tilde{v}^n| \leq |t - s|^{2H}.$$

For the case $k > m$, following the notation of Lemma 6, note that $\mathbb{E}\left[\Delta_1^2\right]$ and $\mathbb{E}\left[\Delta_3^2\right]$ are bounded above by

$$n^{-2H}\min\{n^2|t - s|^2, 1\}.$$

It readily follows that for all $\ell \in \{1, 2, 3\}$

$$\mathbb{E}\left[\Delta_\ell^2\right] \leq |t - s|^{2H}.$$

Moreover, we have

$$\left(t - \frac{k}{n} + \frac{m+1}{n} - s\right)|\tilde{v}^n| \leq \min\{|t - s|, n^{-1}\}n^{1-2H} \leq |t - s|^{2H}.$$

Hence Lemma 6 implies that

$$|\mathbb{P}^n_{s,t} - \mathbb{P}^n_{(m+1)/n,k/n}| \leq 2C_1\left(\Delta_1^2 + (\Delta_1 + \Delta_2 + \Delta_3)\Delta_3 + \Delta_1\Delta_2\right) \\ + \left(t - \frac{k}{n} + \frac{m+1}{n} - s\right)|\tilde{v}^n|,$$

and so by Gaussian chaos

$$\left|\left|\mathbb{P}^n_{s,t} - \mathbb{P}^n_{(m+1)/n,k/n}\right|\right|_{L^2} \leq C_6|t-s|^{2H}.$$

The conclusion now follows from Lemma 8 since $(k-m-1)^{1/2}n^{-2H} \leq |t-s|^{2H}$ for all $H \geq 1/4$.

Proof (*Proof of Theorem* 5) Let $0 \leq s < t \leq 1$. We observe that as $n \to \infty$, it readily follows from Lemmas 6 and 8 that in L^q

$$|\tilde{X}^n_{s,t} - \tilde{Y}^n_{s,t}| \to 0$$

and

$$|\mathbb{A}^n_{s,t} + (t-s)\tilde{v}^n - \mathbb{Y}^n_{s,t}| \to 0.$$

Furthermore, by Gaussian chaos, Lemma 6 implies

$$\sup_{n\geq 1}\mathbb{E}\left[|\tilde{X}^n_{s,t} - \tilde{Y}^n_{s,t}|^q\right] \leq C_q|t-s|^{qH},$$

while Lemma 9 implies

$$\sup_{n\geq 1}\mathbb{E}\left[|\mathbb{A}^n_{s,t} + (t-s)\tilde{v}^n - \mathbb{Y}^n_{s,t}|^q\right] \leq C_q|t-s|^{2qH}.$$

Applying [7, Theorem A.13], and the fact that $\exp(\tilde{Y}^n + \mathbb{Y}^n) \to \tilde{\mathbf{X}}$ in α-Hölder topology in L^q ([7, Theorem 15.42]), completes the proof.

Acknowledgements P.K.F. is partially supported by the European Research Council through CoG-683164 and DFG research unit FOR2402. I.C., affiliated to TU Berlin when this project was commenced, was supported by DFG research unit FOR2402, and is currently supported by a Junior Research Fellowship of St John's College, Oxford. Y.B. thanks Martin Hairer for financial support from his Leverhulme Trust award.

References

1. Hairer, M.: A theory of regularity structures. Invent. Math. **198**(2), 269–504 (2014). https://doi.org/10.1007/s00222-014-0505-4
2. Bruned, Y., Hairer, M., Zambotti, L.: Algebraic renormalisation of regularity structures. ArXiv e-prints (2016)
3. Hairer, M.: The motion of a random string. ArXiv e-prints (2016)
4. Gubinelli, M., Imkeller, P., Perkowski, N.: Paracontrolled distributions and singular PDEs. Forum Math. Pi **3**, e6, 75 (2015). https://doi.org/10.1017/fmp.2015.2
5. Kupiainen, A.: Renormalization group and stochastic PDEs. Ann. Henri Poincaré **17**(3), 497–535 (2016). https://doi.org/10.1007/s00023-015-0408-y
6. Lyons, T.J.: Differential equations driven by rough signals. Rev. Mat. Iberoam. **14**(2), 215–310 (1998). https://doi.org/10.4171/RMI/240
7. Friz, P.K., Victoir, N.B.: Cambridge studies in advanced mathematics. In: Multidimensional stochastic processes as rough paths, vol. 120. Cambridge University Press, Cambridge (2010)
8. Friz, P., Gassiat, P., Lyons, T.: Physical Brownian motion in a magnetic field as a rough path. Trans. Am. Math. Soc. **367**(11), 7939–7955 (2015). https://doi.org/10.1090/S0002-9947-2015-06272-2
9. Flint, G., Hambly, B., Lyons, T.: Discretely sampled signals and the rough Hoff process. Stoch. Process. Appl. **126**(9), 2593–2614 (2016). https://doi.org/10.1016/j.spa.2016.02.011
10. Cannizzaro, G., Friz, P.K., Gassiat, P.: Malliavin calculus for regularity structures: the case of gPAM. J. Funct. Anal. **272**(1), 363–419 (2017). https://doi.org/10.1016/j.jfa.2016.09.024
11. Bruned, Y., Chevyrev, I., Friz, P.K., Preiss, R.: A rough path perspective on renormalization. ArXiv e-prints (2017)

Generalised Weitzenböck Formulae for Differential Operators in Hörmander Form

K. D. Elworthy

Abstract The decomposition of a class of diffusion processes, due to Elworthy–LeJan–Li is described. In particular it applies to processes such as derivative processes coming from stochastic flows. How this decomposition leads automatically to Weitzenböck type formula for related operators acting on sections of associated vector bundles is described in detail clarifying the difference between the action of flows on vector fields and on forms noted recently by Shizan Fang and Dejun Luo. Remarks are made on the possible application to higher order derivative formulae and estimates for heat semigroups, and also to certain diffusions with sub-Riemannian generators using Baudoin's generalised Levi-Civita semi-connection.

Keywords Stochastic analysis · Stochastic flows · Weitzenböck formulae · Diffusion of tensors · Degenerate diffusions · Semi-group domination

Mathematics Subject Classification 58J65 (60G35 60H30 60J60 93E11 53C17)

1 Introduction

It was noted by Fang and Luo, [10], that certain sums of squares of vector fields which give the Hodge–Kodaira Laplacian on differential forms do not do so when applied to vector fields. Here we consider the more general situation of equivariant diffusions on principal bundles and the generalised Weitzenböck formulae they give for the operators they induce on sections of associated vector bundles. This is taken from [8, 9] where more details can be found, but here we clarify the relationship between Weitzenböck formulae for sections of an associated bundle and those of its dual bundle, and correct some formulae in [8, 9].

The resulting formulae for the semi-groups on spaces of sections determined by these operators can theoretically be applied to sections of natural bundles, in

K. D. Elworthy (✉)
Mathematics Institute, University of Warwick, Coventry CV4 7AL, England
e-mail: kde@maths.warwick.ac.uk

A. Eberle et al. (eds.), *Stochastic Partial Differential Equations and Related Fields*, Springer Proceedings in Mathematics & Statistics 229,
https://doi.org/10.1007/978-3-319-74929-7_20

particular to jet spaces to give higher order derivative estimates for diffusion semi-groups, but that has had to be left to a future work.

The usefulness of these representations of the semi-groups depends on the existence of suitable metric semi-connections. In Sect. 5 we note the relevance of Baudoin's generalised Levi-Civita semi-connections, [1], to give immediate applicability of this theory to the sub-elliptic Laplacians transverse to Riemannian foliations with totally geodesic leaves.

2 Diffusion Operators

Let M be a smooth n-dimensional manifold. By a diffusion operator $\mathscr{L}$ on M we will mean a second order differential operator acting on C^2 functions on M which vanishes on constants and is semi-elliptic. The latter concept can be defined in terms of the symbol $\sigma^{\mathscr{L}} : T^*_x M \to TM$ which is defined via $\sigma^{\mathscr{L}}_x : T^*M \to T_x M$ at $x \in M$ by

$$(df)_x \sigma^{\mathscr{L}}_x (dg)_x = \frac{1}{2}\{\mathscr{L}(fg)(x) - f(x)\mathscr{L}(g)(x) - g(x)\mathscr{L}(f)(x)\} \text{ for } f : M \to \mathbf{R}. \tag{1}$$

By definition $\mathscr{L}$ is semi-elliptic if $\sigma^{\mathscr{L}}_x$ is positive semi-definite; it is automatically symmetric. We will always assume for simplicity that the coefficients in local co-ordinates are smooth, and so therefore is the symbol.

If there is a right action $M \times G \to M$ of a group G on M, $(x, g) \mapsto \mathscr{R}_g(x)$ with $\mathscr{R}_g : M \to M$ smooth then $\mathscr{L}$ is said to be *equivariant* if it is invariant under the group action:

$$\mathscr{L}(f \circ \mathscr{R}_g) = \mathscr{L}(f) \circ \mathscr{R}_g \quad \text{for all} \quad g \in G \quad \text{and smooth} \quad f : M \to \mathbf{R}.$$

For each $x \in M$ let $E_x \subset T_x M$ be the image of $\sigma^{\mathscr{L}}$. We will often assume that the disjoint union $E := \bigcup_{x \in M} E_x$ forms a smooth sub-bundle of TM. This is implied by the symbol having constant rank and in turn implies that $\mathscr{L}$ can be written in Hörmander form:

$$\mathscr{L} = \frac{1}{2}\sum_{j=1}^{m} \mathscr{L}_{X^j}\mathscr{L}_{X^j} + \mathscr{L}_A. \tag{2}$$

for some m where $X^1, \ldots, X^m, A$ are smooth vector fields on M and $\mathscr{L}_Z$ denotes Lie differentiation in the direction of a vector field Z.

This is equivalent to the existence of a smooth $X : M \times \mathbf{R}^m \to TM$ with each $X(x) : \mathbf{R}^m \to T_x M$ linear and such that $2\sigma^{\mathscr{L}}_x = X(x)X(x)^*$, as can be seen by taking an orthonormal base $e_1, \ldots, e_m$ for $\mathbf{R}^m$ and setting $X^j(x) = X(x)(e_j)$.

Using the action of Lie differentiation on tensors a choice of Hörmander form determines differential operators on vector fields and differential forms.

Since $\mathscr{L}$ has no zero order term it can be factorised as $\delta^{\mathscr{L}} d$ where $\delta^{\mathscr{L}}$ is a smooth first order differential operator from 1-forms to functions. In case $\mathscr{L}$ has the Hörmander form above we see $\delta^{\mathscr{L}}(\phi)(x) = \sum_{j=1}^{m} \mathscr{L}_{X^j}(\phi(X^j(-))(x) + \phi(A(x))$.

By a *distribution* S on M we will mean a union $S = \bigcup_{x\in M} S_x$ with each S_x a closed linear subspace of $T_x M$.

We will say that $\mathscr{L}$ is *along* S if $\delta^{\mathscr{L}} \phi = 0$ for any C^1 one-form ϕ which vanishes on S. If so then all the vector fields in any Hörmander form representation of $\mathscr{L}$ take values in S provided S is a sub-bundle of TM, and somewhat more generally. We say $\mathscr{L}$ is *cohesive* if the image E of its symbol is a sub-bundle and if it is along E. This holds if and only if it has a Hörmander form all of whose vector fields take values in E.

A Hörmander form representation (2) for $\mathscr{L}$ corresponds to choosing an SDE

$$dx_t = A(x_t)\, dt + X(x_t) \circ dB_t \tag{3}$$

on M with X and A as above and $\{B_t\}_t$ a Brownian motion on $\mathbf{R}^m$. The solutions of this SDE then give an $\mathscr{L}$-diffusion in the sense of [12]. Conversely such an SDE determines a Hörmander form representation.

3 Skew Product Decomposition of Equivariant Diffusions

3.1 Horizontal Lifts, Semi-connections, and the Decomposition

Consider a smooth submersion $\pi : P \to M$ where M is a connected n–dimensional manifold, for example $\mathbf{R}^n$, and P is a smooth manifold. Suppose we have a diffusion operator $\mathscr{B}$ on P *lying over* a diffusion operator $\mathscr{A}$ on M that is

$$\mathscr{B}(f \circ \pi) = \mathscr{A}(f) \circ \pi \quad \text{for any } C^2 \text{ map } f : M \to \mathbf{R}.$$

Assume $\mathscr{A}$ is cohesive.

Theorem 1 ([9]) *There exists a unique, linear, horizontal lift map $h_u : E_{\pi(u)} \to T_u P$ for each $u \in P$, smooth in u, with $T\pi \circ h_u(v) = v$ for $v \in E_{\pi(u)}$ and characterised by*

$$h_u \circ \sigma^{\mathscr{A}}_{\pi(u)} = \sigma^{\mathscr{B}}_u \circ (T_u \pi)^*.$$

If $\mathscr{B}$ is equivariant for a right action of a group G on P then h_- is also equivariant:

$$h_{u.g} = T_u \mathscr{R}_g \circ h_u.$$

There is a unique decomposition of $\mathcal{B}$ into the sum of two diffusion operators:

$$\mathcal{B} = \mathcal{A}^H + \mathcal{B}^V$$

where $\mathcal{B}^V$ is along the vertical tangent bundle $VTP := \ker T\pi$ and $\mathcal{A}^H$ is along the horizontal bundle H, the image of h_- and lies over $\mathcal{A}$.

In the equivariant case both components of B are equivariant.

A horizontal lift map such as h will be called a *semi-connection* over E. In the equivariant case it may be called a *principal semi-connection* over E. This corresponds to the usual notion for (non-linear) connections, e.g. as in [14].

Note that any vector field Z on M which takes values in E, an *E-vector field*, has a horizontal lift

$$u \mapsto Z^H(u) := h_u(Z(u))$$

to an H-vector field on P. If $\mathcal{A}$ has Hörmander form $\frac{1}{2}\sum_{j=1}^m \mathcal{L}_{X^j}\mathcal{L}_{X^j} + \mathcal{L}_A$ then

$$\mathcal{A}^H = \frac{1}{2}\sum_{j=1}^m \mathcal{L}_{(X^j)^H}\mathcal{L}_{(X^j)^H} + \mathcal{L}_{A^H}.$$

This lifted Hörmander form in turn gives an SDE on P whose solutions are $\mathcal{A}^H$-diffusions which project down to the $\mathcal{A}$-diffusion on M given by the Hörmander form we lifted. In general the lifetime of the lift may be less than that of the $\mathcal{A}$-diffusion but it will be the same in the equivariant case by a uniform cover argument, see [3].

In the equivariant case there is the skew-product decomposition for a $\mathcal{B}$-diffusion $u : [0, \zeta) \times \Omega \to P$:

$$u_t = \tilde{x}_t . g_t^{\tilde{x}.} \quad t < \zeta \tag{4}$$

where $\{\tilde{x}_t\}_t$ is the horizontal lift of an $\mathcal{A}$-diffusion $\{x_t\}_t$ on M and $\{g_t^{\tilde{x}.}\}_t$ a continuous semi-martingale on G.

3.2 The Vertical Component $\mathcal{B}^V$

Suppose now that $\pi : P \to M$ is a principal bundle with group G. Let $\mathfrak{g}$ be the Lie algebra of G, and $A_1, \ldots, A_k$ a basis of $\mathfrak{g}$. There are the corresponding fundamental vector fields $A_1^*, \ldots, A_k^*$ on P which trivialise the vertical tangent bundle of P.

A principal semi-connection on P determines and is determined by a *semi-connection one-form* ϖ which for each $u \in P$ gives a linear map

$$\varpi_u : H_u + V_uTP \to \mathfrak{g} \quad \text{s.t. } \varpi_u(A_j^*) = A_j, \ \ j = 1, \ldots, k$$

with $\ker \varpi_u = H_u$.

Theorem 2 *Let $\mathscr{B} = \mathscr{A}^H + \mathscr{B}^V$ be the decomposition of an equivariant $\mathscr{B}$ as above. Then the vertical component has a unique representation*

$$\mathscr{B}^V = \sum \alpha^{i,j} \mathscr{L}_{A_i^*} \mathscr{L}_{A_j^*} + \sum \beta^j \mathscr{L}_{A_j^*} \tag{5}$$

where $\alpha^{i,j}$ and β^j are C^∞ functions on P for $i, j = 1, \ldots, k$.

These functions are given by:

$$\alpha^{i,j} = \varpi^i(\sigma^{\mathscr{B}}(\varpi^j)\,, \quad \beta^j = \delta^{\mathscr{B}}(\varpi^j) \tag{6}$$

where ϖ^ℓ refers to the ℓ'th component of ϖ, extended arbitrarily from $H + VTP$ to TP.

If $\alpha : P \to \mathfrak{g} \otimes \mathfrak{g}$ and $\beta : P \to \mathfrak{g}$ are defined by

$$\alpha(u) = \sum \alpha^{i,j}(u) A_i \otimes A_j \quad \textit{and} \quad \beta(u) = \sum \beta^j A_j$$

then for $g \in G$

$$\alpha(u.g) = \mathrm{ad}(g)^{-1} \otimes \mathrm{ad}(g)^{-1} \alpha(u) \quad \textit{and} \quad \beta(u.g) = \mathrm{ad}(g)^{-1} \beta(u). \tag{7}$$

Equation (7) corrects a misprint on p. 38 of [8] and in Theorem 3.2.1 of [9].

4 Diffusions, and Connections on Associated Bundles: Generalised Weitzenböck Formulae

4.1 Associated Bundles

Let $\pi : P \to M$ be a smooth principal bundle with group a Lie group G. This means there is a right action by G on P as above, the orbits of which are the fibres of π, and which is locally equivariantly trivial, i.e. locally a product G-bundle. Suppose also we have a smooth left action of G on a manifold V, acting by diffeomorphisms, giving a homomorphism $\rho : G \to Diff(V)$. There is then the (possibly weakly) associated bundle $\pi^\rho : F \to M$ where $F = P \times V / \sim$ for the equivalence relation defined by $(u, a) \sim (u.g, \rho(g^{-1})(a)$. Each $u \in P$ determines a diffeomorphism $\bar{u} : V \to F_{\pi(u)}$ by $\bar{u}(a) = [(u, a)]$. If u_1 and u_2 lie in the same fibre of P then $\bar{u_2}^{-1}\bar{u_1} : V \to V$ corresponds to the action of some element of G. Consequently any geometric structure on V preserved by G can be transferred to each fibre of π^ρ. We can consider P as a frame bundle for π^ρ.

Note that a continuous semi-martingale $\{u_t\}_t$ on P together with an element $a \in V$ determines a continuous semi-martingale $\{\bar{u}_t(a)\}_t$ on F. By itself $\{u_t\}_t$ determines diffeomorphisms $\bar{u_t}\bar{u_0}^{-1} : F_{\pi(u_0)} \to F_{\pi(u_t)}$, preserving any structure induced on the fibres. If $\{u_t\}_t$ is a horizontal lift of a diffusion $\{x_t\}_t$ on M via a principal semi-connection, over a sub-bundle E of TM, on P then we write $\bar{u_t}\bar{u_0}^{-1}$ as $\hat{//}_t$. It is parallel translation along $\{x_t\}_t$ for our semi-connection.

When V is a vector space and ρ is a linear representation then $\pi^\rho : F \to M$ is a vector bundle. Using our semi-connection over E any C^1 section $V : M \to F$ of π^ρ can be differentiated in E-directions to give $\hat{\nabla}_u V \in F_z$ for $u \in E_z$. One way to do this is to take a smooth vector field U with values in E such that U(z) = u. Let $\{\eta_t(z)\}_t$ be its integral curve from z, with $\hat{//}_t$ parallel translation along it. Then

$$\hat{\nabla}_u V = \frac{d}{dt}\left(\hat{//}_t^{-1} V(\eta_t(z))\right)|_{t=0}. \tag{8}$$

Similarly if $v : [0, T] \to TM$ is a C^1-vector field along a C^1-curve $\alpha : [0, T] \to M$ so $v(t) \in T_{\alpha(t)}M$ for each t and if $\dot{\alpha}(t) \in E$ for each t we can define the vector field $\frac{\hat{D}}{dt}v$ along α by

$$\frac{\hat{D}}{dt}v := \hat{//}_t \frac{d}{dt} \hat{//}_t^{-1} v(t) \tag{9}$$

where now $\hat{//}_t$ refers to parallel translation along α. Standard stochastic analytic techniques allow parallel translation along continuous semi-martingales α "along E" see: [9] Sect. 9.3. We can therefore define Itô or Stratonovich stochastic derivatives for continuous semi-martingales v along such α in a similar fashion to (9) and denoted by $\hat{D}v_t$.

4.2 Induced Operators on Sections of Associated Bundles

Let $\mathscr{B}$ be an equivariant diffusion operator on P lying over a coherent $\mathscr{A}$.

For P and F as before, the pull back of F over P has a natural trivialisation. On sections of F this is revealed by setting

$$M_\rho(P : V) = \{Z \in C^\infty(P : V) \text{ s.t. } Z(u.g) = \rho(g)^{-1}Z(u) \text{ for } u \in P,\ g \in G\}$$

and defining the bijection $\mathfrak{F}^\rho : M^\rho(P; V) \to C^\infty \Gamma F$, onto the space of C^∞ sections of F by

$$\mathfrak{F}^\rho(Z)(x) = \bar{u}Z(u).$$

Our equivariant $\mathscr{B}$ on P will then define a differential operator $\mathfrak{F}^\rho(\mathscr{B})$ on smooth sections of F by

$$\mathfrak{F}^\rho(\mathscr{B})(\mathfrak{F}^\rho(Z)) = \mathfrak{F}^\rho(\mathscr{B}(Z)) \qquad Z \in M^\rho(P; V) \tag{10}$$

provided we have a G-invariant way to extend $\mathscr{B}$ to act on V-valued functions on P. For V a vector space and ρ a linear representation of G we can take a basis of V and let $\mathscr{B}$ act co-ordinate wise. We will restrict ourselves here to the linear situation, although for completeness we should consider more general jet bundles, and these may not be vector bundles.

Now let $\mathscr{L}$ be an arbitrary equivariant vertical diffusion operator on P. The following is the underlying reason for the existence of Weitzenböck type formulae. As we see later the map λ^ρ can be considered to represent a generalised Weitzenböck term.

Theorem 3 *For any associated vector bundle $\pi^\rho : F \to M$ the induced operator $\mathfrak{F}^\rho(\mathscr{L})$ on sections of F is order zero. It is determined by a smooth section Λ^ρ of $\mathbf{L}(F; F)$, giving $\Lambda^\rho_x \in \mathbf{L}(F_x; F_x)$ for each $x \in M$. In particular there is a smooth equivariant $\lambda^\rho : P \to \mathbf{L}(V; V)$ with*

$$\Lambda^\rho_x[(u, a)] = [(u, \lambda^\rho(u)(a))] \quad u \in P_x \;\; a \in V. \tag{11}$$

If $\mathscr{L}$ has the form as in Eq. (5)

$$\mathscr{L} = \sum \alpha^{i,j} \mathscr{L}_{A_i^*} \mathscr{L}_{A_j^*} + \sum \beta^j \mathscr{L}_{A_j^*}$$

then

$$\lambda^\rho(u) = Comp\ (\rho_* \otimes \rho_*)\alpha(u) - \rho_*(\beta(u)) \tag{12}$$

where $\rho_* : \mathfrak{g} \to \mathbf{L}(V; V)$ is the induced representation of Lie algebras, and

$$Comp : \mathbf{L}(V; V) \otimes \mathbf{L}(V; V) \to \mathbf{L}(V; V)$$

is the composition map.

The negative sign in Eq. (12) corrects a misprint in Eq. (3.19) of [9].

To see why formula (12) holds consider $Z \in M^\rho(P; V)$. Fix $u_0 \in P$. Then

$$\begin{aligned}\mathscr{L}_{A_j^*} Z(u_0) = \frac{d}{ds}\{Z(u_0.\exp sA_j)\}|_{s=0} &= \frac{d}{ds}\{\rho(\exp sA_j)^{-1}(Z(u_0)\}|_{s=0} \\ &= -\rho_*(A_j)Z(u_0).\end{aligned}$$

Iterating this and applying formula (5) gives (12).

Stimulated by the discussion in [10] we next consider the relationship between these generalised Weitzenböck terms on vector bundles and on their dual bundles. For our representation $\rho : G \to GL(V; V)$ as before consider the adjoint representation $\rho^t : G \to GL(V^*)$ induced by it:

$$\rho^t(g)(\ell) = \ell \circ \rho(g^{-1}) = \rho(g^{-1})^*(\ell). \tag{13}$$

There are also the induced maps of Lie algebras $\rho_* : \mathfrak{g} \to \mathbf{L}(V; V)$ and $\rho^t_* : \mathfrak{g} \to \mathbf{L}(V^*; V^*)$ with, for $a \in \mathfrak{g}$ and $\ell \in V^*$,

$$\rho^t_*(a)(\ell) = \frac{d}{ds}\rho^t(\exp(sa))(\ell)|_{s=0} = \frac{d}{ds}\rho(\exp(-sa))^*(\ell)|_{s=0} = -(\rho_*(a))^*(\ell).$$

We have the zero order operators λ^ρ given by Eq. (12) and

$$\lambda^{\rho^t} : B \to \mathbf{L}(V^*; V^*)$$

which, using the symmetry of α, is given by

$$\begin{aligned}\lambda^{\rho^t(u)}(\ell) &= Comp\left((\rho^t_* \otimes \rho^t_*)\alpha(u)\right)(\ell) - \rho^t_*(\beta(u))(\ell)\\ &= \ell \circ \{Comp\,((\rho_* \otimes \rho_*)\alpha(u)) + \rho_*(\beta(u))\}.\end{aligned}$$

Set

$$\tilde{\lambda}^\rho(u) := Comp\,((\rho_* \otimes \rho_*)\alpha(u)) + \rho_*(\beta(u)).$$

Then we have shown

Lemma 1 *For all $\ell \in V^*$ and $u \in P$*

$$\lambda^{\rho^t}(u)(\ell) = \ell \circ \tilde{\lambda}^\rho(u). \tag{14}$$

4.3 Conditioning and the Semigroup on Sections of F^*

Continuing with the same notation and conditions, under suitable bounds on λ^ρ, our equivariant diffusion operator $\mathscr{B}$ on P together with the representation ρ, determines a semi-group $\{P^{\rho^t}_t\}_t$ on some space of measurable sections of F^*. Indeed for $x_0 \in M$ choose $u_0 \in P_{x_0}$ and let $\{u_t\}_t$ be the $\mathscr{B}$-diffusion from u_0. Let $\{v_t\}_t$ be the process induced on F from $[(u_0, a)]$ some $a \in V$,

$$v_t = [(u_t, a)].$$

For ϕ a section of F^*, so $\phi_y \in F^*_y$ for $y \in M$, define

$$P^{\rho^t}_t(\phi)_{x_0}([(u_0, a)] = \mathbf{E}\phi_{x_t}(v_t), \tag{15}$$

if the expectation is defined. Note that the process $\{v_t\}$ is Markovian and depends linearly on $[(u_0, a)] \in F_{x_0}$, so that given suitable bounds we can expect to get semi-groups.

Now take the decomposition $u_t = \tilde{x}_t.g_t^{\tilde{x}.}$ as in Eq. (4). Then

$$v_t = [(\tilde{x}_t.g_t^{\tilde{x}.}, a)] = [(\tilde{x}_t, \rho(g_t^{\tilde{x}.})^{-1}a)].$$

As described in [7] we can take the conditioned process $\mathbf{E}\{v_t|\mathscr{F}_t^{x_0}\}$ where $\{\mathscr{F}_t^{x_0}\}_t$ is the filtration determined by the $\mathscr{A}$-diffusion $x_.$ on M. We see $\mathbf{E}\{v_t|\mathscr{F}_t^{x_0}\} = [(\tilde{x}_t, \mathbf{E}\{\rho(g_t^{\sigma})^{-1}a\})]$ at $\sigma = x_.$ since for any continuous path σ in P the process $g_.^{\sigma}$ is independent of $\mathscr{F}_.^{x_0}$

Now $\{g_t^{\sigma}\}_t$ is a time inhomogeneous diffusion on G with generator $\mathscr{L}_t^{\sigma}$ given at $g \in G$ in terms of the representation (5) of B^V by

$$\sum \alpha^{i,j}(\sigma(t).g)\mathscr{L}_{A_i^*}\mathscr{L}_{A_j^*} + \sum \beta^j(\sigma(t).g)\mathscr{L}_{A_j^*}.$$

It follows, as in the discussion (13), that for $\ell \in V^*$

$$\frac{d}{dt}\mathbf{E}\{\ell(\rho(g_t^{\sigma})^{-1}a)\} = \frac{d}{dt}\mathbf{E}\{(\rho'(g_t^{\sigma})(\ell)\}(a) = \lambda^{\rho'}(\sigma(t))(\mathbf{E}\{\rho'(g_t^{\sigma})(\ell)\})(a).$$

Using (14) this gives

$$\frac{d}{dt}\mathbf{E}\{(\rho(g_t^{\sigma})^{-1}a)\} = \tilde{\lambda}^{\rho}(\sigma(t))\mathbf{E}\{(\rho(g_t^{\sigma})^{-1}a)\}. \tag{16}$$

Define $\tilde{\Lambda}_y^{\rho} \in \mathbf{L}(F_y; F_y)$ for $y \in M$ by $\tilde{\Lambda}_y^{\rho}([(u, a)]) = [(u, \tilde{\lambda}^{\rho}(u)a)]$ for $u \in P_y, a \in V$. We now have:

Proposition 1 *Let $\bar{v}_t = \mathbf{E}\{v_t|\mathscr{F}_t^{x_0}\}$. Then $\bar{v}_t$ satisfies the ordinary differential equation along $\{x_t\}_t$:*

$$\frac{\hat{D}}{dt}\bar{v}_t = \tilde{\Lambda}_{(x_t)}^{\rho}\bar{v}_t.$$

This leads to the semi-group domination result in this context. For this we have to assume that our semi-connection is metric for some Riemannian metric on F. This is equivalent to saying that we have a smooth family of inner products $\{\langle -, -\rangle_y\}_{y\in M}$ on the fibres of F such that, for parallel translation along any suitable path α in M, $\hat{//}_t : F_{\alpha(0)} \to F_{\alpha(t)}$ is isometric.

Theorem 4 *If the semi-connection determined by $\mathscr{B}$ is metric for some metric on F suppose also that, for that metric, we have a bounded continuous $c : M \to \mathbf{R}$ with*

$$\langle v, \tilde{\Lambda}_y^{\rho}(v)\rangle_y \le c(y)||v||_y^2 \quad y \in M, v \in F_y. \tag{17}$$

Let $P_t^{\mathscr{A}+c}, t \ge 0$ be the Feynman–Kac semi-group of $\mathscr{A} + c$, and let $\bar{c} = \sup_y c(y)$. Then for a bounded continuous section ϕ of F^*

$$||P_t^{\rho'}(\phi)||_{x_0} \le P_t^{\mathscr{A}+c}(||\phi.||) \le e^{\bar{c}t}P_t^{\mathscr{A}}(||\phi||.). \tag{18}$$

Proof By definition, for $v_0 = [(u_0, a)] \in F_{x_0}$,

$$|P_t^{\rho'}(\phi)(v_0)| = |\mathbf{E}\phi_{x_t}(v_t)| = |\mathbf{E}\phi_{x_t}(\bar{v}_t)|$$
$$\leq \mathbf{E}\{||\phi||(x_t)||\bar{v}_t||\} \leq \mathbf{E}\{e^{\int_0^t c(x_s)\,ds}||\phi||(x_t)\}|v_0|_{x_0}.$$

The requirement that the semi-connection is metric for some metric on F is serious, e.g. see [7] Example 2C p. 24, and Theorem 5.07.

Let $\mathscr{A} = \frac{1}{2}\sum_{j=1}^m \mathscr{L}_{X^j}\mathscr{L}_{X^j} + \mathscr{L}_A$ be a Hörmander form representation of $\mathscr{A}$. In the decomposition $\mathscr{B} = \mathscr{A}^H + \mathscr{B}^V$, by considering the action of $\mathscr{A}^H$ on $M_{\rho'}(P; V^*)$, and the definition, Eq. (8), of covariant differentiation, we see that the operator $\mathscr{A}^{\rho'}$ induced by $\mathscr{A}^H$ on sections of F^* is given by

$$\mathscr{A}^{\rho'} = \frac{1}{2}\sum_{j=1}^m \hat{\nabla}_{X^j}\hat{\nabla}_{X^j} + \hat{\nabla}_A. \tag{19}$$

The generator $\mathscr{B}^{\rho'}$ of the semigroup $\{P_t^{\rho'}\}_t$ is then given by

$$\mathscr{B}^{\rho'}(\phi) = \mathscr{A}^{\rho'}(\phi) + \phi \circ \tilde{\Lambda}^{\rho}. \tag{20}$$

4.4 The Classical Case: Hodge–Kodaira Operators

Now take P to be the frame bundle $GL(M)$ of TM. It has group $GL(n)$. For our main set of examples suppose we have the SDE (3) for our cohesive $\mathscr{A}$ on M, and let $\{\xi_t\}_t$ denote its solution flow, possibly only partially defined. For a frame $u_0 : \mathbf{R}^n \to T_{x_0}M$ there is a stochastic process $\{T\xi_t\}_t$ on $GL(M)$ starting from u_0 determining a $\mathscr{B}$-diffusion for a certain equivariant operator $\mathscr{B} = \delta\mathscr{A}$ on $GL(M)$ over $\mathscr{A}$. This $\mathscr{B}$ exists without a strong completeness assumption. Using this we obtain a principal semi-connection on $GL(M)$ over E, and so semi-connections on TM and the tensor bundles of M, with covariant derivative to be denoted by $\hat{\nabla}$.

As in Sect. 4.2 we have operators on sections of the associated bundles to $GL(M)$, in particular on differential forms and on vector fields. These can be described in terms of the semi-connection and "Weitzenböck" term Λ^ρ by formulae (19) and (20) for forms and for vector fields by

$$\mathscr{B}^{\rho}(V) = \frac{1}{2}\sum_{j=1}^m \hat{\nabla}_{X^j}\hat{\nabla}_{X^j}V + \hat{\nabla}_A V + \Lambda^\rho(V(-)) \tag{21}$$

where ρ is the standard representation of $GL(n)$ on $\mathbf{R}^n$.

Alternatively they are given by the Hörmander form we have for $\mathscr{A}$ using the action of Lie derivatives on forms or vector fields.

Assume further that M is Riemannian and $\mathscr{A}$ the Laplace–Beltrami operator, so $E = TM$. Assume also that the connection $\hat{\nabla}$ is the Levi-Civita connection ∇ as will hold if we choose a gradient SDE for $\mathscr{A}$ with the drift term A being zero. The term $\sum_{j=1}^{m} \hat{\nabla}_{X^j}\hat{\nabla}_{X^j}$ becomes the trace of the second covariant derivative, the "flat Laplacian", and the zero order term $\tilde{\Lambda}^\rho$ is the negative of half the Weitzenböck curvature: see the Theorem 3.4.7 of [9] (where the calculations are for $\tilde{\lambda}^\rho$ not λ^ρ, so in 3.4.9 the operator is on forms not sections of $\wedge^* TM$). In particular our operator is the Hodge–Kodaira Laplacian on forms: a result going back to [4, 13]. However, as shown in [10], this is not in general true for the operator acting on vector fields. Indeed Eq. (12) gives

$$\lambda^\rho(u) = \text{Comp}\,(\rho_* \otimes \rho_*)\alpha(u) - \rho_*(\beta(u))$$

and from [9] Cor 3.4.9 we see that

$$\text{Comp}(\rho_* \otimes \rho_*)\alpha(u) = \frac{1}{2}\sum \left(u^{-1}\nabla_{\nabla_{u(-)}X^j} X^j\right)$$

while

$$\rho_*(\beta(u)) = -\frac{1}{2}\sum \left(u^{-1}\nabla_{\nabla_{u(-)}X^j} X^j\right) - \frac{1}{2}u^{-1}\text{Ric}^\sharp(u-)$$

in particular we have the "Hodge–Kodaira operator" on vector fields if and only if

$$\sum \left(\nabla_{\nabla X^j} X^j\right) = -\text{Ric}^\sharp. \tag{22}$$

To relate this to the much neater Fang-Luo criterion, using their method, first note that $\sum \nabla_{X^j} X^j = 0$, the connection being Levi-Civita, so for any vector field V

$$0 = \sum\{\nabla_{\nabla_V X^j} X^j + \nabla^2 X^j(V, X^j)\} = \sum\{\nabla_{\nabla_V X^j} X^j + \nabla^2 X^j(X^j, V)\} + \text{Ric}^\sharp(V). \tag{23}$$

Also by integration by parts, for any compactly supported smooth one-form θ,

$$\begin{aligned}\int_M \theta(\sum \operatorname{div} X^j \nabla_V X^j) &= \sum \int_M \operatorname{div} X^j \theta(\nabla_V X^j)\\ &= -\int_M \sum\{\nabla_{X^j}\theta(\nabla_V X^j) + \theta\nabla^2 X^j(X^j, V) + \theta(\nabla_{\nabla_{X^j} V} X^j)\}\\ &= -\int_M \sum \theta\nabla^2 X^j(X^j, V).\end{aligned}$$

Thus, using Eq. (23)

$$\sum \nabla_{\nabla_V X^j} X^j + \text{Ric}^\sharp(V) = -\sum \nabla^2 X^j(X^j, V) = \sum \operatorname{div} X^j \nabla_V X^j = -\sum \operatorname{div} X^j \mathscr{L}_{X^j} V$$

and we obtain the Hodge–Kodaira operator on vector fields iff $\sum \operatorname{div} X^j \mathscr{L}_{X^j} = 0$, as shown more directly in [10].

Note that for Riemannian symmetric spaces, e.g. S^n, there is a natural choice of SDE for which the X^j are Killing vector fields, [7, 9] and so divergence free, fulfilling this criterion.

5 What Semi-connections Arise: A Generalised Levi-Civita Semi-connection

5.1 The Adjoint Connection

The semi-connections over E which arise on GL(M) as above from an SDE for a cohesive $\mathscr{A}$ are precisely the adjoints of metric connections on E obtained by projection of the trivial connection by $X : M \times \mathbf{R}^m \to E$, and called Le Jan-Watanabe connections in [7]. The Weitzenböck terms for forms are derived from the curvature of that connection. When $\mathscr{A}$ is elliptic it is natural to use the Levi-Civita connection, unless some other structure is involved. Both connections are then metric for the same Riemannian structure. For degenerate but cohesive $\mathscr{A}$ the situation is not so clear apart from the special case described below.

A challenge is to see what geometry is involved in the Weitzenböck terms when applying the analogous constructions to jet bundles in order to get higher order derivative estimates for the diffusion semi-groups.

5.2 Baudoin Connections

Interesting examples of hypo-elliptic, and non hypo-elliptic, diffusions arise when M is Riemannian with a Riemannian foliation L for which the metric is bundle like and with totally geodesic leaves, [2, 5, 15]. In this situation there is a unique metric connection $\breve{\nabla}$ on E whose adjoint semi-connection $\hat{\nabla}$ is metric for the given Riemannian metric on M and such that for smooth sections U and V of E we have $P_E(\hat{\nabla}_U V) = \breve{\nabla}_U V$, where $P_E : TM \to E$ is the orthogonal projection. The semi-connection is Baudoin's connection, [1], at $\varepsilon = \frac{1}{2}$. In terms of the Levi-Civita connection ∇ of M, for a section U of E and a vector field V we have

$$\begin{aligned}
\breve{\nabla}_{V(x)} U &= P_E \nabla_{V(x)} U && \text{for } V(x) \in E_x \\
&= P_E \nabla_{V(x)} U + P_E \nabla_{U(x)} (P_L V) && \text{for } V(x) \in T_x L \\
\hat{\nabla}_{U(x)} V &= P_E \nabla_{U(x)} V + P_L [U, V](x) && \text{for } V \text{ a section of } E \\
&= 2 P_E \nabla_{U(x)} V + P_L [U, V](x) && \text{for } V \text{ a section of } L
\end{aligned}$$

where P_L refers to the orthogonal projection onto the foliation sub-bundle of TM.

These assertions follow by using the characterisations of such foliations in Theorems 5.19 and 5.23 of [15]. Using that connection one can immediately apply the results for tensor bundles from [7] described in Sect. 4.2 above, (and also all the analysis on path space in [6, 7]). These connections are induced by SDE's by the general theory, [7], but from the point of view of derivative estimates the precise SDE is not relevant. For more recent work see [11].

Acknowledgements This article is based on the cited joint work with Yves Le Jan and Xue-Mei Li. It was stimulated by correspondence and discussions with Shizan Fang and by his work with Dejun Luo, and Sect. 5 came out of correspondence with Fabrice Baudoin.

References

1. Baudoin, F.: Stochastic analysis on sub-Riemannian manifolds with transverse symmetries. Ann. Probab. **45**(1), 56–81 (2017)
2. Baudoin, F., Bonnefont, M.: Curvature-dimension estimates for the Laplace-Beltrami operator of a totally geodesic foliation. Nonlinear Anal. **126**, 159–169 (2015)
3. Elworthy, K.D.: Stochastic Differential Equations on Manifolds. LMS Lecture Notes Series, vol. 70. Cambridge University Press, Cambridge (1982)
4. Elworthy, K.D.: Stochastic flows on Riemannian manifolds. Diffusion Processes and Related Problems in Analysis, Vol. II (Charlotte, NC, 1990). Progress in Probability, vol. 27, pp. 37–72. Birkhäuser, Boston (1992)
5. Elworthy, D.: Decompositions of diffusion operators and related couplings. Stochastic Analysis and Applications 2014. Springer Proceedings in Mathematics and Statistics, vol. 100, pp. 283–306. Springer, Cham (2014)
6. Elworthy, K.D., Li, X.-M.: Itô maps and analysis on path spaces. Math. Z. **257**(3), 643–706 (2007)
7. Elworthy, K.D., Le Jan, Y., Li, X.-M.: On the Geometry of Diffusion Operators and Stochastic Flows. Lecture Notes in Mathematics, vol. 1720. Springer, Berlin (1999)
8. Elworthy, K.D., Le Jan, Y., Li, X.-M.: Equivariant diffusions on principal bundles. Stochastic Analysis and Related Topics in Kyoto. Advanced Studies in Pure Mathematics, vol. 41, pp. 31–47. The Mathematical Society of Japan, Tokyo (2004)
9. Elworthy, K.D., Le Jan, Y., Li, X.-M.: The Geometry of Filtering. Frontiers in Mathematics. Birkhäuser Verlag, Basel (2010)
10. Fang, S., Luo, D.: A note on Constantin and Iyers representation formula for the NavierStokes equations (2015). arXiv:1508.06387v2 [math.PR]
11. Grong, E., Thalmaier, A.: Curvature-dimension inequalities on sub-Riemannian manifolds obtained from Riemannian foliations: part II. Math. Z. **282**(1–2), 131–164 (2016)
12. Ikeda, N., Watanabe, S.: Stochastic Differential Equations and Diffusion Processes, 2nd edn. North-Holland, Amsterdam (1989)
13. Kusuoka, S.: Degree theorem in certain Wiener Riemannian manifolds. Stochastic Analysis. Proc. Jap.-Fr. Semin., Paris, France, 1987, Lect. Notes Math., vol. 1322, pp. 93–108 (1988)
14. Michor, P.W.: Gauge theory for fiber bundles. Monographs and Textbooks in Physical Science. Lecture Notes, vol. 19. Bibliopolis, Naples (1991)
15. Tondeur, P.: Foliations on Riemannian Manifolds. Universitext, vol. ix. Springer, New York (1988)

On the Rough Gronwall Lemma and Its Applications

Martina Hofmanová

Abstract We present a rough path analog of the classical Gronwall Lemma introduced recently by Deya, Gubinelli, Hofmanová, Tindel in arXiv:1604.00437, [6] and discuss two of its applications. First, it is applied in the framework of rough path driven PDEs in order to establish energy estimates for weak solutions. Second, it is used in order to prove uniqueness for reflected rough differential equations.

Keywords Rough paths · Rough partial differential equations · Reflected rough differential equations · Rough gronwall lemma

2010 Mathematics Subject Classification 60H15 · 35R60 · 35L65

1 Introduction

The theory of rough paths was introduced by Terry Lyons in his seminal work [17]. It can be briefly described as an extension of the classical theory of controlled differential equations which is robust enough to allow for a deterministic treatment of stochastic differential equations. To be more precise and in order to fix the ideas, let us consider a controlled differential equation of the form

$$\mathrm{d}y_t = f(y_t)\mathrm{d}x_t, \tag{1.1}$$

Dedicated to Michael Röckner on the occasion of his 60th birthday.

M. Hofmanová (✉)
Fakultät für Mathematik, Universität Bielefeld, Postfach 100131,
33501 Bielefeld, Germany
e-mail: hofmanova@math.uni-bielefeld.de

M. Hofmanová
Institute of Mathematics, Technical University Berlin, Strasse des 17. Juni 136,
10623 Berlin, Germany

A. Eberle et al. (eds.), *Stochastic Partial Differential Equations and Related Fields*, Springer Proceedings in Mathematics & Statistics 229,
https://doi.org/10.1007/978-3-319-74929-7_21

where the driving signal x possesses only a limited regularity, namely, it does not have finite variation. One requirement for any sensible (deterministic) theory of such equations is certainly robustness: approximating x by smooth paths, say (x^n), and solving the corresponding approximate problems

$$dy_t^n = f(y_t^n)dx_t^n,$$

we expect that solutions y^n remain close, in a suitable sense, to a (unique) solution y of the original problem (1.1). In other words, we look for suitable topologies that render the solution map $x \mapsto y$ continuous, while allowing for driving signals of low regularity. The first positive answer to this problem was given in 1936 by Young [18], who introduced an extension of the Stieltjes integral for paths of finite p-variation with $p < 2$. Roughly speaking, he showed that one can define an integral of y against x such that the mapping

$$V^p([0,T]) \times V^p([0,T]) \to V^p([0,T]), \qquad (x,y) \mapsto \int_0^\cdot y_s \, dx_s$$

is continuous provided $p < 2$, where $V^p([0,T])$ stands for the space of continuous paths with finite p-variation on $[0,T]$.

However, one of the most interesting examples of a driving signal, namely, the Brownian motion, is not covered by this result. Indeed, it can be shown that its sample paths have unbounded p-variation for any $p \leqslant 2$. Moreover, it turns out that the situation is even more delicate: Lyons [16] proved that there exists *no* Banach space $\mathcal{B}$ containing sample paths of Brownian motions such that the map

$$(x,y) \mapsto \int_0^\cdot y_t \, \dot{x}_t \, dt$$

defined on smooth functions extends to a continuous map on $\mathcal{B} \times \mathcal{B}$. The breakthrough by Lyons [17] was then based on the insight that an important part of information on the signal x is missing, due to its low regularity. In particular, he showed that if the driving signal x has finite p-variation for $p \in [2,3)$, the continuity of the solution map to (1.1) can be recovered by enhancing x by a second component X^2, the so-called Lévy's area, which corresponds to the iterated integral

$$X_{st}^2 =: \int_s^t (x_r - x_s) \otimes dx_r.$$

Consequently, the solution map to (1.1), that is $(x, X^2) \mapsto y$, can be shown to be continuous with respect to appropriate topologies.

Remark that the above lines shall be understood as follows: apart from the path x we are given another datum X^2 which satisfies certain analytic and algebraic properties and which plays the role of the iterated integral x against x. Note that for instance in the case of a Brownian motion we still face the same issue as before and the iterated integral cannot be defined by deterministic arguments. However, the striking advantage now is that the only integral that needs to be constructed is the iterated integral of the Brownian motion itself and this can be done via probabilistic arguments. Therefore, we are able to separate the analytic and probabilistic part of solving stochastic differential equations. In the first step, we use probability (or any other available tool for the particular driving signal at hand) to construct the iterated integral. In the second step, we fix one realization of the process and its iterated integral and proceed deterministically.

Since its introduction, the rough path theory has found a large number of applications and tremendous progress has been made in application of rough path ideas to ordinary as well as partial differential equations driven by rough signals. We refer the reader for instance to the works by Friz et al. [4, 5], Gubinelli–Tindel [8, 12, 13], Gubinelli–Imkeller–Perkowski [11], Hairer [14] for a tiny sample of the exponentially growing literature on the subject. In view of these exciting developments it is remarkable that many basic PDE methods have not yet found their rough path analogues. For instance, until recently it was an open problem how to construct (weak) solutions to RPDEs using energy methods.

In [3, 6] we started a long term research programme where these questions will be addressed. One of the aims is to develop a theory applicable to a wide class of RPDEs by following the standard PDE strategies in order to obtain existence and uniqueness results. In [6], we introduced the general framework and developed innovative a priori estimates based on a new rough Gronwall lemma argument. The theory was applied to conservation laws with rough flux. Moreover, these new techniques already proved sufficiently flexible and useful as in [7] we were able to establish uniqueness for reflected rough differential equations, a problem which remained open in the literature as a suitable Gronwall lemma in the context of rough paths was missing. The aim of the present paper is to discuss the main ideas of [6, 7] in simple terms and to make the link between the two applications. It is expected that the framework will find further applications in future.

2 Intrinsic Notion of Solution

2.1 Notation

First of all, let us recall the definition of the increment operator, denoted by δ. If g is a path defined on $[0,T]$ and $s,t\in[0,T]$ then $\delta g_{st}:=g_t-g_s$, if g is a 2-index map defined on $[0,T]^2$ then $\delta g_{sut}:=g_{st}-g_{su}-g_{ut}$. For two quantities a and b the relation $a\lesssim_x b$ means $a\leqslant c_x b$, for a constant c_x depending on a (possibly multidimensional) parameter x.

In the sequel, given an interval I we call a *control on I* (and denote it by ω) any superadditive map on $\Delta_I:=\{(s,t)\in I^2 : s\leqslant t\}$, that is, any map $\omega:\Delta_I\to[0,\infty[$ such that, for all $s\leqslant u\leqslant t$,

$$\omega(s,u)+\omega(u,t)\leqslant\omega(s,t).$$

We will say that a control is *regular* if $\lim_{|t-s|\to 0}\omega(s,t)=0$. Also, given a control ω on an interval $I=[a,b]$, we will use the notation $\omega(I):=\omega(a,b)$. Given a time interval I, a parameter $p>0$, a Banach space E we denote by $\overline{V}_1^p(I;E)$ the space of finite p-variation functions taking values in E. The corresponding seminorm is denoted by

$$\|g\|_{\overline{V}_1^p(I;E)}:=\sup_{(t_i)\in\mathcal{P}(I)}\left(\sum_i|g_{t_i}-g_{t_{i+1}}|^p\right)^{\frac{1}{p}},$$

where $\mathcal{P}(I)$ denotes the set of all partitions of the interval I. If the right hand side is finite then

$$\omega_g(s,t)=\|g\|^p_{\overline{V}_1^p(I;E)}$$

defines a control on I, and we denote by $V_1^p(I;E)$ the set of elements $g\in\overline{V}_1^p(I;E)$ for which ω_g is regular on I. We denote by $\overline{V}_2^p(I;E)$ the set of two-index maps $g:I\times I\to E$ with left and right limits in each of the variables and for which there exists a control ω such that

$$|g_{st}|\leqslant\omega(s,t)^{\frac{1}{p}}$$

for all $s,t\in I$. We also define the space $\overline{V}^p_{2,\mathrm{loc}}(I;E)$ of maps $g:I\times I\to E$ such that there exists a countable covering $\{I_k\}_k$ of I satisfying $g\in\overline{V}_2^p(I_k;E)$ for any k. We write $g\in V_2^p(I;E)$ or $g\in V^p_{2,\mathrm{loc}}(I;E)$ if the control can be chosen regular.

2.2 Rough Drivers

To begin with, let us introduce the notion of a rough path. For a thorough introduction to the theory of rough paths we refer the reader to the monographs [9, 10, 15].

Definition 2.1 Let $d\in\mathbb{N}$, $p\in[2,3)$. A continuous p-rough path is a pair

$$X=(X^1,X^2)\in V_2^p([0,T];\mathbb{R}^d)\times V_2^{p/2}([0,T];\mathbb{R}^{d\times d})\tag{2.1}$$

that satisfies Chen's relation

$$\delta X^2_{sut}=X^1_{su}\otimes X^1_{ut},\qquad s\leqslant u\leqslant t\in[0,T].$$

A rough path X is said to be geometric if it can be obtained as the limit in the p-variation topology given in (2.1) of a sequence of rough paths $X^\varepsilon=(X^{\varepsilon,1},X^{\varepsilon,2})$ explicitly defined as

$$X_{st}^{\varepsilon,1} := \delta x_{st}^{\varepsilon}, \qquad X_{st}^{\varepsilon,2} := \int_s^t \delta x_{su}^{\varepsilon} \otimes \mathrm{d}x_u^{\varepsilon},$$

for some smooth path $x^{\varepsilon} : [0, T] \to \mathbb{R}^d$.

Throughout this paper, we are only concerned with geometric rough paths. A pleasant advantage is a first order chain rule formulae similar to the one known for smooth paths. Recall that this is not true within the Itô stochastic integration theory, where only a (second order) Itô formula is available. However, for the Stratonovich stochastic integral, the first order chain rule holds true. Thus in case of a Brownian motion we employ Stratonovich integration for the construction of the iterated integrals of a geometric rough path, whereas the Itô integral leads to nongeometric setting.

We are now in a position to provide a clear interpretation of the controlled equation (1.1) in the rough path setting.

Definition 2.2 Let $y_{\text{in}} \in \mathbb{R}^N$, let $f : \mathbb{R}^N \to \mathcal{L}(\mathbb{R}^d; \mathbb{R}^N)$ be a differentiable function and let $X = (X^1, X^2)$ be a continuous p-rough path with $p \in [2, 3)$. A path $y \in V_1^p([0, T]; \mathbb{R}^d)$ solves the rough differential equation

$$\mathrm{d}y = f(y)\,\mathrm{d}X, \qquad y_0 = y_{\text{in}},$$

if there exists a 2-index map $y^{\natural} \in V_{2,\text{loc}}^{p/3}([0, T]; \mathbb{R}^N)$ such that for all $s, t \in [0, T]$, we have

$$\delta y_{st} = f(y_s)X_{st}^1 + f_2(y_s)X_{st}^2 + y_{st}^{\natural}, \qquad y_0 = y_{\text{in}}, \tag{2.2}$$

where we have set $f_{2,ij} := f_i' f_j$.

This formulation has an intuitive appeal, however note that more precisely it should be understood as saying that y is a solution of the equation if

$$y_{st}^{\natural} := \delta y_{st} - f(y_s)X_{st}^1 - f_2(y_s)X_{st}^2$$

belongs to $V_{2,\text{loc}}^{p/3}([0, T], \mathbb{R}^N)$. In other words, for every path y one can define $y^{\natural}$ by the above formula, however, this would in general not be a remainder, that is, it would not have the required time regularity.

In the previous example of a rough differential equation we considered a noise term which was irregular in time and its coefficient contained nonlinear but bounded dependence on the solution. Nevertheless, in the PDE theory one is often lead to differential, i.e. unbounded, operators. As a model example, we may think of a rough transport equation

$$\mathrm{d}u = V \cdot \nabla u\,\mathrm{d}X, \qquad u_0 = u_{\text{in}}. \tag{2.3}$$

In order to treat these equations, we introduce the notion of an unbounded rough driver. It can be regarded as an operator valued rough path taking values in a suitable space of unbounded operators.

In what follows, we call a *scale* any sequence $\left(E_n, \|\cdot\|_n\right)_{n\in\mathbb{N}_0}$ of Banach spaces such that E_{n+1} is continuously embedded into E_n. Besides, for $n \in \mathbb{N}_0$ we denote by E_{-n} the topological dual of E_n. Note that it is necessary to distinguish between E_0 and its dual E_{-0} as they are generally different spaces.

Definition 2.3 Let $p \in [2, 3)$ be given. A continuous unbounded p-rough driver with respect to the scale $\left(E_n, \|\cdot\|_n\right)_{n\in\mathbb{N}_0}$, is a pair $A = \left(A^1, A^2\right)$ of 2-index maps such that

$$A^1_{st} \in \mathcal{L}(E_{-n}, E_{-(n+1)}) \text{ for } n \in \{0, 2\}, \qquad A^2_{st} \in \mathcal{L}(E_{-n}, E_{-(n+2)}) \text{ for } n \in \{0, 1\},$$

and there exists a continuous control ω_A on $[0, T]$ such that for every $s, t \in [0, T]$,

$$\begin{aligned} \|A^1_{st}\|^p_{\mathcal{L}(E_{-n}, E_{-(n+1)})} &\leqslant \omega_A(s, t) \qquad \text{for } n \in \{0, 2\}, \\ \|A^2_{st}\|^{p/2}_{\mathcal{L}(E_{-n}, E_{-(n+2)})} &\leqslant \omega_A(s, t) \qquad \text{for } n \in \{0, 1\}, \end{aligned}$$

and, in addition, Chen's relation holds true, that is,

$$\delta A^1_{sut} = 0, \qquad \delta A^2_{sut} = A^1_{ut} A^1_{su}, \qquad \text{for all } \ 0 \leqslant s \leqslant u \leqslant t \leqslant T.$$

It can be checked that under sufficient regularity assumptions on the family of vector fields V, the transport noise in (2.3) defines an unbounded p-rough driver in the scale $W^{n,2}(\mathbb{R}^N)$ by

$$A^1_{st} u := X^{1,k}_{st} V^k \cdot \nabla u, \qquad A^2_{st} u := X^{2,jk}_{st} V^k \cdot \nabla(V^j \cdot \nabla u).$$

Here we employ the Einstein summation convention over repeated indices. Consequently, in analogy to Definition 2.2 we may formulate the notion of weak solution to (2.3) as follows.

Definition 2.4 Let $A = \left(A^1, A^2\right)$ be a continuous unbounded p-rough driver with respect to the scale (E_n). A path $u : [0, T] \to E_{-0}$ is a weak solution to (2.3) provided there exist $u^\natural \in V^{p/3}_{2,\mathrm{loc}}([0, T], E_{-3})$ such that, for every $s, t \in [0, T], s < t$, and every test function $\varphi \in E_3$ it holds

$$(\delta u)_{st}(\varphi) = u_s(\{A^{1,*}_{st} + A^{2,*}_{st}\}\varphi) + u^\natural_{st}(\varphi), \qquad u_0 = u_{\mathrm{in}}. \tag{2.4}$$

The main difficulty in working with the intrinsic formulations (2.2) and (2.4) lies in the remainders $y^\natural$ and $u^\natural$, respectively. Indeed, all the other terms are explicit and do not even contain any time integration. Therefore, estimations of these terms are very straightforward. On the other hand, the only information available so far on the remainders $y^\natural$ and $u^\natural$ are the respective equations (2.2) and (2.4) and they have to be carefully investigated in order to obtain a priori estimates for the solutions y and u.

3 A Priori Estimates for Rough Partial Differential Equations

In view of the possible applications, it is necessary to allow for (deterministic) drift terms. One could then formulate equations in the general form

$$(\delta g)_{st}(\varphi) = (\delta\mu)_{st}(\varphi) + g_s(\{A^{1,*}_{st} + A^{2,*}_{st}\}\varphi) + g^\natural_{st}(\varphi), \tag{3.1}$$

with the drift being for instance another transport term or a second order (possibly nonlinear) elliptic operator, i.e.

$$\mu(\mathrm{d}t) = V_0 \cdot \nabla g \,\mathrm{d}t, \qquad \mu(\mathrm{d}t) = \operatorname{div}(A(x, g, \nabla g)\nabla g)\,\mathrm{d}t.$$

In this case, the rule of thumb is to integrate whenever possible, that is, we do not consider a local approximation by Riemann sums of the drift in (3.1) (as we do for the rough integral) but rather the increments of the integral itself.

The key result establishing the a priori bounds on the remainder $g^\natural$ reads as follows. The proof of its more general form is presented in [6, Theorem 2.5].

Theorem 3.1 *Let I be a subinterval of $[0, T]$. Consider a path $\mu \in \overline{V}^1_1(I; E_{-1})$ satisfying for some control ω_μ and for every $\varphi \in E_1$*

$$|(\delta\mu)_{st}(\varphi)| \leqslant \omega_\mu(s, t)\, \|\varphi\|_{E_1}.$$

Let g be a solution of the equation (3.1) on I such that g is controlled over the whole interval I, that is, $g^\natural \in V_2^{p/3}(I, E_{-3})$. Then there exists a constant $L > 0$ such that if $\omega_A(I) \leqslant L$ then for all $s < t \in I$

$$\|g^\natural_{st}\|_{E_{-3}} \lesssim_{A,I} \|g\|_{L^\infty(s,t;E_{-0})}\, \omega_A(s,t)^{\frac{3}{p}} + \omega_\mu(s,t)\omega_A(s,t)^{\frac{3-p}{p}}. \tag{3.2}$$

Let us explain the application of the above result on the example of a heat equation with transport noise

$$\mathrm{d}u = \Delta u \,\mathrm{d}t + V \cdot \nabla u \,\mathrm{d}X.$$

The ultimate goal is to derive the energy estimates for the solution known from the classical (deterministic) PDE theory. That is, we intend to show that u belongs to $L^\infty(0, T; L^2(\mathbb{R}^N)) \cap L^2(0, T; W^{1,2}(\mathbb{R}^N))$. To this end, one first derives the equation for u^2, which can be done rigorously for instance on the level of smooth approximations. It can be formulated in the scale $E_n = W^{n,\infty}(\mathbb{R}^N)$ as

$$\delta u^2(\varphi)_{st} = -2\int_s^t |\nabla u_r|^2(\varphi)\,\mathrm{d}r - 2\int_s^t (u\nabla u_r)(\nabla\varphi)\,\mathrm{d}r + u_s^2(A_{st}^{1,*}\varphi) + u_s^2(A_{st}^{2,*}\varphi) + u_{st}^{2,\natural}(\varphi). \tag{3.3}$$

Theorem 3.1 then applies with

$$\omega_\mu(s,t) = \int_s^t \|\nabla u_r\|_{L^2}^2\,\mathrm{d}r + \Big(\int_s^t \|\nabla u_r\|_{L^2}^2\,\mathrm{d}r\Big)^{\frac{1}{2}}\Big(\int_s^t \|u_r\|_{L^2}^2\,\mathrm{d}r\Big)^{\frac{1}{2}}.$$

This is the core of our rough Gronwall lemma argument, which is then concluded using the following result, whose proof can be found in [6, Lemma 2.7].

Lemma 3.2 (Rough Gronwall Lemma) *Fix a time horizon $T > 0$ and let $G : [0,T] \to [0,\infty)$ be a path such that for some constants $C, L > 0$, $\kappa \geqslant 1$ and some controls ω_1, ω_2 on $[0,T]$ with ω_1 being regular, one has*

$$\delta G_{st} \leqslant C\Big(\sup_{0\leqslant r\leqslant t} G_r\Big)\,\omega_1(s,t)^{\frac{1}{\kappa}} + \omega_2(s,t),$$

for every $s < t \in [0,T]$ satisfying $\omega_1(s,t) \leqslant L$. Then it holds

$$\sup_{0\leqslant t\leqslant T} G_t \leqslant 2\exp\Big(\frac{\omega_1(0,T)}{\alpha L}\Big)\cdot\Big\{G_0 + \sup_{0\leqslant t\leqslant T}\Big(\omega_2(0,t)\exp\Big(-\frac{\omega_1(0,t)}{\alpha L}\Big)\Big)\Big\},$$

where α is defined as

$$\alpha = \min\Big(1, \frac{1}{L(2Ce^2)^\kappa}\Big).$$

Finally, we have all in hand to derive the desired a priori estimate for a solution to the above heat equation. Indeed, taking the test function $\varphi \equiv 1$ in (3.3) and applying Theorem 3.1 we obtain

$$\begin{aligned}(\delta\|u\|_{L^2}^2)_{st} + 2\int_s^t \|\nabla u_r\|_{L^2}^2\,\mathrm{d}r &\lesssim \sup_{s\leqslant r\leqslant t}\|u_r\|_{L^2}^2\,\omega_A(s,t)^{\frac{1}{p}} + \omega_\mu(s,t)\omega_A(s,t)^{\frac{3-p}{p}}\\ &\lesssim \sup_{s\leqslant r\leqslant t}\|u_r\|_{L^2}^2\Big[\omega_A(s,t)^{\frac{1}{p}} + |t-s|\,\omega_A(s,t)^{\frac{3-p}{p}}\Big] + \int_s^t \|\nabla u_r\|_{L^2}^2\,\mathrm{d}r\,\omega_A(s,t)^{\frac{3-p}{p}}.\end{aligned}$$

Hence Lemma 3.2 applies with

$$G_t := \|u_t\|_{L^2}^2 + 2\int_0^t \|\nabla u_r\|_{L^2}^2\,\mathrm{d}r,\quad \omega_1(s,t) := \omega_A(s,t)^{\frac{\kappa}{p}} + |t-s|^\kappa\omega_A(s,t)^{\frac{(3-p)\kappa}{p}} + \omega_A(s,t)^{\frac{(3-p)\kappa}{p}},$$

$\omega_2(s,t) = 0$, and $\kappa = \min(p, p/(3-p)) \geqslant 1$, and yields

$$\sup_{0\leqslant t\leqslant T}\|u_t\|_{L^2}^2 + \int_0^T \|\nabla u_t\|_{L^2}^2\,\mathrm{d}t \lesssim \exp\Big(\frac{\omega_1(0,T)}{\alpha L}\Big)\|u_0\|_{L^2}^2.$$

With these a priori estimates one can immediately proceed to the proof of existence of a weak solution. It does not rely on the Banach fixed point argument but rather on compactness of suitable approximate solutions. Therefore we are able to separate the proof of existence from the proof of uniqueness, which is needed for many problems of interest and in particular for reflected rough differential equations discussed in the following section.

Let us point out that at the current stage we are not able to treat PDEs with nonlinear noise terms, such as

$$\mathrm{d}u = \Delta u \,\mathrm{d}t + f(u)\,\mathrm{d}X,$$

where f is a sufficiently regular function. Even though such a nonlinear noise can be treated in the finite dimensional framework, for instance as shown in Sect. 4, in the above PDE setting it is an open question how to derive energy estimates from the corresponding variational formulation.

4 Uniqueness for Reflected Rough Differential Equations

Let us present another application of the above rough Gronwall lemma argument from a seemingly rather different field. We are interested in the one-dimensional RDE reflected at 0, which can be described as follows: given a time $T > 0$, a smooth function $f : \mathbb{R} \to \mathcal{L}(\mathbb{R}^d; \mathbb{R})$ and a continuous p-rough path X with $p \in [2, 3)$, find an path $y \in V_1^p([0, T]; \mathbb{R}_{\geqslant 0})$ and an increasing function (the so-called reflection measure) $m \in V_1^1([0, T]; \mathbb{R}_{\geqslant 0})$ that together satisfy

$$\mathrm{d}y_t = f(y_t)\,\mathrm{d}X_t + \mathrm{d}m_t, \qquad y_t\,\mathrm{d}m_t = 0. \tag{4.1}$$

Thus, the idea is to exhibit a path y that somehow follows the dynamics in (1.1), but is also forced to stay nonnegative thanks to the intervention of some regular local time m at 0.

This problem was studied by Aida [1, 2], however, only existence of a solution was established and the uniqueness issue was left open. The main difficulty lies in the lack of regularity of the corresponding Skorokhod map which does not allow to treat the problem via the Banach fixed point theorem. Consequently, it is necessary to study existence and uniqueness separately and a suitable Gronwall lemma argument is needed to establish uniqueness. It turns out that the above introduced framework is well suited for this task: the existence can be proved with significant simplifications and, in addition, uniqueness follows from the rough Gronwall lemma argument by contradiction.

To be more precise, we formulate the problem as follows.

Definition 4.1 Let $y_{\mathrm{in}} > 0$, let $f : \mathbb{R} \to \mathcal{L}(\mathbb{R}^d; \mathbb{R})$ be a differentiable function and let X be a continuous p-rough path with $p \in [2, 3)$. A pair $(y, m) \in V_1^p([0, T];$

$\mathbb{R}_{\geqslant 0}) \times V_1^1([0,T];\mathbb{R}_{\geqslant 0})$ solves the problem (4.1) on $[0,T]$ with initial condition $y_{\rm in}$ if there exists a 2-index map $y^\natural \in V_{2,\rm loc}^{p/3}([0,T];\mathbb{R})$ such that for all $s,t \in [0,T]$, we have

$$\delta y_{st} = f(y_s)X_{st}^1 + f_2(y_s)X_{st}^2 + \delta m_{st} + y_{st}^\natural, \qquad y_0 = y_{\rm in}, \qquad m_t = \int_0^t \mathbf{1}_{\{y_u=0\}} \mathrm{d}m_u, \tag{4.2}$$

where we have set $f_{2,ij} := f_i' f_j$ and $m([0,t]) := m_t$.

With this interpretation in hand, the well-posedness result proved in [7, Theorem 4] reads as follows.

Theorem 4.2 *If* $f \in C_b^3(\mathbb{R};\mathcal{L}(\mathbb{R}^d,\mathbb{R}))$, *then problem* (4.1) *admits a unique solution.*

Sketch of the proof of uniqueness The proof proceeds in several steps, we will only present the main ideas while omitting the technical details. We refer the reader to [7, Theorem 5] for the complete proof.

Let us consider two solutions, say (y,μ) and (z,ν). In order to show that they coincide, we naturally look at the equation satisfied by the difference $y-z$. Denote $Y=(y,z)$ and let $\varphi:\mathbb{R}\to\mathbb{R}_{\geqslant 0}$ be a smooth function. Then a direct computation via Taylor expansion shows for $h(Y) := \varphi(y-z)$ that

$$\delta h(Y)_{st} = H_i(Y_s)X_{st}^{1,i} + H_{2,ij}(Y_s)X_{st}^{2,ij} + \int_s^t \varphi'(y_u - z_u)(\mathrm{d}\mu_u - \mathrm{d}\nu_u) + h_{st}^\natural, \tag{4.3}$$

where $h^\natural$ is a map in $V_2^{p/3}([0,T];\mathbb{R})$ and where we have set for all $Y=(y,z)\in\mathbb{R}^2$

$$H_i(Y) := \varphi'(y-z)(f_i(y) - f_i(z)),$$
$$H_{2,ij}(Y) := \varphi'(y-z)(f_{2,ij}(y) - f_{2,ij}(z)) + \varphi''(y-z)(f_i(y)-f_i(z))(f_j(y)-f_j(z)).$$

Our goal is to deduce a corresponding formula for $\varphi(\xi)=|\xi|$ and to finally show with the help of the rough Gronwall lemma that $|y_t - z_t|$ vanishes for all $t\in[0,T]$. However, since the absolute value function is not smooth at zero, (4.3) does not apply directly and it is necessary to consider a smooth approximation first and then pass to the limit. Namely, we define $\varphi_\varepsilon(\xi) = \sqrt{\varepsilon + |\xi|^2}$ and note that

$$|\varphi_\varepsilon'(\xi)| \leqslant 1\,, \qquad |\varphi_\varepsilon''(\xi)| \leqslant 1/\sqrt{\varepsilon^2 + |\xi|^2}\,, \qquad |\varphi_\varepsilon'''(\xi)| \leqslant 3/(\varepsilon^2 + |\xi|^2).$$

Even though the second and the third derivative blow up as $\varepsilon\to 0$, a detailed computation in the proof of [7, Theorem 5] shows that in order to obtain estimates uniform in ε, it is enough to control the quantity

$$\sup_{\varepsilon\in(0,1)} \sup_{y,z\in\mathbb{R}} \big(|\varphi_\varepsilon'(y-z)| + |y-z||\varphi_\varepsilon''(y-z)| + |y-z|^2|\varphi_\varepsilon'''(y-z)|\big).$$

Since this is indeed finite, we obtain estimates uniform in ε and finally pass to the limit as $\varepsilon \to 0$. This ensures the existence of a limiting remainder $\Phi^{\natural}$ and we get that the path $\Phi(Y) := |y - z|$ satisfies

$$\delta\Phi(Y)_{st} = \Psi_i(Y_s)X^{1,i}_{st} + \Psi_{2,ij}(Y_s)X^{2,ij}_{st} - \omega_M(s,t) + \int_s^t \mathbf{1}_{\{y_u = z_u\}} \mathrm{d}(\mu_u + \nu_u) + \Phi^{\natural}_{st},$$

where

$$\Psi_i(Y) := \operatorname{sgn}(y-z)(f_i(y) - f_i(z)), \qquad \Psi_{2,ij}(Y) := \operatorname{sgn}(y-z)(f_{2,ij}(y) - f_{2,ij}(z))$$

and

$$\omega_M(s,t) := \|\mu\|_{\overline{V}_1^1([s,t])} + \|\nu\|_{\overline{V}_1^1([s,t])}.$$

In addition, further estimations in the spirit of Theorem 3.1 show an apriori estimate for the remainder $\Phi^{\natural}$ of the form

$$|\Phi^{\natural}_{st}| \lesssim \|y - z\|_{L^\infty(s,t)}\, \omega_1(s,t)^{3/p} + \omega_M(s,t)\omega_2(s,t)^{1/p},$$

for some regular controls ω_1, ω_2. Consequently, we are in a position to apply the Rough Gronwall Lemma 3.2 and assert that

$$\sup_{r\in[s,t]} |y_r - z_r| + \omega_M(s,t) \lesssim |y_s - z_s| + \int_s^t \mathbf{1}_{\{y_u = z_u\}} (\mathrm{d}\mu_u + \mathrm{d}\nu_u).$$

Assume now that $[s,t]$ is an interval where $y \neq z$ in (s,t) but $y_s = z_s$. Then

$$\sup_{r\in[s,t]} |y_r - z_r| + \omega_M(s,t) \leqslant 0$$

which implies that $\sup_{r\in[s,t]} |y_r - z_r| = 0$ everywhere so we find a contradiction and such interval cannot exist. This concludes the proof of uniqueness. □

Acknowledgements Financial support by the DFG via Research Unit FOR 2402 is gratefully acknowledged.

References

1. Aida, S.: Rough differential equations containing path-dependent bounded variation terms (2016) arXiv:1608.03083
2. Aida, S.: Reflected rough differential equations. Stochastic Process. Appl. **125**(9), 3570–3595 (2015)
3. Bailleul, I., Gubinelli, M.: Unbounded rough drivers. arXiv:1501.02074 [math], January 2015
4. Caruana, M., Friz, P.K.: Partial differential equations driven by rough paths. J. Differ. Equ. **247**(1), 140–173 (2009)

5. Caruana, M., Friz, P.K., Oberhauser, H.: A (rough) pathwise approach to a class of nonlinear SPDEs. Annales de l'Institut Henri Poincaré / Analyse non linéaire **28**, 27–46 (2011)
6. Deya, A., Gubinelli, M., Hofmanová, M., Tindel, S.: A priori estimates for rough PDEs with application to rough conservation laws. arXiv:1604.00437
7. Deya, A., Gubinelli, M., Hofmanová, M., Tindel, S.: One-dimensional reflected rough differential equations. arXiv:1610.07481
8. Deya, A., Gubinelli, M., Tindel, S.: Non-linear rough heat equations. Probab. Theory Related Fields **153**(1–2), 97–147 (2012)
9. Friz, P.K., Hairer, M.: A Course on Rough Paths: With an Introduction to Regularity Structures. Springer, New York (2014)
10. Friz, P.K., Victoir, N.B.: Multidimensional Stochastic Processes As Rough Paths: Theory and Applications. Cambridge University Press, Cambridge (2010)
11. Gubinelli, M., Imkeller, P., Perkowski, N.: Paracontrolled distributions and singular PDEs. Forum Math. Pi **3:e6**, 75 (2015)
12. Gubinelli, M., Tindel, S.: Rough evolution equations. Ann. Probab. **38**(1), 1–75 (2010)
13. Gubinelli, M., Lejay, A., Tindel, S.: Young integrals and SPDEs. Potent. Anal. **25**, 307–326 (2006)
14. Hairer, M.: A theory of regularity structures. Invent. Math. **198**(2), 269–504 (2014)
15. Lyons, T.J., Caruana, M., Lévy, T.: Differential Equations Driven by Rough Paths. Lecture Notes in Mathematics, vol. 1908. Lectures from the 34th Summer School on Probability Theory held in Saint-Flour, July 6–24, 2004, With an introduction concerning the Summer School by Jean Picard. Springer, Berlin (2007)
16. Lyons, T.: On the nonexistence of path integrals. Proc. R. Soc. London Ser. A **432**(1885), 281–290 (1991)
17. Lyons, Terry J.: Differential equations driven by rough signals. Rev. Mat. Iberoamericana **14**(2), 215–310 (1998)
18. Young, L.C.: An inequality of the Hölder type, connected with Stieltjes integration. Acta Math. **67**(1), 251–282 (1936)

Doubly Damped Stochastic Parallel Translations and Hessian Formulas

Xue-Mei Li

We dedicate this paper to Michael Röckner on the occasion of his 60's birthday.

Abstract We study the Hessian of the solutions of time-independent Schrödinger equations, aiming to obtain as large a class as possible of complete Riemannian manifolds for which the estimate $C(\frac{1}{t} + \frac{d^2}{t^2})$ holds. For this purpose we introduce the doubly damped stochastic parallel transport equation, study them and make exponential estimates on them, deduce a second order Feynman–Kac formula and obtain the desired estimates. Our aim here is to explain the intuition, the basic techniques, and the formulas which might be useful in other studies.

Keywords Heat kernels · Weighted laplacian · Schrödinger operators · Hessian formulas · Hessian estimates

AMS subject classification 60Gxx · 60Hxx · 58J65 · 58J70

1 Introduction

The probability distribution of a Brownian motion or a Brownian bridge are reference measures with which we make L^2 analysis on the space of continuous paths (the Wiener space) and its subspaces of the pinned paths. On the Wiener space, these are Gaussian measures and are well understood. The theory of the probability distributions of Brownian motion and Brownian bridges on more general manifolds is less developed. These include elliptic and semi-elliptic diffusion in an Euclidean

---

X.-M. Li (✉)
Department of Mathematics, Imperial College London, London SW7 2AZ, UK
e-mail: xue-mei.li@Imperial.ac.uk

A. Eberle et al. (eds.), *Stochastic Partial Differential Equations and Related Fields*, Springer Proceedings in Mathematics & Statistics 229,
https://doi.org/10.1007/978-3-319-74929-7_22

space, with non-constant coefficients. For example we would like to describe the tail behaviour of the measure, but how do we describe the set of path far away? Instead, we measure the size of the tails by checking whether a Lipschitz continuous function f is exponentially integrable, or whether $\mathbf{E}(e^{cf^2})$ is finite for a constant c. In fact, a theorem of Herbst states that if a probability measure μ on $\mathbf{R}^n$ satisfies the following logarithmic Sobolev inequality $\mathbf{E}(f^2 \log(f^2)) \le c_0 \mathbf{E}|\nabla f|^2$, then $\mathbf{E}(e^{\epsilon f^2}) \le e^{\frac{\epsilon}{(1-c_0\epsilon)}}$ for any smooth function $f : \mathbf{R}^n \to \mathbf{R}$ with $\|f\|_{L^2(\mu)} = 1$ where ϵ is any number in $(0, 1/c_0)$ and. Also if a mean zero function f is Lipschitz continuous with $\|f\|_{\text{Lip}} = 1$, then $\mathbf{E}(e^{\alpha f}) \le e^{c_0 \alpha^2}$ holds for any number α, On a path or loop space, similar results hold [1]. There are many other applications of Logarithmic Sobolev inequalities, see e.g. M. Ledoux's Saint Flour notes [2] and the reference therein. To obtain such functional inequalities we make use of estimates on the fundamental solutions of Kolmogorov equations.

Let M denote a connected smooth manifold with a complete Riemannian metric g. Denote by (g^{ij}) the inverse of the Riemmanian metric $g = (g_{ij})$. There exists on M a strong Markov process with Markov generator $\frac{1}{2}\Delta$ where Δ is the Laplace–Beltrami operator which in local coordinates takes the form

$$\Delta f(x) = \frac{1}{\sqrt{\det g(x)}} \partial_i \left(\sqrt{\det g}\, g^{ij} \partial_j f \right)(x).$$

This stochastic process is said to be a Brownian motion. If Z is a vector field we denote by L_Z Lie differentiation in the direction of Z so for a real valued function f, $L_Z f = df(Z)$. Observe that any second order elliptic differential operator is of the form $\frac{1}{2}\Delta + L_Z$ where Δ is the Laplacian for the Riemannian metric induced by the operator, so they are generators of Brownian motions (with possibly a non-zero drift). In this article we are mainly concerned with gradient drifts, $Z = 2\nabla h$ where $h : M \to \mathbf{R}$ is a smooth function. In coordinates, the operators are of the form $\sum_{i,j} a_{i,j}(x) \frac{\partial^2}{\partial x_i \partial x_j} + \sum_k \frac{\partial}{\partial x_k}$, which is (locally) elliptic and is in general not strictly elliptic.

Set $\Delta^h = \Delta + 2L_{\nabla h}$, this is called the Bismut-Witten or the weighted Laplacian. With respect to the weighted volume measure $e^{2h}dx$, Δ^h is like a Laplacian. In particular, $\Delta^h = -(d + \delta^h)^2$ where δ^h is the adjoint of d on $L^2(e^{2h}dx)$. All three operators, d, δ^h and Δ^h, extend to acting on differential forms. Then $d + \delta^h$ and all its powers are essentially self-adjoint on C_K^∞, the space of smooth compactly supported differential forms, for the details see [3]. The densities of the probability distributions of the weighted Brownian motion are the weighted heat kernel. These are the fundamental solution to the equation $\frac{\partial}{\partial t} = \frac{1}{2}\Delta^h$. There is also a commutative relation with which one can obtain gradient estimates for the weighted heat kernels under conditions on the Ricci curvature (without involving their derivatives).

We introduce the notations. Let $\text{Ric}_x : T_xM \times T_xM \to \mathbf{R}$ denote the Ricci curvature and let $\mathcal{R}$ denote the curvature tensor. Let $\text{Ric}_x^\sharp : T_xM \to T_xM$ denote the linear map defined by the relation: $\langle \text{Ric}_x^\sharp(u), v\rangle = \text{Ric}_x(u, v)$. One of the novelties is to introduce the symmetrised tensor Θ, see [4],

$$\langle \Theta(v_2)v_1, v_3\rangle = \left(\nabla_{v_3}\mathrm{Ric}^\sharp\right)(v_1, v_2) - \left(\nabla_{v_1}\mathrm{Ric}^\sharp\right)(v_3, v_2) - \left(\nabla_{v_2}\mathrm{Ric}^\sharp\right)(v_1, v_3), \quad (1)$$

where $v_1, v_2, v_3 \in T_{x_0}M$, and to impose growth conditions on a bilinear map Θ^h from $T_{x_0}M \times T_{x_0}M$ to $\mathbf{R}$ instead of imposing conditions on $|\nabla \mathrm{Ric}^\sharp|$. The bilinear form is defined by the formula

$$\Theta^h(v_2)(v_1) = \frac{1}{2}\Theta(v_2, v_1) + \nabla^2(\nabla h)(v_2, v_1) + \mathscr{R}(\nabla h, v_2)(v_1), \qquad v_1, v_2 \in T_{x_0}M.$$

We are particularly interested in adding a zero order potential and consider a time independent Schrödinger equation, which is a parabolic partial differential equation of the form

$$\frac{\partial u}{\partial t} = (\frac{1}{2}\Delta + L_{\nabla h} + V)u, \quad (2)$$

where $u : \mathbf{R}\times M \to \mathbf{R}$ is a real valued function. For simplicity the zero order potential function $V : M \to \mathbf{R}$ will be assumed to be bounded and Hölder continuous and so $\Delta + 2L_{\nabla h} + V$ is essentially self-adjoint on $C_K^\infty \subset L^2(M, e^{2h}dx)$, the space of smooth functions with compact supports.

Our objectives are to obtain global estimates for the Schrödinger semi-group $e^{(\frac{1}{2}\Delta^h+V)t} f$ where f is a bounded measurable function, for its gradient, and for its second order derivatives in terms of the geometric data of the Riemannian manifolds. We will be also interested in such estimates for its fundamental solutions, which we denote by $p^{h,V}(t, x, y,)$, or $P^V(t, x, y)$ if h vanishes identically or $p^h(t, x, y)$ if V vanishes identically, and $p(t, x, y)$ if both h and V vanish. Similar notation, with capital P, e.g. $P_t^{h,V}$, will be used to denote the corresponding semi-groups.

The commutative relation we mentioned earlier is as follows: the differential d and the semi-group $e^{\frac{1}{2}\Delta^h}$ commute on C_K^∞, and consequently $de^{\frac{1}{2}\Delta^h}$ solves the heat equation on differential 1-forms: $\frac{\partial}{\partial t}\phi = \frac{1}{2}\Delta^h\phi$. If $M = \mathbf{R}^n$ this equation on differential 1-forms is an equation on 'vector-valued' functions. Let us denote by $\nabla^{h,*}$ the adjoint of ∇ on $L^2(e^{2h}dx)$, then this equation becomes:

$$\frac{\partial}{\partial t}\phi = \frac{1}{2}\nabla^{h,*}\nabla\phi - \frac{1}{2}\phi(\mathrm{Ric}^\sharp - 2\nabla\nabla h). \quad (3)$$

To see this we observe that, if ∇^* is the adjoint of ∇ on $L^2(dx)$, then there is the Weitzenbock formula $\Delta^h = -\nabla^*\nabla\phi - \mathrm{Ric}^\sharp(\phi) + 2L_{\nabla h}\phi$, $L_{\nabla h}\phi = \nabla\phi(\nabla h)$ $d(\phi(\nabla h) + d\phi(\nabla h, \cdot)$ where $\iota_{\nabla h}$ denotes the interior product. Also, for a differential 1-form ϕ we apply the identity

$$\nabla^{h,*}\phi = \nabla^*\phi - 2\iota_{\nabla h}\phi$$

to see that

$$\nabla^{h,*}\nabla.\phi = \nabla^*\nabla.\phi - 2\iota_{\nabla h}\nabla.\phi.$$

Equation (3) inspired the study of the damped stochastic parallel translation

$$W_t : T_{x_0}M \to T_{x_t(\omega)}M$$

along a path $x_t(\omega)$ which solves the stochastic damped parallel translation equation

$$\frac{DW_t}{dt} = -\frac{1}{2}\mathrm{Ric}^\sharp_{x_t}(W_t) + \nabla_{W_t}\nabla h, \quad W_0 = Id \tag{4}$$

Here Id denotes the identity map on $T_{x_0}M$ and

$$\frac{DW_t}{dt} := //_t(x.(\omega))\frac{d}{dt}\left(//_t^{-1}(x.(\omega))W_t\right)$$

is the covariant derivative along $x_t(\omega)$ and $//_t(x.(\omega)) : T_{x_0}M \to T_{x_t(\omega)}M$ denotes the standard stochastic parallel translation and $//_t^{-1}(x.(\omega)) : T_{x_t(\omega)}M \to T_{x_0}M$ is its inverse.

Stochastic parallel translations along the non-differentiable sample paths of a Brownian motion can be constructed by a stochastic differential equation on the orthonormal frame bundle. This goes back to J. Eells, K. D. Elworthy and P. Malliavin, earlier attempts go back to K. Itô and M. Pinsky. M. Emery and M. Arnaudon studied parallel translations along a general semi-martingale [5]. The damped parallel translation goes back to E. Airault [6]. The damped stochastic parallel translation takes into accounts of the effect of the Ricci curvature along its path and unwind it, leading to the magic well known formula: $de^{\frac{1}{2}\Delta^h t}f(v) = \mathbf{E}df(W_t(v))$ for (x_t) a Brownian motion with the initial value x_0. This holds for all compact manifolds and for more general manifolds.

The global estimates we are after are of the form

$$\left|\nabla^2 p(t, x_0, y_0)\right| \le C\left(\frac{1}{t} + \frac{d^2(x_0, y_0)}{t^2}\right), \quad t \in (0, 1], \ x, y \in M. \tag{5}$$

Such estimates (for $h = 0$, $V = 0$ and for compact manifolds) were obtained in [7] and were generalised to other types of manifolds we refer to the references in [4]. We should remark that adequate care must be taken when generalising estimates from compact manifolds to non-compact manifolds. For example taking a localising sequence of stopping times may not come for free and any technique involving differentiating a stochastic flow with respect to its initial point will likely need the additional assumption the strong 1-completeness [8]. See also [9–13].

Our goal is to establish these estimates for as large a class of manifolds as possible and extend them to the operators $\frac{1}{2}\Delta + L_{\nabla h} + V$. If both V and h vanish, these estimates are relevant for the study of the space of continuous loops and pinned paths using the Brownian bridge measure, e.g. the probability measure induced by a Brownian motion conditioned to return to a point y_0 at time 1. Naturally the Brownian motions with the symmetric drift ∇h, which we refer as an h-Brownian motion, are also candidates for such studies. We recall, that the h-Brownian bridge is a Markov process x_t on $[0, 1)$ with the Markov generator

$$\frac{1}{2}\Delta^h + \nabla \log p(1-t, x, y_0),$$

and $\lim_{t\to 1} x_t = y_0$ where y_0 is the terminal value. Observe that the corresponding damped parallel translation would be

$$\frac{DW_t}{dt} = -\frac{1}{2}\mathrm{Ric}^\sharp_{x_t}(W_t) + \nabla_{W_t}\nabla h + \nabla_{W_t}\nabla \log p(1-t, x_t, y_0), \quad t < 1.$$

Gradient estimates on the semi-group associated with the Brownian bridge will naturally involve the second order derivative of $\log p(1-t, x, y_0)$. It is clear that the small time asymptotics of the Hessian are relevant, and estimates of the type (5) appear to be essential for analysing the Brownian bridge measure and useful for the L^2 analysis of loop spaces. In this paper we explain the main formulas and constructions rom [4] that leads to these estimates.

2 Summary of Results

The following is summary of some results from [4].

(a) We extend estimate (5) to a more general class of manifolds replacing the linear growth condition on $|\nabla \mathrm{Ric}|_{op}$ by a linear growth condition on Θ, Θ being a symmetrised tensor obtained from $\nabla \mathrm{Ric}^\sharp$ after taking into accounts of the effects of ∇h defined (1).

(b) Our proof is based on an elementary Hessian formula which will then lead to an integration by parts type Hessian formula. For these formulas we introduce a doubly damped stochastic parallel transport equation which is defined using Θ. It is natural to call these solutions 'doubly damped stochastic parallel translations'. We denote the solutions by $W_t^{(2)}$, see Lemma 1. We have the formula:

$$\mathrm{Hess}(P_t^h f)(v_2, v_1) = \mathbf{E}\left[\nabla df(W_t(v_2), W_t(v_1))\right] + \mathbf{E}\left[df\left(W_t^{(2)}(v_1, v_2)\right)\right].$$

Here W_t denotes the damped stochastic parallel translation defined by (4). The Second Order Feynman–Kac Formula which does not involve the derivative of f is given in Theorem 1.

(c) Such estimates will be also extended to the symmetric operators $\frac{1}{2}\Delta^h$ and $\frac{1}{2}\Delta^h + V$. Both operators are essentially self-adjoint on C_K^∞. By unitary transformations the drift term and the potential term can be treated almost exchangeably, however the drift and the zero order term do behave differently. For example we assume that h is smooth and pose no direct assumptions on its its growth at the infinity while the zero order term V is only Hölder continuous and is assumed to be bounded. We obtain a second order Feynman–Kac formula, see Theorem 1, the Hölder continuity of V is needed and is used to offset singularities in some of the integrals of the formulas. The modified doubly damped equation involves Θ^h instead of Θ itself.

(d) These estimates are refined for a subclass of manifolds with a pole, for which we make use of and obtain some nice estimates in terms of the semi-classical Brownian bridges, a more careful study of the semi-classical Brownian bridge measure can be found in [14]. See also [15] for generalised Brownian bridges and [16] for gradient estimates.

3 Key Ingredient

Let $X(e)$ be smooth vector field on M given by an isometric embedding $\phi : M \to \mathbf{R}^m$ and so $X(e)$ is the gradient of the real valued function $\langle \phi, e\rangle$ where $e \in \mathbf{R}^m$. If $\{e_i\}$ is an o.n.b. of $\mathbf{R}^m$, this induces a family of vector fields $X_i(x)$ where $X_i(x) = X(x)(e_i)$.

Let $F_t(x, \omega)$ denote the solution to the gradient SDE

$$dx_t = \sum_{i=1}^m X_i(x_t) \circ dB_t^i + \nabla h(x_t)dt = X(x_t) \circ dB_t + \nabla h(x_t)dt$$

where $\circ$ denotes Stratonovich integration and B_t, $B_t = (B_t^1, \ldots, B_t^m)$, is an $\mathbf{R}^m$-valued Brownian motion on a filtered probability space with the usual assumptions. Then $F_t(x_0)$, the solution with the initial value $x_0 \in M$, is a Brownian motion with the initial value x_0.

If x_t is a semi-martingale, the stochastic damped parallel translation $/\!/_t(x_\cdot(\omega))$ along $x_t(\omega)$, which is also denoted by $/\!/_t$, allows us to bring a vector in the tangent space of a solution path at time 0 to its tangent space at time t, to differentiate it there and to bring it back to time t by the inverse parallel translation $/\!/_t^{-1}$. If (x_t) is a Brownian motion with the initial value x_0, the damped stochastic parallel translation W_t along its sample paths, where

$$\frac{DW_t}{dt} = -\frac{1}{2}\mathrm{Ric}_{x_t}^\sharp(W_t) + \nabla_{W_t}\nabla h, \quad W_0 = Id,$$

compensates the effect of the Ricci curvature in Eq. (3) and unwind it, leading to the magic well known formula,

$$de^{\frac{1}{2}\Delta^h t} f(v) = \mathbf{E}\left[df(W_t(v))\right],$$

which holds trivially for compact manifolds and for manifolds with Ric $-$ 2Hess$h$ bounded from below and for more general manifolds.

For the second order derivatives of the fundamental solution of the heat kernel, we ought to differentiate W_t with respect to its initial data, i.e. we differentiate $//_t^{-1}(x.)W_t(x_0)$ which is a map from M to the space of linear maps which we denote by $\mathscr{L}(T_{x_0}M; T_{x_0}M)$.

We introduce the doubly damped stochastic parallel translation equation, whose solution we call doubly damped stochastic parallel transports/translations. Unlike damped parallel translations, the doubly damped ones involve genuine stochastic integrals (unless the curvature vanishes) and it is a challenge to obtain exponential estimates. We also recall that the damped parallel translations are conditional expectations of the spatial derivative of the solution to the gradient SDE. The doubly damped ones are obtained by differentiating the damped parallel translations, followed by taking conditional expectations. The beauty of it is that it satisfies the doubly damped stochastic parallel translation equation:

$$\begin{aligned} Dv_t &= \left(-\frac{1}{2}\text{Ric} + \text{Hess}h\right)^\sharp (v_t)\,dt \\ &\quad + \frac{1}{2}\Theta^h(W_t(v_2))(W_t(v_1))dt + \mathscr{R}(d\{x_t\}, W_t(v_2))W_t(v_1). \end{aligned} \tag{6}$$

We also introduce the notation $d\{x_t\}$, by which we mean integration with respect to the martingale part of $\{x_t\}$, see [17, sect.4.1] for detail. This allows us to give statements on a h-Brownian motion that is independent of its representation as a solution to a specific stochastic differential equation. In particular we may use any of the two canonical representations: (1) $x_t = \pi(u_t)$ where u_t is the solution to the canonical SDE on the orthonormal frame bundle

$$du_t = H(u_t) \circ dB_t + \mathfrak{h}_{u_t}(\nabla h(\pi(u_t))\ dt,$$

where $\mathfrak{h}_u$ denotes the horizontal lift map at a given frame u, and $\pi : OM \to M$ takes a frame, a point of OM, to its base point. (2) x_t is the solution of a gradient SDE. If u_t is the solution to the canonical SDE on the orthonormal frame bundle, then $d\{x_t\}$ is interpreted as $u_t\,dB_t$. If (x_t) is the solution to to a gradient SDE driven by X then $d\{x_t\}$ is interpreted as $X(x_t)\,dB_t$.

Lemma 1 *Suppose that the gradient SDE is strongly 1-complete and suppose that* $v_1, v_2 \in T_{x_t}M$ *and* $x_0 \in M$ *and* (x_t) *is the solution to the gradient SDE. Let* $W_t^{(2)}(v_1, v_2)$ *denote the solution to the following covariant differential equation (the doubly damped stochastic parallel translation equation):*

$$\begin{aligned} Dv_t &= \left(-\frac{1}{2}\text{Ric} + \text{Hess}h\right)^\sharp (v_t)\,dt \\ &\quad + \frac{1}{2}\Theta^h(W_t(v_2))(W_t(v_1))dt + \mathscr{R}(d\{x_t\}, W_t(v_2))W_t(v_1), \\ v_0 &= 0. \end{aligned}$$

Then $W_t^{(2)}(v_1, v_2)$ is the local conditional expectation of $\nabla_{v_2} W_t(v_1)$ with respect to the filtration $\mathscr{F}_t^{x_0} := \sigma\{x_s : s \le t\}$. If furthermore the latter is integrable, then

$$W_t^{(2)}(v_1, v_2) = \mathbf{E}\left\{\nabla_{v_2} W_t(v_1) \,\Big|\, \mathscr{F}_t^{x_0}\right\}.$$

The proof for the lemma consists of stochastic calculus involving $\frac{D}{ds} W_t(j(s))$ where j is a parallel field with $j(0) = v_2$, along the normalised geodesic γ with the initial condition x_0 and the initial velocity $\dot\gamma(0) = v_1$. Observe also that

$$\nabla_{v_2} W_t(v_1) = \frac{D}{ds}|_{s=0} W_t(j(s)).$$

Strong 1-completeness of an SDE is a concept that is weaker than strong completeness, by the latter we mean the existence of a global solution to the SDE which is continuous with respect to the initial value. Let $p = 1, 2, \dots, n$ where n is the dimension of the manifold. Roughly speaking, an SDE is strongly p-complete if for a.s. every ω, and for all t, $F_t(x, \omega)$ is continuous with respect to the initial point x when x is restricted to a sub-manifold of dimension p (or to a smooth C^1 curve if $p = 1$). The first example of an SDE which is complete (i.e. its solution from any initial point has infinite life time) and which is not strongly complete was given by K.D. Elworthy, prior to which it was generally believed that the two problems are equivalent. The concept of strong p-completeness was introduced in [3, 8] where we also give examples of strongly $p - 1$-complete SDEs which are not strongly p-complete and $n > 2$ and $p \le n$. For $n = 1$ completeness is equivalent to strong completeness, similarly for $n = 2$, strong completeness is equivalent to strong completeness. In [18], a non-strongly complete SDE on $\mathbf{R}^2$ is given: it has one single driving linear Brownian motion and is driven a smooth bounded driving vector field. We emphasise that, due to the fact that the exit time of $F_t(x, \omega)$ from a geodesic ball (even one with smooth boundary) is not necessarily continuous with respect to the initial point x, and it is not trivial to solve the strong 1-completeness by localisation. The strong 1-completeness for gradient SDEs was specially studied in [4]. See also the books [19, 20].

Remark 1 If the gradient SDE is strongly 1-complete, $s \mapsto |W_t(\dot\gamma(s))|$ is continuous in L^1 and $\mathbf{E}|T_{\gamma(s)} F_t|$ is finite, we know that for all $f \in BC^1$, $d(P_t^h f)(v_1) = \mathbf{E} df(W_t(v_1))$, [4, 8, 21]. From this we see immediately that

$$|d(P_t f)|_{L_\infty} \le |df|_\infty \, \mathbf{E}\left(e^{\int_0^t \rho^h(x_s) ds}\right),$$

where $\rho^h(x) = \sup_{|v|=1, v \in T_x M}\{-\frac{1}{2}\mathrm{Ric}(v, v) + Hess(h)(v, v)\}$. A more relaxed condition for this to hold can be obtained, but most of the assumptions here will be needed later. If ρ^h is bounded by $-K$ then we see immediately on direction of the characterisation for the Ricci curvature to be bounded below by K, by taking $h = 0$ in the earlier estimate,

$$|d(P_t f)|_{L_\infty} \le |df|_\infty\, e^{-Kt}.$$

We give below the second order analogue. Denote by $T_{x_0}F_t(v_0)$ the derivative flow of $F_t(x)$, it solves the equation

$$dV_t = (\nabla X)_{x_t}(V_t) \circ dB_t + (\nabla^2 h)_{x_t}(V_t)dt,$$
$$V_0 = v_0.$$

Useful moment estimates on the derivatives flows for non-compact manifolds can be found in [22]. Recall that $j(s)$ is a parallel field along the geodesic γ with $\dot\gamma(0) = v_1$. with the initial value $j(0) = v_2$.

Lemma 2 *Suppose that* Ric $-$2Hess(h) *is bounded from below and that the gradient SDE is strongly 1-complete. Suppose also the statements (a) and (b) below hold.*

(a) for every s, $\mathbf{E}|T_{\gamma(s)}F_t|$ *and* $\mathbf{E}|\nabla_{TF_t(\gamma(s))}W_t|$ *are finite.*
(b) $s \mapsto \mathbf{E}\{\frac{D}{ds}W_t(j(s))\big|\, \mathscr{F}_t^{\gamma(s)}\}$ *is continuous in* $L^1(\Omega)$*;*

Then for all $f \in BC^2$,

$$\text{Hess}(P_t^h f)(v_2, v_1) = \mathbf{E}\left[\nabla df(W_t(v_2), W_t(v_1))\right] + \mathbf{E}\left[df\left(W_t^{(2)}(v_1, v_2)\right)\right]. \quad (7)$$

From this lemma we immediately obtain the following estimate:

$$\left|\text{Hess}(P_t^h f)\right|_{L_\infty} \le |\nabla df|_\infty\, \mathbf{E}\left(e^{2\int_0^t \rho^h(x_s)ds}\right) + |df|_{L_\infty}\, \mathbf{E}\left|W_t^{(2)}\right|.$$

It is clear that estimation on $\mathbf{E}\left|W_t^{(2)}\right|$ will be useful, this is given in [4] and which we do not include here. As we shall see, to obtain Hessian estimates of the form (5), we will need to obtain exponential integrability of $\left|W_t^{(2)}\right|^2$, such estimates will be given shortly after we explain why this is so.

The following are the basic assumptions. [**C1**.]

(a) Ric $-$ 2Hess$(h) \ge -K$;
(b) $\sup_{s\le t} \mathbf{E}(\|W_s^{(2)}\|^2) < \infty$;
(c) for all $f \in BC^2(M;\mathbf{R})$, $v_1, v_2 \in T_{x_0}M$, the elementary Hessian formula (7) holds.

For the Schrödinger equation we first set, for $r < t$ and $x_0 \in M$,

$$\mathbb{V}_{t-r,t} = (V(x_{t-r}) - V(x_0))e^{-\int_{t-r}^t [V(x_s)-V(x_0)]ds}. \quad (8)$$

Set also,

$$N_t = \frac{4}{t^2}\int_{\frac{t}{2}}^t \langle X(x_s)dB_s, W_s(v_1)\rangle \int_0^{\frac{t}{2}} \langle X(x_s)dB_s, W_s(v_2)\rangle. \quad (9)$$

Theorem 1 (Second Order Feynman–Kac Formula) *Suppose that* ***C1*** *holds. Let V be a bounded Hölder continuous function. Then for any $f \in \mathscr{B}_b(M;\mathbf{R})$,*

$$\begin{aligned}\mathrm{Hess} P_t^{h,V} f(v_1,v_2) =& e^{-V(x_0)t}\mathbf{E}\left[f(x_t)N_t\right] + e^{-V(x_0)t}\mathbf{E}\left[f(x_t)\frac{2}{t}\int_0^{t/2}\langle X(x_s)dB_s, W_s^{(2)}(v_1,v_2)\rangle\right]\\ &+ e^{-V(x_0)t}\int_0^t \mathbf{E}\left[f(x_t)\frac{2\mathbb{V}_{t-r,t}}{t-r}\int_0^{(t-r)/2}\langle X(x_s)dB_s, W_s^{(2)}(v_1,v_2)\rangle\right]dr\\ &+ e^{-V(x_0)t}\int_0^t \mathbf{E}\left[f(x_t)\mathbb{V}_{t-r,t}N_{t-r}\right]dr.\end{aligned} \tag{10}$$

For $h = V = 0$, a version of the Hessian formula was first given in [21] followed by another in [23]. A version of the Hessian formula for $h \equiv 0$ and $V \neq 0$ was also given in [21], however no proof was given. The doubly damped stochastic parallel translation equations were not present in either papers, nor were any extensive estimates given. Hessian formula and estimates for non-linear potential, on linear space, were given in [24]. A formula for the Laplacian of the semigroup $P_t f$ can be found in [25].

Corollary 1 *We assume $V(x_0) = 0$. Then,*

$$\begin{aligned}&\mathrm{Hess}\, p^{h,V}(t,x_0,y)\\ &= \mathrm{Hess}\, p_t^h(x_0,y) + \int_0^t\int_M V(z)\mathrm{Hess}\, p^h(t-r,x_0,z)p_r^h(z,y)\mathbf{E}[e^{-\int_0^r V(Y_s^{r,z,y})}]dzdr,\end{aligned}$$

where $Y_s^{r,z,y}$ is the h-Brownian bridge with terminal value r, initial value z and terminal value y.

Finally we indicate how to obtain estimates from these formulas. Let us take $V = 0$ for simplicity, so the formula reads:

$$\mathrm{Hess} P_t^h f(v_1,v_2) = \mathbf{E}\left[f(x_t)N_t\right] + \mathbf{E}\left[f(x_t)\frac{2}{t}\int_0^{t/2}\langle X_i(x_s)dB_s, W_s^{(2)}(v_1,v_2)\rangle\right]. \tag{11}$$

We then choose $f(x)$ to be the the fundamental solution $p^h(t,x,y_0)$, so $P_t^h f(x_0,y_0) = p(2t,x_0,y_0)$. In particular,

$$\begin{aligned}\frac{\mathrm{Hess}\, p^h(2t,x_0,y_0)(v_1,v_2)}{p^h(2t,x_0,y_0)} =& \mathbf{E}\left[\frac{p(t,x_t,y_0)}{p(2t,x_0,y_0)}N_t\right]\\ &+ \mathbf{E}\left[\frac{p(t,x_t,y_0)}{p(2t,x_0,y_0)}\frac{2}{t}\int_0^{t/2}\langle X(x_s)dB_s, W_s^{(2)}(v_1,v_2)\rangle\right].\end{aligned} \tag{12}$$

The right hand side can then be estimated. Since $|W_t|$ is bounded by a deterministic function (when ρ^h is bounded above), the first term of the right hand side is easier to estimate. Let us work with the second term,

$$
\begin{aligned}
&\frac{1}{p(2t, x_0, y_0)}\mathbf{E}\left(p(t, x_t, y_0)\frac{2}{t}\int_0^{t/2}\langle X(x_s)dB_s, W_s^{(2)}(v_1, v_2)\rangle\right)\\
&\le \frac{2}{t}\mathbf{E}\left(\frac{p(t, x_t, y_0)}{p(2t, x_0, y_0)}\log\frac{p(t, x_t, y_0)}{p(2t, x_0, y_0)}\right)+\frac{2}{t}\log\mathbf{E}\left(\exp\left(\int_0^{t/2}\langle X_i(x_s)dB_s, W_s^{(2)}(v_1, v_2)\rangle\right)\right)\\
&\le \frac{2}{t}\sup_{y\in M}\log\frac{p(t, y, y_0)}{p(2t, x_0, y_0)}+\frac{2}{t}\log\mathbf{E}\left(\exp\left(\int_0^{t/2}\langle X_i(x_s)dB_s, W_s^{(2)}(v_1, v_2)\rangle\right)\right).
\end{aligned}
$$

This can then be refined by heat kernel estimates and by estimates on $\mathbf{E}\left(e^{\left|W_s^{(2)}(v_1,v_2)\right|^2}\right)$, we illustrate the latter below. The other terms can be treated similarly.

Lemma 3 *Suppose that* $|\rho^h| \le K$, $\|\mathscr{R}_x\| \le \|\mathscr{R}\|_\infty$, *and* $\|\Theta^h\|^2 \le c + \delta r^2$ *for* δ *sufficiently small. Set* $C_1(T, 0) = 1$,

$$
C_1(T, K) = \sup_{0<s\le 3KT}\frac{1}{s}(e^s - 1), \quad \alpha_2(T, K, \|\mathscr{R}\|_\infty) = \frac{1}{49n^2\|\mathscr{R}\|_\infty^2 C_1(T, K)}.
$$

Then there exists a universal constant c such that for unit vectors $v_1, v_2 \in T_{x_0}M$, *and for any* $\alpha \le \alpha_2(T, K, \|\mathscr{R}\|_\infty)$,

$$
\begin{aligned}
\mathbf{E}\exp\left(\alpha\gamma|W_t^{(2)}(v_1, v_2)|^2\right) &\le ce^{2\alpha\gamma}\sqrt{\mathbf{E}\exp\left(4t\,\gamma\alpha\int_0^t e^{3Ks}\|\Theta^h\|_{x_s}^2 ds\right)}\\
&\le ce^{\frac{2\gamma}{49n^2\|\mathscr{R}\|_\infty^2}}\sqrt{\mathbf{E}\exp\left(\frac{4t\gamma}{49n^2\|\mathscr{R}\|_\infty^2 C_1(t, K)}\int_0^t e^{3Ks}\|\Theta^h\|_{x_s}^2 ds\right)}.
\end{aligned}
$$

At this stage we must choose an optimal condition on the growth of $|\Theta^h|$ at the infinity and estimate the exponential integrability of the radial distance function. The condition we imposed is linear growth, with the linear part sufficiently small (or we may compensate the size of the linear part by taking t in a small interval $[0, t_0]$). With these estimates we conclude this paper, and invite the interested reader to consult [4] for technicalities and further results. There we also studied the class of manifolds with a pole using semi-classical bridges. The use of semi-classical bridge for derivatives estimates is novel. See also [14]. Finally we pose the following open question. We know that a bound of the form e^{-Kt} on P_t characterises the lower boundedness of a Ricci curvature. Elworthy asked me whether I can use Eq. (7) and obtain some characterisation for manifolds. Let me make precise a question here.

Open Problem. For $f \in BC^2$, by Lemma 2,

$$
\left|\text{Hess}(P_t^h f)\right|_{L_\infty} \le |\nabla df|_\infty\,\mathbf{E}\left(e^{2\int_0^t \rho^h(x_s)ds}\right) + |df|_{L_\infty}\,\mathbf{E}\left|W_t^{(2)}\right|.
$$

If the Ricci curvature is bounded from below by K, we have

$$
\left|\text{Hess}(P_t^h f)\right|_{L_\infty} \le |\nabla df|_\infty e^{-2Kt} + |df|_{L_\infty}\,\mathbf{E}\left|W_t^{(2)}\right|.
$$

Can we characterise the class of complete Riemannian manifolds, among those whose Ricci curvature is bounded from below by K and whose sectional curvature and symmetrised tensor Θ^h are bounded?

References

1. Aida, S., Masuda, T., Shigekawa, I.: Logarithmic Sobolev inequalities and exponential integrability. J. Funct. Anal. **126**(1), 83–101 (1994)
2. Ledoux, M.: Isoperimetry and Gaussian analysis. Lectures on probability theory and statistics (Saint-Flour, 1994), pp. 165–294, Lecture notes in mathematics, vol. 1648, Springer, Berlin (1996)
3. Li, X.-M.: Stochastic differential equations on non-compact manifolds. University of Warwick Thesis (1992)
4. Li, X.-M.: Hessian formulas and estimates for parabolic schrödinger operators (2016). arXiv:1610.09538
5. Émery, M.: Stochastic calculus in manifolds. With an appendix by P.-A. Meyer. Universitext. Springer, Berlin (1989)
6. Airault, H.: Subordination de processus dans le fibré tangent et formes harmoniques. C. R. Acad. Sci. Paris Sér. A-B. **282**(22):Aiii, A1311–A1314, 1976
7. Malliavin, P., Stroock, D.W.: Short time behavior of the heat kernel and its logarithmic derivatives. J. Differ. Geom. **44**(3), 550–570 (1996)
8. Li, X.-M.: Strong p-completeness of stochastic differential equations and the existence of smooth flows on non-compact manifolds. Probab. Theory Relat. Fields **100**(4), 485–511 (1994)
9. Yau, S.-T., Li, P.: On the upper estimate of the heat kernel of a complete Riemannian manifold. Am. J. Math. **103**(5), 1021–1063 (1981)
10. Hsu, Elton P.: Estimates of derivatives of the heat kernel on a compact Riemannian manifold. Proc. Am. Math. Soc. **127**(12), 3739–3744 (1999)
11. Li, P., Yau, S.-T: On the parabolic kernel of the Schrödinger operator. Acta Math. **156**(3–4), 153–201 (1986)
12. Norris, J.R.: Path integral formulae for heat kernels and their derivatives. Probab. Theory Relat. Fields **94**(4), 525–541 (1993)
13. Shuenn Jyi Sheu: Some estimates of the transition density of a nondegenerate diffusion Markov process. Ann. Probab. **19**(2), 538–561 (1991)
14. Li, X.-M.: On the semi-classical Brownian bridge measure. Electron. Commun. Probab. **22**(38), 1–15 (2017)
15. Li, X.-M.: Generalised Brownian bridges: examples, 2016. arXiv:1612.08716
16. Li, X.-M., Thompson, J. First order Feynman-Kac formula (2016). arXiv:1608.03856
17. Elworthy, K.D., Le Jan, Y., Li, X.-M.: The geometry of filtering. Frontiers in mathematics. Birkhäuser Verlag, Basel (2010)
18. Li, Xue-Mei, Scheutzow, Michael: Lack of strong completeness for stochastic flows. Ann. Probab. **39**(4), 1407–1421 (2011)
19. Elworthy, K.D.: Stochastic differential equations on manifolds. London mathematical society lecture note series, 70. Cambridge University Press, Cambridge-New York (1982)
20. Kunita, H.: Cambridge studies in advanced mathematics. In: Stochastic flows and stochastic differential equations, vol. 24. Cambridge University Press, Cambridge (1990)
21. Elworthy, K.D., Li, X.-M.: Formulae for the derivatives of heat semigroups. J. Funct. Anal. **125**(1), 252–286 (1994)
22. Li, X.-M.: Stochastic differential equations on non-compact manifolds: moment stability and its topological consequences. Probab. Theory Relat. Fields **100**(4), 417–428 (1994)
23. Arnaudon, M., Plank, H., Thalmaier, A.: A Bismut type formula for the Hessian of heat semigroups. C. R. Math. Acad. Sci. Paris **336**(8), 661–666 (2003)

24. Li, X.-M., Zhao, H.Z.: Gradient estimates and the smooth convergence of approximate travelling waves for reaction-diffusion equations. Nonlinearity **9**(2), 459–477 (1996)
25. Elworthy, K.D., Li, X-M.: Bismut type formulae for differential forms. C. R. Acad. Sci. Paris Sér. I Math. **327**(1), 87–92 (1998)

Synchronization, Lyapunov Exponents and Stable Manifolds for Random Dynamical Systems

Michael Scheutzow and Isabell Vorkastner

Abstract During the past decades, the question of existence and properties of a random attractor of a random dynamical system generated by an S(P)DE has received considerable attention, for example by the work of Gess and Röckner. Recently some authors investigated sufficient conditions which guarantee *synchronization*, i.e. existence of a random attractor which is a singleton. It is reasonable to conjecture that synchronization and negativity (or non-positivity) of the top Lyapunov exponent of the system should be closely related since both mean that the system is contracting in some sense. Based on classical results by Ruelle, we formulate positive results in this direction. Finally we provide two very simple but striking examples of one-dimensional monotone random dynamical systems for which 0 is a fixed point. In the first example, the Lyapunov exponent is strictly negative but nevertheless all trajectories starting outside of 0 diverge to ∞ or $-\infty$. In particular, there is no synchronization (not even locally). In the second example (which is just the time reversal of the first), the Lyapunov exponent is strictly positive but nevertheless there is synchronization.

Keywords Synchronization · Lyapunov exponent · Random dynamical system

Mathematics Subject Classification (2010) 37D10 · 37D45 · 37G35 · 37H15

M. Scheutzow (✉) · I. Vorkastner
Institut für Mathematik, MA 7-5, Fakultät II, Technische Universität Berlin,
Straße des 17. Juni 136, 10623 Berlin, Germany
e-mail: ms@math.tu-berlin.de

I. Vorkastner
e-mail: vorkastn@math.tu-berlin.de

A. Eberle et al. (eds.), *Stochastic Partial Differential Equations and Related Fields*, Springer Proceedings in Mathematics & Statistics 229,
https://doi.org/10.1007/978-3-319-74929-7_23

1 Introduction

During the past decades, the question of existence and properties of a random attractor of a random dynamical system generated by an S(P)DE has received considerable attention, see for example [2, 8, 11]. Recently some papers investigated sufficient conditions which guarantee *synchronization*, i.e. existence of a random attractor which is a singleton, see [4, 5, 9, 10, 13]. It is reasonable to conjecture that synchronization and negativity (or non-positivity) of the top Lyapunov exponent of the system should be closely related since both mean that the system is contracting in some sense. A positive result of that kind in the finite dimensional case is [9, Lemma 3.1] which states that (under an ergodicity assumption) negativity of the top Lyapunov exponent plus an integrability assumption on the derivative in a neighborhood of the support of the invariant measure guarantees that for almost every x in the support of the invariant measure, there exists a random neighborhood of x which forms a local stable manifold. In particular, the system contracts locally. In the present paper, we formulate a corresponding result for separable Hilbert spaces. Like [9, Lemma 3.1] the proof is an easy consequence of results by Ruelle [12]. Example 1 in Sect. 4 shows that the result becomes untrue if the integrability assumption on the derivative is dropped. In Example 1 we investigate a simple one-dimensional random dynamical system generated by independent and identically distributed strictly monotone and bijective maps from the real line to itself which fix the point 0. The Lyapunov exponent is strictly negative but nevertheless the point 0 is not even locally asymptotically stable. In fact all trajectories starting outside 0 go to ∞ or $-\infty$ (depending on the sign of the initial condition). In particular, there is no synchronization. The reason for this behaviour is that the random function is very steep outside a very small (random) neighborhood of 0 (even though the derivative at 0 is $1/2$ almost surely).

We also consider the opposite behaviour. Proposition 2 requires that the unstable manifold U of a random fixed point is non-trivial and states that under this condition, synchronization cannot hold. Example 2 shows that replacing the non-triviality of U by positivity of the top Lyapunov exponent does not imply lack of synchronization. In fact, Example 2 is just the time reversal of Example 1.

2 Preliminaries and Notation

Let $(\mathscr{H}, \|\cdot\|)$ be a separable Hilbert space with Borel σ-algebra $\mathscr{B}(\mathscr{H})$ and let $(\Omega, \mathscr{F}, \mathbb{P}, \theta)$ be a metric dynamical system, i.e. $(\Omega, \mathscr{F}, \mathbb{P})$ is a probability space and $(\theta_t)_{t\in\mathbb{T}}$ a group of jointly measurable maps on $(\Omega, \mathscr{F}, \mathbb{P})$ such that $\theta_0 = \mathrm{id}$ with invariant measure $\mathbb{P}$. Here, $\mathbb{T}$ is either $\mathbb{Z}$ or $\mathbb{R}$. We denote the set of non-negative numbers in $\mathbb{T}$ by $\mathbb{T}_+$.

Further, let $\varphi : \mathbb{T}_+ \times \Omega \times \mathscr{H} \to \mathscr{H}$ be jointly measurable, $\varphi_0(\omega, x) = x$, $\varphi_{s+t}(\omega, x) = \varphi_t(\theta_s\omega, \varphi_s(\omega, x))$ for all $x \in \mathscr{H}$, and $x \mapsto \varphi_t(\omega, x)$ continuous, $s, t \in \mathbb{T}_+$

and $\omega \in \Omega$. The collection $(\Omega, \mathscr{F}, \mathbb{P}, \theta, \varphi)$ is then called a *random dynamical system*, see [1] for a comprehensive treatment.

We suppose that there is a family $\left(\mathscr{F}_{s,t}\right)_{-\infty \leq s \leq t \leq \infty}$ of sub-σ algebras of $\mathscr{F}$ such that $\theta_r^{-1}\left(\mathscr{F}_{s,t}\right) = \mathscr{F}_{s+r,t+r}$ for all r, s, t, $\mathscr{F}_{t,u} \subset \mathscr{F}_{s,v}$ whenever $s \leq t \leq u \leq v$, and $\varphi_t(\cdot, x)$ is $\mathscr{F}_{0,t}$-measurable for each $t \in \mathbb{T}_+$. If, additionally, $\mathscr{F}_{s,t}$ and $\mathscr{F}_{u,v}$ are independent whenever $s \leq t \leq u \leq v$, then the collection $(\Omega, \mathscr{F}, \mathbb{P}, \theta, \varphi)$ is called a *white-noise random dynamical system*. We will generally not assume the white-noise property in the following.

Define the *skew product* Θ on $\Omega \times \mathscr{H}$ by $\Theta_t(\omega, x) := (\theta_t\omega, \varphi_t(\omega, x))$ for $t \in \mathbb{T}_+$. We then say that a probability measure ρ on $\Omega \times \mathscr{H}$ is an *invariant measure* for a random dynamical system if its marginal on Ω is $\mathbb{P}$ and $\Theta(t)\rho = \rho$ for all $t \in \mathbb{T}_+$.

Definition 1 A family $\{A(\omega)\}_{\omega \in \Omega}$ of non-empty subsets of $\mathscr{H}$ is called

(i) a *random point* (resp. *random compact set*) if it is $\mathbb{P}$-almost surely a point (resp. compact set) and $\omega \mapsto \inf_{a \in A(\omega)} \|x - a\|$ is $\mathscr{F}$-measurable for each $x \in \mathscr{H}$.
(ii) φ*-invariant* if $\varphi_t(\omega, A(\omega)) = A(\theta_t\omega)$ for all $t \in \mathbb{T}_+$ and almost all $\omega \in \Omega$.

Definition 2 Let $(\Omega, \mathscr{F}, \mathbb{P}, \theta, \varphi)$ be a random dynamical system. A random compact set A is called a *weak attractor* if it satisfies the following properties

(i) A is φ-invariant
(ii) for every compact set $B \subset \mathscr{H}$

$$\lim_{t\to\infty} \sup_{x \in B} \inf_{a \in A(\omega)} \|\varphi_t(\theta_{-t}\omega, x) - a\| = 0 \qquad \text{in probability.}$$

If the convergence in (ii) is even almost sure, then A is called a *pullback attractor*.

Definition 3 Let $(\Omega, \mathscr{F}, \mathbb{P}, \theta, \varphi)$ be a random dynamical system. *Synchronization* occurs if there is a weak attractor $A(\omega)$ being a singleton for $\mathbb{P}$-almost every $\omega \in \Omega$.

Further, we use the following notation. Denote by $B(x, r)$ the open ball centered at $x \in \mathscr{H}$ with radius $r > 0$ and by $\bar{B}(x, r)$ the respective closed ball. We say that a map is of class $C^{1,\delta}$ if its derivative is Hölder continuous of exponent δ. Denote by D the derivative in the state space.

3 Top Lyapunov Exponent and Synchronization

In this section we demonstrate some relations between the sign of the top Lyapunov exponent, stable/unstable submanifolds and synchronization of a random dynamical system φ.

Proposition 1 *Let φ be a discrete or continuous time random dynamical system with state space $\mathscr{H}$ and assume that $\varphi_1(\omega, \cdot) \in C^{1,\delta}$ for some $\delta \in (0, 1)$. Assume further that φ has an invariant measure ρ such that*

$$\int_{\Omega \times \mathscr{H}} \log^+ \|D\varphi_1(\omega, x)\| \, \mathrm{d}\rho(\omega, x) < \infty.$$

and

$$\int_{\Omega \times \mathscr{H}} \log^+ \left(\|\varphi_1(\omega, \cdot + x) - \varphi_1(\omega, x)\|_{C^{1,\delta}(\bar{B}(0,1))}\right) \mathrm{d}\rho(\omega, x) < \infty. \tag{1}$$

Then, the (discrete-time) top Lyapunov exponent

$$\lambda(\omega, x) = \lim_{n \to \infty} \frac{1}{n} \log \|D\varphi_n(\omega, x)\|$$

is defined for ρ-almost all $(\omega, x) \in \Omega \times \mathscr{H}$. Assume that there exists some $\mu < 0$ such that $\lambda(\omega, x) < \mu$ almost everywhere. Then, there exist measurable functions $0 < \alpha(\omega, x) < \beta(\omega, x) < 1$ such that for ρ-almost all (ω, x)

$$S(\omega, x) = \left\{y \in \bar{B}(x, \alpha(\omega, x)) : \|\varphi_n(\omega, y) - \varphi_n(\omega, x)\| \le \beta(\omega, x)\mathrm{e}^{\mu n} \text{ for all } n \ge 0\right\}$$

is a measurable neighborhood of x. We refer to $S(\omega, x)$ as the stable manifold.

Proof By the same construction as in [9, Lemma 3.1], define $M := \Omega \times \mathscr{H}$, $\tilde{\mathscr{F}} := \mathscr{F} \otimes \mathscr{B}(\mathscr{H})$ and $f : M \mapsto M$ given by $f(m) := \Theta_1(\omega, x)$ for $m = (\omega, x) \in M$. Further, set

$$F_m(y) := \varphi_1(\omega, y + x) - \varphi_1(\omega, x) \quad \text{for } m = (\omega, x) \in M$$

and apply [12, Theorem 5.1] with $Q = 0$. Observe that the set D^ν in the proof of [12, Theorem 5.1] is – in our special case – both open and closed in the ball $\bar{B}(0, \alpha(\omega))$ and therefore $\bar{B}(0, \alpha(\omega)) \subset D^\nu$. This implies $\bar{B}(x, \alpha(\omega, x)) \subset S(\omega, x)$ almost everywhere and therefore $\bar{B}(x, \alpha(\omega, x)) = S(\omega, x)$ almost surely. □

Corollary 1 *Under the assumptions of the previous proposition the random dynamical system φ is asymptotically stable, i.e. there exists a deterministic non-empty, open set U in $\mathscr{H}$ such that*

$$\mathbb{P}\left(\lim_{n \to \infty} \operatorname{diam}(\varphi_n(\cdot, U)) = 0\right) > 0.$$

Proof Existence of a neighborhood $B(\omega, \alpha(\omega, x))$ as in Proposition 1 implies local asymptotic stability by [9, Lemma 3.3] (the latter Lemma is formulated and proved only in the finite-dimensional case but the proof in our set-up is almost identical). □

If we further assume that φ is a white-noise random dynamical system, φ has a weak attractor, satisfies an irreducibility condition and contracts on large sets, then synchronization follows, see [9, Theorem 2.14] for exact conditions.

Random attractors with positive top Lyapunov exponent are more difficult to characterize.

Proposition 2 *Assume there exist a φ-invariant random point $A(\omega)$, some $\mu > 0$ and some measurable functions $0 < \alpha(\omega) < \beta(\omega) < 1$ such that*

$$U(\omega) = \big\{x_0 \in \bar{B}(A(\omega), \alpha(\omega)) : \exists (x_n)_{n\in\mathbb{N}} \text{ with } \varphi(\theta_{-n}\omega, x_n) = x_{n-1} \\ \text{and } \|x_n - A(\theta_{-n}\omega)\| \leq \beta(\omega)e^{-\mu n} \text{ for all } n \geq 0\big\} \quad (2)$$

is not trivial (i.e. consists of more than one point) almost surely. Further assume there exists some $x_0(\omega) \in U(\omega) \setminus A(\omega)$ such that $x_n(\omega)$ are random points for $n \geq 0$ where $x_n(\omega)$ are chosen as in (2). Then, the random dynamical system φ does not synchronize.

Proof Suppose φ does synchronize. Then, there exists a weak attractor $\tilde{A}$ being a single random point. By the same arguments as in [9, Lemma 1.3] (stating uniqueness of a weak attractor), A and $\tilde{A}$ have to agree almost surely. Let $(x_n(\omega))_{n\in\mathbb{N}_0}$ be as in the proposition. There exists some $q > 0$ such that

$$\mathbb{P}\left(\|A(\omega) - x_0(\omega)\| > q\right) > \frac{3}{4}.$$

By [6, Proposition 2.15], there exists some compact set K such that $\mathbb{P}\left(A(\omega) \subset K\right) > 3/4$. Define the index set $I(\omega) = \{n \in \mathbb{N}_0 : A(\theta_{-n}\omega) \in K\}$. Then,

$$\mathbb{P}\left(n \in I(\omega)\right) = \mathbb{P}\left(A(\theta_{-n}\omega) \in K\right) > \frac{3}{4}$$

for every $n \in \mathbb{N}_0$ and

$$\lim_{m\to\infty}\Big(\inf_{y\in K}\|x_m(\omega) - y\|1_{\{m\in I(\omega)\}}\Big) = 0$$

Therefore, the set

$$\hat{K}(\omega) := K \cup \{x_m(\omega)\}_{m\in I(\omega)}$$

is a random compact set and hence, by [6, Proposition 2.15], there exists a deterministic compact set $\tilde{K}$ such that $\mathbb{P}(\hat{K}(\omega) \subset \tilde{K}) \geq 3/4$.

Combining these estimates, it follows for each $n \in \mathbb{N}_0$ that

$$\mathbb{P}\left(\sup_{y\in\tilde{K}}\|\varphi_n(\theta_{-n}\omega, y) - A(\omega)\| > q\right) \\ \geq \mathbb{P}\left(\|\varphi_n(\theta_{-n}\omega, x_n(\omega)) - A(\omega)\| > q,\ x_n(\omega) \in \tilde{K}\right) \geq \frac{3}{4} - \mathbb{P}\left(x_n \notin \tilde{K}\right) \\ \geq \frac{3}{4} - \mathbb{P}\left(\hat{K}(\omega) \not\subset \tilde{K}\right) - \mathbb{P}\left(x_n \notin \hat{K}(\omega)\right) \geq \frac{3}{4} - \frac{1}{4} - \frac{1}{4} = \frac{1}{4}.$$

Therefore, there is no synchronization. □

Note that the set $U(\omega)$ is typically the unstable manifold with respect to the invariant measure $\rho(\mathrm{d}\omega, \mathrm{d}x) = \delta_{A(\omega)}(\mathrm{d}x)\,\mathbb{P}(\mathrm{d}\omega)$.

Remark 1 For a finite dimensional space $\mathscr{H}$, the assumption of $x_n(\omega)$ to be random points can be replaced by measurablity of the unstable manifold $U(\omega)$. This condition is a consequence of the stable/unstable manifold theorem [12, Theorem 5.1 and 6.1] due to measurability of φ. Measurability of $U(\omega)$ is sufficient in this case since the selection theorem [3, Theorem III.9, p.67] shows that $x_0(\omega)$ can be chosen to be measurable and $\tilde{K}$ can be replaced by $\{y \in \mathscr{H} : \inf_{z\in K} \|y - z\| \leq 1\}$.

Remark 2 In case of a time-invertible random dynamical system, the unstable manifold $U(\omega)$ can be obtained by choosing a stable manifold of the time-reversed random dynamical system.

More generally, the unstable manifold can be obtained by using [12, Theorem 6.1]. Therefore, let $A(\omega)$ be an φ-invariant random point and define the cocycle $F^n_\omega(y) = \varphi_n(\omega, y + A(\omega)) - A(\theta_n\omega)$. Under similar assumptions as in Proposition 1 with $\rho(\mathrm{d}\omega, \mathrm{d}x) = \delta_{A(\omega)}(\mathrm{d}x)\,\mathbb{P}(\mathrm{d}\omega)$ but supposing a positive (discrete-time) top Lyapunov exponent

$$\lambda(\omega) = \lim_{n\to\infty} \frac{1}{n} \log \|D\varphi_n(\omega, A(\omega))\|$$

such that there exists $\mu > 0$ with $\lambda(\omega) > \mu$ almost everywhere, we can apply the unstable manifold theorem [12, Theorem 6.1]. This theorem shows that there exist measurable functions $0 < \alpha(\omega) < \beta(\omega)$ such that $U(\omega)$ as in (2) is a measurable submanifold of $\bar{B}(A(\omega), \alpha(\omega))$ almost surely. However, this does not exclude the possibility that $U(\omega)$ is a single point, see Example 2.

4 Examples

We provide two examples of independent iterated functions on $\mathbb{R}$. Each of them generates a random dynamical system. The functions will be almost surely strictly increasing, continuous and onto and they will fix 0. In the first example, all trajectories which do not start at 0 converge to ∞ or $-\infty$ almost surely (depending on the sign of the initial condition) in spite of the fact the Lyapunov exponent associated to the equilibrium 0 is strictly negative. In particular, there is no synchronization. The second example just consists of an iteration of the inverses of the functions in the first example (in particular it is also order preserving). In this case the Lyapunov exponent is the negative of the one in the first example and hence strictly positive. From the results about the first example, we immediately obtain that the second example exhibits synchronization, i.e. every compact subset of $\mathbb{R}$ contracts to 0 in probability as $n \to \infty$. Since the convergence in the first example is not only in probability but even almost sure we obtain, that in the second example $\{0\}$ is not only a weak attractor but even a pullback attractor (see [7, Proposition 4.6]).

We will comment on the relation of these examples to the results in the previous section after presenting the examples.

Example 1 Let $(\xi_n)_{n\in\mathbb{N}} > 0$ be independent identically distributed real-valued random variables such that $\mathbb{P}\left(\xi_1 \leq 2^{-k}\right) = 1/(k-1)$ for all $k \geq 2$ and $k \in \mathbb{N}$. Define the function $g : \mathbb{R} \times (0,\infty) \to \mathbb{R}$ by

$$g(z,\xi) = \begin{cases} z/2, & |z| \leq 2\xi \\ z/\xi + \xi - 2, & z > 2\xi \\ z/\xi - \xi + 2, & z < -2\xi \ . \end{cases}$$

Obviously, 0 is a fixed point of $g(\cdot,\xi)$ for each $\xi > 0$ and hence $\rho := \mathbb{P} \otimes \delta_0$ is an invariant measure of the associated discrete time random dynamical system φ given by $\varphi_n(\omega,z) = g\left(\varphi_{n-1}(\omega,z),\xi_n\right)$ for $z \in \mathbb{R}$ and $n \in \mathbb{N}$ with state space $\mathscr{H} = \mathbb{R}$. Clearly, the Lyapunov exponent associated to ρ is $\log(1/2) < 0$. We write $Z_n(\omega) := \varphi_n(\omega,z)$ whenever the initial condition $Z_0 = z$ is clear from the context. We will show that $|Z_n|$ converges to infinity $\mathbb{P}$-almost surely whenever $Z_0 \neq 0$. To see this, observe that the following properties hold for every $m \in \mathbb{N}$:

- $|Z_{m-1}| \geq 1$ implies $|Z_m| \geq 4|Z_{m-1}| - 2 \geq 2|Z_{m-1}|$,
- $|Z_m| < 1$ implies $|Z_{m-1}| \leq 4\xi_m$.

Assume that $|Z_0| > 2^{-k}$ for some $k \in \mathbb{N}$. Then, $|Z_m| > 2^{-k-m}$ for all $m \in \mathbb{N}$ and therefore

$$\begin{aligned}\mathbb{P}\left(|Z_n| < 1\right) &\leq \mathbb{P}\left(\xi_m > 2^{-k-m-1} \text{ for all } 1 \leq m \leq n\right) \\ &= \prod_{m=1}^{n} \frac{k+m-1}{k+m} = \frac{k}{k+n} \xrightarrow{n\to\infty} 0 \ .\end{aligned}$$

Using the first of the two observations above, we obtain $|Z_n| \to \infty$ almost surely whenever $Z_0 \neq 0$.

Example 2 Define the sequence $(\xi_n)_{n\in\mathbb{N}} > 0$ as above and define $f : \mathbb{R} \times (0,\infty) \to \mathbb{R}$ by

$$f(\cdot,\xi) = g^{-1}(\cdot,\xi)$$

for each fixed $\xi > 0$. As mentioned at the beginning of the section, the associated random dynamical system exhibits synchronization in spite of the fact that the top Lyapunov exponenent associated to its invariant measure $\rho = \mathbb{P} \otimes \delta_0$ is strictly positive.

Remark 3 In none of the two examples above the random dynamical system is continuously differentiable in the initial state z. This can easily be mended. Just replace the function g by a function $\tilde{g}$ which is smooth and strictly increasing in its first argument such that $|\tilde{g}(x)| \geq |g(x)|$ for all $x \in \mathbb{R}$ and such that $\tilde{g}(x) = g(x)$ whenever $|x| \notin [\xi, 3\xi]$. Then, the absolute values of the modified trajectories converge to ∞ even faster than for g and in Example 2 the speed of synchronization is even faster after the modification. Note that the change from g to $\tilde{g}$ does not change the Lyapunov exponents.

Let us comment on the relation of the examples to the results in the previous section. Obviously, the random dynamical system φ in Example 1 does not only fail to synchronize but even fails to be asymptotically stable as defined in Corollary 1 (note that in this case asymptotic stability is necessary but not sufficient for synchronization by [9]). Therefore, the assumptions of Proposition 1 cannot hold for this example. Indeed, property (1) fails to hold since

$$\mathbb{E}\left[\log^+ \|\varphi_1\|_{C^1([-1,1])}\right] \geq \mathbb{E}\left[\log^+ \frac{1}{\xi_1}\right] = \infty .$$

The first integrability assumption in Proposition 1 and negativity of the Lyapunov exponent both hold in Example 1 showing that (1) cannot be dropped in Proposition 1.

Actually, the stable manifold of Example 1 is even $\{0\}$. Since the stable manifold of Example 1 and the unstable manifold of Example 2 coincide, Example 2 does not satisfy the assumptions of Proposition 2. In particular, positivity of the top Lyapunov exponent implies neither non-triviality of the unstable manifold nor lack of synchronization.

References

1. Arnold, L.: Random Dynamical Systems. Springer Monographs in Mathematics. Springer, Berlin (1998)
2. Beyn, W.J., Gess, B., Lescot, P., Röckner, M.: The global random attractor for a class of stochastic porous media equations. Commum. Partial Differ. Equ. **36**(3), 446–469 (2011)
3. Castaing, C., Valadier, M.: Convex Analysis and Measurable Multifunctions. Lecture Notes in Mathematics, vol. 580. Springer, Berlin-New York (1977)
4. Chueshov, I., Scheutzow, M.: On the structure of attractors and invariant measures for a class of monotone random systems. Dyn. Syst. **19**(2), 127–144 (2004)
5. Cranston, M., Gess, B., Scheutzow, M.: Weak synchronization for isotropic flows. Discrete Contin. Dyn. Syst. Ser. B **21**(9), 3003–3014 (2016)
6. Crauel, H.: Random Probability Measures on Polish Spaces. Stochastics Monographs. CRC Press (2003)
7. Crauel, H., Dimitroff, G., Scheutzow, M.: Criteria for strong and weak random attractors. J. Dynam. Differ. Equ. **21**(2), 233–247 (2009)
8. Crauel, H., Flandoli, F.: Attractors for random dynamical systems. Probab. Theory Related Fields **100**(3), 365–393 (1994)
9. Flandoli, F., Gess, B., Scheutzow, M.: Synchronization by noise. Probab. Theory Related Fields **168**(3–4), 511–556 (2017)
10. Flandoli, F., Gess, B., Scheutzow, M.: Synchronization by noise for order-preserving random dynamical systems. Ann. Probab. **45**(2), 1325–1350 (2017)
11. Gess, B., Liu, W., Röckner, M.: Random attractors for a class of stochastic partial differential equations driven by general additive noise. J. Differ. Equ. **251**(4–5), 1225–1253 (2011)
12. Ruelle, D.: Characteristic exponents and invariant manifolds in Hilbert space. Ann. Math. **115**(2), 243–290 (1982)
13. Vorkastner, I.: Noise dependent synchronization of a degenerate SDE. Stoch. Dyn. **18**(1), 1850007, 21 (2018)

Nonlinear Fokker–Planck–Kolmogorov Equations for Measures

Stanislav V. Shaposhnikov

Abstract Existence and uniqueness of solutions of the Cauchy problem for nonlinear Fokker–Planck–Kolmogorov equations for measures are investigated. We consider the difficult case when the diffusion matrix depends on the solution. Moreover we give a short survey of the known results connected with these problems.

Keywords Nonlinear Fokker–Planck–Kolmogorov equation · McKean–Vlasov equation · Existence and uniqueness of solutions to the Cauchy problem for nonlinear parabolic equations

AMS subject classification. 35K55 · 35Q84 · 35Q83

1 Introduction

We consider the Cauchy problem for nonlinear Fokker–Planck–Kolmogorov equations for measures:

$$\partial_t \mu = \partial_{x_i} \partial_{x_j} \bigl(a^{ij}(x, t, \mu)\mu\bigr) - \partial_{x_i} \bigl(b^i(x, t, \mu)\mu\bigr), \quad \mu|_{t=0} = \nu, \tag{1}$$

where we assume the summation over repeating indices. The diffusion matrix $A = (a^{ij})$ is symmetric and nonnegative, a solution $\mu = \mu_t(dx)\,dt$ is given by a family of probability measures μ_t on $\mathbb{R}^d$. Such equations describe the evolution of a probability

S. V. Shaposhnikov (✉)
Lomonosov Moscow State University, Moscow, Russia
e-mail: starticle@mail.ru

S. V. Shaposhnikov
National Research University Higher School of Economics, Moscow, Russia

S. V. Shaposhnikov
St.-Tikhon's Orthodox Humanitarian University, Moscow, Russia

A. Eberle et al. (eds.), *Stochastic Partial Differential Equations and Related Fields*, Springer Proceedings in Mathematics & Statistics 229,
https://doi.org/10.1007/978-3-319-74929-7_24

measure under the action of a deterministic or stochastic differential equation and play the crucial role in the theory of diffusion processes. The aim of the present paper is to obtain sufficient conditions for the existence and uniqueness that allow non-smooth and unbounded coefficients. Furthermore, we investigate the difficult situation when the diffusion matrix depends on a solution. It turns out that the uniqueness in this case depends on the metric introduced on the space of measures and also depends on the regularity of the initial condition ν. Below we give an example of the Cauchy problem (1) with $b = 0$ and $A(x, t, \mu) = A(t, \mu)$ such that $|A(t, \mu) - A(t, \sigma)| \le C\|\mu_t - \sigma_t\|_{TV}$ which has two different solutions. We present several new results for the case of a nondegenerate diffusion matrix and recall some known results from [6, 16, 17]. The estimates established in [6] enable us to obtain new uniqueness conditions for equations with a diffusion matrix depending on the solution.

Fokker–Planck–Kolmogorov equations generalize several types of equations important for applications: transport equations, Vlasov equations, linear Fokker–Planck–Kolmogorov equations, and McKean–Vlasov equations. An extensive literature is devoted to each type of equations. Let us mention the classical Kolmogorov paper [13], where he derived linear Fokker–Planck–Kolmogorov equations for the transition probabilities of a diffusion process and the McKean papers [18, 19] on nonlinear parabolic equations. In the general case, such equations and the well-posedness of the martingale problem with Lipschiz coefficients were studied by Funaki [11]. Transport equations, linear Fokker–Planck–Kolmogorov equations, Vlasov equations and Boltzmann equations with Sobolev coefficients are investigated in [8, 9, 14], where the method of renormalized solutions is developed and the existence and uniqueness problems are studied in the space L^p. Many papers (see, e.g., [1, 7, 12]) develop the gradient flow approach. Note also that some existence results for finite- and infinite-dimensional nonlinear continuity equations (the case $A = 0$) with rapidly growing coefficients are given in [3, 4]. Infinite dimensional nonlinear Fokker–Planck–Kolmogorov equations are investigated in [15]. Physical problems leading to the study of nonlinear Fokker–Planck–Kolmogorov equations can be found in [10]. In [5], a survey of results about linear equations is presented.

We use throughout the following notation. Let $C_0^\infty(\Omega)$ denote the class of smooth functions with compact support. By $L_p(\Omega)$ (or $L_p(\mu, \Omega)$ in case of a measure μ) we denote the space of all measurable functions f on Ω such that $|f|^p$ is integrable with respect to Lebesgue measure or μ, respectively. Let $C^\delta(\Omega)$ be the space of Hölder continuous functions of order $\delta \in (0, 1)$. By $\nabla f = (\partial_{x_1} f, \dots, \partial_{x_d} f)$ we denote the gradient of f with respect to the space variable x. We also assume the summation over the repeated indices in terms like $a^{ij}\partial_{x_i}\partial_{x_j}$ and $b^i\partial_{x_i}$.

The paper consists of four sections. The first section is the introduction, the second section contains auxiliary results about linear equations, in the third section we discuss the existence of solutions, and in the fourth section the uniqueness problems are considered.

2 Linear Fokker–Planck–Kolmogorov Equations

Recall several results about linear Fokker–Planck–Kolmogorov equations from [5] that will be applied below to nonlinear equations.

A Borel measure μ on $\mathbb{R}^d \times [0, T]$ is given by a family of probability measures $(\mu_t)_{t\in[0,T]}$ if $\mu_t \geq 0$, $\mu_t(\mathbb{R}^d) = 1$, for every Borel set B the mapping $t \to \mu_t(B)$ is measurable and for every $u \in C_0^\infty(\mathbb{R}^d \times (0, T))$ one has

$$\int_{\mathbb{R}^d\times[0,T]} u\,d\mu = \int_0^T \int_{\mathbb{R}^d} u(x,t)\,\mu_t(dx)\,dt.$$

We write $\mu(dxdt) = \mu_t(dx)\,dt$ or $\mu = \mu_t\,dt$.

Set $L_{A,b} = a^{ij}\partial_{x_i}\partial_{x_j} + b^i\partial_{x_i}$, where $A(x,t) = (a^{ij}(x,t))_{1\leq i,j\leq d}$ is a symmetric and nonnegative matrix, $b(x,t) = (b^i(x,t))_{1\leq i\leq d}$ is a vector field on $\mathbb{R}^d \times [0, T]$, and functions a^{ij}, b^i are Borel measurable. Let ν be a probability measure on $\mathbb{R}^d$.

We say that a measure $\mu = \mu_t\,dt$ given by a family of probability measures μ_t satisfies the Cauchy problem

$$\partial_t\mu = L_{A,b}^*\mu, \quad \mu|_{t=0} = \nu \tag{2}$$

if $a^{ij}, b^i \in L^1(U \times [0, T], \mu)$ for every ball $U \subset \mathbb{R}^d$ and for every $\varphi \in C_0^\infty(\mathbb{R}^d)$ the equality

$$\int_{\mathbb{R}^d} \varphi\,d\mu_t = \int_{\mathbb{R}^d} \varphi\,d\nu + \int_0^t \int_{\mathbb{R}^d} L_{A,b}\varphi\,d\mu_s\,ds$$

holds for almost every $t \in (0, T)$.

We say that $A = (a^{ij})_{1\leq i,j\leq d}$ and $b = (b^i)_{1\leq i\leq d}$ satisfy condition (H) if

(Ha) for every ball U there exist numbers $\lambda = \lambda(U) > 0$ and $\Lambda = \lambda(U) > 0$

$$|A(x,t) - A(y,t)| \leq \Lambda|x - y|, \quad \lambda|\xi|^2 \leq \langle A(x,t)\xi, \xi\rangle \leq \lambda^{-1}|\xi|^2$$

for every $x, y \in U$, $\xi \in \mathbb{R}^d$ and $t \in [0, T]$.

(Hb) for every ball U there exists a number $B(U) > 0$ such that

$$\sup_{(x,t)\in U\times[0,T]} |b(x,t)| \leq B(U).$$

The next result follows immediately form Theorem 6.4.2 in [5].

Proposition 1 *Assume that A and b satisfy* (H). *Then any solution μ to* (2) *has a density ρ with respect to Lebesgue measure on $\mathbb{R}^d \times (0, T)$, ρ is locally Hölder continuous and there exists a number $\delta \in (0, 1)$ such that $\|\rho\|_{C^\delta(U\times J)} \leq C(U, J)$ for every ball $U \subset \mathbb{R}^d$, every interval $J \subset (0, T)$ and some constant $C(U, J)$ that depends only on d and λ, Λ, $B(U)$ from* (H). *Note that $\rho > 0$ according to the Harnack inequality.*

The following theorem (a partial case of Theorem 9.4.8 in [5]) gives sufficient conditions for the existence and uniqueness of a solution μ.

Theorem 1 *Assume that A and b satisfy* (H) *and there exists a function* $V \in C^2(\mathbb{R}^d)$ *such that* $\lim_{|x|\to+\infty} V(x) = +\infty$ *and for some number* $C > 0$ *we have*

$$L_{A,b}V(x,t) \le C + CV(x) \quad \forall (x,t) \in \mathbb{R}^d \times (0,T).$$

Then, for every probability measure ν *on* $\mathbb{R}^d$, *the Cauchy problem* (2) *has a unique solution* $\mu = \mu_t\,dt$, *where* μ_t *is a probability measure for every* $t \in [0,T]$.

The function V is called a Lyapunov function for $L_{A,b}$. In the following proposition (a partial case of Theorem 7.1.1 in [5]) we give an a priori estimate with a Lyapunov function.

Proposition 2 *Assume that* $\mu = \mu_t\,dt$, *where* μ_t *is a family of probability measures, solves the Cauchy problem* (2) *and there exists a function* $V \in C^2(\mathbb{R}^d)$ *such that* $\lim_{|x|\to+\infty} V(x) = +\infty$ *and for all* $(x,t) \in \mathbb{R}^d \times (0,T)$ *we have*

$$L_{A,b}V(x,t) \le C_1(t) + C_2(t)V(x),$$

where C_1, C_2 *are nonnegative continuous functions on* $[0,T]$. *Suppose also that* $V \in L^1(\nu)$. *Then for almost all* $t \in (0,T)$ *we have*

$$\int_{\mathbb{R}^d} V(x)\,d\mu_t \le Q(t) + R(t)\int_{\mathbb{R}^d} V\,d\nu,$$

where

$$R(t) = \exp\Bigl(\int_0^t C_2(s)\,ds\Bigr), \quad Q(t) = R(t)\int_0^t \frac{C_1(s)}{R(s)}\,ds$$

Finally, we present some estimates for distances between different solutions to the Cauchy problem with different coefficients. Assume that $\mu = (\mu_t)_{t\in(0,T)}$ and $\sigma = (\sigma_t)_{t\in(0,T)}$ are two solutions to the Cauchy problem (2) with coefficients A_μ, b_μ and A_σ, b_σ, respectively, and the same initial condition ν.

Let $\mu = \rho_\mu(x,t)\,dxdt$ and $\sigma = \rho_\sigma(x,t)\,dxdt$. Let us introduce the following vector mappings:

$$h_\mu = (h^i_\mu)_{i=1}^d, \quad h_\sigma = (h^i_\sigma)_{i=1}^d, \quad h^i_\mu = b^i_\mu - \sum_{j=1}^d \partial_{x_j} a^{ij}_\mu, \quad h^i_\sigma = b^i_\sigma - \sum_{j=1}^d \partial_{x_j} a^{ij}_\sigma,$$

$$\Phi = \frac{(A_\mu - A_\sigma)\nabla\rho_\sigma}{\rho_\sigma} - (h_\mu - h_\sigma).$$

Let $\|\cdot\|_{TV}$ denote the total variation norm on bounded measures.

Theorem 2 *Suppose throughout that A_μ, b_μ and A_σ, b_σ satisfy* (H) *and there exist a nonnegative function $V \in C^2(\mathbb{R}^d)$ and a number $C \geq 0$ such that*

$$\lim_{|x|\to\infty} V(x) = +\infty, \quad L_{A_\mu,b_\mu} V \leq CV, \quad \frac{\langle \Phi, \nabla V\rangle}{1+V} \in L^1(\mathbb{R}^d \times [0,T], \sigma).$$

Then

$$\|\mu_t - \sigma_t\|_{TV}^2 \leq \int_0^t \int_{\mathbb{R}^d} \left| \frac{A_\mu^{-1/2}(A_\mu - A_\sigma)\nabla \rho_\sigma}{\rho_\sigma} - A_\mu^{-1/2}(h_\mu - h_\sigma)\right|^2 d\sigma_s\, ds. \quad (3)$$

In particular, if $A_\mu = A_\sigma = A$, then

$$\|\mu_t - \sigma_t\|_{TV}^2 \leq \int_0^t \int_{\mathbb{R}^d} |A^{-1/2}(b_\mu - b_\sigma)|^2\, d\sigma_s\, ds.$$

Applying an a priori estimate with a Lyapunov function and estimates of the $L^2(\sigma)$-norm of $\nabla\rho_\sigma/\rho_\sigma$ obtained in [2] (see also [5, Chap. 7]) we obtain the following corollary.

Corollary 1 *Let A_μ, b_μ and A_σ, b_σ satisfy* (H) *with λ and Λ which do not depend on U. Assume also that $|x|^{2m} \in L^1(\nu)$, $\nu = \rho_0\, dx$, $\rho_0 \ln \rho_0 \in L^1(\mathbb{R}^d)$ and*

$$\langle b_\mu(x,t), x\rangle \leq \gamma_1 + \gamma_2|x|^2, \quad |b_\sigma(x,t)| \leq \gamma_3 + \gamma_4|x|^m$$

for some numbers $m, \gamma_i \geq 0$. Then

$$\frac{\lambda}{2}\|\mu_t - \sigma_t\|_{TV}^2 \leq \int_0^t \int_{\mathbb{R}^d} \|A_\mu - A_\sigma\|^2 \frac{|\nabla\rho_\sigma|^2}{\rho_\sigma}\, dx\, ds + \int_0^t \int_{\mathbb{R}^d} |h_\mu - h_\sigma|^2\, d\sigma_s\, ds,$$

where

$$\int_0^t \int_{\mathbb{R}^d} \frac{|\nabla\rho_\sigma|^2}{\rho_\sigma}\, dx \leq C(T)$$

and the number $C(T)$ depends on $T, m, \lambda, \Lambda, \gamma_i, \int |x|^{2m}\, d\nu$, and $\|\rho_0 \ln \rho_0\|_{L^1(\mathbb{R}^d)}$.

The proofs of Theorem 2 and Corollary 1 are given in [6].

3 Nonlinear Fokker–Planck–Kolmogorov Equations: Existence

Let $\mathcal{M}_\tau$ be the set of all measures $\mu = \mu_t\, dt$ on $\mathbb{R}^d \times [0,\tau]$, where $(\mu_t)_{t\in[0,\tau]}$ is a family of probability measures on $\mathbb{R}^d$. Let $\mathcal{M}_0$ be a subset of $\mathcal{M}_\tau$. Assume that for every $\mu \in \mathcal{M}_0$ we are given Borel measurable functions $a^{ij}(x,t,\mu)$ and $b^i(x,t,\mu)$

such that $A(\mu) = (a^{ij}(\mu))$ is a symmetric and nonnegative matrix. Set

$$L_\mu = a^{ij}(x, t, \mu)\partial_{x_i}\partial_{x_j} + b^i(x, t, \mu)\partial_{x_i}.$$

We say that $\mu = \mu_t\,dt$ from $\mathcal{M}_0$ is a solution to the Cauchy problem on $\mathbb{R}^d \times [0, \tau]$

$$\partial_t \mu = L_\mu^* \mu, \quad \mu|_{t=0} = \nu, \tag{4}$$

for the nonlinear Fokker–Planck–Kolmogorov equation if μ is a solution to the Cauchy problem (2) on $\mathbb{R}^d \times [0, \tau]$ for the linear Fokker–Planck–Kolmogorov equation with the operator L_μ.

Let $V \in C^2(\mathbb{R}^d)$, $V \geq 0$ and $\lim\limits_{|x|\to+\infty} V(x) = +\infty$. Let $C^+([0, T])$ denote the space of nonnegative continuous functions on $[0, T]$. For every $\tau \in (0, T]$ and $\alpha \in C^+([0, T])$ we introduce the set $\mathcal{M}_{\alpha,\tau}(V)$ of all measures $\mu = \mu_t\,dt$ on $\mathbb{R}^d \times [0, \tau]$, where $(\mu_t)_{t\in[0,\tau]}$ is a family of probability measures on $\mathbb{R}^d$, such that

$$\int_{\mathbb{R}^d} V(x)\,d\mu_t \leq \alpha(t).$$

for all $t \in [0, \tau]$. Let $\mathcal{M}_\tau(V)$ denote the set of all measures $\mu = \mu_t\,dt$ such that

$$\sup_{t\in[0,\tau]} \int_{\mathbb{R}^d} V(x)\,d\mu_t < \infty.$$

By $\mathcal{L}_\tau(V)$ and $\mathcal{L}_{\alpha,\tau}(V)$ we denote subsets of all measures $\mu = \mu_t\,dt$ from $\mathcal{M}_\tau(V)$ and $\mathcal{M}_{\alpha,\tau}(V)$, respectively, such that $\mu_t = \rho(x, t)\,dt$ for every $t \in (0, T)$.

We suppose that for every $\mu \in \mathcal{L}_T(V)$ there are Borel measurable functions $a^{ij}(x, t, \mu)$ and $b^i(x, t, \mu)$ such that $A(\mu) = (a^{ij}(\mu))$ is a symmetric and nonnegative matrix, and

(NH1) there exist mappings $M_1, M_2\colon C^+([0, T]) \to C^+([0, T])$ such that for every number $\tau \in (0, T]$ and $\alpha \in C^+([0, T])$ we have

$$L_\mu V(x, t) \leq M_1(\alpha)(t) + M_2(\alpha)(t)V(x)$$

for every $(x, t) \in \mathbb{R}^d \times [0, \tau]$ and every $\mu \in \mathcal{L}_{\alpha,\tau}(V)$;

(NH2) for every $\alpha \in C^+([0, T])$, $\tau \in (0, T]$ and every ball $U \subset \mathbb{R}^d$ there exist constants $B = B(\alpha, \tau, U)$, $\lambda = \lambda(\alpha, \tau, U)$ and $\Lambda = \Lambda(\alpha, \tau, U)$ such that

$$|a^{ij}(x, t, \mu) - a^{ij}(x, t, \mu)| \leq \Lambda|x - y|, \quad \lambda|\xi|^2 \leq \langle A(x, t, \mu)\xi, \xi\rangle \leq \lambda^{-1}|\xi|^2,$$

$$|b(x, t, \mu)| \leq B$$

for every $\mu \in \mathcal{L}_{\alpha,\tau}(V)$ and $(x, t) \in U \times [0, \tau]$, $\xi \in \mathbb{R}^d$, moreover if $\{\mu^n\}, \mu \in \mathcal{L}_{\alpha,\tau}(V)$ and $\|\mu_t^n - \mu_t\|_{TV} \to 0$ for almost all $t \in [0, T]$, then

$$\lim_{n\to\infty} a^{ij}(x,t,\mu^n) = a^{ij}(x,t,\mu), \quad \lim_{n\to\infty} b^i(x,t,\mu^n) = b^i(x,t,\mu)$$

for almost every $(x,t) \in \mathbb{R}^d \times [0,\tau]$ with respect to Lebesgue measure.

Theorem 3 *Assume that* (NH1) *and* (NH2) *are fulfilled. Then, for every probability measure ν such that $V \in L^1(\nu)$, there exists $\tau \in (0,T]$ such that on $[0,\tau]$ the Cauchy problem* (4) *has a solution $\mu = \mu_t\,dt$ of class $\mathscr{L}_\tau(V)$. Moreover, if $M_1(\alpha)(t) = g_1(\alpha(t))$, $M_2(\alpha)(t) = g_2(\alpha(t))$, where g_1, g_2 are positive, strictly increasing and continuous functions on $[0,+\infty)$, then on $[0,\tau]$ the Cauchy problem* (4) *has a solution $\mu = \mu_t\,dt$ of class $\mathscr{L}_\tau(V)$ for every $\tau < \min\{\tau^*, T\}$, where*

$$\tau^* = \int_{u_0}^{+\infty} \frac{du}{g_1(u) + ug_2(u)}, \quad u_0 = \int_{\mathbb{R}^d} V\,d\nu.$$

Proof Let us consider the mapping T such that

$$\mu = T(\sigma) \Leftrightarrow \partial_t\mu = L_\sigma\mu, \quad \mu|_{t=0} = \nu.$$

According to Theorem 1 the mapping T is well-defined.

1. We prove that there exist α and $\tau \in (0,T]$ such that $T\colon \mathscr{L}_{\alpha,\tau}(V) \to \mathscr{L}_{\alpha,\tau}(V)$. Let us set $\alpha \equiv 2$. For $\sigma \in \mathscr{L}_{\alpha,\tau}(V)$ we set $\mu = T(\sigma)$. According to Proposition 2 we have

$$\int_{\mathbb{R}^d} V(x)\,d\mu_t \le Q(t) + R(t)\int_{\mathbb{R}^d} V\,d\nu,$$

where

$$R(t) = \exp\Bigl(\int_0^t M_2(\alpha)(s)\,ds\Bigr), \quad Q(t) = R(t)\int_0^t \frac{M_1(\alpha)(s)}{R(s)}\,ds.$$

Note that $\lim_{t\to 0} R(t) = 1$, $\lim_{t\to 0} Q(t) = 0$. Then there is a number $\tau \in (0,T]$ such that $\displaystyle\int_{\mathbb{R}^d} V(x)\,d\mu_t \le 2$ for every $t \in [0,\tau]$. This means that $\mu \in \mathscr{L}_{\alpha,\tau}(V)$.

In the case when $M_1(\alpha)(t) = g_1(\alpha(t))$, $M_2(\alpha)(t) = g_2(\alpha(t))$, we express α from the equation $\alpha(t) = Q(t) + R(t)u_0$. Differentiating we obtain $\alpha' = g_1(\alpha) + \alpha g_2(\alpha)$. For $t \in [0,\tau^*]$ we set

$$\alpha(t) = G^{-1}(t), \quad G(u) = \int_{u_0}^{u} \frac{du}{g_1(u) + ug_2(u)},$$

and $\alpha(t) = \alpha(\tau^*)$ if $t > \tau^*$. Then $\mu \in \mathscr{M}_{\alpha,\tau}(V)$ for every $\tau \in (0,\tau^*)$.

2. Now we assume that $T\colon \mathscr{L}_{\alpha,\tau}(V) \to \mathscr{L}_{\alpha,\tau}(V)$, where α and τ have been defined above. Recall that any solution to the linear Fokker–Planck–Kolmogorov equation with a nondegenerate diffusion matrix has a density with respect to Lebesgue measure. Let $\mu = T(\sigma)$ have a density ρ. According to Proposition 1, for each ball $U \subset \mathbb{R}^d$ and interval $J \subset (0,T)$, there exists a constant $C(U,J)$ depending only on

U, J, d, δ and constants λ, Λ, and B from (NH2) such that

$$\|\rho\|_{C^{\delta}(U\times J)} \le C(U, J).$$

Let us consider the set $\mathcal{K}$ of all nonnegative Hölder continuous functions ρ on $\mathbb{R}^d$ such that $\rho(x,t)\,dx\,dt \in \mathcal{L}_{\alpha,\tau}(V)$ and $\|\rho\|_{C^{\delta}} \le C(U, J)$ for every ball U and interval J. Note that $\mathcal{K}$ is a convex compact set in $L^1(\mathbb{R}^d \times [0, T])$. The convexity is obvious. Let $\{\rho_n\} \subset \mathcal{K}$. One can choose a subsequence ρ_{n_k} such that ρ_{n_k} converges to a function ρ uniformly on every set $U \times J$. Hence $\|\rho\|_{C^{\delta}} \le C(U, J)$ and

$$\int_{\mathbb{R}^d} \rho(x,t)\,dx = 1, \quad \int_{\mathbb{R}^d} V(x)\rho(x,t)\,dx \le \alpha(t).$$

It follows that $\|\rho_{n_k} - \rho\|_{L^1(\mathbb{R}^d\times[0,T])} \to 0$ and $\mathcal{K}$ is a compact set. Finally we should verify that T is a continuous mapping $\mathcal{K} \to \mathcal{K}$. Suppose that measures $\sigma^n = q_n\,dx\,dt$, $\sigma = q\,dx\,dt \in \mathcal{K}$ and $q_n \to q$ in $L^1(\mathbb{R}^d \times [0, T])$. Choosing a subsequence one can assume that $\|q_n(\cdot,t) - q(\cdot,t)\|_{L^1(\mathbb{R}^d)} \to 0$ for almost all $t \in [0, T]$. According to condition (NH2) we have

$$\lim_{n\to\infty} a^{ij}(x,t,\sigma^n) = a^{ij}(x,t,\sigma), \quad \lim_{n\to\infty} b^i(x,t,\sigma^n) = b^i(x,t,\sigma)$$

for almost every $(x,t) \in \mathbb{R}^d \times [0,\tau]$. Let $\mu^n = T(\sigma^n)$ and $\mu^n = \rho_n\,dx\,dt$. Choosing a subsequence again one can assume that $\rho_n \to \rho$ uniformly on every set $U \times J$ for some $\mu = \rho\,dx\,dt \in \mathcal{K}$. Moreover, $\rho_n \to \rho$ in $L^1(\mathbb{R}^d \times [0, T])$. Passing to the limit in the equality

$$\int_{\mathbb{R}^d} \varphi(x)\rho_n(x,t)\,dx = \int_{\mathbb{R}^d} \varphi\,d\nu + \int_0^t \int_{\mathbb{R}^d} \rho_n L_{\sigma_n}\varphi\,dx\,ds,$$

where $\varphi \in C_0^\infty(\mathbb{R}^d)$, we obtain that μ is a solution to the Cauchy problem $\partial_t\mu = L_\sigma^*\mu$, $\mu|_{t=0} = \nu$. According to the uniqueness property for the linear equation we conclude that $\mu_n = T(\sigma_n) \to \mu = T(\sigma)$ in L^1. Thus T is a continuous mapping of the convex compact set $\mathcal{K}$ to itself. By the Schauder fixed point theorem there exists $\mu \in \mathcal{K}$ such that $\mu = T(\mu)$.

The following example shows that condition (NH1) deals with the blow-up phenomenon.

Example 1 Let $d = 1$, $b = 0$, $\nu = \delta_0$ and $a(t,\mu) = 1 + G\Bigl(\int |x|^2\,d\mu_t\Bigr)$. Condition (NH1) with $V(x) = |x|^2/2$ has the following form:

$$a(t,\mu)V''(x) = a(t,\mu) \le 1 + G(\alpha(t)).$$

If μ is a solution to the Cauchy problem $\partial_t\mu = a(t,\mu)\partial_x^2\mu$, $\mu|_{t=0} = \delta_0$, then μ has a density

$$\rho(x,t) = (2\pi h(t))^{-1/2} e^{-x^2/2h(t)}, \quad h(t) = \int_0^t a(s,\mu)\,ds.$$

Moreover, $a(t,\mu) = 1+G(h(t))$ and $h'(t) = 1+G(h(t))$. If $\tau^* = \int_0^\infty \frac{du}{1+G(u)} < \infty$, then the Cauchy problem has no solution on $[0,\tau]$ if $\tau > \tau^*$.

Let us consider an example when conditions (NH1) and (NH2) are fulfilled, but the Cauchy problem has at least two different solutions.

Example 2 Let $d = 1$, $A = a(t,\mu)$, $b = 0$ and $\nu = \delta_0$. Set

$$a(t,\mu) = g(t, l(\mu_t)), \quad l(\mu_t) = \int |x|^2\,d\mu_t, \quad \mu = \mu_t\,dt.$$

It is easy to see that $l(\mu_t^1) = t$ and $l(\mu_t^2) = 4t$ for the measures

$$\mu_t^1 = (2\pi t)^{-1/2} e^{-x^2/2t}\,dx, \quad \mu_t^2 = (8\pi t)^{-1/2} e^{-x^2/8t}\,dx.$$

Assume that g is a function with $1 \le g \le 4$, $g(t,t) = 1$, $g(t,4t) = 4$ if $t > 0$ and for each t the mapping $x \to g(t,x)$ is continuous. Thus conditions (NH1), (NH2) are fulfilled and measures $\mu^1 = \mu_t^1\,dt$ and $\mu^2 = \mu_t^2\,dt$ are two different solutions to the Cauchy problems $\partial_t\mu = a(t,\mu)\partial_x^2\mu$, $\mu|_{t=0} = \delta_0$.

Note that the problem with uniqueness in the above example arises because a is irregular with respect to μ. The next example demonstrates that uniqueness depends on the given metric on the space of measures and also depends on the regularity of the initial condition.

Example 3 Let $d = 1$, $A = a(t,\mu)$, $b = 0$ and $\nu = \delta_0$. Set $\mu^1 = \mu_t^1\,dt$ and $\mu^2 = \mu_t^2\,dt$, where

$$\mu_t^1 = (2\pi t)^{-1/2} e^{-x^2/2t}\,dx, \quad \mu_t^2 = (8\pi t)^{-1/2} e^{-x^2/8t}\,dx.$$

Note that $\|\mu_t^1 - \mu_t^2\|_{TV} = c_0 > 0$ and c_0 does not depend on t. Let

$$a(t,\mu) = 1 + \frac{3}{c_0}\|\mu_t - \mu_t^1\|_{TV}.$$

We have $a(t,\mu^1) = 1$, $a(t,\mu^2) = 4$ and $|a(t,\mu) - a(t,\sigma)| \le \frac{3}{c_0}\|\mu_t - \sigma_t\|_{TV}$. As above, the measures μ^1 and μ^2 are two different solutions to the Cauchy problem with this coefficient a and $\nu = \delta_0$.

Several interesting examples of non-uniqueness are constructed in [17].

Let us mention another existence result established in [16]. Now we do not assume that A is nondegenerate, we only suppose that A is a symmetric and nonnegative matrix. Suppose that

(DH1) condition (NH1) is fulfilled for every μ in $\mathscr{M}_{\alpha,\tau}(V)$,

(DH2) for all $\tau \in (0,T]$, $\alpha \in C^+([0,T])$, $\sigma \in M_{\tau,\alpha}$ and $x \in \mathbb{R}^d$ the mappings $t \mapsto a^{ij}(x,t,\sigma)$ and $t \mapsto b^i(x,t,\sigma)$ are Borel measurable on $[0,\tau]$ and for each closed ball $U \subset \mathbb{R}^d$ the mappings $x \mapsto b^i(x,t,\sigma)$ and $x \mapsto a^{ij}(x,t,\sigma)$ are bounded on U uniformly in $\sigma \in M_{\tau,\alpha}$ and $t \in [0,\tau]$ and continuous on U uniformly in $\sigma \in M_{\tau,\alpha}$ and $t \in [0,\tau]$. Moreover, if a sequence $\mu^n \in M_{\tau,\alpha}$ V-converges to $\mu \in M_{\tau,\alpha}$, which means by definition that, for each function $F \in C(\mathbb{R}^d)$ such that $\lim_{|x|\to\infty} F(x)/V(x) = 0$, one has $\lim\limits_{n\to\infty} \int F\,d\mu_t^n = \int F\,d\mu_t$ for each $t \in [0,\tau]$, then for all $(x,t) \in \mathbb{R}^d \times [0,\tau]$ one has $\lim_{n\to\infty} a^{ij}(x,t,\mu^n) = a^{ij}(x,t,\mu)$, $\lim_{n\to\infty} b^i(x,t,\mu^n) = b^i(x,t,\mu)$.

The following result is Theorem 1 in [16].

Theorem 4 *Assume that* (DH1) *and* (DH2) *are fulfilled. Then, for every probability measure* ν *such that* $V \in L^1(\nu)$, *there exist* $\tau \in (0,T]$ *such that on* $[0,\tau]$ *the Cauchy problem* (4) *has a solution* $\mu = \mu_t\,dt$ *from* $\mathscr{M}_\tau(V)$. *If* $M_1(\alpha)(t) = g_1(\alpha(t))$, $M_2(\alpha)(t) = g_2(\alpha(t))$, *where* g_1, g_2 *are positive strictly increasing and continuous functions on* $[0,+\infty)$, *then on* $[0,\tau]$ *the Cauchy problem* (4) *has a solution* $\mu = \mu_t\,dt$ *of class* $\mathscr{M}_\tau(V)$ *for every* $\tau < \min\{\tau^*, T\}$, *where*

$$\tau^* = \int_{u_0}^{+\infty} \frac{du}{g_1(u) + u g_2(u)}, \quad u_0 = \int V\,d\nu.$$

Note that in [16] the cases $g_1 = 0$ and $g_2 = 0$ are considered separately, but in fact the proof is given for the general case when the time τ^* is given by the above expression. In addition, in [16] the blow up phenomenon is investigated.

4 Nonlinear Fokker–Planck–Kolmogorov Equations: Uniqueness

Set $V(x) = 1 + |x|^{2m}$, where $m \geq 1$. Assume that, for every $\mu \in \mathscr{L}_T(V)$, we are given Borel measurable functions $a^{ij}(x,t,\mu)$, and $b^i(x,t,\mu)$, where $A = (a^{ij})$ is a symmetric and nonnegative matrix. Suppose also that

(UH1) for every $\mu \in \mathscr{L}_T(V)$ there exist numbers $\lambda(\mu) > 0$ and $\Lambda(\mu) > 0$ such that

$$\lambda(\mu)^{-1}|\xi|^2 \leq \langle A(x,t,\mu)\xi,\xi\rangle \leq \lambda(\mu)|\xi|^2,\ |a^{ij}(x,t,\mu) - a^{ij}(y,t,\mu)| \leq \Lambda(\mu)|x-y|$$

for every $x, y, \xi \in \mathbb{R}^d$ and $t \in [0,T]$;

(UH2) for every $\mu \in \mathscr{L}_T(V)$ there exists a number $\gamma(\mu)$ such that

$$\langle b(x,t,\mu), x\rangle \le \gamma(\mu) + \gamma(\mu)|x|^2, \quad |b(x,t,\mu)| \le \gamma(\mu) + \gamma(\mu)|x|^m$$

for every $(x,t) \in \mathbb{R}^d \times [0,T]$;

(UH3) there exist numbers $C_1 > 0$ and $C_2 > 0$ such that for every $\mu, \sigma \in \mathscr{L}_T(V)$

$$|a^{ij}(x,t,\mu) - a^{ij}(x,t,\sigma)| + |\nabla_x a^{ij}(x,t,\mu) - \nabla_x a^{ij}(x,t,\sigma)| \le C_1\|\mu_t - \sigma_t\|_{TV}$$

and $|b^i(x,t,\mu) - b^i(x,t,\sigma)| \le C_2(1+|x|^m)\|\mu_t - \sigma_t\|_{TV}$ for every $(x,t) \in \mathbb{R}^d \times [0,T]$.

Theorem 5 *Suppose* (UH1), (UH2) *and* (UH3) *hold true and ν is a probability measure with a density ρ_0 such that $\rho_0 \ln \rho_0 \in L^1(\mathbb{R}^d)$ and $|x|^{2m} \in L^1(\nu)$. Then, in the class $\mathscr{L}_T(V)$, there exists at most one solution to the Cauchy problem* (4) *with initial condition ν.*

Proof Assume that $\mu = \rho_\mu(x,t)\,dx\,dt$ and $\sigma = \rho_\sigma(x,t)\,dx\,dt$ are two solutions to the Cauchy problem. According to Corollary 1 we have

$$\frac{\lambda}{2}\|\mu_t - \sigma_t\|_{TV}^2 \le \sup_{\mathbb{R}^d\times[0,t]} \|A(\mu) - A(\sigma)\|^2 g(t) + \int_0^t\int_{\mathbb{R}^d} |h_\mu - h_\sigma|^2\,d\sigma_s\,ds,$$

where $g(t) = \int_0^t\int_{\mathbb{R}^d} \frac{|\nabla\rho_\sigma|^2}{\rho_\sigma}\,dx\,ds$. Let $q(t) = \int_0^t\int_{\mathbb{R}^d}(1+|x|^{2m})\,d\sigma_s\,ds$ and $\delta > 0$. Then $\sup_{(0,\delta)}\|\mu_t - \sigma_t\|_{TV}^2 \le C(g(t)+q(t))\sup_{(0,\delta)}\|\mu_t - \sigma_t\|_{TV}^2$, where C depends only on λ, d, C_1 and C_2. One can find $\delta > 0$ such that $|g(t)-g(s)|+|q(t)-q(s)| < 1/(2C)$ whenever $|t-s| < 2\delta$. Then $\sup_{t\in(0,2\delta)}\|\mu_t - \sigma_t\|_{TV}^2 \le 0$ and $\mu_t = \sigma_t$ on $(0,2\delta)$. Applying condition (UH3) we can rewrite the inequality from Corollary 1 in the following form:

$$\frac{\lambda}{2}\|\mu_t - \sigma_t\|_{TV}^2 \le \sup_{\mathbb{R}^d\times[\delta,t]} \|A(\mu) - A(\sigma)\|^2(g(t)-g(\delta)) + \int_\delta^t\int_{\mathbb{R}^d} |h_\mu - h_\sigma|^2\,d\sigma_s\,ds.$$

Applying the above reasoning we obtain that $\mu_t = \sigma_t$ on $(\delta, 3\delta)$. Repeating this procedure we obtain that $\mu_t = \sigma_t$ on $[0,T]$.

The following result from [17] is concerned with the case when the coefficients depend continuously on μ with respect to the metric

$$r_W(\mu,\sigma) = \sup\Bigl\{\int f\,d(\mu-\sigma) \,:\, f \in C_0^\infty(\mathbb{R}^d),\ |\nabla f(x)| \le \sqrt{W(x)}\Bigr\}.$$

Let $V \ge 1$. Assume that, for every $\mu \in \mathscr{M}_T(V)$, we are given Borel measurable functions $(x,t) \to a^{ij}(x,t,\mu)$, and $(x,t) \to b^i(x,t,\mu)$ and $A = (a^{ij})$ is a symmetric and nonnegative matrix. Let us introduce the following assumptions:

(WH1) For each $\mu \in M_T(V)$, there exist constants $\lambda_\mu > 0$ and $\Lambda_\mu > 0$ such that

$$\lambda_\mu^{-1}|\xi|^2 \le \langle A(\mu, x, t)\xi, \xi\rangle \le \lambda_\mu|\xi|^2, \quad |a^{ij}(\mu, x, t) - a^{ij}(\mu, y, t)| \le \Lambda_\mu|x - y|$$

for all $x, y, \xi \in \mathbb{R}^d$, $t \in [0, T]$.

(WH2) For each $\mu \in M_T(V)$ and each $x \in \mathbb{R}^d$, the following quantities are finite:

$$B(\mu, x) = \sup_{t\in[0,T]} \sup_{|x-y|\le 1} |b(\mu, y, t)|,$$

$$\Theta(\mu, x) = \sup_{t\in[0,T]} \sup_{|x-y|\le 1, |x-z|\le 1, y\ne z} \frac{|b(\mu, y, t) - b(\mu, z, t)|}{|y - z|}.$$

Moreover, for some convex function $W \in C^2(\mathbb{R}^d)$ such that $|x|\sqrt{W(x)}V(x)^{-1}$ is a bounded function, $W \ge 1$ and for each measure $\mu \in \mathcal{M}_T(V)$ there exist constants $C_\mu > 0$, $1 > \delta_\mu > 0$ such that

$$L_\mu W(x, t) \le (C_\mu - 2\Theta(\mu, x) - \delta_\mu(1 + |x|^2)^{-1}B^2(\mu, x))W(x).$$

(WH3) There exist numbers $C_1 > 0$ and $C_2 > 0$ such that

$$|A(\mu, x, t) - A(\sigma, x, t)| \le C_1 r_W(\mu_t, \sigma_t),$$

$$|b(\mu, x, t) - b(\sigma, x, t)| \le C_2 V(x)W^{-1/2}(x) r_W(\mu_t, \sigma_t)$$

for all $(x, t) \in \mathbb{R}^d \times [0, T]$ and $\mu, \sigma \in \mathcal{M}_T(V)$.

(WH4) There exists a function $U \in C^2(\mathbb{R}^d)$ with $U \ge 0$ and $\lim_{|x|\to+\infty} U(x) = +\infty$ such that, for each measure $\mu \in \mathcal{M}_T(V)$, there exists a constant $\beta(\mu)$ such that

$$\Bigl(B(\mu, x) + \sqrt{\Theta(\mu, x)}\Bigr) \sup_{|x-y|\le 1} \sqrt{W(y)} + \frac{|\nabla U(x)|^2}{U^2(x)} + \frac{|L_\mu U(x, t)|}{U(x)} \le \beta(\mu)V(x)$$

for all $(x, t) \in \mathbb{R}^d \times [0, T]$.

Theorem 6 *Assume that* (WH1), (WH2), (WH3), *and* (WH4) *hold true. Then, in* $\mathcal{M}_T(V)$, *there exists at most one solution to the Cauchy problem* (4).

We observe that the case when A does not depend on μ is considerably simpler and the corresponding existence and uniqueness results are obtained in [6, 17].

Acknowledgements The author is grateful to Prof. V.I. Bogachev and Prof. M. Röckner for fruitful discussions and valuable remarks. The work was supported by the President Grant MD-207.2017.1, the RFBR grant 17-01-00622, the Simons Foundation, and the SFB 1283 at Bielefeld University.

References

1. Ambrosio, L., Gigli, N., Savaré, G.: Gradient Flows in Metric Spaces and in the Space of Probability Measures. Lectures in Mathematics ETH Zürich. Birkhäuser Verlag, Basel (2005)
2. Bogachev, V.I., Röckner, M., Shaposhnikov, S.V.: Global regularity and bounds for solutions of parabolic equations for probability measures. Teor. Verojatn. Primen. **50**(4), 652–674 (2005) (in Russian); English transl. Theory Probab. Appl. **50**(4), 561–581 (2006)
3. Bogachev, V.I., Röckner, M., Shaposhnikov, S.V.: Nonlinear evolution and transport equations for measures. Doklady Math. **80**(3), 785–789 (2009)
4. Bogachev, V.I., Da Prato, G., Röckner, M., Shaposhnikov, S.V.: Nonlinear evolution equations for measures on infinite dimensional spaces. Stochastic Partial Differential Equations and Applications. Quaderni di Matematica, Dipartimento di Matematica Seconda Universita di Napoli Napoli 25, 51–64 (2010)
5. Bogachev, V.I., Krylov, N.V., Röckner, M., Shaposhnikov, S.V.: Fokker–Planck–Kolmogorov Equations. American Mathematical Society, Providence, Rhode Island (2015)
6. Bogachev, V.I., Röckner, M., Shaposhnikov, S.V.: Distances between transition probabilities of diffusions and applications to nonlinear Fokker–Planck–Kolmogorov equations. J. Funct. Anal. **271**, 1262–1300 (2016)
7. Carrillo, J.A., Difrancesco, M., Figalli, A., Laurent, T., Slepcev, D.: Global-in-time weak measure solutions and finite-time aggregation for non-local interaction equations. Duke Math. J. **156**(2), 229–271 (2011)
8. DiPerna, R.J., Lions, P.L.: Ordinary differential equations, transport theory and Sobolev spaces. Invent. Math. **98**, 511–547 (1989)
9. DiPerna, R.J., Lions, P.L.: On the Fokker–Planck–Boltzmann equation. Commun. Math. Phys. **120**(1), 1–23 (1988)
10. Frank, T.D.: Nonlinear Fokker-Planck Equations. Fundamentals and Applications. Springer, Berlin (2005)
11. Funaki, T.: A certain class of duffusion processes associated with nonlinear parabolic equations. Z. Wahrscheinlichkeitstheorie verw. Geb. **67**, 331–348 (1984)
12. Jordan, R., Kinderlehrer, D., Otto, F.: The variational formulation of the Fokker–Planck equation. SIAM J. Math. Anal. **29**(1), 1–17 (1998)
13. Kolmogorov, A.N.: Über die analytischen Methoden in der Wahrscheinlichkeitsrechnung. Mathematische Annalen **104**, 415–458 (1931)
14. Le Bris, C., Lions, P.L.: Existence and uniqueness of solutions to Fokker–Planck type equations with irregular coefficients. Commun. Partial Differ. Equ. **33**, 1272–1317 (2008)
15. Manita, O.A.: Nonlinear Fokker–Planck–Kolmogorov equations in Hilbert spaces. J. Math. Sci. (New York) **216**(1), 120–135 (2016)
16. Manita, O.A., Shaposhnikov, S.V.: Nonlinear parabolic equations for measures. St. Petersb. Math. J. **25**(1), 43–62 (2014)
17. Manita, O.A., Romanov, M.S., Shaposhnikov, S.V.: On uniqueness of solutions to nonlinear Fokker–Planck–Kolmogorov equations. Nonlinear Anal. Theory Methods Appl. **128**, 199–226 (2015)
18. McKean, H.P.: A class of Markov processes associated with nonlinear parabolic equations. Proc. Natl. Acad. Sci. USA **56**, 1907–1911 (1966)
19. McKean, H.P.: Propagation of chaos for a class of non-linear parabolic equations. In: Lecture Series in Differential Equations, Session, 7, pp. 177–194. Catholic University (1967)

Coupling by Change of Measure, Harnack Inequality and Hypercontractivity

Feng-Yu Wang

Abstract The coupling method is a powerful tool in analysis of stochastic processes. To make the coupling successful before a given time, it is essential that two marginal processes are constructed under different probability measures. We explain the main idea of establishing Harnack inequalities for Markov semigroups using these new type couplings, and apply the coupling and Harnack inequality to the study of hypercontractivity of Markov semigroups.

Keywords Coupling by change of measure · Harnack inequality Hypercontractivity · Degenerate SDEs

1 Coupling Method for Harnack Inequality

In 1887, Carl Gustav Axel Harnack found out the following inequality: for an open domain $D \subset \mathbb{R}^2$ and a compact set $K \subset D$, there exists a constant $C(D, K)$ such that for any positive harmonic function u on D,

$$\sup_K u \leq C(D, K) \inf_K u.$$

This inequality can be reformulated as follows: for any open domain D there exists a locally bounded positive function C on $D \times D$ such that

$$u(x) \leq C(x, y)u(y), \quad x, y \in D$$

holds for all positive harmonic functions u on D. This type of inequality is called Harnack inequality and has been extended and applied to positive solutions of many other elliptic or parabolic PDEs.

In this part, we introduce the main idea of establishing Harnack inequalities for Markov semigroups using the coupling method. Let P_t be a Markov semigroup on

F.-Y. Wang (✉)
Center for Applied Mathematics, Tianjin University, Tianjin 300072, China
e-mail: wangfy@tju.edu.cn

A. Eberle et al. (eds.), *Stochastic Partial Differential Equations and Related Fields*, Springer Proceedings in Mathematics & Statistics 229,
https://doi.org/10.1007/978-3-319-74929-7_25

a Polish space E. Let $\mathscr{B}_b^+(E)$ be the class of all non-negative bounded measurable functions on E. Given $t > 0$ and $x, y \in E$, we aim to compare $P_t f(x)$ and $P_t f(y)$ uniformly in $f \in \mathscr{B}_b^+(E)$.

To apply the coupling method, we assume that the semigroup P_t is associated to a strong Markov process. For fixed $x, y \in E$, we consider the processes $X^x(t)$, $X^y(t)$ on the same probability space starting from x and y respectively such that

$$P_t f(x) = \mathbb{E}[f(X^x(t))], \quad P_t f(y) = \mathbb{E}[f(X^y(t))], \quad t \ge 0, f \in \mathscr{B}_b^+(E). \tag{1}$$

Let $\tau = \inf\{t \ge 0 : X^x(t) = X^y(t)\}$ be the coupling time. By the strong Markov property, we may and do let $X^x(t) = X^y(t)$ for $t \ge \tau$. If $\mathbb{P}(\tau > t) = 0$ then $X^x(t) = X^y(t)$ $\mathbb{P}$-a.s., so that (1) gives

$$P_t f(x) = P_t f(y), \quad f \in \mathscr{B}_b^+(E).$$

This is, however, too strong to be true. Indeed, in general τ is an unbounded random variable such that $\mathbb{P}(\tau > t) > 0$ for $t > 0$. But if $\mathbb{P}(X^x(t) \neq X^y(t)) > 0$, then (1) does not provide any non-trivial comparison of $P_t f(x)$ and $P_t f(y)$ up to a constant independent of f, since, when $X^x(t) \neq X^y(t)$, a function f may be zero at $X^x(t)$ but arbitrarily large at $X^y(t)$. Therefore, to derive the Harnack inequality of P_t using coupling, it seems essential that $\tau \le t$, which is however impossible as explained above. To avoid the contradiction, we will construct the coupling under different probability measures, which is called coupling by change of measure.

From now on, we fix a complete probability space $(\Omega, \mathscr{F}, \mathbb{P})$. We shall define the coupling by change of measure for stochastic processes. Let $\mathscr{L}(X)|_{\mathbb{P}}$ denote the law of a process $X(t)$ under the probability $\mathbb{P}$.

Definition 1.1 Let $X(t)$ and $Y(t)$ be two stochastic processes on E. A stochastic process $(\bar{X}(t), \bar{Y}(t))$ on $E \times E$ is called a coupling by change of measure for $X(t)$ and $Y(t)$ with changed measure $\mathbb{Q}$, if $\mathscr{L}(X)|_{\mathbb{P}} = \mathscr{L}(\bar{X})|_{\mathbb{P}}$ and $\mathbb{Q}$ is a probability measure on $(\Omega, \mathscr{F})$ such that $\mathscr{L}(Y)|_{\mathbb{P}} = \mathscr{L}(\bar{Y})|_{\mathbb{Q}}$. If, in particular, $\mathbb{Q} = \mathbb{P}$, we call $(\bar{X}(t), \bar{Y}(t))$ a coupling for $X(t)$ and $Y(t)$.

In applications, we assume that $\mathbb{Q}$ is absolutely continuous with respect to $\mathbb{P}$. In this case, with a coupling by change of measure satisfying $\bar{X}(T) = \bar{Y}(T)$ $\mathbb{Q}$-a.s for a fixed $T > 0$, one may compare the distributions of $X(T)$ and $Y(T)$ using the density $R := \frac{d\mathbb{Q}}{d\mathbb{P}}$ (we also denote $\mathbb{Q} = R\mathbb{P}$).

The following is a general result on Harnack type inequalities using coupling by change of measure.

Theorem 1.1 *Let P_t be the Markov semigroup and let $x, y \in E, T > 0$ be fixed. Suppose there is a coupling by change of measure $(\bar{X}(t), \bar{Y}(t))_{t\in[0,T]}$ with $\mathbb{Q} := R\mathbb{P}$ such that $\bar{X}(T) = \bar{Y}(T)$ $\mathbb{Q}$-a.s. Then for any $f \in \mathscr{B}^+(E)$,*

$$(P_T f)^p(y) \le \{P_T f^p(x)\}\{\mathbb{E}[R^{p/(p-1)}]\}^{p-1}, \quad p > 1,$$
$$(P_T \log f)(y) \le \log(P_T f)(x) + \mathbb{E}[R \log R].$$

Proof By the definition of coupling by change of measure, we have $P_T f(x) = \mathbb{E}f(\bar{X}(T)), \mathbb{E}[Rf(\bar{Y}(T))] = P_T f(y)$. Combining with $\bar{X}(T) = \bar{Y}(T)$ $\mathbb{Q}$-a.s. and using the Hölder inequality, we obtain

$$\begin{aligned}(P_T f)^p(y) &= \{\mathbb{E}[Rf(Y(T))]\}^p = \{\mathbb{E}[Rf(X(T))]\}^p \\ &\le \{\mathbb{E}[f^p(X(T))]\}\{\mathbb{E}R^{p/(p-1}\}^{p-1} = \{P_T f^p(x)\}\{\mathbb{E}[R^{p/(p-1)}]\}^{p-1}.\end{aligned}$$

Moreover, the Young inequality ([2, Lemma 2.4]) implies

$$\begin{aligned}(P_T \log f)(y) &= \mathbb{E}[R\log f(Y(T))] = \mathbb{E}[R\log f(X(T))] \\ &\le \log \mathbb{E}[f(X(T))] + \mathbb{E}[R\log R] = \log(P_T f)(x) + \mathbb{E}[R\log R].\end{aligned}$$

□

The Harnack inequality with a power $p > 1$ was first found in [16] for diffusion semigroups on manifolds with curvature bounded below using gradient estimates, and was then extended in [1, 2, 18] to unbounded below curvatures using coupling by change of measure. The log-Harnack inequality was introduced in [14, 19] for semi-linear SPDEs and Neumann semigroups on manifolds respectively. Both inequalities have been intensively investigated and applied for many other models, see e.g. [9, 10, 18, 25] for non-linear SPDEs, [12–14, 28] for semi-linear SPDEs, [3, 5, 15, 27] for functional SDEs, [8, 22, 26] for degenerate SDEs, and [6, 24] for SDEs driven by Lévy and fractional noises. We refer to the survey paper [17] and the monograph [21] for more applications of coupling by change of measure and the above type Harnack inequalities.

In the next section, we introduce a general result on the hypercontractivity using coupling and Harnack inequality. Then we apply this result to degenerate SDEs and functional SPDEs in Sects. 3 and 4 respectively.

2 Hypercontractivity Using Coupling and Harnack Inequality

Let $(E, \mathscr{B}, \mu)$ be a probability space, and let P_t be a Markov semigroup on $\mathscr{B}_b(E)$ such that μ is P_t-invariant, i.e. $\mu(P_t f) = \mu(f)$ for $f \in L^1(\mu)$ and $t \ge 0$. P_t is called hypercontractive with respect to the invariant probability measure μ, if $\|P_t\|_{L^2(\mu)\to L^4(\mu)} = 1$ for large enough $t > 0$. By the interpolation theorem, one may replace the operator norm $\|\cdot\|_{L^2(\mu)\to L^4(\mu)}$ by $\|\cdot\|_{L^p(\mu)\to L^q(\mu)}$ for any $q > p > 1$. This property was found by Nelson [11] for the Ornstein-Uhlenbeck semigroup. In general, the hypercontractivity of P_t implies the exponential convergence in entropy, i.e.

$$\mathrm{Ent}_\mu(P_t f) \le c\mathrm{e}^{-\lambda t}\mathrm{Ent}_\mu(f), \quad t \ge 0, f \in \mathscr{B}_b^+(E)$$

holds for some constants $c, \lambda > 0$, where $\mathrm{Ent}_\mu(f) := \mu\big(f \log \frac{f}{\mu(f)}\big)$, see [23] and references therein. According to L. Gross (see e.g. [7]), the hypercontractivity of P_t follows from the log-Sobolev inequality

$$\mu(f^2 \log f^2) - \mu(f^2) \log \mu(f^2) \le C\mu(-fLf), \quad f \in \mathbb{D}(L)$$

for some constant $C > 0$, where $(L, \mathbb{D}(L))$ is the generator of P_t in $L^2(\mu)$. When P_t is symmetric in $L^2(\mu)$, the hypercontractivity and the log-Sobolev inequality are equivalent. However, in the non-symmetric case, the log-Sobolev inequality is essentially stronger than the hypercontractivity, see Sects. 3 and 4 for hypercontractive semigroups for which the log-Sobolev inequality is not available.

We introduce below a general result on hypercontractivity using coupling and Harnack inequality. A process $(X(t), Y(t))$ on $E \times E$ is called a coupling of the Markov process with semigroup P_t, if

$$(P_t f)(X(0)) = \mathbb{E}\big[f(X_t)\big|X(0)\big], \quad (P_t f)(Y(0)) = \mathbb{E}\big[f(Y(t))\big|Y(0)\big], \quad f \in \mathscr{B}_b(E), t \ge 0.$$

Theorem 2.1 ([23]) *Assume that the following three conditions hold for some measurable functions $\rho : E \times E \to (0, \infty)$ and $\phi : [0, \infty) \to (0, \infty)$ with $\lim_{t\to\infty} \phi(t) = 0$:*

(i) There exist two constants $t_0, c_0 > 0$ such that

$$(P_{t_0} f(\xi))^2 \le (P_{t_0} f^2(\eta)) e^{c_0 \rho(\xi,\eta)^2}, \quad f \in \mathscr{B}_b(E), \xi, \eta \in E;$$

(ii) For any $(X(0), Y(0)) \in E \times E$, there exists a coupling $(X(t), Y(t))$ associated to P_t such that

$$\rho(X(t), Y(t)) \le \phi(t)\rho(X(0), Y(0)), \quad t \ge 0;$$

(iii) There exists $\varepsilon > 0$ such that $(\mu \times \mu)(e^{\varepsilon\rho^2}) < \infty$.

Then μ is the unique invariant probability measure and P_t is hypercontractive. Consequently, P_t is compact in $L^2(\mu)$ for large $t > 0$ and is exponentially convergent in entropy.

Proof (*Sketch*) The Harnack inequality implies that P_t has a density with respect to μ, so that besides the exponential convergence in entropy, the hypercontractivity also implies the compactness of P_t in $L^2(\mu)$ for large $t > 0$, see [23] and references therein for details.

According to [27, Proposition 3.1], (i) implies that μ is the unique invariant probability measure for P_{t_0}, and P_{t_0} has a density with respect to μ. It remains to prove $\|P_t\|^4_{L^2(\mu)\to L^4(\mu)} < 2$ for large enough $t > 0$, which implies the hypercontractivity according to [23, Proposition 2.2].

Let $f \in \mathscr{B}_b(E)$ with $\mu(f^2) \le 1$. By (i) and (ii) we have

$$(P_{t+t_0} f(\xi))^2 \le \mathbb{E}[P_{t_0} f(X(t))]^2 \le \mathbb{E}\Big[\{P_{t_0} f^2(Y(t))\}e^{c_0\rho(X(t),Y(t))^2}\Big]$$
$$\le (P_{t_0+t} f^2(\eta))e^{c_0\phi(t)^2\rho(\xi,\eta)^2}, \quad t \ge 0, (\xi,\eta) \in E \times E.$$

Equivalently,

$$(P_{t_0+t} f(\xi))^2 e^{-c_0\phi(t)^2\rho(\xi,\eta)^2} \le P_{t_0+t} f^2(\eta), \quad t \ge 0, (\xi,\eta) \in E \times E.$$

Integrating with respect to $\mu(\mathrm{d}\eta)$ gives

$$(P_{t_0+t} f(\xi))^2 \int_E e^{-c_0\phi(t)^2\rho(\xi,\eta)^2}\mu(\mathrm{d}\eta)$$
$$\le \int_E P_{t_0+t} f^2(\eta)\mu(\mathrm{d}\eta) = \mu(f^2) \le 1, \quad t \ge 0, \xi \in E.$$

Thus,

$$(P_{t_0+t} f(\xi))^4 \le \frac{1}{\big(\int_E \exp[-c_0\phi(t)^2\rho(\xi,\eta)^2]\mu(\mathrm{d}\eta)\big)^2}, \quad \mu(f^2) \le 1, t \ge 0, \xi \in E.$$

Then by Jensen's inequality, for $t \ge 0$

$$\sup_{\mu(f^2)\le 1} \int_E (P_{t+t_0} f(\xi))^4 \mu(\mathrm{d}\xi) \le \int_E \frac{\mu(\mathrm{d}\xi)}{(\int_E \exp[-c_0\phi(t)^2\rho(\xi,\eta)^2]\mu(\mathrm{d}\eta))^2}$$
$$\le \int_E \bigg(\int_E e^{c_0\phi(t)^2\rho(\xi,\eta)^2}\mu(\mathrm{d}\eta)\bigg)^2 \mu(\mathrm{d}\xi) \le \int_{E\times E} e^{2c_0\phi(t)^2\rho(\xi,\eta)^2}\mu(\mathrm{d}\xi)\mu(\mathrm{d}\eta). \tag{2}$$

Since $\lim_{t\to\infty} \phi(t) = 0$, it follows from (iii) that

$$\lim_{t\to\infty} \int_{E\times E} e^{2c_0\phi(t)^2\rho(\xi,\eta)^2}\mu(\mathrm{d}\xi)\mu(\mathrm{d}\eta) = 1.$$

Combining this with (2) we prove $\|P_t\|_{2\to 4}^4 < 2$ for large enough $t > 0$. □

3 Hypercontractivity for Degenerate SDEs

We only consider finite-dimensional stochastic Hamiltonian systems, see [23] for extensions to infinite-dimensions and typical examples.

Consider the following degenerate SDE for $(X(t), Y(t))$ on $\mathbb{R}^m \times \mathbb{R}^d$:

$$\begin{cases} dX(t) = (AX(t) + BY(t))\,dt, \\ dY(t) = Z(X(t), Y(t))dt + \sigma dW(t), \end{cases} \tag{3}$$

where $W(t)$ is a d-dimensional Brownian motion, and

(A1) A is an $m \times m$-matrix, B is a $d \times m$-matrix, σ is a $d \times d$-matrix, such that σ is invertible and $\mathrm{Rank}[B, AB, \cdots, A^{m-1}B] = m$.
(A2) $Z : \mathbb{R}^{m+d} \to \mathbb{R}^d$ is Lipschitz continuous.
(A3) There exist constants $r, \theta > 0$ and $r_0 \in (-\|B\|^{-1}, \|B\|^{-1})$ such that

$$\begin{aligned} &\langle r^2(x - \bar{x}) + rr_0B(y - \bar{y}), A(x - \bar{x}) + B(y - \bar{y})\rangle \\ &+ \langle Z(x, y) - Z(\bar{x}, \bar{y}), y - \bar{y} + rr_0B^*(x - \bar{x})\rangle \\ &\le -\theta(|x - \bar{x}|^2 + |y - \bar{y}|^2), \quad (x, y), (\bar{x}, \bar{y}) \in \mathbb{R}^{m+d}. \end{aligned}$$

Theorem 3.1 ([23]) *Assume* $(A1)$, $(A2)$ *and* $(A3)$. *Let* P_t *be the Markov semigroup associated with* (3). *Then* P_t *has a unique invariant probability measure* μ *and it is hypercontractive. Consequently,* P_t *is compact in* $L^2(\mu)$ *for large* $t > 0$, *and is exponentially convergent in entropy.*

Proof (*Sketch*). Firstly, by $(A1)$ and $(A2)$ we may construct a coupling by change of measure such that Theorem 3.1 gives the following Harnack inequality: for any $t_0 > 0$,

$$(P_{t_0}f)^2(\xi) \le (P_{t_0}f^2(\eta))e^{c_0|\xi-\eta|^2}, \quad f \in \mathscr{B}_b(\mathbb{R}^{m+d}), \xi, \eta \in \mathbb{R}^{m+d}$$

holds for some constant $c_0 > 0$.

Secondly, if $(A3)$ holds then we may find out two constants $c, \lambda > 0$ such that for any two solutions $(X(t), Y(t))$ and $(\mathtt{X}(t), \mathtt{Y}(t))$ of (3),

$$|X(t)-\mathtt{X}(t)|^2+|Y(t)-\mathtt{Y}(t)|^2 \le ce^{-\lambda t}(|X(0)-\mathtt{X}(0)|^2+|Y(0)-\mathtt{Y}(0)|^2), \quad t \ge 0.$$

Finally, if $(A3)$ holds then the standard argument using a Lyapunov condition implies that P_t has an invariant probability measure μ such that $\mu(e^{\varepsilon|\cdot|^2}) < \infty$ for some constant $\varepsilon > 0$.

Therefore, the proof is finished by Theorem 2.1. □

4 Hypercontractivity for Functional SPDEs

We will only consider non-degenerate functional semi-linear SPDEs, see [4] for results on degenerate functional SPDEs and specific examples.

Let $\mathbb{H}$ be a separable Hilbert space. For a fixed constant $r_0 > 0$, consider the path space $\mathscr{C} = C([-r_0, 0]; \mathbb{H})$ equipped with the uniform norm $\|f\|_\infty := \sup_{-r_0 \le \theta \le 0} |f(\theta)|$. For a map $h(\cdot) : [-r_0, \infty) \to \mathbb{H})$, we define its segment functional $h_\cdot : [0, \infty) \to \mathscr{C}$ by letting

$$h_t(\theta) = h(t + \theta), \quad \theta \in [-r_0, 0].$$

Consider the following SPDE on $\mathbb{H}$:

$$\mathrm{d}X(t) = \{AX(t) + b(X_t)\}\mathrm{d}t + \sigma \mathrm{d}W(t), \quad t > 0, \quad X_0 = \xi \in \mathscr{C}, \tag{4}$$

where $W(t)$ is a cylindrical Brownian motion on $\mathbb{H}$; is,

$$W(t) = \sum_{i=1}^{\infty} B_i(t) e_i, \quad t \ge 0$$

for an orthonormal basis $\{e_i\}_{i \ge 1}$ on $\mathbb{H}$ and a sequence of independent one-dimensional Brownian motions $\{B_i(t)\}_{i \ge 1}$. Moreover:

(H1) $(-A, \mathscr{D}(A))$ is a self-adjoint operator on $\mathbb{H}$ with discrete spectrum $0 < \lambda_1 \le \lambda_2 \le \cdots$ counting multiplicities such that $\lambda_i \uparrow \infty$, such that for some constant $\delta \in (0, 1)$,

$$\int_0^1 \|\mathrm{e}^{-t(-A)^{1-\delta}} \sigma\|_{HS}^2 \mathrm{d}t < \infty, \quad t > 0,$$

where $\|\cdot\|_{HS}$ stands for the Hilbert-Schmidt norm.

(H2) $b : \mathscr{C} \to \mathbb{H}$ is such that

$$|b(\xi) - b(\eta)| \le L\|\xi - \eta\|_\infty, \quad \xi, \eta \in \mathscr{C}$$

holds for some constant $L > 0$.

(H3) $(\sigma, \mathbb{D}(\sigma))$ is an invertible linear operator on $\mathbb{H}$, i.e. there exists bounded operator σ^{-1} such that $\sigma^{-1}\mathbb{H} \subset \mathbb{D}(\sigma)$ and $\sigma\sigma^{-1} = I$, the identity operator.

It is easy to see that **(H1)** and **(H2)** imply

$$\int_0^1 \|\mathrm{e}^{tA}\sigma\|_{HS}^{2(1+\varepsilon)} \mathrm{d}t < \infty$$

for some $\varepsilon > 0$. So, according to e.g. [21, Theorem 4.1.3], for any initial point $\xi \in \mathscr{C}$, the equation (4) has a unique continuous mild solution $(X^\xi(t))_{t \ge 0}$. Let $\{X_t^\xi\}_{t \ge 0}$ be the corresponding segment solution. Then the associated Markov semigroup is given by

$$P_t f(\xi) := \mathbb{E} f(X_t^\xi), \quad f \in \mathscr{B}_b(\mathscr{C}), \quad \xi \in \mathscr{C}.$$

Theorem 4.1 ([4]) *Let* **(H1)**–**(H3)** *hold. If* $\lambda := \sup_{s\in(0,\lambda_1]}\left(s - Le^{sr_0}\right) > 0$, *then* P_t *has a unique invariant probability measure and is hypercontractive. Consequently,* P_t *is compact in* $L^2(\mu)$ *for large* $t > 0$ *and is exponentially convergent in entropy.*

Proof (*Sketch*) By constructing a suitable coupling by change of measure in terms of **(H1)** and **(H2)**, we establish the following Harnack inequality according to Theorem 3.1: for any $t_0 > r_0$, there exists a constant $c_0 > 0$ such that (see [21, Theorem 4.2.4]):

$$\left(P_{t_0} f(\eta)\right)^2 \le (P_{t_0} f^2(\xi)) e^{c_0\|\xi-\eta\|_\infty^2}, \quad \xi, \eta \in \mathscr{C}, \ f \in \mathscr{B}_b(\mathscr{C}).$$

Next, by **(H1)** and **(H2)** we have

$$e^{\lambda_1 t}|X^\xi(t) - X^\eta(t)| \le |\xi(0) - \eta(0)| + L\int_0^t e^{\lambda_1 s}\|X_s^\xi - X_s^\eta\|_\infty ds.$$

By Gronwall's inequality this implies

$$\|X_t^\xi - X_t^\eta\|_\infty \le e^{\lambda_1 r_0} e^{-\lambda t}\|\xi - \eta\|_\infty, \quad t \ge 0, \ \xi, \eta \in \mathscr{C}.$$

According to Theorem 2.1, it remains to verify $\mu(e^{\varepsilon\|\cdot\|_\infty^2}) < \infty$ for some constant $\varepsilon > 0$. This can be done by applying an infinite-dimensional Fernique inequality. □

Acknowledgements The author would like to thank the referee for helpful comments and a number of corrections. The work is supported by NNSFC (11771326, 11431014).

References

1. Arnaudon, M., Thalmaier, A., Wang, F.-Y.: Harnack inequality and heat kernel estimates on manifolds with curvature unbounded below. Bull. Sci. Math. **130**, 223–233 (2006)
2. Arnaudon, M., Thalmaier, A., Wang, F.-Y.: Gradient estimates and Harnack inequalities on non-compact Riemannian manifolds. Stoch. Process. Appl. **119**, 3653–3670 (2009)
3. Bao, J., Wang, F.-Y., Yuan, C.: Derivative formula and Harnack inequality for degenerate functional SDEs. Stoch. Dyn. **13**, 1250013, 22 pp (2011)
4. Bao, J., Wang, F.-Y., Yuan, C.: Hypercontractivity for functional stochastic partial differential equations. Comm. Electr. Probab. **20**(9), 1–15 (2015)
5. Es-Sarhir, A., v. Renesse, M.-K., Scheutzow, M.: Harnack inequality for functional SDEs with bounded memory. Electr. Commun. Probab. **14**, 560–565 (2009)
6. Fan, X.-L.: Harnack inequality and derivative formula for SDE driven by fractional Brownian motion. Sci. China-Math. **561**, 515–524 (2013)
7. Gross, L.: Logarithmic Sobolev inequalities. Am. J. Math. **97**, 1061–1083 (1976)
8. Guillin, A., Wang, F.-Y.: Degenerate Fokker-Planck equations: Bismut formula, gradient estimate and Harnack inequality. J. Differ. Equ. **253**, 20–40 (2012)
9. Liu, W.: Harnack inequality and applications for stochastic evolution equations with monotone drifts. J. Evol. Equ. **9**, 747–770 (2009)
10. Liu, W., Wang, F.-Y.: Harnack inequality and strong Feller property for stochastic fast-diffusion equations. J. Math. Anal. Appl. **342**, 651–662 (2008)

11. Nelson, E.: The free Markoff field. J. Funct. Anal. **12**, 211–227 (1973)
12. Ouyang, S.-X.: Harnack inequalities and applications for multivalued stochastic evolution equations. Infin. Dimens. Anal. Quantum Probab. Relat. Top. **14**, 261–278 (2011)
13. Röckner, M., Wang, F.-Y.: Harnack and functional inequalities for generalized Mehler semigroups. J. Funct. Anal. **203**, 237–261 (2003)
14. Röckner, M., Wang, F.-Y.: Log-Harnack inequality for stochastic differential equations in Hilbert spaces and its consequences. Infin. Dimens. Anal. Quant. Probab. Relat. Top. **13**, 27–37 (2010)
15. Shao, J., Wang, F.-Y., Yuan, C.: Harnack inequalities for stochastic (functional) differential equations with non-Lipschitzian coefficients. Electr. J. Probab. **17**, 1–18 (2012)
16. Wang, F.-Y.: Logarithmic Sobolev inequalities on noncompact Riemannian manifolds. Probab. Theory Relat. Fields **109**, 417–424 (1997)
17. Wang, F.-Y.: Dimension-free Harnack inequality and its applications. Front. Math. China **1**, 53–72 (2006)
18. Wang, F.-Y.: Harnack inequality and applications for stochastic generalized porous media equations. Ann. Probab. **35**, 1333–1350 (2007)
19. Wang, F.-Y.: Harnack inequalities on manifolds with boundary and applications. J. Math. Pures Appl. **94**, 304–321 (2010)
20. Wang, F.-Y.: Harnack inequality for SDE with multiplicative noise and extension to Neumann semigroup on non-convex manifolds. Ann. Probab. **39**, 1449–1467 (2011)
21. Wang, F.-Y.: Harnack Inequalities and Applications for Stochastic Partial Differential Equations. Springer, Berlin (2013)
22. Wang, F.-Y.: Derivative formulas and Poincaré inequality for Kohn-Laplacian type semigroups. Sci. China-Math. **59**, 261–280 (2016)
23. Wang, F.-Y.: Hypercontractivity and spplications for stochastic Hamiltonian systems. J. Funct. Anal. **272**, 5360–5383 (2017)
24. Wang, F.-Y., Wang, J.: Harnack inequalities for stochastic equations driven by Lévy noise. J. Math. Anal. Appl. **410**, 513–523 (2014)
25. Wang, F.-Y., Wu, J.-L., Xu, L.: Log-Harnack inequality for stochastic Burgers equations and applications. J. Math. Anal. Appl. **384**, 151–159 (2011)
26. Wang, F.-Y., Xu, L.: Log-Harnack inequality for Gruschin type semigroups. Rev. Matem. Iberoamericana. **30**, 405–418 (2014)
27. Wang, F.-Y., Yuan, C.: Harnack inequalities for functional SDEs with multiplicative noise and applications. Stoch. Process. Appl. **121**, 2692–2710 (2011)
28. Zhang, S.-Q.: Harnack inequality for semilinear SPDEs with multiplicative noise. Statist. Probab. Lett. **83**, 1184–1192 (2013)

Multidimensional Singular Stochastic Differential Equations

Xicheng Zhang

Abstract In this paper we survey recent progress about multidimensional stochastic differential equations with singular drifts and Sobolev diffusion coefficients. Moreover, applications to Navier–Stokes equations and SPDEs are also presented.

Keywords Stochastic flow · Krylov's estimate · Zvonkin's transformation

AMS 2010 Mathematics Subject Classification 60H10 · 60J60

1 Introduction

Consider the following ordinary differential equation in $\mathbb{R}^d$ (abbreviated as ODE)

$$\dot{X}_t = b(X_t), \ X_0 = x \in \mathbb{R}^d, \tag{1}$$

where $b(x) : \mathbb{R}^d \to \mathbb{R}^d$ is a measurable vector field. It is well known that when b is globally Lipschitz continuous, then the above ODE admits a unique solution $X_t = X_t(x)$ so that the map $x \mapsto X_t(x)$ is a homeomorphism from $\mathbb{R}^d$ to $\mathbb{R}^d$. Moreover, the following flow property holds:

$$X_t(X_s(x)) = X_{t+s}(x), \ t, s \geqslant 0,$$

where the left hand side is understood as the composition of the solution as a function of the starting point. The homeomorphism property means that the motion starting from two different points never touches, and the motion can reach any point from

This work is supported by NNSF of China (No. 11731009).

X. Zhang (✉)
School of Mathematics and Statistics, Wuhan University,
Wuhan, Hubei 430072, China
e-mail: XichengZhang@gmail.com

A. Eberle et al. (eds.), *Stochastic Partial Differential Equations and Related Fields*, Springer Proceedings in Mathematics & Statistics 229,
https://doi.org/10.1007/978-3-319-74929-7_26

the starting point; while the flow property means that $T_t f(x) := f(X_t(x))$ forms a semigroup under composition. Notice that these properties would be broken if b is not Lipschitz continuous. For example, when $d = 1$ and $b(x) = \sqrt{|x|}$, then the above ODE has two obvious solutions $X_t(0) = 0$ and $X_t(0) = t^2/4$ with the same starting point 0. Another extremal example is $b(x) = 1_{\mathbb{Q}}(x)$, where $\mathbb{Q}$ is the set of all rational numbers in $\mathbb{R}$. In this case, it is even not expected that the above ODE has an absolutely continuous solution starting from zero. In fact, suppose that X_t satisfies $X_t = \int_0^t 1_{\mathbb{Q}}(X_s)\mathrm{d}s = \gamma\{s \in [0,t] : X_s \in \mathbb{Q}\}$, where γ denotes the Lebesgue measure. Since $t \mapsto X_t$ is increasing and X can not be a constant rational number in any time interval, for given $q \in \mathbb{Q}$, the set $\{s : X_s = q\}$ is at most a single point set. Hence, $X_t \equiv 0$. But this is impossible.

However, the situation has a dramatic change when we consider the following stochastic differential equation (abbreviated as SDE) driven by a Brownian motion:

$$\mathrm{d}X_t = b_t(X_t)\mathrm{d}t + \mathrm{d}W_t,\ X_0 = x \in \mathbb{R}^d, \tag{2}$$

where $b : \mathbb{R}_+ \times \mathbb{R}^d \to \mathbb{R}^d$ is a time-dependent vector field. When $d = 1$ and b is only bounded measurable, Zvonkin [45] first showed the strong well-posedness to the above SDE. In the multidimensional case, Veretennikov [32] proved the same result when b is bounded measurable. By Girsanov's transformation and using the results from PDEs, Krylov and Röckner [23] obtained the existence and uniqueness of strong solutions to SDE (2) with b satisfying

$$\|b\|_{L^q_t(L^p_x)} := \int_0^s \left(\int_{\mathbb{R}^d} |b_t(x)|^p \mathrm{d}x\right)^{q/p} \mathrm{d}t < \infty,\quad \frac{2}{q} + \frac{d}{p} < 1.$$

Later, such results were extended to SDE with Sobolev diffusion coefficients and singular drifts in [36, 38] by using Zvonkin's idea.

Now let us simply describe the method of phase space transformation used in [45]. Consider the simple case of $d = 1$ and time independent drift b. Let $f \in C^2(\mathbb{R})$ and X_t solve equation (2). By Itô's formula, we have

$$\mathrm{d}f(X_t) = (f' \cdot b + \tfrac{1}{2} f'')(X_t)\mathrm{d}t + f'(X_t)\mathrm{d}W_t = f'(X_t)\mathrm{d}W_t,$$

provided that f satisfies ODE $f' \cdot b + \frac{1}{2} f'' = 0$. Notice that one solution of this second order ODE is explicitly given by $f(x) = \int_0^x \mathrm{e}^{-2\int_0^y b(z)\mathrm{d}z}\mathrm{d}y$. Suppose that $b \in L^1(\mathbb{R}) \cap C_b(\mathbb{R})$. It is easy to see that $f : \mathbb{R} \to \mathbb{R}$ is a C^2-diffeomorphism. Let f^{-1} be the inverse of f. If we let $f(X_t) = Y_t$, then

$$\mathrm{d}Y_t = f' \circ f^{-1}(Y_t)\mathrm{d}W_t,\ Y_0 = f(X_0). \tag{3}$$

Thus, we get a new SDE with Lipschitz coefficient $f' \circ f^{-1}$. In particular, we establish a one-to-one correspondence between (2) and (3). Obviously, if $d \geqslant 2$, the same idea requires a better understanding for the associated PDE (Kolmogorov equation). Notice that one dimensional singular SDEs have been systematically studied in [8].

In this survey, we summarize recent results about multidimensional SDEs with singular drifts and Sobolev diffusion coefficients. We also explain some applications to Navier–Stokes equations and stochastic partial differential equations. Other developments about SDEs such as generalized stochastic Lagrangian flows, SDEs driven by stable processes, and singular SDEs in infinite dimensional spaces, will not be discussed in this paper. The interested readers are referred to the references [1, 10, 14, 15, 37, 40, 41] for generalized stochastic Lagrangian flows, and [2, 7, 27, 35, 42] for SDEs driven by Lévy noise, and [11–13] for singular SDEs in infinite dimensional spaces. This paper is organized as follows: In Sect. 2, we present the general settings and useful tools about SDEs. In Sect. 3, we recall multidimensional singular stochastic flows with non-degenerate Sobolev diffusion coefficients, and present an application to Navier–Stokes equations. In Sect. 4, we recall stochastic Hamiltonian flows with singular drifts, and introduce an application to SPDEs.

2 General Settings and Useful Tools

Let $b : \mathbb{R}_+ \times \mathbb{R}^d \to \mathbb{R}^d$ and $\sigma : \mathbb{R}_+ \times \mathbb{R}^d \to \mathbb{R}^d \otimes \mathbb{R}^m$ be two Borel measurable functions, and W an m-dimensional standard Brownian motion defined on some probability space $(\Omega, \mathscr{F}, \mathbb{P})$. For fixed $s \geqslant 0$ and $x \in \mathbb{R}^d$, let $X_{s,t} = X_{s,t}(x)$ solve the following SDE in $\mathbb{R}^d$ starting from x at time s:

$$\mathrm{d}X_{s,t} = b_t(X_{s,t})\mathrm{d}t + \sigma_t(X_{s,t})\mathrm{d}W_t,\ X_{s,s} = x. \tag{4}$$

For any $t \geqslant s \geqslant 0$ and bounded measurable φ, define $P_{s,t}\varphi(x) := \mathbb{E}(\varphi(X_{s,t}(x)))$. Let $\mu^x_{s,t}(\mathrm{d}y)$ be the distribution of $X_{s,t}(x)$. By Itô's formula, one sees that $\mu^x_{s,t}$ solves the following Fokker–Planck equation or Kolmogrov's forward equation:

$$\partial_t \mu^x_{s,t} = \mathscr{L}^*_t \mu^x_{s,t} \Leftrightarrow \partial_t P_{s,t}\varphi = P_{s,t}\mathscr{L}_t\varphi,\ \varphi \in C^\infty_c(\mathbb{R}^d), \tag{5}$$

where $\mathscr{L}^*_t$ is the adjoint operator of $\mathscr{L}_t f := \mathscr{L}^{a,b}_t f := \mathrm{tr}(a \cdot \nabla^2 f) + b \cdot \nabla f$. Here, $\mathrm{tr}(A)$ denotes the trace of the matrix A, $a := \frac{1}{2}\sigma\sigma^*$ with σ^* being the transpose of σ, and $\nabla^2 f$ denotes the Hessian matrix. On the other hand, suppose that the martingale problem for the SDE (4) is well-posed in the sense of Stroock and Varadhan [30]. Then we have the Chapman–Kolmogorov equation $P_{s,t}\varphi = P_{s,r}P_{r,t}\varphi$ for $s \leqslant r \leqslant t$. Under some regularity assumptions on b and σ, the following Kolmogorov's backward equation holds:

$$\partial_s P_{s,t}\varphi = -\mathscr{L}_s P_{s,t}\varphi,\ t \geqslant s \geqslant 0. \tag{6}$$

Kolmogorov's forward and backward equations (5) and (6) will be our cornerstone in the development below.

For $0 \leqslant s < T$ and $p, q \in [1, \infty]$, we introduce the following space for later use:

$$\mathbb{L}^q_p(s, T) := L^q([s, T]; L^p(\mathbb{R}^d))$$

with norm

$$\|f\|_{\mathbb{L}^q_p(s,T)} := \left(\int_s^T \left(\int_{\mathbb{R}^d} |f_t(x)|^p \mathrm{d}x\right)^{q/p} \mathrm{d}t\right)^{1/q},$$

and simply write $\mathbb{L}^p(s, T) := \mathbb{L}^p_p(s, T)$ and $\mathbb{L}^p(T) := \mathbb{L}^p(0, T)$. Next we introduce some useful tools used in the studies of singular SDEs.

Definition 1 (Krylov's estimate [22, Chap. 2]) One says that Krylov's estimate holds for SDE (4) for some $p \geqslant 1$, if for any $T > 0$, there is a constant $C > 0$ such that for all $0 \leqslant s \leqslant T$ and $f \in \mathbb{L}^p(s, T)$,

$$\mathbb{E}^{\mathscr{F}_s}\left(\int_s^T f_r(X_{s,r})\mathrm{d}r\right) \leqslant C\|f\|_{\mathbb{L}^p(s,T)}, \ \mathbb{P} - a.s, \tag{7}$$

where $\mathbb{E}^{\mathscr{F}_s}(\cdot)$ denotes the conditional expectation with respect to the natural filtration $\mathscr{F}_s := \sigma\{W_t : t \leqslant s\}$.

Remark 1 If (7) holds for some $p = q$, then by interpolation, (7) also holds for all $p > q$. In fact, fixing $A \in \mathscr{F}_s$, define an operator $\mathscr{T} : \mathbb{L}^p(s, T) \to L^1(\Omega \times [s, T])$ by

$$(\mathscr{T} f)(t, \omega) := 1_A(\omega) f_t(X_{s,t}(\omega)).$$

Then by (7), we have

$$\|\mathscr{T} f\|_{L^1(\Omega\times[s,T])} \leqslant C\mathbb{P}(A)\|f\|_{\mathbb{L}^q(s,T)}.$$

On the other hand, it is clear that

$$\|\mathscr{T} f\|_{L^1(\Omega\times[s,T])} \leqslant (T - s)\mathbb{P}(A)\|f\|_{\mathbb{L}^\infty(s,T)}.$$

Hence, for any $p \in (q, \infty)$, by the Riesz–Thorin interpolation theorem, we get

$$\|\mathscr{T} f\|_{L^1(\Omega\times[s,T])} \leqslant C^{q/p}(T - s)^{1-q/p}\mathbb{P}(A)\|f\|_{\mathbb{L}^p(s,T)}.$$

Remark 2 Estimate (7) is understood as apriori estimate and has the following consequence: For Lebesgue almost all $r \in [s, T]$, the law of random variable $X_{s,r}(x)$ admits a density $p_{s,r}(x, y)$ with

$$\int_s^T \int_{\mathbb{R}^d} |p_{s,r}(x, y)|^q \mathrm{d}y\mathrm{d}r < \infty, \quad q \in [1, p/(p-1)]. \tag{8}$$

Indeed, define a measure μ on $[s,T]\times\mathbb{R}^d$ by

$$\mu(A) := \mathbb{E}\left(\int_s^T 1_A(r, X_{s,r}(x))\mathrm{d}r\right), \quad A\in\mathscr{B}([s,T]\times\mathbb{R}^d).$$

By (7), one sees that μ is absolutely continuous with respect to the Lebesgue measure, and the density $p_{s,r}(x,y)$ satisfies that for all $f\in\mathbb{L}^p(s,T)$,

$$\int_s^T\int_{\mathbb{R}^d} f_r(y)p_{s,r}(x,y)\mathrm{d}y\mathrm{d}r \leqslant C\|f\|_{\mathbb{L}^p(s,T)}.$$

By duality and interpolation we get (8).

Krylov's estimate (7) has the following useful consequence (see [44, Corollary 4.4]).

Lemma 1 (Khasminskii's type estimate) *Suppose that SDE* (4) *satisfies Krylov's estimate* (7) *for some $p\geqslant 1$. Letting C be the same as in* (7)*, we have*

(i) For each $m\in\mathbb{N}$ and $0\leqslant s\leqslant T$, it holds that for all $f\in\mathbb{L}^p(s,T)$,

$$\mathbb{E}^{\mathscr{F}_s}\left(\int_s^T f_r(X_{s,r})\mathrm{d}r\right)^m \leqslant m!(C\|f\|_{\mathbb{L}^p(s,T)})^m.$$

(ii) For any $\lambda>0$, it holds that $\mathbb{E}\left(\exp\left(\lambda\int_s^T f_r(X_{s,r})\mathrm{d}r\right)\right)<\infty$ for all $f\in\mathbb{L}^p(s,T)$.

Let $U\subset\mathbb{R}^d$ be a bounded C^1-domain. For $p,q,r\in[1,\infty]$ and $T>0$, letting $L^r(T) := L^r([0,T])$, we define

$$\mathbb{W}^{1,q}\big(U;L^p(\Omega;L^r(T))\big) := \Big\{f(x,\omega,r): f,\ \nabla f\in L^q\big(U;L^p(\Omega;L^r(T))\big)\Big\}$$

and

$$\|f\|_{\mathbb{W}^{1,q}(U;L^p(\Omega;L^r(T)))} := \|f\|_{L^q(U;L^p(\Omega;L^r(T)))} + \|\nabla f\|_{L^q(U;L^p(\Omega;L^r(T)))}.$$

The following characterization of weak differentiability of random fields is proven in [34, Theorem 1.1].

Theorem 1 *Let $U\subset\mathbb{R}^d$ be a bounded C^1-domain and $T>0$. Suppose for some $p\in(1,\infty)$ and $q,r\in(1,\infty]$, $f\in L^q\big(U;L^p(\Omega;L^r(T))\big)$. Then $f\in\mathbb{W}^{1,q}\big(U;L^p(\Omega;L^r(T))\big)$ if and only if there exists a nonnegative measurable function $g\in L^q(U)$ such that for Lebesgue-almost all $x,y\in U$,*

$$\|f(x,\cdot)-f(y,\cdot)\|_{L^p(\Omega;L^r(T))} \leqslant |x-y|\big(g(x)+g(y)\big). \tag{9}$$

Moreover, if (9) holds, then for Lebesgue-almost all $x \in U$,

$$\|\partial_i f(x,\cdot)\|_{L^p(\Omega;L^r(T))} \leqslant 2g(x), \quad i = 1, \ldots, d,$$

where $\partial_i f$ is the weak partial derivative of f with respect to the i-th spacial variable.

3 Multidimensional Singular Stochastic Flows

In this section we first consider the nondegenerate case and make the following assumptions about σ:

($\mathbf{H}_K^\alpha$) There exist constants $K \geqslant 1$ and $\alpha \in (0, 1)$ such that for all $(t, x) \in \mathbb{R}_+ \times \mathbb{R}^d$,

$$K^{-1}|\xi| \leqslant |\sigma_t^*(x)\xi| \leqslant K|\xi|, \quad \xi \in \mathbb{R}^d,$$

and for all $t \geqslant 0$ and $x, y \in \mathbb{R}^d$, $\|\sigma_t(x) - \sigma_t(y)\| \leqslant K|x-y|^\alpha$. Here $|\cdot|$ denotes the Euclidean norm and $\|\cdot\|$ the Hilbert-Schmidt norm.

The following result is a combination of [38, Theorem 1], [43, Theorem 1.1] and [35, Theorem 2.1].

Theorem 2 *Assume that σ satisfies ($\mathbf{H}_K^\alpha$) and $\nabla\sigma, b \in \cap_{S<T}\mathbb{L}_p^q(S,T)$ for some $p, q \in (2, \infty)$ with $\frac{d}{p} + \frac{2}{q} < 1$. Then we have the following conclusions:*

(A) *For any $(s, x) \in \mathbb{R}_+ \times \mathbb{R}^d$, there is a unique strong solution denoted by $X_{s,t}(x)$ or $X_{s,t}^{b,\sigma}(x)$ to SDE (4), which has a jointly continuous version with respect to t and x. Moreover, $x \mapsto X_{s,t}(x)$ is a homeomorphism on $\mathbb{R}^d$, $\mathbb{P}$-almost surely.*

(B) *For each $t \geqslant s$ and almost all ω, $x \mapsto X_{s,t}(x, \omega)$ is weakly differentiable. Furthermore, for any $p' \geqslant 1$, the Jacobian matrix $\nabla X_{s,t}(x)$ satisfies*

$$\operatorname{ess.}\sup_{x\in\mathbb{R}^d} \mathbb{E}\left(\sup_{t\in[s,T]} |\nabla X_{s,t}(x)|^{p'}\right) \leqslant C\Big(d, p, q, K, \alpha, p', \|b\|_{\mathbb{L}_p^q(s,T)}, \|\nabla\sigma\|_{\mathbb{L}_p^q(s,T)}\Big),$$

where C is increasing with respect to $\|b\|_{\mathbb{L}_p^q(s,T)}$ and $\|\nabla\sigma\|_{\mathbb{L}_p^q(s,T)}$.

(C) *For each $t \geqslant s$ and $x \in \mathbb{R}^d$, the random variable $\omega \mapsto X_{s,t}(x, \omega)$ is Malliavin differentiable, and for any $p' \geqslant 1$,*

$$\sup_{x\in\mathbb{R}^d} \mathbb{E}\left(\sup_{t\in[s,T]} \|DX_{s,t}(x)\|_{\mathbb{H}}^{p'}\right) < +\infty,$$

where D is the Malliavin derivative.

(D) *For any $f \in C_b^1(\mathbb{R}^d)$, we have the following derivative formula: for Lebesgue-almost all $x \in \mathbb{R}^d$,*

$$\nabla \mathbb{E} f(X_{s,t}(x)) = \frac{1}{t-s} \mathbb{E}\left(f(X_{s,t}(x)) \int_s^t \sigma_r^{-1}(X_{s,r}(x)) \nabla X_{s,r}(x) \mathrm{d}W_r \right),$$

where σ^{-1} is the inverse matrix of σ.

(E) *Assume that $b' \in \mathbb{L}_p^q(S,T)$ with the same p, q as in the assumptions. Let $X_{s,t}^{b,\sigma}(x)$ and $X_{s,t}^{b',\sigma}(x)$ be the solutions to (4) associated with b and b' respectively. Then*

$$\sup_{x \in \mathbb{R}^d} \mathbb{E}\left(\sup_{t \in [s,T]} |X_{s,t}^{b,\sigma}(x) - X_{s,t}^{b',\sigma}(x)|^2 \right) \leqslant C \|b - b'\|^2_{\mathbb{L}_p^q(s,T)},$$

where $C = C\big(d, p, q, K, \alpha, \|b\|_{\mathbb{L}_p^q(s,T)}, \|b'\|_{\mathbb{L}_p^q(s,T)}, \|\nabla \sigma\|_{\mathbb{L}_p^q(s,T)}\big)$.

Let us make the following remarks about Theorem 2. Conclusions **(A)** and **(B)** are essentially contained in [17, 23, 38]. When $\sigma_t(x) = \sigma_t$ and $b_t(x)$ are *bounded*, **(A)**, **(B)** and **(C)** were studied in [25, 26] by using different arguments. We would like to emphasize that in [43], it is required $p = q$ in the case of non-constant $\sigma_t(x)$. Recently, L. Xie told me that K.H. Kim has already established the $L^q(L^p)$-theory of PDEs with varying coefficients in [24]. Thus, without any essential changes, the results in [43] also hold for $p \neq q$ as claimed in [38]. As for the stability estimate **(E)**, it could be used to study numerical solutions of SDEs with singular drifts. For example, consider the following SDE:

$$\mathrm{d}X_t = 1_A(X_t)\mathrm{d}t + \mathrm{d}W_t, \quad X_0 = x,$$

where A is a bounded open subset of $\mathbb{R}^d$. Let $b_n(x) = 1_A * \rho_n(x)$ be the mollifying approximation. By **(E)**, the solution X_t^n of the above SDE corresponding to b_n converges to X_t in L^2. Next, we can approximate X_t^n by Euler's scheme. In this way, one can give a numerical approximation for solutions of singular SDEs.

Below we introduce an application of our results to the probabilistic approach of Navier–Stokes equations. Consider the following classical Navier–Stokes equation in $\mathbb{R}^d$:

$$\partial_t u = \nu \Delta u - (u \cdot \nabla)u + \nabla p, \quad \mathrm{div} u = 0, \quad u_0 = \varphi,$$

where u is the velocity field, ν is the viscosity constant and p is the pressure of the fluid, φ is the initial velocity with vanishing divergence. In [9], Constantin and Iyer provided a probabilistic representation to the above NSE as follows:

$$\begin{cases} X_t(x) = x + \int_0^t u_s(X_s(x))\mathrm{d}s + \sqrt{2\nu} W_t, \\ u_t(x) = \mathbf{P}\mathbb{E}(\nabla^{\mathrm{t}} X_t^{-1} \cdot \varphi(X_t^{-1}))(x), \end{cases} \tag{10}$$

where $X_t^{-1}(x)$ denotes the inverse flow of $x \mapsto X_t(x)$, $\nabla^t X_t^{-1}$ is the transpose of the Jacobian matrix, and $\mathbf{P} = \mathbb{I} - \nabla(-\Delta)^{-1}\mathrm{div}$ is Leray's projection onto the space of all divergence free vector fields.

Recently, in [37, 40], we studied a backward analogue of the stochastic representation (10), that is, for $\nu > 0$ and $s \leqslant t \leqslant 0$,

$$\begin{cases} X_{s,t}(x) = x + \int_s^t u_r(X_{s,r}(x))\mathrm{d}r + \sqrt{2\nu}(W_t - W_s), \\ u_s(x) = \mathbf{P}\mathbb{E}(\nabla^t X_{s,0} \cdot \varphi(X_{s,0}))(x). \end{cases} \tag{11}$$

The advantage of this representation is that the inverse of stochastic flow $x \mapsto X_{s,0}(x)$ does not appear. In this case, $u_s(x)$ solves the following backward Navier–Stokes equation:

$$\partial_s u + \nu\Delta u - (u \cdot \nabla)u + \nabla p = 0, \quad \mathrm{div} u = 0, \quad u_0 = \varphi.$$

Using Theorem 2, we have the following local well-posedness to the system (11) (see [43, Theorem 1.4]).

Theorem 3 *For any $p > d$ and divergence free $\varphi \in \mathbb{W}^{1,p}(\mathbb{R}^d; \mathbb{R}^d)$, where $\mathbb{W}^{1,p}$ is the first order Sobolev space, there exist a time $T < 0$ depending only on $p, d, \nu, \|\varphi\|_{\mathbb{W}^{1,p}}$ and a unique pair (u, X) with $u \in L^\infty([T, 0]; \mathbb{W}^{1,p})$ solving the stochastic system (11). When $d = 2$, T could be arbitrary.*

4 Singular Stochastic Hamiltonian Flows

In this section we consider the following *second order* SDE:

$$\mathrm{d}\dot{X}_t = b_t(X_t, \dot{X}_t)\mathrm{d}t + \sigma_t(X_t, \dot{X}_t)\mathrm{d}W_t, \quad (X_0, \dot{X}_0) = (x, \mathrm{v}) \in \mathbb{R}^{2d},$$

where $b_t(x, \mathrm{v}) : \mathbb{R}_+ \times \mathbb{R}^{2d} \to \mathbb{R}^d$ and $\sigma_t(x, \mathrm{v}) : \mathbb{R}_+ \times \mathbb{R}^{2d} \to \mathbb{R}^d \otimes \mathbb{R}^m$ are two Borel measurable functions, $\dot{X}_t$ denotes the first order derivative of X_t with respect to t. When $\sigma = 0$, the above equation is the classical Newtonian mechanics equation, which describes the motion of a particle. When $\sigma \neq 0$, it means that the motion is perturbed by some random external force. For more backgrounds on the above stochastic Hamiltonian system, we refer to [29, 31], etc. It is noticed that if we let $Z_t := (X_t, \dot{X}_t)$, then Z_t solves the following *one order* (degenerate) SDE:

$$\mathrm{d}Z_t = (\dot{X}_t, b_t(Z_t))\mathrm{d}t + (0, \sigma_t(Z_t)\mathrm{d}W_t), \quad Z_0 = z = (x, \mathrm{v}) \in \mathbb{R}^{2d}, \tag{12}$$

and whose time-dependent infinitesimal generator is given by

$$\mathscr{L}_t^{a,b} f(x, \mathrm{v}) := \mathrm{tr}(a_t \cdot \nabla_{\mathrm{v}}^2 f)(x, \mathrm{v}) + (\mathrm{v} \cdot \nabla_x f)(x, \mathrm{v}) + (b_t \cdot \nabla_{\mathrm{v}} f)(x, \mathrm{v}). \tag{13}$$

Here $a_t = \frac{1}{2}\sigma_t\sigma_t^*$. When b is α-Hölder continuous in x and β-Hölder continuous in v with $\alpha \in (\frac{2}{3}, 1)$ and $\beta \in (0, 1)$, the strong well-posedness of SDE (12) was studied in [5, 33].

The following result is proven in [44, Theorem 1.1].

Theorem 4 *Suppose that for some $K \geqslant 1$ and all $(t, x, \mathrm{v}) \in \mathbb{R}_+ \times \mathbb{R}^{2d}$,*

$$K^{-1}|\xi| \leqslant |\sigma_t^*(x, \mathrm{v})\xi| \leqslant K|\xi|, \quad \forall \xi \in \mathbb{R}^d, \tag{UE}$$

where σ^* denotes the transpose of the matrix σ, and for some $p > 2(2d + 1)$,

$$\kappa_0 := \sup_{s\geqslant 0} \|\nabla\sigma_s\|_p^p + \int_0^\infty \|(\mathbb{I} - \Delta_x)^{1/3} b_s\|_p^p \mathrm{d}s < \infty. \tag{14}$$

Then for any $z = (x, \mathrm{v}) \in \mathbb{R}^{2d}$, SDE (12) admits a unique strong solution $Z_t(z) = (X_t, \dot{X}_t)$ so that $(t, z) \mapsto Z_t(z)$ has a bi-continuous version. Moreover:

(A) *There is a null set N such that for all $\omega \notin N$ and for each $t \geqslant 0$, the map $z \mapsto Z_t(z, \omega)$ is a homeomorphism on $\mathbb{R}^{2d}$.*

(B) *For each $t \geqslant 0$, the map $z \mapsto Z_t(z)$ is weakly differentiable a.s., and for any $q \geqslant 1$ and $T > 0$,*

$$\text{ess.}\sup_z \mathbb{E}\left(\sup_{t\in[0,T]} |\nabla Z_t(z)|^q\right) < \infty,$$

where ∇ denotes the generalized gradient.

(C) *Let $\sigma_t^n(x) = \sigma_t^n * \rho_n(x)$ and $b_t^n(x) := b_t * \rho_n(x)$ be the mollifying approximations of σ and b, where $\rho_n(x) := n^d\rho(nx)$ and ρ is a nonnegative smooth function with compact support and $\int_{\mathbb{R}^d} \rho(x)\mathrm{d}x = 1$. Let Z^n be the corresponding solution of SDE (12) associated with (σ^n, b^n). For any $q \geqslant 1$ and $T > 0$, there exits a constant $C > 0$ only depending on T, K, κ_0, d, p, q such that*

$$\mathbb{E}\left(\sup_{t\in[0,T]} |Z_t^n - Z_t|^q\right) \leqslant C\left(\|b^n - b\|_{\mathbb{L}^p(T)}^q + n^{(\frac{2d}{p}-1)q}\right), \quad n \in \mathbb{N}.$$

Now we introduce the main idea of proving this theorem. As explained in the introduction, the main point is to make a detailed study for the following backward kinetic Fokker–Planck equation:

$$\partial_t u + \mathscr{L}_t^{a,b} u - \lambda u + f = 0, \quad u(T) = 0, \tag{15}$$

where $\lambda > 0$ and $T > 0$. The following result is shown in [44, Theorem 3.2], which is based on the L^p-maximal regularity for the kinetic operator established in [3, 4, 6].

Theorem 5 *Let $p > 1$. Suppose that* **(UE)** *holds, and for some $q \in [p, \infty]$,*

$$\kappa_1 := \|b\|_{\mathbb{L}^p_q(T)} < \infty.$$

(i) For any $f \in \mathbb{L}^p(T)$, there exists a unique solution $u = u^\lambda$ to (15) with

$$\|(\mathbb{I} - \Delta_x)^{1/3} u^\lambda\|_{\mathbb{L}^p(T)} + \|\Delta_{\rm v} u^\lambda\|_{\mathbb{L}^p(T)} \leqslant C\|f\|_{\mathbb{L}^p(T)},$$

where the constant C only depends on d, δ, K, p, q, T and κ_1.

(ii) If in addition, we assume that $p > 4d + 2$ and

$$\kappa_2 := \|(\mathbb{I} - \Delta_x)^{1/3}\sigma\|_{\mathbb{L}^\infty(T)} + \|(\mathbb{I} - \Delta_x)^{1/3} b\|_{\mathbb{L}^p(T)} < \infty,$$

then the unique solution u also enjoys the following regularity

$$\|\nabla_x \nabla_{\rm v} u^\lambda\|_{\mathbb{L}^p(T)} + \|(\mathbb{I} - \Delta_x)^{1/3} \nabla^2_{\rm v} u^\lambda\|_{\mathbb{L}^p(T)} \leqslant C\|(\mathbb{I} - \Delta_x)^{1/3} f\|_{\mathbb{L}^p(T)},$$

where the constant C only depends on d, δ, K, p, T and κ_2.

Now we sketch the proof of Theorem 4. First of all, by (i) and suitable mollifying technique, one can show that Krylov's estimate in Definition 1 holds for SDE (12). For $T > 0$ and $\lambda \geqslant 1$, let $\mathbf{u}^\lambda$ uniquely solve the following PDE:

$$\partial_t \mathbf{u}^\lambda + \mathscr{L}^{a,b}_t \mathbf{u}^\lambda - \lambda \mathbf{u}^\lambda + b = 0, \quad \mathbf{u}^\lambda_T = 0.$$

By (ii) of Theorem 5, there is a constant $C = C(d, p, \kappa_0, K) > 0$ such that

$$\|\nabla \nabla_{\rm v} \mathbf{u}^\lambda\|_{\mathbb{L}^p(T)} \leqslant C\|(\mathbb{I} - \Delta_x)^{1/3} b\|_{\mathbb{L}^p(T)}. \tag{16}$$

Let $\mathrm{H}_t(x, {\rm v}) := {\rm v} + \mathbf{u}^\lambda_t(x, {\rm v})$. One can choose λ large enough so that $\|\nabla_{\rm v} \mathbf{u}^\lambda\|_{\mathbb{L}^\infty(T)} \leqslant 1/2$, and thus,

$$|{\rm v} - {\rm v}'|/2 \leqslant |\mathrm{H}_t(x, {\rm v}) - \mathrm{H}_t(x, {\rm v}')| \leqslant 3|{\rm v} - {\rm v}'|/2. \tag{17}$$

Observing that $\partial_t \mathrm{H} + \mathscr{L}^{a,b}_t \mathrm{H} - \lambda \mathbf{u}^\lambda = 0$, by generalized Itô's formula for $\mathbb{W}^{2,p}$-function (see [22, p. 121, Theorem 1]), we have

$$\mathrm{H}_t(Z_t) = \mathrm{H}_0(Z_0) + \lambda \int_0^t \mathbf{u}^\lambda_s(Z_s) {\rm d}s + \int_0^t \Theta_s(Z_s) {\rm d}W_s,$$

where $\Theta_s(z) := (\nabla_{\rm v} \mathrm{H}_s \cdot \sigma_s)(z)$. From this equation, and by (16), (17) and Khasminskii's type estimate (see Lemma 1), one can show the pathwise uniqueness together with the homeomorphism property. The Sobolev weak differentiability of $Z_t(z)$ in z follows by Theorem 1 and a moment estimate of $\mathbb{E}|Z_t(z) - Z_t(z')|^p$.

Finally, we present an application to SPDEs. Consider the following second order stochastic partial differential equation:

$$\mathrm{d}u = (\mathrm{tr}(a_t \cdot \nabla_{\mathrm{v}}^2 u) - \mathrm{v} \cdot \nabla_x u - \tilde{b}_t \cdot \nabla_{\mathrm{v}} u)\mathrm{d}t - \sigma_t^* \nabla_{\mathrm{v}} u \, \mathrm{d}W_t, \tag{18}$$

where $a_t := \frac{1}{2}\sigma_t\sigma_t^*$ and

$$\tilde{b}_t^i(x, \mathrm{v}) := b_t^i(x, \mathrm{v}) - \sum_{j,k} \sigma_t^{jk}(x, \mathrm{v})\partial_{\mathrm{v}_j}\sigma_t^{ik}(x, \mathrm{v}), \quad i = 1, \dots, d. \tag{19}$$

The following result is a generalization of [18, Theorems 34 and 37].

Theorem 6 *Assume that* **(UE)** *and* (14) *hold. Let* $u_0 \in \cap_{p>1}\mathbb{W}^{1,p}(\mathbb{R}^{2d})$. *We have*

(i) $u_t(z, \omega) := u_0(Z_t^{-1}(z, \omega)) \in L^\infty_{loc}(\mathbb{R}_+; \cap_{p>1} L^p(\Omega; \mathbb{W}^{1,p}(\mathbb{R}^{2d})))$ *solves SPDE* (18) *in the generalized sense that for all* $\varphi \in C_c^\infty(\mathbb{R}^{2d})$,

$$\begin{aligned}\langle u_t, \varphi\rangle = \langle u_0, \varphi\rangle &- \int_0^t [\langle a\nabla_{\mathrm{v}} u, \nabla_{\mathrm{v}}\varphi\rangle - \langle u, \mathrm{v} \cdot \nabla_x \varphi\rangle]\mathrm{d}s \\ &- \int_0^t \langle(\tilde{b} + \mathrm{div} a) \cdot \nabla_{\mathrm{v}} u, \varphi\rangle \mathrm{d}s - \int_0^t \langle \sigma_t^* \nabla_{\mathrm{v}} u, \varphi\rangle \mathrm{d}W_s,\end{aligned}$$

where $\langle u_t, \varphi\rangle := \int_{\mathbb{R}^{2d}} u_t(z)\varphi(z)\mathrm{d}z$.

(ii) If $\mathrm{div}_{\mathrm{v}}\tilde{b} + \sum_{ij} \partial_{\mathrm{v}_i}\partial_{\mathrm{v}_j} a^{ij} \in L^\infty(\mathbb{R}_+ \times \mathbb{R}^{2d})$, *then SPDE* (18) *has a unique solution in the subclass of adapted processes of* $L^\infty_{loc}(\mathbb{R}_+; \cap_{p>1} L^p(\Omega; \mathbb{W}^{1,p}(\mathbb{R}^{2d})))$.

Proof We sketch the proof. Let σ^n and b^n be the mollifying approximations of σ and b. By [28, p. 170, Theorem 2], there is a smooth solution u^n to SPDE (18) associated with σ^n and b^n. Let $Z_t^n(z)$ be the solution of SDE (12) corresponding to σ^n and b^n. By Itô-Ventcel's formula (see [28, p. 29, Theorem 9]), one sees that $u_t^n(Z_t^n(z)) = u_0(z)$. Hence, $u_t^n(z) = u_0((Z_t^n)^{-1}(z))$. Suppose that $\tilde{Z}_{s,t}^n = (\tilde{X}_{s,t}^n, \dot{\tilde{X}}_{s,t}^n)$ solves the following backward SDE:

$$\mathrm{d}\tilde{Z}_{s,t}^n = -(\dot{\tilde{X}}_{s,t}^n, \tilde{b}_s^n(\tilde{Z}_{s,t}^n))\mathrm{d}s - (0, \sigma_s^n(\tilde{Z}_{s,t}^n)\mathrm{d}W_s), \quad \tilde{Z}_{t,t}^n = z,$$

where $\tilde{b}_t^n$ is defined as in (19) with (b^n, σ^n) replacing (b, σ). It is well known that $(Z_t^n)^{-1} = \tilde{Z}_{0,t}^n$ (see [21, p. 265, Lemma 2.2]). Since the above SDE takes the same form as (12), by Theorem 4 and taking weak limits, one can show (i) as in [18] (see also [16, 17, 20] for the non-degenerate case). As for (ii), let $u_0 = 0$ and $u_t^\varepsilon := u_t * \rho_\varepsilon$, where ρ_ε is a family of mollifiers with compact supports. Then by Itô's formula and the commutator estimates in [39, Lemma 4.2], one can show $\lim_{\varepsilon\to 0}\mathbb{E}\|u_t^\varepsilon\|_2^2 = 0$ (see [39] for more details).

References

1. Ambrosio, L.: Transport equation and Cauchy problem for BV vector fields. Invent. Math. **158**(2), 227–260 (2004)
2. Bogachev V.I. , Pilipenko A.Y.: Strong solutions to stochastic equations with Lévy noise and a discontinuous coefficient. Doklady Math. **92**, English transl. No.1, 471–475 (2015)
3. Bouchut, F.: Hypoeliptic regularity in kinetic equations. J. Math. Pures Appl. **81**, 1135–1159 (2002)
4. Bramanti, M., Cupini, G., Lanconelli, E., Priola, E.: Global L^p-estimate for degenerate Ornstein–Uhlenbeck operators. Math Z. **266**, 789–816 (2010)
5. Chaudru de Raynal P. E.: Strong Existence and Uniqueness for Stochastic Differential Equation with Hölder Drift and Degenerate Noise. Annales de l'Institut Henri Poincaré, **53**(1), 259–286 (2017)
6. Chen Z.-Q., Zhang X.: L^p-Maximal Hypoelliptic Regularity of Nonlocal Kinetic Fokker–Planck Operators. J. Math. Pures Appl. (2017). https://doi.org/10.1016/j.matpur.2017.10.003
7. Chen Z.-Q., Song R., Zhang X.: Stochastic Flows for Lévy Processes with Hölder Drifts. Rev. Mat. Iberoam (2018)
8. Cherny A.S., Engelbert H.J.: Singular Stochastic Differential Equations. Lecture Notes in Mathematics, vol. 1858. Springer, Berlin (2005)
9. Constantin, P., Iyer, G.: A stochastic Lagrangian representation of the three-dimensional incompressible Navier–Stokes equations. Commun. Pure Appl. Math. **61**(3), 330–345 (2008)
10. Crippa, G., De Lellis, C.: Estimates and regularity results for the DiPerna–Lions flow. J. Reine Angew. Math. **616**, 15–46 (2008)
11. Da Prato, G., Flandoli, F., Priola, E., Röckner, M.: Strong uniqueness for stochastic evolution equations in Hilbert spaces perturbed by a bounded measurable drift. Ann. Prob. **41**(5), 3306–3344 (2013)
12. Da Prato, G., Flandoli F., Röckner M., Veretennikov A.Yu.: Strong uniqueness for SDEs in Hilbert spaces with nonregular drift. Ann. Prob. **44**(3), 1985–2023 (2016)
13. Da Prato, G., Flandoli, F., Priola, E., Röckner, M.: Strong uniqueness for stochastic evolution equations with unbounded measurable drift term. J. Theor. Prob. **28**(4), 1571–1600 (2015)
14. DiPerna, R.J., Lions, P.-L.: Ordinary differential equations, transport theory and Sobolev spaces. Invent. Math. **98**(3), 511–547 (1989)
15. Fang, S., Luo, D., Thalmaier, A.: Stochastic differential equations with coefficients in Sobolev spaces. J. Funct. Anal. **259**(5), 1129–1168 (2010)
16. Fedrizzi, E., Flandoli, F.: Noise prevents singularities in linear transport equations. J. Funct. Anal. **264**, 1329–1354 (2013)
17. Fedrizzi, E., Flandoli, F.: Hölder flow and differentiability for SDEs with non regular drift. Stoch. Anal. Appl. **31**, 708–736 (2013)
18. Fedrizzi E., Flandoli F., Priola E., Vovelle J.: Regularity of Stochastic Kinetic Equations. http://arXiv.org/pdf/1606.01088v2.pdf
19. Figalli, A.: Existence and uniqueness of martingale solutions for SDEs with rough or degenerate coefficients. J. Funct. Anal. **254**(1), 109–153 (2008)
20. Flandoli, F., Gubinelli, M., Priola, E.: Well-posedness of the transport equation by stochastic perturbation. Invent. Math. **180**(1), 1–53 (2010)
21. Ikeda, N., Watanabe, S.: Stochastic Differential Equations and Diffusion Processes, 2nd edn. North-Holland/Kodanska, Amsterdam (1989)
22. Krylov N.V.: Controlled Diffusion Processes. Translated from the Russian by A.B. Aries. Applications of Mathematics, 14. Springer, New York (1980)
23. Krylov, N.V., Röckner, M.: Strong solutions of stochastic equations with singular time dependent drift. Probab. Theory Relat. Fields **131**, 154–196 (2005)
24. Kim, K.-H.: $L^q(L^p)$-theory of parabolic PDEs with variable coefficients. Bull. Korean Math. Soc. **45**, 169–190 (2008)

25. Menoukeu-Pamen O, Meyer-Brandis T., Nilssen T., Proske F., Zhang T.: A variational approach to the construction and Malliavin differentiability of strong solutions of SDE's. Math. Ann. **357**(2), 761–799 (2013)
26. Mohammed, S.E.A., Nilssen, T., Proske, F.: Sobolev differentiable stochastic flows for SDEs with singular coefficients: applications to the transport equation. Ann. Probab. **43**(3), 1535–1576 (2015)
27. Priola, E.: Pathwise uniqueness for singular SDEs driven by stable processes. Osaka J. Math. **49**, 421–447 (2012)
28. Rozovskii B.L.: Stochastic Evolution Systems: Linear Theory and Applications to Nonlinear Filtering. Math. Appl. (Sov. Ser.), vol. 35. Kluwer Academic Publishers, Dordrecht (1990)
29. Soize C.: The Fokker-Planck Equation for Stochastic Dynamical Systems and its Explicit Steady State Solutions. Ser. Adv. Math. Appl. Sci., vol. 17, World Scientific, Singapore (1994)
30. Stroock, D., Varadhan, S.R.S.: Multidimensional Diffusion Processes. Springer, Berlin (1997)
31. Talay, D.: Stochastic Hamiltonian systems: exponential convergence to the invariant measure and discretization by the implicit Euler scheme. Markov Process. Related Fields **8**, 1–36 (2002)
32. Veretennikov, AJu: On the strong solutions of stochastic differential equations. Theory Probab. Appl. **24**, 354–366 (1979)
33. Wang, F., Zhang, X.: Degenerate SDE with Hölder-Dini drift and non-Lipschitz noise coefficient. SIAM J. Math. Anal. **48**(3), 2189–2226 (2016)
34. Xie, L., Zhang, X.: Sobolev differentiable flows of SDEs with local Sobolev and super-linear growth coefficients. Ann. Probab. **44**(6), 3661–3687 (2016)
35. Xie L., Zhang X.: Ergodicity of Stochastic Differential Equations with Jumps and Singular Coefficients. Preprint
36. Zhang, X.: Strong solutions of SDEs with singular drift and Sobolev diffusion coefficients. Stoch. Proc. Appl. **115**, 1805–1818 (2005)
37. Zhang, X.: Stochastic flows of SDEs with irregular coefficients and stochastic transport equations. Bull. Sci. Math. **134**(4), 340–378 (2010)
38. Zhang, X.: Stochastic homeomorphism flows of SDEs with singular drifts and Sobolev diffusion coefficients. Electron. J. Probab. **16**(38), 1096–1116 (2011)
39. Zhang, X.: Stochastic partial differential equations with unbounded and degenerate coefficients. J. Differ. Equ. **250**, 1924–1966 (2011)
40. Zhang, X.: Well-posedness and large deviation for degenerate SDEs with Sobolev coefficients. Rev. Mat. Iberoam. **29**(1), 25–52 (2013)
41. Zhang X.: Degenerate irregular SDEs with jumps and application to integro-differential equations of Fokker-Planck type. Electron. J. Probab. **18**, Article 55, 1–25 (2013)
42. Zhang, X.: Stochastic differential equations with Sobolev drifts and driven by α-stable processes. Ann. Inst. H. Poincare Probab. Stat. **49**, 1057–1079 (2013)
43. Zhang, X.: Stochastic differential equations with Sobolev diffusion and singular drift. Ann. Appl. Probab. **26**(5), 2697–2732 (2016)
44. Zhang X.: Stochastic Hamiltonian Flows with Singular Coefficients. Sci. China Math. http://engine.scichina.com/doi/10.1007/s11425-017-9127-0
45. Zvonkin A.K.: A Transformation of the Phase Space of a Diffusion Process that Removes the Drift. Mat. Sbornik, **93**(135), No. 1, 129-149 (1974)

Part IV
Dirichlet Forms, Markov Processes and Potential Theory

Invariant, Super and Quasi-martingale Functions of a Markov Process

Lucian Beznea and Iulian Cîmpean

Dedicated to Michael Röckner on the occasion of his sixtieth birthday.

Abstract We identify the linear space spanned by the real-valued excessive functions of a Markov process with the set of those functions which are quasimartingales when we compose them with the process. Applications to semi-Dirichlet forms are given. We provide a unifying result which clarifies the relations between harmonic, co-harmonic, invariant, co-invariant, martingale and co-martingale functions, showing that in the conservative case they are all the same. Finally, using the co-excessive functions, we present a two-step approach to the existence of invariant probability measures.

Keywords Semimartingale · Quasimartingale · Markov process · Invariant function · Invariant measure

Mathematics Subject Classification (2010) 60J45 · 31C05 · 60J40 · 60J25 · 37C40 · 37L40 · 31C25

L. Beznea · I. Cîmpean (✉)
Simion Stoilow Institute of Mathematics of the Romanian Academy,
Research unit No. 2, P.O. Box 1-764, 014700 Bucharest, Romania
e-mail: iulian.cimpean@imar.ro

L. Beznea
e-mail: lucian.beznea@imar.ro

L. Beznea
University of Bucharest, Faculty of Mathematics and Computer Science, and Centre Francophone en Mathématique de Bucarest, Bucharest, Romania

A. Eberle et al. (eds.), *Stochastic Partial Differential Equations and Related Fields*, Springer Proceedings in Mathematics & Statistics 229,
https://doi.org/10.1007/978-3-319-74929-7_27

1 Introduction

Let E be a Lusin topological space endowed with the Borel σ-algebra $\mathscr{B}$ and $X = (\Omega, \mathscr{F}, \mathscr{F}_t, X_t, \mathbb{P}^x, \zeta)$ be a right Markov process with state space E, transition function $(P_t)_{t \geq 0}$: $P_t u(x) = \mathbb{E}^x(u(X_t); t < \zeta)$, $t \geq 0$, $x \in E$.

One of the fundamental connections between potential theory and Markov processes is the relation between excessive functions and (right-continuous) supermartingales; see e.g. [1], Chap. VI, Sect. 10, or [2], Proposition 13.7.1 and Theorem 14.7.1. Similar results hold for (sub)martingales, and together stand as a keystone at the foundations of the so called probabilistic potential theory. For completeness, let us give the precise statement; a short proof is included in Appendix.

Proposition 1 *The following assertions are equivalent for a non-negative real-valued $\mathscr{B}$-measurable function u and $\beta \geq 0$.*

(i) $(e^{-\beta t} u(X_t))_{t \geq 0}$ is a right continuous $\mathscr{F}_t$-supermartingale w.r.t. $\mathbb{P}^x$ for all $x \in E$.
(ii) The function u is β-excessive.

Our first aim is to show that this connection can be extended to the space of differences of excessive functions on the one hand, and to *quasimartingales* on the other hand (cf. Theorem 1 from Sect. 2), with concrete applications to semi-Dirichlet forms (see Theorem 2 below).

Remark 1 Recall the following famous characterization from [3]: *If u is a real-valued $\mathscr{B}$ -measurable function then $u(X)$ is an $\mathscr{F}_t$ -semimartingale w.r.t. all $\mathbb{P}^x$, $x \in E$ if and only if u is locally the difference of two finite 1-excessive functions.*

The main result from Theorem 1 should be regarded as an extension of Proposition 1 and as a refinement of the just mentioned characterization for semimartingales from Remark 1. However, we stress out that our result is not a consequence of the two previously known results.

In Sect. 3 we focus on a special class of (0-)excessive functions called invariant, which were studied in the literature from several slightly different perspectives. Here, our aim is to provide a unifying result which clarifies the relations between harmonic, co-harmonic, invariant, and co-invariant functions, showing that in the Markovian (conservative) case they are all the same. The measurable structure of invariant functions is also involved. We give the results in terms of $L^p(E, m)$-resolvents of operators, where m is assumed sub-invariant, allowing us to drop the strong continuity assumption. In addition, we show that when the resolvent is associated to a right process, then the martingale functions and the co-martingale ones (i.e., martingale w.r.t. to a dual process) also coincide.

The last topic where the existence of (co)excessive functions plays a fundamental role is the problem of existence of invariant probability measures for a fixed Markovian transition function $(P_t)_{t \geq 0}$ on a general measurable space $(E, \mathscr{B})$. Recall that the classical approach is to consider the dual semigroup of $(P_t)_{t \geq 0}$ acting on the space of all probabilities $P(E)$ on E, and to show that it or its integral means, also known as the Krylov-Bogoliubov measures, are relatively compact w.r.t. some convenient topology (metric) on $P(E)$ (e.g. weak topology, (weighted) total variation norm, Wasserstein metric, etc.). In essence, there are two kind of conditions which stand behind the success of this approach: some (Feller) regularity of the semigroup $(P_t)_{t\geq 0}$ (e.g. it maps bounded and continuous (Lipschitz) functions into bounded and continuous (Lipschitz) functions), and the existence of some compact (or *small*) sets which are infinitely often visited by the process; see e.g. [4–10]. Our last aim is to present (in Sect. 4) a result from [11], which offers a new (two-step) approach to the existence of invariant measures (see Theorem 4 below). In very few words, our idea was to first fix a convenient *auxilliary* measure m (with respect to which each P_t respects classes), and then to look at the dual semigroup of $(P_t)_{t\geq 0}$ acting not on measures as before, but on functions. In this way we can employ some weak $L^1(m)$-compactness results for the dual semigroup in order to produce a non-zero and non-negative co-excessive function.

At this point we would like to mention that most of the announced results, which are going to be presented in the next three sections, are exposed with details in [11–13].

The authors had the pleasure to be coauthors of Michael Röckner and part of the results presented in this survey paper were obtained jointly. So, let us conclude this introduction with a

"Happy Birthday, Michael!"

2 Differences of Excessive Functions and Quasimartingales of Markov Processes

Recall that the purpose of this section is to study those real-valued measurable functions u having the property that $u(X)$ is a $\mathbb{P}^x$-*quasimartingale* for all $x \in E$ (in short, "$u(X)$ is a quasimartingale", or "u is a quasimartingale function"). At this point we would like to draw the attention to the fact that in the first part of this section we study quasimartingales with respect to $\mathbb{P}^x$ for all $x \in E$, in particular all the inequalities involved are required to hold pointwise for all $x \in E$. Later on we shall consider semigroups or resolvents on L^p or Dirichlet spaces with respect to some duality measure, and in these situations we will explicitly mention if the desired properties are required to hold almost everywhere or outside some exceptional sets.

For the reader's convenience, let us briefly present some classic facts about quasimartingales in general.

Definition 1 Let $(\Omega, \mathscr{F}, \mathscr{F}_t, \mathbb{P})$ be a filtered probability space satisfying the usual hypotheses. An $\mathscr{F}_t$-adapted, right-continuous integrable process $(Z_t)_{t\geq 0}$ is called $\mathbb{P}$-quasimartingale if

$$Var^{\mathbb{P}}(Z) := \sup_{\tau} \mathbb{E}\{\sum_{i=1}^{n} |\mathbb{E}[Z_{t_i} - Z_{t_{i-1}} | \mathscr{F}_{t_{i-1}}]| + |Z_{t_n}|\} < \infty,$$

where the supremum is taken over all partitions $\tau : 0 = t_0 \le t_1 \le \ldots \le t_n < \infty$.

Quasimartingales played an important role in the development of the theory of semimartingales and stochastic integration, mainly due to M. Rao's theorem according to which any quasimartingale has a unique decomposition as a sum of a local martingale and a predictable process with paths of locally integrable variation. Conversely, one can show that any semimartingale with bounded jumps is locally a quasimartingale. However, to the best of our knowledge, their analytic or potential theoretic aspects have never been investigated or, maybe, brought out to light, before.

We return now to the frame given by a Markov process. Further in this section we deal with a right Markov process $X = (\Omega, \mathscr{F}, \mathscr{F}_t, X_t, \mathbb{P}^x, \zeta)$ with state space E and transition function $(P_t)_{t\ge 0}$. Although we shall not really be concerned with the lifetime formalism, if X has lifetime ξ and cemetery point Δ, then we make the convention $u(\Delta) = 0$ for all functions $u : E \to [-\infty, +\infty]$.

Recall that for $\beta \ge 0$, a $\mathscr{B}$-measurable function $f : E \to [0, \infty]$ is called β-*supermedian* if $P_t^\beta f \le f$ pointwise on $E, t \ge 0$; $(P_t^\beta)_{t\ge 0}$ denotes the β-level of the semigroup of kernels $(P_t)_{t\ge 0}$, $P_t^\beta := e^{-\beta} P_t$. If f is β-supermedian and $\lim_{t\to 0} P_t f = f$ point-wise on E, then it is called β-*excessive*. It is well known that a $\mathscr{B}$-measurable function f is β-excessive if and only if $\alpha U_{\alpha+\beta} f \le f$, $\alpha > 0$, and $\lim_{\alpha\to\infty} \alpha U_\alpha f = f$ point-wise on E, where $\mathscr{U} = (U_\alpha)_{\alpha>0}$ is the resolvent family of the process X, $U_\alpha := \int_0^\infty e^{-\alpha t} P_t dt$. The convex cone of all β-excessive functions is denoted by $E(\mathscr{U}_\beta)$; here $\mathscr{U}_\beta$ denotes the β-level of the resolvent $\mathscr{U}$, $\mathscr{U}_\beta := (U_{\beta+\alpha})_{\alpha>0}$; the *fine topology* is the coarsest topology on E such that all β-excessive functions are continuous, for some $\beta > 0$. If $\beta = 0$ we drop the index β.

Taking into account the strong connection between excessive functions and supermartingales for Markov processes, the following characterization of M. Rao was our source of inspiration: *a real-valued process on a filtered probability space $(\Omega, \mathscr{F}, \mathscr{F}_t, \mathbb{P})$ satisfying the usual hypotheses is a quasimartingale if and only if it is the difference of two positive right-continuous $\mathscr{F}_t$-supermartingales;* see e.g. [14], p. 116.

As a first observation, note that if $u(X)$ is a quasimartingale, then the following two conditions for u are necessary:

(i) $\sup_{t>0} P_t|u| < \infty$ and (ii) u is finely continuous. Indeed, since for each $x \in E$ we have that $\sup_t P_t|u|(x) = \sup_t \mathbb{E}^x|u(X_t)| \le Var^{\mathbb{P}^x}(u(X)) < \infty$, the first assertion is clear. The second one follows by the result from [15] which is stated in the proof of Proposition 1 in the Appendix at the end of the paper.

For a real-valued function u, a partition τ of $\mathbb{R}^+$, $\tau : 0 = t_0 \le t_1 \le \ldots \le t_n < \infty$, and $\alpha > 0$ we set

$$V^{\alpha}(u) := \sup_{\tau} V_{\tau}^{\alpha}(u), \quad V_{\tau}^{\alpha}(u) := \sum_{i=1}^{n} P_{t_{i-1}}^{\alpha} |u - P_{t_i - t_{i-1}}^{\alpha} u| + P_{t_n}^{\alpha} |u|,$$

where the supremum is taken over all finite partitions of $\mathbb{R}_+$.

A sequence $(\tau_n)_{n\geq 1}$ of finite partitions of $\mathbb{R}_+$ is called *admissible* if it is increasing, $\bigcup_{k\geq 1} \tau_k$ is dense in $\mathbb{R}_+$, and if $r \in \bigcup_{k\geq 1} \tau_k$ then $r + \tau_n \subset \bigcup_{k\geq 1} \tau_k$ for all $n \geq 1$.

We can state now our first result, it is a version of Theorem 2.6 from [12].

Theorem 1 *Let u be a real-valued $\mathscr{B}$-measurable function and $\beta \geq 0$ such that $P_t|u| < \infty$ for all t. Then the following assertions are equivalent.*

(i) $(e^{-\beta t} u(X_t))_{t\geq 0}$ is a $\mathbb{P}^x$-quasimartingale for all $x \in E$.

(ii) u is finely continuous and $\sup_n V_{\tau_n}^{\beta}(u) < \infty$ for one (hence all) admissible sequence of partitions $(\tau_n)_n$.

(iii) u is a difference of two real-valued β-excessive functions.

Remark 2 The key idea behind the previous result is that by the Markov property is not hard to show that for all $x \in E$ we have $Var^{\mathbb{P}^x}((e^{-\alpha t} u(X_t)_{t\geq 0}) = V^{\alpha}(u)(x)$, meaning that assertion (i) holds if and only if $V^{\alpha}(u) < \infty$. But $V^{\alpha}(u)$ is a supremum of measurable functions taken over an uncountable set of partitions, hence it may no longer be measurable, which makes it hard to handle in practice. Concerning this measurability issue, Theorem 1, (ii) states that instead of dealing with $V^{\alpha}(u)$, we can work with $\sup_n V_{\tau_n}^{\alpha}(u)$ for any admissible sequence of partitions $(\tau_n)_{n\geq 1}$. This subtile aspect was crucial in order to give criteria to check the quasimartingale nature of $u(X)$; see also Proposition 1 in the next subsection.

2.1 Criteria for Quasimartingale Functions

In this subsection, still following [12], we provide general conditions for u under which $(e^{-\beta t} u(X_t))_{t\geq 0}$ is a quasimartingale, which means that, in particular, $(u(X_t))_{t\geq 0}$ is a semimartingale.

Let us consider that m is a σ-finite sub-invariant measure for $(P_t)_{t\geq 0}$ so that $(P_t)_{t\geq 0}$ extends uniquely to a strongly continuous semigroup of contractions on $L^p(m)$, $1 \leq p < \infty$; $\mathscr{U}$ may as well be extended to a strongly continuous resolvent family of contractions on $L^p(m)$, $1 \leq p < \infty$. The corresponding generators $(\mathsf{L}_p, D(\mathsf{L}_p) \subset L^p(m))$ are defined by

$$D(\mathsf{L}_p) = \{U_\alpha f : f \in L^p(m)\},$$

$$\mathsf{L}_p(U_\alpha f) := \alpha U_\alpha f - f \quad \text{for all } f \in L^p(m), \ 1 \leq p < \infty,$$

with the remark that this definition is independent of $\alpha > 0$.

The corresponding notations for the dual structure are $\widehat{P}_t$ and $(\widehat{\mathsf{L}}_p, D(\widehat{\mathsf{L}}_p))$, and note that the adjoint of L_p is $\widehat{\mathsf{L}}_{p^*}$; $\frac{1}{p} + \frac{1}{p^*} = 1$. Throughout, we denote the standard L^p-norms by $\|\cdot\|_p$, $1 \le p \le \infty$.

We present below the L^p-version of Theorem 1; cf. Proposition 4.2 from [12].

Proposition 2 *The following assertions are equivalent for a $\mathcal{B}$-measurable function $u \in \bigcup_{1\le p\le\infty} L^p(m)$ and $\beta \ge 0$.*

(i) There exists an m-version $\widetilde{u}$ of u such that $(e^{-\beta t}\widetilde{u}(X_t))_{t\ge 0}$ is a $\mathbb{P}^x$-quasimartingale for $x \in E$ m-a.e.
(ii) For an admissible sequence of partitions $(\tau_n)_{n\ge 1}$ of $\mathbb{R}_+$, $\sup_n V^\beta_{\tau_n}(u) < \infty$ m-a.e.
(iii) There exist $u_1, u_2 \in E(\mathcal{U}_\beta)$ finite m-a.e. such that $u = u_1 - u_2$ m-a.e.

Remark 3 Under the assumptions of Proposition 2, if u is finely continuous and one of the equivalent assertions is satisfied then all of the statements hold outside an m-polar set, not only m-a.e., since it is known that an m-negligible finely open set is automatically m-polar; if in addition m is a reference measure then the assertions hold everywhere on E exept a polar set.

Now, we focus our attention on a class of β-quasimartingale functions which arises as a natural extension of $D(\mathsf{L}_p)$. First of all, it is clear that any function $u \in D(\mathsf{L}_p)$, $1 \le p < \infty$, has a representation $u = U_\beta f = U_\beta(f^+) - U_\beta(f^-)$ with $U_\beta(f^\pm) \in E(\mathcal{U}_\beta) \cap L^p(m)$, hence u has a β-quasimartingale version for all $\beta > 0$; moreover, $\|P_t u - u\|_p = \left\| \int_0^t P_s \mathsf{L}_p u ds \right\|_p \le t\|\mathsf{L}_p u\|_p$. The converse is also true, namely if $1 < p < \infty$, $u \in L^p(m)$, and $\|P_t u - u\|_p \le const \cdot t$, $t \ge 0$, then $u \in D(\mathsf{L}_p)$. But this is no longer the case if $p = 1$ (because of the lack of reflexivity of L^1), i.e. $\|P_t u - u\|_1 \le const \cdot t$ does not imply $u \in D(\mathsf{L}_1)$. However, it turns out that this last condition on $L^1(m)$ is yet enough to ensure that u is a β-quasimartingale function. In fact, the following general result holds; see [12], Proposition 4.4 and its proof.

Proposition 3 *Let $1 \le p < \infty$ and suppose $\mathcal{A} \subset \{u \in L^{p^*}_+(m) : \|u\|_{p^*} \le 1\}$, $\widehat{P}_s \mathcal{A} \subset \mathcal{A}$ for all $s \ge 0$, and $E = \bigcup_{f\in\mathcal{A}} \operatorname{supp}(f)$ m-a.e. If $u \in L^p(m)$ satisfies*

$$\sup_{f\in\mathcal{A}} \int_E |P_t u - u| f dm \le const \cdot t \ for\ all\ t \ge 0,$$

then there exists and m-version $\widetilde{u}$ of u such that $(e^{-\beta t}\widetilde{u}(X_t))_{t\ge 0}$ is a $\mathbb{P}^x$-quasimartingale for all $x \in E$ m-a.e. and every $\beta > 0$.

We end this subsection with the following criteria which is not given with respect to a duality measure, but in terms of the associated resolvent $\mathcal{U}$; cf. Proposition 4.1 from [12].

Proposition 4 *Let u be a real-valued $\mathscr{B}$-measurable finely continuous function.*

(i) Assume there exist a constant $\alpha \geq 0$ and a non-negative $\mathscr{B}$-measurable function c such that

$$U_\alpha(|u| + c) < \infty, \quad \limsup_{t\to\infty} P_t^\alpha |u| < \infty, \quad |P_t u - u| \leq ct, t \geq 0,$$

and the functions $t \mapsto P_t(|u| + c)(x)$ are Riemann integrable. Then $(e^{-\alpha t} u(X_t))_{t\geq 0}$ is a $\mathbb{P}^x$-quasimartingale for all $x \in E$.
(ii) Assume there exist a constant $\alpha \geq 0$ and a non-negative $\mathscr{B}$-measurable function c such that

$$|P_t u - u| \leq ct, t \geq 0, \quad \sup_{t\in\mathbb{R}_+} P_t^\alpha(|u| + c) =: b < \infty.$$

Then $(e^{-\beta t} u(X_t))_{t\geq 0}$ is a $\mathbb{P}^x$-quasimartingale for all $x \in E$ and $\beta > \alpha$.
(iii) Assume there exists $x_0 \in E$ such that for some $\alpha \geq 0$

$$U_\alpha(|u|)(x_0) < \infty, \quad U_\alpha(|P_t u - u|)(x_0) \leq const \cdot t, \ t \geq 0.$$

Then $(e^{-\beta t} u(X_t))_{t\geq 0}$ is a $\mathbb{P}^x$-quasimartingale for $\delta_{x_0} \circ U_\beta$-a.e. $x \in E$ and $\beta > \alpha$; if in addition $\mathscr{U}$ is strong Feller and topologically irreducible then the $\mathbb{P}^x$-quasimartingale property holds for q.e. $x \in E$.

2.2 Applications to Semi-Dirichlet Forms

Assume now that the semigroup $(P_t)_{t\geq 0}$ is associated to a semi-Dirichlet form $(\mathscr{E}, \mathscr{F})$ on $L^2(E, m)$, where m is a σ-finite measure on the Lusin measurable space $(E, \mathscr{B})$; as standard references for the theory of (semi-)Dirichlet forms we refer the reader to [16–19], but also [20], Chap. 7. By Corollary 3.4 from [21] there exists a (larger) Lusin topological space E_1 such that $E \subset E_1$, E belongs to $\mathscr{B}_1$ (the σ-algebra of all Borel subsets of E_1), $\mathscr{B} = \mathscr{B}_1|_E$, and $(\mathscr{E}, \mathscr{F})$ regarded as a semi-Dirichlet form on $L^2(E_1, \overline{m})$ is quasi-regular, where $\overline{m}$ is the trivial extension of m to $(E_1, \mathscr{B}_1)$. Consequently, we may consider a right Markov process X with state space E_1 which is associated with the semi-Dirichlet form $(\mathscr{E}, \mathscr{F})$.

If $u \in \mathscr{F}$ then $\widetilde{u}$ denotes a quasi continuous version of u as a function on E_1 which always exists and it is uniquely determined quasi everywhere. Following [22], for a closed set F we define $\mathscr{F}_{b,F} := \{v \in \mathscr{F} : v$ is bounded and $v = 0$ m-a.e. on $E \setminus F\}$.

The next result is a version of Theorem 5.5 from [12], dropping the a priori assumption that the semi-Dirichlet form is quasi-regular.

Theorem 2 *Let $u \in \mathscr{F}$ and assume there exist a nest $(F_n)_{n\geq 1}$ and constants $(c_n)_{n\geq 1}$ such that*

$$\mathscr{E}(u, v) \leq c_n \|v\|_\infty \text{ for all } v \in \mathscr{F}_{b,F_n}.$$

Then $\widetilde{u}(X)$ is a $\mathbb{P}^x$-semimartingale for $x \in E_1$ quasi everywhere.

Remark 4 The previous result has quite a history behind and we take the opportunity to recall some previous achievements on the subject. First of all, without going into details, note that if E is a bounded domain in $\mathbb{R}^d$ (or more generally in an abstract Wiener space) and the condition from Theorem 2 holds for u replaced by the canonical projections, then the conclusion is that the underlying Markov process is a semimartingale. In particular, the semimartingale nature of reflected diffusions on general bounded domains can be studied. This problem dates back to the work of [23], where the authors showed that the reflected Brownian motion on a Lipschitz domain in $\mathbb{R}^d$ is a semimartingale. Later on, this result has been extended to more general domains and diffusions; see [24–28]. A clarifying result has been obtained in [25], showing that the stationary reflecting Brownian motion on a bounded Euclidian domain is a quasimartingale on each compact time interval if and only if the domain is a strong Caccioppoli set. At this point it is worth to emphasize that in the previous sections we studied quasimartingales on the hole positive real semi-axis, not on finite intervals. This slight difference is a crucial one which makes our approach possible and completely different. A complete study of these problems (including Theorem 2 but only in the symmetric case) have been done in a series of papers by M. Fukushima and co-authors (we mention just [22, 29, 30]), with deep applications to BV functions in both finite and infinite dimensions.

All these previous results have been obtained using the same common tools: symmetric Dirichlet forms and Fukushima decomposition. Further applications to the reflection problem in infinite dimensions have been studied in [31, 32], where non-symmetric situations were also considered. In the case of semi-Dirichlet forms, a Fukushima decomposition is not yet known to hold, unless some additional hypotheses are assumed (see e.g. [19]). Here is where our study developed in the previous sections played its role, allowing us to completely avoid Fukushima decomposition or the existence of the dual process. On brief, the idea of proving Theorem 2 is to show that locally, the conditions from Proposition 3 are satisfied, so that $u(X)$ is (pre)locally a semimartingale, and hence a global semimartingale.

Assume that $(\mathscr{E}, \mathscr{F})$ is quasi-regular and that it is *local*, i.e., $\mathscr{E}(u, v) = 0$ for all $u, v \in \mathscr{F}$ with disjoint compact supports. It is well known that the local property is equivalent with the fact that the associated process is a diffusion; see e.g. [18], Chap. V, Theorem 1.5. As in [29], the local property of $\mathscr{E}$ allows us to extend Theorem 2 to the case when u is only locally in the domain of the form, or to even more general situations, as stated in the next result; for details see Sect. 5.1 from [12].

Corollary 1 *Assume that $(\mathscr{E}, \mathscr{F})$ is local. Let u be a real-valued $\mathscr{B}$-measurable finely continuous function and let $(v_k)_k \subset \mathscr{F}$ such that $v_k \xrightarrow[k\to\infty]{} u$ point-wise outside* *an m-polar set and boundedly on each element of a nest* $(F_n)_{n\geq 1}$. *Further, suppose that there exist constants* c_n *such that*

$$|\mathscr{E}(v_k, v)| \leq c_n \|v\|_\infty \quad for\ all\ v \in \mathscr{F}_{b,F_n}.$$

Then $u(X)$ *is a* $\mathbb{P}^x$*-semimartingale for* $x \in E$ *quasi everywhere.*

3 Excessive and Invariant Functions on L^p-spaces

Throughout this section $\mathscr{U} = (U_\alpha)_{\alpha>0}$ is a sub-Markovian resolvent of kernels on E and m is a σ-finite sub-invariant measure, i.e. $m(\alpha U_\alpha f) \leq m(f)$ for all $\alpha > 0$ and non-negative $\mathscr{B}$-measurable functions f; then there exists a second sub-Markovian resolvent of kernels on E denoted by $\widehat{\mathscr{U}} = (\widehat{U}_\alpha)_{\alpha>0}$ which is in *weak duality* with $\mathscr{U}$ w.r.t. M in the sense that $\int_E fU_\alpha g dm = \int_E g\widehat{U}_\alpha f dm$ for all positive $\mathscr{B}$-measurable functions f, g and $\alpha > 0$. Moreover, both resolvents can be extended to contractions on any $L^p(E, m)$-space for all $1 \leq p \leq \infty$, and if they are strongly continuous then we keep the same notations for their generators as in Sect. 2.1. In this part, our attention focuses on a special class of differences of excessive functions (which are in fact harmonic when the resolvent is Markovian). Extending [33], they are defined as follows.

Definition 2 A real-valued $\mathscr{B}$-measurable function $v \in \bigcup_{1\leq p\leq\infty} L^p(E, m)$ is called $\mathscr{U}$*-invariant* provided that $U_\alpha(vf) = vU_\alpha f$ m-a.e. for all bounded and $\mathscr{B}$-measurable functions f and $\alpha > 0$.

A set $A \in \mathscr{B}$ is called $\mathscr{U}$*-invariant* if 1_A is $\mathscr{U}$-invariant; the collection of all $\mathscr{U}$-invariant sets is a σ-algebra.

Remark 5 If $v \geq 0$ is $\mathscr{U}$-invariant, then by [13], Proposition 2.4 there exists $u \in E(\mathscr{U})$ such that $u = v$ m-a.e. If $\alpha U_\alpha 1 = 1$ m-a.e. then for every invariant function v we have that $\alpha U_\alpha v = v$ m-a.e, which is equivalent (if $\mathscr{U}$ is strongly continuous) with v being L_p-harmonic, i.e. $v \in D(\mathsf{L}_p)$ and $\mathsf{L}_p v = 0$.

The following result is a straightforward consequence of the duality between $\mathscr{U}$ and $\widehat{\mathscr{U}}$; for its proof see Proposition 2.24 and Proposition 2.25 from [13].

Proposition 5 *The following assertions hold.*

(i) A function u is $\mathscr{U}$*-invariant if and only if it is* $\widehat{\mathscr{U}}$*-invariant.*

(ii) The set of all $\mathscr{U}$*-invariant functions from* $L^p(E, m)$ *is a vector lattice with respect to the point-wise order relation.*

Let

$$\mathscr{I}_p := \{u \in L^p(E, m) : \alpha U_\alpha u = u \ m\text{-a.e.},\ \alpha > 0\}.$$

The main result here is the next one, and it unifies and extends different more or less known characterizations of invariant functions; cf. Theorem 2.27 and Proposition 2.29 from [13].

Theorem 3 *Let $u \in L^p(E, m)$, $1 \le p < \infty$, and consider the following conditions.*

(i) $\alpha U_\alpha u = u$ m-a.e. for one (and therefore for all) $\alpha > 0$.

(ii) $\alpha \widehat{U}_\alpha u = u$ m-a.e., $\alpha > 0$.

(iii) The function u is $\mathscr{U}$-invariant.

(iv) $U_\alpha u = u U_\alpha 1$ and $\widehat{U}_\alpha u = u \widehat{U}_\alpha 1$ m-a.e. for one (and therefore for all) $\alpha > 0$.

(v) The function u is measurable w.r.t. the σ-algebra of all $\mathscr{U}$-invariant sets.

Then $\mathscr{I}_p$ is a vector lattice w.r.t. the pointwise order relation and (i) $\Leftrightarrow$ (ii) $\Rightarrow$ (iii) $\Leftrightarrow$ (iv) $\Leftrightarrow$ v).

If $\alpha U_\alpha 1 = 1$ or $\alpha \widehat{U}_\alpha 1 = 1$ m-a.e. then assertions (i)–(v) are equivalent.

If $p = \infty$ and $\mathscr{U}$ is m-recurrent (i.e. there exists $0 \le f \in L^1(E, m)$ s.t. $Uf = \infty$ m-a.e.) then the assertions (i)–(v) are equivalent.

Remark 6 Similar characterizations for invariance as in Theorem 3, but in the recurrent case and for functions which are bounded or integrable with bounded negative parts were already investigated in [34]. Of special interest is the situation when the only invariant functions are the constant ones (*irreducibility*) because it entails ergodic properties for the semigroup resp. resolvent; see e.g. [13, 33, 35].

3.1 Martingale Functions with Respect to the Dual Markov Process

Our aim in this subsection is to identify the $\mathscr{U}$-invariant functions with martingale functions and co-martingale ones (i.e., martingales w.r.t some dual process); cf. Corollary 3 below. The convenient frame is that from [36] and we present it here briefly.

Assume that $\mathscr{U} = (U_\alpha)_{\alpha>0}$ is the resolvent of a right process X with state space E and let $\mathscr{T}_0$ be the Lusin topology of E having $\mathscr{B}$ as Borel σ-algebra, and let m be a fixed $\mathscr{U}$-excessive measure. Then by Corollary 2.4 from [36], and using also the result from [21], the following assertions hold: *There exist a larger Lusin measurable space $(\overline{E}, \overline{\mathscr{B}})$, with $E \subset \overline{E}$, $E \in \overline{\mathscr{B}}$, $\mathscr{B} = \overline{\mathscr{B}}|_E$, and two processes $\overline{X}$ and $\widehat{X}$ with common state space $\overline{E}$, such that $\overline{X}$ is a right process on $\overline{E}$ endowed with a convenient Lusin topology having $\overline{\mathscr{B}}$ as Borel σ-algebra (resp. $\widehat{X}$ is a right process w.r.t. to a second Lusin topology on $\overline{E}$, also generating $\overline{\mathscr{B}}$), the restriction of $\overline{X}$ to E is precisely X, and the resolvents of $\overline{X}$ and $\widehat{X}$ are in duality with respect to $\overline{m}$, where $\overline{m}$ is the trivial extension of m to $(E_1, \mathscr{B}_1)$: $\overline{m}(A) := m(A \cap E)$, $A \in \mathscr{B}_1$. In addition, the α-excessive functions, $\alpha > 0$, with respect to $\widehat{X}$ on $\overline{E}$ are precisely the unique extensions by continuity in the fine topology generated by $\widehat{X}$ of the $\widehat{\mathscr{U}_\alpha}$-excessive functions. In particular, the set E is dense in $\overline{E}$ in the fine topology of $\widehat{X}$.*

Note that the strongly continuous resolvent of sub-Markovian contractions induced on $L^p(m)$, $1 \le p < \infty$, by the process $\overline{X}$ (resp. $\widehat{X}$) coincides with $\mathscr{U}$ (resp. $\widehat{\mathscr{U}}$).

Corollary 2 *Let u be function from $L^p(E, m)$, $1 \le p < \infty$. Then the following assertions are equivalent.*

(i) The process $(u(X_t))_{t \ge 0}$ is a martingale w.r.t. $\mathbb{P}^x$ for m-a.e. $x \in E$.
(ii) The process $(u(\widehat{X}_t))_{t \ge 0}$ is a martingale w.r.t. $\widehat{\mathbb{P}}^x$ for m-a.e. $x \in E$.
(iii) The function u is L_p-harmonic, i.e. $u \in D(\mathsf{L}_p)$ and $\mathsf{L}_p u = 0$.
(iv) The function u is $\widehat{\mathsf{L}}_p$-harmonic, i.e. $u \in D(\widehat{\mathsf{L}}_p)$ and $\widehat{\mathsf{L}}_p u = 0$.

Proof The equivalence $(iii) \iff (iv)$ follows by Theorem 3, $i) \iff ii)$, while the equivalence $(i) \iff (iii)$ is a consequence of Proposition 2. □

We make the transition to the next (also the last) section of this paper with an application of Theorem 3 to the existence of invariant probability measures for Markov processes. More precisely, assume that $\mathscr{U}$ is the resolvent of a right Markov process with transition function $(P_t)_{t \ge 0}$. As before, m is a σ-finite sub-invariant measure for $\mathscr{U}$ (and hence for $(P_t)_{t \ge 0}$), while L_1 and $\widehat{\mathsf{L}}_1$ stand for the generator, resp. the co-generator on $L^1(E, m)$.

Corollary 3 *The following assertions are equivalent.*

(i) There exists an invariant probability measure for $(P_t)_{t \ge 0}$ which is absolutely continuous w.r.t. m.
(ii) There exists a non-zero element $\rho \in D(\mathsf{L}_1)$ such that $\mathsf{L}_1 \rho = 0$.

Proof It is well known that a probability measure $\rho \cdot m$ is invariant w.r.t. $(P_t)_{t \ge 0}$ is equivalent with the fact that $\rho \in D(\widehat{\mathsf{L}}_1)$ and $\widehat{\mathsf{L}}_1 \rho = 0$ (see also Lemma 1, (ii) from below). Now, the result follows by Theorem 3.

Remark 7 Regarding the previous result, we point out that if $m(E) < \infty$ and $(P_t)_{t \ge 0}$ is conservative (i.e. $P_t 1 = 1$ m-a.e. for all $t > 0$) then it is clear that m itself is invariant, so that Corollary 3 has got a point only when $m(E) = \infty$. Also, we emphasize that the sub-invariance property of m is an essential assumption. We present a general result on the existence of invariant probability measures in the next section, where we drop the sub-invariance hypothesis.

4 L^1-harmonic Functions and Invariant Probability Measures

Throughout this subsection $(P_t)_{t \ge 0}$ is a measurable Markovian transition function on a measurable space $(E, \mathscr{B})$ and m is an *auxiliary* measure for $(P_t)_{t \ge 0}$, i.e. a finite positive measure such that $m(f) = 0 \Rightarrow m(P_t f) = 0$ for all $t > 0$ and all positive $\mathscr{B}$-measurable functions f. As we previously announced, our final interest concerns the existence of an invariant probability measure for $(P_t)_{t \ge 0}$ which is absolutely continuous with respect to m.

Remark 8 We emphasize once again that in contrast with the previous section, m is not assumed sub-invariant, since otherwise it would be automatically invariant. Also, any invariant measure is clearly auxiliary, but the converse is far from being true. As a matter of fact, the condition on m of being auxiliary is a minimal one: for every finite positive measure μ and $\alpha > 0$ one has that $\mu \circ U_\alpha$ is auxiliary; see e.g. [11, 37].

For the first assertion of the next result we refer to [11], Lemma 2.1, while the second one is a simple consequence of the fact that $P_t 1 = 1$.

Lemma 1 *(i) The adjoint semigroup $(P_t^*)_{t\geq 0}$ on $(L^\infty(m))^*$ maps $L^1(m)$ into itself, and restricted to $L^1(m)$ it becomes a semigroup of positivity preserving operators.*

(ii) A probability measure $\rho \cdot m$ is invariant with respect to $(P_t)_{t\geq 0}$ if and only if ρ is m-co-excessive, i.e. $P_t^*\rho \leq \rho$ m-a.e. for all $t \geq 0$.

Inspired by well known ergodic properties for semigroups and resolvents (see for example [13]), our idea in order to produce co-excessive functions is to apply (not for $(P_t)_{t\geq 0}$ but for its adjoint semigroup) a compactness result in $L^1(m)$ due to [38], saying that an $L^1(m)$-bounded sequence of elements possesses a subsequence whose Cesaro means are almost surely convergent to a limit from $L^1(m)$.

Definition 3 The auxilliary measure m is called *almost invariant* for $(P_t)_{t\geq 0}$ if there exist $\delta \in [0, 1)$ and a set function $\phi : \mathscr{B} \to \mathbb{R}_+$ which is absolutely continuous with respect to m (i.e. $\lim\limits_{m(A)\to 0} \phi(A) = 0$) such that

$$m(P_t 1_A) \leq \delta m(E) + \phi(A) \quad \text{for all } t > 0.$$

Clearly, any positive finite invariant measure is almost invariant. Here is our last main result, a variant of Theorem 2.4 from [11].

Theorem 4 *The following assertions are equivalent.*

(i) There exists a nonzero positive finite invariant measure for $(P_t)_{t\geq 0}$ which is absolutely continuous with respect to m.

(ii) m is almost invariant.

Acknowledgements The first named author acknowledges support from the Romanian National Authority for Scientific Research, project number PN-III-P4-ID-PCE-2016-0372. The second named author acknowledges support from the Romanian National Authority for Scientific Research, project number PN-II-RU-TE-2014-4-0657.

Appendix

Proof of Proposition 1 (i) $\Rightarrow$ (ii). If $(e^{-\beta t}u(X_t))_{t\geq 0}$ is a right-continuous supermartingale then by taking expectations we get that $e^{-\beta t}\mathbb{E}^x u(X_t) \leq \mathbb{E}^x u(X_0)$, hence

u is β-supermedian. Now, by [20], Corollary 1.3.4, showing that $u \in E(\mathscr{U}_\beta)$ reduces to prove that u is finely continuous, which in turns follows by the well known characterization according to which u is finely continuous if and only if $u(X)$ has right continuous trajectories $\mathbb{P}^x$-a.s. for all $x \in E$; see Theorem 4.8 in [15], Chap. II.

(ii) $\Rightarrow$ (i). Since u is β-supermedian, by the Markov property we have for all $0 \le s \le t$

$$\mathbb{E}^x[e^{-\beta(t+s)}u(X_{t+s})|\mathscr{F}_s] = e^{-\beta(t+s)}\mathbb{E}^{X_s}u(X_t) = e^{-\beta(t+s)}P_t u(X_s) \le e^{-\beta s}u(X_s),$$

hence $(e^{-\beta t}u(X_t))_{t\ge 0}$ is an $\mathscr{F}_t$-supermartingale. The right-continuity of the trajectories follows by the fine continuity of u via the previously mentioned characterization. □

References

1. Doob, J.L.: Classical potential theory and its probabilistic counterpart. Springer, Berlin (2001)
2. Le Gall, J.F.: Intégration, probabilités et processus aléatoires. Ecole Normale Supérieure de Paris. Sept 2006
3. Çinlar, E., Jacod, J., Protter, P., Sharpe, M.J.: Semimartingales and Markov processes. Z. Wahrscheinlichkietstheorie verw. Gebiete **54**, 161–219 (1980)
4. Da Prato, G., Zabczyk, J.: Ergodicity for infinite dimensional systems. Cambridge University Press, Cambridge (1996)
5. Hairer, M.: Convergence of Markov processes. Lecture Notes, University of Warwick (2010). http://www.hairer.org/notes/Convergence.pdf
6. Komorowski, T., Peszat, S., Szarek, T.: On ergodicity of some Markov processes. Ann. Probab. **38**, 1401–1443 (2010)
7. Lasota, A., Szarek, T.: Lower bound technique in the theory of a stochastic differential equation. J. Differ. Equ. **231**, 513–533 (2006)
8. Meyn, S.P., Tweedie, R.L.: Markov chains and stochastic stability. Springer, London (1993)
9. Meyn, S.P., Tweedie, R.L.: Stability of markovian processes II: continuous-time processes and sampled chains. Adv. Appl. Probab. **25**, 487–517 (1993)
10. Meyn, S.P., Tweedie, R.L.: Stability of markovian processes III: Foster-Lyapunov criteria for continuous-time processes. Adv. Appl. Probab. **25**, 518–548 (1993)
11. Beznea, L., Cîmpean, I., Röckner, M.: A new approach to the existence of invariant measures for Markovian semigroups. arXiv:1508.06863v3
12. Beznea, L., Cîmpean, I.: Quasimartingales associated to Markov processes. Trans. Am. Math. Soc. (to appear (2017)). arXiv:1702.06282
13. Beznea, L., Cîmpean, I., Röckner, M.L: Irreducible recurrence, ergodicity, and extremality of invariant measures for resolvents. Stoch. Process. Appl. (2017). arXiv:1409.6492v2. https://doi.org/10.1016/j.spa.2017.07.009
14. Protter, P.E.: Stochastic integration and differential equations. Springer, Berlin (2005)
15. Blumenthal, R., Getoor, R.: Markov processes and potential theory. Academic Press, New York (1968)
16. Fukushima, M., Oshima, Y., Takeda, M.: Dirichlet forms and symmetric Markov processes. Walter de Gruyter, Berlin/New York (2011)
17. Ma, Z.M., Overbeck, L., Röckner, M.: Markov processes associated with semi-Dirichlet forms. Osaka J. Math. **32**, 97–119 (1995)
18. Ma, Z.M., Röckner, M.: An introduction to the theory of (non-symmetric) Dirichlet forms. Springer, Berlin (1992)

19. Oshima, Y.: Semi-dirichlet forms and Markov processes. Walter de Gruyter and Co., Berlin (2013)
20. Beznea, L., Boboc, N.: Potential theory and right processes. Springer Series, Mathematics and its applications (572). Kluwer, Dordrecht (2004)
21. Beznea, L., Boboc, N., Röckner, M.: Quasi-regular Dirichlet forms and L^p-resolvents on measurable spaces. Potential Anal. **25**, 269–282 (2006)
22. Fukushima, M.: On semi-martingale characterization of functionals of symmetric Markov processes. Electron. J. Probab. **4**, 1–32 (1999)
23. Bass, R.F., Hsu, P.: The semimartingale structure of reflecting Brownian motion. Proc. Am. Math. Soc. **108**, 1007–1010 (1990)
24. Chen, Z.Q.: On reflecting diffusion processes and Skorokhod decompositions. Probab. Theory Relat. Fields **94**, 281–315 (1993)
25. Chen, Z.Q., Fitzsimmons, P.J., Williams, R.J.: Reflecting Brownian motions: quasimartingales and strong Caccioppoli sets. Potential Anal. **2**, 219–243 (1993)
26. Pardoux, E., Williams, R.J.: Symmetric reflected diffusions. Ann. Inst. H. Poincaré Probab. Stat. **30**, 13–62 (1994)
27. Williams, R.J., Zheng, W.A.: On reflecting Brownian motion - a weak convergence approach. Ann. Inst. Henri Poincaré **26**, 461–488 (1990)
28. Trutnau, G.: Skorokhod decomposition of reflected diffusions on bounded Lipschitz domains with singular non-reflection part. Probab. Theory Relat. Fields **127**, 455–495 (2003)
29. Fukushima, M.: BV functions and distorted Ornstein Uhlenbeck processes over the abstract Wiener space. J. Funct. Anal. **174**, 227–249 (2000)
30. Fukushima, M., Hino, M.: On the space of BV functions and a related stochastic calculus in infinite dimensions. J. Funct. Anal. **183**, 245–268 (2001)
31. Röckner, M., Zhu, R., Zhu, X.: The stochastic reflection problem on an infinite dimensional convex set and BV functions in a Gelfand triple. Ann. Probab. **40**, 1759–1794 (2012)
32. Röckner, M., Zhu, R., Zhu, X.: BV functions in a Gelfand triple for differentiable measure and its applications. Forum Math. **27**, 1657–1687 (2015)
33. Albeverio, S., Kondratiev, Y.G., Röckner, M.: Ergodicity of L^2-semigroups and extremality of Gibbs states. J. Funct. Anal. **144**, 394–423 (1997)
34. Schilling, R.L.: A note on invariant sets. Probab. Math. Statist. **24**, 47–66 (2004)
35. Sturm, K.T.: Analysis on local Dirichlet spaces. I. Recurrence, conservativeness and L^p-Liouville properties. J. Reine Angew. Math. **456**, 173–196 (1994)
36. Beznea, L., Röckner, M.: On the existence of the dual right Markov process and applications. Potential Anal. **42**, 617–627 (2015)
37. Röckner, M., Trutnau, G.: A remark on the generator of a right-continuous Markov process. Infin. Dimens. Anal. Quantum. Probab. Relat. Top. **10**, 633–640 (2007)
38. Komlós, J.: A generalization of a problem of Steinhaus. Acta Math. Acad. Sci. Hungar. **18**, 217–229 (1967)

Mean Value Inequalities for Jump Processes

Zhen-Qing Chen, Takashi Kumagai and Jian Wang

Dedicated to Professor Michael Röckner on the occasion of his 60th birthday.

Abstract Parabolic Harnack inequalities are one of the most important inequalities in analysis and PDEs, partly because they imply Hölder regularity of the solutions of heat equations. Mean value inequalities play an important role in deriving parabolic Harnack inequalities. In this paper, we first survey the recent results obtained in Chen et al. (Stability of heat kernel estimates for symmetric non-local Dirichlet forms, 2016, [15]; Stability of parabolic Harnack inequalities for symmetric non-local Dirichlet forms, 2016, [16]) on the study of stability of heat kernel estimates and parabolic Harnack inequalities for symmetric jump processes on general metric measure spaces. We then establish the L^p-mean value inequalities for all $p \in (0, 2]$ for these processes.

Keywords Symmetric jump process · Heat kernel estimate · Harnack inequality Stability · Mean value inequality

MSC 2010 Primary 31B05 · 60J35 · 60J75 · Secondary 31C25 · 35K08 · 60J45

Z.-Q. Chen
University of Washington, Seattle, WA 98195, USA
e-mail: zqchen@uw.edu

T. Kumagai (✉)
RIMS, Kyoto University, Kyoto 606-8502, Japan
e-mail: kumagai@kurims.kyoto-u.ac.jp

J. Wang
Fujian Normal University, Fuzhou 350007, People's Republic of China
e-mail: jianwang@fjnu.edu.cn

A. Eberle et al. (eds.), *Stochastic Partial Differential Equations and Related Fields*, Springer Proceedings in Mathematics & Statistics 229,
https://doi.org/10.1007/978-3-319-74929-7_28

1 Introduction

Consider a divergence operator $\mathscr{L} = \sum_{i,j=1}^{d} \frac{\partial}{\partial x_i}(a_{ij}(x)\frac{\partial}{\partial x_j})$ acting on functions on $\mathbb{R}^d$, where $(a_{ij}(x))_{i,j=1}^{d}$ is bounded, measurable, and uniform elliptic. In 1964, Moser [28] proved the parabolic Harnack inequalities (PHI(2); see Definition 6 with $\phi(r) = r^2$) for non-negative solutions to the heat equation

$$\frac{\partial u}{\partial t} = \mathscr{L}u. \tag{1}$$

In 1967, Aronson [2] obtained Gaussian type bounds (i.e. (2) with $\mu(B(x, t^{1/2})) = t^{d/2}$ and $d(\cdot, \cdot)$ being the Euclidean metric) for the fundamental solution to (1). These theorems had a profound influence on analysis and differential geometry. An important consequence of the results is that the non-negative solutions to (1) enjoy Hölder regularity (i.e. (16) with $\phi^{-1}(t) = t^{1/2}$). In deriving PHI(2), mean value inequalities (i.e. (18) and (19) without the tail term) play essential roles. In fact, such mean value inequalities were obtained for various linear and non-linear PDEs to derive Harnack inequalities (see, for instance [7, 20, 31, 33]).

There are further significant developments later in the last century. Consider a complete Riemannian manifold M with the Riemannian metric $d(\cdot, \cdot)$ and with the Riemannian measure μ. Let $\mathscr{L}$ be the Laplace–Beltrami operator on M. In 1986, Li-Yau [26] proved the following remarkable fact – if M has non-negative Ricci curvature, then the heat kernel $p_t(x, y)$ enjoys the following estimates

$$\frac{c_1}{\mu(B(x, t^{1/2}))} \exp\Big(-\frac{c_2 d(x, y)^2}{t}\Big) \le p(t, x, y) \le \frac{c_3}{\mu(B(x, t^{1/2}))} \exp\Big(-\frac{c_4 d(x, y)^2}{t}\Big). \tag{2}$$

A few years later, Grigor'yan [21] and Saloff-Coste [30] refined the result and proved that PHI(2) is equivalent to a volume doubling condition (VD; see Definition 1(i)) plus Poincaré inequalities (PI(2); see Definition 8(iii) with $\phi(r) = r^2$). It was also known around 80s that (2) is equivalent to PHI(2), so the following equivalence holds:

$$(2) \Leftrightarrow \mathrm{VD} + \mathrm{PI}(2) \Leftrightarrow \mathrm{PHI}(2). \tag{3}$$

Later, these results were extended to the framework of strongly local Dirichlet forms on metric measure spaces by Sturm [32] and on graphs by Delmotte [17]. One of the important consequence of the equivalence is that (2) and PHI(2) are stable under perturbations, since both VD and PI(2) are stable under the perturbations of rough isometries. Such an equivalence was generalized to the so-called sub-Gaussian heat kernel estimates for symmetric diffusions:

$$\frac{c_1}{\mu(B(x,t^{1/d_w}))}\exp\left(-c_2\Big(\frac{d(x,y)^{d_w}}{t}\Big)^{1/(d_w-1)}\right)$$
$$\le p(t,x,y)\le \frac{c_3}{\mu(B(x,t^{1/d_w}))}\exp\left(-c_4\Big(\frac{d(x,y)^{d_w}}{t}\Big)^{1/(d_w-1)}\right) \tag{4}$$

for some $d_w \ge 2$. When $d_w = 2$, it is just the Aronson Gaussian estimates (2); and when $d_w > 2$, the behaviors of the corresponding diffusions are anomalous. Diffusions on fractals are typical examples that enjoy (4) for some $d_w > 2$. It turns out (see [1, 3, 4, 24]) that there is an inequality $\mathrm{CSA}(d_w)$, a version of the so-called cut-off Sobolev inequality, such that the following equivalence holds:

$$(4) \Leftrightarrow \mathrm{VD} + \mathrm{PI}(d_w) + \mathrm{CSA}(d_w) \Leftrightarrow \mathrm{PHI}(d_w). \tag{5}$$

See Definitions 6 and 8(iii) with $\phi(r) = r^{d_w}$ for definitions of $\mathrm{PHI}(d_w)$ and $\mathrm{PI}(d_w)$, respectively. We will not give the precise definition of $\mathrm{CSA}(d_w)$ (see Definition 4 for the corresponding inequality for symmetric jump processes). Instead, we note that CSA(2) always holds (so that (5) is indeed a generalization of (3)), and that $\mathrm{CSA}(d_w)$ is stable under rough isometries (and, consequently, (4) and $\mathrm{PHI}(d_w)$ are stable under rough isometries).

For symmetric jump processes, the corresponding results have been obtained only recently. Suppose that a metric measure space (M, μ) is an Alhfors d-regular set on $\mathbb{R}^n$; namely, $\mu(B(x,r)) \asymp r^d$ for all $r > 0$, and a regular Dirichlet form $(\mathscr{E}, \mathscr{F})$ on $L^2(M;\mu)$ is defined by

$$\mathscr{E}(f,g) := \int_{M\times M\setminus \Delta} \frac{(f(x)-f(y)(g(x)-g(y))}{|x-y|^{d+\alpha}} c(x,y)\mu(dx)\mu(dy),$$

where $c(\cdot,\cdot)$ is a measurable symmetric function that is bounded between two positive constants and $0<\alpha<2$. The Hunt process X associated with $(\mathscr{E}, \mathscr{F})$ is called a symmetric α-stable-like process on M. It was proved in [12] that the corresponding heat kernel of the Dirichlet form (or equivalently, of X) enjoys the following estimates for all $t > 0$ and $x, y \in M$

$$c_1\Big(t^{d/\alpha}\wedge \frac{t}{|x-y|^{d+\alpha}}\Big) \le p(t,x,y) \le c_2\Big(t^{d/\alpha}\wedge \frac{t}{|x-y|^{d+\alpha}}\Big).$$

In that paper, α-order parabolic Harnack inequalities ($\mathrm{PHI}(\alpha)$; see Definition 6 with $\phi(r) = r^\alpha$) were also proved. In the subsequent paper [13], the results were extended to more general time-scale functions, and in [5] some equivalence criteria were given concerning the heat kernel estimates and parabolic Harnack inequalities for symmetric α-stable-like processes with $0 < \alpha < 2$ on Alhfors regular graphs. In the very recent papers [15, 16], complete equivalences and stability for heat kernel estimates and parabolic Harnack inequalities have been established for symmetric jump processes of variable order on general metric measure spaces. An important ingredient in our approach in these two papers is the L^2 and L^1 mean value inequalities for subharmonic functions of symmetric finite range jump processes.

The aim of this paper is twofold. Firstly, we present the main results obtained in our recent papers [15, 16] on equivalent characterizations of heat kernel estimates and parabolic Harnack inequalities. Secondly, we show that the L^p-mean value inequalities hold not only for $p = 2$ but also for all $p \in (0, 2]$ for a large class of symmetric jump processes. There are done in Sects. 2 and 3, respectively.

2 Stability of Heat Kernel Estimates and Parabolic Harnack Inequalities for Symmetric Non-local Dirichlet Forms

2.1 Setting

Let (M, d) be a locally compact separable metric space, and μ a positive Radon measure on M with full support. The triple (M, d, μ) is called a *metric measure space*. Throughout the paper, we assume for simplicity that $\mu(M) = \infty$. Note that we do not assume M to be connected nor (M, d) to be geodesic.

Let $(\mathscr{E}, \mathscr{F})$ be a regular *Dirichlet form* on $L^2(M; \mu)$ of pure-jump type; namely,

$$\mathscr{E}(f, g) = \int_{M\times M\setminus \Delta} (f(x) - f(y)(g(x) - g(y))\, J(dx, dy), \quad f, g \in \mathscr{F}, \tag{6}$$

where $\Delta := \{(x, x) : x \in M\}$ and $J(\cdot, \cdot)$ is a symmetric Radon measure on $M \times M \setminus \Delta$. In the paper, we will always take the quasi-continuous version for an element of $\mathscr{F}$ (note that since $(\mathscr{E}, \mathscr{F})$ is regular, each function in $\mathscr{F}$ admits a quasi-continuous version). Let $\mathscr{L}$ be the (negative definite) L^2-generator of $(\mathscr{E}, \mathscr{F})$ and $\{P_t\}$ be the associated semigroup on $L^2(M; \mu)$. There exists an μ-symmetric *Hunt process* $X = \{X_t, t \geq 0, \mathbb{P}^x, x \in M \setminus \mathscr{N}\}$ which is associated with the regular Dirichlet form $(\mathscr{E}, \mathscr{F})$ on $L^2(M; \mu)$. Here $\mathscr{N}$ is a properly exceptional set for $(\mathscr{E}, \mathscr{F})$ in that $\mu(\mathscr{N}) = 0$ and $\mathbb{P}^x(X_t \in \mathscr{N}$ for some $t > 0) = 0$ for all $x \in M \setminus \mathscr{N}$. It is known that this Hunt process is uniquely determined up to a properly exceptional set (see [18, Theorem 4.2.8] or [11, Theorem 1.5.1] or [27, Chap. IV, Theorem 6.4]). Furthermore, we can obtain a refined version of $\{P_t\}$ with better regularity properties as follows:

$$P_t f(x) = \mathbb{E}^x f(X_t), \quad x \in M_0 := M \setminus \mathscr{N}$$

for any bounded Borel measurable function f on M.

A measurable function $p(t,x,y):(0,\infty)\times M_0\times M_0\to(0,\infty)$ is called a *heat kernel* associated with $\{P_t\}$ if the following hold:

$$\begin{aligned}\mathbb{E}^x f(X_t) = P_t f(x) &= \int p(t,x,y)f(y)\,\mu(dy), && \forall x\in M_0,\ f\in L^\infty(M,\mu),\\ p(t,x,y) &= p(t,y,x), && \forall t>0,\ x,y\in M_0,\\ p(s+t,x,z) &= \int p(s,x,y)p(t,y,z)\,\mu(dy), && \forall s,t>0,\ x,z\in M_0.\end{aligned}$$

We may extend $p(t,x,y)$ to all $x,y\in M$ by setting $p(t,x,y)=0$ if x or y is outside M_0.

Definition 1 Let $B(x,r)$ be the ball in (M,d) centered at x with radius r, and set

$$V(x,r)=\mu(B(x,r)).$$

(i) We say that (M,d,μ) satisfies the *volume doubling property* (VD) if there exists a constant $C_\mu\geq 1$ so that for all $x\in M$ and $r>0$,

$$V(x,2r)\leq C_\mu V(x,r). \tag{7}$$

(ii) We say that (M,d,μ) satisfies the *reverse volume doubling property* (RVD) if there exist constants $l_\mu, c_\mu>1$ so that for all $x\in M$ and $r>0$,

$$V(x,l_\mu r)\geq c_\mu V(x,r).$$

VD condition (7) is equivalent to the following: there exist positive constants d_2 and $\widetilde{C}_\mu$ so that

$$\frac{V(x,R)}{V(x,r)}\leq \widetilde{C}_\mu\Big(\frac{R}{r}\Big)^{d_2}\quad \text{for all } x\in M \text{ and } 0<r\leq R. \tag{8}$$

It is known that VD implies RVD if M is connected and unbounded (see, for example [22, Proposition 5.1 and Corollary 5.3]).

Let $\mathbb{R}_+:=[0,\infty)$ and $\phi:\mathbb{R}_+\to\mathbb{R}_+$ be a strictly increasing continuous function with $\phi(0)=0$, $\phi(1)=1$ that satisfies the following: there exist constants $c_1,c_2>0$ and $\beta_2\geq\beta_1>0$ such that

$$c_1\Big(\frac{R}{r}\Big)^{\beta_1}\leq\frac{\phi(R)}{\phi(r)}\leq c_2\Big(\frac{R}{r}\Big)^{\beta_2}\quad\text{for all } 0<r\leq R. \tag{9}$$

Definition 2 We say J_ϕ holds if there exists a non-negative symmetric function $J(\cdot,\cdot)$ so that for $\mu\times\mu$-almost all $x,y\in M$,

$$J(dx,dy)=J(x,y)\,\mu(dx)\,\mu(dy), \tag{10}$$

and

$$\frac{c_1}{V(x, d(x, y))\phi(d(x, y))} \le J(x, y) \le \frac{c_2}{V(x, d(x, y))\phi(d(x, y))} \tag{11}$$

for some constants $c_2 \ge c_1 > 0$. We say that $\mathrm{J}_{\phi,\le}$ (resp. $\mathrm{J}_{\phi,\ge}$) if (10) holds and the upper bound (resp. lower bound) in (11) holds.

For a non-local Dirichlet form $(\mathscr{E}, \mathscr{F})$, we define the carré du-Champ operator $\Gamma(f, g)$ for $f, g \in \mathscr{F}$ by

$$\Gamma(f, g)(dx) = \int_{y \in M} (f(x) - f(y))(g(x) - g(y))\, J(dx, dy).$$

Clearly $\mathscr{E}(f, g) = \Gamma(f, g)(M)$. Note that for any $f \in \mathscr{F}_b := \mathscr{F} \cap L^\infty(M, \mu)$, $\Gamma(f, f)$ is the unique Borel measure (called the *energy measure*) on M satisfying

$$\int_M g\, d\Gamma(f, f) = \mathscr{E}(f, fg) - \frac{1}{2}\mathscr{E}(f^2, g), \quad f, g \in \mathscr{F}_b.$$

2.2 *Heat Kernel Estimates*

Definition 3 We say that $\mathrm{HK}(\phi)$ holds if there exists a kernel $p(t, x, y)$ with respect to the measure μ of the semigroup $\{P_t\}$ for $(\mathscr{E}, \mathscr{F})$ so that the following estimates hold for all $t > 0$ and all $x, y \in M_0$,

$$\begin{aligned} c_1\left(\frac{1}{V(x, \phi^{-1}(t))} \wedge \frac{t}{V(x, d(x, y))\phi(d(x, y))}\right) &\le p(t, x, y) \\ &\le c_2\left(\frac{1}{V(x, \phi^{-1}(t))} \wedge \frac{t}{V(x, d(x, y))\phi(d(x, y))}\right), \end{aligned} \tag{12}$$

where $c_1, c_2 > 0$ are constants independent of $x, y \in M_0$ and $t > 0$. Here $\phi^{-1}(t)$ is the inverse function of $t \mapsto \phi(t)$. We say $\mathrm{UHK}(\phi)$ (resp. $\mathrm{LHK}(\phi)$) holds if the upper bound (resp. the lower bound) in (12) holds for $p(t, x, y)$.

Remark 1 (i) We can replace $V(x, d(x, y))$ by $V(y, d(x, y))$ in (12) by modifying the values of c_1 and c_2. Indeed, the following holds (see [15, Remark 1.12]):

$$\frac{1}{V(y, \phi^{-1}(t))} \wedge \frac{t}{V(y, d(x, y))\phi(d(x, y))} \asymp \frac{1}{V(x, \phi^{-1}(t))} \wedge \frac{t}{V(x, d(x, y))\phi(d(x, y))}.$$

Here for two functions f and g, notation $f \asymp g$ means f/g is bounded between two positive constants.

(ii) It follows from [15, Theorem 1.13 and Lemma 5.6] that if HK(ϕ) holds, then the heat kernel $p(t, x, y)$ is Hölder continuous in (x, y) for every $t > 0$, so (12) holds for all $x, y \in M$ and $t > 0$.

In [15], stability of heat kernel estimates has been established for symmetric pure-jump processes on a general metric measure space. Below is the precise statement.

Theorem 1 *Assume that the metric measure space (M, d, μ) satisfies* VD *and* RVD, *and ϕ satisfies* (9)*. Let $(\mathscr{E}, \mathscr{F})$ be a regular (resp. regular and conservative) symmetric Dirichlet form on $L^2(M; \mu)$ of pure-jump type* (6)*. $(\widetilde{\mathscr{E}}, \mathscr{F})$ Let $(\widetilde{\mathscr{E}}, \mathscr{F})$ be another regular (resp. regular and conservative) symmetric Dirichlet form on $L^2(M; \widetilde{\mu})$ of pure-jump type* (6) *with jumping measure $\widetilde{J}(dx, dy)$, and there exists a constant $1 \le c < \infty$ such that for all measurable sets A and B,*

$$c^{-1}\mu(A) \le \widetilde{\mu}(A) \le c\mu(A), \tag{13}$$

$$c^{-1}J(A, B) \le \widetilde{J}(A, B) \le cJ(A, B) \quad \text{when } d(A, B) > 0. \tag{14}$$

Then $(\mathscr{E}, \mathscr{F})$ satisfies HK(ϕ) *(resp.* UHK(ϕ)*) if and only if so does $(\widetilde{\mathscr{E}}, \mathscr{F})$.*

In [15], this theorem is a direct consequence of the stable characterization of HK(ϕ) and UHK(ϕ), which is stable under perturbations (13) and (14). Precise statements will be given in Theorems 2 and 3 below. First we need some definitions.

The following inequality CSJ(ϕ) that controls the energy of cutoff functions, introduced in [15], is a modification of CSA(ϕ) in [1] for strongly local Dirichlet forms as a weaker version of the cut-off Sobolev inequality CS(ϕ) in [3, 4]. In [24], the inequality corresponding to CSJ(ϕ) for strongly local Dirichlet forms is called a generalized capacity inequality.

Definition 4 (i) Let $U \subset V$ be open sets in M with $U \subset \overline{U} \subset V$. A non-negative bounded measurable function φ is said to be a *cutoff function for $U \subset V$* if $\varphi = 1$ on U, $\varphi = 0$ on V^c and $0 \le \varphi \le 1$ on M.

(ii) We say that CSJ(ϕ) holds if there exist constants $c_0 \in (0, 1]$ and $c_1, c_2 > 0$ such that for every $0 < r \le R$, almost all $x \in M$ and any $f \in \mathscr{F}$, there exists a cutoff function $\varphi \in \mathscr{F}_b$ for $B(x, R) \subset B(x, R + r)$ so that the following holds:

$$\int_{B(x,R+(1+c_0)r)} f^2\, d\Gamma(\varphi, \varphi) \le c_1 \int_{U\times U^*} (f(x) - f(y))^2\, J(dx, dy) + \frac{c_2}{\phi(r)} \int_{B(x,R+(1+c_0)r)} f^2\, d\mu,$$

where $U = B(x, R + r)\backslash B(x, R)$ and $U^* = B(x, R + (1 + c_0)r) \setminus B(x, R - c_0 r)$.

Remark 2 As is pointed out in [15, Remark 1.7], under VD, (9) and $\mathrm{J}_{\phi,\le}$, CSJ(ϕ) always holds if $\beta_2 < 2$, where β_2 is the exponent in (9). In particular, CSJ(ϕ) always holds for $\phi(r) = r^\alpha$ with $0 < \alpha < 2$.

For any open set $D \subset M$, $\mathscr{F}_D$ is defined to be the $\mathscr{E}_1$-closure in $\mathscr{F}$ of $\mathscr{F} \cap C_c(D)$, where $\|\cdot\|^2_{\mathscr{E}_1} = \|\cdot\|^2_{\mathscr{E}} + \|\cdot\|^2_2$, and $C_c(D)$ is the space of continuous functions on M with compact support in D. Define

$$\lambda_1(D) = \inf\{\mathscr{E}(f,f) : f \in \mathscr{F}_D \text{ with } \|f\|_2 = 1\},$$

the bottom of the Dirichlet spectrum of $-\mathscr{L}$ on D. For a set $A \subset M$, define its exit time $\tau_A = \inf\{t > 0 : X_t \in A^c\}$.

Definition 5 (i) We say that the *Faber–Krahn inequality* FK(ϕ) holds if there exist constants $c, \nu > 0$ such that for any ball $B(x,r)$ and any open set $D \subset B(x,r)$,

$$\lambda_1(D) \geq \frac{c}{\phi(r)}(V(x,r)/\mu(D))^{\nu}.$$

(ii) We say that E_ϕ holds if there is a constant $c_1 > 1$ such that for all $r > 0$ and all $x \in M_0$,

$$c_1^{-1}\phi(r) \leq \mathbb{E}^x[\tau_{B(x,r)}] \leq c_1\phi(r).$$

We say that $\mathrm{E}_{\phi,\leq}$ (resp. $\mathrm{E}_{\phi,\geq}$) holds if the upper bound (resp. lower bound) in the above display holds for $\mathbb{E}^x[\tau_{B(x,r)}]$.

(iii) We say UHKD(ϕ) holds if there is a constant $c > 0$ such that

$$p(t,x,x) \leq \frac{c}{V(x,\phi^{-1}(t))} \quad \text{for all } t > 0 \text{ and } x \in M_0.$$

(iv) We say $(\mathscr{E},\mathscr{F})$ *is conservative* if its associated Hunt process X has infinite lifetime. This is equivalent to $P_t 1 = 1$ a.e. on M_0 for every $t > 0$.

The following are the main results of [15].

Theorem 2 ([15, Theorem 1.13]) *Assume that the metric measure space (M,d,μ) satisfies* VD *and* RVD, *and ϕ satisfies* (9). *Then the following are equivalent:*

(1) HK(ϕ).
(2) J_ϕ *and* E_ϕ.
(3) J_ϕ *and* CSJ(ϕ).

Theorem 3 ([15, Theorem 1.15]) *Assume that the metric measure space (M,d,μ) satisfies* VD *and* RVD, *and ϕ satisfies* (9). *Then the following are equivalent:*

(1) UHK(ϕ) *and $(\mathscr{E},\mathscr{F})$ is conservative.*
(2) UHKD(ϕ), $\mathrm{J}_{\phi,\leq}$ *and* E_ϕ.
(3) FK(ϕ), $\mathrm{J}_{\phi,\leq}$ *and* CSJ(ϕ).

As is remarked in [15], UHK(ϕ) alone does not imply the conservativeness of the associated Dirichlet form $(\mathscr{E},\mathscr{F})$.

We note that there are two other independent related work around the same time. In [29], stability of discrete-time long range random walks of stable-like jumps is studied on infinite connected locally finite graphs. In [23], stability of stable-like pure-jump processes is studied on metric measure spaces. In both papers, they obtain the stability results under the condition that $\phi(r) = r^\alpha$ and that (M, d, μ) is an Alhfors d-regular set.

2.3 Parabolic Harnack Inequalities

In this subsection, we assume that for each $x \in M$, there is a kernel $J(x, dy)$ so that

$$J(dx, dy) = J(x, dy)\, \mu(dx).$$

Let $Z := \{V_s, X_s\}_{s\geq 0}$ be the space-time process corresponding to X, where $V_s = V_0 - s$. We denote by $\{\widetilde{\mathscr{F}}_s; s \geq 0\}$ the filtration generated by Z satisfying the usual conditions. The law of the space-time process $s \mapsto Z_s$ starting from (t, x) will be denoted by $\mathbb{P}^{(t,x)}$. Define $\tau_D = \inf\{s > 0 : Z_s \notin D\}$ for every open subset D of $[0, \infty) \times M$. A set $A \subset [0, \infty) \times M$ is said to be nearly Borel measurable if for any probability measure μ on $[0, \infty) \times M$, there are Borel measurable subsets A_1, A_2 of $[0, \infty) \times M$ so that $A_1 \subset A \subset A_2$ and that $\mathbb{P}^\mu(Z_t \in A_2 \setminus A_1 \text{ for some } t \geq 0) = 0$. Nearly Borel measurable σ-field is the collection of all nearly Borel measurable subsets of $[0, \infty) \times M$.

Definition 6 (i) We say that a nearly Borel measurable function $u(t, x)$ on $[0, \infty) \times M$ is *parabolic* (or *caloric*) on $D = (a, b) \times B(x_0, r)$ for the process X if there is a properly exceptional set $\mathscr{N}_u$ of the process X so that for every relatively compact open subset U of D, $u(t, x) = \mathbb{E}^{(t,x)} u(Z_{\tau_U})$ for every $(t, x) \in U \cap ([0, \infty) \times (M \setminus \mathscr{N}_u))$.

(ii) A nearly Borel measurable function u on M is said to be *subharmonic* (resp. *harmonic, superharmonic*) in D (with respect to the process X) if for any relatively compact subset $U \subset D$, $t \mapsto u(X_{t\wedge\tau_U})$ is a uniformly integrable submartingale (resp. martingale, supermartingale) under $\mathbb{P}^x$ for q.e. $x \in U$.

(iii) We say that the *parabolic Harnack inequality* PHI(ϕ) holds for the process X, if there exist constants $0 < c_1 < c_2 < c_3 < c_4, 0 < c_5 < 1$ and $c_6 > 0$ such that for every $x_0 \in M$, $t_0 \geq 0$, $R > 0$ and for every non-negative function $u = u(t, x)$ on $[0, \infty) \times M$ that is parabolic on cylinder $Q(t_0, x_0, c_4\phi(R), R) := (t_0, t_0 + c_4\phi(R)) \times B(x_0, R)$,

$$\text{ess sup}_{Q_-} u \leq c_6 \, \text{ess inf}_{Q_+} u, \tag{15}$$

where $Q_- := (t_0 + c_1\phi(R), t_0 + c_2\phi(R)) \times B(x_0, c_5R)$ and $Q_+ := (t_0 + c_3\phi(R), t_0 + c_4\phi(R)) \times B(x_0, c_5R)$.

Note that the above definition of PHI(ϕ) is called a weak parabolic Harnack inequality in [6], in the sense that (15) holds for some $c_1, \ldots, c_5$. The definition of a parabolic Harnack inequality in [6] is (15) valid for any choice of positive constants $c_4 > c_3 > c_2 > c_1 > 0, 0 < c_5 < 1$ with $c_6 = c_6(c_1, \dots, c_5) < \infty$. Since our underlying metric measure space may not be geodesic, we cannot deduce parabolic Harnack inequality from weak parabolic Harnack inequality.

The following stability result for parabolic Harnack inequalities for symmetric pure-jump processes has been obtained in [16].

Theorem 4 *Assume that the metric measure space (M, d, μ) satisfies* VD *and* RVD, *and ϕ satisfies* (9). *Let $(\mathscr{E}, \mathscr{F})$ be a regular Dirichlet form on $L^2(M; \mu)$ of pure-jump type* (6). *Let $(\widetilde{\mathscr{E}}, \mathscr{F})$ be another regular Dirichlet form on $L^2(M; \widetilde{\mu})$ of pure-jump type* (6) *with jumping measure $\widetilde{J}(dx, dy)$ that satisfies* (13) *and* (14). *Then* PHI(ϕ) *holds for $(\mathscr{E}, \mathscr{F})$ if and only if it holds for $(\widetilde{\mathscr{E}}, \mathscr{F})$.*

In fact the above theorem is a direct consequence of the stable characterization of PHI(ϕ) obtained in [16], which is stable under perturbations (13) and (14). A precise statement of the latter will be given below in Theorem 5(7).

Definition 7 (i) We say that the *parabolic Harnack inequality* PHI$^+(\phi)$ holds for the process X, if Definition 6(iii) holds for some constants $c_1 > 0$, $c_k = kc_1$ for $k = 2, 3, 4$, $0 < c_5 < 1$ and $c_6 > 0$.

(ii) We say that the *elliptic Harnack inequality* (EHI) holds for the process X, if there exist constants $c > 0$ and $\delta \in (0, 1)$ such that for every $x_0 \in M$, $r > 0$ and for every non-negative function u on M that is harmonic in $B(x_0, r)$,

$$\text{ess sup}_{B(x_0,\delta r)} h \le c\, \text{ess inf}_{B(x_0,\delta r)} h.$$

(iii) We say that the *parabolic Hölder regularity* PHR(ϕ) holds for the process X, if there exist constants $c > 0$, $\theta \in (0, 1]$ and $\varepsilon \in (0, 1)$ such that for every $x_0 \in M$, $t_0 \ge 0$, $r > 0$ and for every bounded measurable function $u = u(t, x)$ that is caloric in $Q(t_0, x_0, \phi(r), r)$, there is a properly exceptional set $\mathscr{N}_u \supset \mathscr{N}$ so that

$$|u(s, x) - u(t, y)| \le c \left(\frac{\phi^{-1}(|s - t|) + d(x, y)}{r} \right)^{\theta} \text{ess sup}_{[t_0, t_0+\phi(r)]\times M} |u| \tag{16}$$

for every $s, t \in (t_0, t_0 + \phi(\varepsilon r))$ and $x, y \in B(x_0, \varepsilon r) \setminus \mathscr{N}_u$.

(iv) We say that the *elliptic Hölder regularity* (EHR) holds for the process X, if there exist constants $c > 0$, $\theta \in (0, 1]$ and $\varepsilon \in (0, 1)$ such that for every $x_0 \in M$, $r > 0$ and for every bounded measurable function u on M that is harmonic in $B(x_0, r)$, there is a properly exceptional set $\mathscr{N}_u \supset \mathscr{N}$ so that

$$|u(x) - u(y)| \le c \left(\frac{d(x, y)}{r} \right)^{\theta} \text{ess sup}_M |u| \tag{17}$$

for any $x, y \in B(x_0, \varepsilon r) \setminus \mathscr{N}_u$.

Note that in the definition of PHR(ϕ) (resp. EHR) if the inequality (16) (resp. (17)) holds for some $\varepsilon \in (0, 1)$, then it holds for all $\varepsilon \in (0, 1)$ (with possibly different constant c). See [16, Remark 1.13(iv)].

Clearly PHI$^+(\phi) \Longrightarrow$ PHI(ϕ) $\Longrightarrow$ EHI and PHR(ϕ) $\Longrightarrow$ EHR.

In order to discuss stability of parabolic Harnack inequalities, we need some more definitions.

Definition 8 (i) We say that *lower bound near diagonal estimates for Dirichlet heat kernel* (NDL(ϕ)) hold, i.e. there exist $\varepsilon \in (0, 1)$ and $c_1 > 0$ such that for any $x_0 \in M$, $r > 0$, $0 < t \le \phi(\varepsilon r)$ and $B = B(x_0, r)$,

$$p^B(t, x, y) \ge \frac{c_1}{V(x_0, \phi^{-1}(t))}, \quad x, y \in B(x_0, \varepsilon\phi^{-1}(t)) \cap M_0.$$

(ii) We say that the UJS holds if there is a symmetric function $J(x, y)$ so that $J(x, dy) = J(x, y)\,\mu(dy)$, and there is a constant $c > 0$ such that for μ-a.e. $x, y \in M$ with $x \neq y$,

$$J(x, y) \le \frac{c}{V(x, r)} \int_{B(x,r)} J(z, y)\,\mu(dz) \quad \text{for every } 0 < r \le d(x, y)/2.$$

(iii) We say that the *(weak) Poincaré inequality* (PI(ϕ)) holds if there exist constants $c > 0$ and $\kappa \ge 1$ such that for any ball $B_r = B(x, r)$ with $x \in M$ and for any $f \in \mathscr{F}_b$,

$$\int_{B_r} (f - \overline{f}_{B_r})^2\, d\mu \le c\phi(r) \int_{B_{\kappa r} \times B_{\kappa r}} (f(y) - f(x))^2\, J(dx, dy),$$

where $\overline{f}_{B_r} = \frac{1}{\mu(B_r)} \int_{B_r} f\, d\mu$ is the average value of f on B_r.

The following is the main result of [16].

Theorem 5 *Suppose that the metric measure space* (M, d, μ) *satisfies* VD *and* RVD, *and* ϕ *satisfies* (9). *Then the following are equivalent:*

(1) PHI(ϕ).
(2) PHI$^+(\phi)$.
(3) UHK(ϕ), NDL(ϕ) *and* UJS.
(4) NDL(ϕ) *and* UJS.
(5) PHR(ϕ), $\mathrm{E}_{\phi,\le}$ *and* UJS.
(6) EHR, E_ϕ *and* UJS.
(7) PI(ϕ), $\mathrm{J}_{\phi,\le}$, CSJ(ϕ) *and* UJS.

We remark that any of the conditions above implies the conservativeness of the process X. As a corollary of Theorems 2 and 5 (noting that J_ϕ implies UJS), we have the following.

Corollary 1 *Suppose that the metric measure space* (M, d, μ) *satisfies* VD *and* RVD, *and* ϕ *satisfies* (9). *Then*

$$\mathrm{HK}(\phi) \Longleftrightarrow \mathrm{PHI}(\phi) + \mathrm{J}_{\phi,\geq}.$$

Unlike the diffusion case (3), heat kernel estimates and parabolic Harnack inequalities are no longer equivalent for discontinuous Markov processes.

3 L^p-Mean Value Inequality

In this section, we establish L^p-mean value inequality for every $p \in (0, 2]$ for symmetric jump processes. See [8, 9, 25] for the recent study on elliptic Harnack inequalities and mean value inequalities of fractional Laplacian operators.

Definition 9 Let D be an open subset of M. A function f is said to be locally in $\mathscr{F}_D$, denoted as $f \in \mathscr{F}_D^{loc}$, if for every relatively compact subset U of D, there is a function $g \in \mathscr{F}_D$ such that $f = g$ m-a.e. on U. We say that a nearly Borel measurable function u on M is *$\mathscr{E}$-subharmonic* (resp. *$\mathscr{E}$-harmonic, $\mathscr{E}$-superharmonic*) in D if $u \in \mathscr{F}_D^{loc}$ that is locally bounded, and satisfies

$$\int_{U\times V^c} |u(y)|\, J(dx, dy) < \infty$$

for any relatively compact open sets U and V of M with $\bar{U} \subset V \subset \bar{V} \subset D$, and

$$\mathscr{E}(u, \varphi) \leq 0 \quad (\text{resp. } = 0, \geq 0)$$

for any $0 \leq \varphi \in \mathscr{F}_D$.

The following is established in [10, Theorem 2.11 and Lemma 2.3] first for harmonic functions, and then extended in [14, Theorem 2.9] to subharmonic functions.

Theorem 6 *Let D be an open subset of M, and let u be a bounded function. Then u is $\mathscr{E}$-harmonic (resp. $\mathscr{E}$-subharmonic) in D if and only if u is harmonic (resp. subharmonic) in D.*

Following [9, 15], we define the *nonlocal tail* $\mathrm{Tail}(u; x_0, r)$ of a Borel measurable function u on M in the complement of the ball $B(x_0, r)$ by

$$\mathrm{Tail}\,(u; x_0, r) := \phi(r) \int_{B(x_0,r)^c} \frac{|u(z)|}{V(x_0, d(x_0, z))\phi(d(x_0, z))}\, \mu(dz).$$

For simplicity, we denote $B(x_0, r)$ by $B_r(x_0)$. The following L^2-mean value inequality has been obtained in [15, Proposition 4.10].

Proposition 1 (L^2-mean value inequality) *Assume* VD, (9), FK(ϕ), CSJ(ϕ) *and* $\mathrm{J}_{\phi,\leq}$ *hold. For any $x_0 \in M$ and $r > 0$, let u be a bounded $\mathscr{E}$-subharmonic in $B_r(x_0)$. Then there is a constant $c_0 > 0$ independent of x_0 and r so that*

$$\operatorname{ess\,sup}_{B_{r/2}(x_0)} u \leq c_0 \left[\left(\frac{1}{V(x_0, r)} \int_{B_r(x_0)} u^2 \, d\mu \right)^{1/2} + \mathrm{Tail}\,(u; x_0, r/2) \right]. \tag{18}$$

Using Proposition 1, we can establish the following L^p-mean value inequality for every $p \in (0, 2)$ for bounded $\mathscr{E}$-subharmonic functions.

Theorem 7 (L^p-mean value inequality with $p \in (0, 2)$) *Assume that* VD, (9), FK(ϕ), CSJ(ϕ) *and* $\mathrm{J}_{\phi,\leq}$ *hold. For any $x_0 \in M$ and $r > 0$, let u be bounded and $\mathscr{E}$-subharmonic in $B_r(x_0)$ such that $u \geq 0$ on $B_r(x_0)$. Then for any $\sigma \in (0, 1)$ and $p \in (0, 2)$,*

$$\begin{aligned} \operatorname{ess\,sup}_{B_{\sigma r}(x_0)} u \leq & \frac{c_0}{(1-\sigma)^{2(d_2+\beta_2-\beta_1)/p}} \\ & \times \left[\left(\frac{1}{V(x_0, r)} \int_{B_r(x_0)} |u|^p \, d\mu \right)^{1/p} + \mathrm{Tail}\,(u; x_0, r/2) \right], \end{aligned} \tag{19}$$

where β_1, β_2 are the constants in (9), *d_2 is the exponent in* (8) *from* VD, *and $c_0 > 0$ is a constant independent of x_0, σ and r.*

Proof To prove (19), it suffices to consider the case when $\sigma \geq 1/2$. In this case, for any $\sigma \leq t < s \leq 1$ and $z \in B_{tr}(x_0)$, applying Proposition 1 with $B_{(s-t)r}(z)$ playing the role of $B_r(x_0)$, we get that

$$u(z) \leq c_1 \left[\frac{1}{(s-t)^{d_2/2}} \left(\frac{1}{V(x_0, sr)} \int_{B_{sr}(x_0)} u^2 \, d\mu \right)^{1/2} + \mathrm{Tail}\,(u; z, (s-t)r/2) \right],$$

where we have used the facts that $B_{(s-t)r}(z) \subset B_{sr}(x_0)$ for any $z \in B_{tr}(x_0)$, and

$$\frac{V(x_0, sr)}{V(z, (s-t)r)} \leq c' \left(1 + \frac{d(x_0, z) + sr}{(s-t)r} \right)^{d_2} \leq c'' \left(1 + \frac{tr + sr}{(s-t)r} \right)^{d_2} \leq \frac{c'''}{(s-t)^{d_2}},$$

thanks to VD and (9).

Next, by splitting the integration domain of the integral in $\mathrm{Tail}\,(u; z, (s-t)r/2)$ into the sets $B_{r/2}(x_0) \backslash B_{(s-t)r/2}(z)$ and $M \backslash (B_{r/2}(x_0) \cup B_{(s-t)r/2}(z))$, we get that

$$
\begin{aligned}
&\mathrm{Tail}\,(u; z, (s-t)r/2)\\
&=\phi((s-t)r/2)\int_{B_{r/2}(x_0)\setminus B_{(s-t)r/2}(z)} \frac{|u(y)|}{V(z,d(z,y))\phi(d(z,y))}\,\mu(dy)\\
&\quad+\phi((s-t)r/2)\int_{M\setminus(B_{r/2}(x_0)\cup B_{(s-t)r/2}(z))} \frac{|u(y)|}{V(z,d(z,y))\phi(d(z,y))}\,\mu(dy)\\
&\le \int_{B_{r/2}(x_0)\setminus B_{(s-t)r/2}(z)} \frac{|u(y)|}{V(z,d(z,y))}\,\mu(dy)\\
&\quad+\phi((s-t)r/2)\int_{M\setminus(B_{r/2}(x_0)\cup B_{(s-t)r/2}(z))} \frac{|u(y)|}{V(z,d(z,y))\phi(d(z,y))}\,\mu(dy)\\
&\le \frac{c_1}{(s-t)^{d_2}}\frac{1}{V(x_0,r/2)}\int_{B_{r/2}(x_0)}|u|\,d\mu+\frac{c_2}{(s-t)^{d_2+\beta_2-\beta_1}}\mathrm{Tail}\,(u;x_0,r/2)\\
&\le \frac{c_3}{(s-t)^{d_2+\beta_2-\beta_1}}\left[\frac{1}{V(x_0,sr)}\int_{B_{sr}(x_0)}|u|\,d\mu+\mathrm{Tail}\,(u;x_0,r/2)\right]\\
&\le \frac{c_3}{(s-t)^{d_2+\beta_2-\beta_1}}\left[\left(\frac{1}{V(x_0,sr)}\int_{B_{sr}(x_0)}u^2\,d\mu\right)^{1/2}+\mathrm{Tail}\,(u;x_0,r/2)\right],
\end{aligned}
$$

where in the second inequality we have used the following two facts that for any $z \in B_{tr}(x_0)$ and $y \in B_{r/2}(x_0)\setminus B_{(s-t)r/2}(z)$,

$$\frac{V(x_0,r/2)}{V(z,d(z,y))} \le c_4\left(1+\frac{d(x_0,z)+r/2}{d(z,y)}\right)^{d_2} \le \frac{c_5}{(s-t)^{d_2}};$$

for $z \in B_{tr}(x_0)$ and $y \notin B_{r/2}(x_0)\cup B_{(s-t)r/2}(z)$,

$$\frac{V(x_0,d(x_0,y))\phi(d(x_0,y))}{V(z,d(z,y))\phi(d(z,y))} \le \frac{c_6}{(s-t)^{d_2+\beta_2}}$$

and

$$\frac{\phi((s-t)r/2)}{\phi(r/2)} \le c_7(s-t)^{\beta_1},$$

due to VD and (9) again.

Combining both estimates above, we find that for any $1/2 \le t \le s \le 1$,

$$\operatorname{ess\,sup}_{B_{tr}(x_0)} u \le \frac{c_8}{(s-t)^{d_2+\beta_2-\beta_1}}\left[\left(\frac{1}{V(x_0,sr)}\int_{B_{sr}(x_0)}u^2\,d\mu\right)^{1/2}+\mathrm{Tail}\,(u;x_0,r/2)\right].$$

Recall that $u \ge 0$ on $B_r(x_0)$. By VD and the standard Young inequality with exponents $2/(2-p)$ and $2/p$ for $0 < p < 2$, we know that for any $1/2 \le t \le s \le 1$,

$$
\begin{aligned}
&(s-t)^{d_2+\beta_2-\beta_1}\left(\frac{1}{V(x_0, sr)}\int_{B_{sr}(x_0)} u^2\,d\mu\right)^{1/2}\\
&\le c_9(\text{ess sup}_{B_{sr}(x_0)}u)^{(2-p)/2}\frac{1}{(s-t)^{d_2+\beta_2-\beta_1}}\left(\frac{1}{V(x_0, r)}\int_{B_r(x_0)} |u|^p\,d\mu\right)^{1/2}\\
&\le \frac{1}{2}\text{ess sup}_{B_{sr}(x_0)}u+\frac{c_{10}}{(s-t)^{2(d_2+\beta_2-\beta_1)/p}}\left(\frac{1}{V(x_0, r)}\int_{B_r(x_0)} |u|^p\,d\mu\right)^{1/p}.
\end{aligned}
$$

Thus, we have for any $0 < p < 2$ and $1/2 \le t \le s \le 1$,

$$
\begin{aligned}
\text{ess sup}_{B_{tr}(x_0)}u &\le \frac{1}{2}\text{ess sup}_{B_{sr}(x_0)}u\\
&+\frac{c_{11}}{(s-t)^{2(d_2+\beta_2-\beta_1)/p}}\left[\left(\frac{1}{V(x_0, r)}\int_{B_r(x_0)} |u|^p\,d\mu\right)^{1/p}+\text{Tail}\,(u; x_0, r/2)\right].
\end{aligned}
$$

Therefore, the desired assertion (19) now follows from Lemma 1 below. □

The following lemma is taken from [19, Lemma 1.1], which is used in the proof of Theorem 7.

Lemma 1 *Let $f(t)$ be a non-negative bounded function defined for $0 \le T_0 \le t \le T_1$. Suppose that for $T_0 \le t \le s \le T_1$ we have*

$$f(t) \le A(s-t)^{-\alpha} + B + \theta f(s),$$

where A, B, α, θ are non-negative constants, and $\theta < 1$. Then there exists a constant c depending only on α and θ such that for every $T_0 \le r \le R \le T_1$, we have

$$f(r) \le c\Big(A(R-r)^{-\alpha} + B\Big).$$

Proof Consider the sequence $\{t_i; i \ge 0\}$ defined by $t_0 = r$ and $t_{i+1} = t_i + (1-\delta)\delta^i(R-r)$ with $\delta \in (0, 1)$. By iteration

$$f(t_0) \le \theta^k f(t_k) + \left(\frac{A}{(1-\delta)^\alpha}(R-r)^{-\alpha} + B\right)\sum_{i=0}^{k-1}\theta^i\delta^{-i\alpha}.$$

We now choose δ such that $\delta^{-\alpha}\theta < 1$ and let $k \to \infty$, getting the desired assertion holds with $c = (1-\delta)^{-\alpha}(1-\theta\delta^{-\alpha})^{-1}$. □

References

1. Andres, S., Barlow, M.T.: Energy inequalities for cutoff-functions and some applications. J. Reine Angew. Math. **699**, 183–215 (2015)
2. Aronson, D.G.: Bounds on the fundamental solution of a parabolic equation. Bull. Am. Math. Soc. **73**, 890–1896 (1967)
3. Barlow, M.T., Bass, R.F.: Stability of parabolic Harnack inequalities. Trans. Am. Math. Soc. **356**, 1501–1533 (2003)
4. Barlow, M.T., Bass, R.F., Kumagai, T.: Stability of parabolic Harnack inequalities on metric measure spaces. J. Math. Soc. Jpn. **58**, 485–519 (2006)
5. Barlow, M.T., Bass, R.F., Kumagai, T.: Parabolic Harnack inequality and heat kernel estimates for random walks with long range jumps. Math. Z. **261**, 297–320 (2009)
6. Barlow, M.T., Grigor'yan, A., Kumagai, T.: On the equivalence of parabolic Harnack inequalities and heat kernel estimates. J. Math. Soc. Jpn. **64**, 1091–1146 (2012)
7. Björn, A., Björn, J.: Nonlinear Potential Theory on Metric Spaces. EMS Tracts in Mathematics, vol. 17. European Mathematical Society, Zurich (2011)
8. Castro, A.D., Kuusi, T., Palatucci, G.: Nonlocal Harnack inequalities. J. Funct. Anal. **267**, 1807–1836 (2014)
9. Castro, A.D., Kuusi, T., Palatucci, G.: Local behavior of fractional p-minimizers. Annales de l'Institut Henri Poincaré (C) Non Linear Anal. **33**, 1279–1299 (2016)
10. Chen, Z.-Q.: On notions of harmonicity. Proc. Am. Math. Soc. **137**, 3497–3510 (2009)
11. Chen, Z.-Q., Fukushima, M.: Symmetric Markov Processes, Time Change, and Boundary Theory. Princeton University Press, Princeton (2012)
12. Chen, Z.-Q., Kumagai, T.: Heat kernel estimates for stable-like processes on d-sets. Stoch. Process. Appl. **108**, 27–62 (2003)
13. Chen, Z.-Q., Kumagai, T.: Heat kernel estimates for jump processes of mixed types on metric measure spaces. Probab. Theory Relat. Fields **140**, 277–317 (2008)
14. Chen, Z.-Q., Kuwae, K.: On subhamonicity for symmetric Markov processes. J. Math. Soc. Jpn. **64**, 1181–1209 (2012)
15. Chen, Z.-Q., Kumagai, T., Wang, J.: Stability of heat kernel estimates for symmetric non-local Dirichlet forms. Preprint 2016. arXiv:1604.04035
16. Chen, Z.-Q., Kumagai, T., Wang, J.: Stability of parabolic Harnack inequalities for symmetric non-local Dirichlet forms. Preprint 2016. arXiv:1609.07594
17. Delmotte, T.: Parabolic Harnack inequality and estimates of Markov chains on graphs. Rev. Math. Iberoam. **15**, 181–232 (1999)
18. Fukushima, M., Oshima, Y., Takeda, M.: Dirichlet Forms and Symmetric Markov Processes, 2 rev. and ext. edn. de Gruyter, Berlin (2011)
19. Giaquinta, M., Giusti, E.: On the regularity of the minima of variational integrals. Acta Math. **148**, 31–46 (1982)
20. Gilbarg, D., Trudinger, N.S.: Elliptic Partial Differential Equations of Second Order. Reprint of the 1998 edition. Classics in Mathematics, xiv+517 pp. Springer, Berlin (2001)
21. Grigor'yan, A.: The heat equation on noncompact Riemannian manifolds. (in Russian) Matem. Sbornik. **182**, 55-87 (1991); (English transl.) Math. USSR. Sbornik **72**, 47–77 (1992)
22. Grigor'yan, A., Hu, J.: Upper bounds of heat kernels on doubling spaces. Mosco Math. J. **14**, 505–563 (2014)
23. Grigor'yan, A., Hu, E., Hu, J.: Two-sided estimates of heat kernels of jump type Dirichlet forms. Preprint 2016. https://www.math.uni-bielefeld.de/~grigor/gcap.pdf
24. Grigor'yan, A., Hu, J., Lau, K.-S.: Generalized capacity, Harnack inequality and heat kernels on metric spaces. J. Math. Soc. Jpn. **67**, 1485–1549 (2015)
25. Kuusi, T., Mingione, G., Sire, Y.: Nonlocal equations with measure data. Commun. Math. Phys. **337**, 1317–1368 (2015)
26. Li, P., Yau, S.-T.: On the parabolic kernel of the Schrödinger operator. Acta Math. **156**(3–4), 153–201 (1986)

27. Ma, Z.-M., Röckner, M.: Introduction to the Theory of (Non-Symmetric) Dirichlet Forms. Springer, New York (1992)
28. Moser, J.: On Harnack's inequality for parabolic differential equations. Commun. Pure Appl. Math. **17**, 101–134 (1964)
29. Murugan, M., Saloff-Coste, L.: Heat kernel estimates for anomalous heavy-tailed random walks. Preprint 2015. arXiv:1512.02361
30. Saloff-Coste, L.: A note on Poincaré, Sobolev, and Harnack inequalities. Int. Math. Res. Notices **2**, 27–38 (1992)
31. Saloff-Coste, L.: Aspects of Sobolev-type Inequalities. London Mathematical Society Lecture Notes, vol. 289. Cambridge University Press, Cambridge (2002)
32. Sturm, K.-T.: Analysis on local Dirichlet spaces -III. The parabolic Harnack inequality. J. Math. Pures Appl. **75**, 273–297 (1996)
33. Torchinsky, A.: Real-variable Methods in Harmonic Analysis. Academic Press, London (1986)

Positivity Preserving Semigroups and Positivity Preserving Coercive Forms

Xian Chen, Zhi-Ming Ma and Xue Peng

Abstract We recall the idea of Ma and Röckner (Canad. J. Math., 47:817–840, 1995, [13]) and report some further progress along the line of the research initiated by Ma and Röckner (Canad. J. Math., 47:817–840, 1995, [13]). The further progress includes the technique of $h\hat{h}$-transformations for positivity preserving semigroups, and includes our recent results on (σ-finite) distribution flows associated with a given positivity preserving coercive form, which is independent of the choice of h, and equipped with which the canonical cadlag process behaves like a strong Markov process.

Keywords Positivity preserving coercive form · h-associated process $h\hat{h}$-transform · Distribution flow · Revuz correspondence

MSC 2010 Primary 31C25 · 47D03 · Secondary 60J40 · 60J45 · 31C15

1 Introduction

Positivity preserving semigroups appear in various research of mathematics and physics, and have been intensively studied by several authors. In [13], Ma and Röckner studied positivity preserving semigroups which are associated with coercive closed forms (those forms are called positivity preserving coercive forms). They showed that the notion of quasi-regularity characterizes also all the positivity

---

X. Chen
School of Mathematical Sciences, Xiamen University, Xiamen 361005, China
e-mail: chenxian@amss.ac.cn

Z.-M. Ma (✉)
AMSS, University of Chinese Academy of Sciences, Beijing 100190, China
e-mail: mazm@amt.ac.cn

X. Peng
College of Mathematics, Sichuan University, Chengdu 610064, China
e-mail: pengxuemath@scu.edu.cn

A. Eberle et al. (eds.), *Stochastic Partial Differential Equations and Related Fields*, Springer Proceedings in Mathematics & Statistics 229,
https://doi.org/10.1007/978-3-319-74929-7_29

preserving coercive forms which are associated with nice Markov processes via h-transformation, and hence developed probabilistic counterpart for positivity preserving semigroups. In this paper, we are pleased to recall the idea of [13] and report some further progress along the line of the research initiated by [13].

In Sect. 2 we recall the idea and results of [13]. In Sect. 3 we report some results of [10], in which the authors introduced the technique of $h\hat{h}$-transformations, and hence enable us to construct a pair of nice Markov processes h-associated with a quasi-regular positivity preserving form. In Sect. 4 we briefly report part of our recent research [4]. In this research we construct a family of (σ-finite) distribution flows on path space for a given positivity preserving coercive form, which is independent of the choice of h, and equipped with which the canonical cadlag process behaves like a strong Markov process. By this way we can perform a kind of stochastic analysis directly related to a positivity preserving coercive form.

2 Quasi-regular Positivity Preserving Coercive Forms

We start with recalling some idea and results of [13]. Let $(E; \mathscr{B})$ be a measurable space and m be a σ-finite measure on $(E; \mathscr{B})$. We consider a strongly continuous contraction semigroup $(T_t)_{t\geq 0}$ on $L^2(E; m)$. $(T_t)_{t\geq 0}$ is called positivity preserving if for any $f \in L^2(E; m)$ with $f \geq 0$, we have $T_t f \geq 0$ for all $t > 0$. Assume that $(T_t)_{t\geq 0}$ is a holomorphic contraction semigroup, then $(T_t)_{t\geq 0}$ is associated with a coercive closed form $(\mathscr{E}, D(\mathscr{E}))$ (cf. e.g. [12, Corollary I.2.21]). Recall that a bilinear form $(\mathscr{E}, D(\mathscr{E}))$ on $L^2(E; m)$ is called a coercive closed form if its domain $D(\mathscr{E})$ is dense in $L^2(E; m)$ and the following conditions (i) and (ii) are satisfied.

(i) $(\tilde{\mathscr{E}}, D(\mathscr{E}))$ is non-negative definite and closed on $L^2(E; m)$.
(ii) (Sector condition) There exists a constant $K > 0$ (called continuity constant) such that $|\mathscr{E}_1(u, v)| \leq K \mathscr{E}_1(u, u)^{\frac{1}{2}} \mathscr{E}_1(v, v)^{\frac{1}{2}}$ for all $u, v \in D(\mathscr{E})$.

In the above (and henceforth) we write $\tilde{\mathscr{E}}(u, v) := \frac{1}{2}(\mathscr{E}(u, v) + \mathscr{E}(v, u))$, and write $\mathscr{E}_\alpha(u, v) := \mathscr{E}(u, v) + \alpha(u, v)_m$ for $\alpha > 0$.

The notion of positivity preserving coercive form was defined in [13] as follows.

Definition 1 ([13, Definition 1.1]) A coercive closed form $(\mathscr{E}, D(\mathscr{E}))$ on $L^2(E; m)$ is called a positivity preserving coercive form, if for all $u \in D(\mathscr{E})$, it holds that $u^+ \in D(\mathscr{E})$ and $\mathscr{E}(u, u^+) \geq 0$.

The above definition is justified by the following result derived in [13], which extends the first Beurling–Deny criterion for symmetric closed forms (cf. e.g. [16, XIII.50]).

Theorem 1 ([13, Theorem 1.5]) *Let $(\mathscr{E}, D(\mathscr{E}))$ be a coercive closed form on $L^2(E; m)$. Then $(\mathscr{E}, D(\mathscr{E}))$ is a positivity preserving coercive form if and only if its associated semigroup $(T_t)_{t\geq 0}$ is positive preserving.*

After introducing the notion of quasi-regularity for Dirichlet forms by the authors of [1, 2, 12], and introducing the notion of quasi-regularity for semi-Dirichlet forms by the authors of [14], in [13] the authors introduced the notion of quasi-regularity for positivity preserving coercive forms.

Below we assume that E is a Hausdorff topological space, $\mathscr{B} := \mathscr{B}(E)$ is the Borel σ-algebra of E, and m is a σ-finite measure on $(E; \mathscr{B})$. For the notations and terminologies involved in the definition below we refer to e.g. [12, 13].

Definition 2 ([13, Definition 4.9]) A positivity preserving coercive form $(\mathscr{E}, D(\mathscr{E}))$ on $L^2(E; m)$ is called *quasi-regular* if:

(i) There exists an $\mathscr{E}$-nest $(E_k)_{k\in\mathbb{N}}$ consisting of metrizable compact sets.
(ii) There exists an $\tilde{\mathscr{E}}_1^{1/2}$-dense subset of $D(\mathscr{E})$ whose elements have $\mathscr{E}$-quasi-continuous m-versions.
(iii) There exist $u_n \in D(\mathscr{E}), n \geq 1$, having $\mathscr{E}$-quasi-continuous m-versions $\tilde{u}_n, n \geq 1$, and an $\mathscr{E}$-exceptional set $N \subset E$ such that $\{\tilde{u}_n \mid n \geq 1\}$ separates the points of $E \setminus N$.
(iv) There exists a strictly positive $\mathscr{E}$-quasi-continuous α-excessive function $h \in D(\mathscr{E})$ for some $\alpha > 0$.

Note that, comparing with the same notion for Dirichlet forms, condition (iv) is extra. When $(\mathscr{E}, D(\mathscr{E}))$ is a Dirichlet form or semi-Dirichlet form, condition (iv) is a consequence of the conditions (i)–(iii).

We say that $(T_t)_{t\geq 0}$ is sub-Markovian, if $0 \leq f \leq 1$ implies $0 \leq T_t f \leq 1$. When a positivity preserving semigroup $(T_t)_{t\geq 0}$ is not sub-Markovian, then we can not directly associate it with a Markov process. To transfer $(T_t)_{t\geq 0}$ into a sub-Markovian semigroup, in [13] the authors implemented well known h-transformation technique. Let $\alpha > 0$. Recall that a function $h \in L^2(E; m)$ is called α-excessive if $h \geq 0$ and $e^{-\alpha t} T_t h \leq h$ for all $t > 0$. Given a strictly positive α-excessive function h, the conventional h-transform of $(T_t)_{t\geq 0}$ is defined as

$$T_t^h f := h^{-1} e^{-\alpha t} T_t(hf), \;\; \forall\, t \geq 0, \;\; f \in L^2(E; h^2 \cdot m). \tag{1}$$

It is clear that $(T_t^h)_{t\geq 0}$ is a strongly continuous contraction semigroup on $L^2(E; h^2 \cdot m)$ and is sub-Markovian.

By the discussion of [13], we know that if $(T_t)_{t\geq 0}$ is associated with $(\mathscr{E}, D(\mathscr{E}))$, then $(T_t^h)_{t\geq 0}$ is the semigroup associated with the semi-Dirichlet form $(\mathscr{E}_\alpha^h, D(\mathscr{E}^h))$ on $L^2(E; h^2 \cdot m)$, where $(\mathscr{E}^h, D(\mathscr{E}^h))$ is the h-transform of $(\mathscr{E}, D(\mathscr{E}))$, that is (cf. [13, Definition 3.1]),

$$\begin{cases} D(\mathscr{E}^h) := \{u \in L^2(E; h^2 \cdot m) \mid hu \in D(\mathscr{E})\}, \\ \mathscr{E}^h(u, v) := \mathscr{E}(hu, hv) \;\; \text{for } u, v \in D(\mathscr{E}^h). \end{cases} \tag{2}$$

With the preparations of quasi-regularity and h-transformation, the authors in [13] introduced the notion of h-associated processes and hence developed probabilistic counterpart for positivity preserving semigroups.

Let $\mathbf{M} := (\Omega, \mathscr{F}, (\mathscr{F}_t)_{t\geq 0}, (X_t)_{t\geq 0}, (P_x)_{x\in E_\Delta})$ be a right process with state space E and transition semigroup $(P_t)_{t\geq 0}$. Here and henceforth $E_\Delta := E \cup \{\Delta\}$ where Δ is an extra point of E serving as the cemetery of the process. We shall always make the convention that $f(\Delta) = 0$ for any function f originally defined on E. Let $\alpha > 0$ and $h \in D(\mathscr{E})$ be a strictly positive $\mathscr{E}$-quasi-continuous α-excessive function. For the convenience of further discussion we introduce the following notation:

$$Q_t f(x) := h(x)e^{\alpha t} P_t(h^{-1} f)(x) = h(x)E_x[e^{\alpha t}(h^{-1} f)(X_t)], \tag{3}$$

provided that the right hand side makes sense. We shall sometimes use $Q_t(x, \cdot)$ to denote the kernel determined by (3). The concept of h-associated process was introduced in [13] which we restate in the definition below.

Definition 3 (cf. [13, Definition 5.1]) Let $Q_t f$ be defined by (3).

(i) We say that $(\mathscr{E}, D(\mathscr{E}))$ is h-associated with $\mathbf{M}$, or $\mathbf{M}$ is an h-associated process of $(\mathscr{E}, D(\mathscr{E}))$, if $Q_t f$ is an m-version of $T_t f$ for any $f \in L^2(E; m)$.
(ii) We say that $(\mathscr{E}, D(\mathscr{E}))$ is properly h-associated with $\mathbf{M}$, or $\mathbf{M}$ is a properly h-associated process of $(\mathscr{E}, D(\mathscr{E}))$, if in addition $Q_t f$ is an $\mathscr{E}$-quasi-continuous version of $T_t f$ for any $f \in L^2(E; m)$.

The following result was derived in [13].

Theorem 2 ([13, Theorem 5.2]) *Let $h \in D(\mathscr{E})$ be a strictly positive α-excessive function. Then $(\mathscr{E}, D(\mathscr{E}))$ is properly h-associated with an m-special standard process* $\mathbf{M}$ *if and only if $(\mathscr{E}, D(\mathscr{E}))$ is quasi-regular. In this case h is always $\mathscr{E}$-quasi-continuous.*

Remark 1 Suppose that E is a metrizable Lusin space. Let $h \in D(\mathscr{E})$ be a strictly positive α-excessive function. If $(T_t^h)_{t\geq 0}$ is associated with a right process $\mathbf{M}$, then by a result of [7], $(\mathscr{E}^h, D(\mathscr{E}^h))$ is quasi-regular and $\mathbf{M}$ is in fact an m-tight special standard process properly associated with $(\mathscr{E}^h, D(\mathscr{E}^h))$. If in addition h is $\mathscr{E}^h$-quasi-continuous, then $(\mathscr{E}, D(\mathscr{E}))$ is quasi-regular and is properly h-associated with $\mathbf{M}$.

3 $h\hat{h}$-Transform of Positivity Preserving Semigroups

It is known that a quasi-regular (non-symmetric) Dirichlet form is always associated with a pair of nice Markov processes (cf. [12]). Along the line of [13], a natural problem is that: can we construct a pair of nice Markov processes $(\mathbf{M}, \hat{\mathbf{M}})$, which are h-associated with $(\mathscr{E}, D(\mathscr{E}))$ and are in duality in some sense? This problem can not be solved by the conventional h-transform. For example, if $(\mathscr{E}^{\hat{h}}, D(\mathscr{E}^{\hat{h}}))$ is an h-transform of a semi-Dirichlet form with a coexcessive (coexcessive means excessive w.r.t. the dual semigroup) function $\hat{h}$, then the associated semigroup of $(\mathscr{E}^{\hat{h}}, D(\mathscr{E}^{\hat{h}}))$ will in general no longer be sub-Markovian, although the corresponding dual semigroup becomes sub-Markovian.

In this section we report some results of [10], in which the authors solved the above problem by introducing $h\hat{h}$-transformation for positivity preserving semigroups.

Let $(T_t)_{t\geq 0}$ be a positivity preserving semigroup on $L^2(E; m)$ as in the above section. We denote by $(\hat{T}_t)_{t\geq 0}$ the dual semigroup of $(T_t)_{t\geq 0}$ on $L^2(E; m)$. It is known that $(\hat{T}_t)_{t\geq 0}$ is also a positivity preserving semigroup on $L^2(E; m)$.

For given $\alpha > 0$, let h be a strictly positive α-excessive function in $L^2(E; m)$ and $\hat{h}$ be a strictly positive α-coexcessive function in $L^2(E; m)$. We define for $f \in L^\infty(E; m)$ and $t \geq 0$,

$$T_t^h f := h^{-1}e^{-\alpha t}T_t(hf), \quad \hat{T}_t^{\hat{h}} f := \hat{h}^{-1}e^{-\alpha t}\hat{T}_t(\hat{h}f).$$

One finds that T_t^h and $\hat{T}_t^{\hat{h}}$ are contraction operators on $L^\infty(E; h\hat{h}\cdot m)$, and both of them can be (uniquely) continuously extended to be contraction operators on $L^1(E; h\hat{h}\cdot m)$. Applying Riesz–Thorin interpolation theorem, we see that T_t^h and $\hat{T}_t^{\hat{h}}$ can be (uniquely) continuously extended to be contraction operators on $L^2(E; h\hat{h}\cdot m)$. We use still T_t^h and $\hat{T}_t^{\hat{h}}$ to denote their continuous extensions on $L^2(E; h\hat{h}\cdot m)$. The following result was proved in [10].

Theorem 3 ([10, Theorem 2.2]) *$(T_t^h)_{t\geq 0}$ and $(\hat{T}_t^{\hat{h}})_{t\geq 0}$ are a pair of dual sub-Markovian strongly continuous contraction semigroups on $L^2(E; h\hat{h}\cdot m)$.*

By virtue of the above theorem, the authors in [10] solved the problem proposed at the beginning of this section. Below we assume that E is a Hausdorff topological space and $\mathscr{B} := \mathscr{B}(E)$ is the Borel σ-algebra of E.

Theorem 4 ([10, Theorem 3.1]) *Suppose that $(\mathscr{E}, D(\mathscr{E}))$ is a quasi-regular positivity preserving coercive form on $L^2(E; m)$. Then for given $\alpha > 0$, there exist strictly positive α-excessive function $h \in D(\mathscr{E})$, strictly positive α-coexcessive function $\hat{h} \in D(\mathscr{E})$, and a pair of m-tight special standard processes $(\mathbf{M}, \hat{\mathbf{M}})$ satisfying the following properties.*

(i) $\mathbf{M}$ is properly h-associated with $(\mathscr{E}, D(\mathscr{E}))$ and $\hat{\mathbf{M}}$ is properly $\hat{h}$-associated with $(\hat{\mathscr{E}}, D(\mathscr{E}))$. (Here and henceforth $\hat{\mathscr{E}}(u, v) := \mathscr{E}(v, u)$.)
(ii) $\mathbf{M}$ and $\hat{\mathbf{M}}$ are in weak duality w.r.t. $h\hat{h}\cdot m$.
(iii) All semipolar sets of $\mathbf{M}$ and $\hat{\mathbf{M}}$ are $\mathscr{E}$-exceptional.

Remark 2 By Theorem 4, any h-associated process of a positivity preserving coercive form can be studied in the framework of Markov processes in weak duality on $L^2(E; h\hat{h}\cdot m)$, which means many existing results about the weak duality like in [3, Chap. VI], [5, 6] and [6, Chap. 13] can be applied.

If $(T_t)_{t\geq 0}$ is sub-Markovian but $(\hat{T}_t)_{t\geq 0}$ is not sub-Markovian, e.g. $(T_t)_{t\geq 0}$ is associated with a semi-Dirichlet form but not a Dirichlet form. We may perform an h-transform to $(\hat{T}_t)_{t\geq 0}$ only while keeping $(T_t)_{t\geq 0}$ unchanged. A variation of Theorems 3 and 4, which was referred as semi-h-transform, was obtained in [10]. See below.

Theorem 5 ([10, Theorem 2.5]) *Suppose that* $(T_t)_{t\geq 0}$ *is a sub-Markovian strongly continuous contraction semigroup on* $L^2(E;m)$. *Let* $\hat{h}$ *be a strictly positive* α-*coexcessive function in* $L^2(E;m)$. *Then* $(e^{-\alpha t}T_t)_{t\geq 0}$ *and* $(\hat{T}_t^{\hat{h}})_{t\geq 0}$ *are a pair of dual sub-Markovian strongly continuous contraction semigroups on* $L^2(E;\hat{h}\cdot m)$.

Theorem 6 ([10, Theorem 3.3]) *Suppose that* $(\mathscr{E}, D(\mathscr{E}))$ *is a quasi-regular semi-Dirichlet form on* $L^2(E;m)$. *Then for given* $\alpha > 0$, *there exist strictly positive* α-*coexcessive function* $\hat{h} \in D(\mathscr{E})$, *and a pair of* m-*tight special standard processes* $(\mathbf{M}, \hat{\mathbf{M}})$ *satisfying the following properties.*

(i) $\mathbf{M}$ *is properly associated with* $(\mathscr{E}, D(\mathscr{E}))$ *and* $\hat{\mathbf{M}}$ *is properly* $\hat{h}$-*associated with* $(\hat{\mathscr{E}}, D(\mathscr{E}))$.
(ii) $\mathbf{M}$ *and* $\hat{\mathbf{M}}$ *are in weak duality w.r.t.* $\hat{h}\cdot m$.
(iii) All semipolar sets of $\mathbf{M}$ *and* $\hat{\mathbf{M}}$ *are* $\mathscr{E}$-*exceptional.*

4 Distribution Flows Associated with Positivity Preserving Coercive Forms

The notion of h-associated processes introduced in [13] depends on h. Hence a positivity preserving coercive form may have many different h-associated Markov processes. Inspired by the work of pseudo Hunt processes introduced in [15], in our recent research [4] we constructed a family of (σ-finite) distribution flows on path space, which are associated with a given positivity preserving coercive form. The family of distribution flows is independent of the choice of h, and the canonical cadlag process equipped with the distribution flows behaves like a strong Markov process. By this way we can perform a kind of stochastic analysis directly related to a positivity preserving coercive form. In this section we briefly report some of our results obtained in [4].

For simplicity in this section we assume that E is a metrizable Lusin space. As before let $(\mathscr{E}, D(\mathscr{E}))$ be a quasi-regular positivity preserving coercive form on $L^2(E;m)$, and $(T_t)_{t\geq 0}$ be its associated semigroup. We denote by $\mathscr{H}$ the totality of strictly positive $\mathscr{E}$-quasi-continuous $\alpha(\alpha > 0)$-excessive functions in $D(\mathscr{E})$. For given $h \in \mathscr{H}$, let

$$\mathbf{M}^h := (\Omega, \mathscr{F}^h, (\mathscr{F}_t^h)_{t\geq 0}, (X_t)_{t\geq 0}, (P_x^h)_{x\in E_\Delta})$$

with transition semigroup $(P_t^h)_{t\geq 0}$ and life time ζ be a special standard process properly h-associated with $(\mathscr{E}, D(\mathscr{E}))$. Without loss of generality, we may assume that Ω is the Skorohod space over E with cemetery Δ, and $(X_t)_{t\geq 0}$ is the canonical process on Ω, i.e., $X_t(\omega) = \omega_t$ for $\omega \in \Omega$. Denote by $(\mathscr{F}_t^0)_{t\geq 0}$ the natural filtration of $(X_t)_{t\geq 0}$ without augmentation. That is,

$$\mathscr{F}_t^0 := \sigma(X_s, s \in [0,t]), \quad \mathscr{F}_\infty^0 := \sigma(X_s, s \geq 0). \tag{4}$$

Definition 4 ([4, Definition 3.1]) For $t \geq 0, x \in E$, we define a measure $\overline{Q}_{x,t}$ on $\mathscr{F}_t^0$ by setting

$$\overline{Q}_{x,t}(\Lambda) := \overline{Q}_{x,t}(\Lambda; t < \zeta) = h(x)E_x^h\left(\frac{e^{\alpha t}I_\Lambda}{h(X_t)}; t < \zeta\right), \ \forall\, \Lambda \in \mathscr{F}_t^0, \quad (5)$$

where I_Λ is the indicator function of Λ, E_x^h is the expectation related to P_x^h. We call $\overline{Q}_{x,t}$ a *σ-finite distribution up to time t* (in short, distribution up to t), and call $(\overline{Q}_{x,t})_{t\geq 0}$ a *σ-finite distribution flow (in short, distribution flow) associated with* $(\mathscr{E}, D(\mathscr{E}))$.

We emphasize that the above definition is independent of $h \in \mathscr{H}$ in the sense below.

Theorem 7 ([4, Theorem 3.5]) *Let $\overline{Q}_{x,t}$ and $\overline{Q}'_{x,t}$ be distribution flows constructed by (5) with two different $h \in \mathscr{H}$ and $h' \in \mathscr{H}$. Then $\overline{Q}_{x,t}$ and $\overline{Q}'_{x,t}$ are equivalent to each other in the sense that: there exists a Borel set $S \subset E$ such that $(\overline{Q}_{x,t})_{t\geq 0}$ and $(\overline{Q}'_{x,t})_{t\geq 0}$ are identical for all $x \in S$ and $E \setminus S$ is $\mathscr{E}$-exceptional.*

The theorem below explores more features of $\overline{Q}_{x,t}$.

Theorem 8 ([4, Theorem 3.7]) *(i) Let $s < t$. The restriction of $\overline{Q}_{x,t}$ on $\mathscr{F}_s^0 \cap \{s < \zeta\}$ is in general different from $\overline{Q}_{x,s}$.*

(ii) There is in general no measure $\overline{Q}_x$ on $\mathscr{F}_\infty^0$ with the property that the restriction of $\overline{Q}_x$ on $\mathscr{F}_t^0 \cap \{t < \zeta\}$ is equal to $\overline{Q}_{x,t}$.

Remark 3 The measure defined by (5) was first introduced in [15, (3.3.11)] for the dual of a semi-Dirichlet form, and was denoted by $\widehat{\mathbb{P}}_x$ without subscript t. The author of [15] did not notice that his definition of (3.3.11) is in fact dependent on t, and there is in general no single measure $\widehat{\mathbb{P}}_x$ satisfying his definition (3.3.11) simultaneously for different t.

Although there is in general no single measure satisfying (5) simultaneously for different t, but the canonical process $(X_t)_{t\geq 0}$ equipped with the whole family of distribution flows $(\overline{Q}_{x,t})_{t\geq 0}$, $x \in E$, enjoys also strong Markov property. Before showing that we should first augment the natural filtration to make it right continuous and universally measurable, and hence suitable to accommodate stopping times. Note that the distribution P_x^h of the h-associated processes depends on h. For different h, the corresponding P_x^h might be not all equivalent (at least it is not clear for us). Hence we can not directly make use of the augmented filtration $(\mathscr{F}_t^h)_{t\geq 0}$. Thus we have to augment the natural filtration $(\mathscr{F}_t^0)_{t\geq 0}$ with the σ-finite distribution flow $(\overline{Q}_{x,t})_{t\geq 0}$.

We denote by $\mathscr{B}(E_\Delta)$ the Borel sets of E_Δ and by $\mathscr{P}(E_\Delta)$ the collection of all probability measures on $\mathscr{B}(E_\Delta)$. For $\mu \in \mathscr{P}(E_\Delta)$, we define for $t \geq 0$,

$$\overline{Q}_{\mu,t}(\Lambda) := \overline{Q}_{\mu,t}(\Lambda; t < \zeta) = \int_{E_\Delta} \overline{Q}_{x,t}(\Lambda; t < \zeta)\mu(dx), \ \forall\, \Lambda \in \mathscr{F}_t^0, \quad (6)$$

with the convention that $\overline{Q}_{\Delta,t}(\Lambda; t < \zeta) = 0$.

Let $\mathscr{F}^0_{t+} = \bigcap_{s>t} \mathscr{F}^0_s$. For $\mu \in \mathscr{P}(E_\Delta)$ we define

$$\mathscr{M}^\mu_t := \{\Lambda \subset \Omega \mid \exists\, \Lambda' \in \mathscr{F}^0_t,\ \Gamma \in \mathscr{F}^0_{t+},\ \text{s.t.} \quad \Lambda \Delta \Lambda' \subset \Gamma,\ \text{and}\ \overline{Q}_{\mu,T}(\Gamma; T < \zeta) = 0,\ \forall\, T > t\}. \tag{7}$$

We extend $\overline{Q}_{\mu,t}$ to $\mathscr{M}^\mu_t$, denoted again by $\overline{Q}_{\mu,t}$, by setting $\overline{Q}_{\mu,t}(\Lambda) = \overline{Q}_{\mu,t}(\Lambda')$ if Λ and Λ' are related as in (7).

Meanwhile, we define

$$\mathscr{M}_t := \bigcap_{\mu \in \mathscr{P}(E_\Delta)} \mathscr{M}^\mu_t, \quad \mathscr{M}_\infty := \sigma\left(\bigcup_{t \geq 0} \mathscr{M}_t\right). \tag{8}$$

Theorem 9 ([4, Theorems 4.4 and 5.1])

(i) $\{\mathscr{M}_t\} := (\mathscr{M}_t)_{t\geq 0}$ *is a right continuous filtration.*
(ii) Assume that $B \in \mathscr{B}(E_\Delta)$, *then the entrance time* D_B *and the hitting time* σ_B *are* $\{\mathscr{M}_t\}$*-stopping times.*

Below we denote by $\mathscr{T}$ the collection of all the $\{\mathscr{M}_t\}$-stopping times. For $\sigma \in \mathscr{T}$, we set

$$\mathscr{M}_\sigma := \{\Lambda \in \mathscr{M}_\infty \mid \Lambda \cap \{\sigma \leq t\} \in \mathscr{M}_t,\ \forall\, t \geq 0\}. \tag{9}$$

Definition 5 ([4, Definition 5.8]) For $\sigma \in \mathscr{T}$, $x \in E$, we define a measure $\overline{Q}_{x,\sigma}$ on $\mathscr{M}_\sigma$ by setting

$$\overline{Q}_{x,\sigma}(\Lambda) := \overline{Q}_{x,\sigma}(\Lambda; \sigma < \zeta) := h(x) E^h_x\left(\frac{e^{\alpha\sigma} I_\Lambda}{h(X_\sigma)}; \sigma < \zeta\right), \ \forall\, \Lambda \in \mathscr{M}_\sigma, \tag{10}$$

where I_Λ is the indicator function of Λ, E^h_x is the expectation related to P^h_x. We call $\overline{Q}_{x,\sigma}$ a *σ-finite distribution up to time σ* (in short, distribution up to σ), and call $(\overline{Q}_{x,\sigma})_{\sigma \in \mathscr{T}}$ an *expanded σ-finite distribution flow (in short, expanded distribution flow) associated with* $(\mathscr{E}, D(\mathscr{E}))$.

Similar to Theorem 7, we can show that $\overline{Q}_{x,\sigma}$ is independent of the choice of $h \in \mathscr{H}$ (cf. [4, Proposition 5.9]). Now we can state the strong Markov property of the distribution flows. Below for any measurable space $(F, \mathscr{F})$, we shall denote by $p\mathscr{F}$ all the nonnegative $\mathscr{F}$-measurable functions on F.

Theorem 10 ([4, Theorem 5.10]) *Let* $\sigma \in \mathscr{T}$ *and* $\tau \in \mathscr{T}$. *We set*

$$\gamma^* = \sigma + \tau \circ \theta_\sigma \ \text{and}\ \ \gamma = \gamma^*_\zeta := \gamma^* I_{\{\gamma^* < \zeta\}} + \infty I_{\{\gamma^* \geq \zeta\}}.$$

(i) It holds that $\gamma \in \mathscr{T}$ *and* $Y \circ \theta_\sigma \in p\mathscr{M}_\gamma$ *for* $Y \in p\mathscr{M}_\tau$.

(ii) For $\Gamma \in p\mathcal{M}_\sigma$ and $Y \in p\mathcal{M}_\tau$, we have

$$\overline{Q}_{x,\gamma}[\Gamma\,(Y \circ \theta_\sigma); \gamma < \zeta] = \overline{Q}_{x,\sigma}[\Gamma\,\overline{Q}_{X_\sigma,\tau}[Y; \tau < \zeta]; \sigma < \zeta]. \tag{11}$$

In particular, if $\tau = u$ is a constant, then for $\Gamma \in p\mathcal{M}_\sigma$ and $f \in p\mathcal{B}(E)$, we have

$$\overline{Q}_{x,\sigma+u}[\Gamma\, f(X_{\sigma+u}); \sigma + u < \zeta] = \overline{Q}_{x,\sigma}[\Gamma\,(Q_u f)(X_\sigma); \sigma < \zeta]. \tag{12}$$

In the area of Dirichlet forms, positive continuous additive functionals (PCAFs), together with the Revuz correspondence between PCAFs and smooth measures, constitute an active subject and play important roles in stochastic analysis. We derived an analogy in the context of positivity preserving coercive forms. Below are some results in this aspect, for more details see [4].

To handle the problem that there is no single measure on the path space, we introduce the notion of $\mathcal{O}$-measurable ($\mathcal{O}$ stands for the optional σ-field related to the filtration $(\mathcal{M}_t)_{t\geq 0}$) positive continuous additive functionals ($\mathcal{O}$-PCAFs), in which the defining set is an optional set rather than a set in $\mathcal{M}_\infty$.

First, for $\mu \in \mathcal{P}(E_\Delta)$, we define a σ-finite measure $\mathbb{Q}_\mu$ on $\mathcal{O}$ as follows:

$$\mathbb{Q}_\mu(H) := \int_0^{+\infty} \overline{Q}_{\mu,t}(I_H(t, \cdot))dt, \quad \forall\, H \in \mathcal{O}. \tag{13}$$

In particular, for $\mu = \delta_x$ we write $\mathbb{Q}_x := \mathbb{Q}_{\delta_x}$.

We can now define the notion of $\mathcal{O}$-PCAFs.

Definition 6 ([4, Definition 6.2]) (i) An $\overline{\mathbb{R}}_+$-valued optional process $A := (A_t)_{t\geq 0}$ is called an $\mathcal{O}$-measurable positive continuous additive functional ($\mathcal{O}$-PCAF in abbreviation) if there exists a defining set $\Gamma \in \mathcal{O}$ such that:

(a) $I_\Gamma(t, \omega)$ is decreasing and right continuous in t for fixed ω, and $I_\Gamma(t + s, \omega) = 1$ implies $I_\Gamma(s, \theta_t\omega) = 1$;
(b) $\mathbb{Q}_\nu(\Gamma^c) = 0$ for all $\nu \in \mathcal{S}_0$ ($\mathcal{S}_0$: measures of finite energy integral), where $\Gamma^c := [[0, \infty)) \setminus \Gamma$;
(c) Let $\tau_\Gamma(\omega) := \inf\{t \geq 0 \mid (t, \omega) \notin \Gamma\}$ and $\Lambda_\Gamma := \{\omega \mid \tau_\Gamma(\omega) \geq \zeta(\omega)\}$, then $\Lambda_\Gamma = \{\omega \mid \tau_\Gamma(\omega) = \infty\}$.
Furthermore, the restriction of A on Γ, or equivalently, the restriction of A on $\{\tau_\Gamma > 0\}$, possesses the following properties:
(d) A_t is continuous for $0 \leq t < \tau_\Gamma$, $A_0 = 0$, $A_t < \infty$ for $t < \tau_\Gamma \wedge \zeta$, and $A_t = (A_\zeta)_-$ for $t \geq \zeta$;
(e) $A_{t+s}(\omega) = A_t(\omega) + A_s(\theta_t\omega)$ for $t + s < \tau_\Gamma(\omega)$.

(ii) Two $\mathcal{O}$-PCAFs A and A' are said to be $\mathcal{O}$-equivalent if they share a common defining set Γ and their restriction on Γ are identical.

Remark 4 Let A be an $\mathcal{O}$-PCAF and Λ_Γ be specified as in (c) of the above Definition 6. Then for any $h \in \mathcal{H}$, the restriction of A on Λ_Γ is a PCAF of $\mathbf{M}^h$ in the classical sense defined in [9, Sect. 5.1] or [15, Sect. 4.1], with defining set Λ_Γ and some exceptional set N. Conversely, let A^h be a PCAF of $\mathbf{M}^h$ in the classical sense, then applying [4, Theorem 6.5] (see Theorem 11 below), we can construct an $\mathscr{O}$-PCAF A such that the restriction of A on Λ_Γ as a classical PCAF is equivalent to A^h in the classical sense.

Below is the Revuz correspondence between $\mathscr{O}$-PCAFs and the smooth measures.

Theorem 11 ([4, Theorem 6.5]) *(i) For any $\mathscr{O}$-PCAF A, there exists a smooth measure $\mu = \mu_A$, such that for any γ-coexcessive ($\gamma > \alpha$) function $g \in D(\mathscr{E})$ and any bounded function $f \in p\mathscr{B}(E)$, it holds that*

$$\lim_{\beta\to+\infty} \beta(g, U_A^{\beta+\gamma} f)_m = \int_E fg\mu(dx), \tag{14}$$

where

$$U_A^\beta f(x) := h(x)E_x^h\left[\int_0^{+\infty} e^{-(\beta-\alpha)t}\frac{f}{h}(X_t)I_{(t<\zeta)}dA_t\right]. \tag{15}$$

Moreover, if A and B are $\mathscr{O}$-equivalent $\mathscr{O}$-PCAFs, then μ_A and μ_B are identical.

(ii) Conversely, for any smooth measure μ, there exists a unique (in $\mathscr{O}$-equivalent sense) $\mathscr{O}$-PCAF A, such that assertion (14) holds.

We emphasize that the Revuz correspondence defined by (14) and (15) is independent of $h \in \mathscr{H}$ (see Theorem 12 below). To show this we introduce the notion of optional measure $\mathbb{Q}_x^A(\cdot)$ generated by an $\mathscr{O}$-PCAF A, which we believe will have interest by its own and will be useful in the further study of stochastic analysis related to positivity preserving coercive forms.

Definition 7 ([4, Definition 6.9]) Let A be an $\mathscr{O}$-PCAF. For $x \in E$, we define a σ-finite measure $\mathbb{Q}_x^A(\cdot)$ on $\mathscr{O}$ by setting:

$$\mathbb{Q}_x^A(H) := h(x)E_x^h\left[\int_0^\infty I_H(t,\cdot)\frac{e^{\alpha t}I_{\{t<\zeta\}}}{h(X_t)}dA_t\right], \quad \forall\, H \in \mathscr{O}. \tag{16}$$

We call $\mathbb{Q}_x^A(\cdot)$ an optional measure generated by A.

Theorem 12 ([4, Theorem 6.11 and Corollary 6.12]) *(i) The optional measure $\mathbb{Q}_x^A(\cdot)$ defined by (16) is independent of $h \in \mathscr{H}$ in the sense that: let $\mathbb{Q}_x^{'A}(\cdot)$ be defined by (16) with h replaced by another $h' \in \mathscr{H}$, then there exists an $\mathscr{E}$-exceptional set N such that $\mathbb{Q}_x^{'A}(\cdot) = \mathbb{Q}_x^A(\cdot)$ for all $x \in E \setminus N$.*

(ii) The Revuz correspondence specified by (i) and (ii) of Theorem 11 is independent of $h \in \mathscr{H}$.

Remark 5 The proof of Theorem 12 (i) is nontrivial, which employs the structure of optional σ-field and the structure of predictable σ-field (cf. e.g. [11, Theorems 3.17 and 3.21]) in an essential way. See the proof of [4, Theorem 6.11] for more details.

Acknowledgements We thank X.F. Han and W. Sun for permitting us to present our joint work here. Z.M. Ma is grateful to S. Albeverio and M. Röckner for the long standing pleasant collaboration. We are indebted to M. Fukushima who brought our attention to the work of pseudo Hunt processes introduced in [15], which stimulated the research of distribution flows. We are grateful to Y. Oshima who sent us his manuscript [15] which helps our research. We thank Mufa Chen, Zhenqing Chen and Michael Röckner for their comments and discussions. We are grateful to the support of National Center for Mathematics and Interdisciplinary Sciences (NCMIS), and NSFC project (11526214).

References

1. Albeverio, S., Ma, Z.M., Röckner, M.: A Beurling–Deny type structure theorem for Dirichlet forms on general state spaces. In: Albeverio, S., et al. (eds.) Ideas and Methods in Mathematical Analysis, Stochastics, and Applications. Memorial Volume for R. Hϕegh-Krohn, vol. 1. Cambridge University Press, Cambridge (1992). (MR1190494)
2. Albeverio, S., Ma, Z.M., Röckner, M.: Quasi-regular Dirichlet forms and Markov processes. J. Funct. Anal. **111**, 118–154 (1993). (MR1200639)
3. Blumenthal, R.M., Getoor, R.K.: Markov Processes and Potential Theory. Academic Press, New York (1968). (MR0264757)
4. Chen, X., Ma, Z.M., Peng, X.: Distribution flows associated with positivity preserving coercive forms. Preprint http://arxiv.org/abs/1708.06271 (2017)
5. Chen, Z.Q., Fukushima, M., Ying, J.G.: Entrance law, exit system and Lévy system of time-changed processes. Ill. J. Math. **50**, 269–312 (2006). (MR2247830)
6. Chung, K.L., Walsh, J.B.: Markov Processes, Brownian Motion, and Time Symmetry. Springer, New York (2005). (MR2152573)
7. Fitzsimmons, P.Z.: On the quasi-regularity of semi-Dirichlet forms. Potential Anal. **15**, 158–185 (2001). (MR1837263)
8. Fukushima, M.: Almost polar sets and an ergodic theorem. J. Math. Soc. Japan **26**, 17–32 (1974). (MR0350871)
9. Fukushima, M., Oshima, Y., Takeda, M.: Dirichlet Forms and Symmetric Markov Processes. Second revised and extended edition. Walter de Gruyter and Co., Berlin (2011). (MR2778606)
10. Han, X.F., Ma, Z.M., Sun, W.: $h\hat{h}$-transforms of positivity preserving semigroups and associated Markov processes. Acta Math. Sin. (Engl. Ser.) **27**, 369–376 (2011). (MR2754041)
11. He, S.W., Wang, J.G., Yan, J.A.: Semimartingale Theory and Stochastic Calculus. Science Press, Beijing; CRC Press, Boca Raton (1992). (MR1219534)
12. Ma, Z.M., Röckner, M.: Introduction to the Theory of (Non-Symmetric) Dirichlet Forms. Springer, Berlin (1992). (MR1214375)
13. Ma, Z.M., Röckner, M.: Markov processes associated with positivity preserving coercive forms. Canad. J. Math. **47**, 817–840 (1995). (MR1346165)
14. Ma, Z.M., Overbeck, L., Röckner, M.: Markov processes associated with semi-Dirichlet forms. Osaka J. Math. **32**, 97–119 (1995). (MR1323103)
15. Oshima, Y.: Semi-Dirichlet Forms and Markov Processes. Walter de Gruyter and Co., Berlin (2013). (MR3060116)
16. Reed, M., Simon, Y.: Methods of Modern Mathematical Physics IV. Analysis of Operators. Academia Press, New York (1978). (MR0493421)

Some Thoughts and Investigations on Densities of One-Parameter Operator Semi-groups

James Harris and Niels Jacob

Dedicated to Michael Röckner.

Abstract We discuss certain recent ideas on properties of transition densities of Lévy processes.

Keywords Transition density estimates · Lévy processes · Heat kernel bounds

For more than 30 years the second named author has been in close mathematical contact with Michael and although so far they have no joint paper there was and is always a great overlap in their interests, and at the one or another occasion they could cooperate, for example in times when they both were running their departments. The rapid expansion of the probability group in Swansea for example benefits much from his advice. Swansea University has found an appropriate way to honour both the scholar Michael Röckner as well as his support to her Mathematical Department.

This contribution is for the second named author a more personal way to appreciate his friendship with Michael and for this we have chosen a type of survey on the investigations we did in Swansea in the last couple of years with the aim to better understand densities of the kernels of semi-groups generated by pseudo-differential operators with negative definite symbols. A more recent introduction to the general field was published by R. Schilling and co-authors in [5], the monograph [13] may still serve as an introduction. Our survey is much based on more recent work of the second named author with his former and current PhD-students W. Hoh, R. Schilling, V. Knopova, B. Böttcher, K. Evans, S. Landwehr, L. Bray, C. Shen and E. Rhind, and most of all the first author's PhD-thesis [11].

Classical results relate the spectrum of certain differential operators to the geometry of the domain of the functions they are acting upon. In the best studied case

J. Harris (✉) · N. Jacob
Department of Mathematics, Swansea University, Singleton Park, Swansea SA2 8PP, UK
e-mail: jamesrwharris@aol.com

N. Jacob
e-mail: N.Jacob@swansea.ac.uk

A. Eberle et al. (eds.), *Stochastic Partial Differential Equations and Related Fields*, Springer Proceedings in Mathematics & Statistics 229,
https://doi.org/10.1007/978-3-319-74929-7_30

of self-adjoint elliptic operators on a compact manifold it is also since long known that the corresponding heat semi-group "contains" a lot of geometric (and spectral) information. Looking at a differential operator as a pseudo-differential operator it is apparent that the symbol plays a key role, in some sense the analysis of such an operator becomes an analysis of the symbol, i.e. we switch to micro-local analysis. Thus it becomes a natural question whether geometric properties which we can detect from the operator or its associated heat semi-group can already be detected from the symbol.

As an example let us consider the Laplacian - Δ_n on $\mathbb{R}^n$ the symbol of which is $|\xi|^2$. On the Schwartz space $\mathcal{S}(\mathbb{R}^n)$ the Laplacian has a representation as pseudo-differential operator given by

$$-\Delta_n u(x) = (2\pi)^{-\frac{n}{2}} \int_{\mathbb{R}^n} e^{ix\cdot\xi} |\xi|^2 \hat{u}(\xi)\, d\xi \tag{1}$$

and the corresponding heat semi-group, i.e. the Gaussian semi-group $\left(T_t^G\right)_{t\geq 0}$, is given by

$$\begin{aligned} T_t^G u(x) &= \int_{\mathbb{R}^n} \frac{1}{(4\pi t)^{\frac{n}{2}}} e^{-\frac{|x-y|^2}{4t}} u(y)\, dy \\ &= (2\pi)^{-\frac{n}{2}} \int_{\mathbb{R}^n} e^{ix\xi} e^{-t|\xi|^2} \hat{u}(\xi)\, d\xi. \end{aligned} \tag{2}$$

For the heat kernel p_t^G we have

$$\begin{aligned} p_t^G(x-y) &= \frac{1}{(4\pi t)^{\frac{n}{2}}} e^{-\frac{|x-y|^2}{4t}} = (2\pi)^{-\frac{n}{2}} \int_{\mathbb{R}^n} e^{i(x-y)\xi} e^{-t|\xi|^2}\, d\xi \\ &= c_n \lambda^{(n)} \left(B\left(0, \frac{1}{\sqrt{t}}\right)\right) e^{-d_{G,t}^2(x,y)} \end{aligned} \tag{3}$$

where $B(x, r)$ denotes the Euclidean ball with centre x and radius r and $d_{G,t}(x, y) = \frac{1}{2\sqrt{t}}|x - y|$. We see immediately how the geometric property of the symbol (being square of the Euclidean distance) transfers to an interpretation of the heat kernel in geometric terms: the diagonal term $p_t^G(0)$ relates to volume growth, while the off-diagonal term decays exponentially with a certain metric depending on t and the symbol.

Meanwhile for many second order differential operators with non-negative characteristic form analogous results are known, e.g. Hörmander-type operators or certain more general sub-elliptic operators, certain local operators on fractals and many more. These discussions led also to new geometric considerations, for example the emergence of sub-Riemannian geometry. In fact by turning the observation around and starting with certain local operators on Dirichlet spaces it was possible to implement geometric structures on more general metric measure spaces and probabilistic methods became a key tool in these considerations. We are interested in non-local operators generating one-parameter semi-groups, however it has turned out that certain higher order elliptic differential operators fit well into some of our considerations.

Let us fix some terminology and general assumptions first. A continuous function $\psi : \mathbb{R}^n \to \mathbb{C}$ is called negative definite if it admits a Lévy–Khinchin representation

$$\psi(\xi) = c + id \cdot \xi + Q(\xi) + \int_{\mathbb{R}^n \setminus \{0\}} \left(1 - e^{-iy\xi} - \frac{iy \cdot \xi}{1 + |y|^2}\right) \nu(\mathrm{d}y), \tag{4}$$

where $c \geq 0$, $d \in \mathbb{R}^n$, $Q(\cdot)$ is a symmetric positive semi-definite quadratic form on $\mathbb{R}^n$ and the Lévy-measure ν integrates $y \mapsto 1 \wedge |y|^2$. An equivalent characterisation is that $\psi(0) \geq 0$ and $\xi \mapsto e^{-t\psi(\xi)}$ is positive definite in the sense of Bochner for all $t > 0$. In this paper we restrict ourselves to real-valued continuous negative definite functions without a killing term, i.e. $c = 0$, and without a diffusion term, i.e. $Q(\cdot) = 0$. Thus the negative definite functions we are interested in have a representation

$$\psi(\xi) = \int_{\mathbb{R}^n \setminus \{0\}} (1 - \cos y \cdot \xi)\, \nu(\mathrm{d}y), \tag{5}$$

and we assume always in addition

$$e^{-t\psi} \in L^1(\mathbb{R}^n) \tag{6}$$

and

$$\psi(0) = 0 \quad \text{if and only if} \quad \xi = 0. \tag{7}$$

From (5) and (7) follows that $d_\psi(\xi, \eta) := \psi^{\frac{1}{2}}(\xi - \eta)$ is a translation invariant metric on $\mathbb{R}^n$ and we assume that this metric generates the Euclidean topology. We refer to [14] for a more detailed discussion of these conditions as well as for examples.

With a continuous negative definite function as above we can associate a convolution semi-group $\left(\mu_t^\psi\right)_{t \geq 0}$ of probability measures and under our assumptions we have

$$\mu_t^\psi = p_t^\psi \lambda^{(n)} \tag{8}$$

where $\lambda^{(n)}$ denotes the Lebesgue measure and p_t^ψ is given by

$$p_t^\psi(x - y) = \int_{\mathbb{R}^n} e^{i(x-y)\xi} e^{-t\psi(\xi)}\, \mathrm{d}\xi. \tag{9}$$

Writing

$$p_t^\psi(x - y) = p_t^\psi(0) \frac{p_t^\psi(x - y)}{p_t^\psi(0)} \tag{10}$$

we may ask whether $p_t^\psi(0)$ admits an interpretation as a volume term and whether with some metric $\delta_{\psi,t}$ we can write

$$\frac{p_t^{\psi}(x-y)}{p_t^{\psi}(0)} = e^{-\delta^2_{\psi,t}(x,y)}. \tag{11}$$

Already in [16] it was proved that

$$\begin{aligned} p_t^{\psi}(0) &= c_{\psi,n}\int_0^{\infty} \lambda^{(n)}\left(B^{d_\psi}\left(0,\sqrt{\frac{r}{t}}\right)\right)e^{-r}\,\mathrm{d}r \\ &= \tilde{c}_{\psi,n}t\int_0^{\infty} \lambda^{(n)}\left(B^{d_\psi}\left(0,\sqrt{\rho}\right)\right)e^{-t\rho}\,\mathrm{d}\rho. \end{aligned} \tag{12}$$

In [11] it was observed that with the volume function

$$V_{\psi}(r) := \lambda^{(n)}\left(B^{d_\psi}(0,r)\right) \tag{13}$$

and

$$\tilde{V}_{\psi}(r) := V_{\psi}(\sqrt{r}) \tag{14}$$

we can write (12) as

$$p_t(0) = c' t\mathcal{L}\left(\tilde{V}_{\psi}\right)(t) \tag{15}$$

where $\mathcal{L}$ denotes the Laplace transform and $B^{d_\psi}(x,r)$ is the open ball with centre $x \in \mathbb{R}$ and radius $r > 0$ with respect to the metric d_ψ. In the case that the metric measure space $\left(\mathbb{R}^n, d_\psi, \lambda^{(n)}\right)$ has the volume doubling property, i.e. for all $x \in \mathbb{R}^n$ and $r > 0$ we have

$$\lambda^{(n)}\left(B^{d_\psi}(x,2r)\right) \le c\lambda^{(n)}\left(B^{d_\psi}(x,r)\right) \tag{16}$$

we can show, see [14], that

$$p_t^{\psi}(0) \asymp \lambda^{(n)}\left(B^{d_\psi}\left(0,\frac{1}{\sqrt{t}}\right)\right), \tag{17}$$

where $a \asymp b$ means that $0 < c_0 \le \frac{a}{b} \le c_1$. We will return to these considerations soon when dealing with more general semi-groups.

The question whether (11) holds is much more delicate. While there are good reasons to conjecture such a result, so far only for some subclasses proofs are known. Examples show that one has to move away from Lévy processes and allow additive processes. The following results are taken from [7], see also [6].

With ψ as above we define

$$\rho_t := \frac{e^{-t\psi(\cdot)}}{(2\pi)^n p_t^{\psi}(0)}\lambda^{(n)} \tag{18}$$

and

$$\nu_t := \rho_{\frac{1}{t}} = \frac{e^{-\frac{1}{t}\psi(\cdot)}}{(2\pi)^n p_{\frac{1}{t}}(0)} \lambda^{(n)}. \tag{19}$$

It turns out that as a weak limit we have $\nu_t \to \epsilon_0$ as $t \to 0$ and the operators

$$\begin{aligned} S_t u(x) : &= (\nu_t * u)(x) = \int_{\mathbb{R}^n} u(x-y)\, \nu_t(dy) \\ &= (2\pi)^{-\frac{n}{2}} \int_{\mathbb{R}^n} e^{ix\xi} \frac{p_{\frac{1}{t}}(\xi)}{p_{\frac{1}{t}}(0)} \hat{u}(\xi)\, d\xi \end{aligned} \tag{20}$$

define an interesting family of strongly continuous contractions on $L^2(\mathbb{R}^n)$ (or $C_\infty(\mathbb{R}^n)$). It is not hard to see that $(S_t)_{t\geq 0}$ solves the equation

$$\frac{\partial}{\partial t} S_t u(x) + q(x, D) S_t u(x) = 0, \qquad \lim_{t\to 0} S_t u = u, \tag{21}$$

where the latter limit is in $L^2(\mathbb{R}^n)$ or $C_\infty(\mathbb{R}^n)$ and $q(t, D)$ is the pseudo-differential operator with symbol

$$q(t, D) := -\frac{\partial}{\partial t} \ln \frac{p_{\frac{1}{t}}(\xi)}{p_{\frac{1}{t}}(0)} = -\frac{\partial}{\partial t} \ln \sigma_t(\xi). \tag{22}$$

Under the additional and very crucial assumptions that $\xi \mapsto q(t, \xi)$ is a continuous negative definite function we can prove the existence of an additive process $(X_t)_{t\geq 0}$ such that

$$P_{X_t - X_s} = \gamma_{t,s}, \qquad 0 < s < t, \tag{23}$$

where

$$\hat{\gamma}_{t,s}(\xi) = (2\pi)^{-\frac{n}{2}} e^{-\int_s^t q(\tau,\xi)\, d\tau} = (2\pi)^{-\frac{n}{2}} e^{-Q_{t,s}(\xi)}. \tag{24}$$

From our assumptions we can deduce that the distribution $P_{X_t - X_s}$ has a density $\pi_{t,s}$ with respect to the Lebesgue measure and that

$$d_{Q_{t,s}}(\xi, \eta) = Q_{t,s}^{\frac{1}{2}}(\xi - \eta) \tag{25}$$

is for $0 < s < t$ a metric on $\mathbb{R}^n$.

Theorem 1 ([6, 7]) *If for all $\xi \in \mathbb{R}^n$ we have with $0 < \beta_0 \leq \beta_1$ for some $t_0 > 0$ the estimates $\beta_0 q(t_0, s) \leq q(t, \xi) \leq \beta_1 q(t_0, \xi)$ then*

$$\pi_{t,s}(0) \asymp \lambda^{(n)} \left(B^{d_{Q_{s,t}}} \left(0, \sqrt{\frac{\beta_1}{\beta_0}} \right) \right) \tag{26}$$

where we assume that ψ satisfies the conditions for (17) *to hold.* □

Moreover we have

Theorem 2 ([6, 7]) *Let $\psi : \mathbb{R}^n \to \mathbb{R}$ be a continuous negative definite function satisfying the conditions implying* (17) *and assume that $q(t, \xi)$ as defined by* (22) *is with respect to ξ a continuous negative definite function and satisfies $\beta_0 q(t_0, \xi) \leq q(t, \xi) \leq \beta_1 q(t_0, \xi)$ for some $t_0 > 0$ and $0 < \beta_0 \leq \beta_1$. Let $(Y_t^{\psi})_{t\geq 0}$ be the Lévy process associated with ψ with transition density p_t^{ψ}. By*

$$d_{\psi,t}(\xi, \eta) := \sqrt{t\psi(\xi - \eta)} \quad and \quad \delta_{\psi,t}(x, y) = \left(-\ln \sigma_{\frac{1}{t}}(x - y)\right)^{\frac{1}{2}} \tag{27}$$

two metrics are given. Further we can find an additive process $(X_t)_{t\geq 0}$ such that

$$p_t^{\psi}(x - y) \asymp \lambda^{(n)}\left(B^{d_{\psi,t}}(0, 1)\right) e^{-\delta^2_{\psi,\frac{1}{t}}(x,y)} \tag{28}$$

and

$$\pi_{t,0}(x - y) \asymp \lambda^{(n)}\left(B^{\delta_{\psi,\frac{1}{t}}}(0, 1)\right) e^{-d_{\psi,\frac{1}{t}}(x,y)}. \tag{29}$$

where $\pi_{t,0}$ is the density of the distribution of $X_t - X_0$. □

While Theorem 2 is a nice result, it is far a way to solve our general problem.

It turns out that studying more general (translation invariant) operator semi-groups leads for some questions to a clearer picture. Many of the following considerations are taken from [11]. Let $g : [0, \infty] \to [0, \infty]$ be a strictly increasing continuous function with $g(0) = 0$ such that $e^{-tg(|.|^2)} \in L^1(\mathbb{R}^n)$ for all $t > 0$. The function g might be a Bernstein function, but it does not need to be a Bernstein function. On $\mathcal{S}(\mathbb{R}^n)$ we can define the operators

$$T_t u(x) = (2\pi)^{-\frac{n}{2}} \int_{\mathbb{R}^n} e^{ix\xi} e^{-tg(|\xi|^2)} \hat{u}(\xi)\, \mathrm{d}\xi \tag{30}$$

which clearly leads to a semi-group on $L^2(\mathbb{R}^n)$. However, in general we can not assume the existence of a density in $L^1(\mathbb{R}^n)$ such that

$$T_t u = p_{t,g} * u \tag{31}$$

holds. For this we need to assume either that g is a Bernstein function or that $\left\| D^{\alpha} e^{-tg(|\cdot|^2)} \right\|_{L^1} \leq K_{\alpha,t}$ for all $\alpha \in \mathbb{N}_0^n$, $|\alpha| \leq n + 1$. Now we can show.

Proposition 1 ([11]) *Under the assumptions stated above we have*

$$p_{t,g}(0) = (2\pi)^{-\frac{n}{2}} \int_0^{\infty} \lambda^{(n)}\left(\left\{\xi \in \mathbb{R}^n \,\middle|\, g(|\xi|^2) < \frac{r}{t}\right\}\right) e^{-r}\, \mathrm{d}r. \tag{32}$$

Moreover, since

□

$$\lambda^{(n)}\left(\left\{\xi \in \mathbb{R}^n \mid g(|\xi|^2) < \frac{r}{t}\right\}\right) = \frac{\pi^{\frac{n}{2}}}{\Gamma(\frac{n}{2}+1)}\left(g^{-1}\left(\frac{r}{t}\right)\right)^{\frac{n}{2}} \tag{33}$$

we find

$$p_{t,g}(0) = \frac{1}{2^{\frac{n}{2}}\Gamma(\frac{n}{2}+1)}\int_0^{\infty}\left(g^{-1}\left(\frac{r}{t}\right)\right)e^{-r}\,\mathrm{d}r. \tag{34}$$

Assuming with some function $\gamma : [0,\infty) \to [0,\infty)$ such that $\int_1^{\infty}\gamma(r)^s e^{-r}\,\mathrm{d}r < \infty$ that the estimate

$$\frac{g^{-1}(r\rho)}{g^{-1}(\rho)} \le \gamma(r) \tag{35}$$

holds we obtain.

Theorem 3 ([11]) *The following estimates hold*

$$p_{t,g}(0) \asymp \lambda^{(n)}\left(\left\{\xi \in \mathbb{R}^n \mid g(|\xi|^2) < \frac{1}{t}\right\}\right) = \frac{1}{2^{\frac{n}{2}}\Gamma(\frac{n}{2}+1)}\left(g^{-1}\left(\frac{1}{t}\right)\right)^{\frac{n}{2}}. \tag{36}$$

□

Again we can introduce a volume function and we find

$$p_{t,g}(0) = \frac{n}{\Gamma(\frac{n}{2}+1)}\mathcal{L}\left(\frac{r\left(g^{-1}(r^2)\right)^{\frac{n-2}{2}}}{g'(g^{-1}(r^2))}\right)(t) \tag{37}$$

in the case that g is a C^1-function.

These results apply well to semi-groups generated by powers of the Laplacian, i.e. semi-groups with generator $-(-\Delta)^{\frac{s}{2}}$. The interesting case is where s is an even integer, $s = 2m$, i.e. the operator $-(-\Delta)^m$, $m \in \mathbb{N}$, which corresponds to $g(s) = s^m$. For $m > 1$ this function is not a Bernstein function implying that the density of the corresponding semi-group if it exists will not be non-negative. Following the paper [8] in the papers [2, 3] higher order elliptic differential operators in divergence form were investigated as generators of L^2-semi-groups and these results include of course the operator $-(-\Delta)^m$. In order to have L^1-densities $p_{2m,t}$ we need to assume $2m > n$, and then we can give the results in the papers mentioned above some new interpretations. First we find for the diagonal term

$$p_{2m,t}(0) = \Gamma\left(\frac{n}{4m}+1\right)\lambda^{(n)}\left(B\left(0, t^{-\frac{1}{4m}}\right)\right) \tag{38}$$

since

$$\lambda^{(n)}\left(B\left(0, t^{-\frac{1}{4m}}\right)\right) = \frac{\pi^{\frac{n}{2}}}{\Gamma(\frac{n}{2}+1)}t^{-\frac{n}{4m}}.$$

The estimate for the off-diagonal term in [2] or [3] reads as

$$\left|p_{2m,t}(x-y)\right| \leq c_r t^{-\frac{n}{4m}} e^{-\gamma_m \frac{|x-y|^{\frac{2m}{2m-1}}}{rt^{\frac{1}{2m-1}}}} \tag{39}$$

for $r > 0$ and $2m > n$. The estimate fits well to our general conjecture, but reveals even more and it seems to us that this was not observed before: The off-diagonal terms decay exponentially with respect to the metric

$$\delta_{m,t}(x, y) = \frac{1}{t^{\frac{1}{4m-2}}}|x-y|^{\frac{m}{2m-1}}. \tag{40}$$

Thus (39) is of the type we are longing for in the case of symmetric Lévy processes. The interesting observation is that for t_0 fixed $\delta^2_{m,t_0}(x, 0)$ is upto a constant the Legendre transformation of the symbol of the generator! While we can not expect such a result to hold for the off-diagonal term for symmetric stable processes - recall that the decay of the Fourier transform is determined by smoothness, it is very noteworthy and deserves more attention that V. Knopova and R. Schilling in [17] and V. Knopova and A. Kulik in [15] obtained results of this type when the symbol, i.e. the continuous negative definite function associated with a Lévy process, is analytic. Thus it seems that for general translation invariant operator semi-groups on L^2 with generators having very smooth symbols estimates of type

$$|p_t(x-y)| \asymp p_t(0)e^{-\delta^2(x,y)} \tag{41}$$

are possible with δ^2 derived from the symbol of the generator.

We want to extend our considerations into a further direction. Classical analysis of elliptic (pseudo-) differential operators emphasises the role of the principal symbol. However when dealing with pseudo-differential operators with continuous negative definite symbols in some cases a principal symbol is well defined but can not play the role as in the classical elliptic theory, and in other situations we can not even define a principal symbol. Here is an example for the latter case taken from [11]. Consider on $\mathbb{R}^n = \mathbb{R}^{n_1+n_2}$, $\xi \in \mathbb{R}^{n_1}$ and $\eta \in \mathbb{R}^{n_2}$, the two continuous negative definite function ψ_1 and ψ_2 defined by

$$\psi_j(\xi, \eta) = \left(|\xi|^{\alpha_j} + |\eta_j|^{\beta_j}\right)^{s_j} \tag{42}$$

where $0 < \beta_j, \alpha_j \leq 2$ and $0 < s_j < 1$. We choose now $s_1 = s_2$ and $\beta_1 < \alpha_2 < \beta_2 < \alpha_1$, and it is clear that we can not define a proper principal symbol for $\psi(\xi, \eta) = \psi_1(\xi, \eta) + \psi_2(\xi, \eta)$. For the corresponding densities $p_t^{(1)}$, $p_t^{(2)}$ and p_t we find however

$$p_t = p_t^{(1)} * p_t^{(2)} \tag{43}$$

and since we are dealing with probability densities Young's inequality yields

$$\|p_t\|_\infty = p_t(0) \leq p_t^{(1)}(0) \wedge p_t^{(2)}(0). \tag{44}$$

Hence we can in principle determine $T > 0$ such that for $t \leq T$ the term $p_t^{(1)}(0)$ $\left(p_t^2(0)\right)$ gives the sharper estimate. The insight of such a simple example is that in cases we are interested in one shall turn away from a micro-local analysis based on the study of the principal symbol and then apply some perturbation arguments. Instead we shall use an analysis based on the full symbol, see for this idea also in quite a different context [19]. In particular when studying generators with variable coefficients, even in the cases where a symbolic calculus is available, e.g. [4, 12] or [1], a "full symbol analysis" is required.

The last remark brings us to a final problem we want to discuss: determine the shape of the ball

$$B^{d_\psi}(0, r) = \left\{x \in \mathbb{R}^n \mid \psi^{\frac{1}{2}}(x) < r\right\}. \tag{45}$$

A first important question is whether these balls are convex, and in general they are not, see [11, 18] or [14]. Further we may investigate the volume doubling property. When now dealing with x-dependent symbols $\sigma(A)(x, \xi)$ we may ask a question such as: if $\sigma(A)(x_0, \xi)$ induces a metric with volume doubling (convexity of balls, $\cdots$) does the same hold for $\sigma(A)(x_1, \xi)$, $x_0 \neq x_1$. Again, in general one can not expect an affirmative answer and hence typical approaches such as "principal symbol analysis" or "freezing the coefficient analysis" will in general fall short in cases we are interested in. However in [11] uniformity conditions with respect to x were derived which imply that volume doubling for $\sigma(A)(x_0, \xi)$ yields volume doubling for $\sigma(A)(x, \xi)$.

Note that similar problems occur when switching to the state space $\mathbb{Z}^m$ and starting with symbols defined on the torus T^m or on $\mathbb{Z}^m \times T^m$, i.e. when dealing with certain Markov chains, see [9, 10].

The results described here are just the starting point but they make clear that in order to understand Lévy and Lévy-type processes we need much more (geometric) insights in the micro-local analysis of pseudo-differential operators having a continuous negative definite symbol.

References

1. Baldus, F.: Application of the Weyl-Hörmander calculus to generators of Feller semigroups. Math. Nachr. **252**, 3–23 (2003)
2. Barbatis, G., Davies, E.B.: Sharp bounds on heat kernels of higher order uniformly elliptic operators. J. Oper. Theory **36**(1), 179–198 (1996)
3. Barbatis G., Gazzola, F.: Higher order linear parabolic equations. In: Serrin, J.B., Mitidieri, E.L., Radulescu, V.D. (eds.) Recent Trends in Nonlinear Partial Differential Equations I: Evolution Problems. Contemporary Mathematics, vol. 594, pp. 77–97. American Mathematical Society, Providence (2013)
4. Böttcher, B.: A parametrix construction for the fundamental solution of the evolution equation associated with a pseudo-differential operator generating a Markov process. Math. Nachr. **278**, 1235–1241 (2005)

5. Böttcher, B., Schilling, R., Wang, J.: Lévy Matters III: Lévy-Type Processes: Construction, Approximation and Sample Properties. Springer International Publishing, Basel, Switzerland (2013)
6. Bray, L.: Investigations on transition densities of certain classes of stochastic processes. Ph.D-Thesis, Swansea University (2016)
7. Bray, L., Jacob, N.: Some considerations on the structure of transition densities of symmetric Lévy processes, Commum. Stochastic Analysis (to appear)
8. Davies, E.B.: Uniformly elliptic operators with measurable coefficients. J. Funct. Anal. **132**, 141–169 (1995)
9. Evans, K.P., Jacob, N.: Q-matrices as pseudo-differential operators with negative definite symbols. Math. Nachr. **284**, 631–640 (2011)
10. Evans, K.P., Jacob, N.J., Shen, C.: Some Feller semigroups on $C_\infty(\mathbb{R}^n \times \mathbb{Z}^m)$ generated by pseudo-differential operators. Mathematika **61**, 402–413 (2015)
11. Harris, J.: Investigations on metric spaces associated with continuous negative definite functions and bounds for transition densities of certain Lévy processes. Ph.D-Thesis, Swansea University, Swansea (2016)
12. Hoh, W.: A symbolic calculus for pseudo differential operators generating Feller semigroups. Osaka J. Math. **35**, 798–820 (1998)
13. Jacob, N.: Pseudo-Differential Operators and Markov Processes, vol I–III. Imperial College Press, London (2001–2005)
14. Jacob, N., Knopova, V., Landwehr, S., Schiling, R.: A geometric interpretation of the transition density of a symmetric Lévy process. Sci. China Math. 1099–1126 (2012)
15. Knopova, V., Kulik, A.M.: Exact asymptotic for distribution densities of Lévy functionals. Electron. J. Probab. **16**(52), 1394–1433 (2011)
16. Knopova, V., Schilling, R.: A note on the existence of transition probability densities for Lévy processes. Forum Math **25**, 125–149 (2013)
17. Knopova, V., Schilling, R.L.: Transition density estimates for a class of Lévy and Lévy-type processes. J. Theor. Probab. **25**, 144–170 (2012)
18. Landwehr, S.: On the geometry related to jump processes - investigating transition densities of Lévy-type processes. Ph.D-thesis, University of Wales Swansea, Swansea (2010)
19. Maslov, V.P.: Nonstandard characteristics in asymptotic problems. (Engl. Transl.). Russian Math. Surv. **38**(6), 1–42 (1983)

Strong Uniqueness of Dirichlet Operators Related to Stochastic Quantization Under Exponential Interactions in One-Dimensional Infinite Volume

Hiroshi Kawabi

Abstract In this survey paper, we discuss strong uniqueness of Dirichlet operators related to stochastic quantization under exponential (and polynomial) interactions in one-dimensional infinite volume based on joint works with Sergio Albeverio and Michael Röckner (Albeverio et al., J Funct Anal 262:602–638, 2012, [4], Kawabi and Röckner, J Funct Anal 242:486–518, 2007, [11]). We also raise an open problem.

Keywords Strong uniqueness · L^p-uniqueness · Essential self-adjointness Dirichlet operator · Stochastic quantization · Gibbs measure · Path space · SPDE

2010 AMS Classification Numbers 35R15 · 35R60 · 46N50 · 47D07

1 Introduction

Dirichlet form theory on infinite dimensional spaces plays a crucial role in many fields of mathematical physics including Euclidean quantum field theory and statistical mechanics. It is also indispensable in stochastic analysis on path and loop spaces over Riemannian manifolds. From an analytic point of view, it is very important to study L^p-uniqueness of the Dirichlet operator associated with a given Dirichlet form, that is, the question whether or not the Dirichlet operator restricted to some minimal domain uniquely determines a strongly continuous semigroup on the corresponding L^p-space. As is well known, in the case of $p = 2$, this uniqueness is equivalent to essential self-adjointness of the Dirichlet operator. This kind of uniqueness problem on infinite dimensional state spaces has been studied intensively by many authors. We refer to Eberle [6] and references therein for a detailed review. However, it is still understood very insufficiently in the sense that there are several important types of infinite dimensional Dirichlet operators for which it is not known whether uniqueness

H. Kawabi (✉)
Faculty of Science, Department of Mathematics, Okayama University, 3-1-1, Tsushima-Naka, Kita-ku, Okayama 700-8530, Japan
e-mail: kawabi@math.okayama-u.ac.jp

A. Eberle et al. (eds.), *Stochastic Partial Differential Equations and Related Fields*, Springer Proceedings in Mathematics & Statistics 229,
https://doi.org/10.1007/978-3-319-74929-7_31

holds or not. As in Michael Röckner's speech at the conference dinner of the SPDE conference 2016 in Bielefeld, the most prominent example in which essential self-adjointness is not known is the stochastic quantization of $P(\phi)_2$-quantum fields in *infinite volume* in the context of Euclidean quantum field theory. We refer to Albeverio–Ma–Röckner [5] for a concise overview on stochastic quantization. We should also mention that recently there has arisen a renewed interest in singular SPDEs, in connection with Hairer's groundbreaking work on regularity structures [8] and related work by Gubinelli, Imkeller and Perkowski [7]. By using these new theories, Mourrat and Weber [13] constructed a unique strong solution to the stochastic quantization equation associated with the $P(\phi)_2$-quantum fields in infinite volume, and Röckner, R. Zhu and X. Zhu [15] obtained restricted Markov uniqueness for the corresponding Dirichlet operator. Note that essential self-adjointness implies restricted Markov uniqueness. However, the converse does not hold in general.

On the other hand, even in the case of $P(\phi)_1$-quantum fields in infinite volume, essential self-adjointness of the Dirichlet operator has been open for many years and solved by the author and Röckner [11]. Moreover in that paper, it was shown that the corresponding dynamics coincides with the $P(\phi)_1$*-time evolution*, which had been constructed by Iwata [10] as a unique strong solution to the stochastic quantization equation (7) defined on the whole line $\mathbf{R}$.

In this survey paper, we discuss L^p-uniqueness of the Dirichlet operator on an infinite volume path space $C(\mathbf{R}, \mathbf{R}^d)$ with Gibbs measures obtained in [4, 11]. Important examples of the Gibbs measures are $P(\phi)_1$, $\exp(\phi)_1$, and trigonometric quantum fields in infinite volume. In particular, $\exp(\phi)_1$-quantum fields were introduced (for the case where $\mathbf{R}$ occurring in (1) below is replaced by a 2-dimensional Euclidean space-time $\mathbf{R}^2$ and where $d = 1$) by Albeverio and Høegh-Krohn in the early 1970s (cf. [1, 2]). More precisely, we are concerned with Gibbs measures on an infinite volume path space $C(\mathbf{R}, \mathbf{R}^d)$ given by the following formal expression:

$$\mu(dw) = Z^{-1} \exp\Big\{ -\frac{1}{2}\int_{\mathbf{R}} \big((-\Delta_x + m^2)w(x), w(x)\big)_{\mathbf{R}^d} dx \\ - \int_{\mathbf{R}} \Big(\int_{\mathbf{R}^d} e^{(w(x),\xi)_{\mathbf{R}^d}} \nu(d\xi) \Big) dx \Big\} \prod_{x\in\mathbf{R}} dw(x). \tag{1}$$

Here Z is a normalizing constant, $m > 0$ denotes mass, $\Delta_x := d^2/dx^2$, ν is a bounded positive measure on $\mathbf{R}^d$ with compact support, and $\prod_{x\in\mathbf{R}} dw(x)$ stands for a (heuristic) volume measure on the space of maps from $\mathbf{R}$ into $\mathbf{R}^d$. This has the interpretation of a quantized d-dimensional vector field with an interaction of exponential type in the 1-dimensional space-time $\mathbf{R}$, a model which is known as stochastic quantization of the $\exp(\phi)_1$-quantum field model (with weight measure ν).

Furthermore, we also discuss a characterization of the stochastic dynamics corresponding to the above Dirichlet operator. Thanks to a general theory of Albeverio–Röckner [3], the stochastic dynamics constructed through the Dirichlet form approach solves the parabolic SPDE (7) below weakly. However, we obtain something much better, namely existence and uniqueness of a strong solution. We achieve this by first proving pathwise uniqueness for SPDE (7) and then applying the recent work on the Yamada–Watanabe theorem for mild solutions of SPDEs (cf. Ondreját [14]). As a consequence, we have the existence of a unique strong solution to SPDE (7) by using simple and straightforward arguments which do not rely on any finite volume approximations discussed as in [10]. However, as we will mention in Sect. 2, this uniqueness does not imply the L^p-uniqueness of the Dirichlet operator, and vice versa.

The rest of this paper is organized as follows: In Sect. 2, we present the framework and state our strong uniqueness results for both Dirichlet operators and corresponding stochastic dynamics. In Sect. 3, we raise an open problem.

2 Framework and Results

First of all, we introduce some notation and objects we will be working with. We define a weight function $\rho_r \in C^\infty(\mathbf{R}, \mathbf{R})$, $r \in \mathbf{R}$ by $\rho_r(x) := e^{r\chi(x)}$, $x \in \mathbf{R}$, where $\chi \in C^\infty(\mathbf{R}, \mathbf{R})$ is a positive symmetric convex function satisfying $\chi(x) = |x|$ for $|x| \geq 1$. We fix a positive constant $\bar{r}$ sufficiently small. We set $E := L^2(\mathbf{R}, \mathbf{R}^d; \rho_{-2\bar{r}}(x)dx)$. This space is a Hilbert space with its inner product defined by

$$(w, \tilde{w})_E := \int_{\mathbf{R}} \big(w(x), \tilde{w}(x)\big)_{\mathbf{R}^d} \rho_{-2\bar{r}}(x)dx, \quad w, \tilde{w} \in E.$$

Moreover, we set $H := L^2(\mathbf{R}, \mathbf{R}^d)$ and denote by $\|\cdot\|_E$ and $\|\cdot\|_H$ the corresponding norms in E and H, respectively. We regard the dual space E^* of E as $L^2(\mathbf{R}, \mathbf{R}^d; \rho_{2\bar{r}}(x)dx)$. We endow $C(\mathbf{R}, \mathbf{R}^d)$ with the compact uniform topology and introduce a family of Banach spaces

$$\mathscr{C}_r := \big\{w \in C(\mathbf{R}, \mathbf{R}^d) | \lim_{|x|\to\infty} |w(x)|\rho_{-r}(x) < \infty\big\}, \quad r > 0$$

with norms defined by $\|w\|_{r,\infty} := \sup_{x\in\mathbf{R}} |w(x)|\rho_{-r}(x)$, $w \in \mathscr{C}_r$. We also introduce a tempered subspace of $C(\mathbf{R}, \mathbf{R}^d)$ by $\mathscr{C} := \cap_{r>0}\mathscr{C}_r$. We note that $\mathscr{C}$ is a Fréchet space with respect to the system of norms $\{\|\cdot\|_{r,\infty}\}_{r>0}$ and the inclusion $\mathscr{C} \subset E \cap C(\mathbf{R}, \mathbf{R}^d)$ is dense with respect to the topology of E. Let $\mathscr{B}$ be the topological σ-field on $C(\mathbf{R}, \mathbf{R}^d)$. For $T_1 < T_2 \in \mathbf{R}$, we define by $\mathscr{B}_{[T_1,T_2]}$ and $\mathscr{B}_{[T_1,T_2],c}$ the sub-σ-fields of $\mathscr{B}$ generated by $\{w(x); T_1 \leq x \leq T_2\}$ and $\{w(x); x \leq T_1, x \geq T_2\}$, respectively. For $T_1, T_2 \in \mathbf{R}$ and $z_1, z_2 \in \mathbf{R}^d$, let $\mathscr{W}^{z_1,z_2}_{[T_1,T_2]}$ be the path space measure of the Brownian bridge such that $w(T_1) = z_1$, $w(T_2) = z_2$. We sometimes regard this measure as a probability measure on the measurable space $(C(\mathbf{R}, \mathbf{R}^d), \mathscr{B})$ by putting $w(x) = z_1$ for $x \leq T_1$ and $w(x) = z_2$ for $x \geq T_2$.

We now introduce a (U-)Gibbs measure μ on $C(\mathbf{R}, \mathbf{R}^d)$ based on Iwata [9]. Let $U \in C(\mathbf{R}^d, \mathbf{R})$ be a (self-interaction) potential function which can be written as

$$U(z) = U_0(z) + U_1(z), \qquad z \in \mathbf{R}^d,$$

where $U_0 \in C(\mathbf{R}^d, \mathbf{R})$ is convex and $U_1 \in C^1(\mathbf{R}^d, \mathbf{R})$. We impose the following three conditions on U:

(U1): There exists a constant $K_1 \in \mathbf{R}$ such that

$$\big(\widetilde{\nabla} U(z_1) - \widetilde{\nabla} U(z_2), z_1 - z_2\big)_{\mathbf{R}^d} \geq K_1 |z_1 - z_2|^2, \qquad z_1, z_2 \in \mathbf{R}^d,$$

where $\widetilde{\nabla} U(z) := \partial_0 U_0(z) + \nabla U_1(z)$, $z \in \mathbf{R}^d$ and $\partial_0 U_0$ is the minimal section of the subdifferential ∂U_0. (We note that $\widetilde{\nabla} U$ coincides with the usual gradient ∇U provided $U \in C^1(\mathbf{R}^d, \mathbf{R})$.)

(U2): There exist $K_2 > 0$, $R > 0$ and $\alpha > 0$ such that $U_1(z) \geq K_2 |z|^\alpha$, $|z| > R$.

(U3): There exist $K_3, K_4 > 0$ and $0 < \beta < 1 + \frac{\alpha}{2}$ such that

$$|\widetilde{\nabla} U(z)| \leq |\partial_0 U_0(z)| + |\nabla U_1(z)| \leq K_3 \exp(K_4 |z|^\beta), \qquad z \in \mathbf{R}^d.$$

Let $H_U := -\frac{1}{2}\Delta_z + U$ be the Schrödinger operator on $L^2(\mathbf{R}^d, \mathbf{R})$, where $\Delta_z := \sum_{i=1}^d \partial^2/\partial z_i^2$ is the d-dimensional Laplacian. Then condition **(U2)** assures that H_U has purely discrete spectrum and a complete set of eigenfunctions. We denote by $\lambda_0 (> \min U)$ the minimal eigenvalue and by Ω the corresponding normalized eigenfunction in $L^2(\mathbf{R}^d, \mathbf{R})$. This eigenfunction is called ground state and it can be chosen to be strictly positive. Moreover, it has exponential decay at infinity. To be precise, there exist some positive constants D_1, D_2 such that

$$0 < \Omega(z) \leq D_1 \exp\big(-D_2 |z|\, U_{\frac{1}{2}|z|}(z)^{1/2}\big), \quad z \in \mathbf{R}^d, \tag{2}$$

where $U_{\frac{1}{2}|z|}(z) := \inf\{U(y) \mid |y - z| \leq \frac{1}{2}|z|\}$.

For $T_1 < T_2$, and for all $T_1 \leq x_1 < x_2 < \cdots < x_n \leq T_2$, $A_1, A_2, \ldots, A_n \in \mathscr{B}(\mathbf{R}^d)$, we define a cylinder set $A \in \mathscr{B}_{[T_1,T_2]}$ by $A := \{w \in C(\mathbf{R}, \mathbf{R}^d) \mid w(x_1) \in A_1, w(x_2) \in A_2, \ldots, w(x_n) \in A_n\}$. Next, we set

$$\begin{aligned}
\mu(A) &:= \Big(\Omega, e^{-(x_1-T_1)(H_U-\lambda_0)}\big(\mathbf{1}_{A_1} e^{-(x_2-x_1)(H_U-\lambda_0)}\big(\mathbf{1}_{A_2}\cdots \\
&\qquad\qquad e^{-(x_n-x_{n-1})(H_U-\lambda_0)}\big(\mathbf{1}_{A_n} e^{-(T_2-x_n)(H_U-\lambda_0)}\Omega\big)\big)\big)\Big)_{L^2(\mathbf{R}^d,\mathbf{R})} \\
&= e^{\lambda_0(T_2-T_1)} \int_{\mathbf{R}^d} \Omega(z_1) \Big\{ \int_{\mathbf{R}^d} \Omega(z_2) p(T_2 - T_1, z_1, z_2) \\
&\quad \times \Big(\int_{C(\mathbf{R},\mathbf{R}^d)} \mathbf{1}_A(w) \exp\Big(-\int_{T_1}^{T_2} U(w(x))dx\Big) \mathscr{W}^{z_1,z_2}_{[T_1,T_2]}(dw)\Big) dz_2 \Big\} dz_1,
\end{aligned} \tag{3}$$

where $p(t, z_1, z_2)$, $t > 0$, $z_1, z_2 \in \mathbf{R}^d$ is the transition probability density of standard Brownian motion $(B_t)_{t \geq 0}$ on $\mathbf{R}^d$, and we used the Feynman–Kac formula for the second line. Then by recalling that $e^{-tH_U}\Omega = e^{-t\lambda_0}\Omega$, $\|\Omega\|_{L^2(\mathbf{R}^d,\mathbf{R})} = 1$ and by the Markov property of the d-dimensional Brownian motion, (3) defines a consistent family of probability measures, and hence μ can be extended to a probability measure on $C(\mathbf{R}, \mathbf{R}^d)$. We mention that the Gibbs measure μ coincides with the probability law of the $P(\phi)_1$-process associated with the potential U, that is, the stationary solution of the following SDE on $\mathbf{R}^d$:

$$dz_t = \nabla \log \Omega(z_t)dt + dB_t.$$

Carrying out the standard moment estimates of Brownian motion, we see that the Gibbs measure μ is supported on the tempered path space $\mathscr{C}$. Thus we may regard $\mu \in \mathscr{P}(E)$ by identifying it with its image measure under the inclusion map of $\mathscr{C}$ into E. Furthermore, μ satisfies the following DLR-equation:

$$\mathbf{E}^{\mu}\big[\mathbf{1}_A|\mathscr{B}_{[T_1,T_2],c}\big](\xi){=}Z_{[T_1,T_2]}^{-1}(\xi)\int_A \exp\left(-\int_{T_1}^{T_2} U(w(x))dx\right)\mathscr{W}_{[T_1,T_2]}^{\xi(T_1),\xi(T_2)}(dw),$$
$$\mu\text{-a.e. } \xi \in C(\mathbf{R}, \mathbf{R}^d), \text{ for all } A \in \mathscr{B}_{[T_1,T_2]}, T_1 < T_2, \tag{4}$$

where $Z_{[T_1,T_2]}(\xi) := \mathbf{E}^{\mathscr{W}_{[T_1,T_2]}^{\xi(T_1),\xi(T_2)}}[\exp(-\int_{T_1}^{T_2} U(w(x))dx)]$ is a normalizing constant. Although generally there exist other probability measures on $C(\mathbf{R}, \mathbf{R}^d)$ satisfying the DLR-equation (4), we only consider the Gibbs measure μ which has been constructed in (3).

Remark 1 If condition **(U2)** holds with $\alpha > 2$, we obtain the following *Fernique type estimate*:

$$\begin{aligned}
\mathbf{E}^{\mu}\big[\exp(p\|w\|_E^2)\big] &= \sum_{n=0}^{\infty}\frac{p^n}{n!}\mathbf{E}^{\mu}\big[\|w\|_E^{2n}\big] \\
&\le \sum_{n=0}^{\infty}\left(\frac{p}{r}\right)^n\frac{1}{n!}\int_{\mathbf{R}^d}|z|_{\mathbf{R}^d}^{2n}\Omega(z)^2dz \\
&\le D_1^2\sum_{n=0}^{\infty}\left(\frac{p}{r}\right)^n\frac{1}{n!}\int_{\mathbf{R}^d}|z|_{\mathbf{R}^d}^{2n}e^{-c|z|^{1+\alpha/2}}dz \\
&= \frac{C_1S_{d-1}}{(1+\alpha/2)}\sum_{n=0}^{\infty}\left(\frac{p}{r}\right)^n\frac{1}{n!}c^{\frac{1+\alpha/2}{2n+d}}\Gamma\left(\frac{2n+d}{1+\alpha/2}\right) \\
&\le C_2\exp\left\{C_3\left(\frac{p}{r}\right)^{\frac{\alpha+2}{\alpha-2}}\right\}, \quad p > 0,
\end{aligned}$$

where c and C_i $(i = 1, 2, 3)$ are positive constants and S_{d-1} denotes the area of the $(d-1)$-dimensional unit sphere.

Now we are in a position to introduce the pre-Dirichlet form $(\mathscr{E}, \mathscr{F}\mathscr{C}_b^\infty)$. Let $\mathscr{F}\mathscr{C}_b^\infty$ be the space of all smooth cylinder functions on E having the form

$$F(w) = f(\langle w, \varphi_1\rangle, \dots, \langle w, \varphi_n\rangle), \quad w \in E,$$

with $n \in \mathbf{N}$, $f = f(\alpha_1, \dots, \alpha_n) \in C_b^\infty(\mathbf{R}^n, \mathbf{R})$ and $\varphi_1, \dots, \varphi_n \in C_0^\infty(\mathbf{R}, \mathbf{R}^d)$. Here we set $\langle w, \varphi \rangle := \int_{\mathbf{R}} (w(x), \varphi(x))_{\mathbf{R}^d} dx$ if the integral converges absolutely. Note that $\mathscr{F}\mathscr{C}_b^\infty$ is dense in $L^p(\mu)$ for all $p \geq 1$. For $F \in \mathscr{F}\mathscr{C}_b^\infty$, we define the H-Fréchet derivative $D_H F : E \to H$ by

$$D_H F(w) := \sum_{j=1}^n \frac{\partial f}{\partial \alpha_j}(\langle w, \varphi_1 \rangle, \dots, \langle w, \varphi_n \rangle)\varphi_j.$$

Then we consider the pre-Dirichlet form $(\mathscr{E}, \mathscr{F}\mathscr{C}_b^\infty)$ which is given by

$$\mathscr{E}(F, G) = \frac{1}{2}\int_E \big(D_H F(w), D_H G(w)\big)_H \mu(dw), \qquad F, G \in \mathscr{F}\mathscr{C}_b^\infty.$$

Proposition 1 ([4, Proposition 2.7]):

$$\mathscr{E}(F, G) = -\int_E \mathscr{L}_0 F(w) G(w) \mu(dw), \qquad F, G \in \mathscr{F}\mathscr{C}_b^\infty,$$

where $\mathscr{L}_0 F \in L^p(\mu)$, $p \geq 1$, $F \in \mathscr{F}\mathscr{C}_b^\infty$ is given by

$$\begin{aligned}\mathscr{L}_0 F(w) &= \frac{1}{2}\mathrm{Tr}(D_H^2 F(w)) + \frac{1}{2}\langle w, \Delta_x D_H F(w(\cdot))\rangle - \frac{1}{2}\langle (\tilde{\nabla} U)(w(\cdot)), D_H F(w)\rangle \\ &= \frac{1}{2}\sum_{i,j=1}^n \frac{\partial^2 f}{\partial \alpha_i \partial \alpha_j}\big(\langle w, \varphi_1\rangle, \dots, \langle w, \varphi_n\rangle\big)\langle \varphi_i, \varphi_j\rangle \\ &\quad + \frac{1}{2}\sum_{i=1}^n \frac{\partial f}{\partial \alpha_i}\big(\langle w, \varphi_1\rangle, \dots, \langle w, \varphi_n\rangle\big) \cdot \big\{\langle w, \Delta_x \varphi_i\rangle - \langle (\tilde{\nabla} U)(w(\cdot)), \varphi_i\rangle\big\}.\end{aligned}$$

This proposition means that the operator $\mathscr{L}_0$ is the pre-Dirichlet operator which is associated with the pre-Dirichlet form $(\mathscr{E}, \mathscr{F}\mathscr{C}_b^\infty)$. In particular, $(\mathscr{E}, \mathscr{F}\mathscr{C}_b^\infty)$ is closable in $L^2(\mu)$. Let us denote by $\mathscr{D}(\mathscr{E})$ the completion of $\mathscr{F}\mathscr{C}_b^\infty$ with respect to the $\mathscr{E}_1^{1/2}$-norm. By the standard theory of Dirichlet forms, $(\mathscr{E}, \mathscr{D}(\mathscr{E}))$ is a Dirichlet form and the operator $\mathscr{L}_0$ has a self-adjoint extension $(\mathscr{L}_\mu, \mathrm{Dom}(\mathscr{L}_\mu))$, called the Friedrichs extension, corresponding to the Dirichlet form $(\mathscr{E}, \mathscr{D}(\mathscr{E}))$. The semigroup $\{e^{t\mathscr{L}_\mu}\}_{t\geq 0}$ generated by $(\mathscr{L}_\mu, \mathrm{Dom}(\mathscr{L}_\mu))$ is Markovian, i.e., $0 \leq e^{t\mathscr{L}_\mu} F \leq 1$, μ-a.e. whenever $0 \leq F \leq 1$, μ-a.e. Moreover, since $\{e^{t\mathscr{L}_\mu}\}_{t\geq 0}$ is symmetric on $L^2(\mu)$, the Markovian property implies that $\|e^{t\mathscr{L}_\mu} F\|_{L^1(\mu)} \leq \|F\|_{L^1(\mu)}$ holds for $F \in L^2(\mu)$, and $\{e^{t\mathscr{L}_\mu}\}_{t\geq 0}$ can be extended as a family of C_0-semigroup of contractions in $L^p(\mu)$ for all $p \geq 1$.

On the other hand, it is a fundamental question whether the Friedrichs extension is the only closed extension generating a C_0-semigroup on $L^p(\mu)$, $p \geq 1$, which for $p = 2$ is equivalent to the fundamental problem of essential self-adjointness of $\mathscr{L}_0$ in quantum physics. Even if $p = 2$, in general there are many lower bounded self-adjoint extensions $\widetilde{\mathscr{L}}$ of $\mathscr{L}_0$ in $L^2(\mu)$ which therefore generate different symmetric strongly continuous semigroups $\{e^{t\widetilde{\mathscr{L}}}\}_{t\geq 0}$. If, however, we have $L^p(\mu)$-uniqueness of $\mathscr{L}_0$ for some $p \geq 2$, there is hence only one semigroup which is strongly continuous and with generator extending $\mathscr{L}_0$. Consequently, in this case, only one such L^p-, hence only one such L^2-dynamics exists, associated with the Gibbs measure μ.

Before stating our main results of this paper, we recall the notion of "capacity" for the convenience. For an open set $O \subset E$, we define

$$\mathrm{Cap}(O) := \inf\{\mathscr{E}_1(F, F) | \ F \in \mathscr{D}(\mathscr{E}), F \geq 1 \text{ on } O, \ \mu\text{-a.e.}\}$$

and for an arbitrary subset $A \subset E$, we define $\mathrm{Cap}(A) := \inf\{\mathrm{Cap}(O) | \ A \subset O, O$ open$\}$.

The following two theorems are the main results of this survey paper.

Theorem 1 ([11, Theorem 2.4], [4, Theorem 2.8]): (1) *The pre-Dirichlet operator $(\mathscr{L}_0, \mathscr{F}\mathscr{C}_b^\infty)$ is $L^p(\mu)$-unique for all $p \geq 1$, i.e., there exists exactly one C_0-semigroup in $L^p(\mu)$ such that its generator extends $(\mathscr{L}_0, \mathscr{F}\mathscr{C}_b^\infty)$.*

(2) *There exists a diffusion process $\mathbf{M} := (\Theta, \mathscr{F}, \{\mathscr{F}_t\}_{t\geq 0}, \{X_t\}_{t\geq 0}, \{\mathbf{P}_w\}_{w\in E})$ such that the semigroup $\{P_t\}_{t\geq 0}$ generated by the unique (self-adjoint) extension of $(\mathscr{L}_0, \mathscr{F}\mathscr{C}_b^\infty)$ satisfies the following identity for any bounded measurable function $F : E \to \mathbf{R}$, and $t > 0$:*

$$P_t F(w) = \int_\Theta F(X_t(\omega))\mathbf{P}_w(d\omega), \quad \mu\text{-}a.s. \ w \in E. \tag{5}$$

Moreover, $\mathbf{M}$ is the unique diffusion process solving the following "componentwise" SDE:

$$\langle X_t, \varphi\rangle = \langle w, \varphi\rangle + \langle W_t, \varphi\rangle + \frac{1}{2}\int_0^t \big\{\langle X_s, \Delta_x\varphi\rangle - \langle(\widetilde{\nabla}U)(X_s(\cdot)), \varphi\rangle\big\}ds,$$
$$t > 0, \ \varphi \in C_0^\infty(\mathbf{R}, \mathbf{R}^d), \ \mathbf{P}_w\text{-}a.s., \tag{6}$$

for quasi-every $w \in E$ and such that its corresponding semigroup given by (5) consists of locally uniformly bounded (in t) operators on $L^p(\mu)$, $p \geq 1$, where $\{W_t\}_{t\geq 0}$ is an $\{\mathscr{F}_t\}_{t\geq 0}$-adapted H-cylindrical Brownian motion starting at origin defined on $(\Theta, \mathscr{F}, \{\mathscr{F}_t\}_{t\geq 0}, \mathbf{P}_w)$.

Theorem 2 ([4, Theorem 2.9]): *For quasi-every $w \in E$, the parabolic SPDE*

$$\frac{\partial X_t(x)}{\partial t} = \frac{1}{2}\big\{\Delta_x X_t(x) - (\widetilde{\nabla}U)(X_t(x))\big\} + \dot{W}_t(x), \quad x \in \mathbf{R}, \ t > 0 \tag{7}$$

has a unique strong solution $X = \{X_t^w(\cdot)\}_{t\geq 0}$ living in $C([0,\infty), E) \cap C((0,\infty), \mathscr{C}_{\mathcal{T}})$. Namely, there exists a set $S \subset E$ with $\mathrm{Cap}(S) = 0$ such that for any H-cylindrical Brownian motion $\{W_t\}_{t\geq 0}$ starting at origin defined on a filtered probability space $(\Theta, \mathscr{F}, \{\mathscr{F}_t\}_{t\geq 0}, \mathbf{P})$ satisfying the usual conditions and for any initial datum

$w \in E \setminus S$, there exists a unique $\{\mathscr{F}_t\}_{t\ge 0}$-adapted process $X = \{X_t^w(\cdot)\}_{t\ge 0}$ living in $C([0,\infty), E) \cap C((0,\infty), \mathscr{C}_{\bar{r}})$ satisfying (6).

Remark 2 Obviously, the uniqueness result in Theorem 2 implies the (thus weaker) uniqueness stated for the diffusion process $\mathbf{M}$ in Theorem 1. However, it does not imply the $L^p(\mu)$-uniqueness of the Dirichlet operator. This is obvious, since a priori the latter might have extensions which generate non-Markovian semigroups which thus have no probabilistic interpretation as transition probabilities of a process. Therefore, neither of the two uniqueness results in Theorems 1 and 2, i.e., $L^p(\mu)$-uniqueness of the Dirichlet operator and strong uniqueness of the corresponding SPDE respectively, implies the other.

We give three examples which satisfy our conditions **(U1)**, **(U2)** and **(U3)**.

Example 1 ($P(\phi)_1$-quantum fields): We consider the case where the potential function U is written as the following potential function on $\mathbf{R}^d$:

$$U(z) = \sum_{j=0}^{2n} a_j |z|^j, \quad a_{2n} > 0, \ n \in \mathbf{N}.$$

Especially, in the case $U(z) = \frac{m^2}{2}|z|^2$, $m > 0$, the corresponding Gibbs measure μ coincides with the Gaussian measure on $\mathscr{C}$ with mean 0 and covariance operator $(-\Delta_x + m^2)^{-1}$. It is just the (space-time) free field of mass m in terms of Euclidean quantum field theory. A double-well potential $U(z) = a(|z|^4 - |z|^2)$, $a > 0$, is also particularly important from the point of view of physics.

We should mention that the Gibbs measure μ is supported by a smaller subset of $C(\mathbf{R}, \mathbf{R}^d)$ than $\mathscr{C}$. Actually, it holds

$$\mu\left(\left\{ w \,\Big|\, \limsup_{|x|\to\infty} \frac{|w(x)|_{\mathbf{R}^d}}{(\log|x|)^{1/(m+1)}} \le C \right\}\right) = 1 \tag{8}$$

with a suitable constant $C > 0$. See e.g., Rosen–Simon [16] and Lörinczi–Hiroshima–Betz [12]. Following Remark 3 below, we can also show (8) easily.

Example 2 ($exp(\phi)_1$*-quantum fields*): We introduce a Gibbs measure μ with the formal expression (1). Let us consider an exponential type potential function $U : \mathbf{R}^d \to \mathbf{R}$ (with weight ν) given by

$$U(z) = \frac{m^2}{2}|z|^2 + V(z) := \frac{m^2}{2}|z|^2 + \int_{\mathbf{R}^d} e^{(\xi,z)_{\mathbf{R}^d}} \nu(d\xi), \quad z \in \mathbf{R}^d,$$

where ν is a bounded positive measure with $\mathrm{supp}(\nu) \subset \{\xi \in \mathbf{R}^d \,|\, |\xi| \le L\}$ for some $L > 0$. We note that U is a smooth strictly convex function (i.e., $\nabla^2 U \ge m^2$). Hence we can take $K_1 = m^2$, $K_2 = \frac{m^2}{2}$ and $\alpha = 2$. Moreover,

$$|U(z)| \le \frac{m^2}{2}|z|^2 + \nu(\mathbf{R}^d)e^{L|z|} \le \left(\frac{m^2}{2L^2} + \nu(\mathbf{R}^d)\right)e^{2L|z|}, \quad z \in \mathbf{R}^d,$$

and

$$|\nabla U(z)| \le m^2|z| + \int_{\mathbf{R}^d} |\xi| e^{(\xi,z)_{\mathbf{R}^d}} \nu(d\xi) \le \left(\frac{m^2}{L} + L\nu(\mathbf{R}^d)\right)e^{L|z|}, \quad z \in \mathbf{R}^d.$$

Thus we can take $\beta = 1$, which satisfies $\beta < 1 + \frac{\alpha}{2}$ in condition **(U3)**.

Remark 3 Now we consider a simple example of $\exp(\phi)_1$-quantum fields in the case $d = 1$. This example has been discussed in the 2-dimensional space-time case (e.g., $\exp(\phi)_2$-quantum fields) in Albeverio–Høegh-Krohn [2]. Let δ_a be the Dirac measure at $a \in \mathbf{R}$ and we consider $\nu(d\xi) := \frac{1}{2}\big(\delta_{-a}(d\xi) + \delta_a(d\xi)\big)$, $a > 0$. Then the corresponding potential function is $U(z) = \frac{m^2}{2}z^2 + \cosh(az)$, and (2) implies that the Schrödinger operator H_U has a ground state Ω satisfying

$$0 < \Omega(z) \le D_1 \exp\left(-\frac{D_2}{\sqrt{2}}|z|\, e^{\frac{a}{4}|z|}\right), \quad z \in \mathbf{R} \tag{9}$$

for some $D_1, D_2 > 0$. By the translation invariance of the Gibbs measure μ and (9), there exist positive constants M_1 and M_2 such that

$$\begin{aligned} A_T &:= \mu\left(\left\{w \in C(\mathbf{R},\mathbf{R}) \mid |w(T)| > \frac{4}{a}\log\log T\right\}\right) \\ &= \int_{|z| > \frac{4}{a}\log\log T} \Omega(z)^2 dz \\ &\le M_1 \exp\big\{-M_2(\log T)(\log\log T)\big\} = M_1 T^{-M_2 \log\log T} \end{aligned}$$

for T large enough, and it implies $\sum_{T=1}^{\infty} A_T < \infty$. Then the first Borel–Cantelli lemma yields

$$\mu\left(\left\{w \in C(\mathbf{R},\mathbf{R}) \mid \limsup_{|x|\to\infty} \frac{|w(x)|}{\log\log|x|} \le \frac{4}{a}\right\}\right) = 1.$$

This means that μ is supported by a much smaller subset of $C(\mathbf{R},\mathbf{R})$ than $\mathscr{C}$.

Example 3 (*Trigonometric quantum fields*): We consider a trigonometric type potential function $U : \mathbf{R}^d \to \mathbf{R}$ (with weight ν) given by

$$U(z) = \frac{m^2}{2}|z|^2 + V(z) := \frac{m^2}{2}|z|^2 + \int_{\mathbf{R}^d} \cos\big\{(\xi,z)_{\mathbf{R}^d} + \alpha\big\}\nu(d\xi), \quad z \in \mathbf{R}^d,$$

where $\alpha \in \mathbf{R}$, $m > 0$ and ν is a bounded signed measure with finite second absolute moment, i.e.,

$$|\nu|(\mathbf{R}^d) < \infty, \quad K(\nu) := \int_{\mathbf{R}^d} |\xi|^2 |\nu|(d\xi) < \infty.$$

This potential function is smooth, and it can be regarded as a bounded perturbation of a quadratic function. Moreover, $\nabla^2 U \geq m^2 - K(\nu)$ and

$$|\nabla U(z)| \leq m^2 |z| + K(\nu)^{1/2} |\nu|(\mathbf{R}^d)^{1/2}, \quad z \in \mathbf{R}^d.$$

This type of potential functions corresponds to quantum field models with "trigonometric interaction" and has been discussed especially in the 2-dimensional space-time case (cf. [1]).

3 An Open Problem

Finally, we raise an open problem which is concerned with this paper.

If the potential function U is a C^1-function with polynomial growth at infinity, Iwata [10] proves that SPDE (7) has a unique strong solution $X^w = \{X_t^w(\cdot)\}_{t \geq 0}$ living in $C([0, \infty), \mathscr{C})$ for *every* initial datum $w \in \mathscr{C}$. On the other hand, it should be remarked that $(\nabla U)(w(\cdot)) \notin \mathscr{C}$ for $w \in \mathscr{C}$ in the case of $\exp(\phi)_1$-quantum fields. Thus if U has exponential growth at infinity, we cannot apply Iwata's argument directly to solve SPDE (7) in $C([0, \infty), \mathscr{C})$ for *every* initial datum $w \in \mathscr{C}$. Can we overcome this difficulty?

It is natural to think that we can easily construct a unique strong solution to SPDE (7) living in $C([0, \infty), \mathscr{C}_e)$ for *every* initial datum $w \in \mathscr{C}_e$ by only replacing the state space $\mathscr{C}$ by a much smaller tempered subspace $\mathscr{C}_e$ and then by applying Iwata's argument. One might guess that a possible candidate for $\mathscr{C}_e$ is a subspace of $\mathscr{C}$ such that $(\nabla U)(w(\cdot)) \in \mathscr{C}_e$ holds for $w \in \mathscr{C}_e$, which is the space of all paths behaving like

$$|w(x)| \sim \log(\log(\log(\log(\cdots x)))) =: \rho_e(x)^{-1}$$

at infinity. However, we cannot follow all arguments in the papers [4, 10, 11] if we replace $\rho_{-2r}(x)$ by $\rho_e(x)$ because of $\int^{\infty} \rho_e(x)^{-2} dx = \infty$. Hence it seems that this approach is not valid for our problem.

On the other hand, in the case $d = 1$, by applying a comparison theorem for parabolic SPDEs (cf. Shiga [17]), we might construct a unique strong solution to SPDE (7) with exponentially growing drift which lives in $C([0, \infty), \mathscr{C})$ for *every* initial datum $w \in \mathscr{C}$. However, this approach does not work in the case $d \geq 2$.

Hence we still need to find a new approach to tackle this problem. It should be a preliminary step toward a construction of a unique strong solution to the stochastic quantization equation associated with the $\exp(\phi)_2$-quantum fields in *infinite volume*.

Acknowledgements The author was partially supported by JSPS Grant-in-Aid for Scientific Research (C) No. 26400134.

References

1. Albeverio, S., Høegh-Krohn, R.: Uniqueness of the physical vacuum and the Wightman functions in the infinite volume limit for some non-polynomial interactions. Commun. Math. Phys. **30**, 171–200 (1973)
2. Albeverio, S., Høegh-Krohn, R.: The Wightman axioms and the mass gap for strong interactions of exponential type in two-dimensional space-time. J. Funct. Anal. **16**, 39–82 (1974)
3. Albeverio, S., Röckner, M.: Stochastic differential equations in infinite dimensions: solutions via Dirichlet forms. Probab. Theory Relat. Fields **89**, 347–386 (1991)
4. Albeverio, S., Kawabi, H., Röckner, M.: Strong uniqueness for both Dirichlet operators and stochastic dynamics to Gibbs measures on a path space with exponential interactions. J. Funct. Anal. **262**, 602–638 (2012)
5. Albeverio, S., Ma, Z.M., Röckner, M.: Quasi regular Dirichlet forms and the stochastic quantization problem. Festschrift Masatoshi Fukushima, 27–58, Interdisciplinary Mathematical Sciences, vol. 17. World Scientific Publishing, Hackensack (2015)
6. Eberle, A.: Uniqueness and non-uniqueness of singular diffusion operators. Lecture Notes in Mathematics, vol. 1718. Springer, Berlin (1999)
7. Gubinelli, M., Imkeller, P., Perkowski, N.: Paracontrolled distributions and singular PDEs. Forum Math. Pi **3**, e6, 75 p. (2015)
8. Hairer, M.: A theory of regularity structures. Invent. Math. **198**, 269–504 (2014)
9. Iwata, K.: Reversible measures of a $P(\phi)_1$- time evolution. In: Itô, K., Ikeda, N. (eds.), Probabilistic Methods in Mathematical Physics: Proceedings of Taniguchi Symposium, pp. 195–209. Kinokuniya (1985)
10. Iwata, K.: An infinite dimensional stochastic differential equation with state space $C(R)$. Probab. Theory Relat. Fields **74**, 141–518 (1987)
11. Kawabi, H., Röckner, M.: Essential self-adjointness of Dirichlet operators on a path space with Gibbs measures via an SPDE approach. J. Funct. Anal. **242**, 486–518 (2007)
12. Lőrinczi, J., Hiroshima, F., Betz, V.: Feynman-Kac-type theorems and Gibbs measures on path space. De Gruyter Studies in Mathematics, vol. 34. Walter de Gruyter Co., Berlin (2011)
13. Mourrat, J.-C., Weber, H.: Global well-posedness of the dynamic Φ^4 model in the plane. Ann. Probab. **45**, 2398–2476 (2017)
14. Ondreját, M.: Uniqueness for stochastic evolution equations in Banach spaces. Dissertaiones Math. (Rozprawy Mat.) **426** (2004), 63 p
15. Röckner, M., Zhu, R., Zhu, X.: Restricted Markov uniqueness for the stochastic quantization of $P(\Phi)_2$ and its applications. J. Funct. Anal. **272**, 4263–4303 (2017)
16. Rosen, J., Simon, B.: Fluctuations in $P(\phi)_1$ processes. Ann. Probab. **4**, 155–174 (1976)
17. Shiga, T.: Two contrasting properties of solutions for one-dimensional stochastic partial differential equations. Can. J. Math. **46**, 415–437 (1994)

A Probabilistic Proof of the Breakdown of Besov Regularity in L-Shaped Domains

Victoria Knopova and René L. Schilling

Abstract We provide a probabilistic approach in order to investigate the smoothness of the solution to the Poisson and Dirichlet problems in L-shaped domains. In particular, we obtain (probabilistic) integral representations (9), (12)–(14) for the solution. We also recover Grisvard's classic result on the angle-dependent breakdown of the regularity of the solution measured in a Besov scale.

Keywords Brownian motion · Dirichlet problem · Poisson equation · Conformal mapping · Stochastic representation · Besov regularity

MSC 2010 60J65 · 35C15 · 35J05 · 35J25 · 46E35

1 Introduction

Let us consider the (homogeneous) Dirichlet problem

$$\begin{aligned} \Delta f &= 0 \quad \text{in } G, \\ f|_{\partial G} &= h \quad \text{on } \partial G, \end{aligned} \tag{1}$$

where $G \subset \mathbb{R}^d$ is a domain with Lipschitz boundary ∂G and Δ denotes the Laplace operator, i.e. $\Delta = \sum_{i=1}^d \frac{\partial^2}{\partial x_i^2}$. In order to show that there exists a solution to (1) which belongs to some subspace of $L_p(G)$, say, to the Besov space $B^\sigma_{pp}(G)$, $\sigma > 0$, it is necessary that h is an element of the trace space of $B^\sigma_{pp}(G)$ on ∂G; it is well known that the trace space is given by $B^{\sigma-1/p}_{pp}(\partial G)$, see Jerison and Kenig [11, Theorem 3.1],

V. Knopova (✉) · R. L. Schilling
Institut Für Mathematische Stochastik, Fachrichtung Mathematik,
01062 TU Dresden, Dresden, Germany
e-mail: viktoriya.knopova@tu-dresden.de

R. L. Schilling
e-mail: rene.schilling@tu-dresden.de

A. Eberle et al. (eds.), *Stochastic Partial Differential Equations and Related Fields*, Springer Proceedings in Mathematics & Statistics 229,
https://doi.org/10.1007/978-3-319-74929-7_32

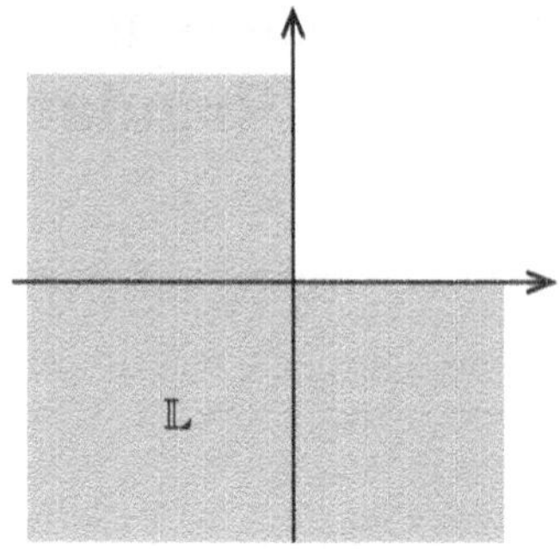

Fig. 1 The L-shaped model domain $\mathbb{L} \subset \mathbb{R}^2$

a more general version can be found in Jonsson and Wallin [12, Chap. VII], and for domains with C^∞-boundary a good reference is Triebel [18, Sects. 3.3.3–4]. The smoothness of the solution f, expressed by the parameter σ in $B^\sigma_{pp}(G)$, is, however, not only determined by the smoothness of h, but also by the geometry of G. It seems that Grisvard [10] is the first author to quantify this in the case when G is a non-convex polygon. Subsequently, partly due to its relevance in scientific computing, this problem attracted a lot of attention; for instance, it was studied by Jerison and Kenig [11], by Dahlke and DeVore [7] in connection with wavelet representations of Besov functions, by Mitrea and Mitrea [13] and Mitrea, Mitrea and Yan [14] in Hölder spaces, to mention but a few references.

In this note we use a probabilistic approach to the problem and we obtain a probabilistic interpretation in the special case when G is an *L-shaped domain* of the form $\mathbb{L} := \mathbb{R}^2 \setminus \{(x, y) : x, y \geq 0\}$, see Fig. 1, and in an L_2-setting. This is the model problem for all non-convex domains with an obtuse interior angle. In this case the Besov space $B^\sigma_{22}(\mathbb{L})$ coincides with the Sobolev–Slobodetskij space $W^\sigma_2(\mathbb{L})$. In particular, we

- give a probabilistic interpretation of the solution to (1) with $G = \mathbb{L}$;
- provide a different proof of the fact that the critical order of smoothness of f is $\sigma < \pi / \frac{3\pi}{2} = \frac{2}{3}$, i.e. even for $h \in C^2_0(\partial\mathbb{L})$ we may have

$$f \in W^{1+\sigma}_{2,\text{loc}}(\mathbb{L}), \quad \sigma < \tfrac{2}{3}, \qquad \text{and} \qquad f \notin W^{1+\sigma}_2(\mathbb{L}), \quad \sigma \geq \tfrac{2}{3}; \tag{2}$$

- apply the "breakdown of regularity" result to the Poisson (or inhomogeneous Dirichlet) problem.

It is clear that this result holds in a more general setting, if we replace the obtuse angle $3\pi/2$ by some $\theta \in (\pi, 2\pi)$.

Results of this type were proved for polygons and in a Hölder space setting by Mitrea and Mitrea [13]. Technically, our proof is close (but different) to that given in [13]—yet our staring idea is different. Dahlke and DeVore [7] proved this regularity result analytically using a wavelet basis for L_p-Besov spaces.

Problem (1) is closely related to the Poisson (or nonhomogeneous Dirichlet) problem

$$\begin{aligned} \Delta F &= g \quad \text{on } G, \\ F|_{\partial G} &= 0 \quad \text{on } \partial G. \end{aligned} \tag{3}$$

If G is bounded and has a C^∞-boundary, the problems (1) and (3) are equivalent. Indeed, in this case for every right-hand side $g \in L_2(G)$ of (3) there exists a unique solution $F \in W_2^2(G)$, see Triebel [18, Theorem 4.3.3]. Denote by N the Newtonian potential on $\mathbb{R}^d$ and define $w := g * N$; clearly, $\Delta w = g$ on G and $w \in W_2^2(G)$. Since the boundary is smooth, there is a continuous linear trace operator $\mathrm{Tr} : W_2^2(G) \to W_p^{3/2}(\partial G)$ as well as a continuous linear extension operator $\mathrm{Ex} : W_2^{3/2}(\partial G) \to W_2^2(G)$, such that $\mathrm{Tr} \circ \mathrm{Ex} = \mathrm{id}$, cf. Triebel [18]. Hence, the function $f := w - F$ solves the inhomogeneous Dirichlet problem (1) with $h = \mathrm{Tr}\, w$ on ∂G. On the other hand, let f be the (unique) solution to (1). Since there exists a continuous linear extension operator from $W_2^{3/2}(\partial G)$ to $W_2^2(G)$ given by $\tilde{h} = \mathrm{Ex}\, h$, we see that the function $F := f - \tilde{h}$ satisfies (3) with $g = \Delta \tilde{h}$.

If the boundary ∂G is Lipschitz the situation is different. It is known, see for example Jerison and Kenig [11, Theorem B] that, in general, on a Lipschitz domain G and for $g \in L_2(G)$ one can only expect that the solution F to (3) belongs to $W_2^{3/2}(G)$; there are counterexamples of domains, for which F cannot be in $W_2^\alpha(G)$ for any $\alpha > 3/2$. Thus, the above procedure does not work in a straightforward way. However, by our strategy we can recover the negative result for this concrete domain, cf. Theorem 2: *If $g \in H_1(\mathbb{R}^2) \cap W_2^1(\mathbb{L})$, then the solution F to* (3) *is not in $W_2^{1+\sigma}(\mathbb{L})$ for any $\sigma \geq 2/3$.* Here $H_1(\mathbb{R}^2) \subset L_1(\mathbb{R}^2)$ is the Hardy space, cf. Stein [17].

If G is unbounded, the solution to (1) might be not unique and, in general, it is only in the local space $W_{2,\mathrm{loc}}^2(G)$ even if ∂G is smooth, cf. Gilbarg and Trudinger [9, Chap. 8]. On the other hand, if the complement G^c is non-empty, if no component of G^c reduces to a single point, and if the boundary value h is bounded and continuous on ∂G, then there exists a unique bounded solution to (1) given by the convolution with the Poisson kernel, see Port and Stone [16, Theorem IV.2.13].

A strong motivation for this type of results comes from numerical analysis and approximation theory, because the exact Besov smoothness of u is very important for computing u and the feasibility of adaptive computational schemes, see Dahlke and DeVore [7], Dahlke, Dahmen and DeVore [6], DeVore [8], Cohen, Dahmen and DeVore [5], Cohen [4]; an application to SPDEs is in Cioika et al. [2, 3]. More precisely—using the set-up and the notation of [5]—let $\{\psi_\lambda,\ \lambda \in \Lambda\}$ be a basis of wavelets on G and assume that the index set Λ is of the form $\Lambda = \bigcup_{i \geq 0} \Lambda_i$ with (usually hierarchical) sets Λ_i of cardinality N_i. By u_{Λ_i} we denote the Galerkin approximation of u in terms of the wavelets $\{\psi_\lambda\}_{\lambda \in \Lambda_i}$ (this amounts to solving a system of linear equations), and by $e_{N_i}(u) := \|u - u_{\Lambda_i}\|_p$ the approximation error in this scheme. Then it is known, cf. [5, (4.2) and (2.35)], that

$$u \in W_p^\sigma(G) \implies e_{N_i}(u) \leq C N_i^{-\sigma/d}, \quad i \geq 1. \tag{4}$$

There is also an adaptive algorithm for choosing the index sets $(\Lambda_i)_{i \geq 1}$. Starting with an initial set Λ_0, this algorithm adaptively generates a sequence of nested sets $(\Lambda_i)_{i \geq 1}$;

roughly speaking, in each iteration step we choose the next set Λ_{i+1} by partitioning the domain of those wavelets ψ_λ, $\lambda \in \Lambda_i$ (i.e. selectively refining the approximation by considering the next generation of wavelets), whose coefficients u_λ make, in an appropriate sense, the largest contribution to the sum $u = \sum_{\lambda\in\Lambda_i} u_\lambda \psi_\lambda$.

Notation. Most of our notation is standard. By $(r, \theta) \in (0, \infty) \times (0, 2\pi]$ we denote polar coordinates in $\mathbb{R}^2$, and $\mathbb{H}$ is the lower half-plane in $\mathbb{R}^2$. We write $f \asymp g$ to say that $cf(t) \le g(t) \le Cf(t)$ for all t and some fixed constants.

2 Setting and the Main Result

Let $B = (B_t^x)_{t\ge 0}$ be a Brownian motion started at a point $x \in G$. Suppose that there exists a conformal mapping $\varphi : G \to \mathbb{H}$, where $\mathbb{H} := \{(x_1, x_2) \in \mathbb{R}^2,\ x_2 \le 0\}$ is the lower half-plane in $\mathbb{R}^2$. Using the conformal invariance of Brownian motion, see e.g. Mörters and Peres [15, p. 202], we can describe the distribution of the Brownian motion inside G in terms of *some* Brownian motion W in $\mathbb{H}$, which is much easier to handle. Conformal invariance of Brownian motion means that there exists a planar Brownian motion $W = (W_t^y)_{t\ge 0}$ with starting point $y \in \mathbb{H}$ such that, under the conformal map $\varphi : G \to \mathbb{H}$ with boundary identification,

$$\left(\varphi(B_t^x)\right)_{0\le t\le \tau_G} \quad \text{has the same law as} \quad \left(W_{\xi(t)}^{\varphi(x)}\right)_{0\le t\le \tau_{\mathbb{H}}}; \tag{5}$$

the time-change ξ is given by $\xi(t) := \int_0^t |\varphi'(B_s^x)|^2\,ds$; in particular, $\xi(\tau_G) = \tau_{\mathbb{H}}$, where $\tau_G = \inf\{t > 0 : B_t^x \in \partial G\}$ and $\tau_{\mathbb{H}} := \inf\{t > 0 :\ W_t^{\varphi(x)} \in \partial\mathbb{H}\}$ are the first exit times from G and $\mathbb{H}$, respectively.

Let us recall some properties of a planar Brownian motion in $\mathbb{H}$ killed upon exiting at the boundary $\partial\mathbb{H} = \{(w_1, w_2) : w_2 = 0\}$. The distribution of the exit position $W_{\tau_{\mathbb{H}}}$ has the transition probability density

$$u \mapsto p_{\mathbb{H}}(w, u) = \frac{1}{\pi}\frac{|w_2|}{|u - w_1|^2 + w_2^2}, \qquad w = (w_1, w_2) \in \mathbb{H}, \tag{6}$$

cf. Bass [1, p. 91]. Recall that a random variable X with values in $\mathbb{R}$ has a Cauchy distribution, $X \sim \mathsf{C}(m, b)$, $m \in \mathbb{R}$, $b > 0$, if it has a transition probability density of the form

$$p(u) = \frac{1}{\pi}\frac{b}{(u - m)^2 + b^2}, \qquad u \in \mathbb{R};$$

if $X \sim \mathsf{C}(m, b)$, then $Z := (X - m)/b \sim \mathsf{C}(0, 1)$. Thus, the probabilistic interpretation of $W_{\tau_{\mathbb{H}}}^w$ is

$$W_{\tau_{\mathbb{H}}}^w \sim Z^w \sim \mathsf{C}(w_1, |w_2|) \quad \text{or} \quad W_{\tau_{\mathbb{H}}}^w \sim \frac{Z - w_1}{|w_2|} \quad \text{where} \quad Z \sim \mathsf{C}(0, 1). \tag{7}$$

This observation allows us to simplify the calculation of functionals Θ of a Brownian motion B on G, killed upon exiting from G, in the following sense:

$$\begin{aligned}\mathbb{E}\Theta(B^x_{\tau_G}) = \mathbb{E}\left(\Theta \circ \varphi^{-1}\right)(\varphi(B^x_{\tau_G})) &= \mathbb{E}\left(\Theta \circ \varphi^{-1}\right)(W^{\varphi(x)}_{\tau_{\mathbb{H}}}) \\ &= \mathbb{E}\left(\Theta \circ \varphi^{-1}\right)\left(\frac{Z - \varphi_1(x)}{|\varphi_2(x)|}\right).\end{aligned} \tag{8}$$

In particular, the formula (8) provides us with a probabilistic representation for the solution f to the Dirichlet problem (1):

$$f(x) = \mathbb{E}h(B^x_{\tau_G}) = \mathbb{E}\left(h \circ \varphi^{-1}\right)(W^{\varphi(x)}_{\tau_{\mathbb{H}}}). \tag{9}$$

Remark 1 The formulae in (8) are very helpful for the numerical calculation of the values $\mathbb{E}\Theta(B^x_{\tau_G})$. In fact, in order to simulate $\Theta(B^x_{\tau_G})$, it is enough to simulate the Cauchy distribution $Z \sim \mathsf{C}(0, 1)$ and then evaluate (8) using the Monte Carlo method.

We will now consider the L-shaped domain $\mathbb{L}$. It is easy to see that the conformal mapping of $\mathbb{L}$ to $\mathbb{H}$ is given by

$$\varphi(z) = \mathrm{e}^{\mathrm{i}\frac{2\pi}{3}} z^{2/3} = r^{2/3} \exp\left(\tfrac{2}{3}\mathrm{i}(\theta + \pi)\right) = \varphi_1(r, \theta) + \mathrm{i}\varphi_2(r, \theta), \tag{10}$$

cf. Fig. 2, where $\theta = \arg z \in (0, 2\pi]$.

The following lemma uses the conformal mapping $\varphi : \mathbb{L} \to \mathbb{H}$ and the conformal invariance of Brownian motion to obtain the distribution of $B^x_{\tau_{\mathbb{L}}}$.

Lemma 1 *Let $\mathbb{L}$ be an L-shaped domain as shown in Fig. 1. The exit position $B_{\tau_{\mathbb{L}}}$ of Brownian motion from $\mathbb{L}$ is a random variable on $\partial\mathbb{L} = \{0\} \times [0, \infty) \cup [0, \infty) \times \{0\}$ which has the following probability distribution:*

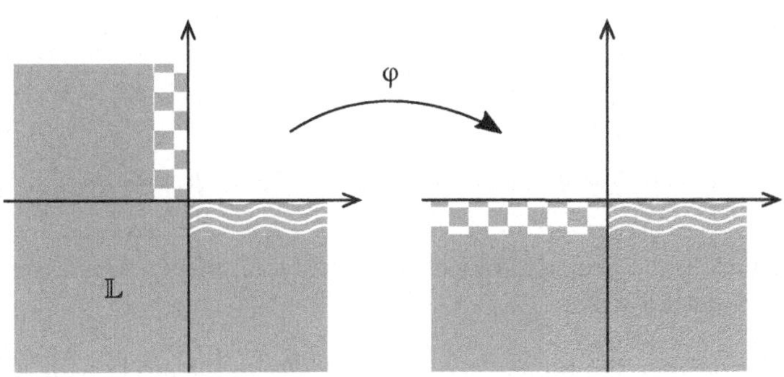

Fig. 2 Conformal mapping from $\mathbb{L}$ to $\mathbb{H}$ and its behaviour at the boundaries

$$\mathbb{P}\left(B^x_{\tau_{\mathbb{L}}} \in \mathrm{d}y\right) = \frac{1}{\pi}\frac{|\varphi_2(x)|}{|\varphi_1(x)+y_2|^2+|\varphi_2(x)|^2}\,\mathrm{d}y_2\,\delta_0(\mathrm{d}y_1) + \frac{1}{\pi}\frac{|\varphi_2(x)|}{|\varphi_1(x)-y_1|^2+|\varphi_2(x)|^2}\,\mathrm{d}y_1\,\delta_0(\mathrm{d}y_2). \tag{11}$$

Lemma 1 provides us with an explicit representation of the solution $f(x)$ to the Dirichlet problem (1) for $G = \mathbb{L}$. Indeed, since (cf. Fig. 2)

$$\left(h \circ \varphi^{-1}\right)(u) = \begin{cases} h(u,0), & u \geq 0, \\ h(0,-u), & u \leq 0, \end{cases}$$

we get

$$f(x) = \int_{\mathbb{R}} f_0(u)\,p_{\mathbb{H}}(\varphi(x),u)\,\mathrm{d}u = \frac{1}{\pi}\int_{\mathbb{R}} f_0(u)\,\frac{|\varphi_2(x)|}{(\varphi_1(x)-u)^2+|\varphi_2(x)|^2}\,\mathrm{d}u, \tag{12}$$

where

$$f_0(u) := h(0,-u)\mathbb{1}_{(-\infty,0)}(u) + h(u,0)\mathbb{1}_{[0,\infty)}(u). \tag{13}$$

After a change of variables, this becomes

$$f(x) = \frac{1}{\pi}\int_{\mathbb{R}} f_0\left(u|\varphi_2(x)| + \varphi_1(x)\right)\frac{\mathrm{d}u}{u^2+1}. \tag{14}$$

If we want to investigate the smoothness of f, it is more convenient to rewrite f in polar coordinates. From the right-hand side of (10) we infer

$$\varphi_1(r,\theta) = r^{2/3}\cos\Phi_\theta \quad \text{and} \quad \varphi_2(r,\theta) = r^{2/3}\sin\Phi_\theta, \tag{15}$$

where we use the shorthand

$$\Phi_\theta := \frac{2}{3}(\pi+\theta).$$

Observe that for $\theta \in (\pi/2, 2\pi]$ we have $\pi < \Phi_\theta \leq 2\pi$, hence $\varphi_2 \leq 0$. This yields

$$f(r,\theta) = \frac{1}{\pi}\int_{\mathbb{R}} f_0\left(r^{2/3}\cos\Phi_\theta - r^{2/3}v\sin\Phi_\theta\right)\frac{\mathrm{d}v}{1+v^2}. \tag{16}$$

Now we turn to the principal objective of this note: the smoothness of f in the Sobolev–Slobodetskij scale.

Theorem 1 *Consider the (homogeneous) Dirichlet problem* (1) *with a boundary term f_0, given by* (17), *and let f denote the solution to* (1).

a. *If* $f_0 \in W_1^2(\mathbb{R}) \cap W_2^2(\mathbb{R})$ *satisfies*

$$\liminf_{\varepsilon \to 0} \int_{|x|>\varepsilon} \frac{f_0'(x)}{x} \, \mathrm{d}x \neq 0, \tag{17}$$

then $f \notin W_2^{1+\sigma}(\mathbb{L})$, *even* $f \notin W_{2,\mathrm{loc}}^{1+\sigma}(\mathbb{L})$, *for any* $\sigma \geq 2/3$.

b. *If* $f_0 \in W_1^2(\mathbb{R}) \cap W_p^1(\mathbb{R})$, *where* $p > \max\{2, 2/(2-3\sigma)\}$, *then* $f \in W_{2,\mathrm{loc}}^{1+\sigma}(\mathbb{L})$ *for all* $\sigma \in (0, 2/3)$.

Remark 2 By the Sobolev embedding theorem we have $W_1^2(\mathbb{R}) \cap W_2^2(\mathbb{R}) \subset C_b(\mathbb{R})$ and $W_1^2(\mathbb{R}) \cap W_p^1(\mathbb{R}) \subset C_b(\mathbb{R})$ if $p > \max\{2, 2/(2-3\sigma)\}$. Hence, the function f given by (14) is the unique bounded solution to (1).

The idea of the proof of Theorem 1 makes essential use of the results by Jerison and Kenig [11] combined with the observation that it is, in fact, enough to show the claim for $\widehat{\mathbb{L}} := \mathbb{L} \cap B(0, 1)$, where $B(0, 1) := \{x \in \mathbb{R}^2 : |x| < 1\}$.

Theorem 1 allows us to prove the negative result for the solution to the Poisson problem, which improves [11, Theorem B]. Recall that $H_1(\mathbb{R}^2) \subset L_1(\mathbb{R}^2)$ is the usual Hardy space, cf. Stein [17].

Theorem 2 *Consider the Poisson (inhomogeneous Dirichlet) problem* (3) *with right-hand side* $g \in H_1(\mathbb{R}^2) \cap W_2^1(\mathbb{L})$ *such that* $f_0(x) := \left((\mathrm{Tr}\, g * N) \circ \varphi^{-1}\right)(x)$ *satisfies* (17), *where* $N(x) = (2\pi)^{-1} \log|x|$ *is the Newton kernel. Then the solution* $F \notin W_2^{1+\sigma}(\mathbb{L})$, *even* $F \notin W_{2,\mathrm{loc}}^{1+\sigma}(\mathbb{L})$, *for any* $\sigma \geq 2/3$.

The proofs of Theorems 1 and 2 are deferred to the next section.

3 Proofs

Proof (*Proof of Lemma* 1) We calculate the characteristic function of $B_{\tau_{\mathbb{L}}}^x$. As before, let $y = (y_1, y_2)$, $x = (x_1, x_2)$ and $\varphi(x) = (\varphi_1(x), \varphi_2(x))$. We have

$$\begin{aligned}
\mathbb{E}\mathrm{e}^{\mathrm{i}\xi \cdot B_{\tau_{\mathbb{L}}}^x} &\overset{(5)}{=} \mathbb{E}\mathrm{e}^{\mathrm{i}\xi \cdot \varphi^{-1}(W_{\tau_{\mathbb{H}}}^{\varphi(x)})} \\
&= \int_{\mathbb{R}^2} \mathrm{e}^{\mathrm{i}\xi \cdot \varphi^{-1}(y)} \, \mathbb{P}(W_{\tau_{\mathbb{H}}}^{\varphi(x)} \in \mathrm{d}y) \\
&\overset{(6)}{=} \frac{1}{\pi} \int_{\mathbb{R}} \mathrm{e}^{\mathrm{i}\xi \cdot \varphi^{-1}(y_1, 0)} \frac{|\varphi_2(x)|}{|\varphi_1(x) - y_1|^2 + |\varphi_2(x)|^2} \mathrm{d}y_1 \\
&= \frac{1}{\pi} \int_{-\infty}^{0} \mathrm{e}^{-\mathrm{i}\xi_2 u} \frac{|\varphi_2(x)|}{|\varphi_1(x) - u|^2 + |\varphi_2(x)|^2} \, du \\
&\quad + \frac{1}{\pi} \int_{0}^{+\infty} \mathrm{e}^{\mathrm{i}\xi_1 u} \frac{|\varphi_2(x)|}{|\varphi_1(x) - u|^2 + |\varphi_2(x)|^2} \, du.
\end{aligned}$$

□

For the proof of Theorem 1 we need some preparations. In order to keep the presentation self-contained, we quote the classical result by Jerison and Kenig [11, Theorem 4.1].

Theorem 3 (Jerison and Kenig) *Let* $\sigma \in (0,1)$, $k \in \mathbb{N}_0$ *and* $p \in [1,\infty]$. *For any function* u *which is harmonic on a bounded domain* Ω, *the following assertions are equivalent:*

a. $f \in B_{pp}^{k+\sigma}(\Omega)$;
b. $\operatorname{dist}(x,\partial\Omega)^{1-\sigma}\,\left|\nabla^{k+1} f\right| + \left|\nabla^k f\right| + |f| \in L_p(\Omega)$.

We will also need the following technical lemma. Recall that $\widehat{\mathbb{L}} = \mathbb{L} \cap B(0,1)$.

Lemma 2 *Suppose that* $f_0 \in W_p^1(\mathbb{R})$ *for some* $p > 2$. *Then* $f \in W_2^1(\widehat{\mathbb{L}})$.

Proof Using the representation (16), the Hölder inequality and a change of variables, we get

$$
\begin{aligned}
&\int_{\pi/2}^{2\pi}\int_0^1 |f(r,\theta)|^2 r\,\mathrm{d}r\,\mathrm{d}\theta\\
&= \frac{3}{2\pi^2}\int_{\pi/2}^{2\pi}\int_0^1 \rho^2\left|\int_{\mathbb{R}} f_0(w)\frac{|\rho\sin\Phi_\theta|}{(w-\rho\cos\Phi_\theta)^2+(\rho\sin\Phi_\theta)^2}\mathrm{d}w\right|^2 \mathrm{d}\rho\,\mathrm{d}\theta\\
&\le C_1\int_{\pi/2}^{2\pi}\int_0^1 \rho^2\left[\left(\int_{\mathbb{R}}|f_0(v)|^p\,\mathrm{d}v\right)^{1/p}\left(\int_{\mathbb{R}}\frac{|\rho\sin\Phi_\theta|^q}{(v^2+|\rho\sin\Phi_\theta|^2)^q}\mathrm{d}v\right)^{1/q}\right]^2 \mathrm{d}\rho\,\mathrm{d}\theta\\
&\le C_2\int_{\pi/2}^{2\pi}\int_0^1 \rho^2\,|\rho\sin\Phi_\theta|^{-2+2/q}\left(\int_{\mathbb{R}}\frac{1}{(w^2+1)^q}\mathrm{d}w\right)^{2/q} \mathrm{d}\rho\,\mathrm{d}\theta\\
&= C_3\int_{\pi/2}^{2\pi}\int_0^1 \rho^{2/q}\,|\sin\Phi_\theta|^{-2+2/q}\,\mathrm{d}\rho\,\mathrm{d}\theta,
\end{aligned}
$$

where $p^{-1}+q^{-1}=1$. Because of $p>2$ we have $-2+2/q>-1$, hence $q<2$. Note that the inequalities $2x/\pi \le \sin x \le x$ for $x\in[0,\pi/2]$ imply

$$
\int_{\pi/2}^{2\pi}|\sin\Phi_\theta|^{-1+\varepsilon}\,\mathrm{d}\theta = \int_0^{\pi}|\sin\varphi|^{-1+\varepsilon}\,\mathrm{d}\varphi = 2\int_0^{\pi/2}|\sin\varphi|^{-1+\varepsilon}\,\mathrm{d}\varphi < \infty.
$$

This shows that $f \in L_2(\widehat{\mathbb{L}})$.

Recall that the partial derivatives of the polar coordinates are

$$
\frac{\partial}{\partial x_1}r = \cos\theta,\quad \frac{\partial}{\partial x_1}\theta = -\frac{\sin\theta}{r},\quad \frac{\partial}{\partial x_1}\Phi_\theta = \frac{2}{3}\frac{\partial}{\partial x_1}\theta = -\frac{2\sin\theta}{3r}. \tag{18}
$$

Therefore, we have for $\theta \in (\pi/2, 2\pi)$

$$\begin{aligned}\frac{\partial}{\partial x_1} f(r,\theta) &= \frac{1}{\pi}\int_{\mathbb{R}} f_0'\left(r^{2/3}\cos\Phi_\theta - vr^{2/3}\sin\Phi_\theta\right)\frac{1}{v^2+1}\times \\ &\quad\times\left[\frac{2\cos\theta}{3r^{1/3}}(\cos\Phi_\theta - v\sin\Phi_\theta)\right. \\ &\quad\left.+r^{2/3}\left(\frac{-2\sin\theta}{3r}\right)(-v\cos\Phi_\theta - \sin\Phi_\theta)\right]\mathrm{d}v \qquad (19)\\ &= \frac{2}{3\pi r^{1/3}}\int_{\mathbb{R}} f_0'\left(r^{2/3}\cos\Phi_\theta - vr^{2/3}\sin\Phi_\theta\right)\frac{1}{v^2+1}\times \\ &\quad\times\left[(\cos\Phi_\theta - v\sin\Phi_\theta)\cos\theta + (v\cos\Phi_\theta + \sin\Phi_\theta)\sin\theta\right]\mathrm{d}v \\ &= \frac{2}{3\pi r^{1/3}}\int_{\mathbb{R}} f_0'\left(r^{2/3}\cos\Phi_\theta - vr^{2/3}\sin\Phi_\theta\right)\frac{K(\theta,v)}{v^2+1}\,\mathrm{d}v,\end{aligned}$$

where

$$K(\theta,v) := \cos\omega_\theta - v\sin\omega_\theta, \qquad (20)$$

and

$$\omega_\theta = \frac{1}{3}(2\pi - \theta). \qquad (21)$$

Note that $\Phi_{\pi/2} = \pi$ and $\omega_{\pi/2} = \pi/2$.

Let us show that the first partial derivatives of f belong to $L_2(\widehat{\mathbb{L}})$. Because of the symmetry of $\widehat{\mathbb{L}}$, is it enough to check this for $\frac{\partial}{\partial x_1} f$.

Using the estimate $|K(\theta,v)|(1+v^2)^{-1} \le C(1+|v|)^{-1}$, a change of variables and the Hölder inequality, we get

$$\begin{aligned}&\int_0^1\int_{\pi/2}^{2\pi}\left|\frac{\partial}{\partial x_1} f(r,\theta)\right|^2 r\,\mathrm{d}\theta\,\mathrm{d}r \\ &= \int_0^1\int_{\pi/2}^{2\pi}\left|\int_{\mathbb{R}}\frac{2}{3\pi r^{1/3}} f_0'\left(r^{2/3}\cos\Phi_\theta - vr^{2/3}\sin\Phi_\theta\right)\frac{K(\theta,v)}{1+v^2}\,\mathrm{d}v\right|^2 r\,\mathrm{d}\theta\,\mathrm{d}r \\ &= \frac{2}{3\pi^2}\int_0^1\int_{\pi/2}^{2\pi}\rho\left|\int_{\mathbb{R}} f_0'(\rho\cos\Phi_\theta - v\rho\sin\Phi_\theta)\frac{K(\theta,v)}{1+v^2}\,\mathrm{d}v\right|^2\mathrm{d}\theta\,\mathrm{d}\rho \\ &\le C_1\int_0^1\int_{\pi/2}^{2\pi}\rho\left(\int_{\mathbb{R}}\frac{|f_0'(w)|}{|\rho\sin\Phi_\theta| + |w-\rho\cos\Phi_\theta|}\mathrm{d}w\right)^2\mathrm{d}\theta\,\mathrm{d}\rho \\ &\le C_2\left(\int_{\mathbb{R}}|f_0'(w)|^p\,\mathrm{d}w\right)^{2/p}\left(\int_{\mathbb{R}}\frac{1}{(1+|w|)^q}\mathrm{d}w\right)^{2/q}\times\end{aligned}$$

$$\times \int_{\pi/2}^{2\pi} \int_0^1 \rho \left(|\rho \sin \Phi_\theta|^{-1+1/q} \right)^2 \mathrm{d}\rho \, \mathrm{d}\theta$$

$$= C_3 \int_{\pi/2}^{2\pi} \int_0^1 |\sin \Phi_\theta|^{-2+2/q} \rho^{-1+2/q} \, \mathrm{d}\rho \, \mathrm{d}\theta < \infty;$$

in the last line we use again that $-2 + 2/q > -1$. □

Proof (*Proof of Theorem* 1) It is enough to consider the set $\widehat{\mathbb{L}}$. We verify that condition b of Theorem 3 holds true. We check whether

$$\operatorname{dist}(0, \cdot)^{1-\sigma} \left| \frac{\partial^2}{\partial x_1^2} f \right| + \left| \frac{\partial}{\partial x_1} f \right| + |f| \quad \text{is in } L_2(\widehat{\mathbb{L}}) \text{ or not.}$$

From Lemma 2 we already know that $\left|\frac{\partial}{\partial x_1} f\right| + |f| \in L_2(\widehat{\mathbb{L}})$. Let us check when

$$\operatorname{dist}(0, \cdot)^{1-\sigma} \left| \frac{\partial^2}{\partial x_1^2} f \right| \in L_2(\widehat{\mathbb{L}}).$$

We will only work out the term $\frac{\partial^2}{\partial x_1^2} f(r, \theta)$ since the calculations for $\frac{\partial^2}{\partial x_1 \partial x_2} f(r, \theta)$ are similar. We have

$$\frac{\partial}{\partial x_1} K(\theta, v) = \frac{\sin \theta}{3r} (-v \cos \omega_\theta - \sin \omega_\theta) =: \frac{\sin \theta}{3r} K^\star(\theta, v),$$

where use that $\frac{\partial}{\partial x_1} \omega_\theta = -\frac{1}{3} \frac{\partial}{\partial x_1} \theta = \frac{\sin \theta}{3r}$ and set

$$K^\star(\theta, v) := -v \cos \omega_\theta - \sin \omega_\theta. \tag{22}$$

Therefore, differentiating $\frac{\partial}{\partial x_1} f$—we use the representation (19)—with respect to x_1 gives

$$\begin{aligned}
\frac{\partial^2}{\partial x_1^2} f(r, \theta) = &- \frac{2 \cos \theta}{9 \pi r^{4/3}} \int_{\mathbb{R}} f_0' \left(r^{2/3} (\cos \Phi_\theta - v \sin \Phi_\theta) \right) \frac{K(\theta, v)}{1 + v^2} \, \mathrm{d}v \\
&+ \frac{4}{9 \pi r^{2/3}} \int_{\mathbb{R}} f_0'' \left(r^{2/3} (\cos \Phi_\theta - v \sin \Phi_\theta) \right) \frac{K^2(\theta, v)}{1 + v^2} \, \mathrm{d}v \\
&+ \frac{2 \sin \theta}{9 \pi r^{4/3}} \int_{\mathbb{R}} f_0' \left(r^{2/3} (\cos \Phi_\theta - v \sin \Phi_\theta) \right) \frac{K^\star(\theta, v)}{1 + v^2} \, \mathrm{d}v.
\end{aligned}$$

Note that

$$\begin{aligned}\int_{\widehat{\mathbb{L}}} \operatorname{dist}(x, \partial\widehat{\mathbb{L}})^{2-2\sigma} \left|\frac{\partial^2}{\partial x_1^2} f(x)\right|^2 \mathrm{d}x \\ = \int_0^1 \int_{\pi/2}^{2\pi} \operatorname{dist}((r,\theta), \partial\widehat{\mathbb{L}})^{2-2\sigma} \left|\frac{\partial^2}{\partial x_1^2} f(r,\theta)\right|^2 r \,\mathrm{d}\theta\, \mathrm{d}r.\end{aligned} \tag{23}$$

Since only the values near the boundary $\Gamma := \partial\widehat{\mathbb{L}} \cap \partial\mathbb{L}$ determine the convergence of the integral, it is enough to check that

$$\mathbb{I} = \int_0^1 \int_{\pi/2}^{2\pi} \operatorname{dist}((r,\theta), \Gamma)^{2-2\sigma} \left|\frac{\partial^2}{\partial x_1^2} f(r,\theta)\right|^2 r \,\mathrm{d}\theta\, \mathrm{d}r \tag{24}$$

is infinite if $\sigma \geq 2/3$ and finite if $\sigma < 2/3$.

We split $\widehat{\mathbb{L}}$ into three parts. For $\delta > 0$ small enough we define, see Fig. 3,

$$\begin{aligned} K_1 &:= \left\{(r,\theta) : 0 < r < 1,\ \frac{\pi}{2} + \delta < \theta < 2\pi - \delta\right\}, \\ K_2 &:= \left\{(r,\theta) : 0 < r < 1,\ \frac{\pi}{2} \leq \theta < \frac{\pi}{2} + \delta\right\}, \\ K_3 &:= \{(r,\theta) : 0 < r < 1,\ 2\pi - \delta < \theta \leq 2\pi\}. \end{aligned}$$

Splitting the integral accordingly, we get

$$\mathbb{I} = \left(\int_{K_1} + \int_{K_2} + \int_{K_3}\right) \operatorname{dist}((r,\theta), \Gamma)^{2-2\sigma} \left|\frac{\partial^2}{\partial x_1^2} f(r,\theta)\right|^2 r \,\mathrm{d}\theta\, \mathrm{d}r;$$

in order to show that $\mathbb{I}$ is infinite if $\sigma \geq 2/3$, it is enough to see that the integral over K_1 is infinite. Noting that in K_1 we have $\operatorname{dist}((r,\theta), \Gamma) \asymp r$, we get

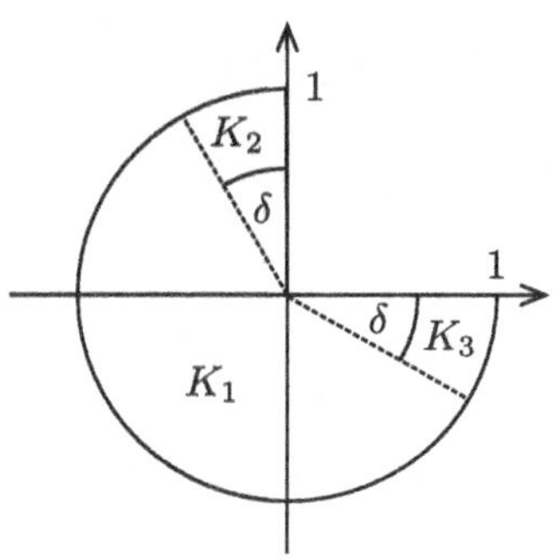

Fig. 3 The set $\widehat{\mathbb{L}}$ is split into three disjoint parts K_1, K_2, K_3

$$
\begin{aligned}
&\int_{K_1} \left| r^{1-\sigma} \tfrac{\partial^2}{\partial x_1^2} f(r,\theta) \right|^2 r \, \mathrm{d}\theta \, \mathrm{d}r \\
&= \int_{K_1} r \left| r^{1-\sigma} \frac{2}{9\pi r^{4/3}} \right|^2 \times \\
&\quad \times \left| \int_{\mathbb{R}} f_0' \left(r^{2/3} (\cos \Phi_\theta - v \sin \Phi_\theta) \right) \frac{K^\star(\theta, v) \sin\theta - K(\theta, v)\cos\theta}{1+v^2} \, \mathrm{d}v \right. \\
&\quad \left. + 2 r^{2/3} \int_{\mathbb{R}} f_0'' \left(r^{2/3} (\cos \Phi_\theta - v \sin \Phi_\theta) \right) \frac{K^2(\theta, v)}{1+v^2} \, \mathrm{d}v \right|^2 \mathrm{d}r \, \mathrm{d}\theta \\
&= \frac{4}{81\pi^2} \int_{K_1} r^{1/3-2\sigma} \left| \int_{\mathbb{R}} f_0' \left(r^{2/3} (\cos \Phi_\theta - v \sin \Phi_\theta) \right) \frac{K^{\star\star}(\theta, v)}{1+v^2} \, \mathrm{d}v \right. \\
&\quad \left. + 2 r^{2/3} \int_{\mathbb{R}} f_0'' \left(r^{2/3} (\cos \Phi_\theta - v \sin \Phi_\theta) \right) \frac{K^2(\theta, v)}{1+v^2} \, \mathrm{d}v \right|^2 \mathrm{d}r \, \mathrm{d}\theta \\
&= \frac{4}{81\pi^2} \int_{K_1} r^{1/3-2\sigma} \left| J(r^{2/3}, \theta) + I(r^{2/3}, \theta) \right|^2 \mathrm{d}r \, \mathrm{d}\theta \\
&= \frac{2}{27\pi^2} \int_{K_1} \rho^{1-3\sigma} \left| J(\rho, \theta) + I(\rho, \theta) \right|^2 \mathrm{d}\rho \, \mathrm{d}\theta,
\end{aligned}
\tag{25}
$$

where we use the following shorthand notation

$$
\begin{aligned}
K^{\star\star}(\theta, v) &:= K^\star(\theta, v) \sin\theta - K(\theta, v) \cos\theta = -v \sin(\theta - \omega_\theta) - \cos(\theta - \omega_\theta), \\
J(\rho, \theta) &:= \int_{\mathbb{R}} f_0' \left(\rho (\cos \Phi_\theta - v \sin \Phi_\theta) \right) \frac{K^{\star\star}(\theta, v)}{1+v^2} \, \mathrm{d}v, \\
I(\rho, \theta) &:= 2\rho \int_{\mathbb{R}} f_0'' \left(\rho (\cos \Phi_\theta - v \sin \Phi_\theta) \right) \frac{K^2(\theta, v)}{1+v^2} \, \mathrm{d}v.
\end{aligned}
$$

Observe that $\theta - \omega_\theta \in (0, 2\pi)$ for $\theta \in (\frac{\pi}{2}, 2\pi)$, and $\theta - \omega_\theta \in (\frac{4\delta}{3}, 2\pi - \frac{4\delta}{3})$ whenever $\theta \in (\frac{\pi}{2} + \delta, 2\pi - \delta)$.

Without loss of generality we may assume that $J(\rho, \theta) + I(\rho, \theta) \not\equiv 0$ on K_1. Let us show that $\lim_{\rho \to 0} |J(\rho, \theta) + I(\rho, \theta)| = C(f_0, \theta) > 0$. This guarantees that we can choose some $K_{11} \subset K_1$ such that

$$
|J(\rho, \theta) + I(\rho, \theta)| \geq C(f_0) > 0 \quad \text{on } K_{11}. \tag{26}
$$

Using the change of variables $x = v\rho$ we get, using dominated convergence,

$$I(\rho,\theta) = 2\int_{\mathbb{R}} f_0''(\rho\cos\Phi_\theta - x\sin\Phi_\theta)\,\frac{(\rho\cos\omega_\theta - x\sin\omega_\theta)^2}{\rho^2+x^2}\,\mathrm{d}x$$

$$\underset{\rho\to 0}{\longrightarrow} 2\sin^2\omega_\theta\int_{\mathbb{R}} f_0''(-x\sin\Phi_\theta)\,\mathrm{d}x = \frac{2\sin^2\omega_\theta}{\sin\Phi_\theta}\int_{\mathbb{R}} f_0''(x)\,\mathrm{d}x = 0,$$

since we assume that $f_0 \in W_1^2(\mathbb{R})$.

For $J(\rho,\theta)$ we have, using the same change of variables,

$$J(\rho,\theta) = -\int_{\mathbb{R}} f_0'(\rho\cos\Phi_\theta - \rho v\sin\Phi_\theta)\,\frac{\cos(\theta-\omega_\theta) + v\sin(\theta-\omega_\theta)}{1+v^2}\,\mathrm{d}v$$

$$= -\int_{\mathbb{R}} f_0'(\rho\cos\Phi_\theta - x\sin\Phi_\theta)\,\frac{\rho\cos(\theta-\omega_\theta) + x\sin(\theta-\omega_\theta)}{\rho^2+x^2}\,\mathrm{d}x$$

$$= -\left(\int_{|x|>\varepsilon} + \int_{|x|\le\varepsilon}\right)(\ldots)\,\mathrm{d}x.$$

The first integral can be treated with the dominated convergence theorem because we have $f_0' \in L_1(\mathbb{R})$ and $\rho(\rho^2+x^2)^{-1} \le \rho x^{-2}$, $x(\rho^2+x^2)^{-1} \le x^{-1}$ are bounded for $|x| > \varepsilon$. Therefore,

$$\lim_{\rho\to 0}\left[-\int_{|x|>\varepsilon}(\ldots)\,\mathrm{d}x\right] = -\sin(\theta-\omega_\theta)\int_{|x|>\varepsilon}\frac{f_0'(-x\sin\Phi_\theta)}{x}\,\mathrm{d}x.$$

Now we estimate the two parts of the second integral. Note that the integral

$$-\int_{|x|\le\varepsilon} f_0'(\rho\cos\Phi_\theta - x\sin\Phi_\theta)\frac{\rho\cos(\theta-\omega_\theta)}{\rho^2+x^2}\,\mathrm{d}x$$

tends to $-\pi\cos(\theta-\omega_\theta)f_0'(0)$ as $\rho\to 0$ since f_0' is continuous by Sobolev embedding. For the second term in this integral we have using a change of variables and the Cauchy–Schwarz inequality,

$$|\sin(\theta-\omega_\theta)|\cdot\left|\int_{|x|\le\varepsilon} f_0'(\rho\cos\Phi_\theta - x\sin\Phi_\theta)\,\frac{x}{x^2+\rho^2}dx\right|$$

$$\le\left|\int_{|w|\le\varepsilon}\left(f_0'(\rho\cos\Phi_\theta - w\sin\Phi_\theta) - f_0'(\rho\cos\Phi_\theta)\right)\frac{w}{\rho^2+w^2}\,\mathrm{d}w\right|$$

$$\le\int_{|w|\le\varepsilon}\int_0^1 |f_0''(\rho\cos\Phi_\theta - rw\sin\Phi_\theta)|\,\mathrm{d}r\,\mathrm{d}w$$

$$
\begin{aligned}
&\le \sqrt{2\varepsilon}\int_0^1\left(\int_{\mathbb{R}}|f_0''(\rho\cos\Phi_\theta - v\sin\Phi_\theta)|^2\,\mathrm{d}v\right)^{1/2}\mathrm{d}r\\
&\le C_1(\theta)\sqrt{\varepsilon}\,\|f_0''\|_2.
\end{aligned}
$$

Altogether we have upon letting $\rho\to 0$ and then $\varepsilon\to 0$, that

$$
\lim_{\rho\to 0} I(\rho,\theta) = 0, \tag{27}
$$

$$
\liminf_{\varepsilon\to 0}\lim_{\rho\to 0} J(\rho,\theta) = -\pi\cos(\theta-\omega_\theta)f_0'(0) + \sin(\omega_\theta-\theta)\liminf_{\varepsilon\to 0}\int_{|x|>\varepsilon}\frac{f_0'(x)}{x}\,\mathrm{d}x. \tag{28}
$$

If the "lim inf" diverges, it is clear that (26) holds, if it converges but is still not equal to 0, we can choose K_{11} in such a way that $\sin(\omega_\theta-\theta)\neq 0$, e.g. taking θ in an open set around $9\pi/8$ the integral over K_1 blows up as $\int_0^1\rho^{1-3\sigma}\,\mathrm{d}\rho=\infty$ for any $\sigma\ge 2/3$.

To show the convergence result, we have to estimate I and J from above. Write

$$
\begin{aligned}
J(\rho,\theta) &= -\int_{\mathbb{R}} f_0'(\rho(\cos\Phi_\theta - v\sin\Phi_\theta))\,\frac{v\sin(\theta-\omega_\theta)}{1+v^2}\,\mathrm{d}v\\
&\quad-\int_{\mathbb{R}} f_0'(\rho(\cos\Phi_\theta - v\sin\Phi_\theta))\,\frac{\cos(\theta-\omega_\theta)}{1+v^2}\,\mathrm{d}v =: J_1(\rho,\theta)+J_2(\rho,\theta).
\end{aligned}
$$

Since $f_0\in W_p^1(\mathbb{R})$, the Hölder inequality and a change of variables give

$$
\begin{aligned}
|J_1(\rho,\theta)| &\le \left(\int_{\mathbb{R}}\left|f_0'(\rho(\cos\Phi_\theta - v\sin\Phi_\theta))\right|^p\mathrm{d}v\right)^{\frac{1}{p}}\left(\int_{\mathbb{R}}\left(\frac{v}{1+v^2}\right)^q\mathrm{d}v\right)^{\frac{1}{q}}\\
&\le c|\rho\sin\Phi_\theta|^{-1/p}
\end{aligned} \tag{29}
$$

for all $\theta\in[\pi/2,2\pi]$ and $\rho>0$. An even simpler calculation yields

$$
|J_2(\rho,\theta)|\le c|\rho\sin\Phi_\theta|^{-1/p} \tag{30}
$$

for all $\theta\in[\pi/2,2\pi]$ and $\rho>0$. Now we estimate $I(\rho,\theta)$. Note that for every $\theta\in[\pi/2,2\pi]$ we have $K^2(\theta,v)/(1+v^2)\le C$. By a change of variables we get

$$
|I(\rho,\theta)|\le\frac{C_1}{|\sin\Phi_\theta|}\int_{\mathbb{R}}\left|f_0''(w+\rho\cos\Phi_\theta)\right|\mathrm{d}w\le\frac{C_2}{|\sin\Phi_\theta|} \tag{31}
$$

for all $\theta\in[\pi/2,2\pi]$ and $\rho>0$. Note that for $\Phi_\theta\in[\pi+2\delta/3,2\pi-2\delta/3]$ it holds that $|\sin\Phi_\theta|>0$. Thus, on K_1 we have

$$
|I(\rho,\theta)+J(\rho,\theta)|\le C\rho^{-1/p},\quad \theta\in[\pi/2+\delta,2\pi-\delta],\ \rho>0, \tag{32}
$$

implying

$$\int_{K_1} \left| r^{1-\sigma} \frac{\partial^2}{\partial x_1^2} f(r,\theta) \right|^2 r \, \mathrm{d}\theta \, \mathrm{d}r \leq C \int_0^1 \rho^{1-3\sigma-2/p} \, \mathrm{d}\rho.$$

The last integral converges if $\sigma \in (0, 2/3)$ and $p > \frac{2}{2-3\sigma}$.

In order to complete the proof of the convergence part, let us show that the integrals over K_2 and K_3 are convergent for all $\sigma \in (0, 1)$.

In K_2 and K_3 we have $\operatorname{dist}((r,\theta), \Gamma) \leq r|\cos\theta|$ and $\operatorname{dist}((r,\theta), \Gamma) \leq r|\sin\theta|$, respectively. We will discuss only K_2 since K_3 can be treated in a similar way. We need to show that

$$\int_{K_2} \left| |r\cos\theta|^{1-\sigma} \frac{\partial^2}{\partial x_1^2} f(r,\theta) \right|^2 r \, \mathrm{d}r \, \mathrm{d}\theta < \infty \quad \text{for all } \sigma \in (0,1). \tag{33}$$

From (29)–(31) we derive that for all $(\rho, \theta) \in \widehat{\mathbb{L}}$

$$|J(\rho,\theta) + I(\rho,\theta)| \leq C\rho^{-\frac{1}{p}} \left(|\sin \Phi_\theta|^{-1} + |\sin \Phi_\theta|^{-\frac{1}{p}} \right) \leq C'\rho^{-\frac{1}{p}} |\sin \Phi_\theta|^{-1}. \tag{34}$$

Now we can use a calculation similar to (25) for K_1 to show that (33) is finite and, therefore, it is enough to show that

$$\int_{\frac{\pi}{2}}^{\frac{\pi}{2}+\delta} \left(\frac{|\cos\theta|^{1-\sigma}}{\sin \Phi_\theta} \right)^2 \mathrm{d}\theta < \infty. \tag{35}$$

Observe that $\lim_{\theta \to \frac{\pi}{2}} \cos\frac{1}{3}(\pi+\theta)/\cos\theta = \frac{1}{3}$, implying

$$\frac{|\cos\theta|^{1-\sigma}}{\sin \Phi_\theta} = \frac{|\cos\theta|^{1-\sigma}}{2\sin\frac{1}{3}(\pi+\theta)\cos\frac{1}{3}(\pi+\theta)} \asymp |\cos\theta|^{-\sigma} \quad \text{as } \theta \to \frac{\pi}{2}.$$

Therefore, it is sufficient to note that for any $\sigma \in (0, 1)$

$$\int_{\pi/2}^{\pi/2+\delta} |\cos\theta|^{-2\sigma} \, \mathrm{d}\theta \asymp \int_0^1 \frac{\mathrm{d}x}{(1-x^2)^\sigma} = \int_0^1 \frac{\mathrm{d}x}{(1-x)^\sigma (1+x)^\sigma} < \infty.$$

Summing up, we have shown that

$$\operatorname{dist}(0,\cdot)^{1-\sigma} \left| \frac{\partial^2}{\partial x_1^2} f \right| \in L_2(\widehat{\mathbb{L}}) \quad \text{resp.} \quad \notin L_2(\widehat{\mathbb{L}}),$$

according to $\sigma \in (0, 2/3)$ or $\sigma \in [2/3, 1)$.

Proof (*Proof of Theorem* 2) Let F be the solution to (3) on $\mathbb{L}$ with source function g, and define $w = g * N$ for the Newtonian potential N on $\mathbb{R}^2$. As we have already mentioned in the introduction, $f := w - F$ is the solution to (1) on $\mathbb{L}$ with the boundary condition $h := \operatorname{Tr} w$ on $\partial\mathbb{L}$. Under the condition $g \in H_1(\mathbb{R}^2) \cap W_2^1(\mathbb{L})$ we have $\Delta w = g$ (cf. Stein [17, Theorem III.3.3, p. 114]), which implies $w \in W_1^3(\mathbb{R}^2) \cap W_2^3(\mathbb{L})$. By the trace theorem we have $h \in W_1^2(\partial\mathbb{L}) \cap W_2^{5/2}(\partial\mathbb{L})$, which in terms of f_0 means $f_0 \in W_1^2(\mathbb{R}) \cap W_2^{5/2}(\mathbb{R})$. The explosion result of Theorem 1 requires $f_0 \in W_1^2(\mathbb{R}) \cap W_2^2(\mathbb{R})$ and (17). The latter is guaranteed by the assumption on the trace in the statement of the theorem. Hence, $f \notin W_{2,\text{loc}}^{1+\sigma}(\mathbb{L})$, $\sigma \geq 2/3$. Since $w \in W_{2,\text{loc}}^2(\mathbb{L})$, this implies that $F \notin W_{2,\text{loc}}^{1+\sigma}(\mathbb{L})$, $\sigma \geq 2/3$. □

Acknowledgements We thank S. Dahlke (Marburg) who pointed out the reference [11], N. Jacob (Swansea) for his suggestions on the representation of Sobolev–Slobodetskij spaces, and A. Bendikov (Wrocław) who told us about the papers [13, 14]. We are grateful to B. Böttcher for drawing the illustrations and commenting on the first draft of this paper. Financial support from NCN grant 2014/14/M/ST1/00600 (Wrocław) for V. Knopova is gratefully acknowledged.

References

1. Bass, R.: Probabilistic Techniques in Analysis. Springer, New York (1995)
2. Cioica, P., Dahlke, S., Kinzel, S., Lindner, F., Raasch, T., Ritter, K., Schilling, R.L.: Spatial Besov regularity for stochastic partial differential equations on Lipschitz domains. Studia Mathematica **207**, 197–234 (2011)
3. Cioica, P., Dahlke, S., Döhring, N., Friedrich, U., Kinzel, S., Lindner, F., Raasch, T., Ritter, K., Schilling, R.L.: On the convergence analysis of spatially adaptive Rothe methods. Found. Comput. Math. **14**, 863–912 (2014)
4. Cohen, A.: Numerical Analysis of Wavelet Methods. Elsevier, Amsterdam (2003)
5. Cohen, A., Dahmen, W., DeVore, R.A.: Adaptive wavelet methods for elliptic operator equations: convergence rates. Math. Comput. **70**, 27–75 (2001)
6. Dahlke, S., Dahmen, W., DeVore, R.A.: Nonlinear approximation and adaptive techniques for solving elliptic operator equations. Wavelet Anal. Appl. **6**, 237–283 (1997)
7. Dahlke, S., DeVore, R.A.: Besov regularity for elliptic boundary value problems. Commun. Partial Differ. Equ. **22**, 1–16 (1997)
8. DeVore, R.A.: Nonlinear approximation. Acta Numerica **7**, 51–150 (1998)
9. Gilbarg, D., Trudinger, N.S.: Elliptic Partial Differential Equations of Second Order. Springer, Berlin (1998)
10. Grisvard, P.: Elliptic Problems in Nonsmooth Domains. Pitman, Boston (1985)
11. Jerison, D., Kenig, C.E.: The inhomogeneous Dirichlet problem in Lipschitz domains. J. Funct. Anal. **130**, 161–219 (1995)
12. Jonsson, A., Wallin, H.: Function Spaces on Subsets of $\mathbb{R}^n$. Harwood Academic, New York (1984)
13. Mitrea, D., Mitrea, I.: On the Besov regularity of conformal maps and layer potentials on nonsmooth domains. J. Funct. Anal. **201**, 380–429 (2003)
14. Mitrea, D., Mitrea, M., Yan, L.: Boundary value problems for the Laplacian in convex and semiconvex domains. J. Funct. Anal. **258**, 2507–2585 (2010)
15. Mörters, P., Peres, Y.: Brownian Motion. Cambridge University Press, Cambridge (2010)
16. Port, S., Stone, C.J.: Brownian Motion and Classical Potential Theory. Academic Press, New York (1978)
17. Stein, E.M.: Harmonic Analysis. Real-Variable Methods, Orthogonality, and Oscillatory Integrals. Princeton University Press, Princeton (NJ) (1993)
18. Triebel, H.: Theory of Function Spaces. Birkhäuser, Basel (1983)

Symmetric Markov Processes with Tightness Property

Masayoshi Takeda

Abstract A symmetric Markov process X is said to be in **Class (T)** if it is irreducible, strong Feller and possesses a tightness property. We give some properties of X in Class (T) and of the semi-group p_t of X: the uniform large deviation principle of X, L^p-independence of growth bounds of p_t, compactness of p_t as an operator in L^2, and boundedness of every eigenfunction of p_t.

Keywords Symmetric Markov process · Dirichlet form · Compactness of semigroup

2010 Mathematics Subject Classification 60J45 · 60G52 · 31C25

1 Introduction

Let E be a locally compact separable metric space and m a positive Radon measure with $\mathrm{supp}[m] = E$. Let $X = (\Omega, X_t, \mathbf{P}_x, \zeta)$ be an m-symmetric Markov process on E, $(p_t f, g)_m = (f, p_t g)_m$, $f, g \in \mathcal{B}_b(E)$. Here ζ is the life time of X and $\{p_t\}_{t\geq 0}$ is the semi-group of X, $p_t f(x) = \mathbf{E}_x(f(X_t))$, $f \in \mathcal{B}_b(E)$. We denote by $\left\{R_\beta\right\}_{\beta\geq 0}$ the resolvent of X, $R_\beta f(x) = \int_0^\infty e^{-\beta t} p_t f(x)dt$, $f \in \mathcal{B}_b(E)$.

The symmetric Markov process is said to be in **Class (T)** if the next three properties hold:

(I) (**Irreducibility**) If a Borel set A is p_t-invariant, i.e., $\int_A p_t 1_{A^c} dm = 0$ for any $t > 0$, then $m(A) = 0$ or $m(A^c) = 0$. Here 1_{A^c} is the indicator function of the complement of the set A.

(II) (**Strong Feller Property**) $p_t(\mathcal{B}_b(E)) \subset C_b(E)$.

(III) (**Tightness**) For any $\varepsilon > 0$, there exists a compact set $K \subset E$ s.t.

M. Takeda (✉)
Tohoku University, Sendai, Japan
e-mail: takeda@math.tohoku.ac.jp

A. Eberle et al. (eds.), *Stochastic Partial Differential Equations and Related Fields*, Springer Proceedings in Mathematics & Statistics 229,
https://doi.org/10.1007/978-3-319-74929-7_33

$$\sup_{x\in E} R_1 1_{K^c}(x) = \sup_{x\in E} \mathbf{E}_x \left(\int_0^\zeta 1_{K^c}(X_t)dt \right) \le \varepsilon. \tag{1}$$

Equation (1) tells us that the two cases occur; X stays on a compact set K for almost all time or explodes fast. More precisely, if X is conservative, $\mathbf{P}_x(\zeta = \infty) = 1$, then the tightness property implies the positive recurrence of X; the measure m turns out to be finite. Moreover, we prove that if, in addition, there exists an increasing sequence $\{K_n\}_{n=1}^\infty$ of compact sets such that the union of $\{K_n\}_{n=1}^\infty$ equals E and each part (or absorbing) process X^{D_n} on D_n ($D_n := K_n^c$) is irreducible, then X possesses the following strong recurrence property: for any positive constant γ, there exists a compact set $K \subset E$ such that

$$\sup_{x\in E} \mathbf{E}_x(\exp(\gamma\sigma_K)) < \infty,$$

where σ_K is the first hitting time of K, $\sigma_K = \inf\{t > 0 : X_t \in K\}$ ([14]). Wu [18] calls this property *uniform hyper-exponential recurrence*, and proves that under this property the uniform large deviation principle holds (Theorem 1). As an example, a one-dimensional diffusion process satisfies the uniform hyper-exponential recurrence and thus the uniform large deviation principle, if both boundaries are *entrance* in Feller's classification of the boundaries.

On the other hand, we see that if X is not conservative, Tightness (III) implies that the lifetime ζ is exponentially integrable: for some $\gamma > 0$

$$\sup_{x\in E} \mathbf{E}_x(\exp(\gamma\zeta)) < \infty.$$

In particular, $\mathbf{P}_x(\zeta = \infty) = 0$ ([14]).

The objective of this note is to survey spectral properties of symmetric Markov processes in Class (T).

2 Remarks on Tightness (III)

We have:

(i) If $R_1 1 \in C_\infty(E)$, then Tightness (III) holds. Indeed, by the maximum principle

$$\sup_{x\in E} R_1 1_{K^c}(x) = \sup_{x\in K^c} R_1 1_{K^c}(x) \le \sup_{x\in K^c} R_1 1(x).$$

(ii) If $m(X) < \infty$ and R_1 is ultracontractive, i.e. $\|R_1\|_{p,\infty} < \infty,\ 1 \le p < \infty$, then

$$\sup_{x\in E} R_1 1_{K^c}(x) \le \|R_1\|_{p,\infty} \cdot m(K^c)^{1/p}$$

and Tightness (III) holds. Here $\| \cdot \|_{p,\infty}$ is the operator norm from $L^p(E; m)$ to $L^\infty(E; m)$.

(iii) If $C_\infty(E)$ is invariant under R_1, i.e. $R_1(C_\infty(E)) \subset C_\infty(E)$, then

$$\text{(III)} \iff R_1 1 \in C_\infty(E).$$

In fact, for a compact set K, take a positive function $g \in C_\infty(E)$ such that $1_K \le g$. We then see from the invariance of $C_\infty(E)$ that $0 \le \lim_{x\to\infty} R_1 1_K(x) \le \lim_{x\to\infty} R_1 g(x) = 0$ and from Tightness (III) that for any $\varepsilon > 0$ there exists a compact set K such that

$$\limsup_{x\to\infty} R_1 1(x) \le \limsup_{x\to\infty} R_1 1_K(x) + \limsup_{x\to\infty} R_1 1_{K^c}(x) \le \sup_{x\in E} R_1 1_{K^c}(x) \le \varepsilon,$$

which implies $R_1 1 \in C_\infty(E)$. Hence, if $C_\infty(E)$ is invariant under R_1 and X is conservative, $p_t 1 = 1$, then X does not have the tightness property, for example, the Ornstein-Uhlenbeck process does not have the tightness property.

(iv) If X is conservative, then Tightness (III) holds if and only if $\{R_1(x, \bullet)\}_{x\in X} (\subset \mathscr{P}(E))$ is tight in the sense of Prohorov. Here $\mathscr{P}(E)$ be the set of probability measures on E equipped with the weak topology.

(v) Let us consider a one-dimensional diffusion process on an open interval $I = (r_1, r_2)$ such that $\mathbf{P}_x(X_{\zeta-} = r_1 \text{ or } r_2,\ \zeta < \infty) = \mathbf{P}_x(\zeta < \infty),\ x \in I$, and $\mathbf{P}_a(\sigma_b < \infty) > 0$ for any $a, b \in I$. The boundary point r_i of I is classified into four classes: *regular, exit, entrance, and natural boundary* ([[6], Chap. 5]):

(a) If r_2 is a regular or exit boundary, then $\lim_{x\to r_2} R_1 1(x) = 0$.
(b) If r_2 is an entrance boundary, then $\lim_{r\to r_2} \sup_{x\in I} R_1 1_{(r,r_2)}(x) = 0$.
(c) If r_2 is a natural boundary, then $\lim_{x\to r_2} R_1 1_{(r,r_2)}(x) = 1$ and thus $\sup_{x\in(r_1,r_2)} R_1 1_{(r,r_2)}(x) = 1$.
Therefore, Tightness (III) is fulfilled if and only if no natural boundaries are present.

3 Uniform Large Deviations

The semi-group $\{p_t\}_{t\ge 0}$ of Sect. 1 can be uniquely extended to a strongly continuous Markovian semi-group $\{T_t\}_{t\ge 0}$ on $L^2(E; m)$. Let $(\mathscr{E}, \mathscr{F})$ be the *Dirichlet form* generated by $\{T_t\}_{t\ge 0}$:

$$\mathscr{E}(u, v) = \lim_{t\to 0} \frac{1}{t}(u - T_t u, v)_m (= (-\mathscr{L} u, v)_m),$$

$$\mathscr{F} = \left\{ u \in L^2(E; m) : \lim_{t\to 0} \frac{1}{t}(u - T_t u, u)_m < \infty \right\}.$$

We define the function $I_{\mathscr{E}}$ on $\mathscr{P}(E)$ by

$$I_{\mathscr{E}}(\nu) = \begin{cases} \mathcal{E}(\sqrt{f}, \sqrt{f}) & \text{if } \nu = f \cdot m, \ \sqrt{f} \in \mathscr{F} \\ \infty & \text{otherwise.} \end{cases}$$

Given $\omega \in \Omega$ with $0 < t < \zeta(\omega)$,

$$L_t(\omega)(A) = \frac{1}{t} \int_0^t 1_A(X_s(\omega))ds \in \mathscr{P}(E)$$

for a Borel set A of E.

Theorem 1 ([14]) *Assume that X is in Class (T).*

(i) For each open set $G \subset \mathscr{P}(E)$

$$\liminf_{t\to\infty} \frac{1}{t} \log \mathbf{P}_x (L_t \in G, t < \zeta) \geq - \inf_{\nu\in G} I_{\mathscr{E}}(\nu). \tag{2}$$

Moreover, if X is conservative, i.e. $p_t 1 = 1$, then

$$\liminf_{t\to\infty} \frac{1}{t} \log \inf_{x\in E} \mathbf{P}_x (L_t \in G) \geq - \inf_{\nu\in G} I_{\mathscr{E}}(\nu). \tag{3}$$

(ii) For each closed set $K \subset \mathscr{P}(E)$

$$\limsup_{t\to\infty} \frac{1}{t} \log \sup_{x\in E} \mathbf{P}_x (L_t \in K, t < \zeta) \leq - \inf_{\nu\in K} I_{\mathscr{E}}(\nu). \tag{4}$$

As remarked in Sect. 2 (iii), the Ornstein-Uhlenbeck process does not possess Tightness (III). Moreover, it is known from [18] that the Ornstein-Uhlenbeck process does not satisfy the uniform large deviation principle, while it satisfies the locally uniform large deviation principle: for any compact set $K \subset E$

$$\liminf_{t\to\infty} \frac{1}{t} \log \inf_{x\in K} \mathbf{P}_x (L_t \in G) \geq - \inf_{\nu\in G} I_{\mathscr{E}}(\nu). \tag{5}$$

It is known from Deuschel and Stroock [3, Theorem 4.2.58, Exercise 4.2.64] that if X is *uniformly ergodic* (cf. Assumption ($\widetilde{\text{U}}$) in [3, pp. 103]), then (3) holds. It is known from Wu [18] that if X satisfies

(H) for any $\gamma > 0$ there exists a compact set K such that $\sup_{x\in E} \mathbf{E}_x(\exp(\gamma\sigma_K)) < \infty$,

then (3) holds. Wu calls the property (H) *uniform hyper-exponential recurrence.*

Let $K \subset E$ be a compact set and denote by D the complement of K, $D = K^c$. Let X^D be the part (absorbing) process on D;

$$X^D = \begin{cases} X_t & t < \tau_D \\ \Delta & t \geq \tau_D, \quad \tau_D = \inf\{t \geq 0 \mid X_t \notin D\}. \end{cases}$$

Let $(\mathscr{E}^D, \mathscr{F}^D)$ be the Dirichlet form generated by X^D and λ_2^D the bottom of the spectrum of X^D:

$$\lambda_2^D = \inf\left\{\mathscr{E}(u,u) : u \in \mathscr{F}^D, \int_D u^2 dm = 1\right\}.$$

We introduce the following condition on X.

(D) there exists a increasing sequence $\{K_n\}_{n=1}^{\infty}$ of compact sets such that $\cup_{n=1}^{\infty} K_n = E$ and X^{D_n}, $D_n = K_n^c$, is irreducible.

Lemma 1 *If X in Class (T) is conservative and satisfies the condition (D), then it has the uniform hyper-exponential recurrence property.*

For the proof of Lemma 1, the next two facts for a conservative symmetric Markov process in Class (T) is crucial ([14]):

1. $\lim_{n\to\infty} \lambda_2^{D_n} = \infty$, for $D_n = K_n^c$ with $K_n \uparrow E$.
2. $\sup_{x\in E} \mathbf{E}_x\left(\exp(\gamma\sigma_{K_n})\right) < \infty$ if and only if $\gamma < \lambda_2^{D_n}$. Here σ_{K_n} is the first hitting time of K_n.

The semi-group $\{p_t\}_{t\geq 0}$ can be extended to a semi-group $\{T_t^p\}_{t\geq 0}$ on $L^p(E;m)$, $1 \leq p \leq \infty$.

Definition 1 (Growth bounds)

$$\lambda_p := -\lim_{t\to\infty} \frac{1}{t} \log \|T_t^p\|_{p,p}, \quad 1 \leq p \leq \infty, \tag{6}$$

where $\|\cdot\|_{p,p}$ is the operator norm from $L^p(E;m)$ to $L^p(E;m)$.

We see that λ_2 equals $\inf\{\mathscr{E}(u,u) \mid u \in \mathscr{F}, \|u\|_2 = 1\}$ and $\lambda_\infty \leq \lambda_p \leq \lambda_2$.

Theorem 2 ([13]) *If X is in Class (T), then λ_p is independent of p, i.e.*

$$\lambda_p = \lambda_2, \quad 1 \leq p \leq \infty.$$

Proof By applying the large deviation principle with $G = K = \mathscr{P}(E)$,

$$\begin{aligned} -\lambda_2 &= -\inf_{\nu\in\mathscr{P}(E)} I_{\mathscr{E}}(\nu) \leq \liminf_{t\to\infty} \frac{1}{t} \log \mathbf{P}_x(t < \zeta) \\ &\leq \limsup_{t\to\infty} \frac{1}{t} \log \sup_{x\in E} \mathbf{P}_x(t < \zeta) = \limsup_{t\to\infty} \frac{1}{t} \log \|p_t\|_{\infty,\infty} = -\lambda_\infty \\ &\leq -\inf_{\nu\in\mathscr{P}(E)} I_{\mathscr{E}}(\nu) = -\lambda_2. \end{aligned}$$

4 Compactness of Symmetric Markov Semigroups

A subset $\mathscr{A} \subset L^1(E;m)$ is said to be *equi-integrable* if

(a) (**tightness**) for any $\varepsilon > 0$ there exist $F \subset \mathcal{B}(E)$ such that $m(F) < \infty$ and $\sup_{v\in\mathscr{A}} \int_{F^c} |v| dm < \varepsilon$.
(b) (**uniform integrability**)

$$\lim_{L\to\infty} \sup_{v\in\mathscr{A}} \int_{\{|v|\geq L\}} |v| dm = 0.$$

For other equivalent definitions of equi-integrablility, see [9, Theorem 16.8]).

Define

$$\mathscr{F}_M = \{u \in \mathscr{F} \mid \mathscr{E}_1(u,u) \leq M\}, \quad M > 0.$$

Lemma 2 *Let* $\mathscr{A}=\{u^2 \mid u \in \mathscr{F}_M\}$*. If* X *is in Class (T), then* $\mathscr{A} \subset L^1(E;m)$ *is equi-integrable.*

Proof For the proof of Lemma 2, the next two facts are used.
• (Stollman-Voigt [12])

$$\int_E u^2 d\mu \leq \|R_1\mu\|_\infty \cdot \mathscr{E}_1(u,u), \; u \in \mathscr{F}. \tag{7}$$

• (Z. -Q. Chen [1]) If X is in Class (T), then $m \in \mathscr{K}_\infty(R_1)$, that is, for any $\varepsilon > 0$ there exists a compact set K and $\delta > 0$ such that for any $B \subset K$ with $m(B) \leq \delta$

$$\|R_1 1_{B\cup K^c}\|_\infty (= \|R_1(1_{B\cup K^c}\cdot m)\|_\infty) < \varepsilon. \tag{8}$$

The tightness (a) follows from (7) because for any $\varepsilon > 0$, there exists a compact set K such that

$$\sup_{u\in\mathscr{F}_M} \int_{K^c} u^2 dm \leq \|R_1 1_{K^c}\|_\infty \cdot M < \varepsilon.$$

Note that $m(\{u^2 \geq L\}) \leq \|u\|_2^2/L \leq M/L \longrightarrow 0, L \to \infty$. Since for $u \in \mathscr{F}_M$ and a compact set $K(\subset E)$

$$\begin{aligned}\int_{\{u^2\geq L\}} u^2 dm &= \int_{\{u^2\geq L\}\cap K} u^2 dm + \int_{\{u^2\geq L\}\cap K^c} u^2 dm \\ &\leq \left(\|R_1 1_{\{u^2\geq L\}\cap K}\|_\infty + \|R_1 1_{K^c}\|_\infty\right)\cdot M.\end{aligned}$$

Tightness (III) and (8) lead us to the uniform integrability of $\mathscr{A}$.

We say that a sequence $\{u_n\}_{n\in\mathbf{N}}$ converges to u *in* m*-measure* if for any $A \in \mathcal{B}(E)$ with $m(A) < \infty$

$$m(\{|u_n - u| > \varepsilon\} \cap A) \to 0 \text{ as } n \to \infty.$$

Let $\{u_n\}_{n\in\mathbf{N}}$ be a sequence in $L^2(E;m)$. We know from Vitali's Theorem (cf. [9, Theorem 16.6]) that $\{u_n\}_{n=1}^\infty$ strongly converges to $u \in L^2(E;m)$ if and only if $\{u_n^2\}_{n=1}^\infty$ is equi-integrable and $u_n \to u$ in m-measure. Hence, we have

Lemma 3 *Suppose X is in Class (T). If $\{u_n\}_{n=1}^\infty \subset \mathscr{F}_M$ converges to u in m-measure, then u is in $L^2(E;m)$ and it converges to u in $L^2(E;m)$.*

Lemma 4 *For $t > 0$ and $f \in L^2(E;m)$*

$$\mathscr{E}(p_t f, p_t f) \le \frac{1}{2et}(f, f)_m.$$

Proof By the spectral theorem

$$\mathscr{E}(p_t f, p_t f) = (-Ae^{tA} f, e^{tA} f)_m = \int_0^\infty \lambda e^{-2t\lambda} d(E_\lambda f, f)_m, \quad \lambda e^{2t\lambda} \le \frac{1}{2et}.$$

Lemma 5 *Let $\{g_n\}_{n=1}^\infty$ be a sequence in $L^2(E;m)$ satisfying that $\|g_n\|_2 + \|g_n\|_\infty \le M < \infty$. Then there exists a subsequence of $\{p_t g_n\}_{n=1}^\infty$ that converges to an $L^2(E;m)$-function.*

Proof Since $\sup_n \|g_n\|_\infty < \infty$, there exists a subsequence $\{g_k\}_{k=1}^\infty$ of $\{g_n\}_{n=1}^\infty$ such that g_k converges to some g in the weak-$\star$ topology $\rho(L^\infty, L^1)$ (Banach-Alaoglu's Theorem). Hence, for any $x \in E$

$$\begin{aligned} p_t g_k(x) &= \int_E p_t(x,y) g_k(y) dm(y) \\ &\longrightarrow \int_E p_t(x,y) g(y) dm(y) = p_t g(x), \quad \forall x \in E. \end{aligned}$$

By Lemma 4

$$\sup_k \mathscr{E}_1(p_t g_k, p_t g_k) \le \left(\frac{1}{2et} + 1\right) \sup_k \|g_k\|_2^2 < \infty.$$

Thus $\{p_t g_k\}_{k=1}^\infty$ converges to $p_t g$ in $L^2(E;m)$ by Lemma 3.

Theorem 3 *For X in Class (T), the operator p_t is compact on $L^2(E;m)$.*

Proof Define

$$g^L(x) = g(x) 1_{\{|g(x)| \le L\}}, \quad L > 0.$$

For $\{g_n\}_{n=1}^\infty \subset L^2(E;m)$ with $\|g_n\|_2 \le 1$, put

$$h_n^L = (p_{t/2} g_n)^L, \quad f_n^L = p_{t/2} h_n^L.$$

Since $\{(p_{t/2}g_n)^2\}$ is equi-integrable,

$$\begin{aligned}\sup_n \|p_t g_n - f_n^L\|_2 &= \sup_n \|p_t g_n - p_{t/2}((p_{t/2}g_n)^L)\|_2 \\ &\le \sup_n \|p_{t/2}g_n - (p_{t/2}g_n)^L\|_2 \longrightarrow 0, \quad L \to \infty.\end{aligned}$$

Hence, for any $\varepsilon > 0$, there exists L such that

$$\sup_n \|p_t g_n - f_n^L\|_2 \le \varepsilon.$$

Since $\sup_n(\|h_n^L\|_2 + \|h_n^L\|_\infty) \le 1 + L$, $\{f_n^L\}_{n=1}^\infty$ has an $L^2(E; m)$-convergent subsequence $\{f_{k(n)}^L\}_{n=1}^\infty$, by Lemma 5, and so

$$\|f_{k(n)}^L - f_{k(m)}^L\|_2 \le \varepsilon, \quad \forall n, m \ge \exists N.$$

Therefore, for $n, m \ge N$

$$\|p_t g_{k(n)} - p_t g_{k(m)}\|_2 \le |p_t g_{k(n)} - f_{k(n)}^L\|_2 + \|f_{k(n)}^L - f_{k(m)}^L\|_2 + \|f_{k(m)}^L - p_t g_{k(m)}\|_2 \le 3\varepsilon.$$

We proved in [15] that if X is in Class (T), then the Dirichlet form $(\mathcal{E}, \mathcal{F})$ generated by X has the ground state φ_0: $\varphi_0 \in \mathcal{F}$, $\|\varphi_0\|_2 = 1$ and $\mathcal{E}(\varphi_0, \varphi_0) = \inf\{\mathcal{E}(u, u) \mid u \in \mathcal{F}, \|u\|_2 = 1\}$. Theorem 3 strengthens this statement.

Theorem 4 *Assume that X is irreducible, strong Feller and transient. If μ is a smooth measure in $\mathcal{K}_\infty(R)$, then $(\mathcal{F}_e, \mathcal{E})$ is compactly embedded in $L^2(E; \mu)$.*

Proof For $\mu \in \mathcal{K}_\infty(R)$ with f-supp$[\mu] = E$, the time-changed process $\check{X}$ by A_t^μ is in Class (T). Let $(\check{\mathcal{E}}, \check{\mathcal{F}})$ is the Dirichlet form generated by $\check{X}$. Then $(\check{\mathcal{F}}, \check{\mathcal{E}}_1 = (\check{\mathcal{E}} + (\ ,\ )_\mu))$ is compactly embedded in $L^2(E; \mu)$. Since $\check{\mathcal{F}} = \mathcal{F}_e$, $\check{\mathcal{E}} = \mathcal{E}$ and

$$\mathcal{E}(u, u) \le \check{\mathcal{E}}_1(u, u) = \mathcal{E}(u, u) + \int_E u^2 d\mu \le (1 + \|R\mu\|_\infty)\mathcal{E}(u, u),$$

$\mathcal{E}$ and $\check{\mathcal{E}}_1$ are equivalent and $(\mathcal{F}_e, \mathcal{E})$ is compactly embedded in $L^2(E; \mu)$.

There exists a strictly positive continuous function g such that $gm \in \mathcal{K}_\infty(R)$. For any $\mu \in \mathcal{K}_\infty(R)$ define $\mu^g = \mu + gm$. Then $(\mathcal{F}_e, \mathcal{E})$ is compactly embedded in $L^2(E; \mu^g)$ and so in $L^2(E; \mu)$.

Note that Theorem 4 is an extension of Theorem 3. Indeed, $(\mathcal{F}, \mathcal{E}_1)$ is a transient regular Dirichlet space and its extended Dirichlet space equals $(\mathcal{F}, \mathcal{E}_1)$. Theorem 4 says that if $m \in \mathcal{K}_\infty(R_1)$, then $(\mathcal{F}, \mathcal{E}_1)$ is compactly embedded in $L^2(E; m)$, which leads us to Theorem 3.

Corollary 1 *Assume that the Markov process X is irreducible, strong Feller. If μ is a smooth measure in $\mathcal{K}_\infty(R_1)$, then $(\mathcal{F}, \mathcal{E}_1)$ is compactly embedded in $L^2(E; \mu)$.*

Consider the Brownian motion on $\mathbb{R}^d$.

Proposition 1 ([17]) *If a domain $D \in \mathbb{R}^d$ satisfies*

$$\lim_{x \in D, |x| \to \infty} |D \cap B(x, 1)| = 0,$$

then the absorbing BM on D is in Class (T), consequently its semi-group is compact. Here $|\cdot|$ means the Lebesgue measure.

Example 1 ([17]) Let D be a domain. The following statements are equivalent:

(i) $1_D dx \in \mathscr{K}_\infty(R_1)$;
(ii) D satisfies $\lim_{|x| \to \infty} |D \cap B(x, 1)| = 0$;
(iii) $(H^1(\mathbb{R}^d), \mathbf{D}_1)$ is compactly embedded in $L^2(D)$.

The equivalence between (ii) and (iii) is known (cf. Edmunds and Evans [4, Lemma 6.22]).

We next consider the symmetric α-stable process on $\mathbb{R}^d$. Let X^V be the subprocess by $\exp(-\int_0^t V(X_s)ds)$.

Example 2 ([17]) Let $V \in \mathscr{K}_{loc}$. The following statements are equivalent:

(i) X^V is in Class (T);
(ii) $\lim_{|x| \to \infty} p_t^V 1(x) = 0$ ($\Longleftrightarrow \lim_{|x| \to \infty} R_1^V 1(x) = 0, \ \forall t > 0$);
(iii) p_t^V is a compact operator on $L^2(\mathbb{R}^d)$.

The equivalence between (ii) and (iii) is shown by Kaleta and Kulczycki [7]. If $\lim_{|x| \to \infty} |\{x \in \mathbb{R}^d \mid V(x) \leq M\} \cap B(x, 1)| = 0$ for any $M > 0$, then X^V is in Class (T) and thus p_t^V is a compact operator on $L^2(\mathbb{R}^d)$. This fact is analytically proved in [8, 10].

5 Boundedness of Eigenfunctions

Suppose that X is in Class (T). Then the semi-group p_t is compact by Theorem 3 and its spectrum are discrete.

Theorem 5 *If X is in Class (T), then every eigenfunction of the operator p_t has a bounded continuous version.*

Proof Let $\varphi \in \mathscr{F}$ be an eigenfunction corresponding to eigenvalue λ. We may suppose $\varphi \in C_\infty(\{K_n\})$, where $\{K_n\}_{n=1}^\infty$ is a compact nest ([5, Theorem 2.1.3]).

Let $D_n = E \setminus K_n$. Then for large n so that $\lambda_2^{D_n} > \lambda$,

$$\varphi(x) = \mathbf{E}_x(e^{\lambda \tau_{D_n}} \varphi(X_{\tau_{D_n}})) \ \ m\text{-a.e.}, \tag{9}$$

where $\lambda_2^{D_n} = \inf\{\mathscr{E}(u, u) \mid u \in \mathscr{F} \cap C_0(D_n), \ \ \|u\|_2 = 1\}$. Indeed, note $\varphi(x) = \mathbf{E}_x(e^{\lambda t} \varphi(X_t))$. Since

$$\int_{D_n} u^2 dm \leq \|R_1 1_{K_n^c}\|_\infty \cdot \mathcal{E}_1(u,u),$$

we have $\lambda_{D_n} \leq 1/\|R_1 1_{K_n^c}\|_\infty - 1 \longrightarrow \infty$ as $n \to \infty$. Since $\sup_{x\in E} \mathbf{E}_x(e^{\lambda\tau_{D_n}}) < \infty$ if and only if $\lambda_{D_n} > \lambda$, we have

$$\begin{aligned}\widetilde{\varphi}(x) := &\ \mathbf{E}_x(e^{\lambda\tau_{D_n}}\varphi(X_{\tau_{D_n}}); \tau_{D_n} < \zeta) = \mathbf{E}_x(e^{\lambda\tau_{D_n}}\varphi(X_{\sigma_{K_n}})) \\ \leq &\ \sup_{x\in K_n} |\varphi(x)| \cdot \sup_{x\in E} \mathbf{E}_x(e^{\lambda\tau_{D_n}}) < \infty.\end{aligned}$$

Put $\varphi' = p_t(\widetilde{\varphi})$. Then φ' is the desired one by the strong Feller property of p_t.

As stated above, the Ornstein-Uhlenbeck process is not in Class (T) and its eigenfunctions are Hermite polynomials, which are unbounded.

To prove the boundedness of ground states of Schrödinger operators, we prove an inequality ([16]): Let μ and ν be positive smooth measures and A^μ and A^ν the corresponding continuous positive additive functionals. Then

$$\sup_{x\in E} \int_E G^\mu(x,y)\nu(dy) \leq \sup_{x\in K} \int_E G^\mu(x,y)\nu(dy) \cdot \sup_{x\in X} \mathbf{E}_x\left(e^{A_\zeta^\mu}\right), \tag{10}$$

where $\int_E G^\mu(x,y)\nu(dy) = \mathbf{E}_x(\int_0^\zeta \exp(A_t^\mu) dA_t^\nu)$ and K is the topological support of ν. If G^μ is a Markov kernel, then the last factor in (10) does not appear, and thus this inequality is regarded as a version of the maximum principle.

Acknowledgements The author would like to thank the referee for helpful comments on this paper.

References

1. Chen, Z.-Q.: Gaugeability and conditional gaugeability. Trans. Am. Math. Soc. **354**, 4639–4679 (2002)
2. Chung, K.L., Zhao, Z.X.: From Brownian Motion to Schrödinger's Equation. Springer, Berlin (1995)
3. Deuschel, J.-D., Stroock, D.W.: Large Deviations. American Mathematical Society, Providence (2001)
4. Edmunds, D.E., Evans, W.D.: Spectral Theory and Differential Operators. Oxford University Press, Oxford (1987)
5. Fukushima, M., Oshima, Y., Takeda, M.: Dirichlet Forms and Symmetric Markov Processes, de Gruyter, 2nd rev. and ext. ed. (2011)
6. Itô, K.: Essentials of Stochastic Processes. American Mathematical Society, Providence (2006)
7. Kaleta, K., Kulczycki, T.: Intrinsic ultracontractivity for Schrödinger operators based on fractional Laplacians. Potential Anal. **33**, 313–339 (2010)
8. Lenz, D., Stollmann, P., Wingert, D.: Compactness of Schrödinger semigroups. Math. Nachr. **283**, 94–103 (2010)
9. Schilling, R. L.: Measures, Integrals and Martingales. Cambridge University Press, Cambridge (2005)

10. Simon, B.: Schrödinger operators with purely discrete spectra. Methods Funct. Anal. Topol. **15**, 61–66 (2009)
11. Simon, B.: Operator Theory, A Comprehensive Course in Analysis, Part 4. American Mathematical Society (2015)
12. Stollmann, P., Voigt, J.: Perturbation of Dirichlet forms by measures. Potential Anal. **5**, 109–138 (1996)
13. Takeda, M.: Lp-independence of spectral bounds of Schrödinger type semigroups. J. Funct. Anal. **252**, 550–565 (2007)
14. Takeda, M.: A tightness property of a symmetric Markov process and the uniform large deviation principle. Proc. Am. Math. Soc. **141**, 4371–4383 (2013)
15. Takeda, M.: A variational formula for Dirichlet forms and existence of ground states. J. Funct. Anal. **266**, 660–675 (2014)
16. Takeda, M.: Criticality and subcriticality of generalized Schrödinger forms. Illinois J. Math. **58**, 251–277 (2014)
17. Takeda, M., Tawara, Y., Tsuchida, K.: Compactness of Markov and Schrödinger semi-groups: a probabilistic approach. Osaka J. Math. **54**, 517–532 (2017)
18. Wu, L.: Some notes on large deviations of Markov processes, Acta Math. Sin. (Engl. Ser.) **16**, 369-394 (2000)

Part V
Applications Including Mathematical Physics

From Non-symmetric Particle Systems to Non-linear PDEs on Fractals

Joe P. Chen, Michael Hinz and Alexander Teplyaev

Abstract We present new results and challenges in obtaining hydrodynamic limits for non-symmetric (weakly asymmetric) particle systems (exclusion processes on pre-fractal graphs) converging to a non-linear heat equation. We discuss a joint density-current law of large numbers and a corresponding large deviation principle.

Keywords (Weakly) Asymmetric exclusion interacting random process Hydrodynamic limit · Nonlinear heat equation · Dirichlet form · Fractal

MSC 2010 Primary 60K35 · Secondary: 05C81 · 28A80 · 31C20 · 35K58 47A07 · 47H05 · 60F05 · 60F10 · 60J25 · 60J27 · 81Q35 · 82B21 · 82C22

Exclusion process on a weighted graph. We consider a locally finite connected (simple and undirected) graph $\Gamma = (V, E)$ with vertex set V and edge set E and endowed with conductances $\mathbf{c} = (c_{xy})_{xy\in E}$ satisfying $c_{xy} > 0$. The pair $(\Gamma, \mathbf{c})$ is called a *weighted graph*. Suppose that $H : [0, T] \times V \to \mathbb{R}$ is a given function with the abbreviated notation $H_t := H(t, \cdot)$. The *weakly asymmetric exclusion process* (WASEP) on $(\Gamma, \mathbf{c})$ associated with H is the Markov chain $(\eta_t)_{t\geq 0}$ on $\{0, 1\}^V$ with time-dependent generator $\mathscr{L}^{\rm EX}_{(\Gamma,\mathbf{c}),H_t}$ defined on functions $f : \{0, 1\}^V \to \mathbb{R}$ by

$$\left(\mathscr{L}^{\rm EX}_{(\Gamma,\mathbf{c}),H_t} f\right)(\eta) = \sum_{xy\in E} c_{xy}\psi_{xy}(H_t, \eta)[f(\eta^{xy}) - f(\eta)],$$

Research supported in part by NSF grants 1262929, 1613025.

J. P. Chen
Department of Mathematics, Colgate University, Hamilton, NY 13346, USA
e-mail: jpchen@colgate.edu

M. Hinz
Fakultät für Mathematik, Universität Bielefeld, Postfach 100131, 33501 Bielefeld, Germany
e-mail: mhinz@math.uni-bielefeld.de

A. Teplyaev (✉)
Department of Mathematics, University of Connecticut, Storrs, CT 06269, USA
e-mail: teplyaev@uconn.edu

A. Eberle et al. (eds.), *Stochastic Partial Differential Equations and Related Fields*, Springer Proceedings in Mathematics & Statistics 229,
https://doi.org/10.1007/978-3-319-74929-7_34

where $\psi_{xy}(H_t,\eta) = \exp\{(\eta(y)-\eta(x))(H_t(x)-H_t(y))\}$ and

$$\eta^{xy}(z) = \begin{cases} \eta(y), & \text{if } z = x, \\ \eta(x), & \text{if } z = y, \\ \eta(z), & \text{otherwise.} \end{cases}$$

We can think of $\eta(z)$ as the occupation variable which returns 1 (resp. 0) when z is occupied with a particle (resp. empty). The configuration η^{xy} is obtained by exchanging the occupation variables $\eta(x)$ and $\eta(y)$ in η. At time t such a transition occurs with rate $c_{xy}\psi_{xy}(H_t,\eta)$, where ψ_{xy} encodes the (weak) asymmetry between the hopping rates from x to y and from y to x. When $H \equiv 0$ we obtain the *symmetric exclusion process* (SEP) on $(\Gamma, \mathbf{c})$.

We also define the *boundary-driven* exclusion process. Declare a nonempty subset $\partial V \subset V$ to be the boundary set. Given the aforementioned exclusion process, we add a birth-and-death process to each boundary point $a \in \partial V$; that is, we consider the Markov chain on $\{0,1\}^V$ generated by

$$\mathscr{L}^{\mathrm{bEX}}_{(\Gamma,\mathbf{c}),H_t} = \mathscr{L}^{\mathrm{EX}}_{(\Gamma,\mathbf{c}),H_t} + \mathscr{L}^{\mathrm{b}}_{\partial V},$$

where for any function $f : \{0,1\}^V \to \mathbb{R}$,

$$(\mathscr{L}^{b}_{\partial V} f)(\eta) = \sum_{a\in\partial V} [\lambda_-(a)\eta(a) + \lambda_+(a)(1-\eta(a))][f(\eta^a) - f(\eta)],$$

with $\lambda_+(a) > 0$ (resp. $\lambda_-(a) > 0$) representing the birth (resp. death) rate at a, and

$$\eta^{a}(z) = \begin{cases} 1-\eta(a), & \text{if } z = a, \\ \eta(z), & \text{otherwise.} \end{cases}$$

We assume that the relative boundary transition rates are bounded away from 0 and ∞, i.e. we assume there exists $\gamma \in [1,\infty)$ such that $\gamma^{-1} \le \frac{\lambda_+(a)}{\lambda_-(a)} \le \gamma$ for all $a \in \partial V$.

Analysis on the Sierpinski gasket and exclusion processes on pre-fractal graphs. We now turn to the Sierpinski gasket (SG). Let a_0, a_1, a_2 be the vertices of an equilateral triangle in $\mathbb{R}^2$, and Γ_0 be the complete graph on the vertex set $V_0 = \{a_0, a_1, a_2\}$. Define the contracting similitude $\Psi_i : \mathbb{R}^2 \to \mathbb{R}^2$, $\Psi_i(x) = \frac{1}{2}(x - a_i) + a_i$ for each $i \in \{0,1,2\}$. For $N \ge 1$, we define $\Gamma_N = (V_N, E_N)$ inductively via the formula $\Gamma_N = \bigcup_{i=0}^2 \Psi_i(\Gamma_{N-1})$. The sequence $(\Gamma_N)_{N\ge0}$ forms a graphical approximation of the SG fractal K, which is the unique compact set satisfying $K = \cup_{i=0}^2 \Psi_i(K)$. By μ we denote the standard self-similar Borel probability measure on K.

A well-known result of Barlow–Perkins [5] states that if for each N we write $(X_t^N)_{t\ge0}$ to denote the natural symmetric random walk on the approximating graph Γ_N, then the sequence of processes $(X^N_{5^N\cdot})_{N\ge0}$ is tight and converges to a generic diffusion process (a "Brownian motion") $(X_t)_{t\ge0}$ on K. The time acceleration factor 5^N can also be observed in the analytic approach of Kigami [40], in the sense that the sequence of graph energies $\mathscr{E}_N(f) = \frac{5^N}{3^N}\sum_{xy\in E_N}[f(x)-f(y)]^2$, $f: V_N \to \mathbb{R}$ is monotone and converges to a limit energy form $(\mathscr{E}, \mathscr{F})$ on K. This form is a resistance form in the sense of Kigami, [41], and in particular, it satisfies the Sobolev embedding $\mathscr{F} \subset C(K)$. Moreover, it defines a strongly local regular Dirichlet form on $L^2(K,\mu)$. This allows to define the standard (Dirichlet) Laplacian Δ by a Gauss-Green formula: given $f \in C(K)$, we say that $u \in \text{dom}\,\Delta$ with $\Delta u = f$ if $\mathscr{E}(u,v) = -\int_K fv\,d\mu$ for all $v \in \mathscr{F}$ with $v|_{V_0} = 0$. Then $(\Delta, \text{dom}\,\Delta)$ is a non-positive self-adjoint operator on $L^2(K,\mu)$, and it is just the $L^2(X,\mu)$-infinitesimal generator of the diffusion $(X_t)_{t\geq 0}$. For more details see [4, 40, 41, 50].

We also need the notions of gradient and divergence. On general weighted graphs they are explained in e.g. [47]. We fix an orientation for each edge $xy \in E_N$ of the approximating graph $\Gamma_N = (V_N, E_N)$, and we may assume that the resulting oriented edge, denoted by $\overrightarrow{xy}$, has initial vertex x and terminal vertex y. By $\ell^2_-(E_N)$ we denote the space of functions $\theta: E_N \to \mathbb{R}$ that are antisymmetric under a change of orientation, i.e. satisfy $\theta(\overrightarrow{yx}) = -\theta(\overrightarrow{xy})$ for all $\overrightarrow{xy} \in E$. On each approximating graph Γ_N we can define the *discrete gradient* $\partial_N : \ell^2(V_N) \to \ell^2_-(E_N)$ by $(\partial_N f)(\overrightarrow{xy}) = \frac{5^N}{3^N}[f(y)-f(x)]$. The continuum analogs of ∂_N on the limiting fractal K can be introduced and studied via Dirichlet form theory, cf. [14, 23–25, 32–36]. There exist a certain *Hilbert bimodule* $\mathscr{H}$ *of generalized* L^2*- vector fields associated with* $(\mathscr{E},\mathscr{F})$ and a derivation $\partial : \mathscr{F} \to \mathscr{H}$ that satisfies $\|\partial f\|^2_{\mathscr{H}} = \mathscr{E}(f)$, $f \in \mathscr{F}$. The adjoint $-\partial^* : \mathscr{H} \to \mathscr{F}^*$ is then defined in the usual way by duality. In particular,

$$\langle \partial f, g\partial h\rangle_{\mathscr{H}} = -\langle f, \partial^*(g\partial h)\rangle_{L^2(K,\mu)} = \int_K g\,d\Gamma(f,h)$$

for all $f, g, h \in \mathscr{F}$, where $\Gamma(f,h)$ denote the *mutual energy measure* of f and h, [22, 48, 50]. The operators ∂ and $-\partial^*$ may be viewed as (abstract) *gradient* and *divergence operators* on K. By construction we have $\lim_{N\to\infty}\|\partial_N f\| = \|\partial f\|_{\mathscr{H}}$ with an appropriate definition of norm involved, However, unlike in the Euclidean setting, we do not have the pointwise convergence $\partial_N f \to \partial f$.

We consider the boundary-driven WASEP $(\eta^N_t)_{t\in[0,T]}$ on Γ_N generated by the operator $5^N\left(\mathscr{L}^{\text{EX}}_{\Gamma_N,H_t} + \mathscr{L}^b_{V_0}\right)$. assuming that $H \in C^1([0,T], \text{dom}\Delta)$, and that the boundary set V_0 and the rates $\{\lambda_\pm(a_i) : i \in \{0,1,2\}\}$ are fixed for all N. See Fig. 1 for a schematic picture.

We are interested in two observables (empirical measures) in the exclusion process: the *empirical density* and the *empirical integrated current*

$$\pi^N_t(A) = \frac{1}{|V_N|}\sum_{x\in V_N}\eta^N_t(x)\mathbf{1}_A(x), \quad A \subset K,$$

$$\mathbf{W}^N_t(B) = \frac{1}{|E_N|}\sum_{\overrightarrow{xy}\in E_N} W^N_t(\overrightarrow{xy})\mathbf{1}_B(\overrightarrow{xy}),$$

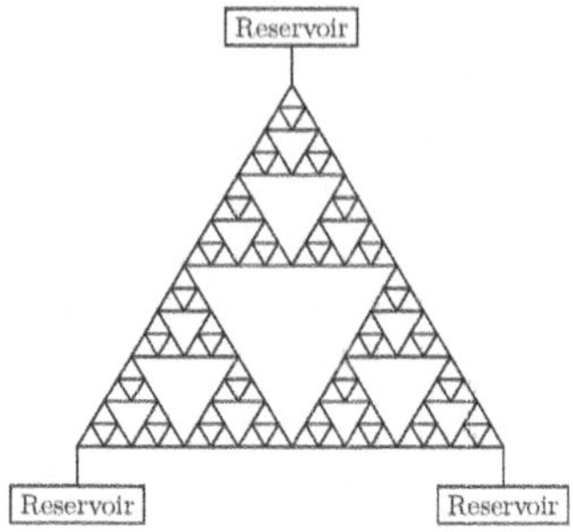

Fig. 1 The pre-Sierpinski gasket graph Γ_4, indicating the three reservoirs which fix the particle densities at the boundary points $\{a_0, a_1, a_2\}$

where B is a subset of oriented edges. Here the *current* $W_t^N(\overrightarrow{xy})$, for each oriented edge $\overrightarrow{xy}$ in E_N, is the net number of particle jumps along $\overrightarrow{xy}$ (i.e. #(jumps from x to y) − #(jumps from y to x)) in the time interval $[0, t]$. We note the mass conservation law $\sum_{y\in V_N:\overrightarrow{xy}\in E_N} W_t^N(\overrightarrow{xy}) = -\left[\eta_t^N(x) - \eta_0^N(x)\right]$ for $x \in V_N\setminus V_0$.

Hydrodynamic limit. The density-current pair $\left(\pi_t^N, \mathbf{W}_t^N\right)$ satisfies a law of large numbers. We fix a macroscopic density ρ_0 satisfying the boundary condition

$$\rho_0(a_i) = \bar{\rho}_i := \frac{\lambda_+(a_i)}{\lambda_+(a_i) + \lambda_-(a_i)}$$

on V_0, and assume that the sequence of initial densities $(\pi_0^N)_{N\geq 1}$ converges weakly to $\rho_0\, d\mu$. We then claim that $(\pi_t^N)_{N\geq 1}$ converges to the weak solution ρ^H of the nonlinear parabolic equation

$$\begin{cases} \partial_t \rho^H(t,x) = \Delta\rho^H(t,x) - \partial^*\left(\chi(\rho^H(t,x))\partial H(t,x)\right) & \text{on } (0,T)\times(K\setminus V_0),\\ \rho^H(0,x) \quad = \rho_0(x), & \text{on } K\setminus V_0,\\ \rho^H(t,a_i) \quad = \bar{\rho}_i & \text{on } (0,T), \end{cases} \tag{1}$$

where $\chi : \mathbb{R} \to [0,1]$ is defined by $\chi(s) := (s(1-s))_+$. The quantity $\chi(\rho^H)$ is the mobility of the exclusion process. In addition, the time derivative of $\mathbf{W}_t^N$ converges to the vector field $\mathbf{J}_t = -\partial\rho_t^H + \chi(\rho_t^H)\partial H_t$, which satisfies the macroscopic continuity equation $\partial_t\rho_t^H + \partial^*\mathbf{J}_t = 0$. An important caveat is that these equations are only interpreted in the weak sense, not in the pointwise sense.

Turning to the formal description, we set

$$C_e(K) := \{f \in C(K) : \text{there exists } \varepsilon > 0 \text{ such that } \varepsilon \leq f \leq 1-\varepsilon\}.$$

Fix $\rho_0 \in C_e(K)$ which satisfies the boundary condition $\rho_0(a_i) = \bar{\rho}_i$. Consider, for every N, the boundary-driven WASEP $(\eta_t^N)_{t\in[0,T]}$ whose initial configurations is η^N, and denote the corresponding law by $\mathbb{P}_{\eta^N}^N$. We assume that $(\eta^N)_{N\geq 0}$ is associated with ρ_0 in the sense of weak convergence, i.e. that

$$\lim_{N\to\infty} \langle f, \pi_0^N \rangle = \langle f, \rho_0 \, d\mu \rangle \quad \text{for all } f \in C(K).$$

Here and below $\langle \cdot, \cdot \rangle$ denotes the dual pairing. Now let $\mathscr{F}_0 = \{f \in \mathscr{F} : f|_{V_0} = 0\}$ and $\mathscr{F}^*$ (resp. $\mathscr{F}_0^*$) be the dual of $\mathscr{F}$ (resp. $\mathscr{F}_0$). A bounded function $\rho^H \in L^2(0, T, \mathscr{F})$ with $\partial_t \rho^H \in L^2(0, T, \mathscr{F}_0^*)$ is said to be a *weak solution of* (1) if for every $t \in [0, T]$ and every $\varphi \in L^2(0, T, \mathscr{F}_0)$,

$$\int_0^t \int_K (\partial_s \rho_s^H) \varphi_s \, d\mu \, ds = - \int_0^t \mathscr{E}(\rho_s^H, \varphi_s) \, ds + \int_0^t \langle \chi(\rho_s^H) \partial H_s, \partial \varphi_s \rangle_{\mathscr{H}} \, ds, \tag{2}$$

if $\rho_0^H = \rho_0$ in $L^2(K, \mu)$ and if $\rho_t^H - h \in \mathscr{F}_0$ for a.e. $t \in (0, T)$, where $h \in \mathscr{F}$ is the unique harmonic function on K with boundary values $\bar{\rho}_i$ on V_0. The specification of the initial condition makes sense, because under the required regularity conditions any such ρ^H will be an element of $C([0, T], L^2(K, \mu))$, see the references mentioned below. There we also comment on the existence and uniqueness of a weak solution ρ^H to (1).

Theorem 1 [Joint density-current large numbers (LLN)] *For every $t \in [0, T]$, $\delta > 0$, $G \in C(K)$, and $F \in \operatorname{dom} \Delta$, we have*

$$\lim_{N\to\infty} \mathbb{P}_{\eta^N}^N \left(\left| \langle \pi_t^N, G \rangle - \int_K G \rho_t^H \, d\mu \right| > \delta \right) = 0$$

$$\lim_{N\to\infty} \mathbb{P}_{\eta^N}^N \left(\left| \langle \mathbf{W}_t^N, \partial_N F \rangle - \int_0^t \langle \mathbf{J}_{H_s}(\rho_s^H), \partial F \rangle_{\mathscr{H}} \, ds \right| > \delta \right) = 0,$$

where ρ^H denotes the unique weak solution of (1), *and for a.e. s the vector field $\mathbf{J}_{H_s}(\rho_s^H)$ is defined by the identity*

$$\langle \mathbf{J}_{H_s}(\rho_s^H), \partial F \rangle_{\mathscr{H}} := \mathscr{E}(\rho_s^H, F) + \langle \chi(\rho_s^H) \partial H_s, \partial F \rangle_{\mathscr{H}}.$$

We also wish to quantify the probability of the (rare) event that a given trajectory deviates from the hydrodynamic solution (1). This is done by proving a large deviations principle (LDP). Fix $\rho_0 \in C_e(K)$. Let

$$\mathscr{F}\mathscr{M}_0 := \{\rho(x) \, d\mu(x) : 0 \le \rho \le 1 \ \mu\text{-a.e.}, \ \rho \in \mathscr{F}\}$$

be the set of positive measures which are absolutely continuous with respect to μ, whose density is bounded by 1 and has finite energy. We will work with $E := D([0, T], \mathscr{F}\mathscr{M}_0 \times \mathscr{H})$, the space of càdlàg paths from $[0, T]$ to $\mathscr{F}\mathscr{M}_0 \times \mathscr{H}$ endowed with the Skorokhod topology.

For $H \in C^1([0, T], \operatorname{dom} \Delta)$, we introduce the functional $J_H := J_{H,T,\rho_0} : E \to \mathbb{R} \cup \{+\infty\}$ as follows. If $(\pi, \mathbf{W}) \notin C([0, T], \mathscr{F}\mathscr{M}_0 \times \mathscr{H})$, set $J_H(\pi, \mathbf{W}) = +\infty$; otherwise set

$$J_H(\pi, \mathbf{W}) = \langle \partial H_T, \mathbf{W}_T \rangle_{\mathscr{H}} - \int_0^T \left\langle \frac{\partial}{\partial t} \partial H_t, \mathbf{W}_t \right\rangle_{\mathscr{H}} dt - \int_0^T \langle \Delta H_t, \pi_t \rangle \, dt$$
$$- \int_0^T \langle \chi(\rho_t) \, \partial H_t, \partial H_t \rangle_{\mathscr{H}} \, dt + \sum_{i=0}^{2} \bar{\rho}_i \int_0^T (\partial^\perp H_t)(a_i) \, dt.$$

Here $\rho_t = \frac{d\pi_t}{d\mu}$ and $\partial^\perp$ denotes the normal derivative as defined in [40, 50]. Put

$$J(\pi, \mathbf{W}) = \sup_{H \in C^1([0,T], \mathrm{dom}\Delta)} J_H(\pi, \mathbf{W}).$$

Let $\mathscr{A}$ be the set of all $(\pi, \mathbf{W}) \in C([0, T], \mathscr{F}\mathscr{M}_0 \times \mathscr{H})$ satisfying $d\pi_0 = \rho_0 \, d\nu$, $\mathbf{W}_0 = 0$, and the *conservation law* $\partial_t \pi_t + \partial^*(\dot{\mathbf{W}}_t) = 0$ in the weak formulation: for every $\varphi \in \mathscr{F}_0$, $\langle \varphi, \pi_t - \pi_0 \rangle = \langle \partial \varphi, \mathbf{W}_t \rangle_{\mathscr{H}}$. We introduce the *dynamical rate function*

$$I(\pi, \mathbf{W}) = \begin{cases} J(\pi, \mathbf{W}), & \text{if } (\pi, \mathbf{W}) \in \mathscr{A}, \\ +\infty, & \text{otherwise.} \end{cases} \tag{3}$$

It turns out that $J(\pi, \mathbf{W})$ can be written in the more symmetric form

$$J(\pi, \mathbf{W}) = \frac{1}{2} \int_0^T \left\langle [\chi(\rho_t)]^{-1} \left(\frac{\partial}{\partial t} \mathbf{W}_t + \partial \rho_t \right), \frac{\partial}{\partial t} \mathbf{W}_t + \partial \rho_t \right\rangle_{\mathscr{H}} dt.$$

Theorem 2 [Joint density-current LDP] *For each closed set $\mathscr{C}$ and each open set $\mathscr{O}$ of E,*

$$\limsup_{N \to \infty} \frac{1}{|V_N|} \log \mathbb{P}^N_{\eta^N} \left((\pi^N, \mathbf{W}^N) \in \mathscr{C} \right) \le - \inf_{(\pi, \mathbf{W}) \in \mathscr{C}} I(\pi, \mathbf{W}),$$
$$\liminf_{N \to \infty} \frac{1}{|V_N|} \log \mathbb{P}^N_{\eta^N} \left((\pi^N, \mathbf{W}^N) \in \mathscr{O} \right) \ge - \inf_{(\pi, \mathbf{W}) \in \mathscr{O}} I(\pi, \mathbf{W}).$$

Outline of proof strategy. Our proof strategy is conceptually aligned with the hydrodynamic limit program originating from [27, 43], which has since been expounded in the monograph [42] and applied to various low-dimensional lattice gas models. In particular we are influenced by the works [9, 10, 13]. For an earlier work on the hydrodynamic limit of a related particle system on SG see [37].

That said, we had to overcome a number of technical obstacles to prove the limit theorems on SG. In a nutshell, the difficulties can be attributed to the well-known fact that on SG (and other fractals), the energy measure is singular to the self-similar (Hausdorff) measure [8, 28, 44]. This has two consequences. On the microscopic level, there is no translational invariance, and one needs new tools (resistance-based energy inequalities) to establish a coarse-graining lemma in order to pass to the scaling limit. On the macroscopic level, one needs to utilize notions of vector calculus and (S)PDEs developed through Dirichlet forms.

For the sake of readability we have divided the proofs of Theorems 1 and 2 into several papers, whose contents are summarized as follows:

The moving particle lemma [17]: we prove an energy inequality in the symmetric exclusion process on any finite weighted graph, bounding the cost of swapping particle configurations at vertices x and y in the exclusion process by the effective resistance distance $R_{\text{eff}}(x, y)$ in the random walk process. The proof is based on the *octopus inequality* of [15], which was key to the positive resolution of Aldous' spectral gap conjecture.

A local ergodic (coarse graining) theorem [16]: we show that on many (strongly) recurrent graphs [6, 51, 52], local functions of the occupation variables η in the (suitably rescaled) exclusion process can be replaced by their macroscopic averages in the scaling limit, with probability superexponentially close to 1. This local ergodic theorem is stated for the conservative version and the boundary-driven version of the exclusion process. It is proved using the aforementioned moving particle lemma, and can be applied to fractal graphs and trees that lack translational invariance.

Evolution equations on resistance spaces [21]: we examine the solvability of evolution PDEs which generalize (1) on spaces which support Kigami's resistance forms [41], using tools from Dirichlet forms and the induced vector analysis.

Hydrodynamic limit of the exclusion process on SG [20]: we utilize the results of the previous three papers, the vector analysis developed in [32–34], along with the established hydrodynamic limit program, to prove Theorems 1 and 2.

Semilinear evolution equations. To verify the existence and uniqueness of a weak solution to problem (1), we first consider the problem

$$\begin{cases} \partial_t w(t,x) = \Delta w(t,x) - \partial^* \left(\chi(w(t,x)+h(x))\partial H(t,x)\right) & \text{on } (0,T)\times(K\backslash V_0), \\ w(0,\cdot) \quad = \rho_0 - h, & \text{on } K\backslash V_0, \\ w(\cdot, a_i) \quad = 0 & \text{on } (0,T), \end{cases} \tag{4}$$

Here $h : K \to \mathbb{R}$ denotes the unique solution $h \in \mathscr{F}$ of the Dirichlet problem

$$\begin{cases} \Delta h \quad = 0 \quad \text{on } K \setminus V_0, \\ h(a_i) = \bar{\rho}_i. \end{cases}$$

We translate (4) into an abstract Cauchy problem and solve it using monotone operator methods [46]. The (Dirichlet) Laplacian Δ may be viewed as a bounded variational operator $\Delta : \mathscr{F}_0 \to \mathscr{F}_0^*$. We consider the nonlinear operator defined by

$$A(t,v) := -\Delta v + \partial^*(\chi(v+h)\partial H_t), \quad v \in \mathscr{F}_0.$$

Recall that $H \in C^1([0,T], \operatorname{dom}\Delta) \subset L^\infty(0,T,\mathscr{F})$. Together with the resistance form properties of $(\mathscr{E}, \mathscr{F})$ this can be used to see that writing $(\mathscr{A}(u))_t := A(t, u_t)$ for a given function $u : (0,T) \to \mathscr{F}_0$, we obtain a bounded and demicontinuous operator $\mathscr{A} : L^2(0,T,\mathscr{F}_0) \to L^2(0,T,\mathscr{F}_0^*)$ which satisfies

$$\lim_{\|u\|_{L^2(0,T,\mathscr{F})}\to\infty} \frac{\langle \mathscr{A}(u), u\rangle}{\|u\|_{L^2(0,T,\mathscr{F})}} = +\infty. \tag{5}$$

Here we use again $\langle\cdot,\cdot\rangle$ to denote the dual pairing in the obvious sense. We now rephrase (4) as the abstract Cauchy problem

$$\begin{cases} \partial_t w_t + A(t, w_t) = 0 & \text{for a.e. } t \in (0,T) \\ w(0) \qquad\qquad = \rho_0 - h. \end{cases} \tag{6}$$

A function $w \in L^2(0,T,\mathscr{F}_0)$ with $\partial_t w \in L^2(0,T,\mathscr{F}_0^*)$ is called a *(strong) solution* to the abstract Cauchy problem (6) if the first identity holds in $\mathscr{F}_0^*$ (for a.e. $t \in (0,T)$) and the second holds in $L^2(K,\mu)$. Note that any $w \in L^2(0,T,\mathscr{F}_0)$ with $\partial_t w \in L^2(0,T,\mathscr{F}_0^*)$ is a member of $C([0,T], L^2(K,\mu))$, so that the second condition makes sense, see [49, Chap. III, Proposition 1.2].

Proposition 1 *There exists a solution w to the abstract Cauchy problem (6).*

This follows from [46, Théorème 2.1] together with certain estimates based on the resistance form properties of $(\mathscr{E},\mathscr{F})$. Given a solution w of (6), the function $\rho^H := w + h$ is an element of $L^2(0,T,\mathscr{F})$ and satisfies $\partial_t \rho^H \in L^2(0,T,\mathscr{F}_0^*)$. We also have

$$\partial_t \rho_t^H = \Delta\rho_t^H - \partial^*\left(\chi(\rho_t^H)\partial H_t\right) \quad \text{for a.e. } t \in (0,T),$$

seen in $\mathscr{F}_0^*$, and this implies the validity of the integral identity (2). Moreover, we observe that $\rho_0^H = \rho_0$ and $\rho^H(t,a_i) = \bar{\rho}_i$ for a.e. $t \in (0,T)$.

Theorem 3 *The function ρ^H is the unique weak solution to (1).*

The uniqueness, due to the Markov property of $(\mathscr{E},\mathscr{F})$, is obtained in [21] by showing that the difference of two solutions has $L^1(K,\mu)$-norm decreasing in time. This argument was already used in [13, Appendix] and [45, Proof of Lemma 7.1].

Perspectives. Theorem 1 may be viewed as a first-principles derivation of a fluid equation on a fractal, subject to external biases on the boundary. The solution of this equation gives the expected trajectory of the macroscopic density and current. Theorem 2 characterizes the fluctuations about this expected trajectory in terms of the large deviations rate function $I(\pi,\mathbf{W})$.

On simple graphs which either are discrete tori or have two boundary points (such as $\mathbb{Z}\cap[-n,n]$), there is an alternative combinatorial approach to deriving the rate function [26]. However it is unclear if this approach generalizes to SG (with the standard 3-point boundary) or other infinite weighted graphs. This explains why we followed the hydrodynamic limit program, which has a more robust analytic flavor.

The rate function can be used to analyze properties of macroscopic fluctuations in boundary-driven diffusive processes on networks, which is a current topic of interest in nonequilibrium statistical mechanics; see the excellent recent review [11]. Some highlights in this area include the universality of the cumulant of the mean long-time current in networks with two boundary points [1], under the hypothesis of the *additivity principle* [12]. It is also of interest to investigate the validity of the additivity principle in closely related boundary-driven particle systems [10]. We hope to investigate these and related problems on SG (and other fractals) in the near future, but keep in mind that there are key differences that make the analysis more difficult than on tori and simple lattices. We list a few open related questions:

- Analytically or numerically characterize the infimum of the rate function $J(\pi, \mathbf{W})$.
- Describe the motion of a tagged particle in the exclusion process.
- Study the aforementioned current cumulant problem.
- Investigate asymmetric exclusion processes and other non-gradient-type particle systems on fractal and other weighted graphs. A key technical step would be to identify energy (spectral gap) inequalities relevant to each process.
- Establish connection to the recent theory of mathematical physics, Dirichlet forms, energy estimates on fractals ([2, 3, 7, 18, 19, 29–31, 38, 39, and references therein]).

References

1. Akkermans, E., Bodineau, T., Derrida, B., Shpielberg, O.: Universal current fluctuations in the symmetric exclusion process and other diffusive systems. EPL **103**(2), 20001 (2013)
2. Alonso-Ruiz, P.: Power dissipation in fractal Feynman-Sierpinski AC circuits. J. Math. Phys. **58**(7), 073503 (2017)
3. Alonso-Ruiz, P., Kelleher, D.J., Teplyaev, A.: Energy and Laplacian on Hanoi-type fractal quantum graphs. J. Phys. A **49**(16), 165206–165236 (2016)
4. Barlow, M.T.: Diffusions on fractals. Lectures on Probability Theory and Statistics (Saint-Flour, 1995). Lecture Notes in Mathematics, vol. 1690, pp. 1–121. Springer, Berlin (1998)
5. Barlow, M.T., Perkins, E.A.: Brownian motion on the Sierpiński gasket. Probab. Theory Relat. Fields **79**(4), 543–623 (1988)
6. Barlow, M.T., Coulhon, T., Kumagai, T.: Characterization of sub-Gaussian heat Kernel estimates on strongly recurrent graphs. Commun. Pure Appl. Math. **58**(12), 1642–1677 (2005)
7. Baudoin, F., Kelleher, D.J.: Differential forms on Dirichlet spaces and Bakry-Emery estimates on metric graphs (2017). arXiv:1604.02520
8. Ben-Bassat, O., Strichartz, R.S., Teplyaev, A.: What is not in the domain of the Laplacian on Sierpinski gasket type fractals. J. Funct. Anal. **166**(2), 197–217 (1999)
9. Bertini, L., De Sole, A., Gabrielli, D., Jona-Lasinio, G., Landim, C.: Large deviations for the boundary driven symmetric simple exclusion process. Math. Phys. Anal. Geom. **6**(3), 231–267 (2003)
10. Bertini, L., De Sole, A., Gabrielli, D., Jona-Lasinio, G., Landim, C.: Large deviations of the empirical current in interacting particle systems. Teor. Ver. Prim. **51**(1), 144–170 (2006)
11. Bertini, L., De Sole, A., Gabrielli, D., Jona-Lasinio, G., Landim, C.: Macroscopic fluctuation theory. Rev. Mod. Phys. **87**, 593–636 (2015)
12. Bodineau, T., Derrida, B.: Current fluctuations in nonequilibrium diffusive systems: an additivity principle. Phys. Rev. Lett. **92**, 180601 (2004)
13. Bodineau, T., Lagouge, M.: Large deviations of the empirical currents for a boundary-driven reaction diffusion model. Ann. Appl. Probab. **22**(6), 2282–2319 (2012)
14. Brzoska, A., Kelleher, D.J., Panzo, H., Teplyaev, A.: Dual graphs and modified Barlow–Bass resistance estimates for repeated barycentric subdivisions. Discret. Contin. Dyn. Syst. (2017). arXiv:1505.03161

15. Caputo, P., Liggett, T.M., Richthammer, T.: Proof of Aldous' spectral gap conjecture. J. Am. Math. Soc. **23**(3), 831–851 (2010)
16. Chen, J.P.: Local ergodicity in the exclusion process on an infinite weighted graph (2017). arXiv:1705.10290
17. Chen, J.P.: The moving particle lemma for the exclusion process on a weighted graph. Electron. Commun. Probab. **22**(47), 1–13 (2017)
18. Chen, J.P., Teplyaev, A.: Singularly continuous spectrum of a self-similar Laplacian on the half-line. J. Math. Phys. **57**(5), 052104–052110 (2016)
19. Chen, J.P., Molchanov, S., Teplyaev, A.: Spectral dimension and Bohr's formula for Schrödinger operators on unbounded fractal spaces. J. Phys. A **48**(39), 395203–395227 (2015)
20. Chen, J.P., Hinz, M., Teplyaev, A.: Hydrodynamic limit of the the boundary-driven exclusion process on the Sierpinski gasket (2018+)
21. Chen, J.P., Hinz, M., Teplyaev, A.: Semi linear evolution equations on resistance spaces (2018+)
22. Chen, Z.Q., Fukushima, M.: Symmetric Markov Processes, Time Change, and Boundary Theory. London Mathematical Society Monographs Series, vol. 35. Princeton University Press, Princeton (2012)
23. Cipriani, F., Sauvageot, J.L.: Derivations as square roots of Dirichlet forms. J. Funct. Anal. **201**(1), 78–120 (2003)
24. Cipriani, F., Guido, D., Isola, T., Sauvageot, J.L.: Integrals and potentials of differential 1-forms on the Sierpinski gasket. Adv. Math. **239**, 128–163 (2013)
25. Cipriani, F., Guido, D., Isola, T., Sauvageot, J.L.: Spectral triples for the Sierpinski gasket. J. Funct. Anal. **266**(8), 4809–4869 (2014)
26. Derrida, B.: Matrix ansatz large deviations of the density in exclusion processes. In: International Congress of Mathematicians, vol. III, pp. 367–382. European Mathematical Society, Zürich (2006)
27. Guo, M.Z., Papanicolaou, G.C., Varadhan, S.R.S.: Nonlinear diffusion limit for a system with nearest neighbor interactions. Commun. Math. Phys. **118**(1), 31–59 (1988)
28. Hino, M.: On singularity of energy measures on self-similar sets. Probab. Theory Relat. Fields **132**(2), 265–290 (2005)
29. Hinz, M.: Sup-norm-closable bilinear forms and Lagrangians. Ann. Mat. Pura Appl. **195**(4), 1021–1054 (2016)
30. Hinz, M., Rogers, L.: Magnetic fields on resistance spaces. J. Fractal Geom. **3**(1), 75–93 (2016)
31. Hinz, M., Teplyaev, A.: Dirac and magnetic Schrödinger operators on fractals. J. Funct. Anal. **265**(11), 2830–2854 (2013)
32. Hinz, M., Teplyaev, A.: Vector analysis on fractals and applications. Fractal Geometry and Dynamical Systems in Pure and Applied Mathematics. II. Fractals in Applied Mathematics. Contemporary Mathematics, vol. 601, pp. 147–163. American Mathematical Society, Providence (2013)
33. Hinz, M., Teplyaev, A.: Local Dirichlet forms, Hodge theory, and the Navier-Stokes equations on topologically one-dimensional fractals. Trans. Am. Math. Soc. **367**(2), 1347–1380 (2015)
34. Hinz, M., Röckner, M., Teplyaev, A.: Vector analysis for Dirichlet forms and quasilinear PDE and SPDE on metric measure spaces. Stoch. Process. Appl. **123**(12), 4373–4406 (2013)
35. Hinz, M., Kelleher, D., Teplyaev, A.: Metric and spectral triples for Dirichlet and resistance forms. J. Noncommutative Geom. **9**(2), 359–390 (2015)
36. Ionescu, M., Rogers, L.G., Teplyaev, A.: Derivations and Dirichlet forms on fractals. J. Funct. Anal. **263**(8), 2141–2169 (2012)
37. Jara, M.: Hydrodynamic limit for a zero-range process in the Sierpinski gasket. Commun. Math. Phys. **288**(2), 773–797 (2009)
38. Kajino, N.: Heat kernel asymptotics for the measurable Riemannian structure on the Sierpinski gasket. Potential Anal. **36**(1), 67–115 (2012)
39. Kelleher, D.J.: Differential forms for fractal subspaces and finite energy coordinates (2017). arXiv:1701.02684
40. Kigami, J.: Analysis on Fractals. Cambridge Tracts in Mathematics, vol. 143. Cambridge University Press, Cambridge (2001)

41. Kigami, J.: Harmonic analysis for resistance forms. J. Funct. Anal. **204**(2), 399–444 (2003)
42. Kipnis, C., Landim, C.: Scaling Limits of Interacting Particle Systems. Grundlehren der Mathematischen Wissenschaften, vol. 320. Springer, Berlin (1999)
43. Kipnis, C., Olla, S., Varadhan, S.R.S.: Hydrodynamics and large deviation for simple exclusion processes. Commun. Pure Appl. Math. **42**(2), 115–137 (1989)
44. Kusuoka, S.: Dirichlet forms on fractals and products of random matrices. Publ. Res. Inst. Math. Sci. **25**(4), 659–680 (1989)
45. Landim, C., Mourragui, M., Sellami, S.: Hydrodynamic limit for a nongradient interacting particle system with stochastic reservoirs. Teor. Ver. Prim. **45**(4), 694–717 (2000)
46. Lions, J.L.: Sur certaines équations paraboliques non linéaires. Bull. Soc. Math. Fr. **93**, 155–175 (1965)
47. Lyons, R., Peres, Y.: Probability on Trees and Networks. Cambridge University Press, Cambridge (2017)
48. Ma, Z.M., Röckner, M.: Introduction to the Theory of (Nonsymmetric) Dirichlet Forms. Universitext. Springer, Berlin (1992)
49. Showalter, R.E.: Monotone Operators in Banach Space and Nonlinear Partial Differential Equations. Mathematical Surveys and Monographs, vol. 49. AMS (1997)
50. Strichartz, R.S.: Differential Equations on Fractals. A Tutorial. Princeton University Press, Princeton (2006)
51. Telcs, A.: Local sub-gaussian estimates on graphs: the strongly recurrent case. Electron. J. Probab. **6**(22), 33 (2001). (electronic)
52. Telcs, A.: Volume and time doubling of graphs and random walks: the strongly recurrent case. Commun. Pure Appl. Math. **54**(8), 975–1018 (2001)

Probabilistic Approach to the Stochastic Burgers Equation

Massimiliano Gubinelli and Nicolas Perkowski

Abstract We review the formulation of the stochastic Burgers equation as a martingale problem. One way of understanding the difficulty in making sense of the equation is to note that it is a stochastic PDE with distributional drift, so we first review how to construct finite-dimensional diffusions with distributional drift. We then present the uniqueness result for the stationary martingale problem of (M. Gubinelli and N. Perkowski, Energy solutions of KPZ are unique. 2015, [18]), but we mainly emphasize the heuristic derivation and also we include a (very simple) extension of (M. Gubinelli and N. Perkowski, Energy solutions of KPZ are unique. 2015, [18]) to a non-stationary regime.

Keywords Stochastic Burgers equation · Martingale problem · Diffusions with distributional drift

1 Introduction

In the past few years there has been a high interest and tremendous progress in so called *singular stochastic PDEs* which involve very irregular noise terms (usually some form of white noise) and nonlinear operations. The presence of the irregular noise prevents the solutions from being smooth functions and therefore the nonlinear operations appearing in the equation are a priori ill-defined and can often only be made sense of with the help of some form of renormalization. The major breakthrough

Financial support by the DFG via CRC 1060 (MG) and Research Unit FOR 2402 (NP) is gratefully acknowledged.

M. Gubinelli
Hausdorff Center for Mathematics & Institute for Applied Mathematics, Universität Bonn, Bonn, Germany
e-mail: gubinelli@iam.uni-bonn.de

N. Perkowski (✉)
Institut für Mathematik, Humboldt–Universität zu Berlin, Berlin, Germany
e-mail: perkowsk@math.hu-berlin.de

A. Eberle et al. (eds.), *Stochastic Partial Differential Equations and Related Fields*, Springer Proceedings in Mathematics & Statistics 229,
https://doi.org/10.1007/978-3-319-74929-7_35

in the understanding of these equations was achieved by Hairer [21] who noted that in a one-dimensional setting rough path integration can be used to make sense of the ill-defined nonlinearities and then used this insight to solve the Kardar-Parisi-Zhang (KPZ) equation for the first time [22] and then proceeded to develop his regularity structures [23] which extend rough paths to higher dimensions and allow to solve a wide class of singular SPDEs. Alternative approaches are paracontrolled distributions [20] and renormalization group techniques [28], but they all have in common that they essentially bypass probability theory and are based on pathwise arguments. A general probabilistic understanding of singular SPDEs still seems out of reach, but recently we have been able to prove uniqueness for the stationary martingale problem for the stochastic Burgers equation [18] (an equivalent formulation of the KPZ equation) which was introduced in 2010 by Gonçalves and Jara [15, 16],[1] see also [19]. The aim of this short note is to give a pedagogical account of the probabilistic approach to Burgers equation. We also slightly extend the result of [18] to a simple non-stationary setting.

One way of understanding the difficulty in finding a probabilistic formulation to singular SPDEs is to note that they are often SPDEs with distributional drift. As an analogy, you may think of the stochastic *ordinary* differential equation

$$\mathrm{d}x_t = b(x_t)\mathrm{d}t + \sqrt{2}\mathrm{d}w_t, \tag{1}$$

where $x\colon \mathbb{R}_+ \to \mathbb{R}$, w is a one-dimensional Brownian motion, and $b \in \mathscr{S}'$, the Schwartz distributions on $\mathbb{R}$. It is a nontrivial problem to even make sense of (1) because x takes values in $\mathbb{R}$ but for a point $z \in \mathbb{R}$ the value $b(z)$ is not defined. One possible solution goes as follows: If there exists a nice antiderivative $B' = b$, then we consider the measure $\mu(\mathrm{d}x) = (\exp(B(x))/Z)\mathrm{d}x$ on $\mathbb{R}$, where $Z > 0$ is a normalization constant. If b itself is a continuous function, then this measure is invariant with respect to the dynamics of (1) and the solution x corresponds to the Dirichlet form $\mathscr{E}(f, g) = \int \partial_x f(x)\partial_x g(x)\mu(\mathrm{d}x)$. But $\mathscr{E}$ makes also sense for distributional b, so in that case the solution to (1) can be simply defined as the Markov process corresponding to $\mathscr{E}$. Among others, the works [29, 30] are based on this approach.

An alternative viewpoint is to consider the martingale problem associated to (1). The infinitesimal generator of x should be $\mathscr{L} = b\partial_x + \partial^2_{xx}$ and we would like to find a sufficiently rich space of test functions φ for which

$$M_t^\varphi = \varphi(x_t) - \varphi(x_0) - \int_0^t \mathscr{L}\varphi(x_s)\mathrm{d}s$$

is a continuous martingale. But now our problem is that the term $b\partial_x\varphi$ appearing in $\mathscr{L}\varphi$ is the product of a distribution and a smooth function. Multiplication with a non-vanishing smooth function does not increase regularity (think of multiplying by 1 which is perfectly smooth), and therefore $\mathscr{L}\varphi$ will only be a distribution and not a function. But then the expression $\int_0^t \mathscr{L}\varphi(x_s)\mathrm{d}s$ still does not make any sense! The

[1]The paper [16] is the revised and published version of [15].

solution is to take non-smooth functions φ as input into the generator because multiplication with a non-smooth function can increase regularity (think of multiplying an irregular function $f > 0$ with $1/f$): If we can solve the equation $\mathscr{L}\varphi = f$ for a given continuous function f, then we can reformulate the martingale problem by requiring that

$$M_t^{\varphi} = \varphi(x_t) - \varphi(x_0) - \int_0^t f(x_s)\mathrm{d}s$$

is a continuous martingale. If we are able to solve the equation $\mathscr{L}\varphi = f$ for all continuous bounded functions f, then we explicitly obtain the domain of the generator $\mathscr{L}$ as $\mathrm{dom}(\mathscr{L}) = \{\varphi : \mathscr{L}\varphi = f \text{ for some } f \in C_b\}$. In that case the distribution of x is uniquely determined through the martingale problem. Of course $\mathscr{L}\varphi = f$ is still not easy to solve because $\mathscr{L}\varphi$ contains the term $b\partial_x\varphi$ which is a product between the distribution b and the non-smooth function $\partial_x\varphi$ and therefore not always well defined. In the one-dimensional time-homogeneous framework that we consider here it is however possible to apply a transformation that removes this ill-behaved term and maps the equation $\mathscr{L}\varphi = f$ to a well posed PDE provided that b is the derivative of a continuous function. This was carried out in the very nice papers [11, 12], where it was also observed that the solution is always a Dirichlet process (sum of a martingale plus a zero quadratic variation term), and that if (b_n) is a sequence of smooth functions converging in $\mathscr{S}'$ to b, then

$$x_t = x_0 + \lim_{n\to\infty} \int_0^t b_n(x_s)\mathrm{d}s + \sqrt{2}w_t. \tag{2}$$

So while $b(x_s)$ at a fixed time s does not make any sense, there is a time-decorrelation effect happening that allows to define the integral $\int_0^t b(x_s)\mathrm{d}s$ using the above limit procedure.

The transformation of the PDE breaks down as soon as we look at multi-dimensional or time-inhomogeneous diffusions. But the philosophy of solving $\mathscr{L}\varphi = f$ to identify the domain of the generator carries over. It is then necessary to deal with PDEs with distributional coefficients, and using paraproducts, rough paths, and paracontrolled distributions respectively, the construction of x could recently be extended to the multi-dimensional setting for comparably regular b (Hölder regularity $> -1/2$) [13], a more irregular one-dimensional setting [8] (Hölder regularity $> -2/3$), and finally the irregular multi-dimensional setting [4] (Hölder regularity $> -2/3$). In all these works the time-homogeneity no longer plays a role. Let us also mention that all these works are concerned with probabilistically weak solutions to the equation. If b is a non-smooth function rather than a distribution, then it is possible to find a unique probabilistically strong solution. This goes back to [33, 35] and very satisfactory results are due to Krylov and Röckner [27]. For a pathwise approach that extends to non-semimartingale driving noises such as fractional Brownian motion see also [5]. A good recent survey on such "regularization by noise" phenomena is [10].

Having gained some understanding of the finite-dimensional case, we can now have a look at the stochastic Burgers equation on the one-dimensional torus $\mathbb{T} = \mathbb{R}/\mathbb{Z}$, $u\colon \mathbb{R}_+ \times \mathbb{T} \to \mathbb{R}$,

$$\mathrm{d}u_t = \Delta u_t \mathrm{d}t + \partial_x u_t^2 \mathrm{d}t + \sqrt{2}\mathrm{d}\partial_x W_t, \tag{3}$$

where $\partial_t W$ is a space-time white noise, that is the centered Gaussian process with covariance $\mathbb{E}[\partial_t W(t,x)\partial_t W(s,y)] = \delta(t-s)\delta(x-y)$. We would like to understand (3) as an infinite-dimensional analogue of (1) for a particular b. It is known that the dynamics of u are invariant under the measure μ, the law of the space white noise on $\mathbb{T}$, which means that for fixed times u_t is a distribution with regularity $C^{-1/2-\varepsilon}$ (i.e. the distributional derivative of a $(1/2-\varepsilon)$ Hölder continuous function) and not a function. But then the square u_t^2 does not make any sense analytically because in general it is impossible to multiply distributions in $C^{-1/2-\varepsilon}$. Let us however simply ignore this problem and try to use the probabilistic structure of u to make sense of u^2. For simplicity we start u in the stationary measure μ. Then u_t^2 is simply the square of the white noise, and this is something that we can explicitly compute. More precisely, let $\rho \in C_c^\infty(\mathbb{R})$ be a nonnegative function with $\int_\mathbb{R} \rho(x)\mathrm{d}x = 1$ and define the convolution

$$(\rho^N * u)(x) := \sum_{k\in\mathbb{Z}} u(N\rho(N(x-k-\cdot))).$$

Then $\rho^N * u$ is a smooth periodic function that converges to u, and in particular $(\rho^N * u)^2$ is well defined. Let us test against some $\varphi \in C^\infty$ and consider $(\rho^N * u)^2(\varphi) = \int_\mathbb{T} (\rho^N * u)^2(x)\varphi(x)\mathrm{d}x$. The first observation we then make is that $\mathbb{E}[\int_\mathbb{T}(\rho^N * u)^2(x)\varphi(x)\mathrm{d}x] = c_N \int_\mathbb{T} \varphi(x)\mathrm{d}x$ with a diverging constant $c_N \to \infty$. So to have any hope to obtain a well-defined limit, we should rather consider the renormalized $((\rho^N * u)^2 - c_N)(\varphi)$. Of course, the constant c_N vanishes under differentiation so we will not see it in the equation that features the term $\partial_x u^2$ and not u^2. Now that we subtracted the mean $\mathbb{E}[(\rho^N * u)^2(\varphi)]$, we would like to show that the variance of the centered random variable stays uniformly bounded in N. This is however not the case, and in particular $((\rho^N * u)^2 - c_N)(\varphi)$ does not converge in $L^2(\mu)$. But $L^2(\mu)$ is a Gaussian Hilbert space and $((\rho^N * u)^2 - c_N)$ lives for all N in the second (homogeneous) chaos. Since for sequences in a fixed Gaussian chaos convergence in probability is equivalent to convergence in L^2 (see [24], Theorem 3.50), $((\rho^N * u)^2 - c_N)(\varphi)$ does not converge in any reasonable sense. Therefore, the square $u^2(\varphi)$ (or rather the renormalized square $u^{\diamond 2}(\varphi) = (u^2 - \infty)(\varphi)$) cannot be declared as a random variable! However, and as far as we are aware it was Assing [1] who first observed this, the renormalized square does make sense as a distribution on $L^2(\mu)$. More precisely, if $F(u)$ is a "nice" random variable (think for example of $F(u) = f(u(\varphi_1), \dots, u(\varphi_n))$ for $f \in C_c^\infty(\mathbb{R})$ and $\varphi_k \in C^\infty(\mathbb{T})$, $k = 1, \dots, n$), then $\mathbb{E}[((\rho^N * u)^2 - c_N)(\varphi)F(u)]$ converges to a limit that we denote with $\mathbb{E}[u^{\diamond 2}(\varphi)F(u)]$ and that does not depend on the mollifier ρ. This means that we cannot evaluate $u \mapsto u^{\diamond 2}(\varphi)$ pointwise, but it makes sense when tested against smooth random variables by evaluating the integral $\int_{C^{-1/2-\varepsilon}} u^{\diamond 2}(\varphi)F(u)\mu(\mathrm{d}u)$. Compare that with a distribution $T \in \mathscr{S}'(\mathbb{R})$ on $\mathbb{R}$ for which the pointwise evaluation $T(x)$

makes no sense, but we can test T against smooth test functions ψ by evaluating the integral $\int_{\mathbb{R}} T(x)\psi(x)\mathrm{d}x$. Of course in finite dimensions the integration is canonically performed with respect to the Lebesgue measure, while in infinite dimensions we have to pick a reference measure which here is the invariant measure μ.

So $u^{\diamond 2}(\varphi)$ is a distribution on the infinite-dimensional space $C^{-1/2-\varepsilon}$ and therefore the stochastic Burgers equation (3) is an infinite-dimensional analog to (1). We would like to use the same tools as in the finite dimensional case, and in particular we would like to make sense of the martingale problem for u. However, while in finite dimensions it is possible to solve the equation $\mathscr{L}\varphi = f$ under rather general conditions, now this would be an infinite-dimensional PDE with a distributional coefficient. There exists a theory of infinite-dimensional PDEs, see for example [3, 6], but at the moment it seems an open problem how to solve such PDEs with distributional coefficients. On the other side, if we simply plug in a smooth function F into the formal generator $\mathscr{L}$, then $\mathscr{L}F$ is only a distribution and not a function and therefore no smooth F can be in the domain of the Burgers generator. Assing [1] avoided this problem by considering a "generalized martingale problem", where the drift term $\int_0^t \partial_x u_s^2 \mathrm{d}s = \int_0^t \partial_x u_s^{\diamond 2} \mathrm{d}s$ never really appears. However, it is still not known whether solutions to the generalized martingale problem exist and are unique. This was the state until 2010, when Gonçalves and Jara [15] introduced a new notion of martingale solution to the stochastic Burgers equation by stepping away from the infinitesimal approach based on the generator and by making use of the same time-decorrelation effect that allowed us to construct $\int_0^t b(x_s)\mathrm{d}s$ in (2) despite the fact that $b(x_s)$ itself makes no sense. In the next chapter we discuss their approach and how to prove existence and uniqueness of solutions.

2 Energy Solutions to the Stochastic Burgers Equation

We would now like to formulate a notion of martingale solution to the stochastic Burgers equation

$$\mathrm{d}u_t = \Delta u_t \mathrm{d}t + \partial_x u_t^2 \mathrm{d}t + \sqrt{2}\mathrm{d}\partial_x W_t.$$

As discussed above the infinitesimal picture is not so clear because the drift is a distribution in an infinite-dimensional space. So the idea of Gonçalves and Jara [16] is to rather translate the formulation (2) to our context. Let us call u a *martingale solution* if $u \in C(\mathbb{R}_+, \mathscr{S}')$, where $\mathscr{S}' = \mathscr{S}'(\mathbb{T})$ are the (Schwartz) distributions on $\mathbb{T}$, and if there exists $\rho \in C_c^\infty(\mathbb{R})$ with $\int_{\mathbb{R}} \rho(x)\mathrm{d}x = 1$ such that with $\rho^N = N\rho(N\cdot)$ the limit

$$\int_0^{\cdot} \partial_x u_s^2 \mathrm{d}s(\varphi) := \lim_{N\to\infty} \int_0^{\cdot} (u_s * \rho^N)^2(-\partial_x \varphi)\mathrm{d}s$$

exists for all $\varphi \in C^\infty(\mathbb{T})$, uniformly on compacts in probability. Note that because of the derivative ∂_x here it is not necessary to renormalize the square by subtracting a large constant. Moreover, we require that for all $\varphi \in C^\infty(\mathbb{T})$ the process

$$M_t(\varphi) := u_t(\varphi) - u_0(\varphi) - \int_0^t u_s(\Delta\varphi)\mathrm{d}s - \int_0^t \partial_x u_s^2 \mathrm{d}s(\varphi), \qquad t \geqslant 0, \tag{4}$$

is a continuous martingale with quadratic variation $\langle M(\varphi)\rangle_t = 2\|\partial_x\varphi\|_{L^2}^2 t$. It can be easily shown that martingale solutions exist, for example by considering Galerkin approximations to the stochastic Burgers equation.

On the other side, martingale solutions are very weak and give us absolutely no control on the nonlinear part of the drift, so it does not seem possible to show that they are unique. To overcome this we should add further conditions to the definition, and Gonçalves and Jara assumed in [16] additionally that u satisfies an *energy estimate*, that is

$$\mathbb{E}\left[\left(\int_s^t \{(u_r * \rho^N)^2(-\partial_x\varphi) - (u_r * \rho^M)^2(-\partial_x\varphi)\}\mathrm{d}r\right)^2\right] \lesssim \frac{(t-s)}{M \wedge N}\|\partial_x\varphi\|_{L^2}^2.$$

The precise form of the estimate is not important, we will only need that it implies $\int_0^\cdot \partial_x u_s^2 \mathrm{d}s \in C(\mathbb{R}_+, \mathscr{S}')$ and that for every test function φ the process $\int_0^\cdot \partial_x u_s^2 \mathrm{d}s(\varphi)$ has zero quadratic variation. Consequently, $u(\varphi)$ is a Dirichlet process (sum of a continuous local martingale and a zero quadratic variation process) and admits an Itô formula [32]. Any martingale solution that satisfies the energy estimate is called an *energy solution*.

This is still a very weak notion of solution, and there does not seem to be any way to directly compare two different energy solutions and to show that they have the same law. To see why the additional structure coming from the energy estimate is nonetheless very useful, let us argue formally for a moment. Let u be a solution to Burgers equation $\partial_t u = \Delta u + \partial_x u^2 + \partial_t\partial_x W$. Then $u = \partial_x h$ for the solution h to the KPZ equation

$$\partial_t h = \Delta h + (\partial_x h)^{\diamond 2} + \partial_t W = \Delta h + (\partial_x h)^2 - \infty + \partial_t W.$$

But the KPZ equation can be linearized through the Cole-Hopf transform [2]: We formally have $h = \log\phi$ for the solution w to the linear stochastic heat equation

$$\partial_t\phi = \Delta\phi + \phi\partial_t W, \tag{5}$$

which is a linear Itô SPDE that can be solved using classical theory [7, 31, 34]. So if we can show that every energy solution u satisfies $u = \partial_x \log\phi$ where ϕ solves (5), then u is unique. And the key difference between energy solutions and martingale solutions is that for energy solutions we have an Itô formula that allows us to perform a change of variables that maps u to a candidate solution to the stochastic heat equation.

So let u be an energy solution, let $\rho \in \mathscr{S}(\mathbb{R})$ be an even function with Fourier transform $\hat\rho = \int_{\mathbb{R}} e^{2\pi i x \cdot}\rho(x)\mathrm{d}x \in C_c^\infty(\mathbb{R})$ and such that $\hat\rho \equiv 1$ on a neighborhood of 0, and define

$$u_t^L := \mathscr{F}_{\mathbb{T}}^{-1}(\hat\rho(L^{-1}\cdot)\mathscr{F}_{\mathbb{T}}u_t) = \rho^L * u_t, \qquad t \geqslant 0 \tag{6}$$

where $\mathscr{F}_{\mathbb{T}}u(k) := \int_{\mathbb{T}} e^{2\pi i k x} u(x)\mathrm{d}x$ respectively $\mathscr{F}_{\mathbb{T}}^{-1}\psi(x) := \sum_{k\in\mathbb{Z}} e^{2\pi i k x}\psi(k)$ denote the Fourier transform (respectively inverse Fourier transform) on the torus. We then integrate u^L by setting

$$h_t^L := \mathscr{F}_{\mathbb{T}}^{-1}(\mathscr{F}_{\mathbb{T}}\Theta\mathscr{F}_{\mathbb{T}}(u_t^L)) = \Theta * u_t^L = (\Theta * \rho^L) * u_t =: \Theta^L * u_t, \qquad t \geqslant 0,$$

where $\mathscr{F}_{\mathbb{T}}\Theta(k) = \mathbb{1}_{k\neq 0}(2\pi i k)^{-1}$ and therefore $\partial_x(\Theta * u) = u - \int_{\mathbb{T}} u(x)\mathrm{d}x$. Since $\int_{\mathbb{T}} u(x)\mathrm{d}x$ is conserved by the stochastic Burgers equation, it suffices to prove uniqueness of $\partial_x(\Theta * u)$. Writing $h_t^L(x) = u_t(\Theta_x^L)$ for $\Theta_x^L(y) := (\Theta * \rho^L)(x-y)$, we get from (4)

$$\mathrm{d}h_t^L(x) = \Delta_x h_t^L(x) + u_t^2(\partial_x\Theta_x^L)\mathrm{d}t + \sqrt{2}\mathrm{d}W_t(\partial_x\Theta_x^L)$$

for a white noise W, and $\mathrm{d}\langle h^L(x)\rangle_t = 2\|\partial_x\Theta_x^L\|_{L^2(\mathbb{T})}^2\mathrm{d}t$. It is not hard to see that $\partial_x\Theta_x^L = \rho_x^L - 1$ for $\rho_x^L(y) = \rho^L(x-y)$. So setting $\phi_t^L(x) := e^{h_t^L(x)}$ we have by the Itô formula for Dirichlet processes [32] and a short computation

$$\begin{aligned}\mathrm{d}\phi_t^L(x) = {} & \Delta_x\phi_t^L(x)\mathrm{d}t + \sqrt{2}\phi_t^L(x)\mathrm{d}W_t(\rho_x^L) + \mathrm{d}R_t^L(x) + K\phi_t^L(x)\mathrm{d}t + \phi_t^L(x)\mathrm{d}Q_t^L \\ & - \sqrt{2}\phi_t^L(x)\mathrm{d}W_t(1) - 2\phi_t^L(x)\mathrm{d}t,\end{aligned}$$

where

$$R_t^L(x) := \int_0^t \phi_s^L(x)\Big\{u_s^2(\partial_x\Theta_x^L) - \Big((u_s^L(x))^2 - \int_{\mathbb{T}}(u_s^L(y))^2\mathrm{d}y\Big) - K\Big\}\mathrm{d}s \tag{7}$$

for a suitable constant K and

$$Q_t^L := \int_0^t \Big\{-\int_{\mathbb{T}}((u_s^L)^2(y) - \|\rho^L\|_{L^2(\mathbb{R})}^2)\mathrm{d}y + 1\Big\}\mathrm{d}s. \tag{8}$$

Proposition 1 *Let $u \in C(\mathbb{R}_+, \mathscr{S}')$ be an energy solution to the stochastic Burgers equation such that*

$$\sup_{x\in\mathbb{T}, t\in[0,T]} \mathbb{E}[|e^{u_t(\Theta_x)}|^2] < \infty. \tag{9}$$

If for R^L and Q^L defined in (7) and (8) respectively and every test function φ the process $R^L(\varphi)$ converges to zero uniformly on compacts in probability and Q^L converge to a zero quadratic variation process Q, then u is unique in law and given by $u_t = \partial_x \log\phi_t + \int_{\mathbb{T}} u_0(y)\mathrm{d}y$, where ϕ is the unique solution to the stochastic heat equation

$$\mathrm{d}\phi_t = \Delta\phi_t\mathrm{d}t + \phi_t\mathrm{d}W_t, \qquad \phi_0(x) = e^{u_0(\Theta_x)}.$$

Proof This is Theorem 2.13 in [18]. Actually we need a slightly stronger convergence of $R^L(\varphi)$ and Q^L than locally uniform, but to simplify the presentation we ignore this here.

Remark 1 The strategy of mapping the energy solution to the linear stochastic heat equation is essentially due to Funaki and Quastel [14], who used it in a different context to study the invariant measure of the KPZ equation. In their approach similar correction terms as R^L and Q^L appeared, and the fact that they were able to deal with them gave us the courage to proceed with the rather long computations that control R^L and Q^L in our setting.

So to obtain the uniqueness of energy solutions we need to verify the assumptions of Proposition 1. Unfortunately we are not able to do this because while the energy condition gives us good control of the Burgers nonlinearity $\int_0^\cdot \partial_x u_s^2 \mathrm{d}s$, it does not allow us to bound general additive functionals $\int_0^\cdot F(u_s)\mathrm{d}s$ such as the ones appearing in the definition of R^L.

To understand which condition to add in order to control such additive functionals, let us recall how this can be (formally) achieved for a Markov process X with values in a Polish space. Assume that X is stationary with initial distribution μ, denote its infinitesimal generator with $\mathscr{L}$, and let $\mathscr{L}^*$ be the adjoint of $\mathscr{L}$ in $L^2(\mu)$. Then $\mathscr{L}$ can be decomposed into a symmetric part $\mathscr{L}_S = (\mathscr{L} + \mathscr{L}^*)/2$ and an antisymmetric part $\mathscr{L}_A = (\mathscr{L} - \mathscr{L}^*)/2$. Moreover, if we reverse time and set $\hat{X}_t = X_{T-t}$ for some fixed $T > 0$, then $(\hat{X}_t)_{t\in[0,T]}$ is a Markov process in its natural filtration, the backward filtration of X, and it has the generator $\mathscr{L}^*$. See Appendix 1 of [25] for the case where X takes values in a countable space. So by Dynkin's formula for $F \in \mathrm{dom}(\mathscr{L}) \cap \mathrm{dom}(\mathscr{L}^*)$ the process

$$M_t^F = F(X_t) - F(X_0) - \int_0^t \mathscr{L}F(X_s)\mathrm{d}s, \qquad t \geqslant 0,$$

is a martingale, and

$$\hat{M}_t^F = F(\hat{X}_t) - F(\hat{X}_0) - \int_0^t \mathscr{L}^* F(\hat{X}_s)\mathrm{d}s, \qquad t \in [0, T],$$

is a martingale in the backward filtration. We add these two formulas and obtain the following decomposition which is reminiscent of the *Lyons-Zheng decomposition* of a Dirichlet process into a sum of forward martingale and a backward martingale:

$$M_t^F + (\hat{M}_T^F - \hat{M}_{T-t}^F) = -2 \int_0^t \mathscr{L}_S F(X_s)\mathrm{d}s,$$

so any additive functional of the form $\int_0^t \mathscr{L}_S F(X_s)\mathrm{d}s$ is the sum of two martingales, one in the forward filtration, the other one in the backward filtration. The predictable quadratic variation of the martingale M^F is

$$\langle M^F \rangle_t = \int_0^t (\mathscr{L}F^2(X_s) - 2F(X_s)\mathscr{L}F(X_s))\mathrm{d}s,$$

and

$$\langle \hat{M}^F \rangle_t = \int_0^t (\mathscr{L}^* F^2(\hat{X}_s) - 2F(\hat{X}_s)\mathscr{L}^* F(\hat{X}_s))\mathrm{d}s.$$

So the Burkholder-Davis-Gundy inequality gives for all $p \geqslant 2$

$$\begin{aligned}
\mathbb{E}\left[\sup_{t\in[0,T]}\left|\int_0^t \mathscr{L}_S F(X_s)\mathrm{d}s\right|^p\right] &\lesssim T^{p/2-1}\int_0^T \mathbb{E}[(\mathscr{L}F^2(X_s) - 2F(X_s)\mathscr{L}F(X_s))^{p/2}]\mathrm{d}s \\
&\quad + T^{p/2-1}\int_0^T \mathbb{E}[(\mathscr{L}^* F^2(\hat{X}_s) - 2F(\hat{X}_s)\mathscr{L}^* F(\hat{X}_s))^{p/2}]\mathrm{d}s \\
&= T^{p/2}\mathbb{E}[(\mathscr{L}F^2(X_0) - 2F(X_0)\mathscr{L}F(X_0))^{p/2}] \\
&\quad + T^{p/2}\mathbb{E}[(\mathscr{L}^* F^2(X_0) - 2F(X_0)\mathscr{L}^* F(X_0))^{p/2}],
\end{aligned}$$

where in the last step we used the stationarity of X. Moreover, if $\mathscr{L}_A$ satisfies Leibniz rule for a first order differential operator, then $\mathscr{L}_A F^2 - 2F\mathscr{L}_A F = 0$ and therefore

$$\mathbb{E}\left[\sup_{t\in[0,T]}\left|\int_0^t \mathscr{L}_S F(X_s)\mathrm{d}s\right|^p\right] \lesssim T^{p/2}\mathbb{E}[(\mathscr{L}_S F^2(X_0) - 2F(X_0)\mathscr{L}_S F(X_0))^{p/2}].$$

That is, we can bound additive functionals of a stationary Markov process X in terms of the symmetric part of the generator $\mathscr{L}_S$ and the invariant measure μ. We call this inequality the *martingale trick*.

For the stochastic Burgers equation this promises to be very powerful, because formally an invariant measure is given by the law of the spatial white noise and the symmetric part of the generator is simply the generator $\mathscr{L}_S$ of the Ornstein-Uhlenbeck process $\mathrm{d}X_t = \Delta X_t \mathrm{d}t + \mathrm{d}\partial_x W_t$, that is the linearized Burgers equation. This also suggests that $\mathscr{L}_A$ is a first order differential operator because it corresponds to the drift $\partial_x u^2$. We thus need a notion of solution to the stochastic Burgers equation which allows us to make the heuristic argumentation above rigorous. This definition was given by Gubinelli and Jara in [19]:

Definition 1 Let u be an energy solution to the stochastic Burgers equation. Then u is called a *forward-backward (FB) solution* if additionally the law of u_t is that of the white noise for all $t \geqslant 0$, and for all $T > 0$ the time-reversed process $\hat{u}_t = u_{T-t}$, $t \in [0, T]$, is an energy solution to

$$\mathrm{d}\hat{u}_t = \Delta\hat{u}_t\mathrm{d}t - \partial_x \hat{u}_t^2 \mathrm{d}t + \sqrt{2}\mathrm{d}\partial_x \hat{W}_t,$$

where $\partial_t \hat{W}$ is a space-time white noise in the backward filtration.

Remark 2 [19] do not define FB-solutions, but they also call their solutions energy solutions. For pedagogical reasons we prefer here to introduce a new terminology for this third notion of solution. Also, the definition in [19] is formulated slightly differently than above, but it is equivalent to our definition.

Of course, we should first verify whether FB-solutions exist before we proceed to discuss their uniqueness. But existence is very easy and can be shown by Galerkin approximation. Also, it is known for a wide class of interacting particle systems that they are relatively compact under rescaling and all limit points are FB-solutions to Burgers equation [9, 16, 17]. A general methodology that allows to prove similar results for many particle systems was developed in [16]. So all that is missing in the convergence proof is the uniqueness of FB-solutions.

Lemma 1 *If u is a FB-solution, then the martingale trick works:*

$$\mathbb{E}\left[\sup_{t\in[0,T]}\left|\int_0^t \mathscr{L}_S F(u_s)\mathrm{d}s\right|^p\right] \lesssim T^{p/2}\mathbb{E}[\mathscr{E}(F(u_0))^{p/2}],$$

where $\mathscr{E}(F(u)) = 2\int_{\mathbb{R}} |\partial_x \mathrm{D}_x F(u)|^2 \mathrm{d}x$ for the Malliavin derivative D which is defined in terms of the law of the white noise.

Proof This is Lemma 2 in [19] or Proposition 3.2 in [18].

Given a FB-solution u we can thus control any additive functional of the form $\int_0^\cdot \mathscr{L}_S F(u_s)\mathrm{d}s$. To bound $\int_0^\cdot G(u_s)\mathrm{d}s$ for a given G we therefore have to solve the Poisson equation $\mathscr{L}_S F = G$. This is an infinite-dimensional partial differential equation and a priori difficult to understand. However, we have to only solve it in $L^2(\mu)$ which has a lot of structure as a Gaussian Hilbert space. In particular we have the chaos decomposition

$$F = \sum_{n=0}^{\infty} W_n(f_n)$$

for all $F \in L^2(\mu)$, where $f_n \in L^2(\mathbb{R}^n)$ and $W_n(f_n)$ is the nth order Wiener-Itô integral over f_n. And the action of the Ornstein-Uhlenbeck generator on $W_n(f_n)$ has a very simple form:

$$\mathscr{L}_S W_n(f_n) = W_n(\Delta f_n),$$

where $\Delta = \partial^2_{x_1} + \cdots + \partial^2_{x_n}$ is the Laplace operator on $\mathbb{R}^n$, see Lemma 3.7 in [18]. This reduces the equation $\mathscr{L}_S F = G$ to the infinite system of uncoupled equations

$$\Delta_n f_n = g_n, \qquad n \in \mathbb{N}_0,$$

where $G = \sum_n W_n(g_n)$.

To test the tools we have developed so far, let us apply them to the Burgers nonlinearity:

Lemma 2 *Let u be a FB-solution. There exists a unique process $\int_0^\cdot u_s^{\diamond 2}\mathrm{d}s \in C(\mathbb{R}_+, \mathscr{S}')$ such that $\partial_x \int_0^\cdot u_s^{\diamond 2}\mathrm{d}s = \int_0^\cdot \partial_x u_s^2 \mathrm{d}s$ and such that for all $T, p > 0$, $\alpha \in (0, 3/4)$, and $\chi, \varphi \in \mathscr{S}$ with $\int_{\mathbb{R}} \chi(x)\mathrm{d}x = 1$ we have with $\chi^N = N\chi(N\cdot)$*

$$\lim_{N\to\infty} \mathbb{E}\left[\left\| \int_0^\cdot u_s^{\diamond 2} \mathrm{d}s(\varphi) - \int_0^\cdot ((u_s * \chi^N)^2 - \|\chi^N\|_{L^2}^2)(\varphi) \mathrm{d}s \right\|_{C^\alpha([0,T],\mathbb{R})}^p\right] = 0.$$

Proof This is a combination of Proposition 3.15 and Corollary 3.17 in [18].

The Hölder regularity of $\int_0^\cdot \partial_x u_s^2 \mathrm{d}s(\varphi)$ for $\varphi \in \mathscr{S}$ is indeed only $3/4 - \varepsilon$ and not better than that. This means that the process $u(\varphi)$ is not a semimartingale. In particular, if we assume for the moment that the stochastic Burgers equation defines a Markov process on a suitable Banach space of distributions, then the map $u \mapsto u(\varphi)$ for $\varphi \in \mathscr{S}$ is not in the domain of its generator. In fact it is an open problem to find any nontrivial function in the domain of the Burgers generator.

Next, we would like to use the martingale trick to prove the convergence of R^L and Q^L that we need in Proposition 1. However, while for the Burgers nonlinearity the additive functional was of the form $\int_0^\cdot G(u_s)\mathrm{d}s$ with a G that has only components in the second Gaussian chaos, the situation is not so simple for R^L because the factor $e^{u(\Theta_x^L)}$ has an infinite chaos decomposition. While in principle it is still possible to write down the chaos decomposition of R^L exlicitly, we prefer to follow another route. In fact there is a general tool for Markov processes which allows to control additive functionals without explicitly solving the Poisson equation, the so called *Kipnis-Varadhan estimate*. It is based on duality and reads

$$\mathbb{E}\left[\sup_{t\in[0,T]}\left|\int_0^t F(u_s)\mathrm{d}s\right|^2\right] \lesssim T\|F\|_{-1}^2$$

for

$$\|F\|_{-1}^2 = \sup_G\{2\mathbb{E}[F(u_0)G(u_0)] + \mathbb{E}[F(u_0)\mathscr{L}_S F(u_0)]\}.$$

Corollary 3.5 in [18] proves that this inequality holds also for FB-solutions, despite the fact that we do not know yet whether they are Markov processes. In fact FB-solutions give us enough flexibility to implement the classical proof from the Markovian setting, as presented for example in [26]. Based on the Kipnis-Varadhan inequality, Gaussian integration by parts, and a lenghty computation we are able to show the following result:

Theorem 1 *Let u be a FB-solution to the stochastic Burgers equation. Then the assumptions of Proposition 1 are satisfied and in particular the law of u is unique.*

Proof This is a combination of Lemmas A.1–A.3 in [18].

Both the assumption of stationarity and the representation of the backward process are only needed to control additive functionals of the FB-solution u and thus to prove the convergence of R^L and Q^L. If we had some other means to prove this convergence, uniqueness would still follow. One interesting situation where this is possible is the following: Let u be an energy solution of the stochastic Burgers equation which satisfies (9), and assume that there exists a FB-solution v such that

$\mathbb{P} \ll \mathbb{P}_{\text{FB}}$, where $\mathbb{P}$ denotes the law of u on $C(\mathbb{R}_+, \mathscr{S}')$ and $\mathbb{P}_{\text{FB}}$ is the law of v. Then the convergence in probability that we required in Proposition 1 holds under $\mathbb{P}_{\text{FB}}$ and thus also under $\mathbb{P}$ and therefore u is still unique in law. Of course, the assumption $\mathbb{P} \ll \mathbb{P}_{\text{FB}}$ is very strong, but it can be verified in some nontrivial situations. For example in [17] some particle systems are studied which start in an initial condition μ^n that has bounded relative entropy $H(\mu^n|\nu)$ with respect to a stationary distribution ν, uniformly in n. Denoting the distribution of the particle system started in π by $\mathbb{P}_\pi$, we get from the Markov property $H(\mathbb{P}_{\mu^n}|\mathbb{P}_\nu) = H(\mu^n|\nu)$. Assume that the rescaled process converges to the law $\mathbb{P}_{\text{FB}}$ of a FB-solution under $\mathbb{P}_\nu$, and to the law of an energy solution $\mathbb{P}$ under $\mathbb{P}_{\mu^n}$. Then

$$H(\mathbb{P}|\mathbb{P}_{\text{FB}}) \leqslant \liminf_{n\to\infty} H(\mu^n|\nu) < \infty,$$

and in particular $\mathbb{P} \ll \mathbb{P}_{\text{FB}}$; here we used that the relative entropy is lower semicontinuous. Therefore, the scaling limits of [17] are still unique in law, even though they are not FB-solutions. Let us summarize this observation:

Theorem 2 *Let u be an energy solution to the stochastic Burgers equation which satisfies* (9) *and let v be an FB-solution. Denote the law of u and v with $\mathbb{P}$ and $\mathbb{P}_{\text{FB}}$, respectively. Then the measure $\mathbb{P}$ is unique. In particular this applies for the non-stationary energy solutions constructed in [17].*

Remark 3 For simplicity we restricted our attention to the equation on $\mathbb{T}$, but everything above extends to the stochastic Burgers equation on $\mathbb{R}$. Also, once we understand the uniqueness of Burgers equation it is not difficult to also prove the uniqueness of its integral, the KPZ equation. For details see [18].

References

1. Assing, S.: A pregenerator for burgers equation forced by conservative noise. Commun. Math. Phys. **225**(3), 611–632 (2002)
2. Bertini, L., Giacomin, G.: Stochastic burgers and KPZ equations from particle systems. Commun. Math. Phys. **183**(3), 571–607 (1997)
3. Bogachev, V.I., Krylov, N.V., Röckner, M., Shaposhnikov, S.V.: Fokker-Planck-Kolmogorov Equations. American Mathematical Society, USA (2015)
4. Cannizzaro, G., Chouk, K.: Multidimensional SDEs with singular drift and universal construction of the polymer measure with white noise potential. arXiv:1501.04751 (2015)
5. Catellier, R., Gubinelli, M.: Averaging along irregular curves and regularisation of ODEs. Stoch. Process. Appl. **126**(8), 2323–2366 (2016)
6. Da Prato, G., Zabczyk, J.: Second Order Partial Differential Equations in Hilbert Spaces. Cambridge University Press, UK (2002)
7. Da Prato, G., Zabczyk, J.: Stochastic Equations in Infinite Dimensions. Cambridge University Press, UK (2014)
8. Delarue, F., Diel, R.: Rough paths and 1d SDE with a time dependent distributional drift: application to polymers. Probab. Theory Relat. Fields **165**(1), 1–63 (2016)
9. Diehl, J., Gubinelli, M., Perkowski, N.: The Kardar-Parisi-Zhang equation as scaling limit of weakly asymmetric interacting Brownian motions. Comm. Math. Phys. **354**(2), 549–589 (2017)

10. Flandoli, F.: Random Perturbation of PDEs and Fluid Dynamic Models: Ecole d'été de Probabilités de Saint-Flour XL-2010. Springer Science and Business Media, Berlin (2011)
11. Flandoli, F., Russo, F., Wolf, J.: Some SDEs with distributional drift. Part I: General calculus. Osaka J. Math. **40**(2), 493–542 (2003)
12. Flandoli, F., Russo, F., Wolf, J.: Some SDEs with distributional drift. Part II: Lyons- Zheng structure, Itô's formula and semimartingale characterization. Random Oper. Stoch. Equ. **12**(2), 145–184 (2004)
13. Flandoli, F., Issoglio, E., Russo, F.: Multidimensional stochastic differential equations with distributional drift. Trans. Am. Math. Soc. **369**, 1665–1688 (2017)
14. Funaki, T., Quastel, J.: KPZ equation, its renormalization and invariant measures. Stoch. Partial Differ. Equ. Anal. Comput. **3**(2), 159–220 (2015)
15. Gonçalves, P., Jara, M.: Universality of KPZ Equation. arXiv:1003.4478 (2010)
16. Gonçalves, P., Jara, M.: Nonlinear fluctuations of weakly asymmetric interacting particle systems. Arch. Ration. Mech. Anal. **212**(2), 597–644 (2014)
17. Gonçalves, P., Jara, M., Sethuraman, S.: A stochastic Burgers equation from a class of microscopic interactions. Ann. Probab. **43**(1), 286–338 (2015)
18. Gubinelli, M., Perkowski, N.: Energy solutions of KPZ are unique. J. Amer. Math. Soc. **31**(2), 427–471 (2018)
19. Gubinelli, M., Jara, M.: Regularization by noise and stochastic Burgers equations. Stoch. Partial Differ. Equ. Anal. Comput. **1**(2), 325–350 (2013)
20. Gubinelli, M., Imkeller, P., Perkowski, N.: Paracontrolled distributions and singular PDEs. Forum Math. Pi **3**(6), 1–75 (2015)
21. Hairer, M.: Rough stochastic PDEs. Commun. Pure Appl. Math. **64**(11), 1547–1585 (2011)
22. Hairer, M.: Solving the KPZ equation. Ann. Math. **178**(2), 559–664 (2013)
23. Hairer, M.: A theory of regularity structures. Invent. Math. **198**(2), 269–504 (2014)
24. Janson, S.: Gaussian Hilbert Spaces. Cambridge University Press, UK (1997)
25. Kipnis, C., Landim, C.: Scaling Limits of Interacting Particle Systems. Springer Science and Business Media, Berlin (2013)
26. Komorowski, T., Landim, C., Olla, S.: Fluctuations in Markov processes: Time Symmetry and Martingale Approximation. Springer, Berlin (2012)
27. Krylov, N.V., Röckner, M.: Strong solutions of stochastic equations with singular time dependent drift. Probab. Theory Relat Fields **131**(2), 154–196 (2005)
28. Kupiainen, Antti: Renormalization group and stochastic PDEs. Ann. Henri Poincaré **17**(3), 497–535 (2016)
29. Mathieu, P.: Zero white noise limit through Dirichlet forms, with application to diffusions in a random medium. Probab. Theory Relat Fields **99**(4), 549–580 (1994)
30. Mathieu, P.: Spectra, exit times and long time asymptotics in the zero-white-noise limit. Stochastics **55**(1–2), 1–20 (1995)
31. Prévôt, C., Röckner, M.: A Concise Course on Stochastic Partial Differential Equations, vol. 1905. Springer, Berlin (2007)
32. Russo, F., Vallois, P.: Elements of Stochastic Calculus via Regularization. Séminaire de Probabilités XL, pp. 147–185. Springer, Berlin (2007)
33. Veretennikov, A.J.: On strong solutions and explicit formulas for solutions of stochastic integral equations. Matematicheskii Sbornik **153**(3), 434–452 (1980)
34. Walsh, J.B.: An Introduction to Stochastic Partial Differential Equations. Ecole d'Eté de Probabilités de Saint Flour XIV-1984, pp. 265–439. Springer, Berlin (1986)
35. Zvonkin, A.K.: A transformation of the phase space of a diffusion process that removes the drift. Math. USSR-Sbornik **22**(1), 129–149 (1974)

Equilibrium States, Phase Transitions and Dynamics in Quantum Anharmonic Crystals

Yuri Kozitsky

Abstract The basic elements of the mathematical theory of states of thermal equilibrium of infinite systems of quantum anharmonic oscillators (quantum crystals) are outlined. The main concept of this theory is to describe the states of finite portions of the whole system (local states) in terms of stochastically positive KMS systems and path measures. The global states are constructed as Gibbs path measures satisfying the corresponding DLR equation. The multiplicity of such measures is then treated as the existence of phase transitions. This effect can be established by analyzing the properties of the Matsubara functions corresponding to the global states. The equilibrium dynamics of finite subsystems can also be described by means of these functions. Then three basic results of this theory are presented and discussed: (a) a sufficient condition for a phase transition to occur at some temperature; (b) a sufficient condition for the suppression of phase transitions at all temperatures (quantum stabilization); (c) a statement showing how the phase transition can affect the local equilibrium dynamics.

Keywords KMS state · Path measure · Stochastic process · Green function

MSC2010 82B10 · 82B20 · 82B26

1 Generalities

In recent years, remarkable progress has been made in the experimental testing of the fundamentals of quantum physics, as well as in developing quantum information theory and basics of quantum computing, see [7]. Due to these advances the elaboration of the mathematical background of quantum theory returned to the circle of actual tasks of applied mathematics. Developing the statistical description of infinite

Y. Kozitsky (✉)
Institute of Mathematics, Maria Curie-Skłodowska University,
20-031 Lublin, Poland
e-mail: jkozi@hektor.umcs.lublin.pl

A. Eberle et al. (eds.), *Stochastic Partial Differential Equations and Related Fields*, Springer Proceedings in Mathematics & Statistics 229,
https://doi.org/10.1007/978-3-319-74929-7_36

systems of interacting quantum particles is one of them. Essential results in this direction were obtained by means of methods developed in stochastic analysis, mostly in the approach in which states of such systems are constructed as probability measures on infinite dimensional path spaces. A substantial part of these results appeared due to Michael Röckner's research activity, see, e.g., [1–5, 12]. The aim of this work is to outline the main aspects of the theory of equilibrium states of quantum anharmonic crystals obtained in [3] in the approach based on path measures.

1.1 The Anharmonic Crystal

An anharmonic oscillator is a mathematical model of a point particle moving in a potential field with multiple minima and sufficient growth at infinity. In the simplest case, the motion is one-dimensional and the potential has two minima (wells) separated by a potential barrier. If the motion is governed by the laws of classical mechanics, the oscillator's states are characterized by a couple $(q, p) \in \mathbb{R}^2$, where q is the displacement of the oscillator from a certain point and p is its momentum (amount of motion). In the states with sufficiently small fixed $|p|$, the particle is confined to one of the wells. This produces a degeneracy – the multiplicity of states (q, p) with the same p and energy E, that is, the multiplicity of q solving the equation

$$H(p,q) := \frac{1}{2m}p^2 + \frac{a}{2}q^2 + V(q) = E. \tag{1}$$

Here $H(p, q)$ is the particle's Hamiltonian in which $m > 0$ is the mass of the particle and the second and third terms constitute the potential energy. If $V \equiv 0$, the oscillator is harmonic (of rigidity $a > 0$), i.e., the third term can be considered as an anharmonic correction to the potential energy. In the quantum case, the particle's states are vectors of unit norm belonging to the complex Hilbert state $L^2(\mathbb{R})$. The displacement and momentum are then unbounded operators defined on $L^2(\mathbb{R})$, satisfying (on a common domain) the following commutation relation

$$[p,q] := pq - qp = -\mathrm{i}, \qquad \mathrm{i} := \sqrt{-1}. \tag{2}$$

In (2), we use the physical units in which the Planck constant is set $\hbar = 1$. Assume now that an infinite system of such particles is arranged into a crystal. That is, each particle is attached to its own crystal site $\ell \in \mathbb{Z}^d$, $d \geq 1$, and performs oscillations in its own copy of $\mathbb{R}$. For a finite $\Lambda \subset \mathbb{Z}^d$, the state space of the particles attached to the sites in Λ is the tensor product of the single-particle spaces, i.e., $\mathcal{H}_\Lambda = L^2(\mathbb{R}^\Lambda)$. The Hamiltonian of this portion of particles is

$$\begin{aligned} H_\Lambda &= \sum_{\ell \in \Lambda} H_\ell + J \sum_{\ell \sim \ell', \Lambda} q_\ell q_{\ell'}, \\ H_\ell &:= \frac{1}{2m} p_\ell^2 + \frac{a}{2} q_\ell^2 + V(q_\ell). \end{aligned} \tag{3}$$

Here H_ℓ is the Hamiltonian of an isolated quantum anharmonic oscillator. The sum in the second term of the first line in (3) is taken over all pairs of $\ell, \ell' \in \Lambda$ satisfying $|\ell - \ell'| = 1$. It describes the interaction between the neighboring oscillators located in Λ with intensity $J > 0$. The anharmonic potential V is assumed to grow at infinity faster than q^2. For simplicity, in this article we take it in the form

$$V(q) = -b_1 q^2 + b_2 q^4, \qquad b_1, b_2 > 0. \tag{4}$$

The Hamiltonian H_Λ in (3) with V as in (4) can be defined as a lower bounded self-adjoint operator in $\mathscr{H}_\Lambda$ such that $\exp(-\beta H_\Lambda)$ is a positive trace-class operator for each $\beta > 0$. Thus, one can set

$$Z_{\beta,\Lambda} = \text{trace} \exp\left(-\beta H_\Lambda\right). \tag{5}$$

The state of thermal equilibrium of the oscillators attached to the sites in Λ (local Gibbs state) is defined as a positive normalized linear functional $\rho_{\beta,\Lambda} : \mathfrak{C}_\Lambda \to \mathbb{C}$ by the following formula

$$\rho_{\beta,\Lambda}(A) = \text{trace}\left[A \exp\left(-\beta H_\Lambda\right)\right] \backslash / Z_{\beta,\Lambda}, \qquad A \in \mathfrak{C}_\Lambda. \tag{6}$$

Here $\beta = 1/k_\mathrm{B} T$, k_B and T are Boltzmann's constant and temperature, respectively, and $\mathfrak{C}_\Lambda$ is the algebra of all bounded linear operators $A : \mathscr{H}_\Lambda \to \mathscr{H}_\Lambda$, called observables. By Høegh-Krohn's theorem [3, p. 72] $\rho_{\beta,\Lambda}$ is uniquely determined by its values on the linear span of products

$$\mathfrak{a}_{t_1}^\Lambda(F_1) \cdots \mathfrak{a}_{t_n}^\Lambda(F_n), \qquad n \in \mathbb{N}, \quad F_1, \ldots, F_n \in \mathfrak{F}_\Lambda, \quad t_1, \ldots, t_n \in \mathbb{R},$$

where $\mathfrak{F}_\Lambda$ is a *complete* family of multiplication operators by bounded measurable functions $F : \mathbb{R}^\Lambda \to \mathbb{C}$, whereas

$$\mathfrak{a}_t^\Lambda(A) := \exp\left(\mathrm{i}t H_\Lambda\right) A \exp\left(-\mathrm{i}t H_\Lambda\right), \qquad A \in \mathfrak{C}_\Lambda.$$

According to [3, Theorem 1.3.6], $\mathfrak{F}_\Lambda$ is complete if it satisfies: (a) for $F_1, F_2 \in \mathfrak{F}_\Lambda$, the point-wise products $F_1 F_2$ is also in $\mathfrak{F}_\Lambda$; (b) the constant function 1 belongs to $\mathfrak{F}_\Lambda$; (c) for each distinct $x_\Lambda, y_\Lambda \in \mathbb{R}^\Lambda$, one finds $F \in \mathfrak{F}_\Lambda$ such that $F(x_\Lambda) \neq F(y_\Lambda)$. Since H_Λ is self-adjoint, the map $A \mapsto \mathfrak{a}_t^\Lambda(A)$ is an isometric automorphism of $\mathfrak{C}_\Lambda$. At the same time, the map $\mathbb{R} \ni t \mapsto \mathfrak{a}_t^\Lambda(A) \in \mathfrak{C}_\Lambda$ is the (time) evolution of the observable

A. The group $\{\mathfrak{a}_t^\Lambda\}_{t\in\mathbb{R}}$ describes the dynamics of the corresponding finite subsystem. The mentioned above Høegh-Krohn theorem implies that $\rho_{\beta,\Lambda}$ is determined by the Green functions

$$G^{\beta,\Lambda}_{F_1,\dots F_n}(t_1,\dots,t_n) := \rho_{\beta,\Lambda}\left[\mathfrak{a}^\Lambda_{t_1}(F_1)\cdots\mathfrak{a}^\Lambda_{t_n}(F_n)\right], \tag{7}$$

with all choices of $F_1,\dots,F_n \in \mathfrak{F}_\Lambda$. Each Green function admits an analytic continuation to the domain

$$\mathscr{D}_{n,\beta} := \{(\zeta_1,\dots,\zeta_n)\in\mathbb{C}^n : 0 < \mathrm{Im}(\zeta_1) < \cdots < \mathrm{Im}(\zeta_n) < \beta\}. \tag{8}$$

Furthermore, see [3, Theorem 1.2.32, p. 78], it can further be continuously extended to the closure $\overline{\mathscr{D}}_{n,\beta}$ of (8). The set

$$\mathscr{D}^{(0)}_{n,\beta} := \{(\zeta_1,\dots,\zeta_n)\in\overline{\mathscr{D}}_{n,\beta} : \mathrm{Re}(\zeta_1) = \cdots = \mathrm{Re}(\zeta_n) = 0\} \tag{9}$$

has the following property: each two continuous functions $f_1, f_2 : \overline{\mathscr{D}}_{n,\beta} \to \mathbb{C}$, analytic on $\mathscr{D}_{n,\beta}$ and equal on $\mathscr{D}^{(0)}_{n,\beta}$, are equal as functions. Then $G^{\beta,\Lambda}_{F_1,\dots F_n}$ is uniquely determined by its restriction to (9), that is, by the Matsubara function

$$\Gamma^{\beta,\Lambda}_{F_1,\dots F_n}(\tau_1,\dots,\tau_n) = G^{\beta,\Lambda}_{F_1,\dots F_n}(\mathrm{i}\tau_1,\dots,\mathrm{i}\tau_n). \tag{10}$$

1.2 The Path Measures

The main ingredient of the technique developed in [3] is the following representation, see [3, Theorem 1.4.5],

$$\Gamma^{\beta,\Lambda}_{F_1,\dots F_n}(\tau_1,\dots,\tau_n) = \int_{\Omega_{\beta,\Lambda}} F_1(\omega_\Lambda(\tau_1))\cdots F_n(\omega_\Lambda(\tau_n))\mu_{\beta,\Lambda}(d\omega_\Lambda). \tag{11}$$

Here $\mu_{\beta,\Lambda}$ is a probability measure on the Banach space $\Omega_{\beta,\Lambda}$ of 'temperature loops', which is

$$\Omega_{\beta,\Lambda} = \{\omega_\Lambda = (\omega_\ell)_{\ell\in\Lambda} : \omega_\ell \in \mathscr{C}_\beta\}, \quad \|\omega_\Lambda\| = \sup_{\ell\in\Lambda}\|\omega_\ell\|_{\mathscr{C}_\beta},$$
$$\mathscr{C}_\beta = \{\phi \in C([0,\beta]\to\mathbb{R}) : \phi(0) = \phi(\beta)\}, \quad \|\phi\|_{\mathscr{C}_\beta} = \sup_{\tau\in[0,\beta]}|\phi(\tau)|.$$

The measure $\mu_{\beta,\Lambda}$ is constructed in the following way. Let

$$H^{\mathrm{har}} = \frac{1}{2m}p^2 + \frac{a}{2}q^2 \tag{12}$$

be the Hamiltonian of a single harmonic oscillator, cf. (1) and (3), which can be defined as an unbounded self-adjoint operator on $L^2(\mathbb{R})$. It has discrete spectrum consisting of nondegenerate eigenvalues

$$E_n^{\text{har}} = (n + 1/2)\Delta^{\text{har}}, \qquad \Delta^{\text{har}} = \sqrt{a/m}, \tag{13}$$

see [3, Proposition 1.1.37, page 41]. Set $Z_\beta^{\text{har}} = \text{trace}\exp(-\beta H^{\text{har}})$, cf. (5), and then

$$S_\beta(\tau, \tau') = \text{trace}\Big(q e^{-|\tau-\tau'|H^{\text{har}}} q e^{-(\beta-|\tau-\tau'|)H^{\text{har}}}\Big)\Big/ Z_\beta^{\text{har}} \tag{14}$$

$$= \Big(e^{-|\tau-\tau'|\Delta^{\text{har}}} + e^{-(\beta-|\tau-\tau'|)\Delta^{\text{har}}}\Big)\Big/ 2\sqrt{ma}\Big(1 - e^{-\beta\Delta^{\text{har}}}\Big), \quad \tau, \tau' \in [0, \beta].$$

By means of the 'propagator' (14) we define a Gaussian measure, χ_β, on $\mathscr{C}_\beta$ by its Fourier transform

$$\int_{\mathscr{C}_\beta} \exp\left(\mathrm{i}\int_0^\beta f(\tau)\phi(\tau)d\tau\right)\chi_\beta(d\phi)$$

$$= \exp\left(-\frac{1}{2}\int_0^\beta\int_0^\beta S_\beta(\tau, \tau')f(\tau)f(\tau')d\tau d\tau'\right), \quad f \in \mathscr{C}_\beta,$$

see [3, pages 99 and 125]. Let $\chi_{\beta,\Lambda}$ be the Gaussian measure on $\Omega_{\beta,\Lambda}$ defined as the product of the corresponding copies of χ_β. Then the path measure in (11) is

$$\mu_{\beta,\Lambda}(d\omega_\Lambda) = \frac{1}{N_{\beta,\Lambda}}\exp\Big(- I_{\beta,\Lambda}(\omega_\Lambda)\Big)\chi_{\beta,\Lambda}(d\omega_\Lambda), \tag{15}$$

where $N_{\beta,\Lambda}$ is the normalization factor and

$$I_{\beta,\Lambda}(\omega_\Lambda) = -J\sum_{\ell\sim\ell',\Lambda}\int_0^\beta \omega_\ell(\tau)\omega_{\ell'}(\tau)d\tau + \sum_{\ell\in\Lambda}\int_0^\beta V(\omega_\ell(\tau))d\tau.$$

Note that by (11), (10) and then by (7) the measure (15) uniquely determines the state (6). That is, the local states (6) can be constructed as Gibbs measures, similarly as in the case of classical anharmonic crystals. Here, however, the classical variable $q_\ell \in \mathbb{R}$ is replaced by a continuous path ω_ℓ, which is an element of an infinite dimensional vector space, $\mathscr{C}_\beta$. Going further in this direction, one can define global Gibbs states of the quantum crystal as the probability measures on the space of tempered configurations $\Omega_\beta^{\mathrm{t}}$ satisfying the Dobrushin-Lanford-Ruelle (DLR) equation, see [3, Chapter 3]. It can be shown, see [3, Theorem 3.3.6] or [11, Theorem 3.1], that the set of all such measures, which we denote by $\mathscr{G}_\beta$, is a nonempty weakly compact simplex with a nonempty extreme boundary $\text{ex}(\mathscr{G}_\beta)$. By virtue of the DLR equation, the set $\mathscr{G}_\beta$ can contain either one or infinitely many elements. Correspondingly, the multiplicity (resp. the uniqueness) of the Gibbs states existing at a given value of the temperature means that $|\mathrm{ex}(\mathcal{G}_\beta)| > 1$ (resp. $|\mathcal{G}_\beta| = 1$). In the physical interpretation, the multiplicity corresponds to a phase transition, cf. [3, Chap. 7].

For a finite $\Lambda \subset \mathbb{Z}^d$, let $\mathfrak{M}_\Lambda$ be the subset of $\mathfrak{C}_\Lambda$ consisting of all multiplication operators by $F \in L^\infty(\mathbb{R}^\Lambda)$. Note that $\mathfrak{M}_\Lambda$ is a maximal C^*-subalgebra of $\mathfrak{C}_\Lambda$. Each such an F can be considered as a function $F : \mathbb{R}^{\mathbb{Z}^d} \to \mathbb{C}$. Set

$$\mathfrak{M} = \bigcup_\Lambda \mathfrak{M}_\Lambda,$$

where the union is taken over all finite $\Lambda \subset \mathbb{Z}^d$. For $F_1, \ldots, F_m \in \mathfrak{M}$ and $\mu \in \mathcal{G}_\beta$, the Matsubara function corresponding to these F_i and μ is

$$\Gamma^\mu_{F_1,\ldots,F_n}(\tau_1, \ldots, \tau_n) = \int_{\Omega^{\mathrm{t}}_\beta} F_1(\omega(\tau_1)) \cdots F_n(\omega(\tau_n))\mu(d\omega), \tag{16}$$

where $\tau_1, \ldots, \tau_n \in [0, \beta]$. Then μ is said to be τ-shift invariant if, for each $\vartheta \in [0, \beta]$, the following holds

$$\Gamma^\mu_{F_1,\ldots,F_n}(\tau_1 + \vartheta, \ldots, \tau_n + \vartheta) = \Gamma^\mu_{F_1,\ldots,F_n}(\tau_1, \ldots, \tau_n), \tag{17}$$

where the addition is modulo β. Let $\mathcal{G}^{\mathrm{phase}}_\beta$ be the subset of $\mathrm{ex}(\mathcal{G}_\beta)$ consisting of all τ-shift invariant measures. Its elements are called *thermodynamic phases* or *states of thermal equilibrium* of the quantum crystal. Each $\mu \in \mathcal{G}^{\mathrm{phase}}_\beta$ is defined by its Matsubara functions (16) corresponding to all possible choices of $n \in \mathbb{N}$ and $F_1, \ldots, F_n \in \mathfrak{M}$, cf. [6]. If $\mathcal{G}_\beta$ is a singleton, then clearly $\mathcal{G}_\beta = \mathcal{G}^{\mathrm{phase}}_\beta$. A state $\mu \in \mathcal{G}^{\mathrm{phase}}_\beta$ is called *translation invariant* if its Matsubara functions are invariant with respect to the shifts of the lattice $\mathbb{Z}^d$.

2 The Results

Now we present three main results concerning the properties of the set $\mathcal{G}^{\mathrm{phase}}_\beta$.

2.1 Phase Transitions and Quantum Stabilization

It can be shown, see [3, Theorem 3.7.4] or [11, Theorem 3.8], that there exist translation invariant $\mu^\pm \in \mathcal{G}^{\mathrm{phase}}_\beta$ such that, for each $\ell \in \mathbb{Z}^d$ and $\mu \in \mathcal{G}^{\mathrm{phase}}_\beta$, the following holds

$$M^- \le M^\mu_\ell \le M^+, \qquad M^- = -M^+, \tag{18}$$

where

$$M_\ell^\mu = \int_{\Omega_\beta^t} \omega_\ell(\tau)\mu(d\omega), \tag{19}$$

and $M^\pm = M_\ell^{\mu^\pm}$. In view of (17), the integral in (19) is independent of τ. By (18) we have that $M^+ = M^- = 0$ whenever $\mathscr{G}_\beta$ is a singleton and $M^+ > 0$ implies that $|\mathscr{G}_\beta^{\text{phase}}| > 1$. Moreover, $M^+ = 0$ is also sufficient for $|\mathscr{G}_\beta| = 1$, see [1, 2]. Assume that the lattice dimension satisfies $d \geq 3$. Set

$$E(p) = \sum_{j=1}^d \left[1 - \cos p_j\right], \qquad \theta(d) = \frac{d}{(2\pi)^d} \int_{(-\pi,\pi]^d} \frac{dp}{E(p)}.$$

It is possible to show that $\theta(d) > 1$ for all $d \geq 3$ and $\theta(d) \to 1^+$ as $d \to +\infty$. For $u \in [0, 1)$, set $t(u) = (\sqrt{u}/2)[\log(1 + \sqrt{u}) - \log(1 - \sqrt{u})]$. Then t is an increasing function and $\lim_{u\to 1^-} t(u) = +\infty$. Let $u(t)$, $t \in \mathbb{R}_+$ be its inverse, which is an increasing function such that $\lim_{t\to+\infty} u(t) = 1$. For b_1, b_2 as in (4) and a as in (12), set

$$\upsilon = \frac{2b_1 - a}{12 b_2}.$$

Note that $\upsilon > 0$ whenever $b_1 > a/2$, and thereby the potential energy in (1) has two wells. Recall that $J > 0$ is the intensity of the interaction of a given pair of oscillators, see (3). Then $\widehat{J} := 2dJ$ is the intensity of the interaction of a given oscillator with all its neighbors. The next statement, cf. [8, Theorem 3.1] or [3, Theorem 6.3.6], gives a sufficient condition for the existing of phase transitions in our model.

Theorem 1 *For $d \geq 3$, assume that $4m\upsilon^2\widehat{J} > \theta(d)$, and hence the equation*

$$4m\upsilon^2\widehat{J}u\,(\beta/4m\upsilon) = \theta(d) \tag{20}$$

has a unique solution, β_*. Then $|\mathscr{G}_\beta^{\text{phase}}| > 1$ whenever $\beta > \beta_*$.

As follows from Theorem 1, the absence of phase transitions, i.e., the fact that $|\mathscr{G}_\beta^{\text{phase}}| = 1$ for all $\beta > 0$ implies $4m\upsilon^2\widehat{J} \leq \theta(d)$. In order to get the corresponding sufficient condition let us turn to the spectral properties of the Hamiltonian H_ℓ given in the second line of (3), which can be defined as a self-adjoint lower bounded operator in $L^2(\mathbb{R})$. By [8, Proposition 4.1] or [3, Theorem 1.1.60], the spectrum of H_ℓ entirely consists of simple eigenvalues E_n, $n \in \mathbb{N}$. The simplicity means that each E_n corresponds to exactly one state, contrary to the classical case where the mentioned degeneracy might occur. By means of the analytic perturbation theory for linear operators it is possible to prove, see [8, Theorem 4.1], that $\Delta := \inf_n(E_{n+1} - E_n)$ is a continuous function of $m \in (0, +\infty)$ such that $m^{2/3}\Delta \to \Delta_0$ as $m \to 0^+$ for some $\Delta_0 > 0$. Then $R_m := m\Delta^2$ is a continuous function of $m \in (0, +\infty)$ such that $R_m \sim m^{-1/3}\Delta_0^2$ as $m \to 0^+$. In the harmonic case (13), we have $R_m^{\text{har}} = a$. By analogy, we call R_m *quantum effective rigidity*, which, however, depends on m as just discussed. The sufficient condition mentioned above is, see [8, Theorem 4.6] or [3, Theorem 7.3.1].

Theorem 2 *Let the parameters introduced above satisfy* $\widehat{J} < R_m$. *Then* $\mathscr{G}_\beta$ *is a singleton for all* $\beta > 0$.

According to Theorem 2 *quantum stabilization* occurs if the interaction intensity is smaller than the effective rigidity, see [2, 4] and Part 2 of [3] for a physical interpretation of this effect. Note that R_m can be made arbitrarily big either by making m small or Δ big (e.g., by making the wells closer to each other). On the other hand, it satisfies $R_m \leq 1/4m\upsilon^2$, see [8, Theorem 4.2] or [3, Theorem 7.1.1]. Therefore, $\widehat{J} < R_m$ implies that $4m\upsilon^2\widehat{J} < 1$, cf. (20).

2.2 Local Dynamics

In this subsection, we show that the dynamics of the oscillators indexed by the elements of a finite Λ can be influenced by the phase transitions described in Theorem 1. To this end, we use the notion of a *stochastically positive KMS system*, see [9]. Such a system is the tuple $(\mathfrak{C}, \mathfrak{B}, \{\mathfrak{a}_t\}_{t\in\mathbb{R}}, \varpi)$, where $\mathfrak{C}$ is a C^*-algebra; $\{\mathfrak{a}_t\}_{t\in\mathbb{R}}$ is a group of automorphisms of $\mathfrak{C}$; $\mathfrak{B}$ is a commutative C^*-subalgebra of $\mathfrak{C}$ such that the algebra generated by $\cup_{t\in\mathbb{R}}\mathfrak{a}_t(\mathfrak{B})$ is $\mathfrak{C}$; ϖ is a faithful state on $\mathfrak{C}$ which is stochastically positive and satisfies the KMS condition with some fixed $\beta > 0$. The latter means that, for each $A, B \in \mathfrak{C}$, there exists a function, $\Phi_{A,B}(z)$, analytic in the strip $\{z \in \mathbb{C} : \mathrm{Im} z \in (0, \beta)\}$ and continuous on its closure, such that $\varpi(A\mathfrak{a}_t(B)) = \Phi_{A,B}(t)$ and $\varpi(\mathfrak{a}_t(B)A) = \Phi_{A,B}(t + \mathrm{i}\beta)$, holding for all $t \in \mathbb{R}$. It can be shown, cf. [9, Theorem 2.1], that, for each collection $A_1, \ldots, A_n$ of the elements of $\mathfrak{C}$, the Green function

$$G^{\varpi}_{A_1,\ldots,A_n}(t_1, \ldots, t_n) := \varpi(\mathfrak{a}_{t_1}(A_1)\cdots\mathfrak{a}_{t_n}(A_n)), \qquad (t_1, \ldots, t_n) \in \mathbb{R}^n, \tag{21}$$

can be continued to a function analytic in the domain defined in (8) and continuous on its closure. The stochastic positivity of ϖ means that, for each collection of positive elements $F_1, \ldots, F_n$ of $\mathfrak{B}$, the function defined in (21) satisfies

$$G^{\varpi}_{F_1,\ldots,F_n}(\mathrm{i}\tau_1, \ldots, \mathrm{i}\tau_n) \geq 0, \qquad 0 \leq \tau_1 \leq \cdots \leq \tau_n \leq \beta.$$

For a finite $\Lambda \subset \mathbb{Z}^d$, let us define

$$\mathfrak{D}_\Lambda = \{Q_{u_\Lambda} = \exp\left(\mathrm{i}\sum_{\ell\in\Lambda} u_\ell q_\ell\right) : u_\Lambda \in \mathbb{Q}^\Lambda\},$$

where $\mathbb{Q}$ stands for the set of rational numbers. Clearly, $\mathfrak{D}_\Lambda \subset \mathfrak{M}_\Lambda$ is countable and complete. The latter follows by the fact that $\mathfrak{D}_\Lambda$ is closed with respect to the point-wise multiplication, contains the unit element and separates the points of $\mathbb{R}^\Lambda$. Let

$\mathfrak{N}_\Lambda$ be the closure (in the norm of $\mathfrak{M}_\Lambda$) of the set of all linear combinations of the elements of $\mathfrak{D}_\Lambda$ with rational coefficients. Then $\mathfrak{N}_\Lambda$ is a separable Banach algebra. Note that $\mathfrak{N}_\Lambda$ is a proper subset of $\mathfrak{M}_\Lambda$, dense in $\mathfrak{M}_\Lambda$ in the σ-weak topology. For the mentioned above states $\mu^\pm \in \mathscr{G}_\beta^{\text{phase}}$, we have, cf. (16), the Matsubara functions $\Gamma^{\mu^\pm}_{F_1,\dots,F_n}$, $F_1, \dots, F_n \in \mathfrak{Q}_\Lambda$. These functions determine two types of dynamics of the considered portion of oscillators.

Theorem 3 *Let $\mathfrak{N}_\Lambda$ and $\mu^\pm \in \mathscr{G}_\beta^{\text{phase}}$ be as just described. Then there exist stochastically positive KMS systems, $(\mathfrak{C}_\pm, \mathfrak{B}_\pm, \{\mathfrak{a}_t^\pm\}_{t\in\mathbb{R}}, \varpi^\pm)$, and injective homomorphisms, $\pi_\pm : \mathfrak{N}_\Lambda \to \mathfrak{B}_\pm$, such that*

$$\Gamma^{\mu^\pm}_{F_1,\dots,F_n}(\tau_1, \dots, \tau_n) = G^{\varpi^\pm}_{\pi_\pm(F_1),\dots,\pi_\pm(F_n)}(\mathrm{i}\tau_1, \dots, \mathrm{i}\tau_n), \quad 0 \le \tau_1 \le \dots \le \tau_n \le \beta, \tag{22}$$

holding for all choices of $F_1, \dots, F_n \in \mathfrak{Q}_\Lambda$.

The proof of this statement readily follows from [6, Theorem 3.1], see also [10]. Its meaning can be seen from the following fact. For $\ell_i \in \Lambda$, let $F_{\ell_i}(q_{\ell_i})$ be real, odd and strictly positive for $q_{\ell_i} > 0$, $i = 1, 2, 3$. Assume also that $|\mathscr{G}_\beta^{\text{phase}}| > 1$, and hence $\mu^+ \neq \mu^-$, see Theorem 1. By the first GKS inequality, see [3, Theorem 3.2.2], it follows that

$$\Gamma^{\mu^+}_{F_{\ell_1},F_{\ell_2},F_{\ell_3}}(\tau_1, \tau_2, \tau_3) > 0, \quad \text{and} \quad \Gamma^{\mu^-}_{F_{\ell_1},F_{\ell_2},F_{\ell_3}}(\tau_1, \tau_2, \tau_3) < 0,$$

for some τ_1, τ_2, τ_3. The second inequality follows from the first one by changing the signs of all ω_ℓ. Then by (22) one obtains that

$$G^{\varpi^+}_{\pi_+(F_{\ell_1}),\pi_+(F_{\ell_2}),\pi_+(F_{\ell_3})} \neq G^{\varpi^-}_{\pi_-(F_{\ell_1}),\pi_-(F_{\ell_2}),\pi_-(F_{\ell_3})},$$

which means that the oscillators in Λ distinguish between the wells in this case, which can be experimentally detected.

Acknowledgements The author was supported by the DFG through the SFB 701 "Spektrale Strukturen and Topologische Methoden in der Mathematik" and by the European Commission under the project STREVCOMS PIRSES-2013-612669.

References

1. Albeverio, S., Kondratiev, Yu., Kozitsky, Yu., Röckner, M.: Small mass implies uniqueness of Gibbs states of a quantum crystal. Commun. Math. Phys. **241**, 69–90 (2003)
2. Albeverio, S., Kondratiev, Yu., Kozitsky, Yu., Röckner, M.: Quantum stabilization in anharmonic crystals. Phys. Rev. Lett. **90**, 170603 (2003). [4 pages]
3. Albeverio, S., Kondratiev, Yu., Kozitsky, Yu., Röckner, M.: The statistical mechanics of quantum lattice systems: A path integral approach. EMS Tracts in Mathematics, vol. 8. European Mathematical Society (EMS), Zürich (2009)

4. Albeverio, S., Kondratiev, Yu., Kozitsky, Yu., Röckner, M.: Phase transitions and quantum effects in anharmonic crystals. Int. J. Mod. Phys. **B 26**, 1250063 (2012). [32 pages]
5. Albeverio, S., Kondratiev, Yu., Pasurek, T., Röckner, M.: Euclidean Gibbs measures on loop lattices: existence and a priori estimates. Ann. Probab. **32**, 153–190 (2004)
6. Birke, L., Fröhlich, J.: KMS, ect. Rev. Math. Phys. **14**, 829–871 (2002)
7. Blanchard, Ph, Fröhlich, J. (eds.): The message of quantum science—Attempts towards a synthesis. Lecture Notes in Physics, vol. 899. Springer, Berlin (2015)
8. Kargol, A., Kondratiev, Yu., Kozitsky, Yu.: Phase transitions and quantum stabilization in quantum anharmonic crystals. Rev. Math. Phys. **20**, 529–595 (2008)
9. Klein, A., Landau, L.: Stochastic processes associated with KMS states. J. Func. Anal. **42**, 368–428 (1981)
10. Kozitsky, Yu.: Equilibrium dynamics and phase transitions in quantum anharmonic crystals. In: Berche, B., Bogolyubov, N. Jr., Folk. R., Holovatch, Y. (eds.) Statistical Physics: Modern Trends and Applications, pp. 87-94. AIP (2009)
11. Kozitsky, Yu., Pasurek, T.: Euclidean Gibbs measures of interacting quantum anharmonic oscillators. J. Stat. Phys. **127**, 985–1047 (2007)
12. Röckner, M.: PDE approach to invariant and Gibbs measures with applications. Infin. Dimens. Harm. Anal. **III**, 233–247 (2005)

On Continuous Coding

Xing Liu and Bogusław Zegarliński

Abstract We review some results, ideas and open problems related to a continuous coding based on Kolmogorov representation theorem.

Keywords Kolmogorov representation theorem · K-basis · Continous coding and decoding

Mathematics Subject Classification 68Pxx · 65Dxx · 41-XX

1 Kolmogorov Representation Theorem

Let $\mathbb{B}$ and $\mathbb{A}$ be a Banach space of (some) functions $f : X \to Z$ and $g : Y \to Z$, respectively, where X, Y, Z are Polish spaces. Let $\mathscr{O}$ be a space of functions $\xi : Y \to Z$. A set of functions $\boldsymbol{\xi} \equiv \{\xi_j \in \mathscr{O}\}_{j \in J}$, with a countable set J, is called a reference basis (for the space $\mathbb{B}$ relative to $\mathbb{A}$) iff $\forall f \in \mathbb{B} \quad \exists\, \mathbf{g} \equiv \{g_j \in \mathbb{A} | j \in J\}$

$$f = \sum_{j \in J} g_j(\xi_j) \equiv \mathbf{g} \bullet \xi$$

We remark that similar notions can be discussed in case when the above spaces have a structure of (noncommutative) algebra or a group. The usefulness of the above representation could be justified if elements of the classes of functions $\mathbb{A}$ and $\mathscr{O}$ are somehow "simpler" or have some desired "nice" properties.

In a linear space a general representation is achieved by a use of a notion of a basis - the Hamel basis if a representation of finite type is allowed or otherwise

X. Liu
Hande Finmaker, Shenzhen, China
e-mail: xing.liu11@outlook.com

B. Zegarliński (✉)
Imperial College London, London, UK
e-mail: b.zegarlinski@imperial.ac.uk

A. Eberle et al. (eds.), *Stochastic Partial Differential Equations and Related Fields*, Springer Proceedings in Mathematics & Statistics 229,
https://doi.org/10.1007/978-3-319-74929-7_37

the Schauder basis which allows a representation of each element of a (topological) vector space by a convergent linear combination of vectors from a possible countable infinite (linearly independent) family. One well known example of a linear basis for a space of square integrable functions on n-dimensional cube is provided by trigonometric functions periodic along a discrete set of directions, that is a set of the form $\{\cos(\varepsilon\mathbf{n}\cdot\mathbf{x}), \sin(\varepsilon\mathbf{k}\cdot\mathbf{x})\}$, defined with some $\varepsilon \in (0,\infty)$ and $\mathbf{k} \equiv (k_1, .., k_n) \in \mathbb{Z}^n$. By Stone–Weierstrass theorem finite combinations of such functions form a dense set in the space of continuous functions on a compact set. Representation by such functions naturally appear in analysis of certain PDEs.

Another interesting example where representation in terms of composition of functions of one variable and one dimensional projection is provided by the following Hilberts identity (1909): For any $k \in \mathbb{N}$ and $n \in \mathbb{N}$, there exist $N \in \mathbb{N}$ and positive rational numbers a_i, and integers η_{ij}, such that

$$(x_1^2 + \cdots + x_n^2)^k = \sum_{i=1,\dots,N} a_i(\eta_{i1}x_1 + \cdots + \eta_{in}x_n)^{2k}.$$

As a consequence of this identity one can deduce a remarkable theorem of Hilbert-Waring saying that, for any $l \in \mathbb{N}$, every nonnegative number $m \in \mathbb{N}$ can be represented as a sum of at most $N \equiv N(l)$ terms of $2l$th powers of nonnegative integers. (One could expect some interesting application of Hilbert's identity to the probability associated to the measures with radial symmetry.)

More generally one can pursue a question of approximation by the functions of the following form

$$f(x) = \sum_{i=1,\dots,k} g_i(a_i \cdot x)$$

where g_i are some functions of one variable and a_i, $i = 1, \dots, k$, are some (pairwise different) vectors. One can show that any polynomial functions admits such representation [4] (and refs therein) and thus the set of such functions is dense in the space of continuous functions on a given compact set in an Euclidean space. [Note that such representation may be not unique on a dense set in the space of continuous functions. In [4] the authors introduce the following notion of nonuniqueness: a function $f(x)$ has strongly nonunique representations if there are two sets of directions $\{(a_i)\}_{i=1,\dots,k}$, $\{(\alpha_j)\}_{j=1,\dots,l}$, all distinct from each other, such that

$$f(x) = \sum_{i=1,\dots,k} g_i(a_i \cdot x) = \sum_{j=1,\dots,l} h_j(\alpha_j \cdot x)$$

for some functions of one variable g_i and h_i. They show that a function has strongly nonunique representations if and only if it is a polynomial.]

In connection to the above representations, one can mention the following well known metrics in the space of probability measures on $\mathbb{R}^n$. First, the distance of Fourier transforms of the measures given by [8]

$$\lambda(\mu_1, \mu_2) \equiv = \min_{T>0} \max \left\{ max_{||t|| \leq T} |\mu_1 e^{t \cdot x} - \mu_2 e^{t \cdot x}|, \frac{1}{T} \right\}.$$

Second, the Wasserstein distance on the space of probability measures with finite pth moment, given by

$$W_p(\mu_1, \mu_2) \equiv \left(\sup_{f,g} \{ \mu_1 f - \mu_2 g : f(x) - g(y) \leq d(x, y)^p \} \right)^{\frac{1}{p}}$$

where d denotes a distance on the underlying space.

In the first case in [8] one can give estimates on the distance provided one knows that the marginals on a given number of directions are the same. In the second case, [12] claims the following bound for the $p = 1$ Wasserstein distance $W_{\mathbb{R}^n}(\mu, \nu)$ of probability measures μ and ν on $\mathbb{R}^n$, $n \geq 2$, by the $p = 1$ Wasserstein distance $W_{\mathbb{R}}(\mu_a, \nu_a)$ of one dimensional marginals μ_a, ν_a on the set $\{x \cdot a\}$

$$W_{\mathbb{R}^n}(\mu, \nu) \leq C \sup_{|a|=1} W_{\mathbb{R}}(\mu_a, \nu_a).$$

with a constant $C \in (0, \infty)$ dependent on the dimension of the space, but not on the measures.

If composition of (more general) functions is allowed into the game, one can obtain more compact representations of multivariate functions. In connection to the Hilbert's XIII problem Kolmogorov (1957) proved the following remarkable fact about representation of continuous functions on $n \in \mathbb{N}$-dimensional cube $\mathbb{I}^n$:

Theorem 1 (Kolmogorov Superposition Theorem)
There exist continuous (increasing) functions $\{\xi_{q,p} : p = 1, \ldots, 2n + 1 \text{ and } q = 1, \ldots, n\}$, of one variable such that every continuous functions $f : \mathbb{I}^n \to \mathbb{R}$ can be represented as follows

$$f = \mathbf{g} \bullet \xi$$

with vectors of continuous functions $\xi_p \equiv \sum_q \xi_{q,p}$, $p = 1, \ldots, 2n + 1$ and $\mathbf{g} \equiv \{\frac{1}{2n+1} g_p\}_{q=1,\ldots,n}$ of one variable dependent on f and the choice of the functions $\xi_{q,p}$, $(p = 1, \ldots, 2n + 1,\ q = 1, \ldots, n)$.

Later Lorenz (1966) proved [10] the following elegant representation theorem:

Theorem 2 (Kolmogorov Superposition Theorem: Lorentz' version)
There exist $\lambda_j \in (0, 1)$, $\sum_{j=1,\ldots,n} \lambda_j \leq 1$ and strictly increasing continuous functions of one variable Φ_p, $p = 1, \ldots, 2n + 1$, such that for any $f \in \mathscr{C}(\mathbb{I}_n)$ one has

$$f(x) = \mathbf{g} \bullet \boldsymbol{\xi} \tag{1}$$

with a reference basis $\boldsymbol{\xi} \equiv \{\xi_j \in \mathscr{C}(\mathbb{I}_n, \mathbb{R})\}_{j=1,\dots,2n+1}$ *independent of* f *and given by*

$$\xi_j(x) = \sum_{i=1,\dots,n} \lambda_i \Phi_j(x_i)$$

and with $\mathbf{g}_j \equiv \frac{1}{2n+1} g \equiv \frac{1}{2n+1} g_{f,\xi} \in \mathscr{C}(\mathbb{R})$ *dependent on* f *[as well as the choice of the basis* $\boldsymbol{\xi}$*].*

It is known that at best the inner functions Φ_p can be Lipschitz continuous; see [9] and references therein. Moreover the numbers λ_j and the inner functions can be chosen to be computable (in the sense of theoretical computing analysis), [1] and refs therein. Similar representation theorem is known for **bounded** continuous functions on $\mathbb{R}^n$ instead of the cube, [3]. Sprecher (see also [2]) introduced particularly simple form of the reference basis where

$$\sum_{j=1,\dots,n} \lambda_j \Phi_p(x_j) = \sum_{j=1,\dots,n} \lambda_j \varphi(x_j + ap) + bp$$

with some constants $a, b \in (0, \infty)$ dependent on $n \in \mathbb{N}$.

Let Φ be the set of all non-decreasing continuous functions from the closed unit interval $\mathbb{I} = [0, 1]$ to itself, (which is a closed subset of $C[0, 1]$).

Theorem 3 (Kolmogorov Superposition Theorem: Kahane's version [5–7])
Let $\lambda_1, \lambda_2, \dots, \lambda_n$ *be pairwise different strictly positive reals adding up to one. Then, for a generic* $(2n+1)$*-tuple* $(\phi_1, \phi_2, \dots, \phi_{2n+1}) \in \Phi^{2n+1}$*, every continuous real-valued function* f *on the unit n-cube In can be represented as*

$$f = \mathbf{g} \bullet \xi$$

where $\mathbf{g}_j \equiv \frac{1}{2n+1} g$ *is a continuous function on* $\mathbb{I}$ *(depending on* f *and* ξ*) and* $\xi_p(x) \equiv \sum_{q=1}^n \lambda_q \phi_p(x_q)$*.*

The term generic in this theorem means that a set of such $2n + 1$ tuples for which the above representation holds is of 2nd category. Therefore such the set could be a support of a nice probability measure. It is an interesting problem if one could give a constructive characterisation of such the measure, so one could randomly sample possible reference basis. Also it is an interesting problem how to characterise connected sets of reference basis. For example one can show, [9], that the set of Lorentz' type with fixed functions ϕ_p is not connected. As a consequence it is impossible to talk about continuous deformation of the reference basis in the given class.

2 Continuous Coding

Given the same reference basis on the input and exit of a transmission channel, instead of passing over the full multivariate function, we need only to transmit a one dimensional signal. For example the simplest situation when one may like to transmit two dimensional function would appear in a situation of 3D printing. Here a shape of printed body is determined by its boundary which is a two dimensional manifold and such the manifold can be represented as a union of graphs of functions of two variables. A text printed with multicolour font can be encoded, first by assigning to it a discontinuous (piecewise constant) surface with high corresponding to enumeration of colours. Then we modify it on the boundaries of connected subsets (corresponding to constant value of a colour) to obtain a continuous function to which Kolmogorov representation could be applied. (Of course we could also deform the corresponding surface to a complicated manifold if one would like to hide deeper the information.) In this way we obtain a coding which is not based on conventional notion of alphabet and essentially there is no grammar which may help one with decoding.

We mention that the outer function can be non unique, which can be convenient in practice, [9] (see also refs therein):

Theorem 4
Let X be a compact separable metric space. Suppose $\xi_q : X \to \mathbb{R}, q = 1, \ldots, 2n+1$ is a reference basis such that $Y := \bigcup_1^{2n+1} Y_q := \bigcup_1^{2n+1} \xi_q[X]$ is not connected in $\mathbb{R}$, then for any $f \in C(X)$, there exist infinitely many $g \in C(\mathbb{R})$ such that

$$f = g \bullet \xi.$$

[One can also show that there are problems with positivity preservation property for semigroups, [9].]

If we are given a one dimensional signal g representing a function f with a use of a reference basis $\xi_j, j = 1, \ldots, 2n+1$, we can try to guess the function using some other reference basis $\widetilde{\xi}_k, k = 1, \ldots, 2m+1$, (we may choose $m \neq n$ which reflects a fact that we do not know what is a dimension of the original underlying space). In this way we get a function

$$\widetilde{f}(x) = \tfrac{1}{2m+1} \sum_{k=1,\ldots,2m+1} g(\widetilde{\xi}_k)$$

An example, which besides other things illustrates how conventional coding can be embedded into the functional coding, can be provided by a reference basis of Lorentz' type with permutation of the constants λ_j. More generally, any bijection of a set $\mathscr{A} \subset \mathbb{R}_+^n$ of positive algebraically independent n-tuples into itself will provide a way to generate new reference basis of Lorenz' type.

Next given a homeomorphism $T : \mathbb{I}_n \to \mathbb{I}_n$ and a reference basis ξ we can define a new reference basis $\widetilde{\xi}_j := \xi_j \circ T, j = 1, \ldots, 2n+1$. Note that in general the reference basis obtained in this way may be of more complicated type.

Under suitable conditions, one can also generate new basis by projection and extension (see [9] Theorems 3.2.6 and 3.2.7).

Remark 1 In view of basis extension theorem one could ask the following question. Given a continuous bounded function on $\mathbb{I}_n$ one can define cylinder functions on higher dimensional spaces and have a sequence of representations

$$f(x) = \frac{1}{2m+1} \sum_{j=1}^{m} g_{m,\xi^{(m)}}(\xi_j^{(m)}(x))$$

Can one find a suitable sequence of reference basis and corresponding outer functions such that both g_m, $\xi^{(m)}$ and $\xi_j^{(m)}(x)$ converge, and *(for any cylinder function)* one has a universal representation of the following form

$$f(x) = \int g_\infty(\xi_s(x))ds.$$

Decoding Error. If we decode a signal $\mathbf{g}$ (obtained with a given key ξ), with the wrong key $\widetilde{\xi}$, what is the possible error between $f \equiv \mathbf{g} \bullet \xi$ and $\widetilde{f} \equiv \mathbf{g} \bullet \widetilde{\xi}$? One can show that the error can be maximised when measured in L_p-norm.

To this end we consider a special class of homeomorphisms to transform a reference basis. Let T be a homeomorphism on $\mathbb{I}_n$ and let μ be a probability measure absolutely continuous with respect to the Lebesgue measure restricted to $\mathbb{I}_n$. We say T admits a measure decomposition with respect to μ, iff there exist μ-measurable sets A, B such that

$$\mathbb{I}_n = A \cup T[A] \cup B, \quad with\, \mu(B) = 0\, and\, A \cap T[A] = \emptyset.$$

One has the following result [9].

Theorem 5
Let μ be a measure absolutely continuous with respect to the Lebesgue measure, and T be a homeomorphism on $\mathbb{I}_n$ which admits a measure decomposition with respect to μ. Given reference basis ξ, let $\widetilde{\xi} \equiv \xi \circ T$. Then

$$\sup_{\substack{f \in \mathscr{C}(\mathbb{I}_n) \\ 0 \le f \le 1}} \|\mathbf{g}_f \bullet \xi - \mathbf{g}_f \bullet \widetilde{\xi}\|_{L_p(\mu)} = 1$$

where $f = \mathbf{g}_f \bullet \xi$.

To discuss further coding-decoding problems it is useful to get a bit deeper into the integration theory. First we notice the following simple relation for an integral of a function $f = \mathbf{g} \bullet \xi$, $\mathbf{g}_j \equiv \frac{1}{2dim+1} g_{f,\xi}$, $j = 1, \ldots, 2dim + 1$, with a probability measure μ on $(\mathbb{I}_n, \Sigma)$

$$\int_{\mathbb{I}_n} f d\mu = \int_{\mathbb{R}} g_{f,\xi} dF_\xi \tag{2}$$

where

$$F_\xi(s) \equiv \frac{1}{2n+1} \sum_{j=1,\ldots,2n+1} \mu(\{\xi_j \leq s\})$$

is called a distribution functions of a (Lorentz' type) reference basis ξ. Similar formulas hold with another reference basis $\eta \equiv \{\eta_j\}_{j=1,\ldots,2n+1}$. Next we note that

$$\int_{\mathbb{R}} g_{f,\eta} dF_\eta = \int_{\mathbb{R}} g_{f,\eta} \circ T_{\xi,\eta} dF_\xi$$

with a transportation map $T_{\xi,\eta}$, which in case of continuous distributions F_ξ and F_η is given by

$$T_{\xi,\eta} \equiv F_\eta^{-1} \circ F_\xi$$

Thus, at least in the integral sense to get the transition for external functions associated to different reference basis' it is enough to know the distribution functions of the basis' and we have

$$g_{f,\xi} = g_{f,\eta} \circ T_{\xi,\eta}. \tag{3}$$

Public Key Cryptography. At this point we would like to note that one way for a Public Key Cryptography to be realised would be by sending a reference key ξ with a monotone function F which is not easy to invert except if one knows that $F = F_1 \circ .. \circ F_k$ where each factor has a known inverse. (For example choosing F_i, $i = 1, \ldots, k$, as a part of a trivial word of some combinatorial group with a functional realisation, see e.g. [13].)

Kolmogorov-Fourier Transform. We remark that one can study a transition between representations in different basis using the following Kolmogorov-Fourier transform formally given by

$$f(x) = \int \mathcal{K}_\xi(q, x)\hat{g}_\xi(q)dq$$

with a kernel

$$\mathcal{K}_\xi(q, x) \equiv \tfrac{1}{2n+1} \sum_{j=1}^{2n+1} e^{iq\xi_j(x)}$$

jointly continuous in q and x. By Kolmogorov representation theorem, there exists a continuous function k such that

$$\mathcal{K}_\xi(q, x) = k_{q,\xi,\eta} \bullet \eta$$

with another reference basis η. Applying formally to k_q the Kolmogorov-Fourier representation one arrives at

$$f(x) = \int \mathcal{K}_\eta(q, x)\hat{g}_\eta(q)dq$$

with the following formal representation of Fourier transform of exterior function with respect to the new reference basis η

$$\hat{g}_\eta(q) \equiv \int \hat{k}_{q,\xi,\eta},(p)\hat{g}_\xi(p)dp.$$

Transportation and Wasserstein Distance. Let X and Z be a compact separable metric space (possibly of different dimensions) with a reference basis' $\xi_j : X \to \mathbb{R}$, $1 \leq j \leq 2n$ and $\xi_k : Z \to \mathbb{R}$, $1 \leq k \leq 2m$, $n, m \in \mathbb{N}$, respectively. Let μ and ν be a probability measure on (X, Σ_X) and (Z, Σ_Z), respectively. Given a (jointly) continuous function $c : X \times Z \to \mathbb{R}$, one can introduce the following cost functional

$$W(\mu, \nu) \equiv \inf\left\{\int c(x, z)\Pi(dx, dz) \mid \Pi_{|X} = \mu \, and \, \Pi_{|Z} = \nu\right\} \tag{4}$$

For example one could chose X to be a Riemaniann submanifold of Z and, given a metric $d(\cdot, \cdot)$ on Z, one could choose $c(x, z) \equiv d^r(\iota(x), z)$ with $r \in (0, \infty)$ and ι denoting continuous embedding of X into Z. One can apply representation theory to the function $c : X \times Z \to \mathbb{R}$, [9], to get

$$c(x, z) = C(\bullet\xi(x), \bullet\eta(z)),$$

where the *bullets* indicate composition separately with respect to each basis'. Then

$$W(\mu, \nu) = \inf\left\{\int C(s, t)dF_{\xi,\eta}(ds, dt)\right\}$$

where inf is taken over joint distributions of $\xi(x)$ and $\eta(z)$ with respect to $\{\Pi(dx, dz) \mid \Pi_{|X} = \mu \, and \, \Pi_{|Z} = \nu\}$. We remark also that W can also be represented as follows

$$W(\mu, \nu) = \sup\left(\mu(f) - \nu(\tilde{f}) | \, f(x) - \tilde{f}(z) \leq c(x, z), \; x \in X, z \in Z\right)$$

$$= \sup\left(\int_{\mathbb{R}} g_\xi dF_\xi - \int_{\mathbb{R}} \tilde{g}_\eta dF_\eta\right)$$

where sup is over continuous functions such that $f = \mathbf{g} \bullet \xi$ and $\tilde{f} = \tilde{\mathbf{g}} \bullet \eta$. We remark that Lip property of the reference basis is achievable, [9] and references therein, so the external functions g_ξ and $\tilde{g}_\eta$ belong to the Lip class. Therefore in case $r = 1$ and

$X = Z$ one has the following relation, [9],

$$W(\mu,\nu) \leq \sup_{\tilde{g},g\in Lip}\left(\int_{\mathbb{R}} g dF_\xi - \int_{\mathbb{R}} \tilde{g} dF_\eta\right)$$

Given transportation maps T such that $\mu = \nu \circ T^{-1}$ and the cost function W one is interested in its minimum over the transportation maps. Moreover one would like to find a path of the measures ν_τ and minimizing cost function transport maps T_τ such that $\nu_{\tau=0} = \mu$ and $\nu_{\tau=1} = \mu$. In our case the interesting question is if one can find a (continuous) path in the space of reference basis' so we could have a correspondence on every level. It particular this could be helpful for discussion of approximative decoding procedures. We remark, [9], that the set of Lorentz basis' with fixed functions ξ_p but variable constants λ_q, described earlier, is not connected.

References

1. Brattka, V.: A Computable Kolmogorov Superposition Theorem. In: Blanck, J., Brattka, V., Hertling, P., Weihrauch, K. (eds.) Computability and Complexity in Analysis, Vol. 272 of Informatik Berichte, FernUniversität Hagen, pp. 7–22 (2000). http://cca-net.de/vasco/publications/kolmogorov.pdf
2. Braun, J., Griebel, M.: On a constructive proof of Kolmogorov's superposition theorem. Constr. Approx. **30**, 653 (2009). https://doi.org/10.1007/s00365-009-9054-2
3. Demko, S.: A superposition theorem for bounded continuous functions. Proc. Amer. Math. Soc. **66**, 75–78 (1977)
4. Diaconis, P., Shahshahani, M.: On nonlinear functions of linear combinations. SIAM J. Sci. Stat. Comput. **5**, 175–191 (1984). https://doi.org/10.1137/0905013
5. Kahane, J.-P.: Sur le theoréme de superposition de Kolmogorov. J. Approx. Theory **13**, 229–234 (1975)
6. Kahane, J.-P.: Sur le treizieme probleme de Hilbert, le theoreme de superposition de Kolmogorov et les sommes algebriques d'arcs croissants. In: Petridis, N., Pichorides, S., Varopoulos, N. (eds.) Harmonic Analysis Iraklion 1978, Vol. 781, pp. 76–101 (1980). https://doi.org/10.1007/bfb0097649
7. Kahane, J.-P.: Le 13ème problème de Hilbert: un carrefour de l'algèbre, de l'analyse et de la géométrie. Cahiers du sminaire d'histoire des mathmatiques **3**, 1–25 (1982)
8. Klebanov, L.B., Rachev, S.T.: Proximity of probability measures with common marginals in a finite number of directions. Lecture Notes-Monograph Series, vol. 28, pp. 162–174 (1996)
9. Liu, X.: Kolmogorov Superposition Theorem and Its Applications, Ph.D. Thesis. Imperial College, London (2015)
10. Lorentz, G.G.: Approximation of Functions. Chelsea Publishing Company, London (1966)
11. Nathanson, M.B.: Additive Number Theory: The Classical Basses. Springer, Berlin (1996)
12. Zurkowski, V.: http://mathoverflow.net/revisions/215993/2
13. Zegarlinski, B.: Crystallographic Groups of Hörmander Fields, in the Special Issue in Honour of Alexander Grigor'yan, Sci. j. Volgograd State Univ. Math. Phys. (2017). arXiv:hal-01160736

Infinite-Dimensional Stochastic Differential Equations with Symmetry

Hirofumi Osada

Abstract We review recent progress in the study of infinite-dimensional stochastic differential equations with symmetry. This paper contains examples arising from random matrix theory.

Keywords Random matrices · Infinitely many particle systems · Interacting Brownian motions · Dirichlet forms · Logarithmic potentials

AMC2010 60H110 · 60J60 · 60K35 · 60B20 · 15B52

1 Introduction

We consider $(\mathbb{R}^d)^{\mathbb{N}}$-valued infinite-dimensional stochastic differential equations (ISDEs) of $\mathbf{X} = (X^i)_{i\in\mathbb{N}}$ of the form

$$dX_t^i = \sigma(X_t^i, \mathbf{X}_t^{i\diamond})dB_t^i + b(X_t^i, \mathbf{X}_t^{i\diamond})dt \quad (i \in \mathbb{N}). \tag{1}$$

Here $\mathbf{B} = (B^i)_{i\in\mathbb{N}}$ is $(\mathbb{R}^d)^{\mathbb{N}}$-valued standard Brownian motion. For $\mathbf{X} = (X^i)_{i\in\mathbb{N}}$ we set $\mathbf{X}^{i\diamond} = (X^j)_{j\in\mathbb{N}\setminus\{i\}}$. Coefficients $\sigma(x, \mathbf{y})$ and $b(x, \mathbf{y})$ are defined on a subset $\mathbf{S_0}$ of $\mathbb{R}^d \times (\mathbb{R}^d)^{\mathbb{N}}$ independent of $i \in \mathbb{N}$. By definition $\sigma(x, \mathbf{y})$ is $\mathbb{R}^{d^2}$-valued, and $b(x, \mathbf{y})$ is $\mathbb{R}^d$-valued. We assume that $\sigma(x, \mathbf{y})$ and $b(x, \mathbf{y})$ are symmetric in $\mathbf{y} = (y_i)_{i\in\mathbb{N}}$ for each $x \in \mathbb{R}^d$. Therefore, the set of (1) is referred to as "ISDEs with symmetry". In the present article, we review recent results in this regard. Using a Dirichlet form technique and an analysis on tail σ-fields of configuration spaces, we prove the existence and pathwise uniqueness of strong solutions of the ISDEs of (1). We emphasize that the coefficients are defined only on a *thin* subset in $(\mathbb{R}^d)^{\mathbb{N}}$ and the state space of the solution $\mathbf{X}$ is in this subset. Solving the ISDEs of (1) includes identifying such a subset.

H. Osada (✉)
Department of Mathematics, Kyushu University, Nishi-ku, Fukuoka 819-0395, Japan
e-mail: osada@math.kyushu-u.ac.jp

A. Eberle et al. (eds.), *Stochastic Partial Differential Equations and Related Fields*, Springer Proceedings in Mathematics & Statistics 229,
https://doi.org/10.1007/978-3-319-74929-7_38

Let $\mathsf{S} = \{\mathsf{s} = \sum_i \delta_{s_i};\ \mathsf{s}(K) < \infty$ for any compact $K \subset \mathbb{R}^d\}$ be the configuration space over $\mathbb{R}^d$. S is a Polish space equipped with the vague topology. With the symmetry of $\sigma(x, \mathbf{y})$ and $b(x, \mathbf{y})$ in $\mathbf{y}$, we regard σ and b as functions on $\mathbb{R}^d \times \mathsf{S}$. We denote these by the same symbol such that $\sigma(x, \mathbf{y}) = \sigma(x, \mathsf{y})$ and $b(x, \mathbf{y}) = b(x, \mathsf{y})$, where $\mathsf{y} = \sum_i \delta_{y_i}$ for $\mathbf{y} = (y_i)$. Then we rewrite the ISDEs of (1) for $\mathbf{X} = (X^i)_{i\in\mathbb{N}}$ as:

$$dX_t^i = \sigma(X_t^i, \mathsf{X}_t^{i\diamond})dB_t^i + b(X_t^i, \mathsf{X}_t^{i\diamond})dt \quad (i \in \mathbb{N}). \tag{2}$$

Here $\mathsf{X}^{i\diamond} = \sum_{j\neq i}^{\infty} \delta_{X^j}$, which is the S-valued process $\{\sum_{j\neq i}^{\infty} \delta_{X_t^j}\}$. $\mathbf{X}$ is called the *labeled* dynamics, and the associated *unlabeled* dynamics X is given by $\mathsf{X} = \sum_{i\in\mathbb{N}} \delta_{X^i}$.

If σ is a unit matrix and b is given by a pair interaction $\Psi(x - y)$, (1) becomes

$$dX_t^i = dB_t^i + \frac{\beta}{2} \sum_{j=1, j\neq i}^{\infty} \nabla\Psi(X_t^i - X_t^j)dt. \tag{3}$$

Here $\beta > 0$ is inverse temperature. Having Ψ of Ruelle class and $\Psi \in C_0^3(\mathbb{R}^d)$, Lang [10, 11] then solved (3), Fritz [2] constructed non-equilibrium solutions for $d \leq 4$, and Tanemura [25] provided solutions for hard core Brownian balls. The stochastic dynamics $\mathbf{X}$ given by the solution of (3) are called the interacting Brownian motions (IBM).

These solutions are strong solutions in the sense that $\mathbf{X}$ are functionals of the given Brownian motions $\mathbf{B}$ and initial starting points $\mathbf{s}$. The method used in these studies are based on the classic Itô scheme. Hence, if Ψ is of long range such as a polynomial decay, then it is difficult to apply this scheme. Tsai [26] solved (3) for the Dyson model. He used very cleverly a specific monotonicity of the logarithmic potential and its one-dimensional structure. As for the weak solution, we present a robust method based on the Dirichlet form technique from [15]. We present a general theory to give μ-pathwise unique strong solutions applicable to the logarithmic interaction from [20].

Thus, our demonstration is divided into two steps. In the first step, we obtain weak solutions of ISDEs (1). That is, we construct solutions $(\mathbf{X}, \mathbf{B})$ satisfying (1) (see Sects. 2–4). In the second step, we prove the existence of strong solutions and the μ-pathwise uniqueness. For this, we perform a fine analysis of the tail σ-field of S (see Sects. 5 and 6). In Sect. 7, we give ISDEs arising from random matrix theory. In Sect. 8, we present the algebraic construction of the dynamics, and the coincidence of the algebraic dynamics with solutions of ISDEs.

2 Unlabeled Dynamics: Quasi-Gibbs Property

We next construct a natural μ-reversible unlabeled diffusion, where μ is a point process. The key point is the quasi-Gibbs property of μ, which we proceed to describe.

Let $S_r = \{x \in \mathbb{R}^d;\ |x| < r\}$. Let $\pi_r, \pi_r^c : \mathsf{S} \to \mathsf{S}$ be projections such that $\pi_r(\mathsf{s}) = \mathsf{s}(\cdot \cap S_r)$, $\pi_r^c(\mathsf{s}) = \mathsf{s}(\cdot \cap S_r^c)$. For a point process μ, we set

$$\mu_{r,\mathsf{t}}^m(\cdot) = \mu(\pi_r(\mathsf{s}) \in \cdot | \mathsf{s}(S_r) = m, \pi_r^c(\mathsf{s}) = \pi_r^c(\mathsf{t}))$$

Let $\Phi : \mathbb{R}^d \to \mathbb{R} \cup \{\infty\}$ and $\Psi : (\mathbb{R}^d)^2 \to \mathbb{R} \cup \{\infty\}$ be potentials. We set

$$\mathscr{H}_r = \sum_{s_i \in S_r} \Phi(s_i) + \sum_{s_i, s_j \in S_r, i<j} \Psi(s_i, s_j).$$

A point process μ is called a canonical Gibbs measure if μ satisfies Dobrushin–Lanford–Ruelle (DLR) equation, that is, for μ-a.s. $\mathsf{t} = \sum_j \delta_{t_j}$

$$\mu_{r,\mathsf{t}}^m = c_{r,\mathsf{t}}^m e^{-\mathscr{H}_r - \sum_{x_i \in S_r, t_j \in S_r^c} \Psi(x_i, t_j)} d\Lambda_r^m. \tag{4}$$

Here $\Lambda_r^m = \Lambda(\cdot | \mathsf{s}(S_r) = m)$ and Λ_r is the Poisson PP with intensity $1_{S_r} dx$.

Point processes appear in random matrix theory in the form sine, Airy, Bessel, and Ginibre point processes having logarithmic potentials

$$\Psi(x, y) = -\beta \log |x - y|.$$

However, the DLR equation (4) does not make sense for a logarithmic potential. Hence we introduce the notion of quasi-Gibbs measures:

Definition 1 μ is (Φ, Ψ)-quasi-Gibbs measure if $\exists\, c_{r,\mathsf{t}}^m$ such that

$$c_{r,\mathsf{t}}^{m\ -1} e^{-\mathscr{H}_r} d\Lambda_r^m \le \mu_{r,\mathsf{t}}^m \le c_{r,\mathsf{t}}^m e^{-\mathscr{H}_r} d\Lambda_r^m$$

By definition a canonical Gibbs measure is a quasi-Gibbs measure. We refer to [16, 17] for a sufficient condition for quasi-Gibbs property. We assume:
(A1) μ is a quasi-Gibbs measure with upper semi-continuous (Φ, Ψ). Furthermore, $a(x, \mathsf{s}) = \sigma^t(x, \mathsf{s})\sigma(x, \mathsf{s})$ is bounded and uniformly elliptic.
(A2) There exists a $1 < p \le \infty$ such that the k-point correlation function ρ^k of μ is in $L^p_{\mathrm{loc}}((\mathbb{R}^d)^m)$ for each $k \in \mathbb{N}$.

For a given point process μ we introduce a Dirichlet form such that

$$\mathscr{E}^\mu(f, g) = \int_{\mathsf{S}} \mathbb{D}[f, g] d\mu, \quad \mathbb{D}[f, g] = \frac{1}{2} \sum_i a(s_i, \mathsf{s}^{i\diamond}) \frac{\partial \check{f}}{\partial s_i} \cdot \frac{\partial \check{g}}{\partial s_i}. \tag{5}$$

Here we set $\mathsf{s}^{i\diamond} = \sum_{j \ne i} \delta_{s_j}$ for $\mathsf{s} = \sum_i \delta_{s_i}$, and $f(\mathsf{s}) = \check{f}(s_1, s_2, \ldots)$, where $\check{f}$ is symmetric in $(s_1, s_2, \ldots)$. Note that $\mathbb{D}[f, g]$ is a function of s by construction.

A function f defined on the configuration space S is called local if f is $\sigma[\pi_r]$-measurable for some $r \in \mathbb{N}$. f is called smooth if $\check{f}$ is smooth. Let $\mathscr{D}_\circ$ be the set of local, smooth functions on S. We set $\mathscr{D}_\circ^\mu = \{f \in \mathscr{D}_\circ \cap L^2(\mu);\ \mathscr{E}^\mu(f, f) < \infty\}$.

Theorem 1 ([13, 16, 20]) *(i) Assume (A1). Then $(\mathscr{E}^{\mu}, \mathscr{D}^{\mu}_{\circ})$ is closable on $L^2(\mu)$. (ii) Assume (A1) and (A2). Then there exists a diffusion $\mathsf{X}_t = \sum_i \delta_{X_t^i}$ associated with the closure $(\mathscr{E}^{\mu}, \mathscr{D}^{\mu})$ of $(\mathscr{E}^{\mu}, \mathscr{D}^{\mu}_{\circ})$ on $L^2(\mu)$.*

The local boundedness of the correlation functions is used for the quasi-regularity of the Dirichlet form. Once quasi-regularity is established, the existence of μ-reversible diffusion is immediate from the general theory [3, 12].

Unlabeled dynamics are also obtained in [1, 27] with a different frame work. It is now proved these are the same dynamics as in [21, 22]. We remark that ergodicity of unlabeled dynamics with grand canonical Gibbs measures with small enough activity constant is obtained in [1].

3 Labeled Dynamics: A Scheme of Dirichlet Spaces

We next lift the unlabeled dynamics X in Theorem 1 to a labeled dynamics $\mathbf{X} = (X^i)_{i\in\mathbb{N}}$ solving (1). For this we present a natural scheme of Dirichlet spaces describing the labeled dynamics $\mathbf{X}$. We assume a pair of mild assumptions:
(A3) $\{X^i\}$ do not collide with each other (non-collision)
(A4) each tagged particle X^i never explode (non-explosion)

Let $\mathsf{S}_{s,i} = \{\mathsf{s} \in \mathsf{S}\,;\ \mathsf{s}(\{x\}) = 0 \text{ for all } x \in S,\ \ \mathsf{s}(S) = \infty\}$. Then (A3) is equivalent to $\mathrm{Cap}^{\mu}(\mathsf{S}_{s,i}^c) = 0$. (A4) follows from $\rho^1(x) = O(e^{|x|^{\alpha}})$, $\alpha < 2$.

We call $\mathfrak{u}$ the unlabeling map if $\mathfrak{u}((s_i)) = \sum_i \delta_{s_i}$. We call $\mathfrak{l}$ a label if $\mathfrak{l}$ is defined for μ-a.s.s, and $\mathfrak{u}\circ\mathfrak{l}(\mathsf{s}) = \mathsf{s}$. For a unlabeled dynamics satisfying (A3) and (A4), the particles can keep the initial label $\mathfrak{l}(\mathsf{s})$. Thus we can construct a map $\mathfrak{l}_{\text{path}}$ to $C([0,\infty); (\mathbb{R}^d)^{\mathbb{N}})$ such that $\{\mathfrak{u}(\mathfrak{l}_{\text{path}}(\mathsf{X}_t))\}_{t\in[0,\infty)} = \mathsf{X}$. Hence we obtain:

Theorem 2 ([14]) *Assume (A1)–(A4). Then there exists a labeled dynamics $\mathbf{X} = (X^i)_{i\in\mathbb{N}}$ such that $\mathsf{X} = \sum_{i\in\mathbb{N}} \delta_{X^i}$ and that $\mathbf{X}_0 = \mathfrak{l}(\mathsf{X}_0)$.*

Remark that $(\mathbb{R}^d)^{\mathbb{N}}$ has no good measures. Then no Dirichlet forms on $(\mathbb{R}^d)^{\mathbb{N}}$ associated with the labeled dynamics $\mathbf{X}$ exist. We hence introduce the scheme of spaces $(\mathbb{R}^d)^m \times \mathsf{S}$ with Campbell measures $\mu^{[m]}$ such that $d\mu^{[m]} = \rho^m(\mathbf{x}_m)\mu_{\mathbf{x}_m}(d\mathsf{s})d\mathbf{x}_m$, where ρ^m is the m-point correlation function of μ and $\mu_{\mathbf{x}_m}$ is the reduced Palm measure conditioned at $\mathbf{x}_m$. For $a = \sigma^t\sigma$, let $\mathbb{D}^{[m]}$ be the square field on $(\mathbb{R}^d)^m \times \mathsf{S}$ defined similarly as $\mathbb{D}$ on S given by (5) in Sect. 2. Let

$$\mathscr{E}^{[m]}(f,g) = \int_{(\mathbb{R}^d)^m\times\mathsf{S}} \mathbb{D}^{[m]}[f,g]d\mu^{[m]}.$$

Let $\mathscr{D}_{\circ}^{[1],\mu} = \{f \in C_0^{\infty}(\mathbb{R}^d)\otimes\mathscr{D}_{\circ};\ \mathscr{E}^{[m]}(f,f) < \infty,\ f \in L^2(\mu^{[m]})\}$.

Theorem 3 ([14]) *Assume (A1) and (A2). Then $(\mathscr{E}^{[m]}, \mathscr{D}_{\circ}^{[1],\mu})$ is closable on $L^2(\mu^{[m]})$, and its closures is quasi-regular. Hence the associated diffusion $(\mathbf{X}^{[m]}, \mathsf{X}^{[m]*})$ exists. Here we write $(\mathbf{X}^{[m]}, \mathsf{X}^{[m]*}) = (X^{[m],1}, \dots, X^{[m],m}, \sum_{i=m+1}^{\infty} \delta_{X^{[m],i}})$.*

Let $(\mathcal{E}^{\mu}, \mathcal{D}^{\mu}, L^2(\mu))$ be the original Dirichlet form. Let $\mathsf{X} = \sum_{i=1}^{\infty} \delta_{X^i}$ be the associated unlabeled diffusion. We fix a label $\mathfrak{l}$. Let $\mathbf{X} = (X^i)_{i\in\mathbb{N}}$ be the labeled dynamics given by $\mathfrak{l}$. We set $(\mathbf{X}^m, \mathsf{X}^{m*}) = (X^1, \ldots, X^m, \sum_{i=m+1}^{\infty} \delta_{X^i})$.

Theorem 4 ([14]) *Assume (A1)–(A4). Assume* $(\mathbf{X}^{[m]}, \mathsf{X}^{[m]*})$ *and* $(\mathbf{X}^m, \mathsf{X}^{m*})$ *start at the same initial point. Then* $(\mathbf{X}^{[m]}, \mathsf{X}^{[m]*}) = (\mathbf{X}^m, \mathsf{X}^{m*})$ *in distribution for each* $m \in \mathbb{N}$.

Instead of the huge space $(\mathbb{R}^d)^{\mathbb{N}}$, we use a scheme of countably infinite *good* infinite-dimensional spaces $\{(\mathbb{R}^d)^m \times \mathsf{S}\}_{m\in\{0\}\cup\mathbb{N}}$. Using the diffusion X on the original unlabeled space S, we construct a scheme of the *coupled* diffusions $(\mathbf{X}^m, \mathsf{X}^{m*})$ on $(\mathbb{R}^d)^m \times \mathsf{S}$ associated with the scheme of Dirichlet spaces $(\mathcal{E}^{[m]}, L^2(\mu^{[m]}))$ on $(\mathbb{R}^d)^m \times \mathsf{S}$. This construction is key for the ISDE-representation below.

4 ISDE-Representation: Logarithmic Derivative

We pursue the ISDE describing the labeled dynamics $\mathbf{X}$ obtained in Theorem 2. The key notion for this is the logarithmic derivative of μ introduced below.

Definition 2 ([15]) Let ∇_x be the nabla on $\mathbb{R}^d$. $\mathsf{d}^{\mu} \in L^1_{\mathrm{loc}}(\mathbb{R}^d \times \mathsf{S}, \mu^{[1]})^d$ is called the *logarithmic derivative* of μ if, for all $f \in C_0^{\infty}(\mathbb{R}^d) \otimes L^{\infty}(\mathsf{S})$,

$$\int_{\mathbb{R}^d\times\mathsf{S}} \nabla_x f d\mu^{[1]} = -\int_{\mathbb{R}^d\times\mathsf{S}} f \mathsf{d}^{\mu} d\mu^{[1]}. \tag{6}$$

Let $a(x, \mathsf{s}) = \sigma^t(x, \mathsf{s})\sigma(x, \mathsf{s})$ as before. We set $\nabla_x a$ such that $\nabla_x a = \sum_{j=1}^{d} \frac{\partial}{\partial x_j} a_{ij}$. We introduce a "geometric" differential equation on $\mathsf{d}^{\mu}(x, \mathsf{s}) =: \nabla_x \log \mu^{[1]}(x, \mathsf{s})$:

$$\nabla_x a(x, \mathsf{s}) + a(x, \mathsf{s})\nabla_x \log \mu^{[1]}(x, \mathsf{s}) = 2b(x, \mathsf{s}). \tag{7}$$

(A5) μ has a logarithmic derivative d^{μ}.
(A6) The logarithmic derivative d^{μ} satisfies (7).

Theorem 5 ([15]) *Assume (A1)–(A6). Then there exists an* $\mathsf{S}_0 \subset \mathsf{S}$ *such that* $\mu(\mathsf{S}_0) = 1$ *and that, for each* $\mathbf{s} \in \mathfrak{u}^{-1}(\mathsf{S}_0)$*, ISDE* (1) *has a solution* $(\mathbf{X}, \mathbf{B})$ *satisfying* $\mathbf{X}_0 = \mathbf{s}$ *and* $\mathbf{X}_t \in \mathfrak{u}^{-1}(\mathsf{S}_0)$ *for all t.*

From the coupling in Theorem 4 and Fukushima decomposition (Itô formula), we prove that $\mathbf{X} = (X^i)_{i\in\mathbb{N}}$ satisfies the ISDEs of (1). We use the m-labeled process $(\mathbf{X}^m, \mathsf{X}^{m*})$, to apply Itô formula to coordinate functions $x_1, \ldots, x_m$.

5 Strong Solutions of ISDEs and Pathwise Uniqueness

We lift the weak solutions $(\mathbf{X}, \mathbf{B})$ to pathwise unique strong solutions. For this purpose, we introduce a scheme consisting of *an infinite system of finite-dimensional SDEs with consistency* (IFC), and perform an analysis of the tail σ-field of the path space $W((\mathbb{R}^d)^{\mathbb{N}}) = C([0, T]; (\mathbb{R}^d)^{\mathbb{N}})$. The key idea is the following interpretations:

- a single ISDE $\Longleftrightarrow$ a scheme of IFC.
- the tail σ-field of $W((\mathbb{R}^d)^{\mathbb{N}}) \Longleftrightarrow$ the boundary condition of the ISDEs.

The method is robust and may be applied to many other models. We consider *non-Markovian* ISDEs because the argument is general. Let $(W_{\text{sol}}, \mathbf{S_0}, \{\sigma^i\}, \{b^i\})$ be

W_{sol} F a Borel subset of $W((\mathbb{R}^d)^{\mathbb{N}})$	(space of solutions of the ISDEs),
$\mathbf{S_0}$: a Borel subset of $(\mathbb{R}^d)^{\mathbb{N}}$	(initial starting points of the ISDEs),
$\sigma^i, b^i : W_{\text{sol}} \to W((\mathbb{R}^d)^{\mathbb{N}})$	(coefficients of the ISDEs).

We consider the ISDEs on $(\mathbb{R}^d)^{\mathbb{N}}$ of the form: $\mathbf{X} = \{(X_t^i)_{i\in\mathbb{N}}\}_{t\in[0,T]} \in W((\mathbb{R}^d)^{\mathbb{N}})$

$$dX_t^i = \sigma^i(\mathbf{X})_t dB_t^i + b^i(\mathbf{X})_t dt \quad (i \in \mathbb{N}),$$
$$\mathbf{X}_0 = \mathbf{s} = (s_i)_{i\in\mathbb{N}} \in \mathbf{S_0}, \quad \mathbf{X} \in W_{\text{sol}}. \tag{8}$$

Note that (1) is a special case of (8). We assume:
(P1) The ISDEs (8) has a weak solution $(\mathbf{X}, \mathbf{B})$. (not a strong solution!)
From a weak solution $(\mathbf{X}, \mathbf{B})$, we define a new SDE of $\mathbf{Y}^m = (Y_t^{m,i})_{i=1}^m$ such that

$$dY_t^{m,i} = \sigma^i(\mathbf{Y}^m, \mathbf{X}^{m*})_t dB_t^i + b^i(\mathbf{Y}^m, \mathbf{X}^{m*})_t dt \quad (i = 1, \dots, m),$$
$$\mathbf{Y}_0^m = (s_1, \dots, s_m), \quad (\mathbf{Y}^m, \mathbf{X}^{m*}) \in W_{\text{sol}} \tag{9}$$

for each $\mathbf{X} \in W_{\text{sol}}^{\mathbf{s}}$, $\mathbf{s} = (s_i)_{i=1}^{\infty} \in \mathbf{S_0}$, and $m \in \mathbb{N}$. Here $\mathbf{X}^{m*} := (X^n)_{n>m}$ is interpreted as a part of the coefficients of the SDE (9) and $W_{\text{sol}}^{\mathbf{s}} = \{\mathbf{X} \in W_{\text{sol}}; \mathbf{X}_0 = \mathbf{s}\}$. Indeed, we regard (9) as *finite-dimensional* SDEs of $\mathbf{Y}^m$. Equation (9) become automatically time-inhomogeneous SDEs. We have therefore obtained a scheme of finite-dimensional SDEs of $\{\mathbf{Y}^m\}_{m\in\mathbb{N}}$. We assume:

(P2) The SDE (9) has a unique, strong solution for each $\mathbf{s} \in \mathbf{S_0}$, $\mathbf{X} \in W_{\text{sol}}^{\mathbf{s}}$, and $m \in \mathbb{N}$.

Let $\bar{P}_{\mathbf{s}}$ be the distribution of solution $(\mathbf{X}, \mathbf{B})$ on $W_{\text{sol}}^{\mathbf{s}} \times W((\mathbb{R}^d)^{\mathbb{N}})$. Let

$$\bar{P}_{\mathbf{s},\mathbf{B}} = \bar{P}_{\mathbf{s}}(\mathbf{X} \in \cdot|\mathbf{B}) \text{ and } P_{\text{Br}}^{\infty} = \bar{P}_{\mathbf{s}}(\mathbf{B} \in \cdot),$$
$$\mathscr{T}_{\text{path}}((\mathbb{R}^d)^{\mathbb{N}}) = \cap_{m=1}^{\infty} \sigma[\mathbf{X}^{m*}],$$
$$\mathscr{T}_{\text{path}}^{(1)}[\bar{P}_{\mathbf{s},\mathbf{B}}] = \{\mathbf{A} \in \mathscr{T}_{\text{path}}((\mathbb{R}^d)^{\mathbb{N}})\, ;\ \bar{P}_{\mathbf{s},\mathbf{B}}(\mathbf{A}) = 1\}.$$

(P3) $\mathscr{T}_{\text{path}}((\mathbb{R}^d)^{\mathbb{N}})$ is $\bar{P}_{\mathbf{s},\mathbf{B}}$-trivial for each $\mathbf{s} \in \mathbf{S_0}$ and P_{Br}^{∞}-a.s. $\mathbf{B}$.

Theorem 6 ([20]) *Assume (P1)–(P3). Then*
(i) $\mathbf{X}$ *is a strong solution of the ISDEs* (8) *for each* $\mathbf{s} \in \mathbf{S}_0$.
(ii) Let $\mathbf{X}_\mathbf{s}$ *and* $\mathbf{X}'_\mathbf{s}$ *be strong solutions of the ISDEs* (8) *starting at* $\mathbf{s} \in \mathbf{S}_0$ *defined on the same space of Brownian motions* $\mathbf{B}$. *Then* $\mathbf{X}_\mathbf{s} = \mathbf{X}'_\mathbf{s}$ *for* P_{Br}^{∞}*-a.s.* $\mathbf{B}$ *if and only if* $\mathscr{T}_{\mathrm{path}}^{(1)}[\bar{P}_{\mathbf{s},\mathbf{B}}] = \mathscr{T}_{\mathrm{path}}^{(1)}[\bar{P}'_{\mathbf{s},\mathbf{B}}]$ *for* P_{Br}^{∞}*-a.s.* $\mathbf{B}$.

Idea of the proof (i): Let $(\mathbf{X}, \mathbf{B})$ be a weak solution of ISDE given by (P1), and fix it. Let $\mathbf{Y}^m$ be the unique strong solution of (9) given by (P2). By construction $\mathbf{Y}^m$ is $\sigma[\mathbf{B}] \bigvee \sigma[\mathbf{X}^{m*}]$-measurable. Because the solution (9) is unique, we see that $\mathbf{Y}^m = \mathbf{X}^m := (X^1, \dots, X^m)$. Let $\mathbf{Y}$ be the limit $\mathbf{Y} = \lim_{m\to\infty} \mathbf{Y}^m$. Then $\mathbf{Y} = \mathbf{X}$ and $\mathbf{Y}$ is $\sigma[\mathbf{B}] \bigvee \mathscr{T}_{\mathrm{path}}((\mathbb{R}^d)^{\mathbb{N}})$-measurable. Because $\mathscr{T}_{\mathrm{path}}((\mathbb{R}^d)^{\mathbb{N}})$ is $\bar{P}_{\mathbf{s},\mathbf{B}}$-trivial by (P3), $\mathbf{Y}$ depends only on $\mathbf{s}$ and $\mathbf{B}$. This means $\mathbf{Y} = \mathbf{X}$ is a strong solution. □

In [20] we introduce a notion of IFC solution, with which we generalize Theorem 6.

6 Tail Triviality: Application to Interacting Brownian Motions

We return to the Markovian-type ISDEs of (1). We assume (A1)–(A6). We apply Theorem 6 to ISDEs of (1) by checking (P1)–(P3). (P1) follows from Theorem 5. Controlling the capacity of $(\mathscr{E}^\mu, \mathscr{D}^\mu, L^2(\mu))$ we obtain (P2). Because $\mathbf{X} \in W_{\mathrm{sol}}$ and W_{sol} is a nice subset of $W((\mathbb{R}^d)^{\mathbb{N}})$, we can assume (P2) for the solution of (1). Dirichlet form theory proves that $\mathbf{X}$ stays in W_{sol}. Indeed, such a condition is reduced to a calculation of capacity related to the unlabeled Dirichlet space [3, 12]. Roughly speaking, (P2) is satisfied if $\nabla_x^m \mathsf{d}^\mu \in \mathscr{D}_{\mathrm{loc}}^{\mu^{[1]}}$ for a suitable m, see [20, Sects. 8 and 9].

Theorem 7 ([20]) *Assume (Q1)–(Q3) below. Then (P3) holds.*
(Q1) μ *is tail trivial. That is,* $\mu(A) \in \{0, 1\}$ *for all* $A \in \mathscr{T}(\mathsf{S}) := \cap_{r=1}^{\infty} \sigma[\pi_r^c]$.
(Q2) $P_\mu \circ \mathsf{X}_t^{-1} \prec \mu$ *for all* t. *(absolute continuity condition).*
(Q3) $P_\mu(\cap_{r=1}^{\infty}\{\mathsf{m}_r(\mathsf{X}) < \infty\}) = 1$,
where $\mathsf{m}_r = \inf\{m \in \mathbb{N};\ X^i \in C([0, T]; S_r^c)$ *for* $m < \forall i \in \mathbb{N}\}$ *for* $\mathsf{X} = \sum_{i\in\mathbb{N}} \delta_{X^i}$.

Remark 1 (i) Determinantal point processes satisfy (Q1) (see [18]).
(ii) (Q2) is obvious because the unlabeled dynamics is μ-reversible.
(iii) (Q3) is satisfied if the one-point correlation function ρ^1 satisfies $\rho^1(x) = O(e^{|x|^\alpha})$ $(|x| \to \infty)$ for some $\alpha < 2$.

Let $\mathbf{T} = \{\mathbf{t} = (t_1, \dots, t_m)\,;\ t_i \in [0, T], m \in \mathbb{N}\}$ and $\mathbf{X}_\mathbf{t}^{n*} = (\mathbf{X}_{t_1}^{n*}, \dots, \mathbf{X}_{t_m}^{n*})$. Let

$$\tilde{\mathscr{T}}_{\mathrm{path}}(\mathsf{S}) = \bigvee_{\mathbf{t}\in\mathbf{T}} \bigcap_{r=1}^{\infty} \sigma[\pi_r^c(\mathsf{X}_\mathbf{t})], \quad \tilde{\mathscr{T}}_{\mathrm{path}}((\mathbb{R}^d)^{\mathbb{N}}) = \bigvee_{\mathbf{t}\in\mathbf{T}} \bigcap_{n=1}^{\infty} \sigma[\mathbf{X}_\mathbf{t}^{n*}].$$

Hence by definition, $\tilde{\mathscr{T}}_{\text{path}}(\mathsf{S})$ is the cylindrical tail σ-field of the unlabeled path space and $\tilde{\mathscr{T}}_{\text{path}}((\mathbb{R}^d)^{\mathbb{N}})$ is the cylindrical tail σ-field of the labeled path space $W((\mathbb{R}^d)^{\mathbb{N}})$. We deduce the triviality of $\mathscr{T}_{\text{path}}((\mathbb{R}^d)^{\mathbb{N}})$ from that of S. We do this step-by-step following the scheme:

$$\begin{array}{cccccccc} \mathscr{T}(\mathsf{S}) & \xrightarrow[\text{(Q1), (Q2)}]{\text{(Step I)}} & \tilde{\mathscr{T}}_{\text{path}}(\mathsf{S}) & \xrightarrow[\text{(Q3)}]{\text{(Step II)}} & \tilde{\mathscr{T}}_{\text{path}}((\mathbb{R}^d)^{\mathbb{N}}) & \xrightarrow[\text{(IFC)}]{\text{(Step III)}} & \mathscr{T}_{\text{path}}((\mathbb{R}^d)^{\mathbb{N}}) \\ \mu & & \mathsf{P}_{\mu} & & \mathbf{P}_{\mu^{\mathfrak{l}}} = \int \bar{P}_{\mathbf{s}}(\mathbf{X} \in \cdot) d\mu^{\mathfrak{l}} & & \bar{P}_{\mathbf{s},\mathbf{B}} \text{ a.s. } (\mathbf{s}, \mathbf{B}). \end{array}$$

We denote by $\mathscr{B}(\mathsf{S})^{\mu}$ the completion of the σ-field $\mathscr{B}(\mathsf{S})$ with respect to μ.

Definition 3 For $\mathbf{s} \in (\mathbb{R}^d)^{\mathbb{N}}$, we set $\mathbf{X}_{\mathbf{s}} = \{\mathbf{X}_{\mathbf{s},t}\}_{t\in[0,\infty)}$ such that $\mathbf{X}_{\mathbf{s},0} = \mathbf{s}$. We set

$$\{\mathbf{X}_{\mathbf{s}}\}_{\mathbf{s}\in\mathfrak{l}(\mathsf{H})} = \{\{\mathbf{X}_{\mathbf{s},t}\}_{t\in[0,\infty)}\}_{\mathbf{s}\in\mathfrak{l}(\mathsf{H})}.$$

(i) We call $(\{\mathbf{X}_{\mathbf{s}}\}_{\mathbf{s}\in\mathfrak{l}(\mathsf{H})}, \mathbf{B})$ a μ-solution of (1) if $\mathsf{H} \in \mathscr{B}(\mathsf{S})^{\mu}$ satisfying $\mu(\mathsf{H}) = 1$ and $\mathfrak{u}^{-1}(\mathsf{H}) \subset \mathbf{S}_0$, and if $(\mathbf{X}_{\mathbf{s}}, \mathbf{B})$ is a solution of ISDE (1) for each $\mathbf{s} \in \mathfrak{l}(\mathsf{H})$.
(ii) We call $(\{\mathbf{X}_{\mathbf{s}}\}_{\mathbf{s}\in\mathfrak{l}(\mathsf{H})}, \mathbf{B})$ a μ-strong solution if it is a μ-solution such that $(\mathbf{X}_{\mathbf{s}}, \mathbf{B})$ is a strong solution for each $\mathbf{s} \in \mathfrak{l}(\mathsf{H})$.

Definition 4 We say that the μ-strong uniqueness holds if the following holds.
(i) The μ-uniqueness in law holds. That is, $\mathbf{X}_{\mathbf{s}} = \mathbf{X}'_{\mathbf{s}}$ in law for each $\mathbf{s} \in \mathfrak{l}(\mathsf{H} \cap \mathsf{H}')$ for any pair of μ-solutions $(\{\mathbf{X}_{\mathbf{s}}\}_{\mathbf{s}\in\mathfrak{l}(\mathsf{H})}, \mathbf{B})$ and $(\{\mathbf{X}'_{\mathbf{s}}\}_{\mathbf{s}\in\mathfrak{l}(\mathsf{H}')}, \mathbf{B}')$ satisfying (Q2).
(ii) A μ-solution $(\{\mathbf{X}_{\mathbf{s}}\}_{\mathbf{s}\in\mathfrak{l}(\mathsf{H})}, \mathbf{B})$ satisfying (Q2) is a μ-strong solution $(\{\mathbf{X}_{\mathbf{s}}\}_{\mathbf{s}\in\mathfrak{l}(\mathsf{H}')}, \mathbf{B})$ for some $\mathsf{H}' \subset \mathsf{H}$.
(iii) The μ-pathwise uniqueness holds. That is, $P^{\mathbf{B}}(\mathbf{X}_{\mathbf{s}} = \mathbf{X}'_{\mathbf{s}}) = 1$ for each $\mathbf{s} \in \mathfrak{l}(\mathsf{H} \cap \mathsf{H}')$, where $\{\mathbf{X}_{\mathbf{s}}\}_{\mathbf{s}\in\mathfrak{l}(\mathsf{H})}$ and $\{\mathbf{X}'_{\mathbf{s}}\}_{\mathbf{s}\in\mathfrak{l}(\mathsf{H}')}$ are any pair of μ-strong solutions defined for the same Brownian motion $\mathbf{B}$ satisfying (Q2).

Theorem 8 ([20]) *Make the same assumptions as for Theorem 5. Assume (P2), (Q1), and (Q3). Then* (1) *has a μ-strong solution $\mathbf{X}$ such that the associated unlabeled dynamics X is μ-reversibile, and the μ-strong uniqueness holds.*

7 Examples Arising from Random Matrix Theory

The first three examples are particle systems in $\mathbb{R}$ ($[0, \infty)$ for Bessel), whereas the last example is in $\mathbb{R}^2$. All examples have logarithmic interaction potential.
sine, Airy, and Bessel IBM [5, 15, 19, 26]**:** Let $d = 1$.

$$dX_t^i = dB_t^i + \frac{\beta}{2}\lim_{r\to\infty}\sum_{|X_t^i - X_t^j|<r,\ j\neq i}\frac{1}{X_t^i - X_t^j}dt, \qquad \text{(Dyson model, sine)}$$

$$dX_t^i = dB_t^i + \frac{\beta}{2}\lim_{r\to\infty}\{(\sum_{j\neq i,\ |X_t^j|<r}\frac{1}{X_t^i - X_t^j}) - \frac{1}{\pi}\int_{-r}^{0}\frac{\sqrt{-x}}{-x}dx\}dt, \qquad \text{(Airy)}$$

$$dX_t^i = dB_t^i + \frac{a}{2X_t^i}dt + \frac{\beta}{2}\sum_{j\neq i}^{\infty}\frac{1}{X_t^i - X_t^j}dt \quad a \geq 1. \qquad \text{(Bessel)}$$

The equilibrium states of these dynamics are sine, Airy, and Bessel point processes ($\beta = 1, 2, 4$ (sine, Airy), $\beta = 2$ (Bessel)). These point processes correspond to bulk, soft edge, and hard edge scaling limits respectively. The relationships to inverse temperature are: $\beta = 1 \Rightarrow$GOE, $\beta = 2 \Rightarrow$GUE, and $\beta = 4 \Rightarrow$GSE, respectively.
Ginibre IBM [15]**:** Let $d = 2$ and $\beta = 2$. We consider two ISDEs.

$$dX_t^i = dB_t^i + \lim_{r\to\infty}\sum_{|X_t^i - X_t^j|<r,\ j\neq i}\frac{X_t^i - X_t^j}{|X_t^i - X_t^j|^2}dt, \tag{10}$$

$$dX_t^i = dB_t^i - X_t^i + \lim_{r\to\infty}\sum_{|X_t^j|<r,\ j\neq i}\frac{X_t^i - X_t^j}{|X_t^i - X_t^j|^2}dt. \tag{11}$$

The equilibrium state μ_{gin} is the Ginibre point process, which has various rigidities such as small variance [24], number rigidity [4], and dichotomy in its reduced Palm measures [23]. The drift coefficients are equal on and tangential to the support of μ_{gin}, yielding the coincidence of the solutions of (10) and (11). This dynamical rigidity reflects rigidity of μ_{gin}.

8 Algebraic Construction and Finite Particle Approxiations

An algebraic construction is known for stochastic dynamics related to point processes appearing in random matrix theory in $\mathbb{R}$ with $\beta = 2$, which is given by space-time correlation functions e.g. [6–8]. For example, as for the Airy$_2$ point process, the multi-time, moment generating function is

$$\mathbf{E}\big[\exp\{\sum_{m=1}^{M}\langle f_{t_m}, \mathsf{X}_{t_m}\rangle\}\big] = \operatorname*{Det}_{(s,t)\in\mathbf{t}^2,\ (x,y)\in\mathbb{R}^2}\Big[\delta_{st}\delta(x-y) + \mathbf{K}_{\mathrm{Airy}}(s,x;t,y)\chi_t(y)\Big].$$

Here $\mathbf{t} = \{t_1, \ldots, t_M\}$, $\chi_t = e^{f_t} - 1$, and $\mathbf{K}_{\mathrm{Ai}}$ is the extended Airy kernel

$$\mathbf{K}_{\mathrm{Ai}}(s,x;t,y)=\begin{cases}\int_0^{\infty}du\, e^{-u(t-s)/2}\mathrm{Ai}(u+x)\mathrm{Ai}(u+y), & t\geq s\\ -\int_{-\infty}^{0}du\, e^{-u(t-s)/2}\mathrm{Ai}(u+x)\mathrm{Ai}(u+y), & t<s.\end{cases}$$

Theorem 9 ([21, 22]) *The algebraic construction and the ISDEs define the same stochastic dynamics for* $sine_2$*,* $Airy_2$*, and* $Bessel_2$*.*

By algebraic method, the finite particle approximation for sine_2, Airy_2, and Bessel_2 is proved [22]. By analytic method, the same is proved for these point processes with $\beta = 1, 2, 4$ and also the Ginibre point process [9]. The latter approach is robust and valid for many other examples.

Acknowledgements H.O. is supported in part by a Grant-in-Aid for Scenic Research (KIBAN-A, No.24244010; KIBAN-A, No.16H02149; KIBAN-S, No.16H06338) from the Japan Society for the Promotion of Science.

References

1. Albeverio, S., Kondratiev, YuG, Röckner, M.: Analysis and geometry on configuration spaces: the Gibbsian case. J. Funct. Anal. **157**, 242–291 (1998)
2. Fritz, J.: Gradient dynamics of infinite point systems. Ann. Probab. **15**, 478–514 (1987)
3. Fukushima, M., Oshima, Y., Takeda, M.: Dirichlet Forms And Symmetric Markov Processes. De Gruyter Studies in Mathematics, vol. 19, 2nd edn. Walter de Gruyter & Co., Berlin (2011)
4. Ghosh, S., Peres, Y.: Rigidity and tolerance in point processes: Gaussian zeroes and Ginibre eigenvalues. Duke Math. **166**, 1789–1858 (2017)
5. Honda, R., Osada, H.: Infinite-dimensional stochastic differential equations related to Bessel random point fields. Stoch. Process. Appl. **125**, 3801–3822 (2015)
6. Johansson, K.: Non-intersecting paths, random tilings and random matrices. Probab. Theory Relat. Fields **123**, 225–280 (2002)
7. Katori, M., Tanemura, H.: Noncolliding Brownian motion and determinantal processes. J. Stat. Phys. **129**, 1233–1277 (2007)
8. Katori, M., Tanemura, H.: Markov property of determinantal processes with extended sine, Airy, and Bessel kernels. Markov Process. Relat. Fields **17**, 541–580 (2011)
9. Kawamoto, Y., Osada, H.: Finite-particle approximations for interacting Brownian particles with logarithmic potentials, J. Math. Soc. Jpn. (to appear). arXiv:1607.06922
10. Lang, R.: Unendlich-dimensionale Wienerprocesse mit Wechselwirkung I. Z. Wahrschverw. Gebiete **38**, 55–72 (1977)
11. Lang, R.: Unendlich-dimensionale Wienerprocesse mit Wechselwirkung II. Z. Wahrschverw. Gebiete **39**, 277–299 (1978)
12. Ma, Z.-M., Röckner, M.: Introduction to the Theory of (Non-symmetric) Dirichlet Forms Universitext. Springer, Berlin (1992)
13. Osada, H.: Dirichlet form approach to infinite-dimensional Wiener processes with singular interactions. Commun. Math. Phys. **176**, 117–131 (1996)
14. Osada, H.: Tagged particle processes and their non-explosion criteria. J. Math. Soc. Jpn. **62**, 867–894 (2010)
15. Osada, H.: Infinite-dimensional stochastic differential equations related to random matrices. Probab. Theory Relat. Fields **153**, 471–509 (2012)

16. Osada, H.: Interacting Brownian motions in infinite dimensions with logarithmic interaction potentials. Ann. Probab. **41**, 1–49 (2013)
17. Osada, H.: Interacting Brownian motions in infinite dimensions with logarithmic interaction potentials II: Airy random point field. Stoch. Process. Appl. **123**, 813–838 (2013)
18. Osada, H., Osada, S.: Discrete approximations of determinantal point processes on continuous spaces: tree representations and tail triviality. J. Stat. Phys. **170**(2), 421–435 (2018). https://doi.org/10.1007/s10955-017-1928-2
19. Osada, H., Tanemura, H.: Infinite-dimensional stochastic differential equations related to Airy random point fields. arXiv:1408.0632
20. Osada, H., Tanemura, H.: Infinite-dimensional stochastic differential equations and tail σ-fields. arXiv:1412.8674
21. Osada, H., Tanemura, H.: Cores of Dirichlet forms related to random matrix theory. Proc. Jpn. Acad. Ser. A Math. Sci. **90**, 145–150 (2014)
22. Osada, H., Tanemura, H.: Strong Markov property of determinantal processes with extended kernels. Stochastic Process. Appl. **126**, 186–208 (2016)
23. Osada, H., Shirai, T.: Absolute continuity and singularity of Palm measures of the Ginibre point process. Probab. Theory Relat. Fields **165**, 725–770 (2016)
24. Shirai, T.: Large deviations for the Fermion point process associated with the exponential kernel. J. Stat. Phys. **123**, 615–629 (2006)
25. Tanemura, H.: A system of infinitely many mutually reflecting Brownian balls in $\mathbb{R}^d$. Probab. Theory Relat. Fields **104**, 399–426 (1996)
26. Tsai, Li-Cheng: Infinite dimensional stochastic differential equations for Dyson's model. Probab. Theory Relat. Fields **166**, 801–850 (2015)
27. Yoshida, M.: W., Construction of infinite-dimensional interacting diffusion processes through Dirichlet forms. Probab. Theory Relat. Fields **106**, 265–297 (1996)

Recent Progress on the Dirichlet Forms Associated with Stochastic Quantization Problems

Rongchan Zhu and Xiangchan Zhu

Abstract In this paper we present recent progress on the Dirichlet forms associated with stochastic quantization problems obtained in Röckner et al. (J Funct Anal, 272(10):4263–4303, 2017, [23]), Röckner et al. (Commun Math Phys, 352(3):1061–1090, 2017, [24]), Zhu and Zhu (Dirichlet form associated with the Φ_3^4 model, 2017, [27]). In the two dimensional case we have obtained the equivalence of the two notions of solutions, the restricted Markov uniqueness and the uniqueness of martingale problem. In the three dimensional case we construct the Dirichlet form associated with the dynamical Φ_3^4 model obtained in Catellier and Chouk (Paracontrolled distributions and the 3-dimensional stochastic quantization equation, [6]), Hairer (Invent Math, 198:269–504, 2014, [14]), Mourrat and Weber (Global well-posedness of the dynamic Φ_3^4 model on the torus, [20].

Keywords ϕ_3^4 model · Dirichlet form · Regularity structures · Paracontrolled distributions · Space-time white noise · Renormalisation

2010 Mathematics Subject Classification AMS 60H15 · 82C28

1 Introduction

In this paper we analyze stochastic quantization equations on $\mathbb{T}^d$, $d = 2, 3$ and on $\mathbb{R}^2$: Let $H = L^2(\mathbb{T}^d)$ or $L^2(\mathbb{R}^2)$ and consider

R. Zhu (✉)
Department of Mathematics, Beijing Institute of Technology, Beijing 100081, China
e-mail: zhurongchan@126.com

R. Zhu · X. Zhu
Department of Mathematics, University of Bielefeld, 33615 Bielefeld, Germany
e-mail: zhuxiangchan@126.com

X. Zhu
School of Science, Beijing Jiaotong University, Beijing 100044, China

A. Eberle et al. (eds.), *Stochastic Partial Differential Equations and Related Fields*, Springer Proceedings in Mathematics & Statistics 229,
https://doi.org/10.1007/978-3-319-74929-7_39

$$\begin{aligned} dX &= (AX - :X^3:)dt + dW(t), \\ X(0) &= z, \end{aligned} \tag{1}$$

where $A : D(A) \subset H \to H$ is the linear operator

$$A\phi = \Delta\phi - \phi,$$

where $:X^3:$ means the renormalization of X^3. W is a cylindrical $\mathscr{F}_t$-Wiener process defined on a probability space $(\Omega, \mathscr{F}, P)$ with a normal filtration $(\mathscr{F}_t)_{t\geq 0}$.

This equation arises in stochastic quantization of Euclidean quantum field theory. Heuristically, (1) has an invariant measure ν defined as

$$\nu(d\phi) = ce^{-\frac{1}{2}\int :\phi^4: dx}\mu(d\phi),$$

where c is a normalization constant and μ is the Gaussian free field. ν is called the Φ_d^4-quantum field. There have been many approaches to the problem of giving a meaning to the above heuristic measure for the two dimensional case and the three dimensional case (see [8, 9, 13] and references therein). In [21] Parisi and Wu proposed a program for Euclidean quantum field theory of getting Gibbs states of classical statistical mechanics as limiting distributions of stochastic processes, especially as solutions to non-linear stochastic differential equations. Then one can use the stochastic differential equations to study the properties of the Gibbs states. This procedure is called stochastic field quantization (see [16]). The Φ_d^4 model is the simplest non-trivial Euclidean quantum field (see [9] and the reference therein). The issue of the stochastic quantization of the Φ_d^4, $d = 2, 3$ model is to solve the Eq. (1) and to prove that the invariant measure is the limit of the time marginals as $t \to \infty$, which means the marginals converge to the Euclidean quantum field.

When $d = 2$: In [1] weak solutions to (1) have been constructed by using the Dirichlet form approach in the finite and infinite volume case. In [18] the stationary solution to (1) has also been considered in their general theory of martingale solutions for stochastic partial differential equations; In [7] Da Prato and Debussche define the Wick powers of solutions to the stochastic heat equation in the paths space and study a shifted equation instead of (1) in the finite volume case. In [19] the authors also consider the shifted equation and obtain global existence and uniqueness of the solution directly from every starting point both in the finite and infinite volume case.

When $d = 3$: Local existence and uniqueness of the solutions to (1) has been obtained in [14] by Hairer's theory of regularity structures and in [6] by paracontrolled distributions proposed by Imkeller, Gubinelli, Perkowski in [12]. Global existence and uniqueness of the solutions to (1) has been obtained in [20] recently.

As a result of these new development in the SPDE theory for the stochastic quantization problem, we can also study the properties of the associated Dirichlet forms such as restricted Markov uniqueness in [23], the properties of the Φ_2^4 field in [24] and can also construct the Dirichlet forms associated to the dynamical Φ_3^4 model in [27]. In this paper we give a survey on these results.

This paper is organized as follows: In Sect. 2 we survey recent results on the associated Dirichlet form and the solutions to the shifted equation in the two dimensional case. We obtain Markov uniqueness in the restricted sense and uniqueness of the martingale solutions (probabilistically weak solution) to (1). In Sect. 3 we give the main results for construction of the Dirichlet form associated with the solution obtained in [20].

Besov Spaces

In the following we recall the definitions of Besov spaces. The space of real valued infinitely differentiable functions of compact support is denoted by $\mathscr{D}(\mathbb{R}^d)$ or $\mathscr{D}$. The space of Schwartz functions is denoted by $\mathscr{S}(\mathbb{R}^d)$. Its dual, the space of tempered distributions, is denoted by $\mathscr{S}'(\mathbb{R}^d)$. The Fourier transform and the inverse Fourier transform are denoted by $\mathscr{F}$ and $\mathscr{F}^{-1}$, respectively.

Let (χ, θ) be a dyadic partition of unity, and for the existence of dyadic partitions of unity we refer to [5, Proposition 2.10]. The Littlewood–Paley blocks are now defined as $\Delta_{-1}u = \mathscr{F}^{-1}(\chi\mathscr{F}u)$, $\Delta_j u = \mathscr{F}^{-1}(\theta(2^{-j}\cdot)\mathscr{F}u)$, $j \geq 1$.

For $\alpha \in \mathbb{R}$, $p, q \in [1, \infty]$, $u \in \mathscr{D}$ we define $\|u\|_{B^\alpha_{p,q}} := (\sum_{j\geq -1}(2^{j\alpha}\|\Delta_j u\|_{L^p})^q)^{1/q}$, with the usual interpretation as l^∞ norm in case $q = \infty$. The Besov space $B^\alpha_{p,q}$ consists of the completion of $\mathscr{D}$ with respect to this norm and the Hölder–Besov space $\mathscr{C}^\alpha$ is given by $\mathscr{C}^\alpha(\mathbb{R}^d) = B^\alpha_{\infty,\infty}(\mathbb{R}^d)$. We use $\|\cdot\|_\alpha$ to denote $\|\cdot\|_{\mathscr{C}^\alpha}$. We point out that everything above and everything that follows can be applied to distributions on the torus (see [25, 26]).

2 Two Dimensional Case

In this section we consider (1) on the torus $\mathbb{T}^2$.

2.1 Wick Power

In the following we define the Wick powers. First we define Wick powers on $L^2(\mathscr{S}'(\mathbb{T}^2), \mu)$ for $\mu := N(0, C)$ with $C := \frac{1}{2}(-\Delta + 1)^{-1}$.

Wick Powers on $L^2(\mathscr{S}'(\mathbb{T}^2), \mu)$

Now we define the Wick powers by using approximations: for $\phi \in \mathscr{S}'(\mathbb{T}^2)$ define $\phi_\varepsilon := \rho_\varepsilon * \phi$, with ρ_ε an approximate delta function on $\mathbb{R}^2$ given by $\rho_\varepsilon = \varepsilon^{-2}\rho(\frac{\cdot}{\varepsilon}) \in \mathscr{D}$, $\int \rho = 1$. Here the convolution means that we view ϕ as a periodic distribution in $\mathscr{S}'(\mathbb{R}^2)$ and do convolution on $\mathbb{R}^2$. For every $n \in \mathbb{N}$ we set

$$:\phi_\varepsilon^n:_C \;:= c_\varepsilon^{n/2} P_n(c_\varepsilon^{-1/2}\phi_\varepsilon),$$

where P_n, $n = 0, 1, \ldots$, are the Hermite polynomials defined by the formula $P_n(x) = \sum_{j=0}^{[n/2]}(-1)^j \frac{n!}{(n-2j)!j!2^j}x^{n-2j}$, and $c_\varepsilon = \int \phi_\varepsilon^2 \mu(d\phi) = \frac{1}{2}\int\int \bar{G}(x-y)\rho_\varepsilon(y)dy\rho_\varepsilon$

$(x)dx$. Then $: \phi_\varepsilon^n :_C \in \mathscr{H}_n$, with $\mathscr{H}_n$ being the nth Wiener chaos. Here and in the following $\bar{G}$ is the Green function associated with $-A$ on $\mathbb{T}^2$.

A direct calculation yields the following:

Lemma 1 ([23, Lemma 3.1]) *Let $\alpha < 0$, $n \in \mathbb{N}$ and $p > 1$. $: \phi_\varepsilon^n :_C$ converges to some element in $L^p(\mathscr{S}'(\mathbb{T}^2), \mu; \mathscr{C}^\alpha)$. This limit is called the nth Wick power of ϕ with respect to the covariance C and denoted by $: \phi^n :_C$.*

Wick Powers on a Fixed Probability Space

Now we follow the idea from [7, 19] to define the Wick powers of the solutions to the stochastic heat equation in the paths space. We fix a probability space $(\Omega, \mathscr{F}, P)$ and W is a cylindrical Wiener process on $L^2(\mathbb{T}^2)$. In the following we set $Z(t) = \int_0^t e^{(t-s)A} dW(s)$ and let $Z_\varepsilon := \rho_\varepsilon * Z$. For $t > 0$, define

$$: Z_\varepsilon^n(t) :_C \ := c_\varepsilon^{\frac{n}{2}} P_n \left(c_\varepsilon^{-\frac{1}{2}} Z_\varepsilon(t) \right).$$

Then we have the following.

Lemma 2 ([23, Lemma 3.4]) *For $\alpha < 0$, $p > 1$, $n \in \mathbb{N}$, $: Z_\varepsilon^n :_C$ converges in $L^p(\Omega, C((0, T]; \mathscr{C}^\alpha))$. Here $C((0, T]; \mathscr{C}^\alpha)$ is equipped with the norm $\sup_{t \in [0,T]} t^{\rho/2} \| \cdot \|_\alpha$ for $\rho > 0$. The limit is called nth Wick powers of $Z(t)$ with respect to the covariance C and denoted by $: Z^n(t) :_C$.*

Now following the technique in [19] we combine the initial value part with the Wick powers. We set $V(t) = e^{tA} z$, $z \in \mathscr{C}^\alpha$ for $\alpha < 0$ and

$$\bar{Z}(t) = Z(t) + V(t), \quad : \bar{Z}^n(t) :_C = \sum_{k=0}^{n} C_n^k V(t)^{n-k} : Z^k(t) :_C .$$

In [23, Lemma 3.6] we give relations between the above two different Wick powers.

2.2 Relations Between the Two Solutions

First we recall some basic results related to Dirichlet forms from [1].

Solutions Given by Dirichlet Forms

Let $H = L^2(\mathbb{T}^2)$ and let $-\Delta + I$ be the generator of the following quadratic form on $H : (u, v) \mapsto \int_{\mathbb{T}^2} \langle \nabla u, \nabla v \rangle_{\mathbb{R}^d} dx + \int_{\mathbb{T}^2} uv dx$ with $u, v \in \{g \in L^2(\mathbb{T}^2) | \nabla g \in L^2(\mathbb{T}^2)\}$ (where ∇ is in the sense of distributions). Let $\{e_k | k \in \mathbb{Z}^2\} \subset C^\infty(\mathbb{T}^2)$ be the (orthonormal) eigenbasis of $-\Delta + I$ in H and $\{\lambda_k | k \in \mathbb{Z}^2\} \subset (0, \infty)$ the corresponding eigenvalues. Define for $s \in \mathbb{R}$, $H^s := \{u \in \mathscr{S}'(\mathbb{T}^2) | \sum_{k \in \mathbb{Z}^2} \lambda_k^s {}_{\mathscr{S}'}\langle u, e_k \rangle_{\mathscr{S}}^2 < \infty\}$, equipped with the inner product $\langle u, v \rangle_{H^s} := \sum_{k \in \mathbb{Z}^2} \lambda_k^s {}_{\mathscr{S}'}\langle u, e_k \rangle_{\mathscr{S}} {}_{\mathscr{S}'}\langle v, e_k \rangle_{\mathscr{S}}$. For $s \geq 0$ ${}_{H^s}\langle \cdot, \cdot \rangle_{H^{-s}}$ denotes the dualization between H^s and its dual space H^{-s}. Let $E = H^{-1-\varepsilon}$, $E^* = H^{1+\varepsilon}$ for some $\varepsilon > 0$. We denote their Borel σ-algebras by $\mathscr{B}(E)$, $\mathscr{B}(E^*)$ respectively. Define

$$\mathscr{F}C_b^\infty := \{u : u(z) = f({}_{E^*}\langle l_1, z\rangle_E, {}_{E^*}\langle l_2, z\rangle_E, \ldots, {}_{E^*}\langle l_m, z\rangle_E), z \in E,$$
$$l_1, l_2, \ldots, l_m \in E^*, m \in \mathbb{N}, f \in C_b^\infty(\mathbb{R}^m)\}.$$

Define for $u \in \mathscr{F}C_b^\infty$ and $l \in H$, $\frac{\partial u}{\partial l}(z) := \frac{d}{ds}u(z + sl)|_{s=0}, z \in E$. Let Du denote the H-derivative of $u \in \mathscr{F}C_b^\infty$, i.e. the map from E to H such that $\langle Du(z), l\rangle = \frac{\partial u}{\partial l}(z)$ for all $l \in H, z \in E$. By [1] we easily deduce that the form

$$\mathscr{E}(u, v) := \frac{1}{2}\int_E \langle Du, Dv\rangle_H d\nu, \quad u, v \in \mathscr{F}C_b^\infty$$

is closable and its closure $(\mathscr{E}, D(\mathscr{E}))$ is a quasi-regular Dirichlet form on $L^2(E; \nu)$ in the sense of [17]. Here ν is the measure defined in the introduction. By [1, Theorem 3.6] we know that there exists a (Markov) diffusion process $M = (\Omega, \mathscr{F}, (\mathscr{M}_t)_{t\geq 0}, (X(t))_{t\geq 0}, (P^z)_{z\in E})$ on E *properly associated with* $(\mathscr{E}, D(\mathscr{E}))$. By [1, Theorem 6.10] we have the following Fukushima decomposition for $X(t)$ under P^z.

Theorem 1 *There exist a map* $W : \Omega \to C([0, \infty); E)$ *and a* properly $\mathscr{E}$*-exceptional set* $S \subset E$*, i.e.* $\nu(S) = 0$ *and* $P^z[X(t) \in E \setminus S, \forall t \geq 0] = 1$ *for* $z \in E\backslash S$*, such that* $\forall z \in E\backslash S$ *under* P^z*,* W *is an* $\mathscr{M}_t$*- cylindrical Wiener process and the sample paths of the associated process* $M = (\Omega, \mathscr{F}, (\mathscr{M}_t)_{t\geq 0}, (X(t))_{t\geq 0}, (P^z)_{z\in E})$ *on* E *satisfy the following: for* $l \in H^{2+s}$*,* $s > 0$

$$\begin{aligned} {}_{E^*}\langle l, X(t) - X(0)\rangle_E = \int_0^t \langle l, dW(r)\rangle + \int_0^t \Big[{}_{H^{-s-2}}\langle - : X(r)^3 :, l\rangle_{H^{2+s}} \\ + {}_{H^s}\langle \Delta l - l, X(r)\rangle_{H^{-s}} \Big] dr \quad \forall t \geq 0\ P^z\text{--a.s.} \end{aligned} \tag{2}$$

Moreover, ν *is an invariant measure for* M *in the sense that* $\int P_t u d\nu = \int u d\nu$ *for* $u \in L^2(E; \nu) \cap \mathscr{B}_b(E)$*. Here* P_t *is the semigroup for* M*.*

Solutions Given by the Shifted Equation

Now we fix a stochastic basis $(\Omega, \mathscr{F}, (\mathscr{F}_t)_{t\geq 0}, P)$ and on it a cylindrical Wiener process W in $L^2(\mathbb{T}^2)$. Define $\bar{Z}(t) = \int_0^t e^{(t-s)A} dW(s) + e^{tA}z$ with $z \in \mathscr{C}^\alpha$ for $\alpha < 0$. Now we consider the following equation:

$$Y(t) = e^{tA}y - \int_0^t e^{(t-s)A} \sum_{l=0}^3 C_3^l Y(s)^l : \bar{Z}(s)^{3-l} : ds. \tag{3}$$

Global existence and uniqueness of the solutions to (3) have been obtained in [19].

Relations Between the Two Solutions

In the following we discuss the relations between M constructed above and the shifted equation. For W constructed in Theorem 1 define $\bar{Z}(t) := \int_0^t e^{(t-s)A} dW(s) + e^{tA}X(0)$.

Theorem 2 (cf. [23, Theorem 3.9]) *Let $\alpha \in (-\frac{1}{3}, 0)$, $-\alpha < \beta < \alpha + 2$. There exists a properly $\mathscr{E}$-exceptional set $S_2 \subset E$ in the sense of Theorem 1 such that for every $z \in \mathscr{C}^\alpha \setminus S_2$ under P^z, $Y := X - \bar{Z} \in C([0, T]; \mathscr{C}^\beta)$ is a solution to* (3) *with $y = 0$. Moreover, $P^z[X(t) \in \mathscr{C}^\alpha \setminus S_2, \forall t \geq 0] = 1$ for $z \in \mathscr{C}^\alpha \setminus S_2$.*

By using solutions given by Dirichlet form theory we also obtain that ν is an invariant measure of the solution to $X_0 = Y_0 + \bar{Z}$, where Y_0 is the unique solution to (3) with $y = 0$.

Theorem 3 (cf. [23, Theorem 3.10]) *ν is an invariant measure of the solution $X_0 = Y_0 + \bar{Z}$, where Y_0 is the unique solution to* (3) *with $y = 0$.*

2.3 Markov Uniqueness in the Restricted Sense

By [17, Chap. 4, Sect. 4b] it follows that there is a point separating countable $\mathbb{Q}$-vector space $D \subset \mathscr{F}C_b^\infty$ such that $D \subset D(L(\mathscr{E}))$. Let $\mathscr{E}^{q.r.}$ be the set of all quasi-regular Dirichlet forms $(\tilde{\mathscr{E}}, D(\tilde{\mathscr{E}}))$ (cf. [17]) on $L^2(E; \nu)$ such that $D \subset D(L(\tilde{\mathscr{E}}))$ and $\tilde{\mathscr{E}} = \mathscr{E}$ on $D \times D$. Here for a Dirichlet form $(\tilde{\mathscr{E}}, D(\tilde{\mathscr{E}}))$ we denote its generator by $(L(\tilde{\mathscr{E}}), D(L(\tilde{\mathscr{E}})))$. In the following we consider the martingale problem in the sense of [2] to (1):

Definition 1 A ν-special standard process $M = (\Omega, \mathscr{F}, (\mathscr{M}_t)_{t\geq 0}, (X_t)_{t\geq 0}, (P^z))$ in the sense of [17, Chap. IV] with state space E is said to solve the martingale problem for $(L(\mathscr{E}), D)$ if for all $u \in D$, $u(X(t)) - u(X(0)) - \int_0^t L(\mathscr{E})u(X(s))ds$, $t \geq 0$, is an $\mathscr{M}_t$-martingale under P^ν.

To explain the uniqueness result below we also introduce the following concept: Two strong Markov processes M and M' with state space E and transition semigroups $(p_t)_{t>0}$ and $(p'_t)_{t>0}$ are called *ν-equivalent* if there exists $S \in \mathscr{B}(E)$ such that (i) $\nu(E\backslash S) = 0$, (ii) $P^z[X(t) \in S, \forall t \geq 0] = P'^z[X'(t) \in S, \forall t \geq 0] = 1, z \in S$, (iii) $p_t f(z) = p'_t f(z)$ for all $f \in \mathscr{B}_b(E), t > 0$ and $z \in S$.

Combining Theorems 2 and 3, we obtain Markov uniqueness in the restricted sense for $(L(\mathscr{E}), D)$ (see part (ii)) and the uniqueness of martingale (probabilistically weak) solutions to (1) if the solution has ν as an invariant measure (see part (i)):

Theorem 4 (cf. [23, Theorem 3.12])

(i) There exists (up to ν-equivalence) exactly one ν-special standard process M with state space E solving the martingale problem for $(L(\mathscr{E}), D)$ and satisfying $P^z(X \in C([0, \infty); E)) = 1$ for ν-a.e. $z \in E$ and having ν as an invariant measure.

(ii) $\sharp\mathscr{E}^{q.r.} = 1$. Moreover, there exists (up to ν-equivalence) exactly one ν-special standard process M with state space E associated with a Dirichlet form $(\mathscr{E}, D(\mathscr{E}))$ solving the martingale problem for $(L(\mathscr{E}), D)$.

Remark 1 (i) For the stationary case, we refer to [23, Sect. 3.5]. In this case, we can obtain a probabilistically strong solution.

(ii) All the results above can also be obtained in the infinite volume case (see [23, Sect. 4]), including the equivalence of the two notions of solutions, the Φ_2^4 field being an invariant measure of the shifted equation, the restricted Markov uniqueness, the uniqueness of martingale problem. Instead of Besov space, we have to consider weighted Besov space as state space.

2.4 Ergodicity

It is natural to ask whether this invariant measure ν is the unique invariant measure for X_0 obtained in Theorem 2. If ν is the unique invariant measure for X_0, then ν is the limiting distribution of the stochastic process X_0. This problem is main point in the stochastic field quantization as we mentioned above in the Φ_2^4 model on the torus. This problem has been studied in [3] and the references therein. It is proved in [3] that the stochastic quantization of a Guerra–Rosen–Simon Gibbs state on $\mathscr{S}'(\mathbb{R}^2)$ in infinite volume with polynomial interaction is ergodic if the Gibbs state is a pure phase. This result also holds for the finite volume case if one takes Dirichlet boundary conditions. Moreover, by [22] we know that ν constructed above with A changed to a Dirichlet Laplacian on a bounded domain is a pure phase, which implies that the stochastic quantization of the Gibbs state is ergodic. However, the idea in [22] and the results in [3] cannot be applied for the torus. In this case we don't know whether ν is a pure phase. We also emphasize that it is not obvious that ν is a pure phase even if ν is absolutely continuous with respect to μ. In this case, the zero set of $\frac{d\nu}{d\mu}$, i.e. $\{\frac{d\nu}{d\mu}=0\}$, which is hard to analyze analytically, may divide the state space into different irreducible components, which immediately implies non-ergodicity, i.e. the existence of two invariant measures. In [24] we study this problem using the techniques from SPDE theory. We analyze the shifted equation directly and obtain that ν is the unique invariant measure of X_0, where X_0 is the solution considered in Theorem 2.

Theorem 5 (cf. [24, Theorem 1.1]) *ν is the unique invariant probability measure for the process X_0. Moreover, the associated semigroup P_t converges to ν weakly in $\mathscr{C}^\alpha$, as t goes to ∞.*

Remark 2 As in [7, 23], one can replace the term $-:X^3:$ by any Wick polynomial of odd degree with negative leading coefficient and obtain the same results in an analogous way.

Remark 3 By [9] we know that for polynomials $\phi^4-\lambda\phi^2$ with λ large enough, the quantum fields in the infinite volume case may have different phases. We expect finding two different Gibbs state ν_1, ν_2 in this case such that they have similar property as in [9, Corollary 12.2.4]. If this is true, we can also obtain that these two states correspond to two different invariant measures for X_0 in the infinite volume case.

However, so far one only knows one state in the infinite volume case obtained in [9, Chap. 11] satisfying the property in [9, Corollary 12.2.4].

As a consequence of Theorem 5 we can give a characterization of ν in terms of its density under translation:

Theorem 6 (cf. [24, Theorem 1.4]) *ν is the unique probability measure such that the following hold*

(i) *ν is absolutely continuous with respect to μ with $\frac{d\nu}{d\mu} \in L^p(\mathscr{S}'(\mathbb{T}^2), \mu)$ for some $p > 1$;*
(ii) *("quasi-invariance") $\frac{d\nu(z+tk)}{d\nu(z)} = a_{tk}(z) = a^0_{tk}(z)a^1_{tk}(z)$ for $z \in H_2^{-1-\varepsilon}$, $k \in C^\infty(\mathbb{T}^2)$, $t > 0$ with*

$$a^0_{tk}(z) = \exp\left[-t\langle(-\Delta+1)k, z\rangle - \frac{1}{2}t^2\langle(-\Delta+1)k, k\rangle\right]$$

and

$$a^1_{tk}(z) = \exp\left[-\frac{1}{2}\sum_{i=0}^{3} C_4^i t^{4-i} : z^i : (k^{4-i})\right].$$

Here $H_2^{-1-\varepsilon}$ for some $\varepsilon > 0$ is defined in Sect. 2.2 and in the following $\langle\cdot,\cdot\rangle$ means the dualization between the elements in $C^\infty(\mathbb{T}^2)$ and $H_2^{-1-\varepsilon}$, respectively. $: z^3 :$ is a fixed version of the Wick power we defined in Sect. 2.2.

Remark 4 (ii) in Theorem 6 can be replaced by the condition that $\int Luv d\nu = \int Lvu d\nu$ for $u, v \in \mathscr{F}C_b^\infty$, where $Lu(z) = \frac{1}{2}\mathrm{Tr}(D^2u)(z) + \langle z, ADu\rangle - \langle : z^3 :, Du(z)\rangle$ for $z \in H_2^{-1-\varepsilon}$ and $\mathscr{F}C_b^\infty$ is defined in Sect. 2.2. (cf. [24, Theorem 1.5]).

Moreover, we can prove that ν is an extreme point of the following convex set.

Corollary 1 (cf. [24, Corollary 1.7]) *ν is an extreme point of the convex set $\mathscr{M}^a$, which denotes the set of all probability measures on $\mathscr{S}'(\mathbb{T}^2)$ satisfying (ii) in Theorem 6.*

Remark 5 (i) The condition (ii) in Corollary 1 can also be replaced by the condition in Remark 4.

(ii) By [3, Theorem 3.3] we know that ν being an extreme point of the convex set $\mathscr{M}^a$ is equivalent to ν being $C^\infty(\mathbb{T}^2)$-ergodic, which is also equivalent to the maximal Dirichlet form $(\mathscr{E}_\nu, D(\mathscr{E}_\nu))$ being irreducible. For the definition of the maximal Dirichlet form $(\mathscr{E}_\nu, D(\mathscr{E}_\nu))$, we refer to [3, Sect. 3].

(iii) Since the irreducibility is so crucial we recall here some characterizations of it in terms of the semigroup $(T_t)_{t>0}$ and generator $(L, D(L))$ of $(\mathscr{E}_\nu, D(\mathscr{E}_\nu))$. The following are equivalent:

1. $(\mathscr{E}_\nu, D(\mathscr{E}_\nu))$ is irreducible.

2. $(T_t)_{t>0}$ is irreducible, i.e., if $g \in L^2(\nu)$ such that $T_t(gf) = gT_tf$ for all $t > 0$, $f \in L^2(\nu)$, then $g = const.$
3. If $g \in L^2(\nu)$ such that $T_t g = g$ for all $t > 0$, then $g = const.$
4. $\int (T_t g - \int g d\nu)^2 d\nu \to_{t\to\infty} 0$ for all $g \in L^2(\nu)$.
5. If $u \in D(L)$ with $Lu = 0$, then $u = const.$

Here we emphasize that we don't know whether the maximal Dirichlet form is the same as the minimal Dirichlet form defined in the proof of Theorem 1.4 below, which is the issue of the Markov uniqueness problem.

3 Three Dimensional Case

Consider the Φ^4_3 field ν constructed in [8]. In the three dimensional case, we cannot directly obtain that the bilinear form:

$$\bar{\mathscr{E}}(u, v) := \frac{1}{2} \int_E \langle Du, Dv \rangle_{L^2} d\nu, \quad u, v \in \mathscr{F}C_b^\infty$$

is closable, since the measure ν is more singular and may not be quasi-invariant along smooth direction (see [4]). Instead we construct the Dirichlet form associated to the solution Φ constructed by [6, 14, 20].

Choose $E = H^{-\frac{1}{2}-\varepsilon}$ with $\varepsilon > 0$ and $H = L^2(\mathbb{T}^3)$. Then we have the following relations:

$$E^* \subset H^* \simeq H \subset E.$$

We recall that the solution $\Phi \in C([0, \infty); \mathscr{C}^{-\alpha})$ to (1) with initial value $z \in \mathscr{C}^{-\alpha}$ for $\frac{1}{2} < \alpha < \frac{1}{2} + \varepsilon$. By [15] we have $(\Omega, \mathscr{F}, (\Phi(t))_{t\geq 0}, (P^z)_{z\in\mathscr{C}^{-\alpha}})$ is a Markov process on $\mathscr{C}^{-\alpha}$ for $\frac{1}{2} < \alpha < \frac{1}{2} + \varepsilon$. It is easy to obtain that $(\Phi(t))_{t\geq 0}$ is a Feller process on $\mathscr{C}^{-\alpha}$. By [17, Chap. IV, Theorem 6.7] and $P^z(\Phi \in C([0, \infty); \mathscr{C}^{-\alpha})) = 1$ for $z \in \mathscr{C}^{-\alpha}$ we know that the Feller process will satisfy the corresponding strong Markov property on $\mathscr{C}^{-\alpha}$. Now we would like to extend Φ to a process Φ' with state space E in such a way that each $z \in E \backslash \mathscr{C}^{-\alpha}$ is a trap for Φ' (see [17, p. 118]). For notation's simplicity we still use $(\Omega, \mathscr{F}, (\Phi(t))_{t\geq 0}, (P^z)_{z\in E})$ to denote Φ'. $(\Omega, \mathscr{F}, (\Phi(t))_{t\geq 0}, (P^z)_{z\in E})$ is a continuous strong Markov process with state space E. Define $\mathscr{F}C_b^\infty$ as in Sect. 2.2. In [27] we have obtained the Dirichlet form associated with Φ:

Theorem 7 (cf. [27, Theorems 1.1 and 1.2]) *There exists a quasi-regular Dirichlet form $(\mathscr{E}, D(\mathscr{E}))$ associated with Φ. Moreover, $\mathscr{F}C_b^\infty \subset D(\mathscr{E})$ and the Dirichlet form $(\mathscr{E}, D(\mathscr{E}))$ coincides with $\int |Df|^2 d\nu$ for $f \in \mathscr{F}C_b^\infty$.*

As a consequence, we obtain the bilinear form is closable, which we cannot directly obtain as we mentioned before:

Theorem 8 (cf. [27, Theorem 1.3]) *The bilinear form* $\bar{\mathscr{E}}(u, v) = \int \langle Du, Dv \rangle d\nu$, $u, v \in \mathscr{F}C_b^\infty$ *is closable and the closure* $(\bar{\mathscr{E}}, D(\bar{\mathscr{E}}))$ *is a quasi-regular Dirichlet form.*

Remark 6 From Theorem 8 we know that there exists another Markov process which admits μ as an invariant measure. Is this Markov process the same as the solution Φ to (1)? In Dirichlet form theory it corresponds to the problem of the relations between the domains of the Dirichlet forms $D(\mathscr{E})$ and $D(\bar{\mathscr{E}})$. In the two dimensional case, they are the same (corresponding to restricted Markov uniqueness in Theorem 4). In the three dimensional case we do not know the answer until now, since we do not know in which direction the integration by parts formula holds.

References

1. Albeverio, S., Röckner, M.: Stochastic differential equations in infinite dimensions: solutions via Dirichlet forms. Probab. Theory Relat. Field **89**, 347–386 (1991)
2. Albeverio, S., Röckner, M.: Dirichlet form methods for uniqueness of martingale problems and applications. Stochastic Analysis (Ithaca, NY, 1993). Proceedings of Symposia in Pure Mathematics, vol. 57, pp. 513–528. American Mathematical Socoiety, Providence (1995)
3. Albeverio, S., Kondratiev, Y.G., Röckner, M.: Ergodicity for the stochastic dynamics of quasi-invariant measures with applications to Gibbs States. J. Funct. Anal. **149**(2), 415–469 (1997)
4. Albeverio, S., Liang, S., Zegarlinski, B.: Remark on the integration by parts formula for the Φ_3^4-quantum field model. Infin. Dimens. Anal. Quantum Probab. Relat. Top. **9**(1), 149–154 (2006)
5. Bahouri, H., Chemin, J.-Y., Danchin, R.: Fourier Analysis and Nonlinear Partial Differential Equations. Grundlehren der Mathematischen Wissenschaften [Fundamental Principles of Mathematical Sciences], vol. 343. Springer, Heidelberg (2011)
6. Catellier, R., Chouk, K.: Paracontrolled distributions and the 3-dimensional stochastic quantization equation. arXiv:1310.6869
7. Da Prato, G., Debussche, A.: Strong solutions to the stochastic quantization equations. Ann. Probab. **31**(4), 1900–1916 (2003)
8. Feldman, J.: The $\lambda\Phi_3^4$ field theory in a finite volume. Commun. Math. Phys. **37**, 93–120 (1974)
9. Glimm, J., Jaffe, A.: Quantum Physics: A Functional Integral Point of View. Springer, New York, Heidelberg, Berlin (1986)
10. Glimm, J., Jaffe, A.: Quantum Physics: A Functional Integral Point of View, 2nd edn. Springer, New York (1987)
11. Gubinelli, M.: Controlling rough paths. J. Funct. Anal. **216**(1), 86–140 (2004)
12. Gubinelli, M., Imkeller, P., Perkowski, N.: Paracontrolled distributions and singular PDEs. arXiv:1210.2684
13. Guerra, F., Rosen, J., Simon, B.: The $P(\Phi)_2$ Euclidean quantum field theory as classical statistical mechanics. Ann. Math. **101**, 111–259 (1975)
14. Hairer, M.: A theory of regularity structures. Invent. Math. **198**, 269–504 (2014)
15. Hairer, M., Matetski, K.: Discretisations of rough stochastic PDEs. arXiv:1511.06937v1
16. Jona-Lasinio, G., Mitter, P.K.: On the stochastic quantization of field theory. Commun. Math. Phys. **101**(3), 409–436 (1985)
17. Ma, Z.M., Röckner, M.: Introduction to the Theory of (Non-Symmetric) Dirichlet Forms. Springer, Berlin (1992)
18. Mikulevicius, R., Rozovskii, B.: Martingale problems for stochasic PDE's. Stochastic Partial Differential Equations: Six Perspectives. Mathematical Surveys and Monographs, vol. 64, pp. 243–325. American Mathematical Society, Providence (1999)

19. Mourrat, J.-C., Weber, H.: Global well-posedness of the dynamic Φ^4 model in the plane. arXiv:1501.06191v1
20. Mourrat, J.-C., Weber, H.: Global well-posedness of the dynamic Φ_3^4 model on the torus. arXiv:1601.01234
21. Parisi, G., Wu, Y.S.: Perturbation theory without gauge fixing. Sci. Sinica **24**(4), 483–496 (1981)
22. Röckner, M.: Specifications and Martin boundaries for $P(\Phi)_2$-random fields. Commun. Math. Phys. **106**, 105–135 (1986)
23. Röckner, M., Zhu, R., Zhu, X.: Restricted Markov unqiueness for the stochastic quantization of $P(\phi)_2$ and its applications (2015). arXiv:1511.08030; J. Funct. Anal. **272**(10), 4263–4303 (2017)
24. Röckner, M., Zhu, R., Zhu, X.: Ergodicity for the stochastic quantization problems on the 2D-torus (2016). arXiv:1606.02102; Commun. Math. Phys. **352**(3), 1061–1090 (2017)
25. Sickel, W.: Periodic spaces and relations to strong summability of multiple Fourier series. Math. Nachr. **124**, 15–44 (1985)
26. Stein, E.M., Weiss, G.L.: Introduction to Fourier Analysis on Euclidean Spaces. Princeton University Press, Princeton (1971)
27. Zhu, R., Zhu, X.: Dirichlet form associated with the Φ_3^4 model (2017). arXiv:1703.09987

Author Index

A. Eberle et al. (eds.), *Stochastic Partial Differential Equations and Related Fields*, Springer Proceedings in Mathematics & Statistics 229,
https://doi.org/10.1007/978-3-319-74929-7

The manufacturer's authorised representative in the EU is Springer Nature Customer Service Centre GmbH, Europaplatz 3, 69115 Heidelberg, Germany. If you have any concerns regarding our products, please contact ProductSafety@springernature.com

Printed and bound by CPI Group (UK) Ltd, Croydon, CR0 4YY

22/07/2026

02174179-0006